SEMICONDUCTOR MANUFACTURING TECHNOLOGY

MICHAEL QUIRK
STAKTEK CORPORATION

JULIAN SERDA
ADVANCED MICRO DEVICES

Pearson Education International

SEMICONDUCTOR MANUFACTURING TECHNOLOGY

MICHAEL QUIRK
STAKTEK CORPORATION

JULIAN SERDA
ADVANCED MICRO DEVICES

Upper Saddle River, New Jersey
Columbus, Ohio

Library of Congress Cataloging-in-Publication Data

Quirk, Michael
   Semiconductor manufacturing technology / Michael Quirk, Julian Serda.
     p. cm.
   ISBN 0-13-081520-9
   1. Electronic industries. 2. Semiconductors—Design and construction. I. Serda, Julian.
II. Title.

TK7836.Q27 2001

00-041653

Vice President and Publisher: Dave Garza
Editor in Chief: Stephen Helba
Assistant Vice President and Publisher: Charles E. Stewart, Jr.
Production Editor: Alexandrina Benedicto Wolf
Production Coordination: Lithokraft II
Design Coordinator: Robin G. Chukes
Cover Designer: Linda Fares
Cover Image: John Foxx
Production Manager: Matthew Ottenweller
Marketing Manager: Barbara Rose

This book was set in Times Roman by Lithokraft II. It was printed and bound by R. R. Donnelley & Sons Company. The cover was printed by Phoenix Color Corp.

Copyright © 2001 by Prentice-Hall, Inc., Upper Saddle River, New Jersey 07458. All rights reserved. Printed in the United States of America. This publication is protected by Copyright and permission should be obtained from the publisher prior to any prohibited reproduction, storage in a retrieval system, or transmission in any form or by any means, electronic, mechanical, photocopying, recording, or likewise. For information regarding permission(s), write to: Rights and Permissions Department.

10 9 8 7 6 5 4 3
ISBN 0-13-081520-9

CONTENTS

CHAPTER 1 INTRODUCTION TO THE SEMICONDUCTOR INDUSTRY
OBJECTIVES 1
INTRODUCTION 2
DEVELOPMENT OF AN INDUSTRY 2
  Industry Roots 2
  The Solid State 3
CIRCUIT INTEGRATION 4
  Integration Eras 5
IC FABRICATION 6
  Wafer Fab 7
  Stages of IC Fabrication 7
SEMICONDUCTOR TRENDS 9
  Increase in Chip Performance 9
  Increase in Chip Reliability 12
  Reduction in Chip Price 12
THE ELECTRONIC ERA 14
  The 1950s: Transistor Technology 14
  The 1960s: Process Technology 14
  The 1970s: Competition 14
  The 1980s: Automation 15
  The 1990s: Volume Production 15
CAREERS IN SEMICONDUCTOR MANUFACTURING 16
  Technician 16
  Job Descriptions 18
SUMMARY 19
KEY TERMS 19
REVIEW QUESTIONS 20
SELECTED INDUSTRY WEB SITES 20
REFERENCES 20

CHAPTER 2 CHARACTERISTICS OF SEMICONDUCTOR MATERIALS
OBJECTIVES 21
INTRODUCTION 21
ATOMIC STRUCTURE 22
  Electrons 22
THE PERIODIC TABLE 24
  Ionic Bonds 26
  Covalent Bonds 28
CLASSIFYING MATERIALS 28
  Conductors 29
  Insulators 30
  Semiconductors 32
SILICON 33
  Pure Silicon 33
  Why Silicon? 33
  Doped Silicon 34
  pn Junctions 39
ALTERNATIVE SEMICONDUCTOR MATERIALS 39
  Gallium Arsenide (GaAs) 40
SUMMARY 41
KEY TERMS 41
REVIEW QUESTIONS 42
REFERENCES 42

CHAPTER 3 DEVICE TECHNOLOGIES
OBJECTIVES 43
INTRODUCTION 43
CIRCUIT TYPES 44
  Analog Circuits 44
  Digital Circuits 44
PASSIVE COMPONENT STRUCTURES 45
  IC Resistor Structures 45
  IC Capacitor Structures 46
ACTIVE COMPONENT STRUCTURES 46
  The pn Junction Diode 46
  The Bipolar Junction Transistor 49
  Schottky Diode 51
  Bipolar IC Technology 52
  CMOS IC Technology 52

Enhancement and Depletion-Mode MOSFETs 58
LATCHUP IN CMOS DEVICES 59
INTEGRATED CIRCUIT PRODUCTS 59
  Linear IC Product Types 60
  Digital IC Product Types 60
SUMMARY 62
KEY TERMS 62
REVIEW QUESTIONS 63
IC MANUFACTURERS' WEB SITES 64
REFERENCES 65

CHAPTER 4 SILICON AND WAFER PREPARATION

OBJECTIVES 67
INTRODUCTION 67
SEMICONDUCTOR-GRADE SILICON 68
CRYSTAL STRUCTURE 68
  Amorphous Materials 69
  Unit Cells 69
  Polycrystal and Monocrystal Structures 70
CRYSTAL ORIENTATION 71
MONOCRYSTAL SILICON GROWTH 72
  CZ Method 72
  Float-Zone Method 75
  Reasons for Larger Ingot Diameters 76
CRYSTAL DEFECTS IN SILICON 78
  Point Defects 78
  Dislocations 79
  Gross Defects 80
WAFER PREPARATION 80
  Shaping Operations 81
  Wafer Slicing 82
  Wafer Lapping and Edge Contour 82
  Etching 83
  Polishing 83
  Cleaning 84
  Wafer Evaluation 84
  Packaging 84
QUALITY MEASURES 84
  Physical Dimensions 85
  Flatness 86
  Microroughness 86
  Oxygen Content 86
  Crystal Defects 87

  Particles 87
  Bulk Resistivity 87
EPITAXIAL LAYER 87
SUMMARY 88
KEY TERMS 88
REVIEW QUESTIONS 89
SELECTED INDUSTRY WEB SITES 90
REFERENCES 90

CHAPTER 5 CHEMICALS IN SEMICONDUCTOR FABRICATION

OBJECTIVES 91
INTRODUCTION 91
STATES OF MATTER 91
PROPERTIES OF MATERIALS 92
  Chemical Properties for Semiconductor Manufacturing 93
PROCESS CHEMICALS 99
  Liquids 99
  Gases 103
SUMMARY 109
KEY TERMS 109
REVIEW QUESTIONS 110
CHEMICAL SUPPLIERS' WEB SITES 110
REFERENCES 111

CHAPTER 6 CONTAMINATION CONTROL IN WAFER FABS

OBJECTIVES 113
INTRODUCTION 113
  Clean Background 114
TYPES OF CONTAMINATION 114
  Particles 115
  Metallic Impurities 116
  Organic Contamination 117
  Native Oxides 118
  Electrostatic Discharge 119
SOURCES AND CONTROL OF CONTAMINATION 120
  Air 120
  Humans 121
  Facility 123
  Water 126
  Process Chemicals 130

Production Equipment 131
Workstation Design 131
WAFER WET CLEANING 135
Wet-Cleaning Overview 135
Wet-Clean Equipment 138
Alternatives to RCA Clean 142
SUMMARY 143
KEY TERMS 143
REVIEW QUESTIONS 144
CHEMICAL AND EQUIPMENT SUPPLIERS' WEB SITES 145
REFERENCES 146

CHAPTER 7 METROLOGY AND DEFECT INSPECTION

OBJECTIVES 149
INTRODUCTION 149
IC METROLOGY 150
Measurement Equipment 151
Yield 151
Data Management 152
QUALITY MEASURES 152
Film Thickness 153
Film Stress 158
Refractive Index 158
Dopant Concentration 159
Unpatterned Surface Defects 161
Patterned Surface Defects 164
Critical Dimension (CD) 165
Step Coverage 167
Overlay Registration 167
Capacitance-Voltage (C-V) Test 168
Contact Angle 171
ANALYTICAL EQUIPMENT 171
Secondary-Ion Mass Spectrometry (SIMS) 171
Atomic Force Microscope (AFM) 173
Auger Electron Spectroscopy (AES) 174
X-Ray Photoelectron Spectroscopy (XPS) 174
Transmission Electron Microscope (TEM) 175
Energy- and Wavelength-Dispersive Spectrometer (EDX and WDX) 176
Focused Ion Beam (FIB) 176
SUMMARY 177
KEY TERMS 178

REVIEW QUESTIONS 178
METROLOGY EQUIPMENT SUPPLIERS' WEB SITES 179
REFERENCES 180

CHAPTER 8 GAS CONTROL IN PROCESS CHAMBERS

OBJECTIVES 181
INTRODUCTION 181
VACUUM 183
Vacuum Ranges 183
Mean Free Path 184
VACUUM PUMPS 184
Roughing Pump 185
High Vacuum Pump 186
Vacuum in Integrated Tools 188
PROCESS CHAMBER GAS FLOW 189
Mass Flow Controllers 189
RESIDUAL GAS ANALYZER (RGA) 190
RGA Basics 190
RGA as Real-Time Monitor 191
PLASMA 192
Glow Discharge 193
PROCESS CHAMBER CONTAMINATION 195
SUMMARY 196
KEY TERMS 196
REVIEW QUESTIONS 196
VACUUM EQUIPMENT SUPPLIERS' WEB SITES 197
REFERENCES 197

CHAPTER 9 IC FABRICATION PROCESS OVERVIEW

OBJECTIVES 199
INTRODUCTION 199
CMOS PROCESS FLOW 199
Overview of Areas in a Wafer Fab 200
CMOS MANUFACTURING STEPS 205
1. Twin Well Process 205
2. Shallow Trench Isolation Process 207
3. Poly Gate Structural Process 210
4. Lightly Doped Drain (LDD) Implant Process 211
5. Sidewall Spacer Formation 212
6. Source/Drain (S/D) Implant Processes 213

7. Contact Formation 214
8. Local Interconnect (LI) Process 214
9. Via-1 and Plug-1 Formation 216
10. Metal-1 Interconnect Formation 217
11. Via-2 and Plug-1 Formation 218
12. Metal-2 Interconnect Formation 219
13. Metal-3 to Pad Etch and Alloy 220
14. Parametric Testing 221

SUMMARY 222
KEY TERMS 223
REVIEW QUESTIONS 223
REFERENCES 224

CHAPTER 10 OXIDATION

OBJECTIVES 225
INTRODUCTION 225
OXIDE FILM 226
    Nature of Oxide Film 226
    Uses of Oxide Film 227
THERMAL OXIDATION GROWTH 231
    Chemical Reaction for Oxidation 231
    Oxidation Growth Model 232
FURNACE EQUIPMENT 239
HORIZONTAL VERSUS VERTICAL FURNACES 240
    Vertical Furnace 241
    Fast Ramp Vertical Furnace 245
    Rapid Thermal Processor 246
OXIDATION PROCESS 248
    Pre Oxidation Cleaning 248
    Oxidation Process Recipe 249
QUALITY MEASUREMENTS 250
OXIDATION TROUBLESHOOTING 252
SUMMARY 252
KEY TERMS 253
REVIEW QUESTIONS 253
FURNACE AND RTP EQUIPMENT SUPPLIERS' WEB SITES 254
REFERENCES 255

CHAPTER 11 DEPOSITION

OBJECTIVES 257
INTRODUCTION 257
    Film Layering Terminology 258
FILM DEPOSITION 260
    Thin-Film Characteristics 260
    Film Growth 263
    Film Deposition Techniques 264
CHEMICAL VAPOR DEPOSITION 265
    CVD Chemical Processes 265
    CVD Reaction 266
CVD DEPOSITION SYSTEMS 269
    CVD Equipment Design 270
    APCVD (Atmospheric Pressure CVD) 271
    LPCVD (Low Pressure CVD) 273
    Plasma-Assisted CVD 277
DIELECTRICS AND PERFORMANCE 282
    Dielectric Constant 282
    Device Isolation 286
SPIN-ON-DIELECTRICS 287
    Spin-On-Glass (SOG) 287
    Spin-On-Dielectric (SOD) 288
EPITAXY 289
    Epitaxy Growth Methods 290
CVD QUALITY MEASURES 292
CVD TROUBLESHOOTING 292
SUMMARY 294
KEY TERMS 295
REVIEW QUESTIONS 295
DEPOSITION EQUIPMENT SUPPLIERS' WEB SITES 296
REFERENCES 296

CHAPTER 12 METALLIZATION

OBJECTIVES 299
INTRODUCTION 300
TYPES OF METALS 302
    Aluminum 302
    Aluminum-Copper Alloys 305
    Copper 305
    Barrier Metals 307
    Silicides 309
    Metal Plugs 312
METAL DEPOSITION SYSTEMS 313
    Evaporation 313
    Sputtering 314
    Metal CVD 320
    Copper Electroplate 323
METALLIZATION SCHEMES 325
    Traditional Aluminum Structure 325

Copper Damascene Structure 326
METALLIZATION QUALITY MEASURES 329
METALLIZATION TROUBLESHOOTING 330
SUMMARY 331
KEY TERMS 331
REVIEW QUESTIONS 332
METALLIZATION EQUIPMENT AND MATERIALS SUPPLIERS' WEB SITES 332
REFERENCES 333

CHAPTER 13 PHOTOLITHOGRAPHY: VAPOR PRIME TO SOFT BAKE

OBJECTIVES 335
INTRODUCTION 336
   Photolithography Concepts 336
PHOTOLITHOGRAPHY PROCESSES 339
   Negative Lithography 339
   Positive Lithography 340
EIGHT BASIC STEPS OF PHOTOLITHOGRAPHY 342
   Step 1: Vapor Prime 343
   Step 2: Spin Coat 343
   Step 3: Soft Bake 344
   Step 4: Alignment and Exposure 344
   Step 5: Post-Exposure Bake (PEB) 344
   Step 6: Develop 344
   Step 7: Hard Bake 344
   Step 8: Develop Inspect 345
VAPOR PRIME 345
   Wafer Cleaning 345
   Dehydration Bake 346
   Wafer Priming 346
SPIN COAT 348
   Photoresist 348
   Photoresist Physical Properties 349
   Conventional I-Line Photoresists 351
   Deep UV (DUV) Photoresists 354
   Photoresist Dispensing Methods 357
SOFT BAKE 360
   Soft Bake Equipment 360
   Process Characterization 361
PHOTORESIST QUALITY MEASURES 362
PHOTORESIST TROUBLESHOOTING 363
SUMMARY 364
KEY TERMS 364
REVIEW QUESTIONS 364
PHOTORESIST MATERIALS AND EQUIPMENT SUPPLIERS' WEB SITES 365
REFERENCES 366

CHAPTER 14 PHOTOLITHOGRAPHY: ALIGNMENT AND EXPOSURE

OBJECTIVES 367
INTRODUCTION 367
   Importance of Alignment and Exposure 368
OPTICAL LITHOGRAPHY 370
   Light 370
   Exposure Sources 372
   Optics 376
   Resolution 385
PHOTOLITHOGRAPHY EQUIPMENT 388
   Contact Aligner 388
   Proximity Aligner 389
   Scanning Projection Aligner 390
   Step-and-Repeat Aligner (Stepper) 390
   Step-and-Scan System 393
   Reticles 394
   Optical Enhancement Techniques 398
   Alignment 400
   Environmental Conditions 404
   Comparison of Photo Tools 405
MIX AND MATCH 406
ALIGNMENT AND EXPOSURE QUALITY MEASURES 407
ALIGNMENT AND EXPOSURE TROUBLESHOOTING 408
SUMMARY 409
KEY TERMS 410
REVIEW QUESTIONS 410
PHOTORESIST MATERIALS AND EQUIPMENT SUPPLIERS' WEB SITES 411
REFERENCES 412

CHAPTER 15 PHOTOLITHOGRAPHY: PHOTORESIST DEVELOPMENT AND ADVANCED LITHOGRAPHY

OBJECTIVES 413
INTRODUCTION 413
    Advanced Lithography 414
POST-EXPOSURE BAKE 414
    DUV Post-Exposure Bake (PEB) 415
    Conventional I-Line PEB 416
DEVELOP 416
    Negative Resist 417
    Positive Resist 418
    Development Methods 419
    Resist Development Parameters 421
HARD BAKE 422
DEVELOP INSPECT 422
ADVANCED LITHOGRAPHY 424
    Next-Generation Lithography 424
    Advanced Resist Processing 428
DEVELOP QUALITY MEASURES 429
DEVELOP TROUBLESHOOTING 431
SUMMARY 432
KEY TERMS 433
REVIEW QUESTIONS 433
PHOTOLITHOGRAPHY MATERIALS AND EQUIPMENT SUPPLIERS' WEB SITES 433
REFERENCES 434

CHAPTER 16 ETCH

OBJECTIVES 435
INTRODUCTION 435
    Etch Processes 436
ETCH PARAMETERS 437
    Etch Rate 437
    Etch Profile 438
    Etch Bias 439
    Selectivity 440
    Uniformity 441
    Residues 441
    Polymer Formation 442
    Plasma-Induced Damage 442
    Particle Contamination 443
DRY ETCH 443
    Etching Action 444
    Potential Distribution 444
PLASMA ETCH REACTORS 446
    Barrel Plasma Etcher 446
    Parallel Plate (Planar) Reactor 447
    Downstream Etch Systems 448
    Triode Planar Reactor 448
    Ion Beam Milling 448
    Reactive Ion Etch (RIE) 450
    High-Density Plasma Etchers 450
    Etch System Review 453
    Endpoint Detection 455
    Vacuum for Etch Chambers 456
DRY ETCH APPLICATIONS 456
    Dielectric Dry Etch 457
    Silicon Dry Etch 459
    Metal Dry Etch 462
WET ETCH 464
    Types of Wet Etch 465
HISTORICAL PERSPECTIVE 466
PHOTORESIST REMOVAL 466
    Plasma Ashing 466
ETCH INSPECTION 469
ETCH INSPECTION QUALITY MEASURES 469
DRY ETCH TROUBLESHOOTING 470
SUMMARY 471
KEY TERMS 472
REVIEW QUESTIONS 472
ETCH EQUIPMENT SUPPLIERS' WEB SITES 473
REFERENCES 474

CHAPTER 17 ION IMPLANT

OBJECTIVES 475
INTRODUCTION 475
    Doped Regions 477
DIFFUSION 479
    Diffusion Principles 479
    Diffusion Process 481
ION IMPLANTATION 482
    Overview 483
    Ion Implant Parameters 484
ION IMPLANTERS 488
    Ion Source 488
    Extraction and Ion Analyzer 490
    Acceleration Column 491

Scanning System 494
Process Chamber 498
Annealing 499
Channeling 501
Particles 502
ION IMPLANT TRENDS IN PROCESS INTEGRATION 503
Deep Buried Layers 503
Retrograde Wells 503
Punchthrough Stoppers 504
Threshold Voltage Adjustment 504
Lightly Doped Drain 505
Source/Drain Implants 505
Polysilicon Gate 506
Trench Capacitor 506
Ultrashallow Junctions 506
Silicon-On-Insulator (SOI) 507
ION IMPLANT QUALITY MEASURES 508
ION IMPLANT TROUBLESHOOTING 509
SUMMARY 510
KEY TERMS 511
REVIEW QUESTIONS 512
ION IMPLANTER EQUIPMENT SUPPLIERS' WEB SITES 513
REFERENCES 513

CHAPTER 18 CHEMICAL MECHANICAL PLANARIZATION

OBJECTIVES 515
INTRODUCTION 515
TRADITIONAL PLANARIZATION 518
Etchback 518
Glass Reflow 519
Spin-On Films 520
CHEMICAL MECHANICAL PLANARIZATION 520
CMP Planarity 521
Advantages of CMP 522
CMP Mechanisms 523
CMP Slurry and Pad 526
CMP Equipment 529
CMP Clean 532
CMP Equipment Manufacturers 534
CMP APPLICATIONS 535
STI Oxide Polish 535

LI Oxide Polish 535
LI Tungsten Polish 536
ILD Oxide Polish 536
Tungsten Plug Polish 536
Dual-Damascene Copper Polish 537
CMP QUALITY MEASURES 538
CMP TROUBLESHOOTING 540
SUMMARY 541
KEY TERMS 541
REVIEW QUESTIONS 542
CMP EQUIPMENT SUPPLIERS' WEB SITES 542
REFERENCES 543

CHAPTER 19 WAFER TEST

OBJECTIVES 545
INTRODUCTION 545
IC Electrical Tests 546
WAFER TEST 547
In-Line Parametric Test 547
Wafer Sort 555
Yield 560
Wafer Sort Yield Models 563
TEST QUALITY MEASURES 565
TEST TROUBLESHOOTING 567
SUMMARY 567
KEY TERMS 568
REVIEW QUESTIONS 568
TEST AND PROBER EQUIPMENT SUPPLIERS' WEB SITES 569
REFERENCES 569

CHAPTER 20 ASSEMBLY AND PACKAGING

OBJECTIVES 571
INTRODUCTION 571
Packaging Levels 573
TRADITIONAL ASSEMBLY 574
Backgrind 574
Die Separation 575
Die Attach 575
Wirebonding 577
TRADITIONAL PACKAGING 580
Plastic Packaging 581
Ceramic Packaging 584
Final Test 586

ADVANCED ASSEMBLY AND PACKAGING 586
- Flip Chip 587
- Ball Grid Array (BGA) 589
- Chip on Board (COB) 590
- Tape Automated Bonding (TAB) 591
- Multichip Module (MCM) 591
- Chip Scale Packaging (CSP) 592
- Wafer-Level Packaging 592

ASSEMBLY AND PACKAGING QUALITY MEASURES 596
IC PACKAGING TROUBLESHOOTING 597
SUMMARY 597
KEY TERMS 598
REVIEW QUESTIONS 598
ASSEMBLY AND PACKAGING SUPPLIERS' WEB SITES 599
REFERENCES 600

APPENDICES

- APPENDIX A CHEMICALS AND SAFETY 601
- APPENDIX B CONTAMINATION CONTROLS IN CLEANROOMS 610
- APPENDIX C UNITS 614
- APPENDIX D COLOR AS A FUNCTION OF OXIDE THICKNESS 617
- APPENDIX E OVERVIEW OF PHOTORESIST CHEMISTRY 618
- APPENDIX F ETCH CHEMISTRY 622

GLOSSARY 624

INDEX 648

PREFACE

This text started with a simple premise: as instructors, we need to teach relevant microchip technology to students and employees in semiconductor manufacturing. Unfortunately, in the semiconductor industry, changes in technology are measured in months, not years. Our challenge was to write a relevant book that would not be outdated by the time it was published. With that in mind, we researched the material and applied ourselves to writing the chapters and creating the artwork. Following the aggressive pace of Moore's law, the technical material in our book is at most only 18 to 24 months old. This permits us to keep abreast of the changing technology nodes swirling through the semiconductor industry.

This text is written for students in two-year and four-year technology programs at community colleges and universities. The text will also be a practical reference as well as a standard text in corporate and technical training classes. Students are expected to have an understanding of high school chemistry, physics, and math. Chapters are organized around the broad technologies applicable to semiconductor manufacturing.

ORGANIZATION OF THE TEXT

Our goal is to accomplish three objectives:

1. Help technology students grasp the fundamental technologies used in manufacturing semiconductor devices.
2. Present some of the many challenges in microchip fabrication.
3. Instill in the reader an appreciation of the conceptual simplicity of semiconductor manufacturing.

All fundamental technical information relevant to semiconductor manufacturing is first presented in Chapters 1 to 8. Chapter 9 presents a process model overview with a general flowchart that links the major areas in a wafer fab. Chapters 10 to 19 cover each of the major processes in the fab. Finally, Chapter 20 provides an overview of the back-end process for IC assembly and packaging. The content in the process chapters (Chapters 10 to 20) addresses critical process technology, followed by the various equipment designs needed to support this technology. Each process chapter concludes with a summary of quality measures and troubleshooting issues to familiarize the student with the practical, day-to-day challenges encountered during wafer fabrication.

The latest technologies for sub-0.25 µm processing are covered in detail. This includes chemical mechanical planarization (CMP), shallow trench isolation (STI), chemically amplified deep UV photoresists, step-and-scan systems, copper metallization with dual damascene, and the widespread move to process integration with cluster tools. Throughout the text, we explain all process and equipment technology in light of the long history of change in the industry. Early tools and processes are described to clarify the development of current technology. In some cases, the linkage between the latest equipment and earlier tools is obvious, while in other instances the change is dramatic.

Professors, students, and other readers of this book can send comments or questions about this text to the authors at the following website: http://www.smtbook.com. We look forward to any exchange of information that can help advance semiconductor manufacturing education.

ACKNOWLEDGMENTS

We would like to thank the following reviewers for their valuable feedback: Bruce Bothwell, Chemeketa Community College; Mike D'Elia, Advanced Micro Devices; Chris Dennis and David Hata, Portland Community College; John Fowler, Arizona State University; Lance Kinney; Bassam Matar, Maricopa Community College; John Nistler, Advanced Micro Devices; Val Shires, Gwinnett Technical Institute; and Joseph White, Texas A&M University.

We acknowledge the support of the following colleagues, friends, and new acquaintances for their technical inputs and encouragement throughout the project.

Technical Reviewers and Editors:

- Mike D'Elia, Member of Technical Staff, Advanced Micro Devices—a tremendous inspiration to us—reviewed the process flow, oxidation and deposition chapters and was the first SMT instructor to use selected chapters.
- Mike Hillis, Module Shift Manager, Advanced Micro Devices, provided encouragement and reviewed and edited the photolithography sections.
- Steve Hymes, Ph.D., Applications Manager, Ashland Specialty Chemicals, who seemed more excited than us at times, reviewed and edited the CMP and metallization chapters.
- Rick Jarvis, Senior Member Technical Staff, Advanced Micro Devices, on assignment to International SEMATECH, with great insight to the industry, reviewed and edited the contamination, etch, and deposition sections.
- Larry Knipp, President, Nykar Inc., reviewed and edited the RF and plasma sections.
- Jeremy Lansford, Module Manager, Advanced Micro Devices, very carefully reviewed and edited the CMP section, then met with us to ensure his suggestions were understood.
- Federico Miller, Chemistry Department, Austin Community College, provided advice and encouragement throughout the project and reviewed all the chemistry related information.
- Sergei Postnikov, Ph.D., Photolithography Development, Motorola, reviewed the photolithography sections and provided encouragement throughout the project.
- Alan Romriell, Senior Test Engineer, Advanced Micro Devices, reviewed the test chapter.
- Andy Shurtleff, President, Liquid Asset Services, an expert in ultra-pure water systems, reviewed and edited the section on ultra-pure water.
- Darby Webster, Senior Manager, Manufacturing Engineering, Staktek Corp., reviewed the assembly and packaging chapter and gave his support throughout the project.
- Joseph W. Wiseman, Member of Technical Staff, Advanced Micro Devices, a professional in his engineering approach and also an etch instructor, reviewed and edited the etch section.

Computer-Aided Illustrators and Writers:

- Thuat Duong, CAD specialist, Solectron, created many of the drawings used in the assembly and packaging section.
- Matt Hanson, Computer Graphics Artist, provided graphic support for the metrology section.
- Josh Hix, Technical Writer, a full-time engineering student at Georgia Tech University, produced most of the glossary.
- Trent Romriell, Computer Graphics Artist, supported us with many drawings in the beginning chapters.
- James Samara, Computer Graphics Artist, provided graphics support for the metrology chapter.
- Mark Serda, Computer Graphics Artist, provided graphics support for the gas controls chapter.
- Edgard Torres, Computer Graphic Artist, provided 3-dimensional models of silicon lattice structures.

PREFACE **xiii**

Providers of Logistics, Technical Information, Graphics, and Photographs:

- Shannon Bacon, Continuous Improvement Facilitator, Advanced Micro Devices, made arrangements for us to meet with subject matter experts at AMD's facilities.
- Tony Denboer, Integrated Circuit Engineering Corporation, provided SEM micrographs of actual integrated circuit structures.
- Jim DeVries, Facilities Operations Manager, Advanced Micro Devices, directed the photo shoot of some AMD facilities.
- Charlie Earnhart, Vice-President, Staktek Corp., for his support throughout the project.
- James A. Elliott, Technical Instructor, Varian Semiconductor Equipment, assisted the writers with obtaining permission to use technical documentation and illustrations of Varian ion implanters.
- Robert Emmons, Learning and Development Manager, Advanced Micro Devices, allowed Julian to adjust his work hours so he could devote more quality time to the book.
- Kathleen Fitts, Global Quality Programs Manager, Applied Materials, supported the project with various diagrams and photographs.
- Dahlia Hernandez-Fridge, Executive Assistant to the President, CEDRA Corp., researched technical information.
- Van Pham, Process Development Engineer, Staktek Corporation, provided input regarding etch.
- Richard Rogoff, Major Accounts Manager, ASML, provided excellent diagrams and photographs of step-and-scan systems.
- Tim Roy, Director, Staktek Corp., for his support throughout the project.
- Sharon Shaw, Librarian, Advanced Micro Devices, showed us what the word "resourceful" means by always being able to locate a reference for us and for handling the logistics.
- Michael Sullivan, MEGASYSTEM Manager, Air Products, allowed us to photograph several bulk gas distribution sites.
- Chuck Tully, Austin Technical Training Center Manager, Applied Materials, who was always there when we needed him, provided technical information and graphics on electrostatic chuck and DPS etch equipment.
- Phil Ware, Director and General Manager of Marketing, Semiconductor Equipment Division, Canon USA, Inc., provided contacts, graphics, and photographs of several generations of photolithography tools.
- Scott A. Williams, Central Regional Account Manager, SpeedFam-IPEC, arranged to have photographs taken and provided diagrams of CMP equipment.

Michael Quirk
Julian Serda

This text is dedicated to our families, who endured so much during this project:

Monique, Justin, and Morgan Quirk

Dorothy, Mark, Ruben, and Elena Serda

CHAPTER 1
INTRODUCTION TO THE SEMICONDUCTOR INDUSTRY

Society witnessed a technology revolution during the twentieth century with the change from products derived from mechanical technology to those centered on electronic technology. Digital compact disc players replaced record players, and automotive engines are now controlled by electronic ignition systems. Electronic computers quickly perform tasks in nearly every aspect of society, promoting efficient use of our resources. Given the breadth of changes brought by electronic technology, the revolution has just begun.

The semiconductor industry has been at the center of this technology revolution. The principal building material, the *semiconductor*, is the main ingredient of the electronic products found throughout society. Semiconductor products are manufactured by people with diverse technical skills: designers who create new designs based on customer needs, engineers who specify improved requirements for the equipment and process, and technicians who fabricate the semiconductor products in automated factories. All the while, the growing semiconductor market continues to demand more performance at lower cost.

OBJECTIVES

After studying the material in this chapter, you will be able to:

1. Describe the current economic state and the technical roots of the semiconductor industry.
2. Explain what an integrated circuit (IC) is and list the five circuit integration eras.
3. Describe a wafer, including how it is layered, and describe the essential aspects of the five stages of wafer fabrication.
4. State and discuss the three major trends associated with improvement in wafer fabrication.
5. Explain what a critical dimension (CD) is and how Moore's law predicts future wafer fabrication improvement.
6. Describe the different eras of electronics since the invention of the transistor up to modern wafer fabrication.
7. Discuss different career paths in the semiconductor industry.

Microprocessor Chip
(Photo courtesy of Advanced Micro Devices)

Microprocessor Chip
(Photo courtesy of Intel Corporation)

INTRODUCTION

The basic semiconductor material from which electronic devices are made comes in the form of round thin crystalline disks called *wafers*. Semiconductor products are produced from wafers in wafer fabrication (*wafer fab*) factories, and are referred to as *microchips*, or *chips*.

The technology of semiconductor fabrication is complex, requiring many specialized process steps, materials, equipment, and supplier industries. Once the microchips are fabricated, they are packaged into the various electronic and mechanical assemblies required for the numerous product applications. Examples of such applications are automotive electronics, electronic commerce, personal computers, and mobile cellular communications.

Worldwide sales of microchips are expected to exceed $200 billion in 2001.[1] Today, semiconductors account for 30% to 40% of the cost for a personal computer, and about $100 worth of semiconductors are found in each cellular telephone. Every automobile has approximately $140 worth of microchips, an amount that is increasing as cars become "smarter."

The semiconductor industry is actually a subset of a larger entity—the high-tech industry. Fabricating the microchips produces electronic hardware that is coupled with software to integrate and control the chip functions. The high-tech industry encompasses all the hardware and software technology found in all semiconductor applications (See Figure 1.1).

The high-tech industry in the United States is big. In the mid-1990s, the high-tech industry contributed 27% of the U.S. economy, compared to 14% for residential housing and 4% for the automotive industry. Why has the semiconductor industry become so large? A major factor has been the industry's ability to consistently increase semiconductor product performance while decreasing the price. The ability to meet the market demands for high performance at low cost can be directly traced to the technology improvements regularly made in product design and manufacturing throughout the history of the industry.

DEVELOPMENT OF AN INDUSTRY

The technology used to fabricate semiconductor devices developed out of many different technological inventions from early pioneers in the electronics field.

Industry Roots

The developmental roots of the semiconductor industry grew out of technologies developed in the first half of the twentieth century.[2] Key technical knowledge was gained from a web of industry and academia: vacuum tube electronics, wireless communications, mechanical tabulators, and solid-state physics. These industries were the fabric of the high-tech industry at the time.

The triode vacuum tube for amplifying electrical signals was invented by Lee De Forest in 1906. It was developed out of earlier vacuum tube work by John Fleming and Thomas Edison. A triode consists of three elements: two electrodes and a grid separated in an evacuated glass enclosure. A vacuum is necessary to keep the elements from burning up and also to facilitate the transfer of electrons between the electrodes. De Forest patented and named his vacuum tube invention the audion because he thought there was some potential for amplifying and reproducing sound. He

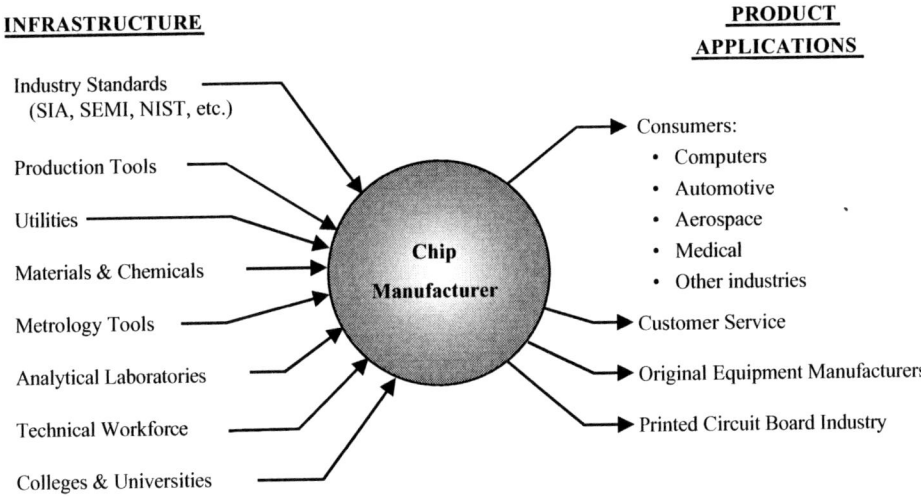

FIGURE 1.1 The Semiconductor Industry

was right, and the vacuum tube became the primary electronic device for the growth of modern radio, television, and general electronics up until the 1950s.

A relatively new material, a crystal known as silicon, was used in the early 1900s to convert wireless communications signals from alternating current to direct current. The term "semiconductor" was first adopted in Germany to cover this class of material.[3] However, to truly develop the power of semiconductor technology required additional research by physical chemists and physicists around the world. This research was needed in order to understand the quantum theory of electron behavior before a full explanation of semiconductor properties was available. This substantial work extended over several decades and beyond World War II.

There was a market need to electronically calculate numbers. The first mechanical computer was the Hollerith mechanical tabulating machine, invented by Herman Hollerith, to tabulate the 1890 census figures in an amazing six weeks. This tabulator was driven by an electric motor and based on punched cards (punch cards remained the common input method for computers up through the 1970s). Vacuum tubes were used to develop the first electronic computer, the ENIAC (Electronic Numeric Integrator and Calculator) at the University of Pennsylvania during World War II. The ENIAC weighed 50 tons, required 3,000 ft^2 of floor space, needed 19,000 vacuum tubes, and used the equivalent electrical power of 160 lighthouses.

A major drawback to the ENIAC, besides its bulky size, was the number of problems associated with vacuum tubes. Tubes were large, unreliable and consumed large amounts of electric power. Tubes had a limited lifetime due to burnout. Vacuum tubes clearly were not the optimum technology to produce small, reliable electronic products that were needed for the rapidly developing electronic markets.

The Solid State

There was a concerted effort at Bell Telephone Laboratories after World War II to study the properties of solid silicon and germanium semiconductor crystals. The scientists conducting the research sensed the need to replace vacuum tubes and the benefits from using a solid semiconductor material in place of vacuum tubes.

The modern-day semiconductor industry was born with the invention of the *solid-state transistor* at Bell Telephone Laboratories on December 16, 1947, by William Shockley, John Bardeen, and Walter Brattain.[4] The transistor, a name taken from the two terms *transconductance* and *varistor*,[5] offered the same electrical functions as a vacuum tube, but with the distinct advantages of a solid state: miniscule size, no vacuum, reliable, lightweight, minimal heat production, and low power consumption. The three scientists were awarded the 1956 Nobel Prize in physics for their invention. This discovery launched the modern semiconductor industry based on solid-state materials and technology.

The semiconductor industry started growing rapidly in the 1950s to commercialize transistor technology based on silicon. Many of the early pioneers started in northern California, in the area

Vacuum Tubes

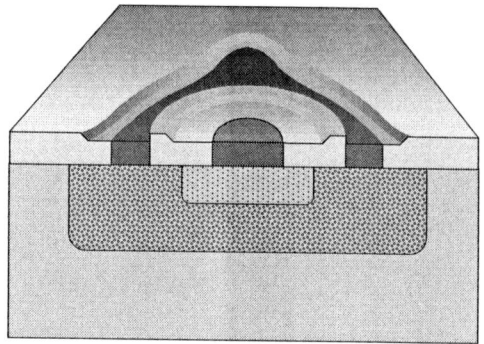

FIGURE 1.2 The First Planar Transistor

now known as Silicon Valley. The first commercial planar transistor was fabricated at Fairchild Semiconductor in Palo Alto, California, in 1957. It had a layer of aluminum interconnect material that was deposited on top of the silicon wafer to connect the different parts of the transistor (see Figure 1.2). A natural oxide layer thermally grown from the silicon was used to insulate the aluminum conductor. The use of layers was an important development in semiconductors and is the reason for the term *planar technology*.

CIRCUIT INTEGRATION

The semiconductor name derives from the material's ability to sometimes be a conductor to electricity and other times act as a nonconductor to electricity. The earliest semiconductor material was germanium, built into a single chip with one function (referred to as a discrete component). Nowadays, greater than 85% of all microchips are made from silicon semiconductor material. For this reason, we will emphasize silicon in this book.

An important step forward in the semiconductor industry was the integration of multiple electronic components on one silicon substrate. Referred to as an *integrated circuit,* or *IC,* it was coinvented independently by Robert Noyce at Fairchild Semiconductor and Jack Kilby at Texas Instruments in 1959. On the silicon surface of an IC are manufactured the many different semiconductor *devices,* such as transistors, diodes, resistors, and capacitors, which are connected into a circuit to define how the chip will function. The expression *IC* is often meant to describe the chip and all its components.

The first IC, developed in July 1958 by Jack Kilby at a Texas Instruments facility in Dallas, Texas, was made using a slice of germanium semiconductor material as the substrate.[6] The invention combined a transistor and other components on the germanium while also using the natural resistance of the germanium as a resistor. The devices were connected with individual wires.

Robert Noyce at Fairchild Semiconductor also invented the concept of an IC by extending the idea of how to interconnect the different components on the planar silicon material. His idea was to use an aluminum metal conductor on the silicon surface to interconnect the different transistors while using a layer of oxide grown from the silicon as an insulator to separate the metal conductor from the silicon devices. This was the first practical structure of an IC as a single-structure silicon chip.[7]

Circuit integration has increased dramatically since the initial IC. Because all the components are integrated on the silicon substrate, the IC has developed into a cost-effective and reliable way to produce and interconnect many components. The ability to integrate many different components on an IC has spurred engineers to design ever more complex electronic circuits to meet new customer needs.

Introduction to the Semiconductor Industry **5**

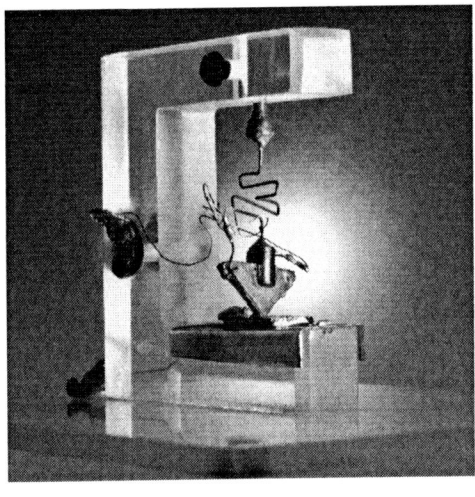

The First Transistor from Bell Labs
(Photo courtesy of Lucent Technologies, Bell Labs Innovations)

Jack Kilby's First Integrated Circuit
(Photo courtesy of Texas Instruments, Inc.)

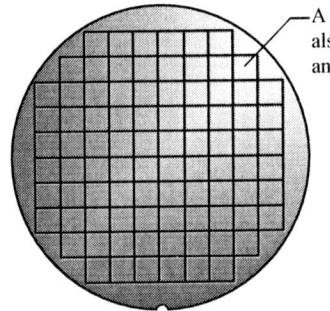

A single integrated circuit, also known as a die, chip, and microchip

FIGURE 1.3 Top View of Wafer with Chips

Integration Eras

We can roughly categorize the integration era by the number of components integrated on a chip. This is a useful method of organizing the development of the semiconductor industry since 1960 to the present (see Table 1.1).

TABLE 1.1 Circuit Integration of Semiconductors

Circuit Integration	Semiconductor Industry Time Period	Number of Components per Chip
No integration (discrete components)	Prior to 1960	1
Small scale integration (SSI)	Early 1960s	2 to 50
Medium scale integration (MSI)	1960s to Early 1970s	50 to 5,000
Large scale integration (LSI)	Early 1970s to Late 1970s	5,000 to 100,000
Very large scale integration (VLSI)	Late 1970s to Late 1980s	100,000 to 1,000,000
Ultra large scale integration (ULSI)	1990s to present	> 1,000,000

Circuit integration continues today as the number of devices on a chip is continually increased. A significant challenge for circuit integration is the capability of the semiconductor manufacturing process to improve the fabrication technology to produce highly integrated ULSI chips at acceptable costs. To this end, the semiconductor industry has become highly standardized, with most manufacturers using similar manufacturing process and equipment technology. The key to market success is the ability of a company to deliver the right product at the right time. Industry

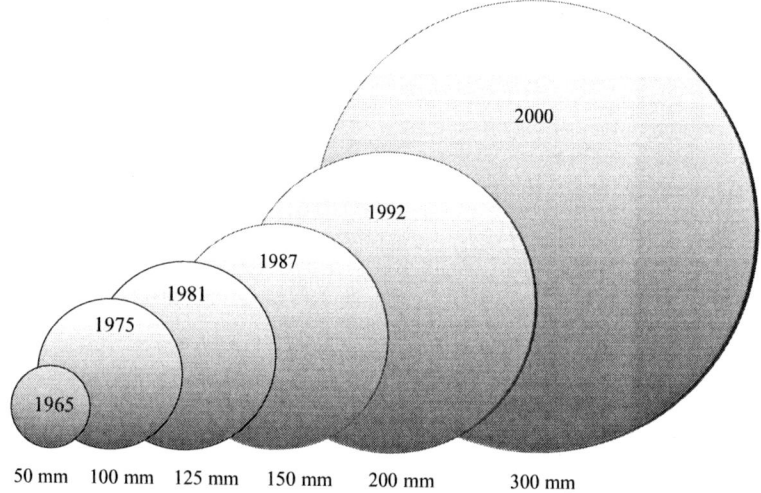

FIGURE 1.4 Evolution of Wafer Size

consortiums, such as International SEMATECH and the Semiconductor Industry Association (SIA), have been created to assist chip producers to be competitive in the world market of high-performance ICs.

IC FABRICATION

Tens or hundreds of identical chips are fabricated simultaneously on a silicon *wafer* (see Figure 1.3 on page 5). The number of chips on a wafer varies depending on the type of product and the size of each chip. The chip size changes depending on the level of integration on the chip.

The chip is also referred to as a *die* (singular and plural for chips or ICs), while the silicon wafer is often called the *substrate*. The wafer diameter has evolved over the years from an initial diameter of less than 1″[8] to the present common diameter of 8″ (200 mm), which is undergoing a change to 12″, or 300 mm (see Figure 1.4). The cost of fabricating ICs drops dramatically if there are more chips on a wafer due to benefits from the economies of scale (more chips are produced from the same effort).

Fabrication of the semiconductor devices occurs only in the first few microns of the silicon near the wafer surface. The bulk of the silicon wafer thickness is to give the wafer adequate rigidity during processing. Once the devices are fabricated in the silicon, then layers of metal circuitry are placed on the silicon to interconnect between the devices and the various electronic signals outside the chip (see Figure 1.5). The concept and materials for interconnecting a modern day IC are remarkably similar to the initial planar transistor first commercialized at Fairchild Semiconductor in 1957. The major difference is that today's chips are much more complex.

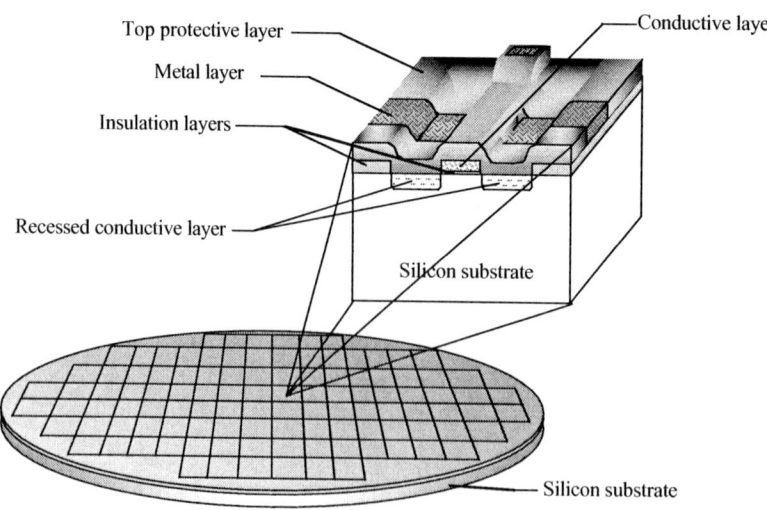

FIGURE 1.5 Devices and Layers from a Silicon Chip

Wafer Fab

Early wafer fabs were simple, with operators manually processing wafers through the different operations. An essential aspect of wafer fabs is that as wafers become more densely integrated, the permissible level of contamination decreases dramatically. Contamination that can damage wafers and cause them not to function properly evolves from many sources: humans, materials, water, air, and equipment. Modern wafer fabs have become specialized facilities that provide the clean manufacturing environment and specialized equipment to produce wafers with minimal defects from contamination. This includes limited human exposure, ultrapure chemicals and utilities, and special wafer handling procedures needed to fabricate ICs in the ULSI era.

A wafer typically requires two to three months of processing to complete the 450 or more process steps in a wafer fab. At the end of the fabrication process, the individual chips are separated from the entire wafer and then prepared for packaging into the final product.

Stages of IC Fabrication

Fabrication of microchips involves five general stages of manufacturing (see Figure 1.6):

- ◆ Wafer preparation
- ◆ Wafer fabrication
- ◆ Wafer test/sort
- ◆ Assembly and packaging
- ◆ Final test

These five stages are interdependent, with a large infrastructure within the semiconductor company and a support network of industries that provide specialized chemicals and equipment. Companies that operate only in an individual stage (such as a chip company that only fabricates chips) have to meet industrywide standards to ensure that the final microchip meets performance objectives.

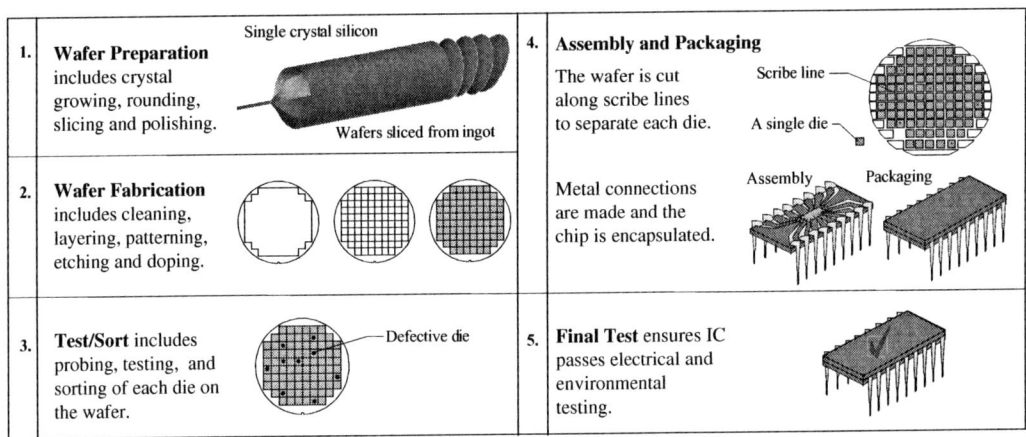

FIGURE 1.6 Stages of IC Fabrication

Wafer Preparation ■ In the first stage, *silicon* is mined from sand and purified. It undergoes specialized processing to create a silicon ingot with the appropriate diameter (see Figure 1.7 on page 8). The silicon ingot is then sliced into thin wafers that are used to fabricate the microchips. Wafers are prepared to special specifications for criteria such as flatness and contamination.

Since the 1980s, most companies that fabricate microchips purchase their wafers from suppliers who specialize in crystal growth and wafer preparation. The industry also produces wafers made from germanium or compound semiconductor materials. These are specialized applications, with the majority of all semiconductor wafers made from silicon.

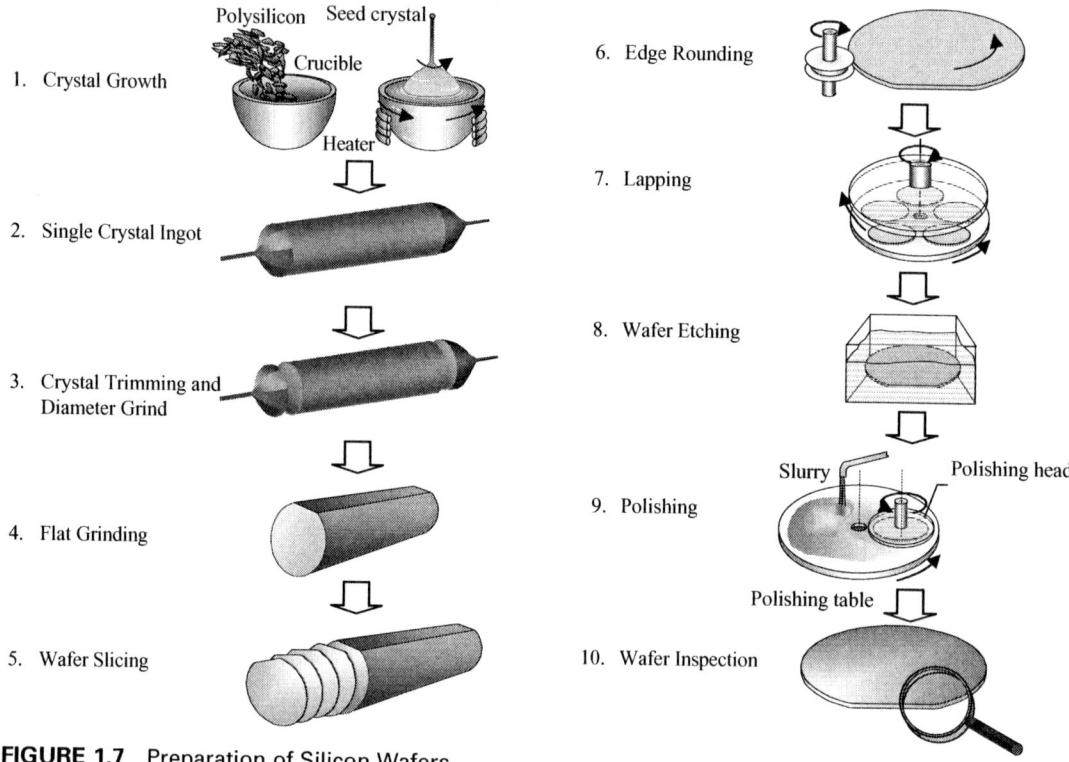

FIGURE 1.7 Preparation of Silicon Wafers
(*Note:* Terms in Figure 1.7 are explained in Chapter 4.)

Wafer Fabrication ■ The fabrication of microchips from wafers is the second stage, referred to as *wafer fabrication*. The bare silicon wafers arrive in the wafer fab, then they go through a variety of cleaning, layering, patterning, etching, and doping steps. The completed wafers have a full complement of integrated circuits permanently etched into each silicon wafer. Other names for wafer fabrication are *microchip fabrication, chip fabrication,* and *chip fab.*

Companies that fabricate chips are both merchant and captive producers. *Merchant chip suppliers* fabricate chips to sell on the open market, such as a chip manufacturer that produces memory chips for customers. *Captive chip producers* fabricate chips to be used in-house in their company's products. An example of a captive producer is a company that makes computers and also fabricates chips for their computers. Some chip makers fabricate chips for in-house use and also sell chips on the open market, while others will fabricate specialty chips and purchase other chips on the open market.

Another type of chip maker is the *fabless company.* This type of company designs chips for a particular market, such as a video microchip, while another chip maker fabricates the chips. Finally,

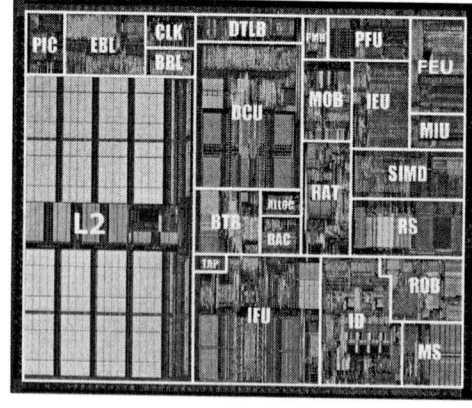

ULSI Chip
(Photo courtesy of Intel Corporation)

another type of semiconductor manufacturer, the *foundry,* produces chips only for other companies. Semiconductor foundries have become more common since the 1980s, with about 10% of all chips now fabricated at foundries. A major reason for the increase in fabless chip producers and foundries is the high cost of constructing and maintaining a wafer fab. Currently, a wafer fab for high-performance ICs costs around $1.5 to $3.0 billion, with about 75% of the total cost for equipment.

Wafer fabrication of ICs involves the interaction of many complicated process steps using automated equipment to produce the hundreds of millions of devices on one ULSI wafer. Because of the complexity associated with fabricating high-performance ICs, the semiconductor industry has always been at the leading edge of equipment design and manufacturing technology. This innovation has fostered continual improvements in wafer fabrication.

Wafer Test/Sort ■ At the completion of wafer fabrication, the wafer is then sent to the test/sort area where individual die are probed and electrically tested. Defective die are marked for sorting later into acceptable and unacceptable chips. Die that fail the wafer test will not be sent to customers, while die that pass the wafer test will continue through the process.

Assembly and Packaging ■ After wafer test/sort, the wafer enters assembly and packaging to enclose the individual chips in a protective package. The backside of the wafer undergoes a backgrind to reduce the thickness of the substrate. A thick membrane of acetate tape is attached to the backside of each wafer, then the front of the wafer is sliced along the scribe lines with a diamond-tip saw blade to separate the die from each wafer. The sticky acetate holds the silicon chips to keep them from falling off. At the assembly plant, the good die have wire bonds added or have bumps applied to form the attachment to the assembly package. Soon after, the die is hermetically sealed in a plastic or ceramic package. The actual form of the final package varies depending on the type of chip and its application (see Figure 1.8 on page 10). See Chapter 20 for more information on assembly and packaging.

Final Test ■ To ensure a chip's functionality, each packaged integrated circuit is tested to meet the manufacturer's electrical and environmental specifications. After the final test, the chips are delivered to the customer for assembly into the specific application, such as mounting memory components on a circuit board for a personal computer.

SEMICONDUCTOR TRENDS

The rapid technology changes needed to design and fabricate ICs with ULSI integration leads to the continual introduction of new equipment and processes. The semiconductor industry introduces new fabrication technology into the wafer fab every 18 to 24 months. Wafer fabrication technology changes are driven by the customers' needs. Customers require faster, more reliable, and lower cost chips. To achieve this, chip manufacturers have learned to reduce the size of components on a chip. This increases chip speed and reduces power consumption. Chips undergo extensive testing and analysis to verify long-term reliability. The number of chips on a wafer are increased to reduce cost.

There are three major trends associated with improvements in microchip technology:

- Increase in chip performance
- Increase in chip reliability
- Reduction in chip cost

As the different process areas of wafer fabrication are studied in this text, the introduction of new technologies will be discussed based on their contribution toward these trends.

Increase in Chip Performance

The performance of semiconductor microchips has dramatically improved since the SSI era of the early 1960s. A common way to judge chip performance is speed. *Chip speed* will improve if devices are made smaller and packed closer together on the chip because the electrical signal moving through the circuit has less distance to travel. Another way to increase speed is to use materials that improve the passage of the electrical signal through the device and circuitry on the chip surface. We will study these new materials in later chapters. Microprocessor chip performance is also judged by the number of instructions the chip can perform, as measured in million instructions per second (MIPS). A faster microchip that can perform more instructions per second is beneficial to the customer.

Critical Dimension (CD) ■ The physical dimension of a feature on a chip is referred to as the *feature size*. Another term to describe feature size is *circuit geometry*. Of special note is the minimum feature size on a wafer, known as the *critical dimension,* or *CD*. We will refer throughout the text to the CD as the criteria that defines the level of fabrication complexity (i.e., if you possess the capability to fabricate the CD on a wafer, then you can fabricate all other feature sizes because the dimension is larger and therefore easier to produce). For instance, if the smallest linewidth on a chip is 0.18 μm, then this dimension is the CD (see Figure 1.9). Device CDs have continually shrunk since the beginning of semiconductor fabrication, starting with a CD of about 125 μm in the early 1950s and currently at 0.18 μm and less.[9] The semiconductor industry uses the term *technology nodes* to describe the applicable CD in use during wafer fabrication. The actual and projected industry technology nodes for CDs from 1 μm and below are shown in Table 1.2.

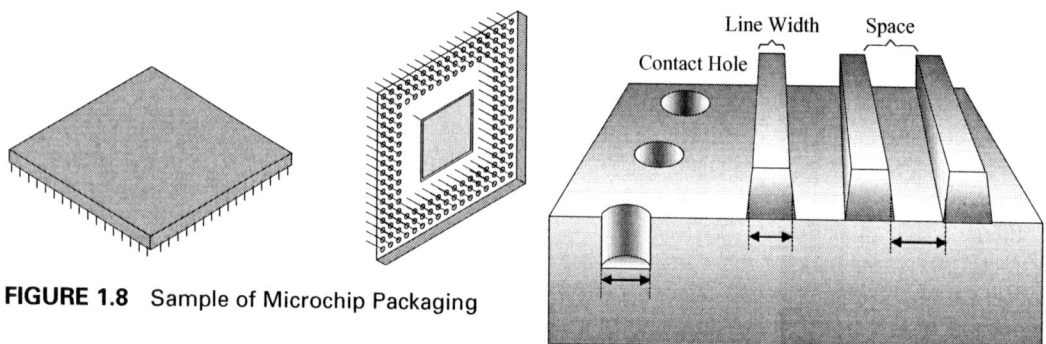

FIGURE 1.8 Sample of Microchip Packaging

FIGURE 1.9 Critical Dimension

TABLE 1.2 Past & Future Technology Nodes for Feature Critical Dimension (CD)

	1988	1992	1995	1997	1999	2001	2002	2005
CD (μm)	1.0	0.5	0.35	0.25	0.18	0.15	0.13	0.10

Adapted from *1999 Roadmap: Solutions and Caveats,* Solid State Technology, May, 2000, p. 77.

The coordinated shrinking of device dimensions on the chip is referred to as *scaling*. It is not acceptable to only reduce one feature on a chip. This would be like designing a large pickup truck with very small tires. Normally, a large vehicle for heavy loads requires large tires. A small car with a light load is acceptable with small tires. The same goes for chip design. All dimensions must be reduced simultaneously, or scaled, in order to optimize electrical performance. For advanced semiconductor manufacturing, device scaling occurs for dimensions in both the vertical and lateral directions.

Components Per Chip ■ Reducing the feature size on a chip permits more components to be fabricated on the silicon wafer. For microprocessors, the number of transistors on the chip surface can illustrate the increased chip integration permitted by the reduction in CD. Since the number of transistors on a chip has dramatically increased over the years of wafer fabrication, there has been an improvement in chip performance (see Figure 1.10).

Moore's Law. In 1964, Gordon Moore, one of the early pioneers in the semiconductor industry and a founder of Intel, predicted that the number of transistors on a chip would double roughly every year.[10] This prediction is known in the industry as *Moore's law* (it was later modified

Introduction to the Semiconductor Industry 11

Wafer Fab
(Photo courtesy of Advanced Micro Devices-Dresden Copyright by S. Doering)

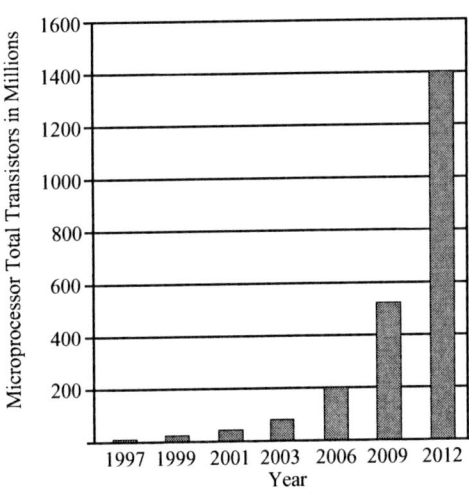

FIGURE 1.10 Increase in Total Transistors/Chip
Redrawn from Semiconductor Industry Association, *The National Technology Roadmap for Semiconductors*, 1997

in 1975 to predict doubling every 18 months). A key contributor to Moore's law is the ability to fabricate wafers with a reduction in the device feature size to achieve a new CD and an increase in the number of transistors on a chip with the introduction of each new product generation. Moore's law for transistors on microprocessors, shown in Figure 1.11 has been surprisingly accurate.

Size reduction in semiconductor devices as predicted by Moore's law is important because it leads to smaller microchip packaging. Smaller microchips fit into small volumes, which leads to smaller commercial products. Semiconductors are the enabling technology for the development of portable electronic products such as the palmtop computer and cellular telephones (see Figure 1.12 on page 12).

Power Consumption ■ Another important aspect of chip performance is the power consumption of the device during operation. Whereas vacuum tubes were power hungry, semiconductor devices consume substantially less power. As device miniaturization proceeds, there is a corresponding reduction in power consumption.[11] Although the number of transistors per chip is rapidly increasing, the chip power consumption grows at a much slower rate (see Figure 1.13 on p. 12). This has become a key performance parameter given the market growth for portable electronic products.

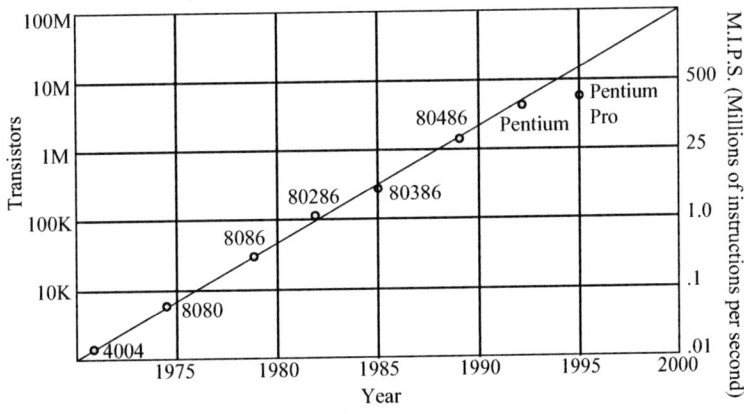

FIGURE 1.11 Moore's Law for Microprocessors
(Used with permission from *Proceedings of the IEEE*, January, 1998, © 1998 IEEE)

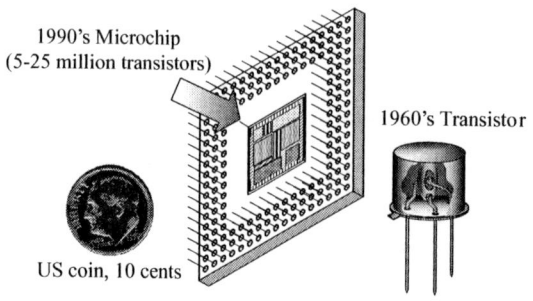

FIGURE 1.12 Size Comparison of Early and Modern Semiconductors

Increase in Chip Reliability

Chip reliability addresses the ability of the chip to function as intended over its expected life. Advances in technology have improved the product reliability of chips (see Figure 1.14). For example, contaminants are controlled through strict application of cleanroom procedures in areas such as particle-free air and control of chemical purity. Fabrication processes are constantly analyzed to improve device reliability. Wafers are monitored and microchips tested to verify acceptable performance. This translates into products that function with low failures during operation.

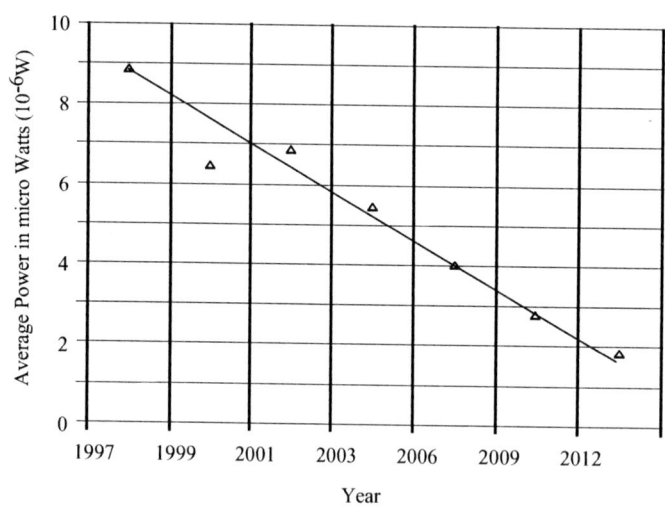

FIGURE 1.13 Reduction in Chip Power Consumption per IC
Redrawn from Semiconductor Industry Association, *The National Technology Roadmap for Semiconductors*, 1997

Reduction in Chip Price

The price of semiconductor microchips has steadily decreased (see Figure 1.15). During the nearly 50 years prior to 1996, the price of semiconductor microchips has decreased by a phenomenal 100 million times.[12] For example, in 1958 a single silicon transistor that was of poor quality cost around $10. Today, $10 might buy a memory chip with over 20 million transistors, an equal number of other components, and the necessary interconnections to make a useful chip.

There are several reasons for the decreasing prices for semiconductor chips. As previously explained, factors such as reducing the feature size and increasing the wafer diameter to put more chips on a wafer are contributors to price reduction. The cost benefits from this decrease in CD are substantial. For example, if a semiconductor manufacturer decreased the CD on an 8-inch wafer from 0.35 μm to 0.25 μm in 1997, they were able to increase the number of chips per wafer from 150 to 275. In other words, they produced almost twice as many chips per wafer at nearly the same manufacturing cost. This cost benefit was achieved because the reduced CD permitted millions of circuit lines to be compressed into a smaller area.

Another reason for price reductions has been the large market growth for semiconductor products. This growth requires increased volume from chip fabrication companies, which introduces economies of scale for manufacturing. The introduction of manufacturing improvements to

Introduction to the Semiconductor Industry **13**

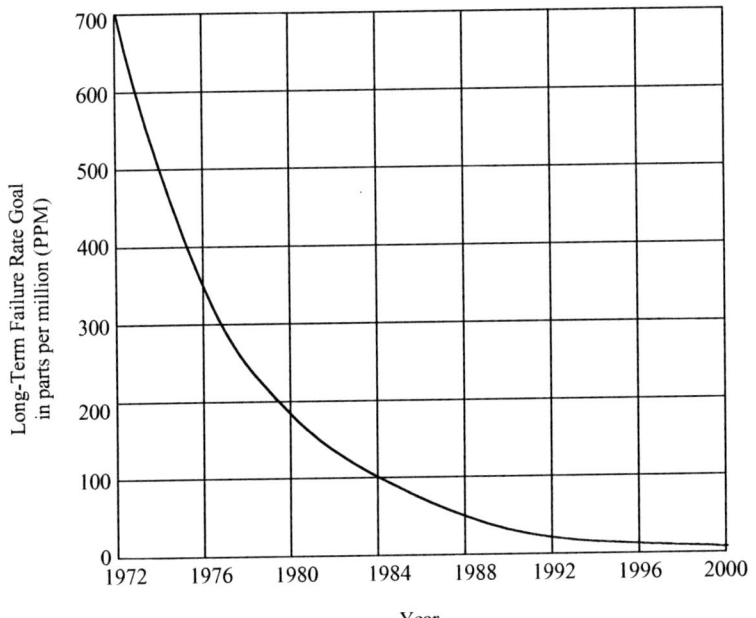

FIGURE 1.14 Reliability Improvement of Chips

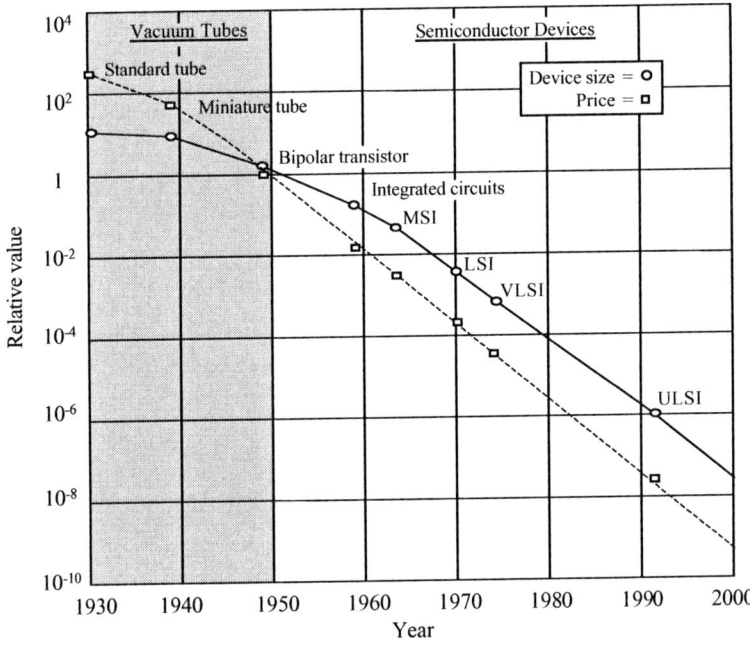

FIGURE 1.15 Price Decrease of Semiconductor Chips Redrawn from C. Chang and S. Sze, *ULSI Technology*, (New York: McGraw-Hill, 1996), p. xxiii.

the equipment and process used to fabricate microchips has also reduced costs. Work teams have an important role in improving manufacturing technology and cost. They work firsthand with the chip fabrication equipment and contribute to improving manufacturing productivity. The efforts of manufacturing teams make a direct difference in lowering the cost to manufacture chips.

THE ELECTRONIC ERA

The semiconductor industry has undergone intensive growth since the invention of the first transistor at Bell Labs in 1947. It is beneficial to review the major industry changes since the beginning of solid state electronics to gain insight from the foundations of modern wafer fabrication.

The 1950s: Transistor Technology

The 1950s were a rich era in the development of different types of semiconductor transistor technology. Many different types of transistors were developed at Bell Labs and the early semiconductor efforts in northern California and Texas. The first transistors were developed from germanium semiconductor material. In 1954 the first manufacturable silicon transistor was made by Gordon Teal at Texas Instruments.[13]

With the invention of the IC in 1959, the industry was poised to solidify transistor technology as the replacement for vacuum tubes and to develop new market applications. The basic elements needed to commercialize the transistor were in place. The semiconductor industry was growing in high technology centers such as Silicon Valley in northern California; Route 128 near Boston, Massachusetts; and Texas Instruments in Dallas, Texas. This concentration of ideas and technology was important for the development of semiconductor technology and the new manufacturing techniques necessary to move wafer fabrication out of the laboratory processes of the 1950s into production processes of the 1960s. The growing dominance of semiconductors and silicon in all aspects of society gave credence to the expression, "the silicon age."

The 1960s: Process Technology

In the 1960s, the semiconductor industry entered an era oriented toward solving the basic problems of producing semiconductor ICs. This was the beginning of integration and the SSI era. Semiconductor manufacturers were proliferating. The market for semiconductor ICs was growing rapidly, with sales exceeding $1 billion in 1961. The subsequent growth of the many chip manufacturers served to increase IC performance and reduce cost. Because of process commonality across semiconductor companies, special supplier industries developed to provide the chemicals and equipment needed for wafer fabrication.

There was a proliferation of engineers wanting to exploit the new semiconductor technology, which led to the formation of many new high-tech companies in the 1960s. Signetics was founded in 1961 by a group of Fairchild engineers to commercialize ICs. Robert Noyce, Gordon Moore, and Andrew Grove left Fairchild to found Intel in 1968. Jerry Sanders and other scientists, also from Fairchild, founded Advanced Micro Devices in 1969. There were also some large captive supplier companies in the IC business, such as IBM and Digital Equipment Corporation, and midsize companies such as Hewlett-Packard. The exploding demand for semiconductor products helped the many small start-up companies overcome the hurdles of constant price reductions, rapid company growth, and financial challenges of starting new companies.

The 1970s: Competition

The early 1970s was the period of medium-scale integration (MSI) for chip design. The manufacturing processes were largely manual operations based on batch processing. A typical wafer fab was fortunate to start a new product with a yield of 5 or 10% (yield is the percent of acceptable die produced at the completion of the process) and after an intense effort possibly raise it to 30%.[14]

Microprocessors were invented separately in the early 1970s at both Texas Instruments and Intel, and with its wide acceptance in market applications came the need for more chip integration. LSI-level integration remained only a few years and then was quickly replaced by VLSI, the integration standard by the end of the 1970s.

The rapidly changing semiconductor industry became chaotic. Much of the equipment and many of the processes used in the wafer fabs were developed by the same companies that manufactured the semiconductor devices, with few industrywide standards. This created inefficiencies for manufacturers and suppliers. Asia emerged as a formidable competitor, with Japan as a powerhouse of semiconductor innovation and manufacturing. Asian electronic giants challenged

American dominance in semiconductors. By 1979, Japan had captured more than 40% of the world's demand for memory microchips.[15]

As the demand for more complex chips grew, equipment technology changed from the manual tools of the 1960s to semiautomatic operation with buttons for operator control and onboard solid-state controllers. Equipment suppliers also encountered stiff competition from Japan and Asia for key semiconductor tools.

The cost of constructing a wafer fab became extremely expensive. Control of wafer contamination levels became critical for shrinking device feature sizes, requiring special wafer fab cleanliness standards that exceeded nearly every other industry. Special purity requirements were needed for water, air, and the many chemicals and gases used during fabrication. By the end of the 1970s, the cost of a wafer fab was around $30 million and rising. Market demand and the need to continually replace equipment to maintain the advancing technology encouraged the construction of new fabs.

The industry attempted to form collaborative ventures to address the disorder. In 1970, SEMI (Semiconductor Equipment and Materials Institute) was founded to standardize and promote equipment, materials, and services in the industry. The Semiconductor Industry Association (SIA) was started in 1977, with leadership by Robert Noyce. The goal was to have more industry coordination about mutual problems created by rapid growth.

The 1980s: Automation

By the 1980s, the widespread acceptance of solid-state electronics in society was established, driving the growth of chip fabrication in the semiconductor industry. This was the VLSI era, with a high degree of component integration on a chip. The growth of the personal computer industry fueled hardware and software demand. At the same time, the industry faced competitive pressures by Japanese IC manufacturers that expanded their manufacturing capacity while at the same time driving yield and quality to unprecedented levels. U.S. semiconductor companies were clearly shaken by these developments and reached near panic because of their inability to compete. By the mid-1980s, Japan almost totally dominated the fast growing and technologically demanding DRAM (dynamic random access memory) market segment.[16]

The semiconductor industry formed SEMATECH in 1987, under the guidance of the Department of Defense, with Robert Noyce as its first Chief Executive Officer. The goal was to develop industrywide policies with respect to the specification and evaluation of manufacturing equipment. These organizations were formed, in part, because Japanese competition threatened the very existence of U.S. semiconductor manufacturers.[17] Almost simultaneously SEMI/SEMATECH was formed of U.S. equipment and materials suppliers in support of SEMATECH's mission. The names of the consortiums would be changed in 1999 to International SEMATECH and the Semiconductor Industry Suppliers Association, respectively.

Improvements in semiconductor equipment, manufacturing efficiency, and product quality were emphasized in U.S. companies. Fabrication tools were automated to include all significant wafer processing steps. The intent was to largely remove the operator from the process, especially since the human was a major source of contamination in the cleanroom.

The industry was able to maintain the predicted growth of Moore's law by means of ongoing chip design changes and reduction in chip feature sizes. These design changes presented manufacturing challenges that led to the development of sophisticated processes. The equipment and cleanroom controls necessary to fabricate the new ICs escalated the cost of building wafer fabs, which rose to nearly $1 billion by the late 1980s (see Figure 1.16 on page 16). To overcome this high cost of entry into new technology with new fabs, many U.S. semiconductor companies entered into joint venture agreements with Japanese and European chip manufacturers to share the cost of constructing wafer fabs.

The 1990s: Volume Production

Feature size dimensions shrunk below 1 μm for production chips during the late 1980s and early 1990s. By the end of the 1990s, the minimum feature size was at 0.18 μm. Microchip design experts predict that at some point there will be a physical limitation to further reductions in feature size. At this time, there is no barrier anticipated until the CD goes well below 0.1 μm. CDs

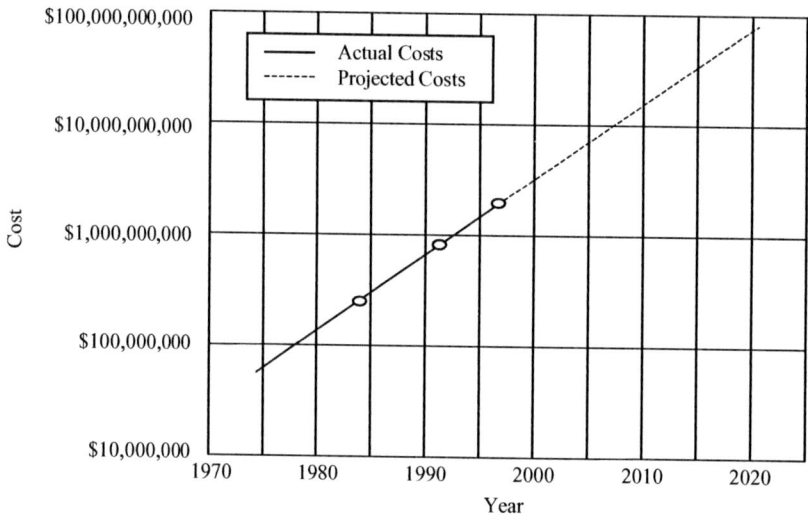

FIGURE 1.16 Start-Up Cost of Wafer Fabs (Used with permission from *Proceedings of IEEE*, January, 1998, © 1998 IEEE)

continue to shrink because of the cost pressures for manufacturing efficiency resulting from putting more chips on the same wafer.

Submicron geometry processing ushered in the ULSI era of integration, with high-performance ICs containing 10 million transistors or more. The two types of chips with the most advanced technology are microprocessors and memory chips. Highly integrated chips require multiple layers of circuitry to interconnect the devices (up to eight layers and increasing) and up to 450 or more process steps to fabricate the chips.

The semiconductor industry has become increasingly competitive in the 1990s. To survive in the worldwide chip market, it is critical for manufacturers to produce complex, high quality chips in record time. Delivery of the right microchip to the market has a small window of opportunity based on customers and their need for advanced technology. Semiconductor manufacturers try to be the first (or nearly the first) with the new technology. If a company misses this product window, then it risks spending substantial money to develop the chip technology without realizing any sales potential.

Semiconductor equipment is highly automated. Advanced material handling systems move wafers between workstations with no required human intervention. Expert software systems control nearly all equipment functions, including troubleshooting diagnostics. Technicians and engineers intervene to download product recipes into the equipment software database and interpret software diagnostic commands to take corrective action for equipment repair.

CAREERS IN SEMICONDUCTOR MANUFACTURING

The career paths in semiconductor manufacturing are grouped into three main areas: technician, engineering, and management. The choice of a particular career path generally depends on an individual's technical knowledge, education, and personal goals. Examples of different career paths are shown in Figure 1.17.

Technician

The technician has the technical skills to operate, troubleshoot and maintain the advanced equipment used to fabricate microchips in a production environment. Equipment operation requires knowledge to set up and operate the automated equipment so that it functions properly in the manufacturing process. Most equipment operation is done through software interfaces to the computer-controlled equipment. Troubleshooting equipment problems requires expertise in equipment systems (e.g., electronics, mechanical systems, motors, etc.) with the ability to interpret software diagnostic routines. Equipment maintenance involves extensive training and knowledge of electrical, mechanical, chemical, and computer equipment systems. An equipment technician often performs the troubleshooting and maintenance of wafer fab equipment.

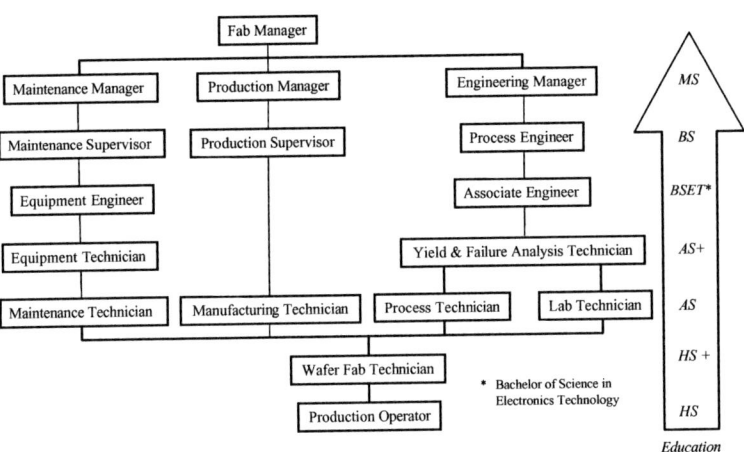

FIGURE 1.17
Career Paths in the Semiconductor Industry

Technicians must be capable of performing their technical role as a manufacturing team member, while contributing to the productivity objectives of the wafer fab (see Figure 1.18). Technicians are knowledgeable about safety procedures for chemicals and equipment. As equipment automation continues toward more integrated equipment across different operations, the skills of the technician in the wafer fabs must also advance.

Technicians in a wafer fab are referred to by different names at the various chip manufacturers, such as a wafer fab technician (WFT), self sustaining technician (SST), manufacturing association (MA), or manufacturing technician (MT). Technicians are also used in the semiconductor industry, such as for field service representatives, to install and service sophisticated wafer fab equipment. The skills required of technicians include computer usage, knowledge of equipment procedures and support, analytical skills to solve problems for process improvement, and critical thinking skills for decision-making. Due to the advanced sophistication of wafer fab equipment, it is clear moving wafers and operating equipment are not the only things people do in a wafer fab. Furthermore, humans create too many problems with respect to wafer contamination, manufacturing repeatability, and inefficiency.

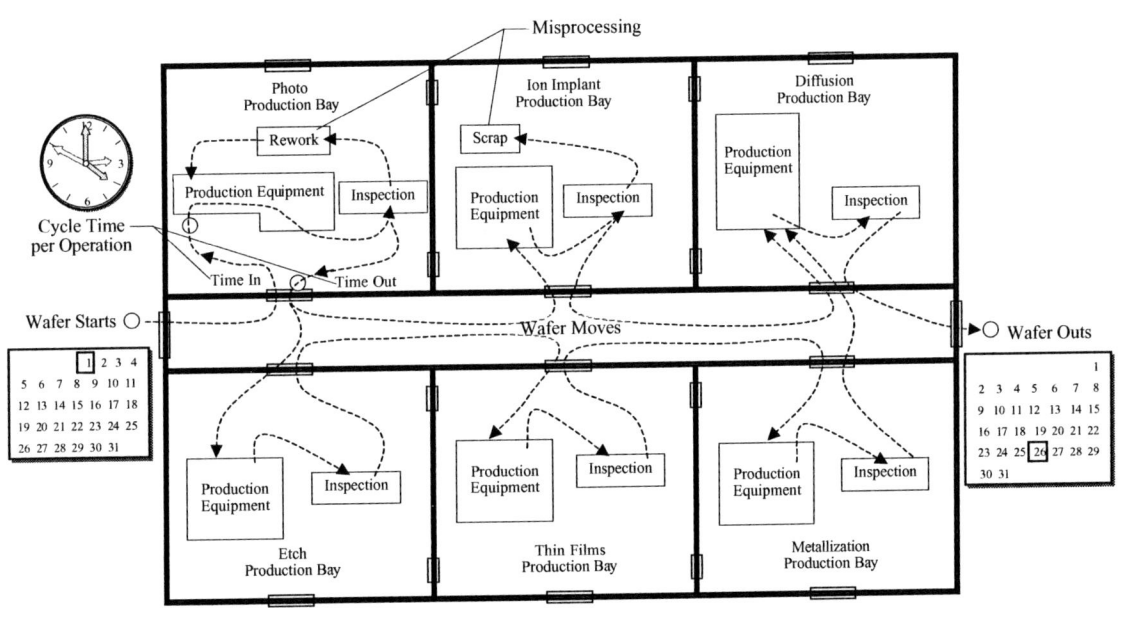

FIGURE 1.18 Productivity Measurements in a Wafer Fab

A. Equipment Technician in a Wafer Fab

B. Technician in a Wafer Fab
(Photos courtesy of Advanced Micro Devices)

Job Descriptions

Within each of the three major career paths, there are various options available for different jobs. A description of some different technician, engineer and management positions in the wafer fab is given.

- ◆ Wafer Fab Technician: The wafer fab technician is responsible for operating the wafer fab equipment. The wafer fab technician performs some equipment maintenance and basic troubleshooting of process and equipment. Semiconductor companies often prefer high school or some college (e.g., a certificate). In some cases, a two-year associate of applied science degree may be desirable.

- ◆ Equipment Technician: The equipment technician troubleshoots and maintains advanced equipment systems so that the equipment performs correctly during wafer manufacturing. This position typically requires at least a two-year associate of applied science degree and extensive hands-on training.

- ◆ Equipment Engineer: The equipment engineer specifies the equipment design parameters and optimizes equipment performance for wafer production. This career typically requires at least a four-year engineering degree.

- ◆ Process Technician: The process technician supports the production equipment and process engineering teams by troubleshooting process-related problems. A two-year associate of applied science degree or four-year engineering technology degree is helpful; however, good writing, math, science, and analytical thinking skills are necessary.

- ◆ Process Engineer: The process engineer analyzes the performance of the fabrication process and equipment to determine optimum parameter settings. This position typically requires a four-year engineering or science degree.

- ◆ Field Service Representative: A field service representative in a wafer fab installs the fabrication equipment. The field service representative also performs equipment maintenance, diagnostics, and repairs to ensure the equipment is production-worthy. This position requires at least a two-year associate degree or four-year engineering technology degree.

- ◆ Lab Technician: Lab technicians can work in development laboratories setting up and running experiments. This job skill usually requires some experience and people who are self-motivated.

- ◆ Yield/Failure Analysis Technician: These technicians perform the tasks associated with defect analysis, such as preparing the materials to be analyzed and operating analytical equipment to determine the source of problems that arise during wafer fabrication.
- ◆ Yield Improvement Engineer: The yield improvement engineer collects and analyzes yield and test data to improve wafer fabrication performance. This position often requires a four-year science or engineering degree.
- ◆ Facilities Technician: Facilities technicians support the facility equipment and utilities required in a wafer fab, including chemical management and cleanroom utilities.
- ◆ Facilities Engineer: This engineer provides engineering design support for the wafer fab utility infrastructure for chemicals, clean air and equipment that are used in the facility. The position typically requires a four-year engineering degree.
- ◆ Supervisor/Manager: Supervisory and management positions in a wafer fab combine technical skills with people management skills to achieve the organizational goals of the company. These positions often will require either a four-year science, engineering, or management degree or an additional two years of college after the associate degree.

SUMMARY

The semiconductor industry is a major part of the high tech industry in the U.S. The invention of the solid-state transistor launched the semiconductor industry in 1947, followed by the introduction of silicon and the development of the integrated circuit (IC). The IC combines multiple components on one chip to increase chip performance and lower cost. There are IC integration eras, with the most recent era being ULSI (ultra large scale integration). IC fabrication occurs on a wafer, with the devices on the top of the silicon and layers of circuitry deposited on the substrate. There are five stages of microchip fabrication: wafer preparation, wafer fabrication, wafer test/sort, assembly and packaging and final test. The three major trends for improvement in microchip technology are: increase in performance, increase in reliability, and reduction in cost. Chip speed is important for performance and is improved by reducing the critical dimension (CD), or minimum feature size, on a chip. Moore's law predicts that the number of components on a chip will double every 18-24 months, which has remained relatively true since the 1960s. The reliability of chips has improved while the price of chips has substantially decreased since 1947. The semiconductor industry has become increasingly competitive and the wafer fab technician plays an important role in a manufacturer's ability to fabricate microchips.

KEY TERMS

semiconductor
wafers
wafer fab
microchips (chips)
solid-state transistor
planar technology
integrated circuit (IC)
devices
wafer
die
substrate
wafer fab
fab
silicon
wafer fabrication

microchip fabrication
chip fabrication
chip fab
merchant chip suppliers
captive chip producers
fabless company
wafer sort
packaging
chip speed
feature size
circuit geometry
critical dimension (CD)
technology node
scaling
Moore's law
wafer fab technician (WFT)

REVIEW QUESTIONS

1. How big is the high-tech industry relative to the U.S. economy? What has been a major factor in helping the semiconductor industry become so large?
2. List four different industries from the first half of the twentieth century that contributed to the development of the semiconductor industry.
3. When, where, and by whom was the solid-state transistor invented?
4. What is an integrated circuit (IC)? When was it invented, and by whom?
5. List the five integration eras, provide the time period for each era, and give the number of components on a chip in each era.
6. What is a wafer? What is a substrate? What is a die?
7. Give a short description of a wafer fab.
8. List the five major stages of IC fabrication, and give a short description of each stage.
9. Describe merchant chip supplier, captive chip producer, fabless company, and foundry.
10. List the three major trends associated with improvement in microchip fabrication technology, and give a short description of each trend.
11. What is the chip critical dimension (CD)? Why is this dimension important?
12. Describe scaling and its importance in chip design.
13. What is Moore's law and what does it predict? Has this law been basically true?
14. By what factor have chip prices reduced since 1947? Give two reasons for this change.
15. Provide a short description of how the electronic era developed from 1950 through 2000.
16. Describe the responsibilities of a wafer fab technician and an equipment technician.
17. List and give a short description of eight different job descriptions in the semiconductor industry.

SELECTED INDUSTRY WEB SITES

Advanced Micro Dvices — http://www.amd.com
AT&T Tech. History — http://akpublic.research.att.com/
Semiconductor International Magazine — htt;p://www.semiconductor.net/
Fairchild Semiconductor — http://www.fairchildsemi.com/
IBM Microelectronics — http://www.chips.ibm.com/
Intel — http://www.intel.com/
International SEMATECH — http://www.sematech.org/public/index.htm
MATEC, Maricopa Advanced Technology Education Center — http://matec.org/
Motorola Semiconductor — http://mot-sps.com/
Mitsubishi — http://www.mmc-sil.com/
National Semiconductor — http://www.national.com/
NIST, National Institute of Standards and Technology — http://www.nist.gov/
Selete, Semiconductor Leading Edge Technologies, Inc. — http://ww.selete.co.jp
SEMI, Semiconductor Equipment and Materials International — http://www.semi.org/
Semiconductor Search Engine — http://www.semiseek.com/
SIA, Semiconductor Industry Association — http://www.semichips.org/
SISA, Semiconductor Industry Suppliers Association — http://www.sisa.org/
Solid State Technology Magazine — http://sst.pennwellnet.com/home/home.cfm
Texas Instruments — http://www.ti.com/

REFERENCES

1. *Chip Scale Review,* Volume 3, Number 4 (July-August 1999): p. 25.
2. Historical information generally taken from F. Seitz and N. Einspruch, *Electronic Genie: The Tangled History of Silicon* (Urbana: University of Illinois Press, 1998), p. 52.
3. F. Seitz and N. Einspruch, *Electronic Genie,* p. 52.
4. P. Bondyopadhyay, "In the Beginning," *Proceedings of the IEEE* 86, no. 1 (January 1998): p. 63.
5. J. Pierce, "The Naming of the Transistor," *Proceedings of the IEEE* 86, no. 1 (January 1998): p. 37.
6. Texas Instruments, *The Chip That Jack Built Changed the World,* Texas Instruments, (September 1997).
7. G. Moore, "The Role of Fairchild in Silicon Technology in the Early Days of Silicon Valley,'" *Proceedings of the IEEE* 86, no. 1 (January 1998): p. 59.
8. Ibid., p. 54.
9. Ibid., p. 56.
10. G. Moore, "Cramming More Components onto Integrated Circuits," *Proceedings of the IEEE* 86, no. 1 (January 1998): p. 84. Reprinted from *Electronics,* April 19, 1965.
11. C. Chang and S. Sze, *ULSI Technology* (New York: McGraw-Hill, 1996), p. xxi.
12. Ibid., p. xix
13. I. Ross, "The Invention of the Transistor," *Proceedings of the IEEE* 86, no. 1 (January 1998): p. 17.
14. F. Seitz and N. Einspruch, *Electronic Genie,* p. 94.
15. C. Melliar-Smith, M. Borrus, D. Haggan, T. Lowrey, A. Vincentelli, W. Troutman, "The Transistor: An Invention Becomes a Big Business," *Proceedings of the IEEE* 86, no. 1 (January 1998): p. 93.
16. Ibid., p. 94.
17. F. Seitz and N. Einspruch, *Electronic Genie,* p. 238.

CHAPTER 2
CHARACTERISTICS OF SEMICONDUCTOR MATERIALS

Understanding the structure and bonding of atoms provides the key to knowing how silicon performs its vital role as a semiconductor. Once unlocked, this knowledge provides the foundation for understanding how the simplicity of a semiconductor device can be a part of the complex world of the microchip.

OBJECTIVES

After studying the material in this chapter, you will be able to:

1. Describe the atom, including the valence shell, band theory, and ions.
2. Interpret the periodic table of the elements with regards to main-group elements, and explain how ionic and covalent bonds are formed.
3. State the three classes of materials, and describe each one with regards to current flow.
4. Explain resistivity, resistance, and capacitance, and discuss their importance to wafer fabrication.
5. Describe pure silicon and give four reasons why it is the most common semiconductor material.
6. Explain doping and discuss how the trivalent and pentavalent dopant elements make silicon a useful semiconductor material.
7. Discuss the difference between p-type (acceptor) silicon and n-type (donor) silicon, describe how silicon resistivity changes with the addition of a dopant, and explain the pn junction.
8. Discuss alternative semiconductor materials, with emphasis on gallium arsenide.

INTRODUCTION

The most important semiconductor material in use today is silicon. Devices are fabricated in silicon and function as a semiconductor because of the unique properties of this material. In this chapter, we will first review some physical and chemical concepts associated with the individual atom. We then will use this information to study properties of silicon material to learn its unique physical and chemical structure and how it permits a semiconductor to function. This chapter will establish basic material and semiconductor concepts that will be used throughout the different process steps in wafer fabrication.

ATOMIC STRUCTURE

Matter is everything that has shape, size, and occupies space in the universe. If we take a material, such as a grain of sand, and continually divide it into smaller and smaller particles, we eventually arrive at the smallest particle, an atom. In the atomic model, atoms consist of three types of particles: the neutral neutrons, the positively charged protons that together make up the nucleus of the atom, and the negatively charged electrons that orbit the nucleus (see Figure 2.1). The number of protons in the atom equals the number of electrons, which makes the atom electrically neutral.

An *element* is a substance formed from one type of atom and is the simplest substance with unique physical and chemical properties. A *molecule* consists of two or more atoms in a structure that are chemically bound together, behaving as an independent unit. A *compound* is a substance composed of two or more atoms (similar or dissimilar) that are chemically bound together and have formed a new substance which has different properties from its individual component atoms.

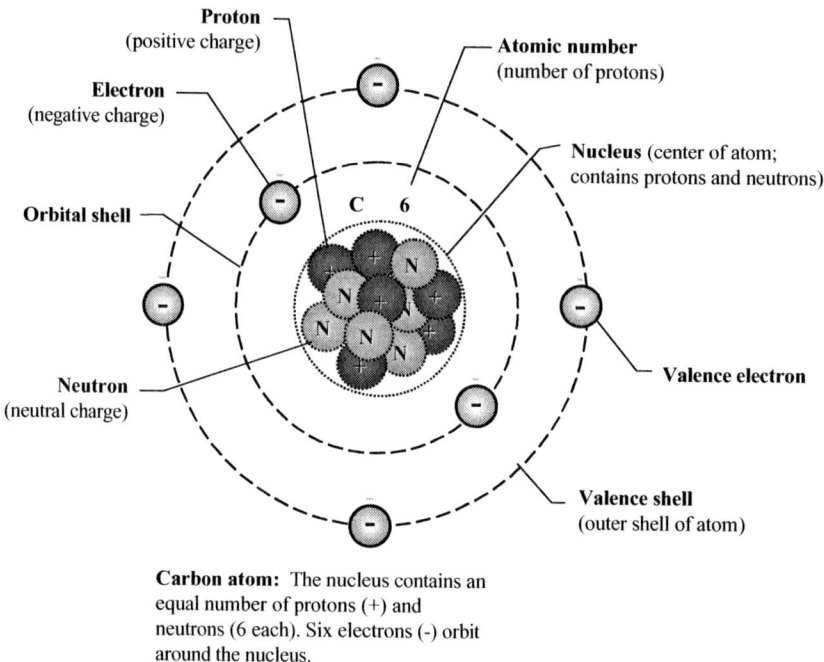

Carbon atom: The nucleus contains an equal number of protons (+) and neutrons (6 each). Six electrons (-) orbit around the nucleus.

FIGURE 2.1 Elementary Model of the Carbon Atom

Electrons

The basic law of attraction and repulsion that occurs at the microscopic level states that like charges repel and unlike charges attract. Orbiting negative electrons are distributed in a space cloud around the nucleus and held in place by their attraction toward the positive nucleus. Individual electrons occupy orbitals within seven distinct shells. In a hydrogen atom, for example, each shell corresponds to a specific energy level and is identified using the letters K through Q (see Figure 2.2). For all other atoms, referred to as many-electron atoms, the orbital energies within each shell are not the same.

Electron Energy ■ The unit of energy at the atomic level is the electron volt (eV). It represents the kinetic energy (energy associated with motion) gained by an electron in passing from a point of low potential to a point 1 volt higher in potential. One electron volt is equal to 1.6×10^{-19} joules of energy. The electron volt is used to describe electron energy in different processes of semiconductor manufacturing (e.g., ion implant energy levels, as described in Chapter 17).

Valence Shells ■ The outermost electron shell for a given atom is the *valence shell*. Valence electrons are found in the valence shell and have considerable influence on the chemical and physical properties of the atom and exist at the highest energy level for a given atom. The number of

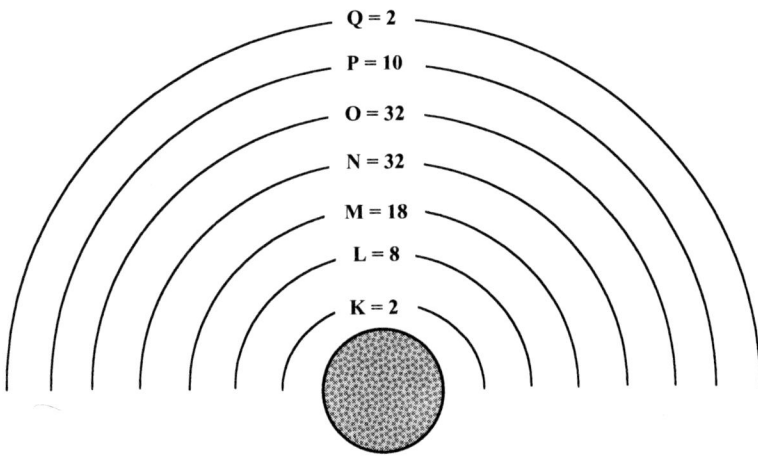

FIGURE 2.2 Electron Shells in Atoms

electrons in a valence shell varies depending on the shell level, K to Q. Atoms with one electron in a valence shell that permits eight electrons easily give up this electron. On the other hand, atoms with seven electrons in a valence shell have an affinity for electrons and readily accept one electron to fill the valence shell. The electrons and valence shells for sodium and chlorine are shown in Figure 2.3.

Energy-Band Theory of Solids ■ Energy-band theory explains how electrons change orbital levels in a solid material. The valence electrons are in the valence band. There is a band gap between the valence band and the conduction band (see Figure 2.4 on page 24). The band gap has very high energy levels in some materials that creates a forbidden gap (usually > 2 eV). These materials are referred to as *insulators* because it is hard to move an electron from the valence band to the conduction band. Other materials, called *conductors,* have a valence band that overlaps the conduction band, requiring very little energy for electrons to move into the conduction band. A third class of materials has a band-gap energy level found somewhere between insulators and conductors. These materials are referred to as *semiconductors*. The forbidden gap of semiconductors is moderate. The band-gap energy of silicon is 1.11 eV.

Ions ■ Ions are formed when an atom gains or loses one or more electrons. An atom becomes positively charged if it loses an electron and negatively charged if it gains an electron. Ions with opposite charges are mutually attracted and can form a chemical bond as an ionic compound.

An example of an ionic compound is common table salt, sodium chloride (see Figure 2.5 on page 24). A sodium atom, which is a metal, is initially neutral because it has the same number of

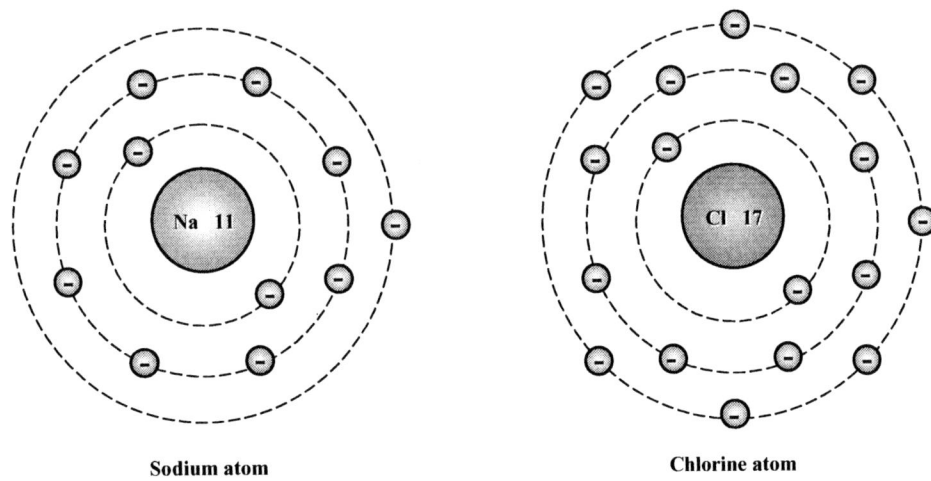

FIGURE 2.3 Electron Shells for Sodium and Chlorine Atoms

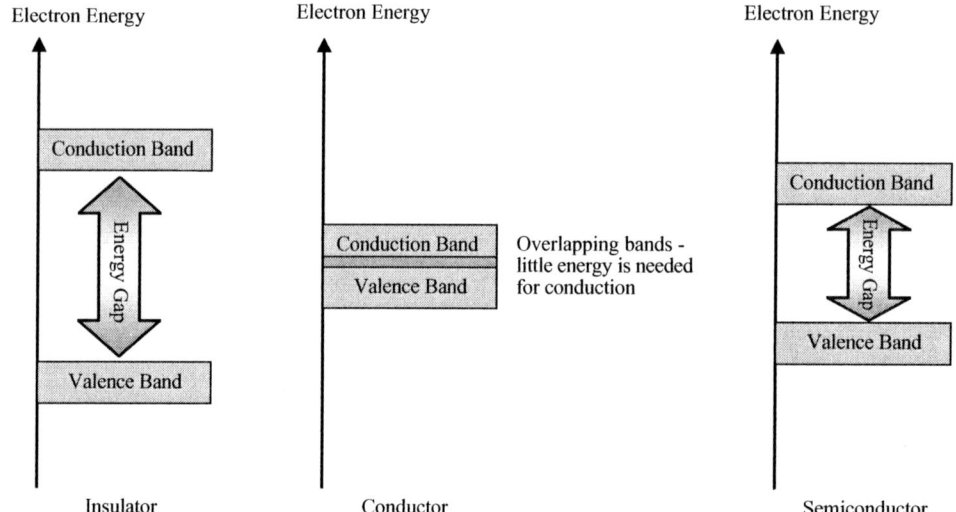

FIGURE 2.4 Energy Band Gaps

protons as electrons. If it loses one electron, it becomes positively charged as Na$^+$ and is referred to as a *cation*.[1] A chlorine atom, which is a nonmetal, gains an electron and becomes negatively charged as Cl$^-$, referred to as the *anion*. The Na$^+$ and Cl$^-$ are attracted to one another due to opposite charges and form an ionic compound.

Many gases exist as molecules. *Ionization* is the removal of electrons from atoms, creating positively charged atoms or molecules. Ionization is used in many process areas in semiconductor fabrication. Once gas particles are charged by ionization, the gas flow and motion of atoms can be controlled by the use of electrostatic and magnetic fields.

THE PERIODIC TABLE

The periodic table lists all elements known at this time (see Figure 2.6). It is organized in such a manner because of certain periodic patterns of behavior among the elements.

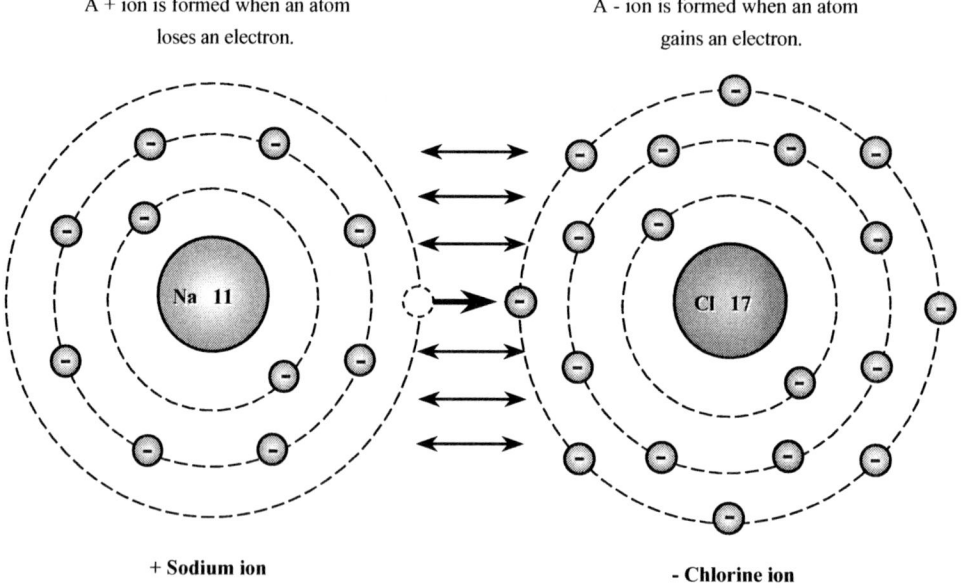

FIGURE 2.5 Sodium Chloride

FIGURE 2.6 The Periodic Table of the Elements

The periodic table has a box for each element (see Figure 2.7). Each element box on the periodic table provides information about the element, including the following:

Atomic symbol: The symbol for each element (e.g., *Si* is silicon).

Atomic number: Equals the number of protons in the nucleus. All atoms of a particular element have the same atomic number (and therefore the same number of electrons).

Atomic mass number: The sum of the protons and neutrons in an atom. Isotopes of the same element have the same number of protons but a different number of neutrons; therefore, isotopes have a different mass number.

Atomic weight: Also referred to as atomic mass. This figure is the average of the masses of the naturally occurring isotopes of an element and is weighted according to the abundance of these isotopes. One *atomic mass unit* (*amu*) is a relative measure of mass and is exactly equal to $1/12$ the mass of a carbon-12 atom.

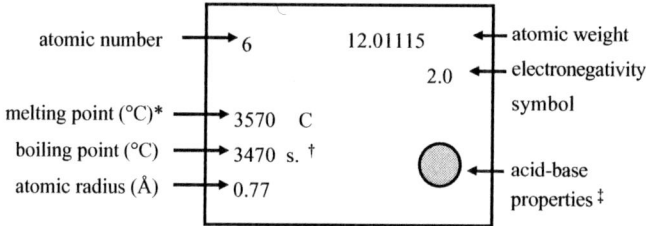

* Based on carbon-12. ( ) indicates the most stable or best-known isotope.

† s. indicates sublimation

‡ For representative oxides of group. Oxide is acidic if color is red, basic if color is blue, and amphoteric if both colors are shown. Intensity of color indicates relative strength.

FIGURE 2.7 Element Box of the Periodic Table

We will be concerned primarily with the main-group elements of the periodic table found in the columns with group numbers from IA through VIIIA. In the periodic table, the group number is given for each column and represents the number of valence-shell electrons. For instance, Boron (B) is in Group IIIA (found in column IIIA) and therefore has three valence electrons. The group number characteristics of elements commonly used in the semiconductor industry are outlined in Table 2.1.

Ionic Bonds

An ionic bond is formed when valence-shell electrons are transferred from the atoms of one element to another. Unstable atoms easily form ionic bonds. An example is sodium chloride. Sodium (Na) is located in Group IA, indicating there is one electron in its valence shell. Na is an unstable element that is highly corrosive. It is carefully controlled in wafer fabrication to avoid device contamination (see Chapter 6). Chlorine (Cl) is found in Group VIIA and has seven valence-shell electrons. This atom lacks one electron for a fully occupied valence shell; therefore, it also is unstable. Because of their instability, these two elements (Na and Cl) have an affinity for each other. Na readily gives up its valence-shell electron to Cl, forming an ionic bond (see Figure 2.8).

In the formation of ionic bonds, the transfer of electrons is such a common process that we define two terms that apply specifically to this type of bond:

Oxidation: loss of electrons

Reduction: gain of electrons

Thus, in the formation of NaCl, the neutral Na atom is oxidized into a positive Na^+ ion when it loses an electron. A neutral Cl atom is reduced into a negative Cl^- ion by gaining an electron.

TABLE 2.1 Group Number Characteristics of Commonly Used Chemical Elements in Wafer Fabrication

Group Number	Characteristics
IA	• 1 valence electron that is easily given up; low electronegativity • Highly unstable • Very reactive; explosive • Form ionic bonds • Prefer not to use the metals in this group due to contamination issues
IIA	• 2 valence electrons • Somewhat unstable • Quite reactive • Prefer not to use metals in this group
IIIA	• 3 valence electrons • Dopant elements (primarily B) added to semiconductor material • Common interconnect conductor material (Al)
IVA	• 4 valence electrons • Semiconductor materials • Form covalent bonds
VA	• 5 valence electrons • Dopant elements (primarily P and As) added to semiconductor material
VIA	• 6 valence electrons
VIIA	• 7 valence electrons; readily accept electrons; high electronegativity • Corrosive • Very reactive • Form ionic bonds • Useful in some semiconductor applications; used as etching and cleaning compounds
VIIIA	• 8 valence electrons • Stable; nonreactive • Inert gas • Safe to use in some aspects of semiconductor manufacturing
IB	• Best metal conductors • Cu is replacing Al as primary interconnect conductor material
IVB – VIB	• Refractory (high melting temperature) metals commonly used in semiconductor manufacturing to improve metallization (especially Ti, W, Mo, Ta, and Cr) • React well with silicon to form stable compound with good electrical characteristics

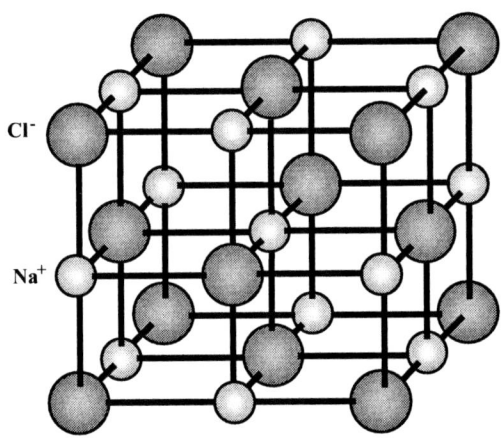

FIGURE 2.8 Ionic Bond for NaCl

Covalent Bonds

Another form of chemical bond is a covalent bond, where atoms of different elements share valence-shell electrons. The atoms share these valence electrons in order to attain a full valence shell and become stable. An example of a covalent bond is hydrogen chloride (HCl), where molecules consist of one hydrogen (H) atom forming a covalent bond with one chlorine (Cl) atom to form HCl (see Figure 2.9).

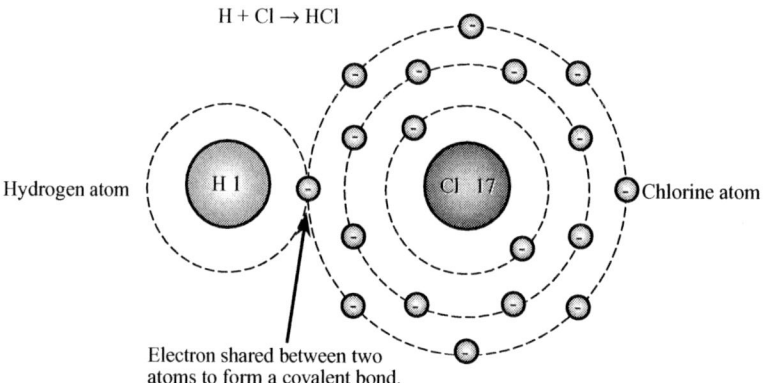

FIGURE 2.9 Covalent Bond for HCl

CLASSIFYING MATERIALS

Electrical current is the movement of charge carriers (electrons in metals, electrons or holes in semiconductors) from one point to another in a material under the influence of applied electric fields.[2] The unit for current is amperes, or amps (see Figure 2.10). A major way of classifying electronic materials is based on how current flows through the material.

There are three different classes of materials based on the flow of electric current through these materials:

- Conductors
- Insulators
- Semiconductors

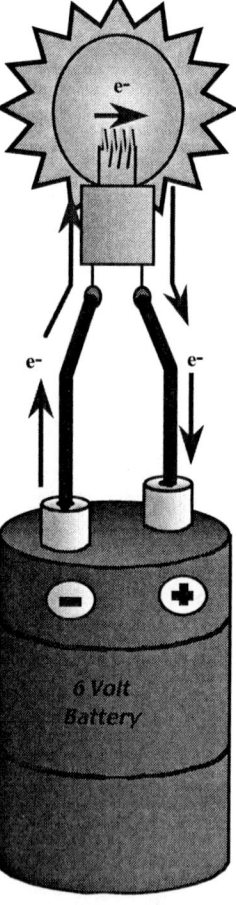

Copper wires provide connections that allow electrons to flow from the - terminal, through the filament inside the lamp, and back into the + terminal of the battery.

FIGURE 2.10 Electrical Current Flow
(Used with permission from Advanced Micro Devices)

Conductors

A material is a *conductor* if electrons readily flow through the material as electrical current. A good conductor has high current-carrying capability, referred to as current density. A conductor generally has few valence electrons in its outermost shell that are loosely bound and easily given up by the atom. Metals typically have this valence-shell configuration. Common conductor metals used in semiconductor manufacturing are aluminum, used for interconnect wiring between devices, or tungsten, used as the material for forming electrical interconnections between metal layers.

An example of a good metal conductor is copper (Cu), which recently has been introduced as a replacement for aluminum metallization for interconnecting the different devices on a microchip during wafer fabrication. Cu has twenty-nine electrons, with one electron in the valence shell that is relatively far from the nucleus. This electron is easily freed from the atom and is available to conduct electricity (see Figure 2.11).

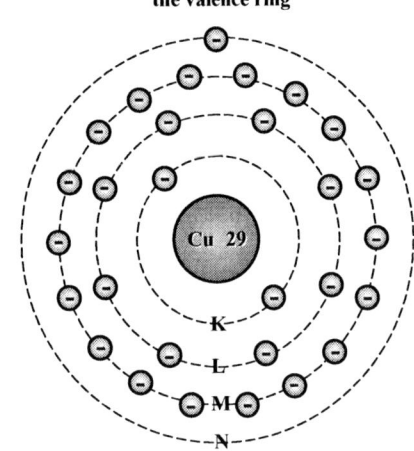

FIGURE 2.11 Flow of Free Electrons in Copper

Conductors require little energy to move a valence electron from its valence band into the conduction band. This electron movement leaves a gap in the covalent bond of the atom, which is referred to as a hole. For every conduction-band electron there must exist a valence-band hole. Valence electrons can jump into the position of the hole, which promotes electron movement and conduction. This condition is referred to as the *electron-hole pair*.

Very quickly after becoming a free electron in the conduction band, an electron gives up its energy and falls into a hole in the covalent bond of the valence band. This process is known as *recombination*. The time period from when an electron moves from the conduction band until recombination is known as the *lifetime* of the electron-hole pair. Thermal energy leads to the continual creation of electron-hole pairs followed by their recombination.

Conductivity and Resistivity ■ The property of a material to conduct electricity is its *conductivity*. Materials are usually characterized by their *resistivity* (ρ), or resistance to current flow. A material with lower resistivity has better conductivity. These two properties of resistivity and conductivity depend only on the material, not its geometry, and are related by the following formula:

$$\rho = \frac{1}{C}$$

where, ρ = resistivity in ohms-centimeter (Ω-cm)
C = conductivity

Resistivity of materials is important in semiconductor products because of its effects on the electrical operation of an integrated circuit. Resistivity can be deliberately designed into an IC in the form of an electronic component called the resistor. In this case the resistor serves as a desired control for current in a particular part of an electronic circuit. In some cases, resistivity is an

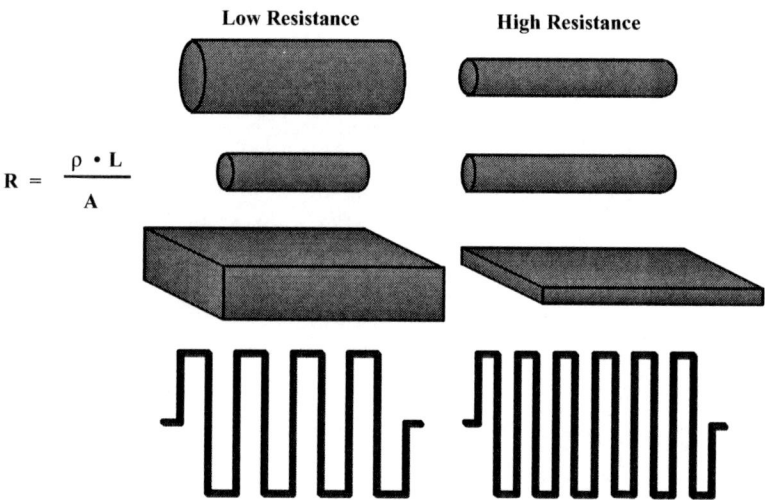

FIGURE 2.12 How Wiring Sizes Affect Resistance

undesirable characteristic of materials that introduce too much resistance to the flow of current, resulting in power losses and overheating within the IC.

Resistance. *Resistance* is the opposition to current flow and is accompanied by the dissipation of heat. A material with high resistance has high opposition to current flow. Resistance depends on both resistivity and the physical geometry of the material and is calculated by:

$$R = \frac{\rho l}{area}$$

where, R = resistance of the conductor material, in ohms
ρ = resistivity of the conductor material, in ohms-cm
l = length of the conductor, in cm
area = cross-sectional area of conductor, in cm^2

Reduced feature sizes in wafer fabrication makes resistance an important parameter. Smaller geometries cause increased resistance in interconnect lines, which is undesirable due to increased heat losses. Lower resistance is why copper is replacing aluminum as the primary interconnect wiring material. Another example of the application of resistance in semiconductor manufacturing is the sheet resistance measurement to control the thickness of conductive films (see Chapter 7). Since resistance is dependent on geometry, the sheet resistance of a thin film changes depending on the film's thickness.

Insulators

An insulator is a material that has a high resistance to current flow. Another term for an insulator is a *dielectric*. Insulators have no loosely bound electrons in the valence shell available for electrical conduction, with a high energy gap to keep valence band electrons out of the conduction band. Examples of insulators in everyday life are rubber, plastics, glass, and ceramics. Some insulator materials in semiconductor manufacturing are silicon dioxide (SiO_2), silicon nitride (Si_3N_4), and polyimide (a plastic material).

Purified and deionized (DI) water is a good example of an insulator, with a resistivity of about 18×10^6 ohms-cm, or 18 megohms-cm. There are insufficient free electrons in DI water to be able to sustain current flow from a small battery. However, the conductivity of an insulator can be improved by adding an impurity to it (see Figure 2.13). In the case of water, we can add common table salt. The salt dissociates in water into its basic ionic elements of sodium (Na^+) and chlorine (Cl^-), forming an electrolyte (a conducting solution). The net effect of these charged atoms is the same as having free electrons in a piece of copper wire—if there are sufficient charged atoms, current flow can be sustained. As we shall learn later, adding an impurity to alter an insulator material's conductivity is an important aspect of semiconductor technology.

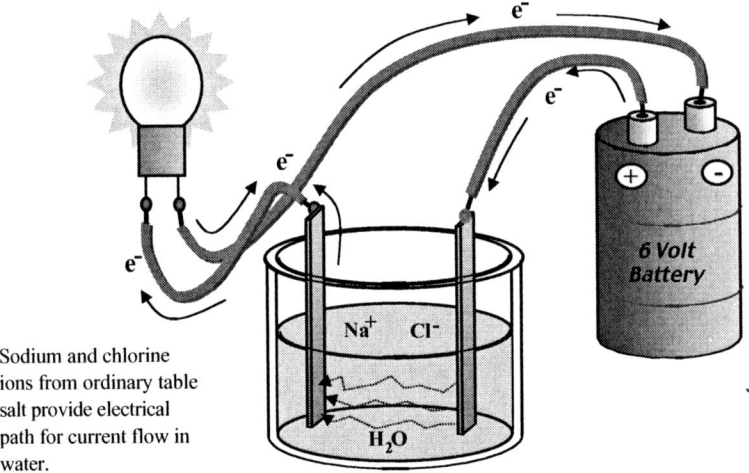

Sodium and chlorine ions from ordinary table salt provide electrical path for current flow in water.

FIGURE 2.13 Adding an Impurity to Water to Improve its Conductivity

Capacitance ■ *Capacitance* is the storage of electrical charge on two conductive plates separated by a dielectric material (see Figure 2.14). The unit of measure for capacitance is farads and is usually expressed as picofarads (10^{-12} farads) for capacitance in integrated circuit structures. The amount of charge that can be stored by a capacitor varies depending on certain physical characteristics. These characteristics include the area of the plates, the spacing between the plates, and the quality of the insulation material between the plates, better known as the *dielectric constant, k* (farads/cm). The k value of air is 1, while that of glass is between 4 and 7.

We can use a simple circuit to explain how a capacitor holds an electrical charge (see Figure 2.15 on page 32). In this circuit, a 1.5 V battery is connected to a capacitor. When the switch is closed, electrons from the left metal plate are attracted to the positive (+) side of the battery. At the same time, electrons from the negative side (-) of the battery flow to the right plate to balance the deficiency of electrons from the left side. The initial current stops due to the dielectric material between the plates. The result, however, is a difference in potential between the two plates. An electrostatic field exists between the charged plates.

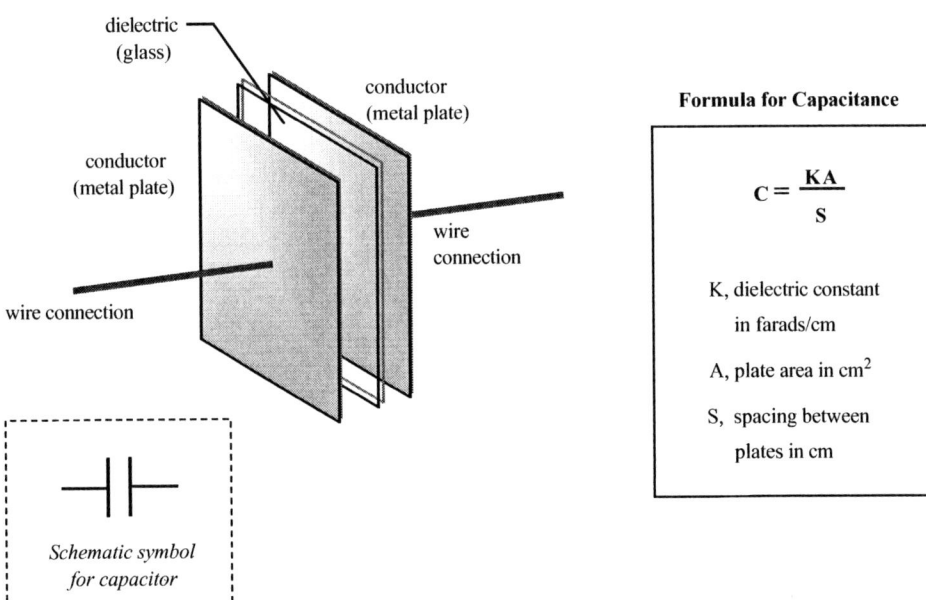

Formula for Capacitance

$$C = \frac{KA}{S}$$

K, dielectric constant in farads/cm

A, plate area in cm^2

S, spacing between plates in cm

FIGURE 2.14 Basic Capacitor Structure

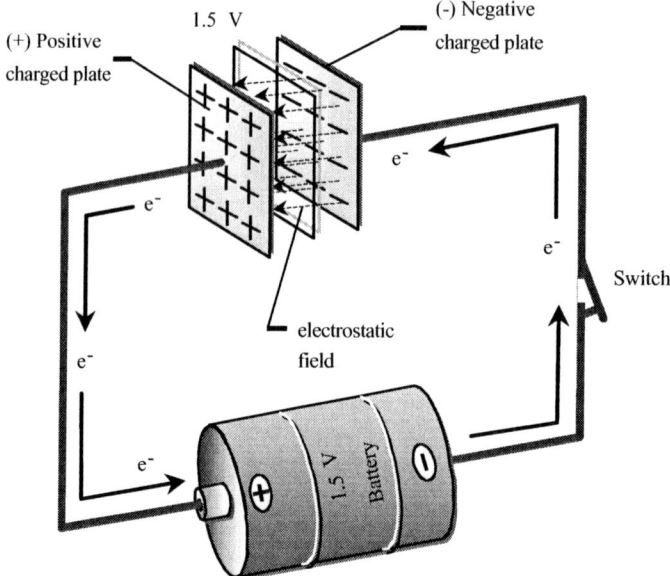

FIGURE 2.15 Battery Charges a Capacitor

If the battery and circuit are removed from the capacitor, the left plate is positively charged and the right plate is negatively charged (see Figure 2.16). The overall difference in potential is equal to the original battery value and remains as long as there is no other way for electrons to flow to the opposite side of the capacitor (in reality, some leakage does occur through the dielectric). Understanding these concepts regarding how a capacitor works is crucial for understanding how a field-effect transistor (FET) works (see Chapter 3).

Dielectric Constant. The dielectric material is a key component of capacitors. This material can change the capacitance of a capacitor by more effectively concentrating the electric field occurring between the two conductors. This dielectric constant, k, has become an important parameter in semiconductor performance. As current flows through adjacent metal conductors that interconnect the devices on the chip (referred to as wiring), it is desirable to have a low-k dielectric material to minimize capacitance losses (see Figure 2.17).

Semiconductors

The third type of material is *semiconductors*. These materials are special because they can function as either conductors or insulators. A semiconductor material has a small energy gap (i.e., 1.11 eV for silicon) that is a value between an insulator (>2 eV) and a conductor. This energy gap permits the electrons to jump from the valence band into the conduction band when energy is supplied. This

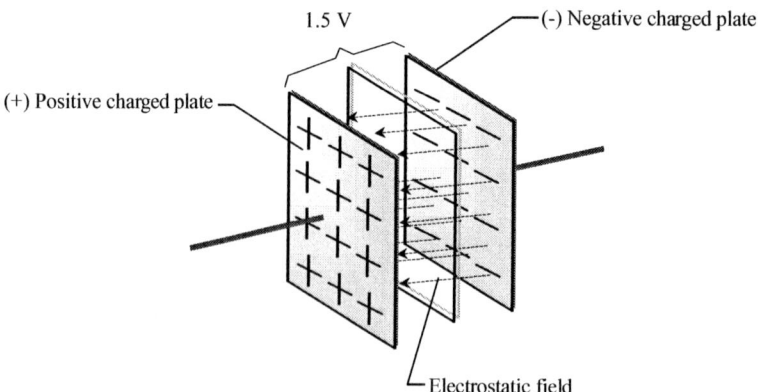

FIGURE 2.16 Capacitor Holds a Charge

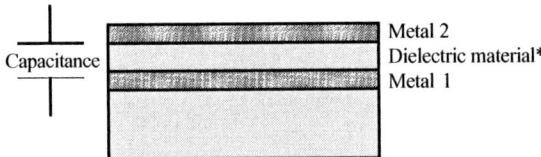

Low k dielectric reduces capacitance between metal layers.

FIGURE 2.17 Low-*k* Dielectric Material

action occurs when a semiconductor is heated, thus improving conductivity with an increase in temperature (which is the opposite condition for a conductor).

The most important semiconductor material in wafer fabrication is silicon, used as the semiconductor wafer substrate in over 85% of all chips.[3] This text will only address silicon because of its predominant use in the semiconductor industry. A short review of some of the other semiconductor materials and their applications is given at the end of this chapter.

SILICON

Silicon is an elemental semiconductor material because it has four valence-shell electrons, along with the other elements located in Group IVA of the periodic table (see Figure 2.18 on page 34). The number of valence-shell electrons in silicon places it midway between the valence-shell condition of a good conductor (one valence electron) and an insulator (eight valence electrons).

Silicon is not found pure in nature. It must be refined and purified to become pure for semiconductor manufacturing (see Chapter 4). It usually occurs as silica (silicon dioxide, or SiO_2) and other silicates. Silica is sand and is a primary ingredient of glass. Other forms of SiO_2 are rock crystal, quartz, agate, and opal.

The melting point of silicon is 1412°C. Silicon is a hard and brittle material that fractures easily if deformed, much like a piece of glass. It can be polished to a mirrorlike finish. Silicon exhibits many of the same characteristics as metals and at the same time exhibits characteristics of nonmetals. This is why silicon is classified as a semiconductor, midway between the conductors (metals) and insulators (nonmetals) on the periodic table.

Pure Silicon

Pure silicon is referred to as *intrinsic silicon,* with no contaminants or impurities from other substances. The silicon atoms in pure silicon bond together through covalent bonds to share electrons and complete their valence shells (see Figure 2.19 on page 34).

Many of silicon's properties result from this strong covalent bonding. The covalent bonds in pure silicon hold the atoms together to form a solid, electrically stable material that is an insulator. Pure silicon is a poor conductor because all valence shells are fully occupied in covalent bonds. In this pure form, silicon is not useful as a semiconductor.

The solid material formed when two or more silicon atoms bond together in this set, repeatable pattern, as the one shown in Figure 2.19 on page 34, is referred to as a *crystal*. A crystal is a smooth, glassy solid that forms a lattice structure with a three-dimensional shape. An example of a crystal material is window glass. We will study more about crystals in Chapter 4 when we discuss silicon wafer preparation.

Why Silicon?

Germanium was the first material used for semiconductors in the 1940s and early 1950s, but as we learned in Chapter 1, it was soon replaced by silicon. Why was silicon chosen as the predominant semiconductor material? There are four main reasons for the selection of silicon as the primary semiconductor material:

- Abundance of silicon
- Higher melting temperature for wider processing range

◆ Wider temperature range of operation
◆ Natural growth of silicon dioxide

Semiconductors

Group IVA	
C, Carbon	6
Si, Silicon	14
Ge, Germanium	32
Sn, Tin	50
Pb, Lead	82

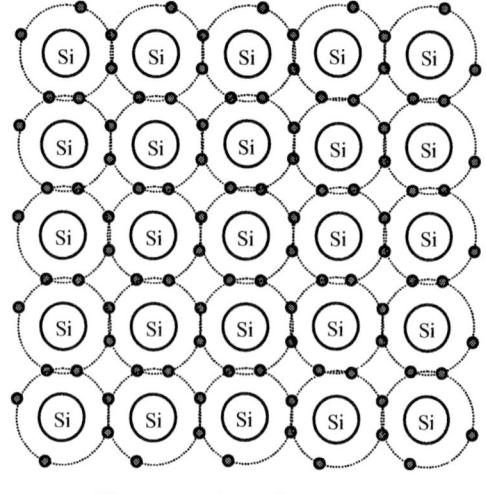

Silicon atoms share valence electrons to form insulator-like bonds.

FIGURE 2.18 Group IVA Elemental Semiconductors

FIGURE 2.19 Covalent Bonding of Pure Silicon

Silicon is the second most abundant element on earth and makes up about 25% of the earth's crust. If processed properly, silicon can be refined into ample quantities of the very pure form necessary for semiconductor fabrication which leads to lower costs. Silicon's melting temperature of 1412°C is much higher than the melting temperature for germanium which is 937°C. This higher melting temperature permits silicon to withstand high-temperature processing. Another advantage to using silicon is that a semiconductor device made from silicon can function over a wider temperature range than germanium, which increases the semiconductor's applications and reliability.

Finally, an important reason for using silicon as a semiconductor material is the ability to naturally grow *silicon dioxide* (SiO_2) on the silicon surface (see Figure 2.20). SiO_2 is a high-quality, stable electrical insulator material that also serves as a good chemical barrier to protect silicon from external contaminants.[4] Electrical stability is important to avoid leakage between adjacent conductors in an IC. The ability to grow stable, thin SiO_2 material is fundamental to the fabrication of high-performance metal-oxide semiconductor (MOS) devices. SiO_2 has mechanical properties similar to silicon, which allows high temperature processing without excessive wafer warpage.

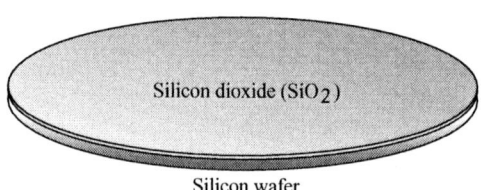

FIGURE 2.20 SiO_2 on Silicon Wafer

Doped Silicon

Silicon in its pure state has little use in semiconductor technology. However, the structure of silicon can be altered to greatly enhance its conductivity by adding small amounts of other elements to the material through a process known as doping. *Doping* is the process of adding certain elements to pure silicon to improve the conductivity of the semiconductor (see Figure 2.21). For example, the electrical resistivity (ρ) of pure silicon is approximately 2.5×10^5 ohms-cm. If only one in

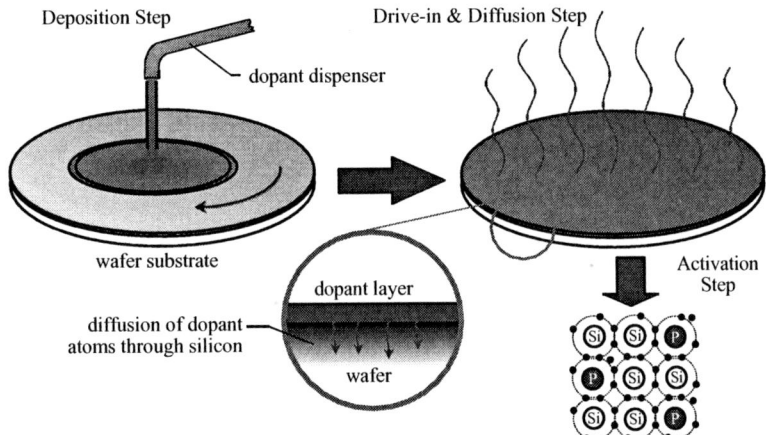

FIGURE 2.21 Doping of Silicon
(Used with permission from Advanced Micro Devices)

every one million silicon atoms is replaced by one atom of arsenic, resistivity will drop to 0.2 ohms-cm.[5] This is an improvement in conductivity of 1,250,000 times.

The elements added during doping are referred to as *dopants* or *impurities* because the silicon is no longer pure. In other words, we dope the silicon with an impurity so that it will conduct electricity. The more impurity added then the higher the conductivity (or the lower the resistivity). Note that we use the term impurity to indicate that we have added another element to the silicon. We intentionally add the impurity to increase the conductivity of silicon. Doped silicon is also known as *extrinsic silicon*.

The concept of doping the silicon with an impurity to improve electrical conductivity is a critical aspect of semiconductor fabrication. If we can introduce impurities that alter the silicon's resistivity and then control when the silicon performs as an insulator or as a conductor, then we have the essence of solid-state technology.

Dopant Materials ■ Silicon is located in Group IVA of the periodic table and has four valence electrons. Elements from the two adjacent groups are commonly used for doping: Group IIIA and Group VA (see Figure 2.22). Group IIIA elements are referred to as *trivalent* because of three valence electrons, whereas elements from group VA are *pentavalent* because of five valence electrons. The trivalent dopant increases the number of holes (positive doping or p-type), whereas the pentavalent will increase the number of free electrons (negative doping or n-type).

Acceptor Impurities		Semiconductor		Donor Impurities	
Group III (p-type)		**Group IV**		**Group V (n-type)**	
Boron	**5**	Carbon	6	Nitrogen	7
Aluminum	13	**Silicon**	**14**	**Phosphorus**	**15**
Gallium	31	Germanium	32	**Arsenic**	**33**
Indium	49	Tin	50	**Antimony**	**51**

* *Elements underlined are the most commonly used in silicon-based IC manufacturing.*

FIGURE 2.22 Silicon Dopants

When trivalent dopant atoms are added to silicon, the resulting material is called a *p-type silicon*. The trivalent dopants are referred to as *acceptors* (they accept an extra mobile electron), with the most common acceptor element being boron. When a pentavalent element is added to pure silicon, the resulting material is referred to as an *n-type silicon*. Pentavalent dopants are referred to as *donors* (they donate an extra mobile electron), and are typically either phosphorus, arsenic, or antimony. The different doping elements are listed in Table 2.2.

TABLE 2.2 Dopant Elements Commonly Used in Semiconductor Manufacturing

Trivalent Dopants (p-type, positive doping, acceptor)	Pentavalent Dopants (n-type, negative doping, donor)
Boron (B)	Phosphorus (P)
	Arsenic (As)
	Antimony (Sb)

n-Type Silicon. For an n-type silicon, there are more electrons than valence-band holes. This is shown in Figure 2.23 for silicon with a pentavalent phosphorous dopant atom.

The silicon atoms will form covalent bonds with the donor phosphorous atom, each sharing one phosphorous electron. However, the fifth phosphorous electron is not bound to any of the surrounding silicon atoms. Because of this, the fifth phosphorous electron needs little energy to break away and enter the conduction band. For n-type silicon, conduction-band free electrons are the majority carriers, of which the material has an abundance. There are also many fewer minority carrier valence-band holes.

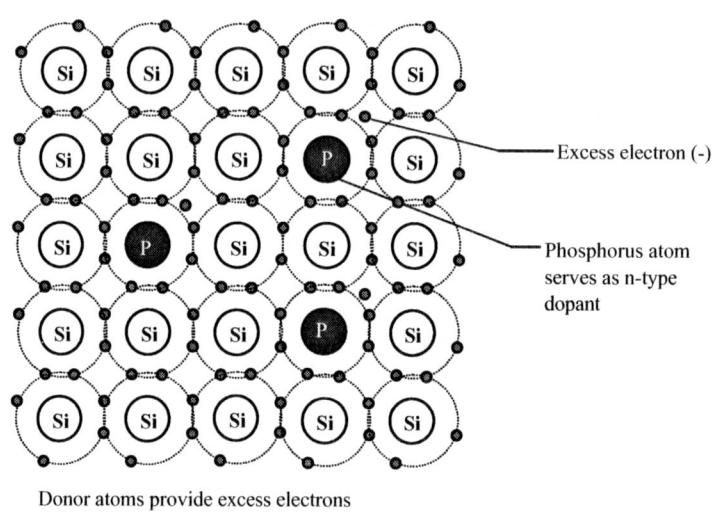

Donor atoms provide excess electrons to form n-type silicon.

FIGURE 2.23 Electrons in n-Type Silicon with Phosphorous Dopant

One electron in the conduction band is not significant for conduction. However, when we dope silicon with a dopant, we add literally millions of dopant atoms, producing many electrons that are not part of the covalent bonds. There is a high amount of movement between electrons and holes. Negative electrons are attracted to the positive holes. Electrons enter the conduction band relatively easily. If a voltage is applied to this material, the electrons can be made to flow through the material as electric current (see Figure 2.24).

Note that the doped silicon is still electrically neutral (this is true for n-type or p-type silicon). In the case of n-type silicon, this is because each phosphorous atom still has the same number of protons as electrons, as do the silicon atoms. Thus the overall number of protons and electrons in the semiconductor is still equal and the result is a net charge of zero. What is not equal is that there are many more conduction-band electrons (majority carriers) than there are valence-band holes (minority carriers).

p-Type Silicon. In the p-type silicon shown in Figure 2.25, the boron atom is a p-type acceptor dopant that forms a covalent bond with four adjacent silicon atoms. The boron acceptor atoms provide a deficiency of electrons caused by the lack of a fourth electron in the boron

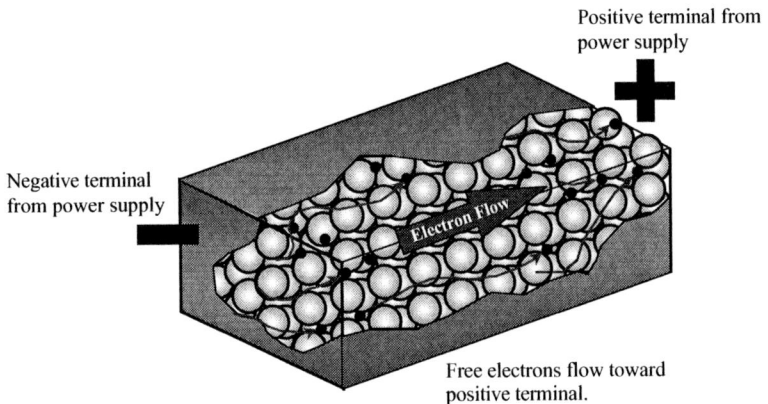

FIGURE 2.24 Flow of Free Electrons in n-Type Silicon

atom, thus creating the p-type silicon. There is an excess of holes (absence of electrons) in the valence shell.

Since there are many more valence-band holes than conduction-band electrons, the holes are the major current carriers in p-type material. The holes are referred to as *majority carriers* and electrons are the *minority carriers*. If a dc voltage is applied across a p-type semiconductor, then the large number of holes in the material will attract electrons from the negative terminal of the voltage source into the p-type semiconductor. This is current flow in a p-type material (see Figure 2.26 on page 38). The holes appear to move because each time an electron moves into a hole it serves to create a hole in its previous position. The holes seemingly move in the opposite direction as the electrons.

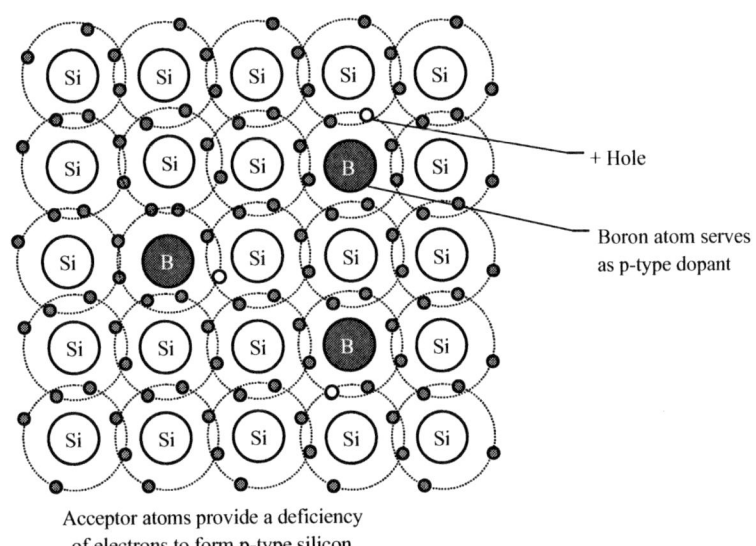

FIGURE 2.25 Holes in p-Type Silicon with Boron Dopant

Resistivity of Doped Silicon. We obtain precise control of silicon's resistivity by introducing dopants into the silicon crystal structure. The concentration of the dopant atom placed in the silicon crystal defines how well the material will conduct electrical current. Pure silicon has a resistivity of about 250,000 Ω-cm and is an insulator. Compare this to copper, a good conductor, with a resistivity of 1.7 $\mu\Omega$-cm (0.0000017 Ω-cm). By adding the proper type and concentration of dopants to pure silicon, the doped silicon's resistivity is decreased and conductivity is improved (see Figure 2.27 on page 38). For a given resistivity there is less concentration of n-type dopants than p-type dopants. This is because it takes less energy to move an electron than to move a hole.

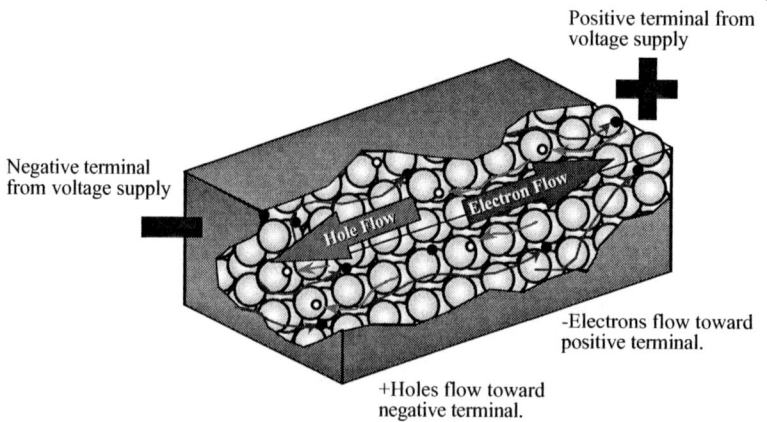

FIGURE 2.26 Flow of +Holes in p-Type Silicon

It only takes a small amount of dopant (from as little as 0.000001% to 0.1%) to make silicon a useful conductor. This is important for the fabrication of semiconductor devices on wafers. The amount of dopant in the silicon, or concentration, is carefully controlled during semiconductor fabrication to attain a precise resistivity. The ion implantation process for adding dopants to silicon will be covered in Chapter 17.

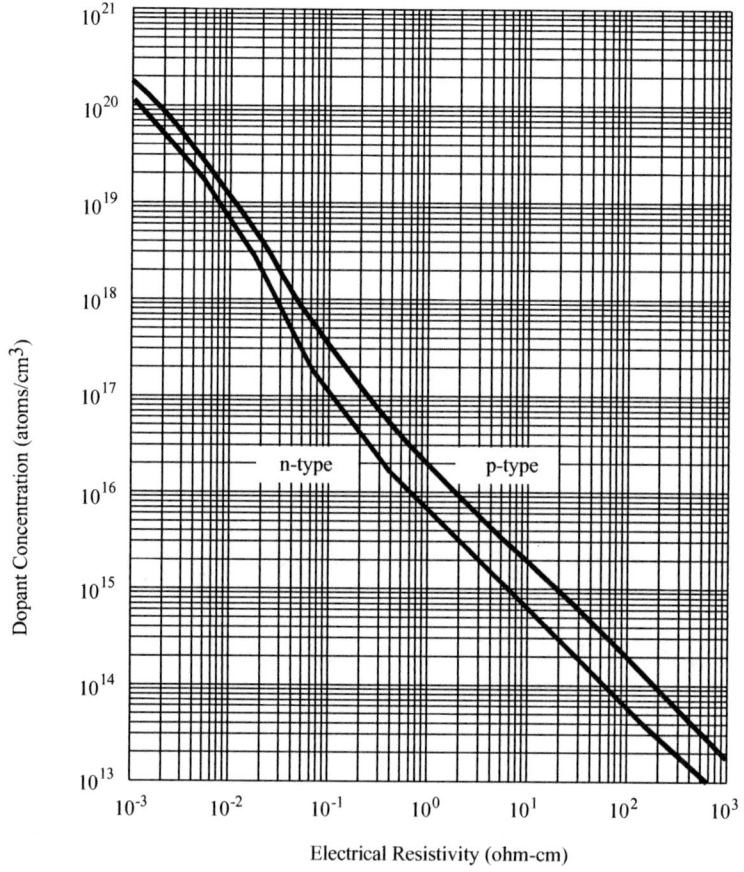

FIGURE 2.27 Silicon Resistivity Versus Dopant Concentration
Redrawn from S. Ghandi, *VLSI Fabrication Principles: Silicon and Gallium Arsenide,* John Wiley & Sons, 1994.

pn Junctions

We dope pure silicon with pentavalent or trivalent elements to obtain either an n-type or p-type semiconductor. The type and concentration of the dopant element determines whether conduction occurs by hole or by electron and determines the final value of the silicon's resistivity. The number of carriers, and therefore the resistivity, is determined by the net donor and acceptor atom concentration in the silicon crystal. Because of this, an n-dopant can be implanted or diffused into a p-region to convert that region into an n-region (or vice versa). To achieve this, the concentration of the n-type donor dopant must exceed that of the p-type acceptor dopants in the region.

The ability to incorporate both n-type and p-type regions in the silicon crystal is beneficial because semiconductor devices require both of these regions in order to function as a useful electronic device. It is the junction between the n-type and p-type regions that is important and creates the useful characteristics of silicon as a semiconductor. This junction is referred to as a *pn junction* (see Figure 2.28).

The pn junction is the essence of solid-state electronics and is the basis for how semiconductor wafers can achieve their unique ability to act as an insulator and a conductor, depending on the bias voltage applied to the junction. The specifics of how a pn junction functions as a useful electronic device are discussed in Chapter 3. The creation of a pn junction is almost universally done in wafer fabrication with ion implantation (see Chapter 17).

Remember that a pn junction is formed between two pieces of essentially the same material. The p-type and the n-type materials scarcely differ except for a minute amount of a dopant. The n-type material has extra mobile electrons from the donor impurity, whereas the p-type material has extra mobile holes. It is deceptive to speak of one material meeting the other.[6] The junction is so intimate that the n-type and p-type materials are formed from one continuous solid. The silicon crystal with a pn junction continues to look like and in most ways behave like the solid crystal material that it is.

The depth and definition of this junction is critical in semiconductor manufacturing. As device critical dimensions are reduced, the ability to precisely control both the pn junction in silicon (e.g., junction depth) and the concentration of dopants is becoming a major challenge for semiconductor chip fabrication.

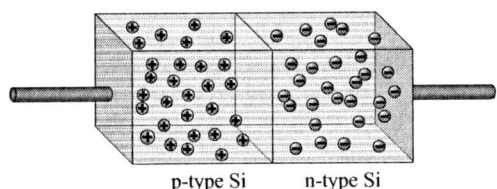

FIGURE 2.28 Cross Section of Planar pn Junction

ALTERNATIVE SEMICONDUCTOR MATERIALS

Germanium and silicon are the two elemental semiconductor materials from Group IVA with four valence electrons. We have learned that germanium was the first semiconductor material used for transistor fabrication and was replaced by silicon in the 1950s for process and performance reasons.

There are alternative semiconductor materials used for specific market applications. These are primarily the *compound semiconductors*. A major class of compound semiconductors are formed from Group IIIA and Group VA of the periodic table (often referred to as III-V compounds). An example is gallium arsenide (GaAs). In addition, other semiconductor compounds come from elements found in Groups IIA and VIA, and are referred to as II-VI compounds.

The need for alternative semiconductor materials is driven by the growth of markets that require IC performance beyond that capable by silicon semiconductors. The key IC performance factor is speed. Wireless and high-speed digital communications, space applications, and consumer markets such as the automotive industry, are developing special niche semiconductor markets that are centered on high speed that can support higher signal frequencies. In addition, III-V compound semiconductors are used for light-emitting diodes (LEDs).

GaAs is the most common compound III-V semiconductor and will be discussed in the next section. There are variations of this compound, such as gallium nitride (GaN) for use in the fabrication of blue semiconductor lasers and LEDs. Another compound semiconductor is silicon germanium (SiGe), which has been researched for more than two decades and may compete against GaAs for market applications.

There is also some growth in II-VI semiconductors, with their market revenue expected to be $179 million in 2001.[7] The two principal II-VI materials are cadmium telluride (CdTe) and zinc sellenide (ZnSe). CdTe semiconductors are used primarily in infrared (IR) detection systems. ZnSe is another II-VI compound material used to fabricate blue LEDs.

Gallium Arsenide (GaAs)

Gallium arsenide (GaAs) is the most common III-V compound semiconductor material and is formed by combining gallium (Ga) from Group IIIA with the element arsenic (As) from Group VA. GaAs has greater electron mobility than silicon, so the majority carriers move faster than in silicon. There are also some attributes of GaAs semiconductor material that reduce parasitic capacitance (see Chapter 3) and signal losses. These result in ICs that are faster than those made with silicon.[8] The improved signal speed of GaAs devices permits them to react to high-frequency microwave signals and accurately convert them into electrical signals for communication systems. Semiconductors based on silicon are too slow to react to the microwave frequency. For these reasons, products for wireless and high-speed digital communications and high-speed optoelectronics devices are made from GaAs and other compound semiconductors.

An advantage of GaAs is the increased resistivity of the material, with values up to 10^8 Ω-cm. This makes it easier to isolate semiconductor devices fabricated in the GaAs substrate without loss in electrical performance. GaAs devices also exhibit a higher radiation hardness than silicon, which makes this material attractive for military and space applications.

A summary of some different semiconductor material properties is shown in Table 2.3. The properties of SiO_2 have been included for comparison.

TABLE 2.3 Comparison of Some Physical Properties for Semiconductor Materials

Property	Si	Ge	GaAs	SiO_2
Melting point (°C)	1412	937	1238	1700 (approx.)
Atomic Weight	28.09	72.60	144.63	60.08
Atomic Density (atoms/cm^3)	4.99×10^{22}	4.42×10^{22}	2.21×10^{22}	2.3×10^{22}
Energy Band Gap (eV)	1.11	0.67	1.40	8 (approx.)

A. Grove, *Physics and Technology of Semiconductor Devices,* John Wiley and Sons, 1967.

The primary disadvantage of GaAs semiconductor material is the lack of a natural oxide. This feature hinders the development of standard MOS devices that require the ability to grow a surface dielectric. Another problem with GaAs is the fragility of the material, which makes wafer handling a major issue in wafer fabrication. The cost of GaAs is as much as ten times higher than silicon because of the relative scarcity of gallium and the energy used in its purification process. A final point to note is that the extreme toxicity of arsenic requires special controls for the equipment, process, and waste disposal facilities. These precautions have a significant effect on the manufacturing cost of GaAs semiconductors.

SUMMARY

All matter is made up of atoms. The number of valence-shell electrons in an atom is one of several factors that influences its chemical and physical properties and its ability to bond with other atoms. The energy-band theory of solids explains how electrons change orbital levels between the valence band and conduction band to be an insulator, conductor or semiconductor. Chemical elements are described in the periodic table, which is used to explain how ionic and covalent bonds form and how certain elements are used in semiconductor manufacturing. All material is classified as either a conductor, which conducts current, an insulator, which does not pass current, or semiconductor, which conducts current under certain conditions. Resistivity is a material property, while resistance depends on both resistivity and the geometry of the material. A capacitor uses a dielectric to store an electric charge.

Silicon is the most common semiconductor material because of its abundance, high melting temperature, wide temperature range, and ability to oxidize. Oxidation permits masking of the silicon surface and the ability to dope silicon with impurities to make it conductive. Silicon is doped with trivalent (p-type acceptor dopant) and pentavalent (n-type donor dopant) elements. Introducing a specific concentration of dopants into the silicon crystal permits the precise control of silicon's resistivity. When p-type and n-type silicon form a junction in the silicon crystal, then silicon can act as either an insulator or a semiconductor. There are alternative semiconductor materials, primarily the compound semiconductors, of which gallium arsenide is the most common.

KEY TERMS

element
molecule
compound
valence shell
insulators
conductors
semiconductors
cation
anion
ionization
atomic mass unit (amu)
electron-hole pair
recombination
lifetime
conductivity
resistivity
resistance
capacitance

silicon
intrinsic silicon
crystal
silicon dioxide
doping
dopants (or impurities)
extrinsic silicon
trivalent
pentavalent
p-type silicon
acceptors
n-type silicon
donors
majority carriers
minority carriers
pn junction
compound semiconductors
gallium arsenide

REVIEW QUESTIONS

1. What is matter?
2. Describe the model for atomic structure, including the proton, neutron, and electron, and state the charge of each.
3. Describe an element, a molecule and a compound.
4. What is a valence electron and why are valence electrons important for forming compounds?
5. How does energy-band theory explain the difference among insulators, conductors, and semiconductors? How does the energy gap change for each material?
6. How is an ion formed? What is a cation and an anion?
7. What does the group number on the periodic table represent in terms of valence electrons? What is the difference between the atomic number and the atomic mass number?
8. Define the atomic mass unit (amu).
9. How is an ionic bond formed? How is a covalent bond formed? What is oxidation and reduction and how are these processes related to the formation of a bond?
10. State the three classes of materials and explain how electric current flows in each one.
11. What is resistivity? What is conductivity? How is resistivity related to conductivity?
12. What is resistance? State the resistance formula.
13. What is another term for an insulator material?
14. What is capacitance? What is the dielectric constant and why is this concept important to semiconductor manufacturing?
15. Name the most common semiconductor material and give four reasons why it is so commonly used.
16. What is the difference between intrinsic and extrinsic silicon?
17. Describe silicon dioxide. Discuss why the oxidation of silicon is important.
18. What is doping? Why is doping important for semiconductor silicon?
19. Give one example of a trivalent doping element and three examples of a pentavalent dopant element. From what group number of the periodic table is each of these elements?
20. Describe the n-type doping of silicon with phosphorus.
21. n-type silicon has what type of majority and minority carriers? Give the majority and minority carriers for p-type silicon.
22. What characteristics determine how well silicon will conduct electricity?
23. Describe a pn junction and explain why it is important for a semiconductor.
24. Compound semiconductors come from what group numbers of the periodic table?
25. What is the advantage of gallium arsenide over silicon?
26. What is the primary disadvantage of gallium arsenide over silicon?

REFERENCES

1. Basic chemistry concepts referenced from M. Silberberg, *Chemistry, The Molecular Nature of Matter and Change,* (St. Louis: Mosby, 1996).
2. S. Wolf and R. Tauber, *Silicon Processing for the VLSI Era, Vol. 1—Process Technology* (Sunset Beach: Lattice Press, 1986), p. 118.
3. *Solid State Technology* (January 1998): p. 18.
4. F. Seitz and N. Einspruch, *Electronic Genie: The Tangled History of Silicon,* Urbana, (University of Illinois Press, 1998), p. 175.
5. G. Anner, *Planar Processing Primer,* (New York: Van Nostrand Reinhold, 1990), p. 57.
6. A. Holden, *Conductors and Semiconductors,* (Bell Telephone Laboratories, 1964), p. 109.
7. "Growth Seen for II-VI Materials," *Semiconductor International* (March 1997): p. 57.
8. S. Ghandhi, *VLSI Fabrication Principles: Silicon and Gallium Arsenide,* 2nd ed., (New York: Wiley, 1994), p. 2.

CHAPTER 3
DEVICE TECHNOLOGIES

There are many different types of semiconductor devices, each meeting a functional need for users. This chapter introduces various electronic components and explains how they are constructed in silicon for use in the evolution of specific IC technologies.

OBJECTIVES

After studying the material in this chapter, you will be able to:

1. Identify differences between analog and digital devices and passive and active components. Explain the effects of parasitic structures in passive components.
2. Describe the pn junction, discuss why it is important, and explain reverse and forward biasing.
3. State the characteristics of bipolar technology and the bipolar junction transistor in terms of function, biasing, structure, and applications.
4. Explain the basic characteristics of CMOS technology, including the field-effect transistor, biasing, and the CMOS inverter.
5. Explain the difference between enhancement mode and depletion mode MOSFETs.
6. Explain the effects of parasitic transistors and the implications for CMOS latchup.
7. Give examples of IC products and state some applications of each.

INTRODUCTION

Electronic devices used in microchips are constructed in the silicon substrate. Common microchip devices include resistors, capacitors, fuses, diodes, and transistors. Their integration on a silicon substrate is the basis for IC wafer fabrication technology.

How a particular electronic device is formed in silicon is referred to as a *structure*. There are literally thousands of different semiconductor device structures and we will evaluate only a few of them. In this chapter, we discuss the actual formation of the devices in order to understand how they can perform in their many applications. Also covered in this chapter is a review of the various classifications of IC products.

Components on Printed Circuit Board

CIRCUIT TYPES

Circuits constructed from electronic components can be classified into two basic types: analog and digital circuits. Each circuit type has applications where it is most beneficial, as illustrated on page 30.

Analog Circuits

In electronics technology an *analog circuit* is one in which the electrical data varies continuously over a range of voltage, current, or power values. Analog circuits can be designed to operate with direct current (DC), alternating current (AC), or a combination of the two, pulsating DC. Some examples of electronics products that operate primarily as analog (also known as *linear*) circuits are radio transmitters and receivers, audio recording and playback systems, X-ray machines, and automotive ignition systems. The magnitude of the input and output signals, however, may not always be set at any predetermined level. For example, when scanning radio stations on an AM/FM radio receiver, not all radio signals have the same signal strength. As a result, the volume control may have to be adjusted according to the strength of the incoming signal. A volume control on a radio, a manual thermostat on a wall heater, the light control for a household light fixture—these are all examples of common analog devices.

Digital Circuits

Digital circuits have operating signals that vary about two distinct voltage levels—a high and a low. A logic high represents a binary bit = 1 and a logic low represents a binary bit = 0. Digital circuits are associated with *logic devices* such as computers and calculators. Other applications of digital logic devices include clocks, handheld computer games, and barcode readers. Digital devices can be used to measure and/or control events that require either an on/off type of command or that can be controlled in discrete incremental changes that approximate the actions of a linear circuit. That is why today it is sometimes difficult to tell the difference between an analog device and a digital system. Exactly what voltage levels a high and a low are depends on a specific *device technology*. Here are two examples of these logic voltage levels:

Logic Family	High State = 1	Low State = 0
TTL	5 VDC	0.0 VDC
CMOS	3.5 VDC	0.0 VDC

Note: VDC is volts DC.

PASSIVE COMPONENT STRUCTURES

You learned in Chapter 2 about the differences in materials categorized as insulators, conductors, or semiconductors. Now you will begin to see how these basic materials are used in the construction of some very basic electronic components—for example, a resistor and a capacitor. These components are referred to as passive components for various reasons. *Passive components* can conduct electrical current regardless of how the component is connected across a voltage supply. A resistor, for example, will conduct current equally as well regardless of which end of the resistor is connected to the positive or negative terminal of a battery.

IC Resistor Structures

An IC resistor can be created out of metal films or from doped polysilicon or by diffusion of dopant impurities into specific areas of the substrate (see Figure 3.1).[1] The resistors are microstructures and, therefore, occupy very small areas of the substrate. The connection of the resistors to the chip circuitry is formed from conductor metals such as aluminum or tungsten in the form of contacts (see Chapter 12).

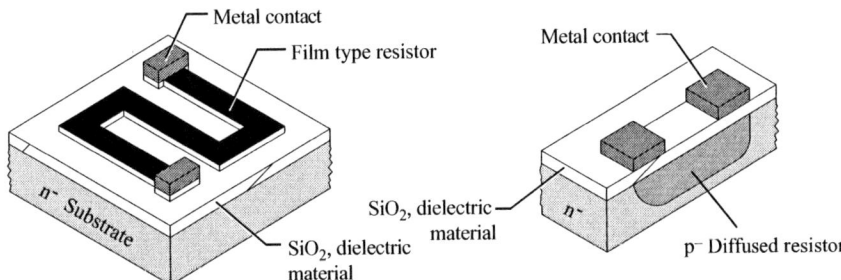

FIGURE 3.1 Examples of Resistor Structures in ICs

Parasitic Resistance Structures ■ *Parasitic resistance* is an unwanted resistance found in the design of components of an IC. It may exist in structures simply for reasons such as the size, shape, material type, dopant species, and quantity of dopant in the material. Parasitic resistance is an undesirable condition because it reduces the operational performance of IC devices. Figure 3.2 illustrates the locations of parasitic resistances in a transistor.

Parasitic resistances are cumulative, which means that the overall effect of these resistors in series is greater than a single resistance alone. The effect of parasitic resistances in IC devices becomes a detriment to the ability to reduce device feature sizes on chips. With higher circuit density comes higher resistances and overall degradation of electrical performance. Designers are aware of this problem and automatically calculate resistance losses into their design equations as well as select low-resistance metals for contacts and design process specifications to reduce the bulk resistance within active devices.

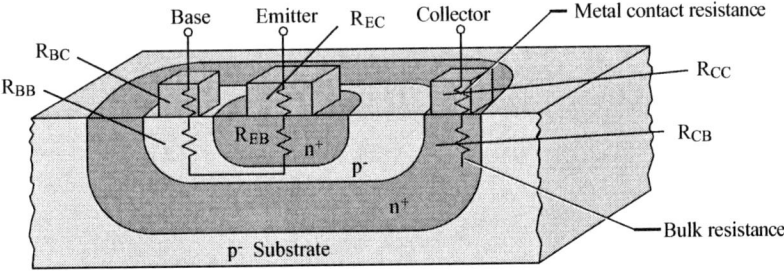

FIGURE 3.2 Cross Section of Parasitic Resistances in a Transistor

IC Capacitor Structures

You may recall in Chapter 2 that a simple capacitor is formed when two conductors are separated by a dielectric (insulator) material. In microchip manufacturing the dielectric material is usually silicon dioxide (glass, SiO_2), more commonly referred as simply *oxide*. *Planar capacitors,* which are built laterally on the substrate, may be formed from metallic films, doped polysilicon, or diffused areas of the silicon substrate. In general, capacitors are formed on substrates using four basic techniques (see Figure 3.3).

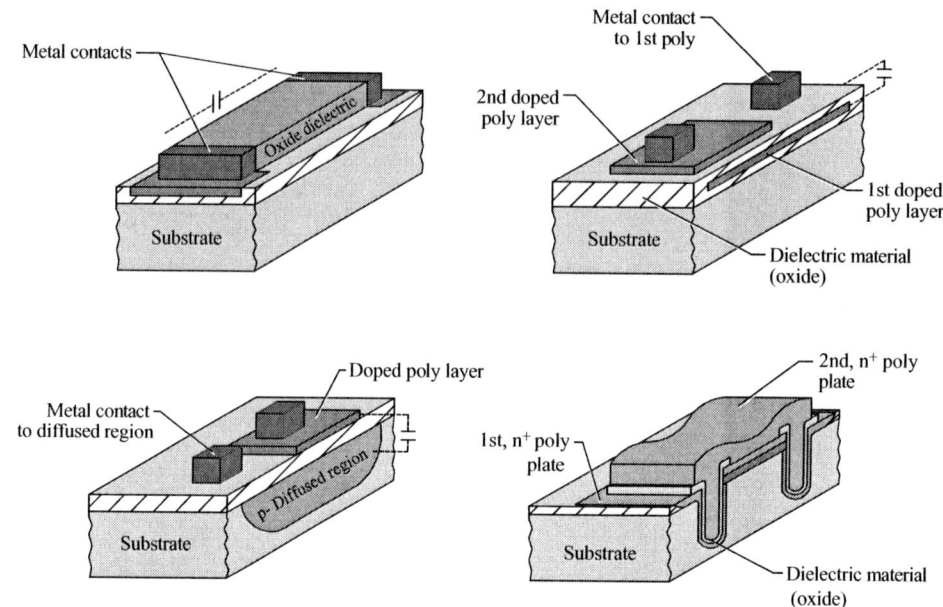

FIGURE 3.3 Examples of Capacitor Structures in ICs

Parasitic Capacitance Structures ■ Quite frequently capacitance occurs unintentionally as a result of the way materials are constructed on the substrate. This condition is referred to as *parasitic capacitance.*[2] An example of this would be two adjacent metal conductors with a dielectric material between them. This type of capacitance is undesirable for IC performance. In fact, parasitic capacitance interferes with the ability of electronic circuits to operate at faster switching speeds. In some cases parasitic capacitance may create instability in circuits, cause parasitic oscillations, and even create short-circuit paths for AC signals where they are not needed. Figure 3.4 illustrates the location of parasitic capacitance in between the electrodes of bipolar junction transistors and field-effect transistors.

ACTIVE COMPONENT STRUCTURES

Active components, such as diodes and transistors, offer some very different electronic control attributes that are not characteristic of passive components. They can be used to control the direction of current flow. Furthermore, active components can amplify small signals and can be used to create more complicated circuits, such as current and voltage regulators, oscillators, and logic gates. Active components require definite polarity (+ or -) when connecting these devices to power supplies. Active components utilize the flow of both electrons and holes in their operation.

The pn Junction Diode

A *pn junction diode* is formed any time there is a region of n-type semiconductor adjacent to a region of p-type semiconductor. A pn junction may be deliberately designed as a functional part of a larger integrated circuit or it may exist as a nonfunctioning diode in another part of the integrated

Device Technologies 47

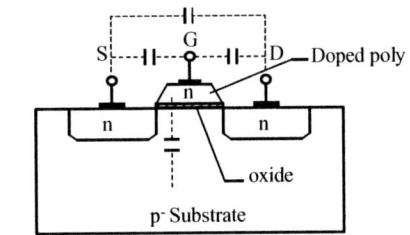
Field-effect transistor

...ance in Transistors

...because they are fundamental to the operation of both
...ransistors.[3]
...gle crystal of semiconductor material, e.g. silicon. As
...bstrate is doped heavily with donor dopant, such as
...an n-type silicon region. On the other side, acceptor
...-type silicon region. The actual manufacturing proce-
...pical metal contact materials used to connect the diode
...titanium, or copper.

...liode
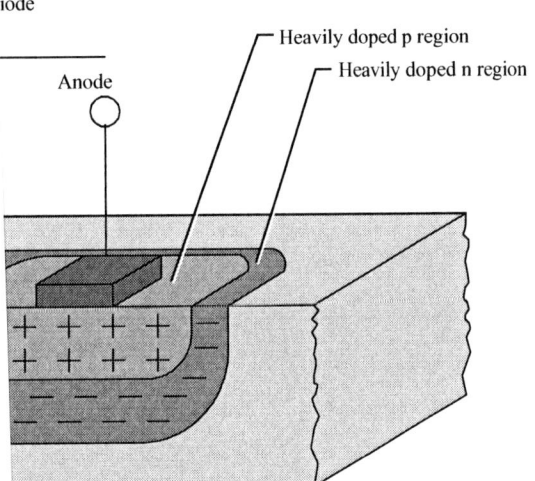

...he pn Junction Diode

...is shown in Figure 3.6 on page 48. Holes (+) make
...while electrons (-) are the majority carriers in the
...are electrically neutral; however, some holes from
...electrons from the n region diffuse into the p region.
...ction of the diode. This recombination process has
...jority current carriers. The p region has fewer holes
...leted of electrons. Thus, the term *carrier-depletion*
...ea. The depletion of holes from the p region creates
...s of electrons from the n region results in a positive
...produce corresponding electric fields that oppose
...ite side of the silicon crystal.
...s across the depletion region is to create a potential
...across the junction. This results in a *barrier voltage* that has to be overcome before the
diode can be operated.

Reverse Biasing the pn Junction Diode ■ Assume a voltage or bias supply (see the battery symbol in Figure 3.7) is connected across the pn junction as shown in Figure 3.7 on page 48. This bias

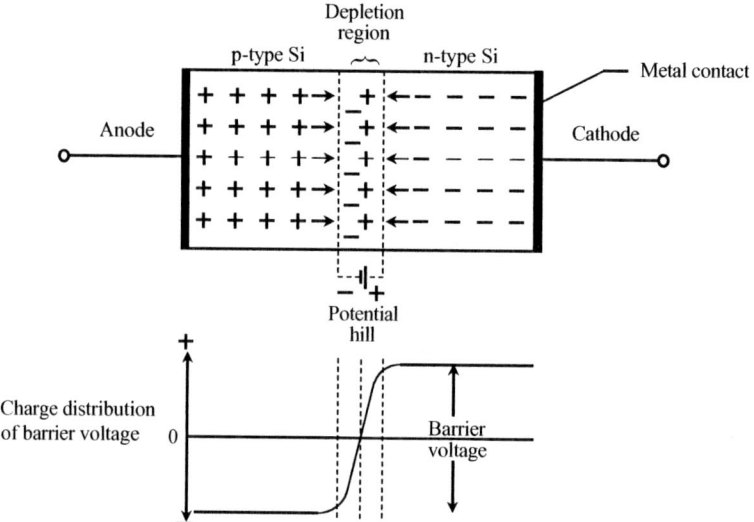

FIGURE 3.6 Open-Circuit Condition of a pn Junction Diode

configuration, called *reverse bias,* results in little or no conduction through the diode. The lamp is placed in the circuit as an indicator of current flow. Any appreciable current through the device can be attributed to minority current carriers in the diode.

Forward Biasing the pn Junction Diode ■ Figure 3.8 illustrates the effects of applying a *forward bias* to a pn junction diode. In this circuit electrons in the n region are repelled away from the negative terminal of the bias supply. Additional electrons are injected from the negative terminal to fill holes left vacant by the electrons in the n region. Similarly, holes in the p region are repelled away from the positive terminal of the bias supply. Holes are supplied from the positive side of the bias supply to balance the electrons supplied from the opposite side of the bias supply. Holes recombine with electrons at the junction and overcome the barrier voltage, greatly reducing its action as a barrier to current flow. Current flow will continue through the circuit as long as the bias supply can maintain a constant injection of holes and electrons through the diode.

Holes and electrons are drawn toward the junction to overcome the potential hill, thus, allowing current to flow through the circuit. Note that in Figure 3.8 solid arrows illustrate the direction of hole flow and broken arrows depict the direction of electron flow. Solid arrows also indicate conventional current flow.

The electrical characteristics of the diode are represented by the current-versus-voltage curve (see Figure 3.9). The graph in Figure 3.9 represents the forward and reverse bias characteristics of a typical silicon diode. Note the breakdown voltage at the knee of the forward bias curve. This voltage is characteristic of silicon diodes and is generally in the range of 0.6 to 0.8 volts. The reverse bias curve shows a point when the diode will conduct in the opposite direction when the junction voltage is exceeded.

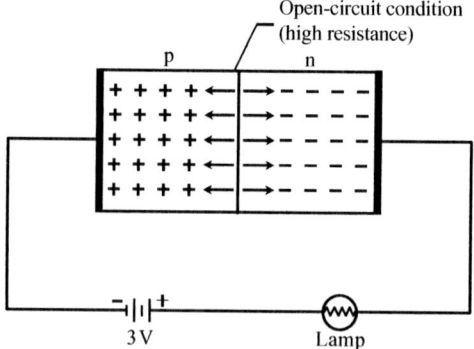

FIGURE 3.7 Reverse-Biased pn Junction Diode. Holes and electrons are drawn away from the pn junction, essentially creating an open-circuit condition.

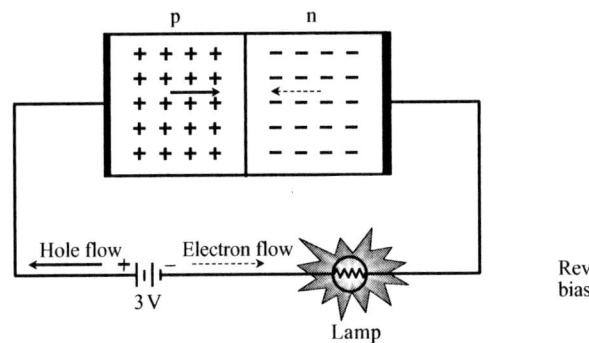

FIGURE 3.8 Forward-Biased pn Junction Diode

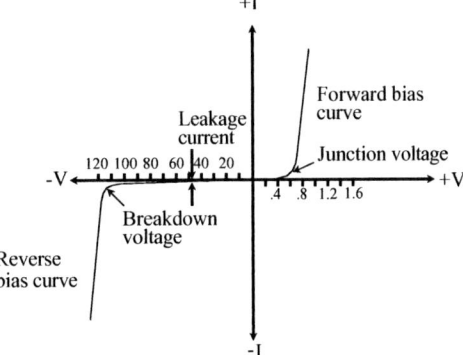

FIGURE 3.9 Forward and Reverse Electrical Characteristics of a Silicon Diode

The Bipolar Junction Transistor

The *bipolar junction transistor* (BJT) has three electrodes and two pn junctions. The entire transistor is constructed from a single semiconductor substrate. There are two variations of the bipolar junction transistor—*npn* and *pnp*. The simplified structures and schematic symbols of these transistors are shown in Figure 3.10. The electrodes are labeled *emitter, base,* and *collector*. The emitter arrows indicate the direction of hole flow through the respective transistor.

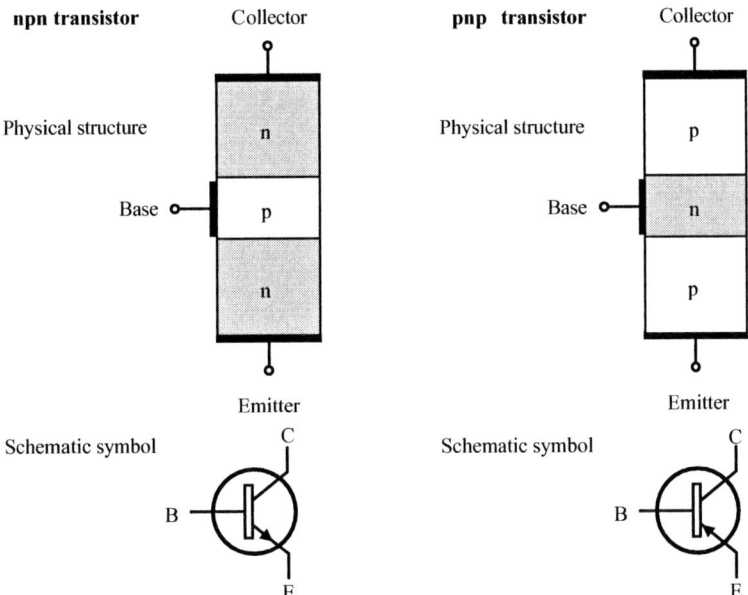

FIGURE 3.10 Two Types of Bipolar Transistors

Biasing the npn Transistor for Conduction Mode ■ The biasing scheme for proper operation of the npn transistor is shown in Figure 3.11 on page 50. Note the polarity of the two bias supplies. The emitter-base junction is forward-biased by a smaller bias supply, i.e., a simple 1.5-volt cell. The collector bias supply, a 3-volt battery, is connected in reverse polarity relative to the n-type collector. The emitter is shown as the reference point (ground) for all voltages. In this circuit configuration, the base serves as the input to the transistor and the collector is the output. The lamp serves the purpose as an output load and as an indicator of current flow through the collector.

There can be no conduction through the collector of the transistor without a completed circuit path through the emitter-base (E-B) junction. This nonconducting mode is shown in Figure 3.11 on page 50. As long as switch S_1 remains open, the pn junction of the emitter and base remains in a nonconducting mode (open circuit). This condition will not allow for current flow from emitter to

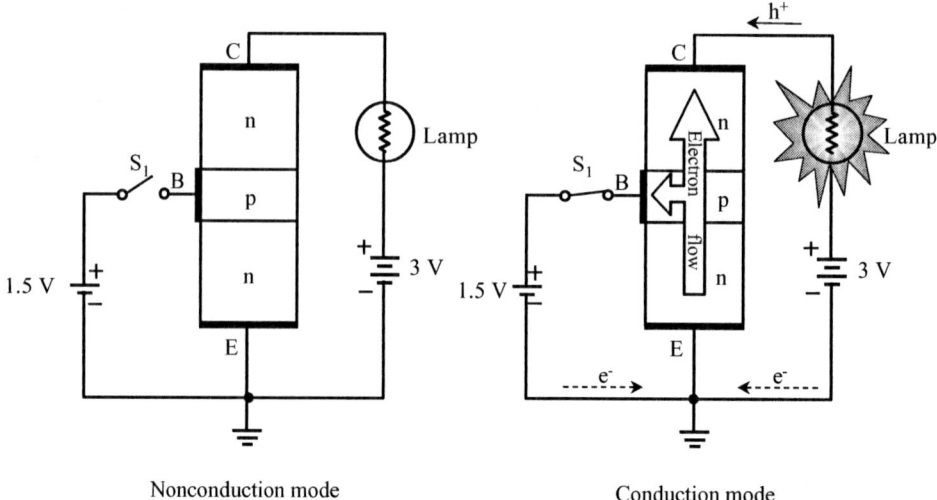

FIGURE 3.11 npn Transistor Biasing Circuit

collector. Closing switch S_1 forward-biases the E-B junction (the same as a diode). Conduction through the E-B junction immediately reduces the barrier and allows electrons to be injected from the negative side of the 3-volt battery through the E-B region and into the collector. At the same time holes are injected from the positive side of the collector battery through the lamp and then into the top of the collector to recombine with electrons. The lamp now indicates current flow through the output section of the transistor. This condition will continue until a change in S_1 or one of the circuit elements is disconnected.

Note that in Figure 3.11 solid arrows illustrate the direction of hole flow and broken arrows depict the direction of electron flow. Solid arrows also indicate conventional current flow.

Biasing the pnp Transistor for Conduction Mode ■ The operation of the pnp transistor is similar to the npn except that the bias supply connections are reversed. Conduction of a pnp transistor requires the same conditions as with the npn: first, the E-B junction must be forward biased (See Figure 3.12), and second, the collector supply voltage, V_{CC}, must be reversed relative to the collector. Upon initial conduction through the E-B junction, holes are injected from the positive side of the 1.5-volt supply through the E-B junction and into the collector. Electrons from the 3-volt battery leave the negative side of the battery, flow through the lamp, and enter the top of the collector to recombine with holes.

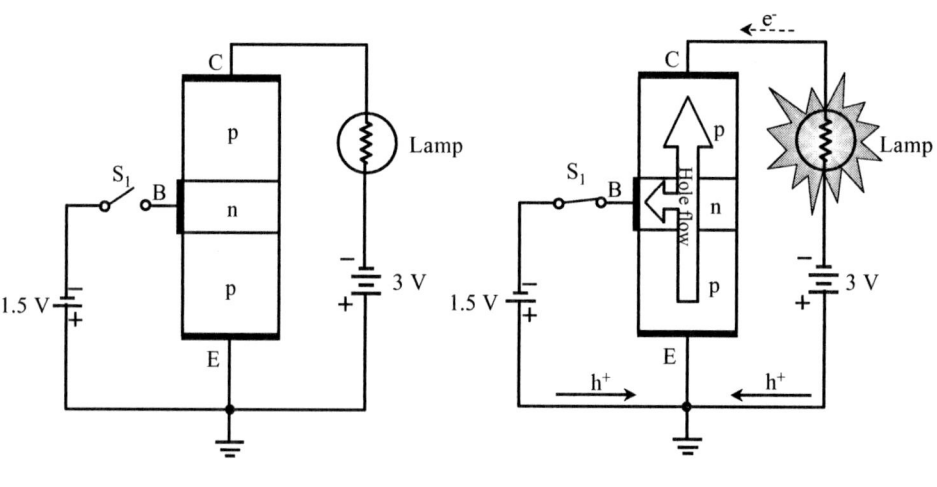

FIGURE 3.12 pnp Transistor Biasing Circuit

Structure of a Bipolar Junction Transistor ■ Figure 3.13 illustrates the structural features of the cross section of an npn bipolar junction transistor. Note the differences in the amount of dopant concentration in each of the electrodes. The emitter (E) and collector (C) are both heavily doped with n-type dopant, such as arsenic or phosphorus. The base (B) is lightly doped with boron, a p-type dopant. With fewer current carriers in the base, the current drawn through the base is significantly smaller than the current drawn through the collector. This difference accounts for the *gain* (amplification) in current from the input to the output of the transistor. This feature is the major difference between the transistor and the diode. The transistor is able to amplify small input signals literally hundreds of times to drive output devices such as speakers, motors, lamps, relays, and other electromechanical devices. The BJT is a *current-driven current amplifier* device. Thus, bipolar transistors are commonly used in radios, tape recorders, automotive electronics, aircraft control systems, biomedical instrumentation, robotics, and wherever there is a need for high power control.

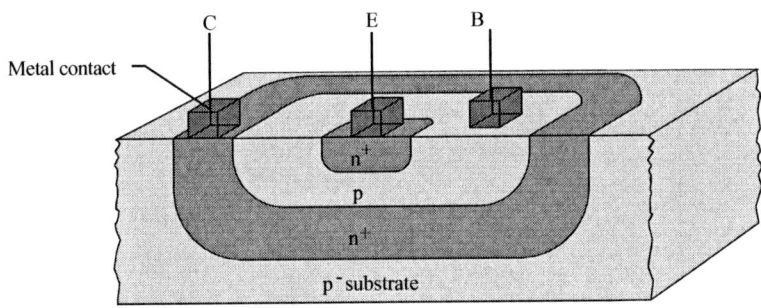

FIGURE 3.13 Cross Section of an npn BJT

Schottky Diode

The *Schottky diode* is formed when metal is brought in contact with lightly doped n-type semiconductor material. The resulting device operates just like an ordinary pn junction diode—low resistance when forward-biased and high resistance when reverse-biased. The forward junction voltage drop of the silicon Schottky diode (0.3 to 0.5 volts) is nearly half that of the silicon pn junction diode (0.6 to 0.8 volts). The Schottky diode's biggest advantage is that its conduction is due entirely to electrons, which results in faster switching time from on to off.[4] Figure 3.14 shows the schematic symbol and structured cross section of the Schottky diode.

The discovery of the Schottky diode has helped extend the usefulness of bipolar IC technologies into the twenty-first century. The Schottky diode concept has been utilized in the development of faster and more power efficient bipolar integrated circuits.

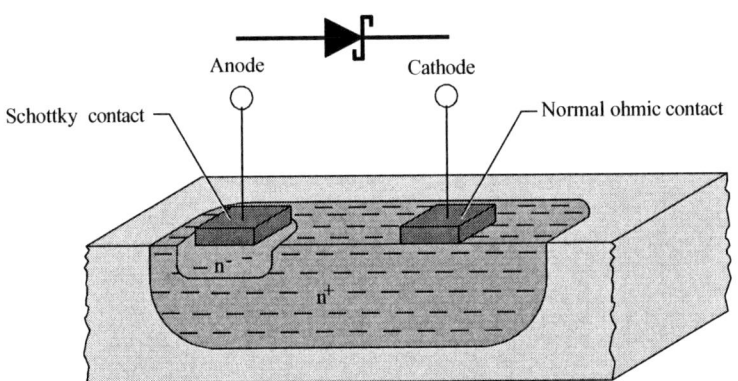

FIGURE 3.14 Schematic Symbol and Structural Cross Section of the Schottky Diode

Bipolar IC Technology

Diodes and bipolar transistors—along with supporting components of resistors, capacitors, insulators, and conductors—are used in the development of a breed of integrated circuits called *bipolar technology*. Bipolar technology was the first of many IC technologies that was used in the production of analog and digital integrated circuits. For many years bipolar devices have been noted for fast speeds, durability, and power-controlling abilities. The biggest drawback, however, has been their high power consumption. This, of course, meant higher costs for operating these devices in terms of electrical utilities and batteries for portable equipment. Some families of the bipolar technology era have since become obsolete. Other bipolar manufacturers have managed to maintain their business based on a continued need for bipolar devices for use in high-power applications. A list of the bipolar logic families is shown in Table 3.1. Further information about bipolar technology can be obtained by consulting the reference list at the end of this chapter.

TABLE 3.1 Bipolar Logic Families

Bipolar Logic Family	Abbreviation
Direct-Coupled Transistor Logic	DCTL**
Resistor-Transistor Logic	RTL**
Resistor-Capacitor-Transistor Logic	RCTL**
Diode-Transistor Logic	DTL**
Transistor-Transistor Logic*	TTL**
Schottky TTL Logic*	STTL†
Emitter-Coupled Logic*	ECL†

*Some forms of TTL, STTL, and ECL still in use through were 2000. **From G. Deboo and C. Burrous, *Integrated Circuits and Semiconductor Devices: Theory and Application,* 2nd ed. (New York: McGraw-Hill, 1977), p. 192. †From A. Sedra and K. Smith, *Microelectric Circuits* (Oxford: Oxford University Press, 1998), p. 1187, 1196.

CMOS IC Technology

When energy conservation became an international concern in the late 1970s, the semiconductor industry responded with the *field-effect transistor* (*FET*), which benefits from more compact and power-efficient electronic devices. Although early experiments with FETs date back to the 1930s, the first mass-produced FETs became available in the 1960s. Several versions of the FET have been used since then. Today the most popular IC technology, *CMOS* (*complementary metal-oxide semiconductor*), revolves around the improvements that have been made in FET design and manufacturing. The remainder of this text concentrates on the development and manufacturing processes related to CMOS devices.

The Field-Effect Transistor ■ The development of the field-effect transistor essentially created a new era in the history of the semiconductor industry. As opposed to the current-amplifying bipolar junction transistor, the FET is a *voltage-amplifying device.* The only similarities between the BJT and the FET are in the types of materials used to build the transistors and the number of electrodes. Both have three electrodes and both types are constructed of a single crystalline substrate. The greatest advantage of the FET is its low voltage and low power requirements. Whereas the BJT requires input current at the base to turn on the transistor, the FET turns on as a result of an electric field created when an input voltage is applied to the *gate*—thus the name field-effect transistor.

The FET has seen applications as an amplifier in linear/analog circuits and as a switching component in digital electronics. Its high input resistance and moderate gain characteristics make it an excellent device for use in instrumentation and communications. Its low power consumption and compactness make it extremely suitable for the ever-shrinking dimensions of VLSI and ULSI technology.

There are two basic types of FETs: the junction (*JFET*) and the metal-oxide (*MOSFET*) semiconductor. The major difference between these two types is that the gate on the MOSFET, which is the input to the FET, is insulated by a thin dielectric (silicon dioxide, referred to as *gate oxide*) from the other two electrodes of the transistor. The gate of the JFET actually forms a physical pn junction with the other electrodes of the transistor. JFETs are used extensively in GaAs integrated circuits. When metal gates are used in GaAs JFETs, the term *MESFET* is used. Because of the popularity of MOSFETs in silicon VLSI applications, the remainder of our discussion of FETs will focus on the MOSFET. Consult the reference list at the end of this chapter for further information on the JFET.

MOSFETs. MOSFETs gained wide acceptance in logic applications in the 1970s and have been the mainstay transistor in IC products ever since. There are two categories of MOSFETs: *nMOS* (*n-channel*) and *pMOS* (*p-channel*). Each type is distinguishable by the majority current carriers in each device. Figure 3.15 shows some common schematic symbols and physical cross sections of the two types of MOSFETs.

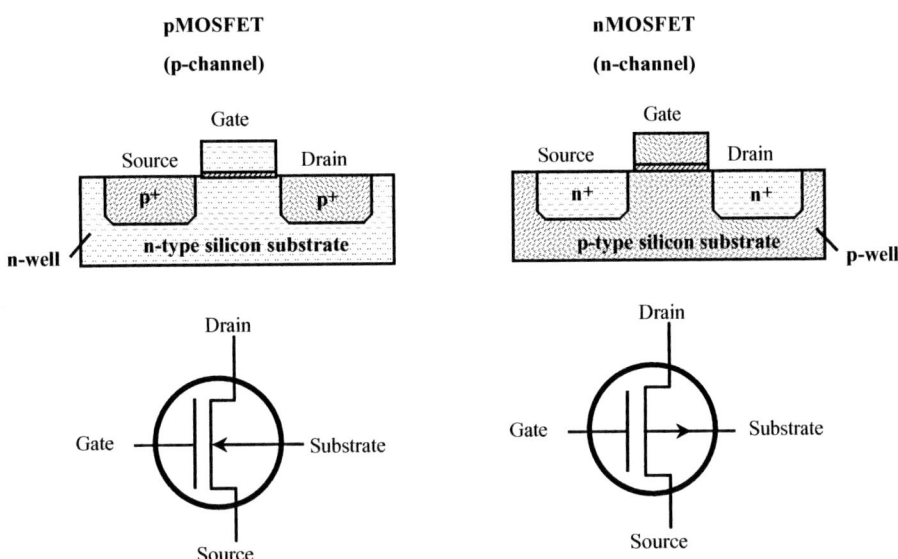

FIGURE 3.15 Two Types of MOSFETs

Each MOSFET has an input electrode called the gate. The term *metal oxide* refers to the material the gate is made of. However, the "metal" descriptor in MOS derives from the early days of MOS technology and is no longer true. The most popular material used in the formation of gates for MOSFETs is *polysilicon,* a polycrystalline silicon material that is deposited on the substrate during IC fabrication (see Chapter 11). The polysilicon, however, must be doped with one of the common p- or n-type dopants to give the material its conductive characteristics.

The *source* and the *drain* electrodes are heavily doped, respectively, with n-type or p-type dopant depending on the category of transistor being manufactured. The supply of majority current carriers comes from these two electrodes. The *nMOSFET* uses electrons as the majority carriers; therefore, the channel is n-type. The *pMOSFET,* then, has a p-channel formed by holes from the source and drain. When in the nonconduction mode, the channel is an open circuit consisting of an opposing doped region called the *well.* The *n-channel MOSFET* is built inside a *p-well,* while the *p-channel MOSFET* is built inside an *n-well.* When in the conduction mode, opposing carriers in the upper part of the well move away from the gate oxide interface and a channel of majority current carriers forms across from the source to the drain to complete the circuit (very similar to closing a switch).

Biasing the nMOSFET for Conduction Mode. Figure 3.16 on page 54 illustrates the biasing scheme for operating an n-channel MOSFET. The lamp in the circuit is an indicator of output current and normal operation of the transistor. At the moment there is no conduction in the n-channel MOSFET in Figure 3.16 because there is no input voltage applied to the gate. The condition of the channel is that of an open circuit. The majority of carriers in the area immediately beneath the gate oxide and between the source and drain electrodes are holes. At the moment the source and the p-well are at the same potential. As a pn junction there is no forward bias applied, so the pn junction remains in a nonconducting mode.

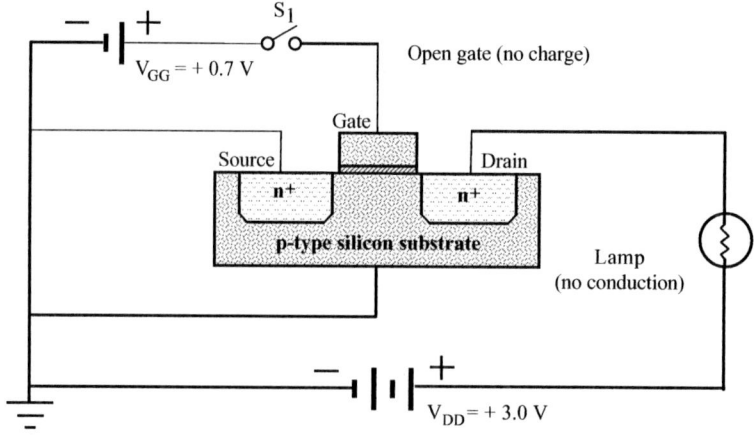

FIGURE 3.16 Biasing Circuit for an nMOS Transistor

When the switch, S_1, is closed as in Figure 3.17, certain conditions begin to happen. The small bias voltage of 0.7 V places a charge equal to the bias supply directly across the gate and source electrodes. An electrostatic charge is created with positive holes on the gate and electrons flowing up from the source to the gate oxide interface. At the same time an electric field is created by the positive charges at the gate. This field opposes the electric field of the holes in the upper part of the p-well. These opposing fields force the holes in the upper part of the p-well to move away from the gate oxide interface. As a result, electrons move from the source and drain to fill the gaps left vacant by the holes. The electric field from the positive charge on the gate continues to attract and hold the electrons in the channel area, thus closing the gap between the source and the drain with only electrons. Now, the electrons form a continuous crystalline structure of n-type silicon. Electrons from the 3-volt battery flow freely from the negative terminal through the source, to the n-channel, into the drain, and then through the lamp and back to the positive side of the battery. This condition will remain the same until any part of the input or output circuit is changed.

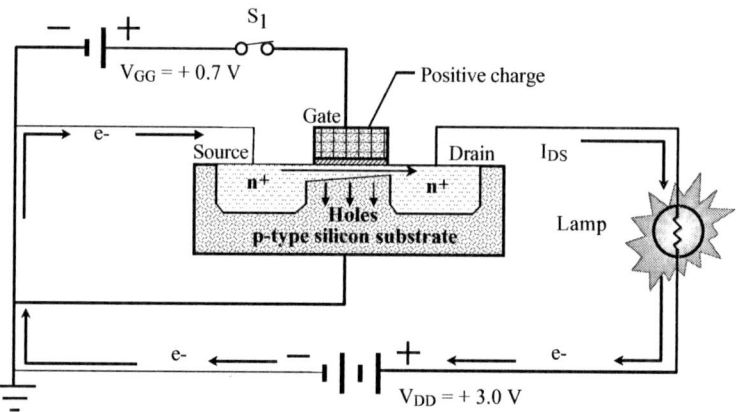

FIGURE 3.17 nMOS Transistor in Conduction Mode. Closing S_1 allows V_{GG} to place a positive charge on the gate of the transistor that causes an n-channel to form; therefore, allowing electrons to flow from the source to the drain.

Increasing the voltage at the gate increases all the electrical activities just mentioned. The electrostatic charge on the gate increases, which increases the electric field strength and forces the holes in the p-well to move further away from the gate-oxide interface. This results in an increase in the size of the n-channel and the number of electrons flowing through the channel. The net effect is for more drain current to flow, thus increasing the power delivered to the load (increased brilliance of the indicator lamp).

Figure 3.18 provides an example of a graphical representation of the electrical characteristics of an n-channel MOSFET. The graph represents drain current, I_{DS}, as a function of V_{DS} and V_{GS}. Each curve labeled V_{GS} represents a specific setting of the gate-source voltage. Using the curves it is possible to determine the value of I_{DS} for a given value of V_{GS} and V_{DS}. For example, assume $V_{GS} = 4$ V and $V_{DS} = 2$ V. The resulting drain current, $I_{DS} = 0.28$ mA. The *threshold voltage* is the lowest attainable V_{GS} value that will turn on the FET. This value is usually less than 1.0 V and is dependent on several variables, including structural sizes and doping characteristics of the electrodes and p-well.

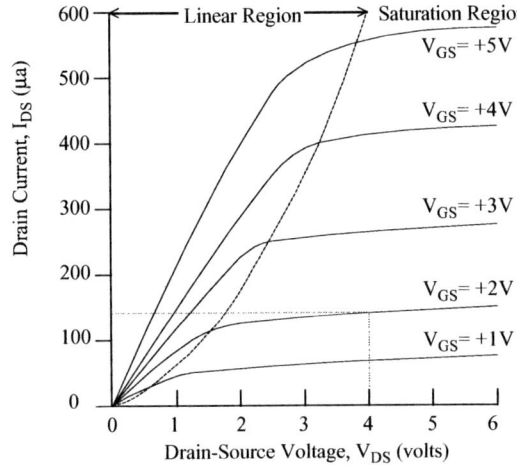

FIGURE 3.18 Example of Characteristics Curves of an n-channel MOSFET

Biasing the pMOSFET for Conduction Mode. Figure 3.19 illustrates the biasing circuit for a p-channel MOSFET. The operation of the p-channel transistor is similar to the operation of the n-channel type. The only difference is that the majority current carriers in the p-channel MOSFET are holes and the bias supplies are reversed. The other major difference between these two transistors is in terms of performance. The speed of the p-channel transistor is slower than that of its counterpart, primarily due to the holes, which move slower than electrons.

Closing S_1, allows V_{GG} to place a negative charge on the gate of the transistor that causes a p-channel to form, therefore allowing holes to flow from the source to the drain (see Figure 3.20 on page 56).

CMOS Technology ■ Manufacturers of MOSFET-based integrated circuits for many years concentrated on the development and manufacturing of products based solely on n-channel MOSFET technology. Whereas some semiconductor manufacturers produced both discrete nMOS and pMOS transistors, few, if any, manufactured integrated circuits made primarily with pMOS transistors.

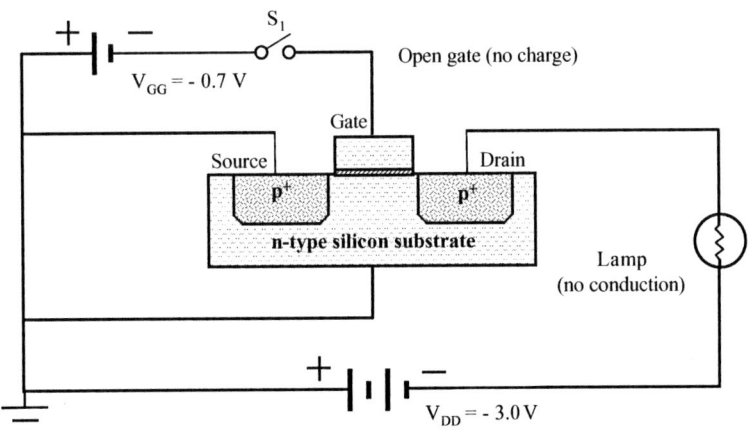

FIGURE 3.19 Biasing Circuit for a p-Channel MOSFET

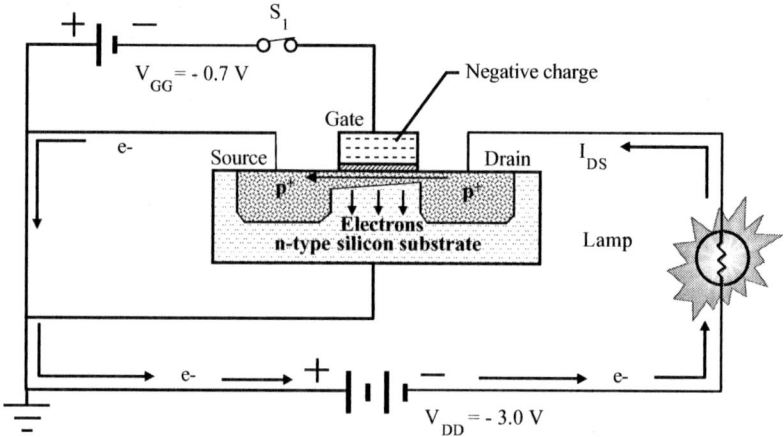

FIGURE 3.20 pMOS Transistor in Conduction Mode

Discrete pMOS transistors have served many usable functions in specific electronic applications, but in general, nMOS IC devices outperformed the pMOS technology. Thus, nMOS became the choice of most IC manufacturers.

CMOS incorporates both nMOS and pMOS transistors in the same integrated circuit. The combination of power efficiency, design scaling techniques, and improved manufacturing processes have made CMOS one of the most popular device technologies since the early 1980s. The term *scaling* has been commonly used to describe the process of *shrinking* the overall dimensions and operating voltages of existing ICs to attain improved operating performance as well as compactness. All dimensions and voltages must shrink in unison (thus the source of the word scaling) through the use of a design model employed by the IC designer during the circuit design and layout stage.

A schematic diagram of a simple CMOS inverter circuit is shown in Figure 3.21. The schematic shows the gates of both transistors connected together. These gates serve as a single input to the inverter. The output of the inverter is taken out of the two drains, which are tied together. The source of the n-channel transistor is grounded, while the source of the p-channel transistor is connected to the bias supply, V_{DD}. A signal applied to the input of the CMOS is inverted at the output, as shown in Figure 3.21. Normal operation has the n-channel operating during the positive transition of the input signal and the p-channel operating during the negative transition. The efficiency of the CMOS inverter circuit occurs during the transition period when the input signal is at zero. There is no power consumed by the transistors when the input signal is at zero. nMOS, TTL, and ECL circuits differ from CMOS in that those logic families all dissipate power even with no signal applied. This fact is one major reason why CMOS is currently the preferred IC technology for use in the manufacture of portable electronics products such as calculators, clocks, cellular telephones, and notebook computers.

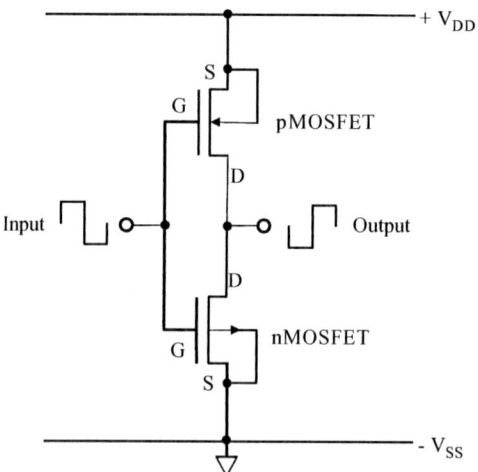

FIGURE 3.21 Schematic of a CMOS Inverter

The physical structure of a simple CMOS inverter is shown in the top view in Figure 3.22 and cross-sectional view in Figure 3.23. The n-substrate serves as the n-well for the p-channel transistor. A separate p-well must be created for the n-channel transistor. This well is created by a selective doping process called *ion implantation*. Isolation regions separate transistors from each other and from other transistors (not shown). The isolation regions are referred to as *field oxide,* which is manufactured from silicon dioxide (also called *glass*). The field oxide insulates transistors from each other in order to prevent the flow of undesirable leakage current between transistors. The actual electrical connections between the two transistors are made possible with metal deposited over the insulated transistors. Later, a patterning process is used to define the exact locations where the metal connections are to be made, followed by an etching process to remove the excess metal. Further details of the CMOS manufacturing process are beyond the scope of this chapter. Chapter 9 focuses on the specific physical and chemical requirements and process flow in CMOS IC manufacturing.

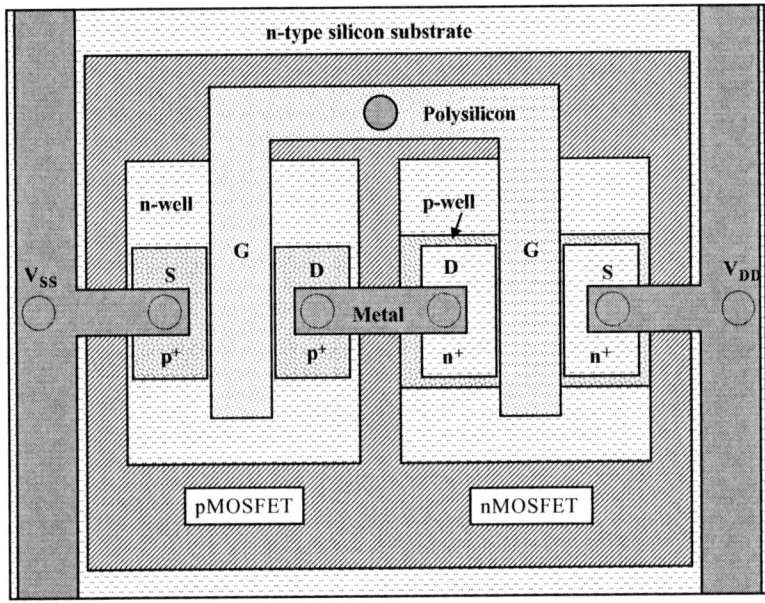

FIGURE 3.22 Top View of CMOS Inverter

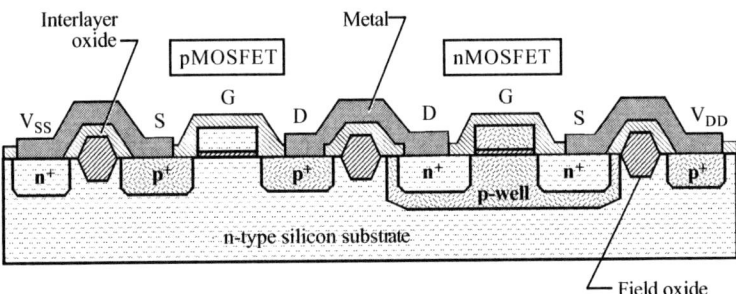

FIGURE 3.23 Cross-section of CMOS Inverter

BiCMOS Technology ■ *BiCMOS technology* makes use of the best features of both CMOS and bipolar technology in the same IC device. BiCMOS incorporates the low-power, high-density CMOS structures with the high-current drive capabilities of TTL or ECL device structures. Applications of BiCMOS products can be found wherever the need for complex digital control of high-power loads is desired. In which case, *digital/analog* (D/A) converter chips may be used to provide the analog drive signals that are used to control electromechanical equipment. On the instrumentation side, *analog/digital* (A/D) chips may be used to measure the outcome of analog drive signal. Figure 3.24 shows a basic example of BiCMOS chips used in an instrumentation and control application. Other applications for BiCMOS include automotive electronics, aerospace, robotics, and industrial equipment.

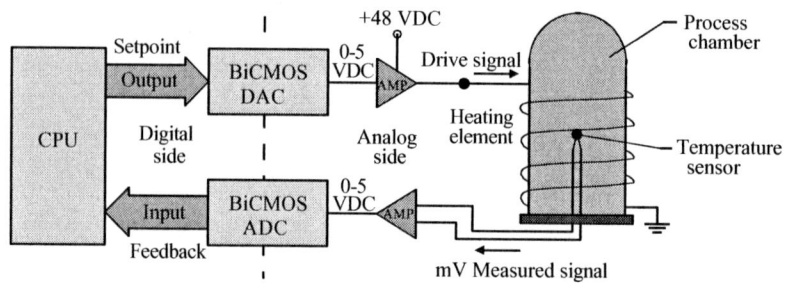

FIGURE 3.24 BiCMOS Chips used in the Control of a Simple Heating System

Figure 3.25 shows additional bipolar transistors added to the CMOS inverter mentioned earlier in Figure 3.21 on page 56. The bipolar transistors (Q3, Q4) provide greater current drive than the CMOS inverter (Q1, Q2) is capable of.

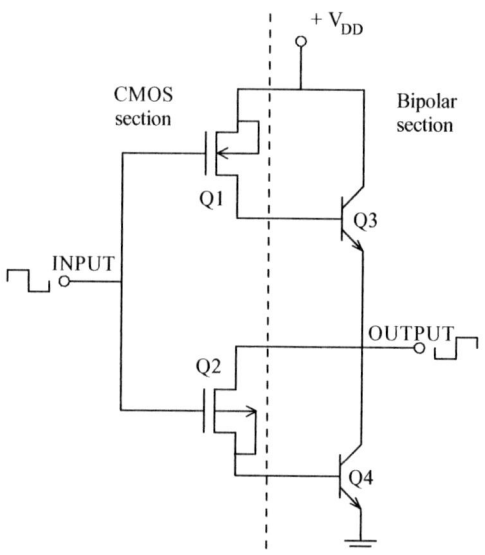

FIGURE 3.25 Simple BiCMOS Inverter
Redrawn from H. Lin, J. Ho, R. Iyer, and K. Kwong, "Complementary MOS-Bipolar Transistor Structure," *IEEE Transactions Electron Devices,* ED-16, 11 Nov. 1969, p. 945–951.

Enhancement and Depletion-Mode MOSFETs

Another way to categorize FETs is in terms of *enhancement mode* and *depletion mode*. Up to now all n-channel and p-channel MOSFETs discussed in this chapter have been enhancement-mode transistors. Enhancement-mode MOSFETs are the most commonly used transistors in the industry. The difference between the enhancement-mode and depletion-mode transistors is shown in Figure 3.26.

In general, the major difference between enhancement- and depletion-mode transistors is in the way the channels are doped. While the enhancement-mode channels are doped opposite in polarity to the source and drain regions, the depletion-mode channels are doped the same as their respective source and drain regions. This setup creates open and closed circuit conditions, respectively, for the enhancement- and depletion-mode transistors. Consequently, the enhancement-mode transistor is regarded as *normally-off* and the depletion mode as *normally-on*.

The enhancement-mode transistor works very well in digital logic applications and requires only a single polarity input signal (V_{GS}) to operate the FET. On the other hand, the depletion-mode transistor is partly turned on by the already existing closed channel. The input voltage can swing in one direction to increase the current through the channel or reverse in the opposite direction to decrease the current through the channel. The depletion-mode transistor will turn off completely if the input voltage to the gate is further increased in the opposite direction. Both types of transistors can be designed into practical applications in both analog and digital circuit applications in either discrete component applications or in integrated circuit applications.

MOSFET Type	Mode	Standby Condition	V_{GG} Switching Requirements	Physical Structure
nMOS	Enhancement	Off	+	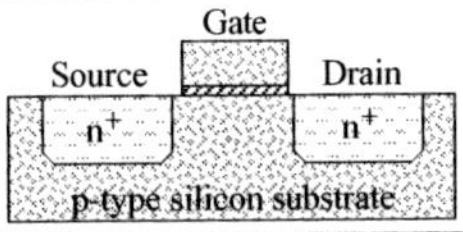 Gate, Source, Drain, n^+, n^+, p-type silicon substrate
nMOS	Depletion	On	−	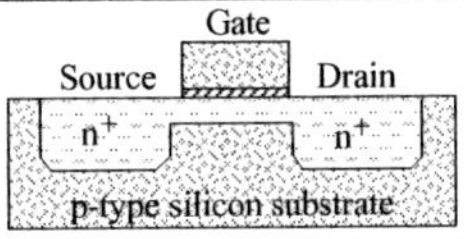 Gate, Source, Drain, n^+, n^+, p-type silicon substrate
pMOS	Enhancement	Off	−	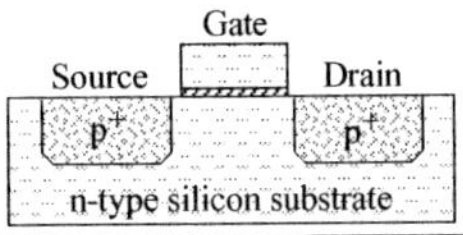 Gate, Source, Drain, p^+, p^+, n-type silicon substrate
pMOS	Depletion	On	+	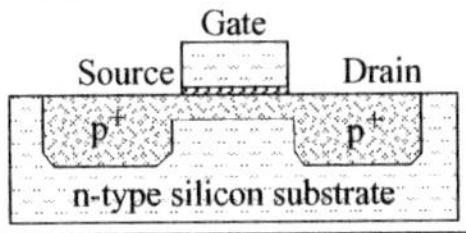 Gate, Source, Drain, p^+, p^+, n-type silicon substrate

FIGURE 3.26 Comparison of Enhancement- and Depletion-Mode MOSFETS

LATCHUP IN CMOS DEVICES

Just as undesirable parasitic resistances and capacitances exist in semiconductor devices, sometimes the pn junctions in CMOS devices can produce *parasitic transistors* that can create a *latchup* condition in CMOS ICs that causes transistors to unintentionally turn on. Figure 3.27 illustrates the parasitic transistors in a CMOS inverter structure. Complementary junction transistors are formed as a result of normal manufacturing of MOSFETs in a CMOS structure. Given certain operating conditions it is possible for the parasitic junction transistors to turn on and create a low resistance path for current to flow across the CMOS structure.[5] The transistors become latched, thus preventing any further control of the MOSFETs in the CMOS device.

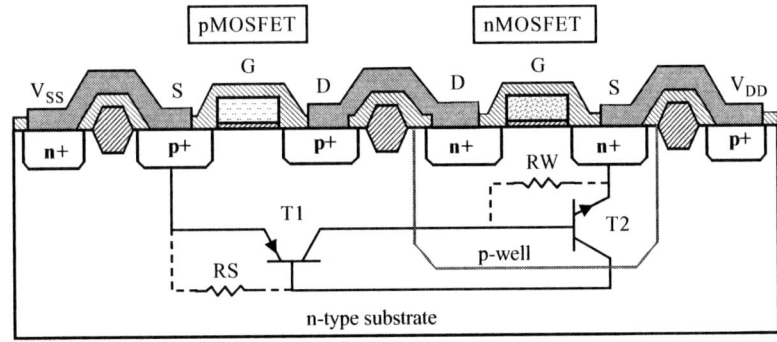

FIGURE 3.27 Parasitic Junction Transistors within a CMOS Structure

Indeed, the latchup phenomenon is a very complex concept. Any further discussion is beyond the scope of this chapter. Suffice it to say that knowing about its existence will help you understand the design and manufacturing steps that are taken to prevent this condition. Later chapters describe several manufacturing techniques used in the prevention of latchup, namely creating isolation barriers between transistors, placing an epitaxial layer explained in Chapter 11 between the substrate and the CMOS structures, and using ion implantation to create retrograde wells explained in Chapter 17.

INTEGRATED CIRCUIT PRODUCTS

Semiconductor products can be found in most electrical and electronic apparatuses throughout the world. The most familiar applications of ICs are found in the personal computer (PC). Such a computer would employ a microprocessor unit (MPU), read-only memory (ROM), random-access memory (RAM), communications interface chips, video graphics controller chips, disk drive controller chips, and more. For the most part, these chips are built using CMOS technology, followed by BiCMOS and bipolar technologies.

The types of ICs used vary depending on customer needs, environment, cost, power requirements, and other specifications. Most ICs, as was mentioned earlier, fall into two basic categories —analog or digital. The following sections list the most common types of ICs found in most consumer and industrial applications.

Linear IC Product Types

The *linear IC* family consists of devices that are designed to operate primarily in analog circuit applications. Linear ICs are typically used in audio systems, radio communications, industrial controls and instrumentation, aerospace, and in automotive electronics. The following are a few examples of linear ICs.

Operational Amplifier ■ The *op-amp* is a high-gain and high-input impedance amplifier that can be adapted to a variety of electronic control applications. It can be used as an amplifier, oscillator, voltage regulator, voltage limiter, sample-and-hold amp, voltage rectifier, integrator, and differentiator.

Voltage Regulator ■ The *voltage regulator* is a single-chip device containing diodes, transistors, resistors (VR) and capacitors for the sole purpose of regulating the voltage delivered to a load. VRs can be used in any electronic subassembly or system where a constant voltage source is needed to maintain a continuous voltage over a specific range of load resistances. Some applications include computers, peripherals, and various types of instrumentation.

Stepper Motor Driver ■ The *stepper motor driver* (*SMD*) is a bipolar IC used for controlling stepper motors. An SMD can be used any place where a stepper motor must be accurately controlled, such as in laser printers, photocopiers, scanners, and robotics, as well as aerospace, automotive, and industrial applications.

Digital IC Product Types

The digital IC family includes devices that operate with binary (1s and 0s) bits of data signals, such as those used in calculators, computers, digital pagers, cellular phones, and many others similar applications.

Volatile Memory ■ *Volatile memory* is a semiconductor device that allows data to be stored and changed as desired. Data in the volatile memory is lost when the power is turned off. Volatile type memory devices are used in computers, calculators, and appliances as well as automotive, aerospace, medical, military, and industrial equipment—in any application where logic instructions must be stored and changed as needed by the user. All memory devices contain millions of individual memory cells. The probability of defects increases with device density; therefore, redundant memory cells are built into memory chips. Semiconductor fuses are built into the memory cells that are deliberately blown during the initial product test to disable defective cells.

RAM. A *random-access memory* device is one that can be accessed to read its stored data or it can be erased and new data rewritten into it. RAM chips do not need to be removed from the printed circuit board to change the data held in storage within its memory cells. Its contents can be easily changed in the normal operations of the logic system.

DRAM. The *dynamic RAM* is the most common and least expensive of the RAM family. The term *dynamic* refers to the refresh voltage that must be applied regularly to the storage capacitors to retain the data. DRAMs require more power to operate the capacitors.

SRAM. The *static RAM* uses flip-flops as data storage registers. SRAMs do not require refreshing; thus, they use less power than the DRAMs. The data is also lost when the power is removed.

MPU or CPU. *Microprocessor units* (also known as *central processing units*), are complex logic ICs that are capable of executing instructions programmed into a separate or internal ROM. The MPU is capable of decision making and of performing math functions. Microprocessors are used in computers, calculators, and appliances as well as automotive, aerospace, medical, military, and industrial equipment—in any application where a controlling function is needed or where computer functions are a requirement.

Nonvolatile Memory ■ The *nonvolatile memory* is a semiconductor device designed to store digital data in the form of an electrical charge. The charge remains in storage even after the power is turned off. Nonvolatile type memory devices are used in computers, calculators, and appliances as well as automotive, aerospace, medical, military, and industrial equipment—in any application where logic instructions must be stored for later recall.

ROM. A *read-only memory* IC is nonvolatile and is programmed directly in the IC manufacturing process. It is also referred to as a *mask-programmable* ROM. The mask sets used during manufacturing contain code patterns for a specific ROM. ROMs are very expensive to produce. ROMs were the earliest version of nonvolatile memory devices. The consumer can read electronically what has been coded into the ROM but cannot change its contents.

PROM. A *programmable read-only memory* is an IC that can be reprogrammed in the field and is less expensive than the mask-programmable ROM. This tool applies large voltage pulses that allow memory cells to be changed as needed. The data programmed into this memory cannot be changed during normal operation of a computer. There are two types of *field-programmable* proms —the EPROM and the EEPROM.[6]

EPROM. An *erasable PROM* can be erased and reprogrammed in the field. The chip must be removed from the circuit board and erased by exposure to UV light. The EPROM can then be reprogrammed. The EPROM is an improvement over its two predecessors; however, any change in the stored code requires complete erasure of the PROM followed by reprogramming.

EEPROM. The *electrically erasable PROM* IC can be electrically erased and reprogrammed in the field without having to remove it from the circuit board. The EEPROM is the most convenient form of ROM. The device operates more like a random-access memory in that it can be erased and rewritten without the use of any special equipment. EPROMs have many useful applications in the computer and peripheral industry.

Flash RAM. Flash memory is a type of nonvolatile memory that can be erased and reprogrammed—similar to electrically-erasable programmable read-only memory (EEPROM), but faster to update. Flash memory is often used to store the operating code such as the Basic Input/Output System (BIOS) in a personal computer. Flash memory is used in computers, digital cellular phones, digital cameras, embedded controllers, and others products.

ASIC. *Application-specific ICs* are fully custom-designed and manufactured to meet an individual customer's needs. The ASIC device may include existing logic circuitry with new design features as requested by the customer. ASIC chips are also nonvolatile. All logic functions of the IC are determined by specific masks that have been designed to achieve unique structural features given in the specific ASIC design specifications. While ASIC devices provide the exact product to meet the user's needs, the cost of manufacturing them is high.

PLD. *Programmable logic devices* are ICs that utilize a wide variety of logic components. The actual logic function implemented is determined by the user, using some form of design software to specify the state of the internal programming points.[7] During the programming process, higher than normal voltages are applied to specific semiconductor fuses. The fuses heat up and vaporize leaving behind a desirable logic circuit configuration as programmed by the user.

PAL. A *programmable array logic* IC contains a network of programmable logic gates used to create a custom logic circuit. The PAL has an array of input AND gates that drive an array of hard-wired output OR gates. The AND gates can be field-programmed as required, but the output OR gates are fixed.[8]

PLA. A *programmable logic array* differs from the PAL in that both the input AND gates and the output OR gates are programmable. PLAs are available in both mask-programmed and field-programmed versions.

MPGA. A *mask-programmable gate array* chip can be customized to meet the functional needs of individual customers. Initially during the manufacturing process all gate array products passing through a specific manufacturing facility may contain the same basic number and configuration of transistor circuitry. It is during the interconnect layers of the manufacturing process that specific masking steps are used to define the function of thousands of transistors within individual chips. The mask sets are designed in cooperation with customers to define the proper interconnects for a specific gate-array application.[9]

FPGA. A *field-programmable gate array* is an IC device in which the final logic product can be defined in the field without the use of an IC manufacturing facility.[10] Programmable logic switches within the FPGA can be activated to create connections between desired logic circuits. The convenience of customizing in the field and its low manufacturing cost make the FPGA a viable alternative to the MPGA. However, the MPGA device operates faster than the FPGA since all interconnects are hard-wired internally during the IC manufacturing process.

SUMMARY

Electronic components are assembled into analog or digital circuits. Passive components, such as resistors and capacitors, conduct current without specific polarity requirements and are constructed on silicon substrates. These components have parasitic losses that are detrimental to IC performance. Active components, such as transistors and diodes, control the direction of current flow. The pn junction diode can form an open circuit (reversed biased) or closed circuit (forward biased). The bipolar junction transistor (BJT) can be biased to conduct current and also amplify input signals. The structure of the BJT is based on n-type or p-type dopant regions that form two pn junctions per transistor as either an npn or pnp. The Schottky diode has faster switching times, which has helped bipolar IC technology extend its applications in to the twenty-first century.

CMOS IC technology is more power efficient than bipolar and is one of the most popular device technologies with the most common transistor being the MOSFET as either an nMOS or pMOS. The MOSFET has three electrodes, the gate, source and drain that are attached to a biasing circuit for conduction mode. The minimum voltage to turn on a MOSFET is the threshold voltage. An example of CMOS technology is the inverter circuit. BiCMOS technology combines both CMOS and bipolar technology. MOSFETs can be in enhancement (normally-off) or depletion (normally-on) mode for both analog and digital applications. Parasitic transistor effects can occur in pn junctions that create a latchup condition that causes the transistor to turn on with no further control. There is a wide range of IC products, including analog and digital.

KEY TERMS

- structure
- analog circuit
- linear circuit
- digital circuit
- logic devices
- device technology
- passive components
- parasitic resistance
- planar capacitors
- parasitic capacitance
- active components
- pn junction diode
- carrier-depletion region
- barrier voltage
- reverse bias
- forward bias
- bipolar junction transistor (BJT)
- npn
- pnp
- emitter
- base
- collector
- current-driven current amplifier
- well
- n-channel MOSFET
- p-well
- p-channel MOSFET
- n-well
- scaling
- shrinking
- field oxide
- glass
- BiCMOS technology
- digital/analog (D/A)
- analog/digital (A/D)
- enhancement mode
- depletion mode
- normally-off
- normally-on
- parasitic transistors
- latchup
- op amp
- voltage regulator (VR)
- stepper motor driver (SMD)
- volatile memory
- random-access memory (RAM)
- gain
- Schottky diode
- bipolar technology
- field-effect transistor (FET)
- CMOS (complementary metal-oxide semiconductor)
- voltage-amplifying device
- gate
- JFET
- MOSFET gate oxide
- MESFET
- nMOS (n-channel)
- pMOS (p-channel)
- polysilicon
- source electrodes
- drain electrodes
- nMOSFET
- pMOSFET
- dynamic RAM (DRAM)
- static RAM (SRAM)
- microprocessor units (MPU)
- central processing units (CPU)
- nonvolatile memory
- read-only memory (ROM)
- programmable read-only memory (PROM)
- field-programmable PROM
- erasable PROM
- electrically erasable PROM
- flash RAM
- application specific IC
- programmable logic devices (PLD)
- programmable array logic (PAL)
- programmable logic array (PLA)
- mask-programmable gate array (MPGA)
- field-programmable gate array (FPGA)

REVIEW QUESTIONS

1. What is an analog circuit? Provide two examples of applications.
2. What is a digital circuit? Provide two examples of applications.
3. What is a passive component? Give two examples of this type of component.
4. List three ways to make a resistor on a single crystalline substrate.
5. What is parasitic resistance? Why is it undesirable in ICs?
6. What is a planar capacitor? Describe four techniques for building this component on a silicon substrate.
7. What is parasitic capacitance? What problems can this condition create in ICs?
8. What is an active component? Give two examples of this type of component.
9. When is a pn junction diode formed? Why is this junction important for an IC?
10. What is the barrier voltage of a pn junction, and how is it formed?
11. Explain what happens when a pn junction is reversed biased.
12. Explain what happens when a pn junction is forward biased.
13. A bipolar junction transistor (BJT) has how many electrodes, junctions, and variations? What are the names of the electrodes? What are the names of the variations?
14. Refer to Figure 3.11. What happens when switch S1 is open? What happens to the E-B junction when switch S1 is closed? When does conduction occur and why?
15. Describe the dopant concentrations in the emitter, collector, and base in an npn bipolar junction transistor.
16. What type of amplifier device is the BJT? How does this affect the applications in terms of power requirements?
17. How is the Schottky diode formed? What is its biggest advantage?
18. What are some notable characteristics of bipolar technology? What is biggest drawback to bipolar technology?
19. What are the benefits of the field-effect transistor (FET)?
20. What is the most popular IC technology today?
21. Is the FET a voltage-amplifying or current-amplifying device?
22. What is the greatest advantage of the FET?
23. How does the FET turn on? What voltage level must be applied to turn on the FET and where should the voltage be applied in the FET?
24. Why is the FET suitable for VLSI and ULSI technology?
25. What are the two basic types of FETs? What is the major difference between them?
26. What are the two categories of MOSFETs? How are they distinguishable from one another?
27. What is the most popular conductor material used in the formation of gates for MOSFETs? How is this material made a conductor?
28. What purpose does the source and drain serve in a MOSFET?
29. What is the majority carrier for a nMOSFET, and what type of channel does it have?
30. What is the majority carrier for a pMOSFET, and what type of channel does it have?
31. What is the well in a MOSFET? Explain what happens in the well region in the conduction mode.
32. If an nMOSFET is in an open-circuit mode, explain what is happening in the three electrode regions (i.e., the source, the drain and the channel below the gate).
33. When switch S1 is closed on the nMOSFET transistor as shown in Figure 3.17, explain what happens with respect to holes and electrons in the electrode regions.
34. What is the threshold voltage in a FET?
35. What is the major difference in performance between a pMOSFET and nMOSFET?
36. What does scaling mean with regards to IC design?
37. What is the efficiency of a CMOS inverter circuit with respect to power consumption?
38. If a p-substrate is used for a CMOS inverter, how is the n-well created?
39. What is the purpose of the field oxide? What is another name for silicon dioxide?
40. What two IC technologies are used in BiCMOS?
41. What could a digital/analog (D/A) converter chip be used for? What could an analog/digital (A/D) chip be used for?
42. Explain the difference between an enhancement-mode transistor and depletion-mode transistor with regards to their standby condition.
43. Which type of MOSFET, enhancement-mode or depletion-mode, is the most commonly used in the semiconductor industry?
44. What is latchup in CMOS devices? What undesirable condition can cause this?
45. Describe three manufacturing techniques used to prevent latchup. What is a linear IC and when would it be used?
46. What is an op amp?
47. What is a voltage regulator?
48. What is a stepper motor driver?
49. What is volatile memory?
50. Describe RAM, DRAM, and SRAM memory.
51. What is MPU?
52. What is nonvolatile memory?
53. Describe the differences among ROM, PROM, EPROM, and EEPROM memory. What is flash RAM, and what are some applications?
54. What is ASIC?
55. What is PLD?
56. What are PAL and PLA?
57. What is an MPGA and a FPGA?

IC MANUFACTURERS' WEB SITES

Actel	http://www.actel.com/
Advanced Micro Devices	http://www.amd.com/
Altera	http://www.altera.com/
Analog Devices	http://www.analog.com/
AT&T Tech. History	http://www.akpublic.research.att.com/
Burr Brown	http://www.bbrown.com/
Cirrus Logic	http://www.cirrus.com/
Cypress Semiconductor	http://www.cypress.com/
Dallas Semiconductor	http://www.dalsemi.com/
Fairchild Semiconductor	http://www.fairchildsemi.com/
Fujita Laboratory	http://www.fujita3.iis.u-tokyo.ac.jp/
Fujitsu	http://www.fujitsu.com/
General Semiconductor	http://www.gensemi.com/
Hitachi Semiconductor	http://semiconductor.hitachi.com/
IBM Microelectronics	http://www.chips.ibm.com/
Intel	http://www.intel.com/
International SEMATECH	http://www.sematech.org/public/index.htm
Intersil Corporation	http://www.intersil.com/
Lattice Semiconductor	http://www.latticesemi.com/
LSI Logic	http://www.lsilogic.com/
Micron Semiconductor	http://www.micron.com/mti/
Mitsubishi Silicon Amer.	http://www.mmc-sil.com/
Motorola Semiconductor	http://mot-sps.com/
National Semiconductor	http://www.national.com/
NEC Semiconductor	http://www.nec.com/semiconductors/
Philips Semiconductor	http://www-us2.semiconductors.philips.com/
QuickLogic	http://www.quicklogic.com/
Rockwell International	http://www.rockwell.com/
Samsung Semiconductor	http://www.usa.samsungsemi.com/
SEMATECH	http://www.sematech.org/public/index.htm
ST Microelectronics	http://www.st.com/stonline/index.shtml
Texas Instruments	http://www.ti.com/
Vishay Siliconix	http://www.vishay.com/brands/siliconix/
Xicor	http://www.xicor.com/
Xilinx	http://www.xilinx.com/
Zilog	http://zilog.com/

Note: Web site addresses are subject to change. The Web site addresses listed here were current at the time of publication.

REFERENCES

1. S. Wolf, *Silicon Processing for the VLSI Era,* vol. 2 of *Process Integration,* IC Resistor Fabrication, (Sunset Beach, CA: Lattice Press, 1990), Appendix A.
2. R. Jacob Baker, H. Li, and D. Boyce, chap. 7 in *CMOS Circuit Design, Layout and Simulation,* (IEEE Press, 1998).
3. A. Sedra and K. Smith, *Microelectronic Circuits,* (New York: Oxford University Press, 1998), p. 138.
4. Ibid., p. 197.
5. R. Jacob Baker, H. Li, and D. Boyce, *Circuit Design,* p. 212–215.
6. S. Brown et al., *Field-Programmable Gate Arrays,* (Norwall, MA: Kluwer Academic Publishers, 1992), p. 2.

7. R. Seals and G. Whapshott, *Programmable Logic PLDs and FPGAs,* (New York: MacMillan Press Ltd., 1997), p. 1.

8. J. Carter, *Digital Designing With Programmable Logic Devices,* (Upper Saddle River, NJ: Prentice-Hall, 1997), p. 17.

9. M. Smith, *Application-Specific Integrated Circuits,* (Reading, MA: Addison-Wesley Longman, 1997), p. 4–15.

10. S. Brown et al., *Field Programmable Gate Arrays,* (Norwell, MA: Kluwer Academic Publishers, 1992), p. 1.

CHAPTER 4

SILICON AND WAFER PREPARATION

The primary semiconductor material used to fabricate microchips is silicon, the most important material in the semiconductor industry. In order for silicon to be acceptable for the fabrication of semiconductor devices, a special purity grade is used that meets stringent material and physical conditions.

The typical U.S. semiconductor manufacturer does not produce silicon wafers. The production of the silicon material and its preparation into wafers occurs in a highly specialized factory dedicated to the production of silicon wafers. These wafers are then supplied to wafer fabrication facilities for processing into various types of microchips by semiconductor manufacturers.

The quality of the final microchip produced in a wafer fab depends directly on the quality of the starting silicon wafer. If defects exist in the incoming wafer, then there will surely be defects in the microchip. An understanding of the silicon wafer and its preparation process will assist us in comprehending the silicon wafer's importance to the complete microchip process.

OBJECTIVES

After studying the material in this chapter, you will be able to:

1. Describe how raw silicon is refined into semiconductor-grade silicon.
2. Explain the crystal structure and growth method for producing monocrystalline silicon.
3. Discuss the major defects in silicon crystal.
4. Outline and describe the basic process steps for wafer preparation, starting from a silicon ingot and finishing with a wafer.
5. State and discuss seven quality measures for wafer suppliers.
6. Explain what epitaxy is and why it is important for wafers.

INTRODUCTION

To exploit silicon's desirable properties as a semiconductor, the natural silica must be purified into an extremely clean silicon material. Pure silicon is required to minimize microdefects at the atomic level of the silicon that are detrimental to semiconductor performance. Once pure silicon is obtained, it is then transformed into the physical wafer with the desired crystal orientation, proper amount of dopants, and physical dimensions necessary for semiconductor wafer fabrication.

SEMICONDUCTOR-GRADE SILICON

The highly refined silicon used for wafer fabrication is termed *semiconductor-grade silicon* (SGS), which is also sometimes referred to as electronic-grade silicon. Obtaining SGS from natural silicon with the acceptable purity level needed for semiconductor device performance is a multistep process (see Table 4.1).[1] There are alternative methods for obtaining SGS, but the following description is the predominant production process.

TABLE 4.1 Steps to Obtaining Semiconductor-Grade Silicon (SGS)

Step	Description of Process	Reaction
1	Produce metallurgical-grade silicon (MGS) by heating silica with carbon.	$SiC\ (s) + SiO_2\ (s) \rightarrow Si\ (l) + SiO(g) + CO\ (g)$
2	Purify MG silicon through a chemical reaction to produce a silicon-bearing gas of trichlorosilane ($SiHCl_3$).	$Si\ (s) + 3HCl\ (g) \rightarrow SiHCl_3\ (g) + H_2\ (g) + heat$
3	Using the Siemens process, $SiHCl_3$ and H_2 react to produce pure semiconductor-grade silicon (SGS).	$2SiHCl_3\ (g) + 2H_2\ (g) \rightarrow 2Si\ (s) + 6HCl\ (g)$

The first step to achieving SG silicon is to produce a metallurgical-grade silicon (MGS) by heating silica (SiO_2), a pure sand, with carbon in a reducing-gas atmosphere.

$$SiC\ (solid) + SiO_2\ (solid) \rightarrow Si\ (liquid) + SiO\ (gas) + CO\ (gas)$$

The resulting metallurgical-grade silicon on the right side of the reaction is 98% pure. Because MG silicon has unacceptable levels of contaminants, it is of no use for semiconductor fabrication. This metallurgical-grade silicon is then crushed and purified through a chemical reaction that produces a silicon-bearing gas of trichlorosilane ($SiHCl_3$).

$$Si\ (solid) + 3HCl\ (gas) \rightarrow SiHCl_3\ (gas) + H_2\ (gas) + heat$$

The silicon-bearing gas of $SiHCl_3$ undergoes additional chemical processing and reduction in hydrogen to produce 99.9999999% (nine nines) pure, semiconductor-grade silicon (SGS).[2] This rate equation is,

$$2SiHCl_3\ (gas) + 2H_2\ (gas) \rightarrow 2Si\ (solid) + 6HCl\ (gas)$$

The pure SGS is produced in a process referred to as the Siemens process.[3] The $SiHCl_3$ gas and hydrogen are injected into the Siemens reactor (See Figure 4.1) and chemically react on heated rods of ultrapure silicon (the rod temperature is about 1100°C). After several days, the process is complete and the rods of deposited SGS are cut up into smaller pieces for growth into the silicon crystal.

Semiconductor-grade silicon has the ultrahigh purity required for semiconductor manufacturing, with less than two parts per million (ppm) of carbon and less than one part per billion (ppb) for elements from Groups III and V (the critical doping elements) of the periodic table.[4] However, the silicon produced by the Siemens process does not have atoms arranged in a desirable crystal order. We will now analyze crystal structure to understand the correct atomic order for semiconductor-grade silicon.

CRYSTAL STRUCTURE

It is not only critical that the semiconductor-grade silicon used to fabricate semiconductor devices have ultrahigh purity, but it must also have a near-perfect crystal structure. This condition is necessary to avoid electrical and mechanical defects that are detrimental to device performance.

A *crystal* is a solid material that has an ordered, repeatable three-dimensional pattern over a long range of many atoms. Figure 4.2 shows how atoms are related in a crystal structure, referred to as the *crystal lattice*. The crystal lattice represents the repeatable order within a crystal at the atomic level of the internal structure (atomic order), even though the surfaces of the crystal may be abraded or rough. An example is the uneven surface of common beach sand, which has an internal crystalline structure with atomic order. Our goal is to obtain a specific crystal lattice necessary for the specific requirements for SG silicon used to fabricate wafers.

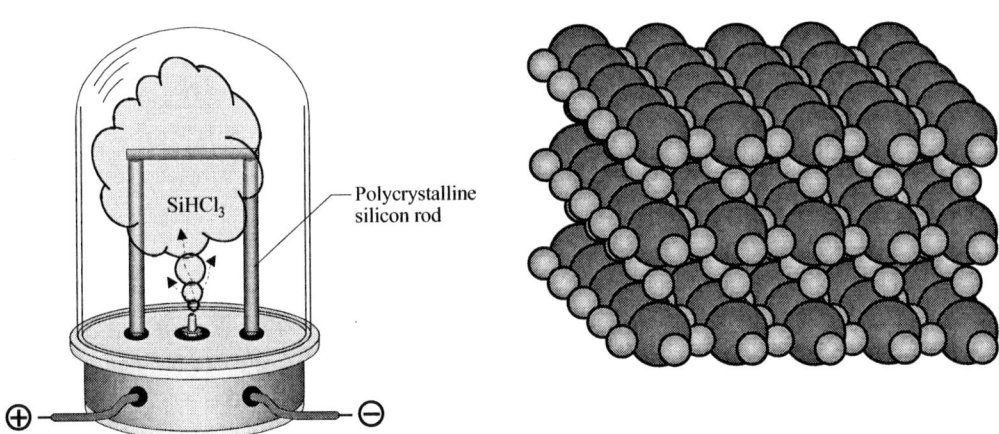

FIGURE 4.1 Siemens Reactor for SG Silicon   **FIGURE 4.2** Atomic Order of a Crystal Structure

Amorphous Materials

Amorphous materials are noncrystalline solids that lack a repetitive structure and demonstrate structural disorder at the atomic level (see Figure 4.3). Amorphous materials are the opposite of crystalline materials. Plastic is an example of an amorphous material. Amorphous silicon would be an unacceptable structure for the silicon used in wafers to fabricate semiconductor devices. This is because many of the electrical and mechanical aspects of a device occur at the atomic level in the silicon, which requires order and predictability to yield repeatable results from chip to chip.

Uni Cells

The most fundamental entity for the long-range order found in the atomic pattern of a crystal material is the *unit cell*. A unit cell is the simplest arrangement of atoms that, when repeated in a three-dimensional framework, gives the crystal structure. We need to study unit cells in order to understand how silicon replicates itself in its crystal structure. A two-dimensional analogy to a unit cell can be seen in a checkerboard or a section of a tiled floor. A three-dimensional analogy would be a neat arrangement of a child's building blocks. An illustration of a unit cell in a three-dimensional structure is shown in Figure 4.4.

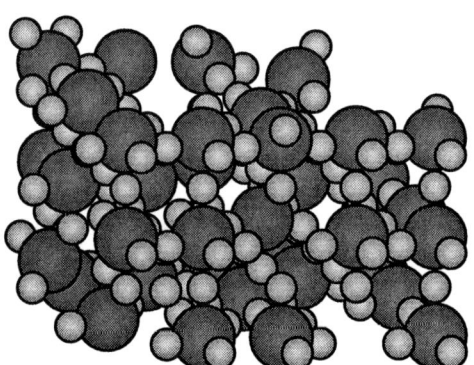
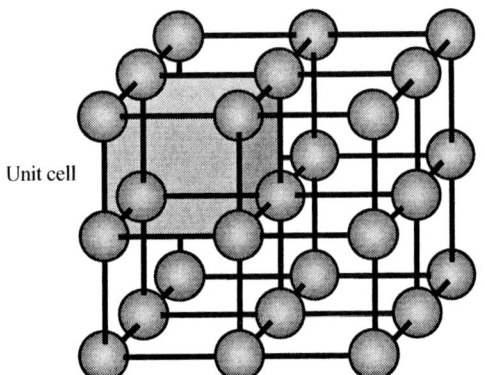

FIGURE 4.3 Amorphous Atomic Structure   **FIGURE 4.4** Unit Cell in 3-D Structure

Since crystal structures are identical in three dimensions, unit cells have a framework structure, such as a cube. There are seven possible crystal systems in nature.[5] The crystal system relevant to silicon technology is the cubic system. We will only consider one of the basic crystal structures because it is applicable to silicon: *face-centered cubic,* or *FCC*. The FCC unit cell is shown in Figure 4.5.

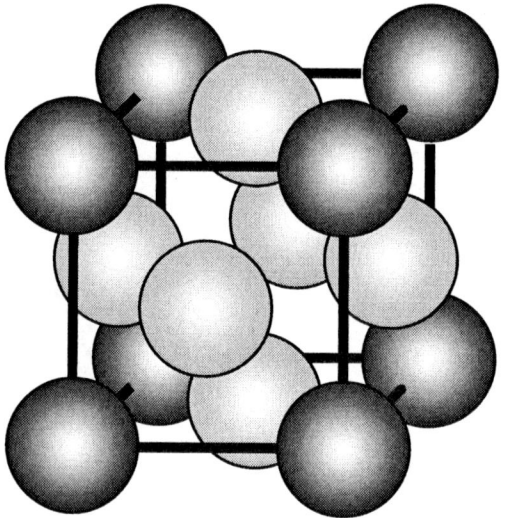

FIGURE 4.5 Face-Centered Cubic (FCC) Unit Cell

Within a crystal structure, unit cells are tightly packed and, therefore, share atoms. Sharing atoms is important because this is how unit cells build up into a cohesive crystal lattice structure. In an FCC unit cell, each corner shares an atom with eight other unit cells, while each face shares with one other unit cell. For an FCC unit cell, there is the equivalent of four shared atoms.

For a silicon crystal, the unit cell is a variation of the FCC known as the *FCC diamond structure* (see Figure 4.6). Atoms are shared as with the FCC unit cell, plus there are four complete atoms situated inside the cubic structure. For a silicon unit cell, there are a total of eight complete atoms, with four shared and four unshared.

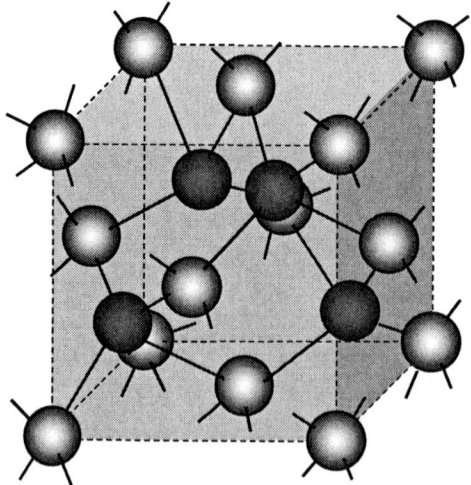

FIGURE 4.6 Silicon Unit Cell: FCC Diamond Structure

Polycrystal and Monocrystal Structures

Now that the unit cell is defined for a crystal structure, we can define how the unit cells are organized. If the unit cells are not in a regular arrangement, then the material is a *polycrystal.* The silicon produced from the semiconductor-grade purification process is polycrystalline silicon. Another name for polycrystalline silicon material is *polysilicon.* An analogy for a polycrystal structure is a pile of bricks. Each individual brick represents the unit cell and the pile is polycrystalline because there is no repetitive arrangement (see Figure 4.7).

If the unit cells are neatly arranged in a three-dimensional, repeatable manner, then the crystal structure is *monocrystal.* Another term often used for monocrystal is *single crystal.* For our brick analogy, if the bricks are now stacked neatly in repeating rows next to one another, then the bricks represent a monocrystal. A monocrystal structure is also shown in Figure 4.7.

Semiconductor wafer processing requires a pure, monocrystalline silicon structure (single crystal). This is because the repeatable unit cells of a monocrystal structure provide the desirable electrical and mechanical properties necessary for silicon wafer processing and performance. Unacceptable crystal structure and defects influence the formation of microdefects that affect wafer processing (see the following section on crystal defects in silicon).

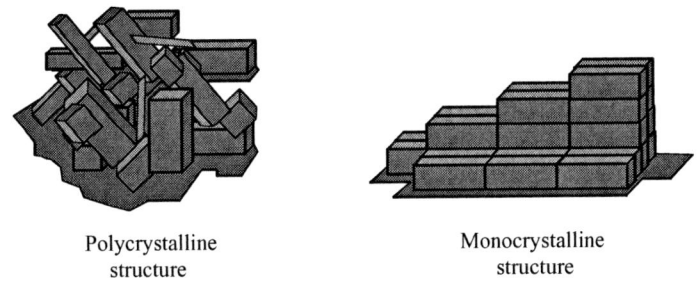

FIGURE 4.7 Polycrystalline and Monocrystalline Structures

CRYSTAL ORIENTATION

Before studying the production process to obtain a monocrystal structure for silicon, let's first examine how unit cells are oriented in silicon. Unit cell orientation is important because it defines how the silicon crystal structure is physically aligned with the silicon wafer. Different crystal orientations change the chemical, electrical, and mechanical properties of the silicon, which affects the manufacturing process conditions and final device performance.

To define the silicon unit cell orientation, we need a coordinate system. In crystals, the coordinate system has three axes, x, y, and z, shown in Figure 4.8. We will arbitrarily assign the number 1 at an equal distance along each axis, while the center point is given the value of 0. These are referred to as unit values. If the crystal is a monocrystal, then all unit cells are aligned in a repetitive manner along these three coordinate axes.

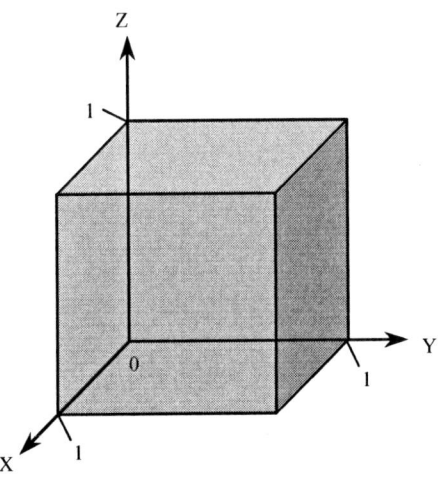

FIGURE 4.8 Axes of Orientation for Unit Cells

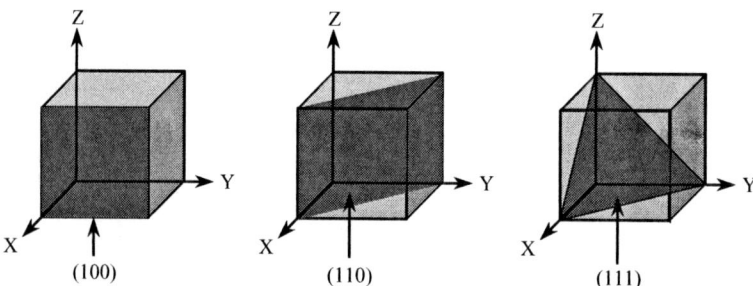

FIGURE 4.9 Miller Indices of Crystal Planes

The orientation of planes in the silicon crystal is described by a set of numbers known as *Miller indices*.[6] It is beyond the scope of this book to completely describe the Miller notation system for crystal planes and their directions. In the Miller system of notation, parentheses ( ) are used to denote a specific plane, whereas brackets < > denote groups of equivalent directions.[7]

From our standpoint, what is important is to recognize that each set of Miller indices enclosed in a parenthesis specifies a unique plane in the crystal structure. For semiconductor fabrication, the Miller notations applicable to the most common crystal planes on a wafer are (100), (110), and (111). These three crystal planes are shown in Figure 4.9 on page 72. They are obtained in the silicon crystal by maintaining precise control of orientation during the crystal growth process (described in the next section). You interpret each Miller notation based on where the plane intersects the axes.

The (100) crystal plane is parallel to the y-z plane and intersects the x-axis at the unit value of one. The (110) intersects the x- and y-axes only, whereas the (111) intersects the x-, y-, and z-axes. Wafers with a (100) crystal plane orientation are most common for fabricating MOS devices. The reason for this is that the surface state condition for (100) silicon is more conducive toward controlling the threshold voltage required to turn MOS devices on and off.[8] The (111) crystal plane orientation has a tighter packing density at the atomic level, making it easier to grow. These are the least expensive crystals to grow and are often used for bipolar devices.[9] Gallium arsenide (GaAs) technology also uses wafers with (100)-oriented planes.

MONOCRYSTAL SILICON GROWTH

Crystal growth is the process of converting the polysilicon chunks of semiconductor-grade silicon into a large monocrystal of silicon. The grown silicon monocrystal is referred to as an *ingot*. The most common technique today for growing a silicon monocrystal ingot to be used in wafer fabrication is the *Czochralski (CZ) method,* which is named for the inventor who developed the process in the early 1900s.

CZ Method

The Czochralski (CZ) growth of a silicon monocrystal involves the transformation of molten SG silicon liquid into a solid silicon ingot that has the correct crystal orientation and is doped either as n-type or p-type. Over 85% of all silicon crystals are grown according to the CZ method.[10]

A piece of monocrystalline silicon having the desired crystal orientation is used as a starter seed to grow the silicon ingot that is a replica of the original seed crystal. To achieve crystal growth with the CZ method, precise conditions are controlled at the contact interface between molten silicon and a single monocrystal seed of silicon. These conditions permit thin films of silicon to accurately replicate the seed crystal structure and grow into a large ingot. This is done with equipment known as a *CZ crystal puller*.

CZ Crystal Puller ■ To grow a silicon ingot, chunks of SG silicon are placed in a fused silica (amorphous quartz) crucible holder, along with small amounts of dopant to create either n-type or p-type silicon. The crucible is large, with crucible diameters for a 300-mm wafer at 32 inches or larger. These crucibles must hold from 150 to 300 kg of silicon.[11] An alternative approach for larger diameter wafers is to use granular polysilicon in the crucible, which permits the gradual

Silicon Ingot Grown by CZ Method
(Photo courtesy of Kayex Corporation)

introduction of the silicon during the meltdown to minimize stresses on the large crucible during the melting process.[12] The crucible is positioned in the crystal puller, which is where the silicon crystal ingot is grown (see Figure 4.10).

The crucible of silicon is heated in the furnace of the puller using either resistance heaters or RF (radio frequency) heating coils. Resistance heaters are used for larger diameter ingots. As the silicon is heated, it turns to liquid and is referred to as the *melt*. A seed of a perfect silicon crystal is attached to a pull mechanism to start the structure of the new crystal. The seed is placed at the surface of the melt and is slowly drawn away while rotating in a direction that is opposite to the rotation of the crucible.

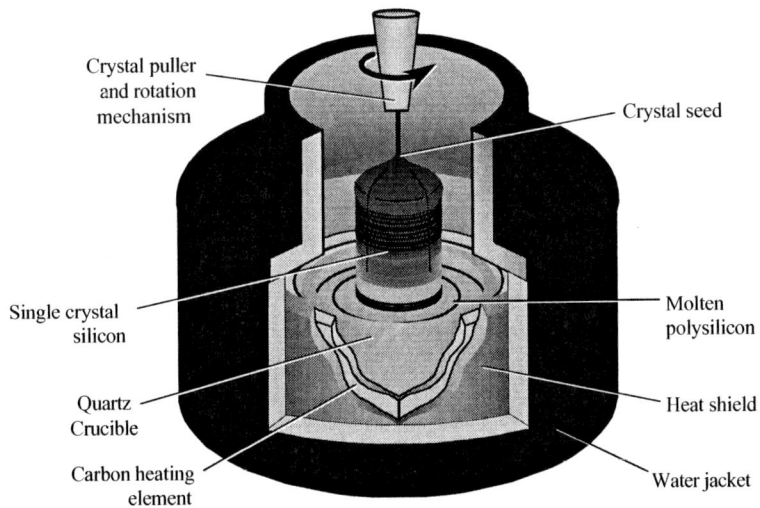

FIGURE 4.10 CZ Crystal Puller

CZ Crystal Puller
(Photo courtesy of Kayex Corporation)

As the single seed is drawn away from the melt during the pull process, liquid from the melt is raised with it by surface tension. The interface at the single seed dissipates heat and solidifies downward toward the melt. As the seed is steadily pulled out of the melt and rotated, a single crystal grows with the same crystal orientation as the seed. While the seed is rotated, the crucible is also rotated. Different ingot growth patterns result depending on the speed of the seed and crucible rotation and their respective directions.

The objective of the pull process is to precisely replicate the seed structure while obtaining dopant uniformity, achieving the correct ingot diameter, and limiting introduction of impurities into the silicon. The two main parameters that affect the pull process are the pull rate and crystal rotation.[13] The pull starts at a relatively high speed and then slows down considerably. This action forms a neck in the ingot, since the diameter of the growing crystal is directly related to the rate of pull.

A goal of the crystal growth process is to have uniform (or homogeneous), large-diameter crystals. This result gives more predictable crystal properties with respect to parameters such as resistivity. A technique developed to achieve this with the CZ process is to use a magnetic field around the silicon melt to stabilize the crystal during its growth.[14] This condition is referred to as *magnetic CZ (MCZ)*.

Doping ■ Dopant material is added to the melt in the crystal puller in order to obtain the desired electrical resistivity in the finished crystal. The resistivity of pure silicon is approximately 2.5×10^5 Ω-cm. The most common dopants used in crystal growth are trivalent boron to create p-type silicon or pentavalent phosphorus to create n-type silicon. Dopant concentration ranges in the silicon are indicated through letter designations and superscripts, as shown in Table 4.2. Usually raw dopants are not added directly to the melt because the dopant amounts are extremely small. It is typical to add the dopant in the form of a highly doped powder of crushed silicon.[15]

TABLE 4.2 Dopant Concentration Nomenclature in Silicon

Dopant	Material Type	Concentration (Atoms/cm³)			
		< 10^{14}	10^{14} to 10^{16}	10^{16} to 10^{19}	>10^{19}
		(Very Lightly Doped)	(Lightly Doped)	(Doped)	(Heavily Doped)
Pentavalent	n	n^{--}	n^-	n	n^+
Trivalent	p	p^{--}	p^-	p	p^+

Impurity Control ■ Impurity control is important during crystal growth since unacceptable impurities affect the device performance. One impurity that must be controlled but that can also be beneficial is oxygen. The primary source of oxygen in the CZ method results from the dissolution of the crucible during crystal growth.[16]

Small amounts of oxygen in the ingot are desirable because oxygen can serve as an internal getter to tie up metallic contaminants that are introduced into the wafer during the fabrication process. Note that the term *getter* is used to describe any process that immobilizes or ties up impurities. Most of the oxygen in the silicon crystal near the wafer surface is removed during the many heating processes that take place during wafer fabrication, leaving higher oxygen concentrations deeper in the wafer. This oxygen then serves as the getter that attracts the contaminants away from the devices located on the surface. The amount of oxygen required to serve as a getter is continually being reduced as the device fabrication process is improved to reduce the sources of contaminants.[17]

Float-Zone Method

An alternative crystal growth method is the *float-zone* method, which produces a silicon monocrystal ingot with significantly lower oxygen content.[18] The float-zone method was developed in the 1950s and produces the purest bulk silicon single crystals known to date. A schematic of the float-zone process is shown in Figure 4.11.

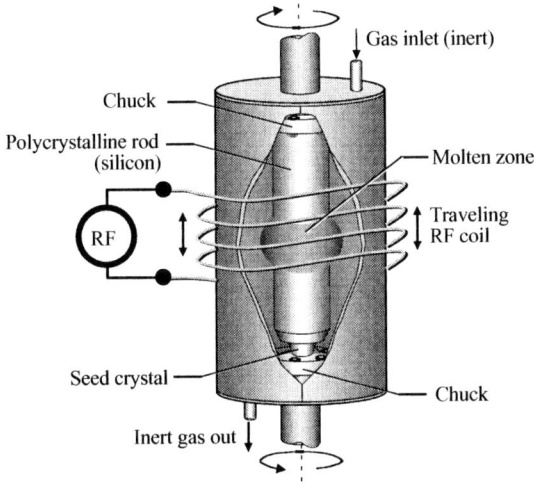

FIGURE 4.11 Float Zone Crystal Growth

The float-zone method of growing a monocrystal ingot of silicon starts with a bar of doped polysilicon that has been cast in a mold. A monocrystal seed is attached to one end of the bar and then placed in the grower. An RF (radio frequency) coil applies heat at the contact area of the bar and seed. Heating of the polysilicon bar is the most important aspect of the float-zone process, as each section of the bar is molten only for about 30 minutes before it solidifies again at the monocrystal interface.[19] This monocrystal growth process progresses along the bar as the heater is moved along its axis.

The float-zone process typically makes smaller-diameter wafers than the CZ process and was producing predominantly 125-mm wafers at the turn of the century. By not using a crucible, the float-zone process results in high-purity silicon with lower oxygen content.

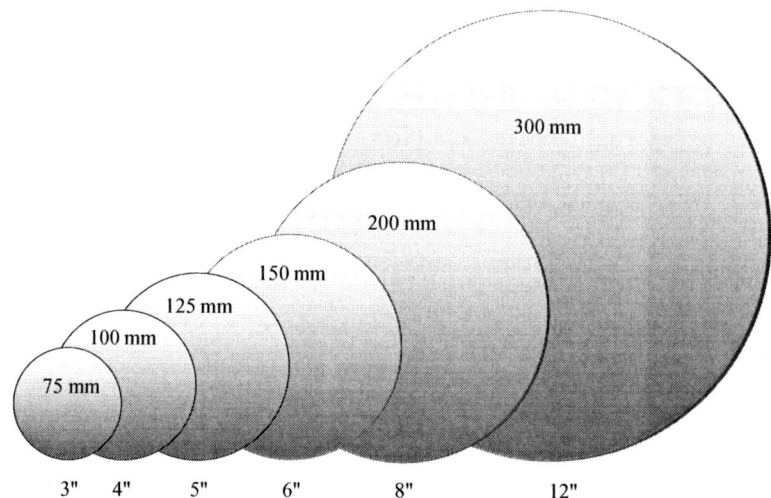

FIGURE 4.12 Wafer Diameter Trends

Reasons For Larger Ingot Diameters

Silicon ingot diameters have increased from the early diameters of less than 25 mm in the 1950s to the most recent diameter of 300 mm. The historical trend for larger semiconductor wafer diameters is shown in Figure 4.12.

There are wafer fabrication facilities still in use that produce wafers with diameters of 75 mm, 100 mm, 125 mm, and 150 mm. The most common practice is to build a new fab to introduce a new wafer diameter, since upgrading a facility to a larger wafer diameter can cost hundreds of millions of dollars. The semiconductor industry began to transition to 300-mm wafers around the year 2000. There are also early evaluations underway to understand the possibility of increasing wafer diameters to 400 mm. Table 4.3 highlights different attributes of various wafer sizes.

TABLE 4.3 Wafer Dimensions & Attributes

Diameter (mm)	Thickness (μm)	Area (cm^2)	Weight (grams/lbs)	Weight/25 Wafers (lbs)
150	675 ± 20	176.71	28 / 0.06	1.5
200	725 ± 20	314.16	53.08 / 0.12	3
300	775 ± 20	706.86	127.64 / 0.28	7
400	825 ± 20	1256.64	241.56 / 0.53	13

From H. Huff, R. Goodall, R. Nilson, and S. Griffiths, "Thermal Processing Issues for 300 mm Silicon Wafers: Challenges and Opportunities," *ULSI Science and Techonology* (New Jersey: The Electrochemical Society, 1997), p. 139.

Larger ingot diameters bring new challenges to achieving correct crystal growth and maintaining correct process control during ingot growth. Ingots with a 300-mm diameter are about 1 meter long and require between 150 to 300 kg of SG silicon in the crucible for the melt. With the increased complexity needed to manufacture the silicon ingot, what is the main reason to keep increasing the diameter? The answer lies in the cost benefits to the wafer fabrication process due to increasing the wafer diameter.

Larger-diameter wafers have a greater surface area for chips. For 300-mm wafers, there is 2.25 times the surface area than that available on a 200-mm wafer (see Figure 4.13). This increase translates into more chips that can be produced from a single wafer.

An analogy would be driving a car 200 miles from city A to city B. If the driver travels alone, then this establishes a certain cost for making the trip (e.g., fuel, wear on the vehicle, and so on). However, if a passenger goes with the driver, then the driver and the passenger make the trip for less cost. If three passengers make the trip with the driver, then there are even more savings per passenger. This approach to efficiency is called *economies of scale*.

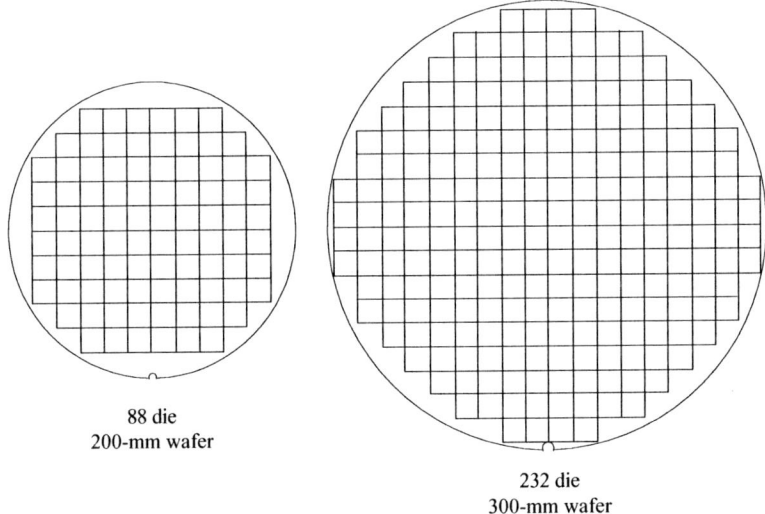

FIGURE 4.13 Increase in Number of Chips on Larger Wafer Diameter

Larger wafers means more chips per wafer, leading to improved equipment productivity due to economies of scale for less handling per chip and less processing time per chip. It is estimated that conversion to a 300-mm wafer diameter will reduce the fabrication cost per chip by approximately 30% through equipment utilization improvements.[20] Another benefit from larger-diameter wafers is that fewer chips are near the edge of the wafer, which translates into higher production yield. There is also a benefit from tool repeatability from chip to chip since more chips are exposed to the same process conditions.

The total industry cost to convert the standard wafer diameter from 200 mm to 300 mm has been estimated at $13 billion to $15 billion. No one firm or country has the resources to make this change alone. Global 300-mm standards have been developed and 23 semiconductor companies worldwide have joined two separate consortia to coordinate the conversion to 300-mm wafers. The two consortia are the International 300 mm Initiative (I300I), which includes the U.S., Europe, Korea and Taiwan, and the Semiconductor Leading Edge Technology (Selete), which represents ten IC companies from Japan.[21] Ultimately, cost savings will be the major factor that drives the industry toward a larger wafer diameter. A summary of dimensional specifications for 300-mm wafers is provided in Table 4.4.

TABLE 4.4 Developmental Specifications for 300-mm Wafer Dimensions and Orientation Requirements

Parameter	Units	Nominal	Some Typical Tolerances
Diameter	mm	300.00	± 0.20
Thickness (center point)	μm	775	± 25
Warp (max)	μm	100	
Nine-Point Thickness Variation (max)	μm	10	
Notch Depth	mm	1.00	+ 0.25, -0.00
Notch Angle	Degree	90	+5, -1
Back Surface Finish		Bright Etched/Polished	
Edge Profile Surface Finish		Polished	
FQA (Fixed Quality Area—radius permitted on the wafer surface)	mm	147	

From H. Huff, R. Goodall, R. Nilson, and S. Griffiths, "Thermal Processing Issues for 300 mm Silicon Wafers: Challenges and Opportunities," *ULSI Science and Technology* (New Jersey: The Electrochemical Society, 1997), p. 139.

CRYSTAL DEFECTS IN SILICON

Semiconductors require a nearly perfect silicon crystal structure in order to function optimally for advanced ICs. A *crystal defect* is any interruption in the repetitive nature of the unit cell crystal structure. Another term used for a crystal defect is *microdefect*. It is essentially impossible to grow or process silicon without making a defect. However, progress has been made on growing nearly defect-free silicon with a very low defect density.

Defect density is a commonly used term in crystal growth and wafer fabrication. It is defined as the number of defects per cm^2 of wafer surface that may occur during processing due to all kinds of causes. Reducing defect density is a critical aspect for increasing wafer yield (see Figure 4.14). Yield is the percentage of good chips produced out of the total chips on a wafer.

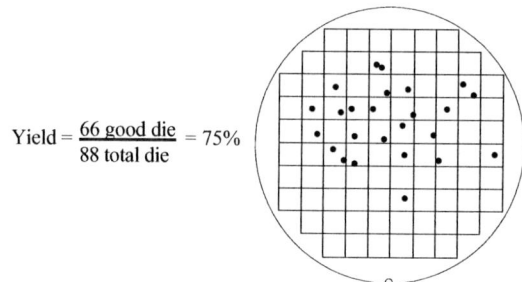

FIGURE 4.14 Yield of a Wafer

It is important to study silicon crystal defects because they have a damaging effect on the electrical properties of the semiconductor. These problems includes SiO_2 dielectric breakdown and leakage current failures. As device geometries shrink and active gate areas increase due to more transistors on a chip, the probability of a defect occurring in a sensitive area of the chip increases. Such a defect can negatively affect the device yield for advanced ICs.

Crystal defects can occur during crystal growth and the subsequent processing of the silicon ingot and wafers. Some crystal defects occur due to surface damage, including cracks and surface defects from mechanical operations on the wafer. There are three general forms of crystal defects in silicon:

1. Point defects: Localized crystal defect at the atomic level
2. Dislocations: Displaced unit cells
3. Gross defects: Defects in crystal structure

Point Defects

A *point defect* occurs at a particular location in the crystal lattice. Three types of point defects are shown in Figure 4.15. The most basic type of point defect is a *vacancy* (also referred to as a *void*).

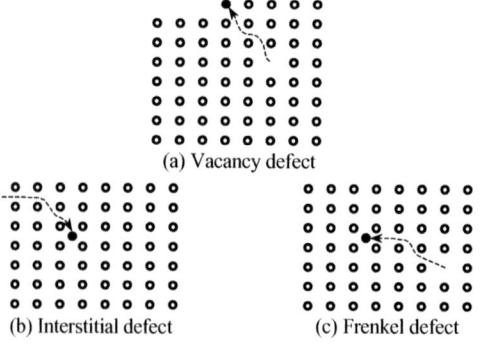

FIGURE 4.15 Point Defects
Redrawn from S. Ghandi, *VLSI Fabrication Principles: Silicon and Gallium Arsenide,* 2nd ed. (New York: Wiley, 1994), p. 23.

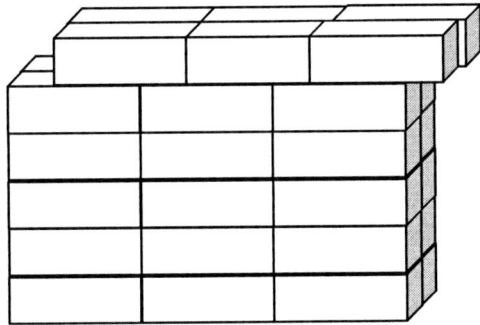

FIGURE 4.16 Dislocations

This defect occurs when an atom is removed from its lattice site and moves to the surface of the crystal. Another type of point defect is an *interstitial,* which occurs when an atom becomes located in one of the many voids within the crystal structure. When an atom leaves its lattice site and takes up position in a void, it creates a vacancy-interstitial pair, or *Frenkel defect.*

Point defects in semiconductor silicon have become increasingly important as device technology becomes more complex. Crystal growth conditions that affect the creation of point defects during wafer growth are the growth rate (how fast the crystal is pulled) and the thermal gradient (the temperature difference between the melt and the solid crystal) at the crystal-melt interface.[22] The silicon crystal cooling rate is controlled, which effectively reduces the defect generation. High-temperature processing during semiconductor manufacturing is also a source of point defects.

Another type of point defect is due to chemical impurities that are introduced into the lattice. Impurities are introduced intentionally or unintentionally during fabrication. They can take up locations at sites ordinarily occupied by atoms, known as *substitutional impurities,* or locate themselves in voids as interstitial impurities.

The most significant impurities not introduced intentionally into the silicon are oxygen and carbon. It is important in VLSI and ULSI device fabrication to control these impurities. Oxygen enters the silicon melt via dissolution of the crucible during the CZ process, as discussed earlier in this chapter. Most of this oxygen is removed, but less than 5% remains in the growing crystal. Oxygen present in silicon can create crystal defects such as dislocations (see the next section). The presence of donor electrons from oxygen impurities in the silicon lattice can also affect the electrical parameters of a pn junction. Carbon can act as a nucleation site for crystalline defect formation in silicon.

Dislocations

Unit cells form a repetitive structure in a monocrystal. If unit cells are displaced, this condition is known as a *dislocation* (see Figure 4.16). An analogy would be a neat stack of bricks that has a group of bricks displaced along a row. A form of dislocation is *stacking faults* which is due to layer stacking errors. Dislocations can be introduced at any stage of the crystal growth or wafer fabrication process. However, dislocations occurring after crystal growth are often associated with mechanical stress to the wafer, such as that caused by uneven heating and cooling or excessive force applied to the wafer.

In some instances, dislocation faults are induced following the thermal oxidation of the silicon wafer surface during device fabrication (see Chapter 10).[23] The result is referred to as *oxygen-induced stacking faults* (*OISF*). OISF defects appear as shallow, saucer-shaped depressions in the lattice. These defects can be detected through X-ray analysis or surface etching. Specialized thermal treatments (referred to as annealing) or gettering can be used to minimize stacking faults and dislocations in silicon crystal structures.

Crystal growth dislocations have been reduced for large-diameter wafers through use of a neck-down procedure during ingot growth.[24] This procedure consists of necking down the cross section of the single seed at the beginning of the pull so that the dislocations grow out to the surface, and then growing several centimeters at sufficiently high speed so that high-vacancy densities remove the edge dislocations. However, as described above, dislocations can still occur from other

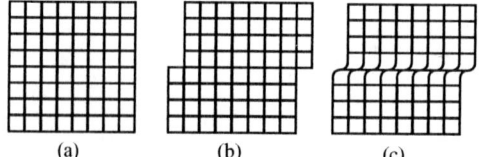

FIGURE 4.17 Crystal Slip
Redrawn from S. Ghandi. *VLSI Fabrication Principles: Silicon and Gallium Arsenide,* 2nd ed. (New York: Wiley, 1994), p. 49.

sources during the wafer fabrication process, such as from wafer edge chipping or high temperature processing.

Gross Defects

Gross defects are related to the structure of the crystal and typically occur during crystal growth. A *slip* is a gross defect in a crystal that occurs when there is slippage of the crystal along one or more crystal planes (see Figure 4.17).

Another gross defect is *twin planes* (or *twinning*), where the crystal grows in two different directions from the same plane (see Figure 4.18). The source of twin planes could be thermal or mechanical shock during the growth process. The crystal on either side of the defect may be perfect. Crystals with either slip or twin planes are unacceptable for semiconductor manufacturing.

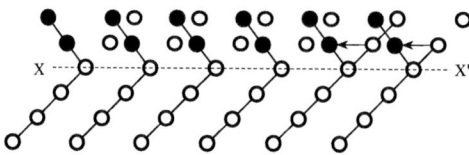

FIGURE 4.18 Crystal Twin Planes
Redrawn from S. Ghandi. *VLSI Fabrication Principles: Silicon and Gallium Arsenide,* 2nd ed. (New York: Wiley, 1994), p. 55.

WAFER PREPARATION

Silicon is a hard, brittle material and of little use for semiconductor manufacturing in its form as an ingot after crystal growth. The cylindrical, single-crystal ingot (also called a *boule*) must undergo a series of process steps to transform it into wafers that meet stringent specifications for semiconductor manufacturing. These wafer preparation steps include machining operations, chemical operations, surface polishing, and quality measures. The basic process flow for wafer preparation is shown in Figure 4.19.

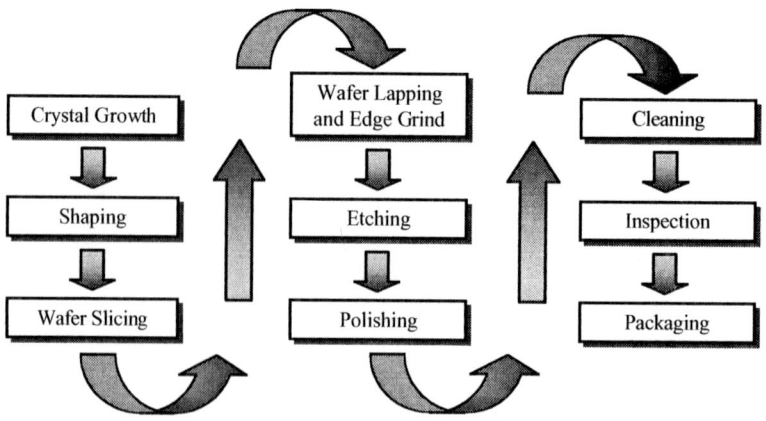

FIGURE 4.19 Basic Process Steps for Wafer Preparation

Because of the continued advances in chip design and fabrication requirements, the wafer preparation process must be capable of delivering wafers that meet tighter specification requirements. Wafer preparation requirements include the wafer geometry (diameter, flatness, and warpage), surface perfection (roughness and light scattering), and cleanliness (sources of particles). These wafer specifications address issues such as dimensional control for robotic material handling or surface conditions required for critical process steps during IC fabrication.

Shaping Operations

After growth of the ingot in the crystal puller, the shaping operations are the first process steps. Shaping operations include all preliminary steps to prepare the monocrystal silicon ingot prior to slicing into wafers.

End Removal ■ The first operation is to remove the two ends of the ingot. The ends are referred to as the *seed end* (where the seed was located) and the *tang end* (opposite to the seed end). Once the ends are removed (cropped off), a four-point probe resistivity check is done to confirm the proper dopant uniformity throughout the ingot (see Chapter 7 for the resistivity measurement procedure).

Diameter Grinding ■ The next operation is *diameter grinding* to create the precise diameter of the material. Ingots are grown slightly oversize to permit this grinding step, since diameter control and roundness cannot be adequately controlled during crystal growth. Precise diameter control is critical for semiconductor manufacturing given the automated wafer-handling steps in wafer fabs. Figure 4.20 illustrates the diameter-grinding process.

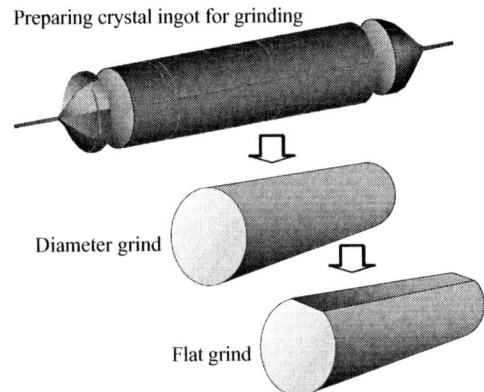

FIGURE 4.20 Ingot Diameter Grind

Wafer Flat or Notch ■ The semiconductor industry has traditionally placed *flats* on the ingot to identify the crystal structure and to orient the wafer. The primary flat orients the wafer to the crystal structure, as shown in Figure 4.21 on page 82. There is a secondary flat which identifies the orientation and conductivity type of the wafer.

Wafer flats have been replaced in the U.S. with a *notch* for wafers 200 mm and larger. Notched wafers have information about the wafer laser-scribed in a small region on the wafer. The location and depth of the laser-scribing initially created some fear of contamination collecting in the small laser marks. For 300-mm wafers, a standard has been approved for laser-scribing all wafers on the back surface in an exclusion zone near the wafer's edge.[25] The notch and laser scribe are illustrated in Figure 4.22 on page 82.

For 300-mm wafers, the exclusion zone is outside an area known as the fixed-quality area (FQA), which is where chips are permissible on the wafer surface. The exclusion zone is currently 3 mm, but may be reduced to 2 mm at some time in the future.

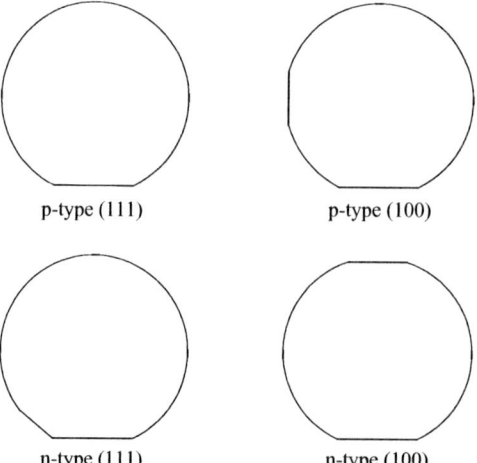

FIGURE 4.21 Wafer Identifying Flats

Wafer Slicing

Once the shaping operations are complete, the ingot is ready to undergo *wafer slicing*. This is the first major process step after ingot growth. For wafers up to 200 mm in diameter, slicing is done with an internal diameter saw that has a diamond cutting edge, as seen in Figure 4.23. The internal diameter saw is used because the cutting edge is more stable to yield a flat cutting surface.

For 300-mm wafers, internal diameter saws are not desirable for slicing wafers due to the larger diameter. Wafer slicing for 300-mm ingots is currently being done with wire saws.[26] Wire saws yield more wafer slices per inch of crystal than conventional ID saws due to thinner kerf (saw blade thickness) losses associated with using a slurry-coated wire instead of a diamond-coated blade.[27] Wire sawing reduces mechanical damage to the wafer surface during the slicing process, but there are still concerns about inadequate wafer-flatness control with sawing.[28] Equipment development is ongoing for wafer slicing of 300-mm and larger wafers to obtain precise wafer dimensions.

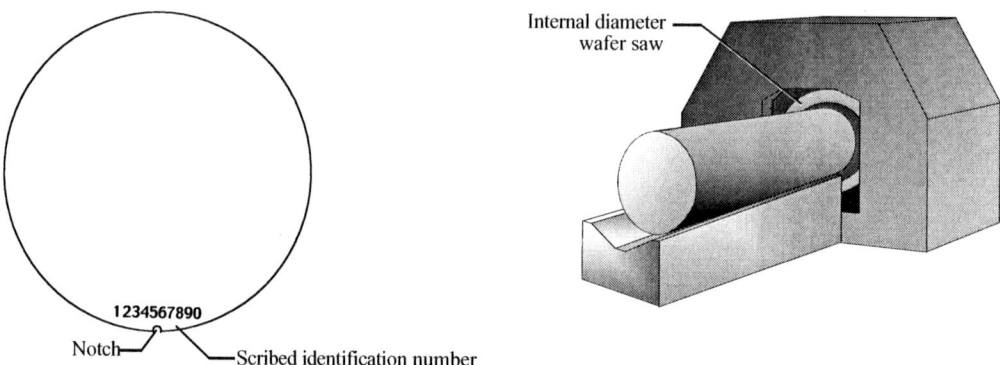

FIGURE 4.22 Wafer Notch and Laser Scribe   FIGURE 4.23 Internal Diameter Saw

The thickness of the wafer is carefully controlled during the slicing process. The 300-mm wafer is currently specified at a thickness of 775 ± 25 microns. Thicker wafers are better able to withstand thermal and mechanical shocks introduced during high temperature processing in the semiconductor fabrication process.

Wafer Lapping and Edge Contour

Following wafer slice, there is traditionally a mechanical, two-sided *lapping* operation to remove damage left by slicing and to achieve a high degree of parallelism and flatness of the wafer. Lapping is performed under rotational pressure with pads and an abrasive slurry mixture that typically consists of alumina or silicon carbide and glycerine. Flatness is a critical wafer parameter for many process steps in the wafer fabrication process.

A *polished wafer edge* finish (also called *edge grind*) is applied to the wafer to contour a smooth radius on the edge of the wafer (see Figure 4.24). This step may takes place either before or after lapping. Cracks and small crevices at the edge of the wafer create mechanical stress in the wafer that activates crystal dislocations, especially during thermal process steps that occur in wafer fabrication. Crevices can also be the source of unwanted contamination buildup as well as particulate flaking during fabrication. A smooth edge radius is important to minimize these concerns.[29] Furthermore, chipped edges are a source of edge dislocation growth during thermal cycles in the wafer fabrication process.

Etching

Shaping the wafer leaves the surface and edges damaged and contaminated. The depth of the wafer damage depends on the particular process at a manufacturer, but it is usually on the order of several microns deep. To remove the damaged surface from the wafer, wafer suppliers use a technique known as wafer etch or chemical etching. *Wafer etch* is a process involving the chemical removal of selective surfaces of a material (see Figure 4.25). Wafers undergo a wet chemical etch process to remove the damage and contamination on the wafer. Typically about 20 microns of silicon wafer surface is removed during the etch process to ensure that all damage is removed.[30] The etching is done with either acid or alkaline chemicals, depending on the process in place at the wafer manufacturer. The subject of etching will be covered in detail in Chapter 16.

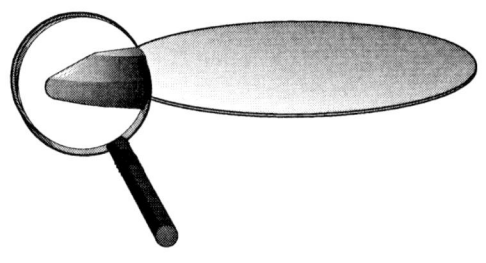

FIGURE 4.24 Polished Wafer Edge

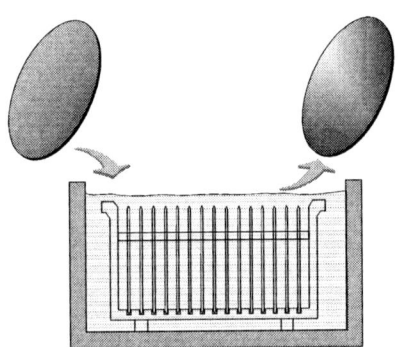

FIGURE 4.25 Chemical Etch of Wafer Surface to Remove Damage
Redrawn from M. S. Bawa, E. F. Petro, H. m. Grimes, "Fracture Strength of Large Diameter Silicon Wafers," *Semiconductor International* (November 1995), p. 115.

Polishing

The final step in the wafer preparation is a *chemical mechanical planarization* (*CMP*) to achieve a smooth wafer surface with a high degree of flatness. CMP is also referred to as polish (see Chapter 18 for a discussion of CMP). This CMP step has traditionally been done only on the topside of the wafer for 200-mm and earlier wafers, leaving the chemically-etched surface on the backside. This etched backside left a relatively rough surface, being approximately three times rougher then a surface processed with CMP.[31] Its purpose was to give a rougher surface for handling devices. However, there has been concern about the ability of an etched surface to meet the flatness requirements for deep submicron photolithography, plus the possibility of introducing particulate contamination into the wafer fabrication process.

For 300-mm wafers, double-sided polishing (DSP) with CMP is the final major manufacturing step. The planetary motion of the wafer between the polishing plates produces flat and parallel surfaces while improving surface roughness (see Figure 4.26 on page 84). Since this is the final wafer preparation step, flatness across the large wafer diameter can be readily maintained. The polished back surface also permits the wafer manufacturer to characterize wafer cleanliness prior to delivery to the wafer fab. The final wafer surface has a mirrorlike finish on both sides of the wafer.

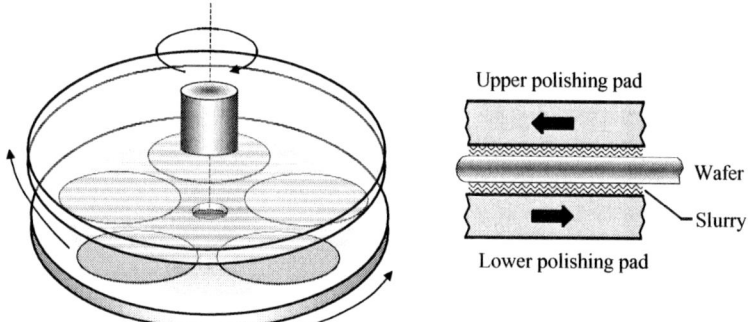

FIGURE 4.26 Double-Sided Wafer Polish

Cleaning

Semiconductor wafers must be cleaned to achieve an ultraclean state of cleanliness prior to shipment to wafer fabrication facilities. Cleaning specifications have undergone considerable development in the last few years, to the extent that wafers should be free of particles and contamination. The different types of wafer contamination, the appropriate cleaning steps, and the process procedures for cleaning are discussed in Chapter 6.

Wafer Evaluation

Prior to packaging the wafer, it is inspected for quality criteria that are specified by the customer. Standard quality measures are discussed later in this chapter. The most critical criteria relate to surface defects such as particulate contamination and stains.

Packaging

Wafer suppliers must carefully package wafers for shipment to the wafer fabrication facilities. Considerable effort can be lost if the wafers are damaged from transit or from the packaging material. The wafers are stacked in a plastic cassette or "boat" that has grooved slots to hold the wafers. A fluorocarbon resin material such as Teflon is often used as the cassette material to minimize particle generation. In addition, the Teflon is made conductive so that it does not generate an electrostatic discharge. All equipment and operators are grounded to drain off any charge buildup that could attract particles.

Once filled with wafers, the cassette is then placed in a nitrogen-filled, sealed canister that prevents oxidation and other contamination during transit. When the wafers are received at the wafer fab, they are transferred to other standardized cassettes used for transport and handling during wafer processing inside the fabrication facility. The shipping container is designed to minimize the need for wafer handling. One shipping concept under development is a 25-wafer container referred to as a front-opening shipping box (FOSB) which would interface with automated material handling systems in the wafer fab.[32]

QUALITY MEASURES

Silicon wafer suppliers produce wafers to requirements that are becoming more stringent, which represents improved control over wafer quality. A sampling of some wafer specifications and their rate of improvement relative to the critical dimension on a wafer is shown in Table 4.5.

For wafer quality measures, uniformity throughout the wafer is critical. Important silicon wafer quality requirements are:

- Physical dimensions
- Flatness
- Microroughness
- Oxygen content
- Crystal defects
- Particles
- Bulk resistivity

The suppliers of silicon wafers must control their wafer quality by performing ingot and wafer quality inspections to demonstrate that quality specifications are met. Defective wafers shipped unknowingly to a semiconductor manufacturing wafer fab facility would be catastrophic.

TABLE 4.5 Improving Silicon Wafer Requirements

	Year (Critical Dimension)			
	1995 (0.35 μm)	1998 (0.25 μm)	2000 (0.18 μm)	2004 (0.13 μm)
Wafer diameter (mm)	200	200	300	300
Site flatness[A] (μm)	0.23	0.17	0.12	0.08
Site size (mm × mm)	(22 × 22)	(26 × 32)	26 × 32	26 × 36
Microroughness[B] of front surface (RMS)[C] (nm)	0.2	0.15	0.1	0.1
Oxygen content (ppm)[D]	≤ 24 ± 2	≤ 23 ± 2	≤ 23 ± 1.5	≤ 22 ± 1.5
Bulk microdefects[E] (defects/cm^2)	≤ 5000	≤ 1000	≤ 500	≤ 100
Particles per unit area (#/cm^2)	0.17	0.13	0.075	0.055
Epilayer[F] thickness (± % uniformity) (μm)	3.0 (± 5%)	2.0 (± 3%)	1.4 (± 2%)	1.0 (± 2%)

Adapted from K.-M. Kim, "Bigger and Better CZ Silicon Crystals," *Solid State Technology* (November 1996), p. 71.

Notes:
A: Flatness is the linear thickness variation across the wafer or a site on a wafer (see the following).
B: See the following for a description of microroughness.
C: RMS is a method for determining the best estimate of group of measurements—in this case, the surface finish measurements (see the following). It is calculated by taking the root-mean-square (square root of the average of all measurements squared). Surface finish measurements are obtained by measuring the highest point relative to the lowest point on a surface.
D: ppm is part per million.
E: Bulk microdefects represent all defects within a square centimeter.
F: See the following section for a description of epilayer.

Physical Dimensions

Wafers must be physically dimensioned in order to meet the requirements for device fabrication and to accommodate the automated wafer handling equipment used in the wafer fab. The wafer physical dimensions controlled and inspected during wafer preparation include measurements such as diameter, thickness, location and size of orientation, flat (or notch), and wafer deformation. Figure 4.27 shows an example of wafer deformation. The most likely source of wafer deformation is the slicing process.

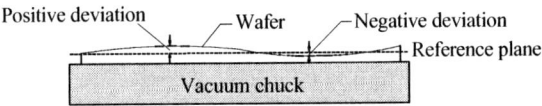

FIGURE 4.27 Wafer Deformation

Flatness

Flatness is one of the most critical wafer parameters specified, primarily because the process of photolithography (Chapters 13 to 15) is highly sensitive to local-site flatness. *Wafer flatness* is the linear thickness variation across the wafer. It is measured by determining the distance that the front surface of the wafer deviates from a specified reference plane. For a wafer, the reference plane is theoretically defined by the back wafer surface if it were held perfectly flat, such as when pulled down by a vacuum onto a clean, flat surface (see Figure 4.28). Note that the front surface of the wafer is the wafer side where chips will be placed.

Flatness can be specified as site flatness for a specific site on the wafer or specified as global flatness, which is the total wafer flatness across the fixed-quality area (FQA) of the-wafer surface. The FQA does not include the exclusion zone on the periphery of the wafer surface. A larger area for measuring flatness is more difficult to control than a smaller area. Typical flatness specifications for site flatness are shown in Table 4.5 on page 85.

FIGURE 4.28 Flatness of Wafer Front Surface

Microroughness

Microroughness is the small-scale deviation of the actual surface from a nominal plane surface, with many small, closely spaced peaks and valleys.[33] It is an indication of the surface texture on the wafer. Surface microroughness measures the maximum and minimum height deviation on the wafer surface and is specified in units of nanometers (10^{-9} meters). The roughness criteria is usually specified as a root-mean-square (RMS), which is the square root of the average of the square of all deviation measurements from a defined plane. This is a general statistical technique for determining the most probable measurement. Measurement of wafer surface microroughness is done using one of several types of optical surface profile instruments.

Surface microroughness is important to control for wafer fabrication because it is known to have a negative effect leading to the breakdown of very thin dielectric films that are placed on the wafer during device fabrication.[34] Wafers undergo a lapping step followed by an etching step to remove surface microroughness. Typical specification values for microroughness are shown in Table 4.5 on page 85.

Oxygen Content

Control of the oxygen level and uniformity in the silicon ingot is important and becoming increasingly difficult with larger crystal diameters. Small amounts of oxygen can have beneficial gettering effects, which can tie up contaminants in the wafer, as noted previously in this chapter. However, excessive oxygen in the silicon ingot can affect the mechanical and electrical properties of the silicon. For instance, oxygen can lead to an increase in the leakage current at the pn junction and also to increased MOS leakage.[35]

Oxygen content in silicon is verified by using cross-sectioning, which permits composition analysis of the silicon crystal structure. A representative piece of the silicon is potted in epoxy and then ground and polished flat to expose the grain structure of the solid silicon. A chemical etchant is used to darken and highlight certain elements for identification. Once the sample is prepared, special microscopes such as transmission electron microscopy (TEM) are used to characterize the crystal structure. The use of TEM is discussed in Chapter 7. The oxygen content is currently controlled in silicon wafers within a range of 24 to 33 ppm.[36] Typical specifications are shown in Table 4.5 on page 85.

Crystal Defects

It is necessary to control the silicon in order to minimize the various crystal defects previously discussed in this chapter. As shown in Table 4.5 on page 85, the current requirement is for the number of crystal defects per square centimeter to be less than 1000. Cross-sectioning is the technique used to control bulk microdefects in the crystal.

Particles

The number of surface particles are controlled on a wafer to minimize yield loss during wafer fabrication. The primary means of reducing particles is to minimize the generation of particles during wafer processing, followed by effective cleaning steps to remove particles. A typical wafer-cleanliness specification is to have less than 0.13 particles per square centimeter (1300 particles per square meter) of wafer surface area on a 200-mm wafer. This particle size measured is greater than or equal to 0.08 micron.

Bulk Resistivity

The resistivity of the bulk silicon ingot depends on the density of the dopant material that was added to the silicon melt prior to crystal growth. Recall that the most common dopant materials are either boron to create p-type silicon or phosphorus to make n-type silicon. The result of adding these Group IIIA or VA dopants to silicon is to decrease its resistivity due to increased carrier mobility.

It is important to achieve uniform resistivity throughout the bulk silicon. During the actual crystal growth process, there is a radial temperature gradient that reaches a maximum at the center of the ingot and decreases toward the outer edge. This radial temperature gradient produces variations in the radial doping concentration of the ingot.[37]

The bulk silicon ingot is checked for correct resistivity and uniformity after the ends of the ingot have been removed. The four-point probe measurement tool is used to measure wafer resistivity. This measurement tool is described in Chapter 7.

EPITAXIAL LAYER

In some cases, it is desirable to have wafers with a very pure silicon surface of the same crystal structure as the substrate wafer (monocrystalline), yet retain specific control over the doping type or concentration. This condition is achieved by depositing an *epitaxial layer* on the surface of the silicon (referred to as an *epilayer*). Epitaxial is a combination of two Greek words, *epi,* meaning "upon", and *taxis,* meaning "ordered."

In epitaxial silicon, the base wafer is used as a seed crystal to grow a thin layer of silicon on the wafer. The crystal structure of the new epitaxial layer will duplicate that of the wafer. Since the substrate wafer is monocrystal, the epitaxial layer is monocrystal. Furthermore, the dopant of the epitaxial layer can be n-type or p-type and is independent of the initial wafer's dopant type. For example, it is possible to grow a p-type epilayer on a p-type wafer with a lower concentration of electrically active dopant in the epilayer than what is found on the wafer (see Figure 4.29).

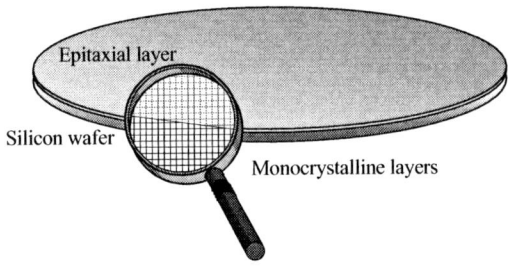

FIGURE 4.29 Formation of Epitaxial Silicon Layers

The original reason for the development of silicon epitaxy was to attain improved performance in bipolar transistors and integrated circuits. Epitaxy permitted the growth of a lightly doped epilayer over a heavily doped substrate, which optimized for a high breakdown voltage in the pn junction while permitting low collector resistance, yielding high device operating speeds at moderate currents. Epitaxy has become important in CMOS ICs because it minimizes latchup problems as device geometries continue to shrink. An epitaxial layer is generally free of contaminants (such as no oxygen particles, which is not true for CZ-grown silicon).

The thickness of the epitaxial layer can vary, with a typical thickness of 0.5 to 5 μm for high speed digital circuit applications, and an epilayer as thick as 50 to 100 μm for silicon power devices.[38] The methods for depositing an epitaxial layer on silicon will be discussed under the topic of deposition in Chapter 11.

SUMMARY

Naturally occurring silicon is used to produce ultra-pure, semiconductor-grade polycrystalline silicon. Silicon is a crystal with a repetitive FCC diamond unit cell structure at the atomic level. The crystal orientation is defined by Miller indices, with (100) most common for MOS devices. Polycrystalline silicon is converted into the monocrystal (single-crystal) ingot necessary for wafer fabrication through crystal growth using the CZ method and equipment referred to as a crystal puller. Dopant material is added to the liquid silicon during the CZ process to achieve proper doping levels. Undesirable impurities are tightly controlled in order to grow pure silicon. An alternative process, the float-zone method, produces silicon with very low oxygen content.

Ingot diameters have been increasing over the years to place more chips on a wafer and lower cost through economies of scale. The major crystal defects controlled during growth are point defects, dislocations, and gross defects. Ingots undergo a multistep process to make acceptable wafers, including diameter grinding, wafer notch, wafer slicing, wafer lapping, edge contour, etching, polishing, cleaning, inspection, and packaging.

Wafer suppliers have specific quality measures to control the wafer quality, for characteristics such as physical dimension, flatness, microroughness, oxygen content, crystal defects, particles, and bulk resistivity. An epitaxial layer is sometimes grown on the silicon surface to achieve a very pure silicon layer with the same crystal structure yet have precise control of resistivity by the controlled addition of dopants.

KEY TERMS

semiconductor-grade silicon
crystal
crystal lattice
amorphous materials
unit cell
face-centered cubic (FCC)
FCC diamond structure
polycrystal
polysilicon
monocrystal
single crystal
Miller indices
crystal growth
ingot
Czochralski (CZ) method
CZ crystal puller
melt
magnetic CZ (MCZ)
getter

float-zone method
economies of scale
crystal defect
microdefect
yield defect density
point defect
vacancy (or void)
interstitial
Frenkel defect
substitutional impurities
dislocation
stacking faults
oxygen-induced stacking faults
gross defect
slip
twin planes (twinning)
boule
seed end
tang end

diameter grinding

flats

notch

fixed-quality area (FQA)

wafer slicing

lapping

polished wafer edge (edge grind)

wafer etch

chemical mechanical planarization (CMP)

double-sided polishing (DSP)

wafer flatness

microroughness

epitaxial layer (epilayer)

REVIEW QUESTIONS

1. List the three steps to obtaining semiconductor-grade silicon. How pure is semiconductor-grade silicon?
2. Name the process for producing pure semiconductor-grade silicon?
3. What is a crystal? What is a crystal lattice?
4. Describe an amorphous material. Why is this unacceptable for the silicon used in wafers?
5. Define a unit cell. What type of a unit cell does silicon form?
6. How many complete atoms are in the silicon unit cell? How many atoms are shared and how many are unshared in the silicon unit cell?
7. Describe a polycrystal. What is another name for polycrystalline silicon?
8. Describe a monocrystal. What is another name for a monocrystal?
9. Why is it necessary to have monocrystal silicon for wafer fabrication?
10. What are Miller indices, and what do they indicate?
11. Draw a picture of the following three crystal planes: (100), (110) and (111).
12. Which crystal plane orientation is most common for MOS? Which is most common for bipolar?
13. Define crystal growth.
14. What is the CZ method for crystal growth?
15. Describe the silicon seed for the CZ method and how it is used.
16. What is the name of the main equipment used in the CZ method?
17. Describe a crystal puller. Specifically, describe the crucible holder, seed, melt, pull process and ingot growth.
18. What is the objective of the pull process and what are the two main parameters that affect the pull process in the CZ method?
19. What is the purpose of magnetic CZ?
20. Why is dopant material added to the melt during the CZ method?
21. How can small amounts of oxygen in the ingot be beneficial?
22. Why is the float-zone method desirable to grow silicon crystal?
23. Describe the float-zone process.
24. What is the main reason for the continued increase in wafer diameter?
25. Give three benefits to using larger-diameter wafers.
26. What is a crystal defect?
27. Define defect density. How is defect density related to wafer yield?
28. What are the three general forms of crystal defects in silicon?
29. List and describe three types of point defects.
30. Which crystal growth parameters can affect the occurence of point defects.
31. What is a dislocation? Explain a stacking fault.
32. What is an oxidation-induced stacking fault?
33. Explain how crystal growth dislocations have been reduced for large-diameter wafers.
34. What is a gross defect? Explain the difference between crystal slip and crystal twin planes.
35. What is the purpose of diameter grinding of the ingot?
36. Describe or draw a picture of the four types of wafer flats. What have the wafer flats been replaced with on 200-mm diameter and larger wafers?
37. How is wafer slicing done on wafers up to 200 mm in diameter?
38. What is the purpose of wafer lapping?
39. Why is a polished wafer edge (edge grind) done to the wafer? Give three reasons why this grinding benefits the wafer quality.
40. Why is chemical etching done on the wafer?
41. Why is chemical mechanical planarization (CMP) done on the wafer surface?
42. How are wafers packaged for shipment to the wafer fab?
43. List seven wafer quality requirements for a silicon wafer.
44. What is wafer flatness, and how is it measured?
45. What is surface microroughness, and why is it important to control?
46. What is an epitaxial layer, and why is it used on wafers?

SELECTED INDUSTRY WEB SITES

GT Equipment Technologies, Inc.	http://www.gtequipment.com/
International SEMATECH	http://www.sematech.org/public/index.htm
Kayex CZ Crystal Growers	http://www.kayex.com/
Mitsubishi	http://www.munc-sil.com/
NIST, National Institute of Standards and Technology	http://www.nist.gov/
Selete, Semiconductor Leading Edge Technologies, Inc.	http://www.selete.co.jp
SEMI, Semiconductor Equipment and Materials International	http://www.semi.org/
Semiconductor International Magazine	http://semiconductor.net/
Semiconductor Search Engine	http://www.semiseek.com/
SIA, Semiconduictor Industry Association	http://www.semichips.org/
SISA, Semiconductor Industry Suppliers Association	http://www.sisa.org/
Solid State Technology Magazine	http://sst.pennwellnet.com/home/home.cfm
Wafer World, Inc.	http://www.waferworld.com/

REFERENCES

1. C. Pearce, *Crystal Growth and Wafer Preparation,* VLSI Technology, 2nd ed., ed. S. Sze (Boston: McGraw-Hill, 1988), p. 11.
2. S. Ghandhi, *VLSI Fabrication Principles: Silicon and Gallium Arsenide,* 2nd ed., (New York: Wiley, 1994), p. 102.
3. K. Bachmann, *The Materials Science of Microelectronics* (New York: VCH Publishers, 1995), p. 102.
4. Society of Chemical Engineers of Japan, eds., *Introduction to VLSI Process Engineering* (New York: Chapman and Hall, 1993), p. 148.
5. L. Van Vlack, *Elements of Materials Science and Engineering,* 4th ed., (Reading: Addison-Wesley, 1980), p. 73.
6. S. Ghandhi, *VLSI Fabrication Principles,* p. 16.
7. S. Wolf and R. Tauber, *Silicon Processing for the VLSI Era,* vol. 1 of *Process Technology* (Sunset Beach: Lattice Press, 1986), p. 2.
8. S. Ghandhi, *VLSI Fabrication Principles,* p. 21.
9. Ibid.
10. S. Takasu and W. Zulehner, "Silicon Crystal Growth," *Semiconductor Silicon Proceedings* 94–10 (Pennington: The Electrochemical Society, 1994): p. 55.
11. K.-M. Kim, "Bigger and Better CZ Silicon Crystals," *Solid State Techonology* (November 1996), p. 73.
12. K. Bachmann, *Materials Science of Microelectronics,* p. 248.
13. S. Sze, *VLSI Technology,* 2nd ed. (New York: McGraw-Hill, 1988), p. 26.
14. K. Bachmann, *Materials Science of Microelectronics,* p. 270.
15. S. Ghandhi, *VLSI Fabrication Principles,* p. 120.
16. K. Bachmann, *Materials Science of Microelectronics,* p. 248.
17. K.-M. Kim, "Bigger and Better CZ Silicon Crystals," p. 74.
18. Ibid., p. 74.
19. H. Wolfgang, "High Purity Silicon IV," *International Symposium,* (Pennington, NJ: The Electrochemical Society, 1996): p. 5.
20. H. Huff et al., "Thermal Processing Issues for 300 mm Silicon Wafers: Challenges and Opportunities," *ULSI Science and Technology: Proceedings of the Electrochemical Society,* 97-3: p. 136.
21. Ibid., p. 136.
22. K.-M. Kim, "Bigger and Better CZ Silicon Crystals," p. 74.
23. B. El-Kareh, *Fundamentals of Semiconductor Processing Technologies,* (Boston: Kluwer Academic Publishers, 1995), p. 68.
24. A. Takao, "Innovative Silicon Crystal Growth and Wafering Techonologies," *ULSI Science and Technology: Proceedings of the Electrochemical Society,* 97-3: p. 123.
25. S. Brunkhorst and D. Sloat, "The Impact of the 300-mm Transition on Silicon Wafer Suppliers," *Solid State Technology* (January 1998): p. 90.
26. Ibid.
27. G. Fisher, "Challenge for 300mm Polished Wafer Manufacturers," *Semiconductor International* (September 1998): p. 98.
28. A. Takao, *Innovative Silicon Crystal Growth,* p. 124.
29. H. Huff, et al., "Thermal Processing Issues," p. 140.
30. S. Sze, *VLSI Technology,* p. 40.
31. S. Brunkhorst and D. Sloat, "Impact of the 300-mm Transition," p. 88.
32. G. Fisher, "Challenges," p. 102.
33. W. Runyan and T. Shaffner, *Semiconductor Measurements and Instrumentation,* 2nd ed., (New York: McGraw-Hill, 1997), p. 214.
34. W. Bullis, "Microroughness of Silicon Wafers," *Proceedings of Semiconductor Silicon* 94-10 (Pennington, NJ: The Electrochemical Society, 1994): p. 1156.
35. J. Vanhellemont et al., "On the Electrical Activity of Oxygen-Related Extended Defects in Silicon," *Semiconductor Silicon Proceedings* 94–10 (Pennington, NJ) The Electrochemical Society, 1994): p. 670.
36. K.-M. Kim, "Bigger and Better CZ Silicon Crystals," p. 73.
37. S. Ghandhi, *VLSI Fabrication,* p. 258.
38. S. Ghandhi, *VLSI Fabrication,* p. 258.

CHAPTER 5

CHEMICALS IN SEMICONDUCTOR FABRICATION

The fabrication of semiconductors is largely a chemical process. Chemical manufacturers obtain raw chemicals from nature, refine the chemical makeup of these materials to achieve ultrahigh purity, and transport them to the fabrication area, where these chemicals are introduced to the surface of the wafer. Chemical reactions alter the silicon to create semiconductors and then apply many layers of interconnections. These chemical reactions form the basis for semiconductor device fabrication and microchip performance.

The semiconductor industry is undergoing rapid changes to introduce fundamentally new materials and processes into wafer fabrication. The goal is to achieve improved microchip performance and productivity. Examples of new technologies are insulators with a low dielectric constant (see Chapter 11) and copper metallurgy (see Chapter 12). Chip design and process improvements create the ongoing need for new and improved chemicals.

OBJECTIVES

After studying the material in this chapter, you will be able to:

1. Identify and discuss the four states of matter.
2. Describe the important chemical properties relevant to semiconductor manufacturing.
3. State how the different process chemicals are categorized and used in a wafer fab.
4. Explain how an acid, a base, and a solvent are used in chip manufacturing.
5. State whether a gas is a bulk or specialty gas and explain how each type of gas is delivered and used in wafer fabrication.

INTRODUCTION

Semiconductor manufacturing makes use of large quantities of chemicals to fabricate wafers. There is also a high usage of chemicals for cleaning wafers and the processing tools used in the manufacturing processes. Chemicals used in wafer fabrication are referred to as *process chemicals*. They exist in various chemical states and are controlled to strict purity requirements. To understand process chemicals, we will first review some concepts in basic chemistry.

STATES OF MATTER

All matter in the universe exists in three basic states: solid, liquid, or gas. There is an additional fourth state that is not widely understood, known as plasma. We can describe each state by the way the substance fills a container (see Figure 5.1 on page 92).

A *solid* has its own fixed shape and will not conform to the shape of the container. A *liquid* conforms to the container shape. It will fill the container to the extent of the liquid's volume, and has a surface. A *gas* also conforms to the container shape, but it fills the entire container volume

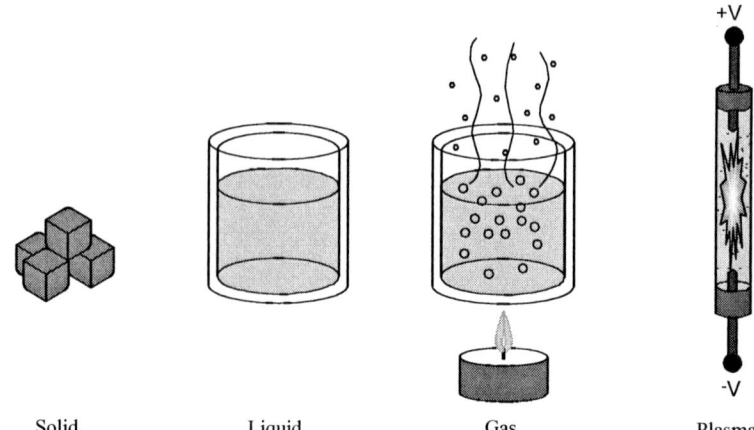

Solid   Liquid   Gas   Plasma

FIGURE 5.1 Physical States of Matter

without forming a surface. Gas particles are small and freely moving. Some gases, like hydrogen and oxygen, are reactive and easily form stable chemical compounds with other gases or elements. Other gases, like helium and argon, are inert. An *inert gas* (also called a *noble gas*) does not readily form compounds. Inert gases are widely used in semiconductor manufacturing because they do not react with other chemicals.

A fourth state of matter is *plasma*. A plasma state exists when there is a high-energy collection of ionized molecules or atoms. This state is considerably different than the three basic states of matter. Examples of a plasma state are a star, a fluorescent light, and neon signs. Plasma states can be induced in certain gases through exposure to high-energy electric fields. As we shall see in later chapters, plasma states are widely used in semiconductor manufacturing.

Many substances can exist in any one of the three basic physical states. For example, water undergoes changes of state depending on the temperature and pressure of its surroundings. As temperature increases, solid water (ice) melts to a liquid water, and then boils into a gas state (water vapor). Many substances behave in this same manner.

PROPERTIES OF MATERIALS

We can understand how to use materials to our benefit in semiconductor manufacturing by investigating their properties. *Properties* are the characteristics of a material that describe its unique identity. For instance, a solid material can have different properties, such as a rigid solid (iron), a soft solid (wax) at room temperature (25°C, 77°F) and a flexible solid (lead). Identifying properties for materials used in semiconductor manufacturing is important for understanding how to properly fabricate the silicon wafer to build chips.

There are two types of material properties: physical and chemical. *Physical properties* of a material are those that reflect a material by itself without its interacting with another substance. Some of these physical properties are the melting point, boiling point, resistivity, and density. *Chemical properties* of a material result from an interaction or transformation with another substance. Examples of chemical properties are flammability, reactivity, and corrosiveness. A *chemical reaction* occurs when a material is converted into a different material that has a different composition and properties. The starting material is called a *reactant,* and the resultant material is called a *product*.

An example of a chemical reaction is the combustion of hydrogen in the presence of oxygen. Both chemicals exist as gases in their natural state. Hydrogen will undergo a chemical reaction with oxygen when the temperature of hydrogen exceeds 600°C. The result is an explosive heat reaction with water vapor as the byproduct. The chemical change is:

$$2 \text{ parts of hydrogen (gas)} + \text{oxygen (gas)} \xrightarrow{\text{Heat}} \text{water (liquid)} + \text{energy}$$

Chemical Properties for Semiconductor Manufacturing

There are many different types of chemicals and materials used to fabricate silicon microchips. Manufacturers of advanced ICs use new materials to improve chip performance and reduce device feature sizes. Some chemical properties are important to understanding how existing and new semiconductor process materials perform. These properties are:

- Temperature
- Pressure and vacuum
- Condensation
- Vapor pressure
- Sublimation and deposition
- Density
- Surface tension
- Thermal expansion
- Stress

Temperature ■ *Temperature* is a measure of how hot or cold a substance is relative to another substance. Thus it is a measure of the average kinetic or thermal energy (molecular or atomic motion) of a substance. The transfer of energy between objects at different temperatures is *heat*. Wafer fabrication requires extensive processing at high temperatures for reasons such as using heat to affect chemical reactions (e.g., changing the rate of a reaction) or annealing the silicon crystal structure to rearrange atoms.

Three temperature scales exist: Fahrenheit in °F, Celsius in °C (commonly called centigrade), and Kelvin in K. Figure 5.2 illustrates how these three scales are related.

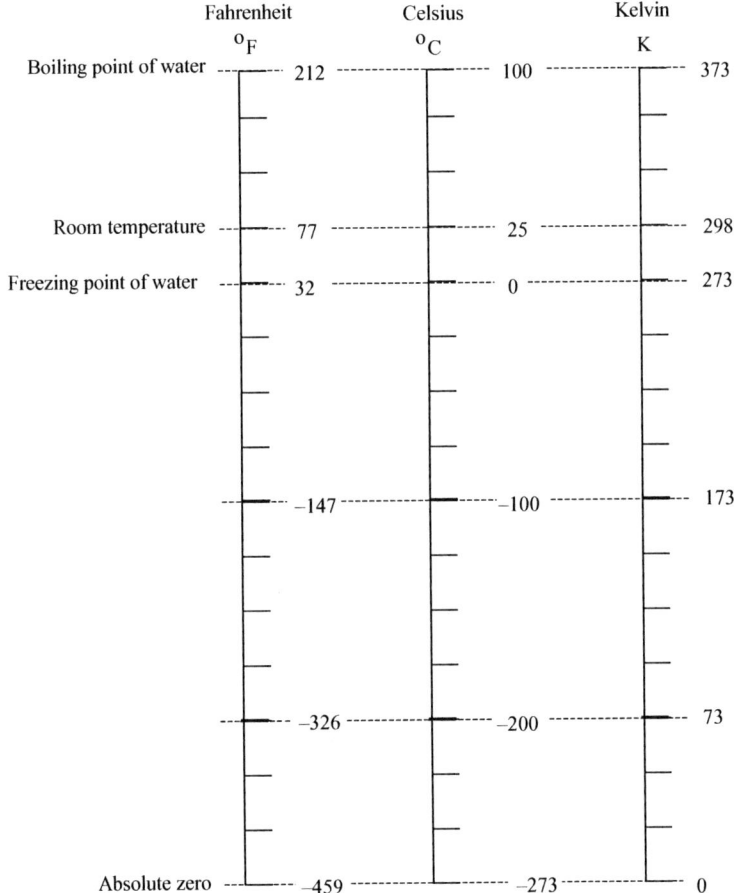

FIGURE 5.2 Temperature Scales

The most common temperature scale in scientific work is the *Celsius* scale. This scale is based on changes in the physical state of water: 0°C is set at the water's freezing point and 100°C at its boiling point when measured at standard atmospheric pressure at sea level. To convert between Fahrenheit and Celsius,

$$°F = \tfrac{9}{5}\,°C + 32$$
$$°C = \tfrac{5}{9}[°F - 32]$$

The *Kelvin* scale, or absolute scale, is the base unit of temperature for the metric unit system. The temperature 0 K (absolute zero) is −273°C (more precisely, −273.15°C), which is the lowest temperature attainable and the theoretical temperature at which all atomic motion would cease. To convert between Kelvin and Celsius,

$$K = °C + 273$$
$$°C = K - 273$$

Pressure and Vacuum of a Gas ■ Gas fills the entire volume of a container and exerts uniform pressure on its walls. *Pressure* (P) is defined as the force exerted per unit area against a surface:

$$\text{Pressure} = \frac{\text{Force}}{\text{Area}} \quad \text{(pounds per square inch, or psi)}$$

The pressure of any gas is dependent on the number of gas molecules present, the temperature, and the volume of the chamber. If there are more gas molecules striking the sides of the container per unit time, then pressure increases (see Figure 5.3). This is also evident when blowing up a balloon.

The English unit of pressure common in the United States is pounds per square inch (psi), specified as either psia (absolute pressure, which includes atmospheric pressure of 14.7 psi) or psig (gauge pressure). These two gauges are illustrated in Figure 5.4. Different units of pressure at standard atmosphere and temperature (sea level and 23°C) are shown in Table 5.1.

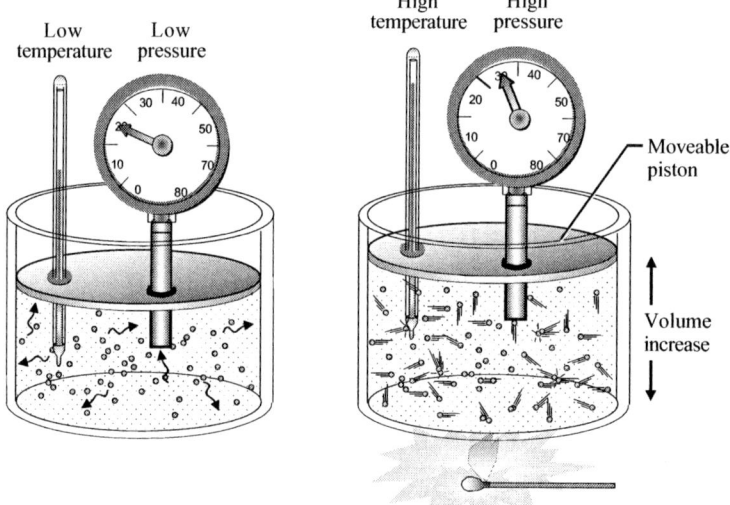

FIGURE 5.3 Pressure Against a Container Wall

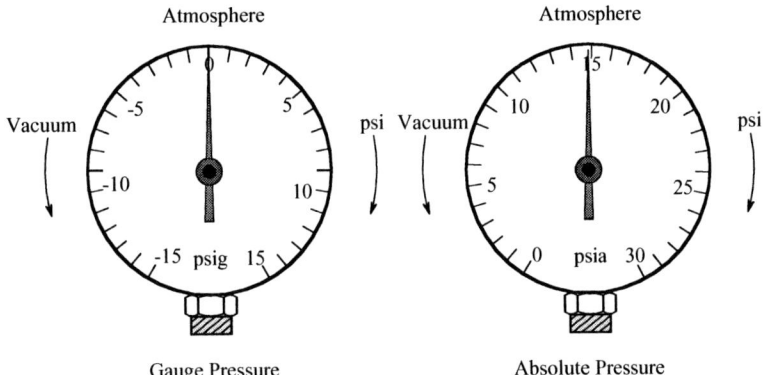

FIGURE 5.4 Gauge Pressure (PSIG) Versus Absolute Pressure (PSIA)

TABLE 5.1 Units of Pressure at Sea Level and 23°C

Unit of Pressure	Sea Level and 23°C
Psig (gauge)	0 psi
Psia (absolute)	14.7 psi
Atmosphere	14.7 psi
Inches of mercury	29.92 inches
Millimeters of mercury	760 mm
Torr	760 torr
mtorr	760,000 mtorr
Bar	1.013 bar
millibar	1013 mbar
Pascal	101,325 pascal

Pressure is a widely used property in semiconductor manufacturing. Chemicals and gases flow from regions of higher pressure to lower pressure. Some fabrication processes occur at atmospheric pressure, while some require higher pressures. Other processes occur in chambers at less than atmospheric pressure (vacuum).

Vacuum. If the gas pressure in a container is less than 14.7 psi, then a vacuum exists. *Vacuum* is the removal of gas molecules (e.g., air, moisture, and gas residues) in a closed container to achieve a pressure less than atmosphere. Many semiconductor operations occur in vacuum conditions. Vacuum is commonly measured in units of torr. A *torr* is defined as the equivalent of one mm of mercury in a pressure-measuring device known as a barometer (see Figure 5.5 on page 96).

At one atmosphere, the weight of the air will push down on the mercury in the bowl and cause it to rise 760 mm (760 torr or 29.92 inches) in the vacuum of the column. The amount of mercury rise is proportional to the pressure bearing down on the mercury in the dish. Less pressure (or vacuum) makes the mercury rise less. For the sensitivity required to measure and control vacuum conditions in many semiconductor processes, the units of millitorr are required.

Condensation and Vaporization ■ The process of a gas changing into a liquid is called *condensation*. When water vapor cools, a mist appears as the particles form tiny droplets of liquid that then collect into a bulk sample with a single surface. The opposite process, changing from a liquid into a gas, is called *vaporization*.

Liquids and gases interact with materials in different ways. *Absorption* is the taking up of a liquid or a gas into the bulk of another material, as in the dissolving of a gas in a liquid. *Adsorption* is the condensing of a gas or a liquid on the surface of a solid. The adsorbed molecules actually stick to the surface through a chemical bond or through a weaker bond of physical attraction.

Vapor Pressure ■ *Vapor pressure* is the pressure exerted by vapor in a closed container when there is an equilibrium condition between the rate of vaporization and rate of condensation. Vapor pressure is shown in Figure 5.6 on page 96.

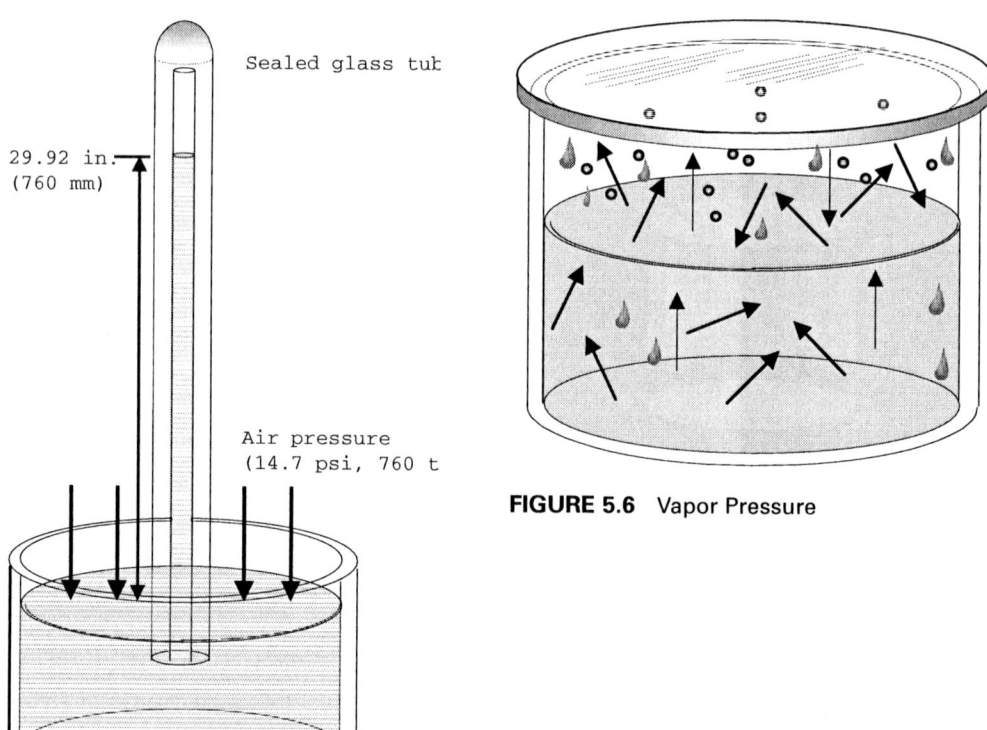

FIGURE 5.6 Vapor Pressure

FIGURE 5.5 Barometer at Atmospheric Pressure

High vapor pressure materials are *volatile* (tend to become a gas). A material with a high vapor pressure will release vapors more easily when exposed to vacuum conditions. Examples are solvents, perfumes, and lotions. These materials typically exhibit odors that are readily detectable through human nasal passages.

Sublimation and Deposition ■ A solid can change directly to a gas through a process termed *sublimation*. Have you ever put an ice tray in the freezer and left it there for six months? The ice cubes are smaller because the ice sublimes. Other examples of sublimation are dry ice (shown in Figure 5.7) and moth balls.

Dry ice (CO_2)

FIGURE 5.7 Sublimation

The opposite process, changing from a gas into a solid, is called *deposition* (see Figure 5.8). This process is how ice forms on a cold window from the deposition of water vapor. As we shall see in a later chapter, deposition is an important process in semiconductor manufacturing.

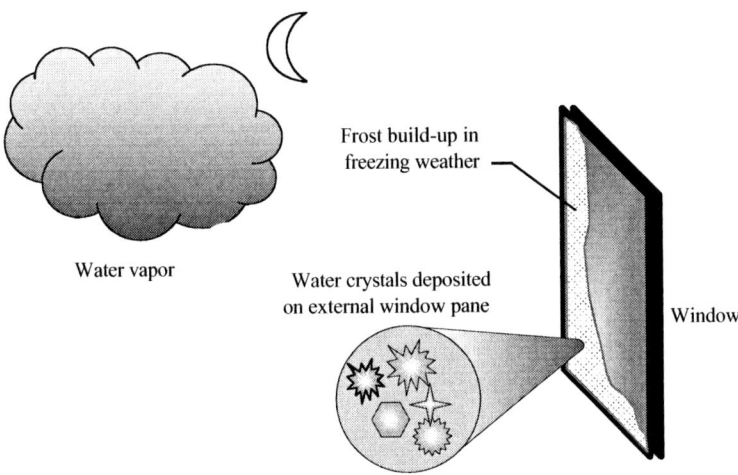

FIGURE 5.8 Deposition

Density ■ The *density* of a substance is defined as its mass (or weight) divided by its volume:

$$\text{Density} = \frac{\text{Mass}}{\text{Volume}} \text{ (grams/cm}^3\text{)}$$

A dense material is thought of as being heavy. If two objects have the same volume and one object is heavier than the other is, then it is denser (see Figure 5.9). A sponge has a lower density than an equal volume of steel. We know this because the sponge weighs less than the steel.

Table 5.2 on page 98 gives the density for different substances at standard temperature and pressure.[1] Water is the standard, with a density of 1 gram/cm^3. The densities of other substances are often expressed as a ratio of their density to that of a comparable volume of water. For instance, silicon has a density of 2.3. Interpret this number to mean that a piece of silicon occupying a volume of one cm^3 will weigh 2.3 grams.

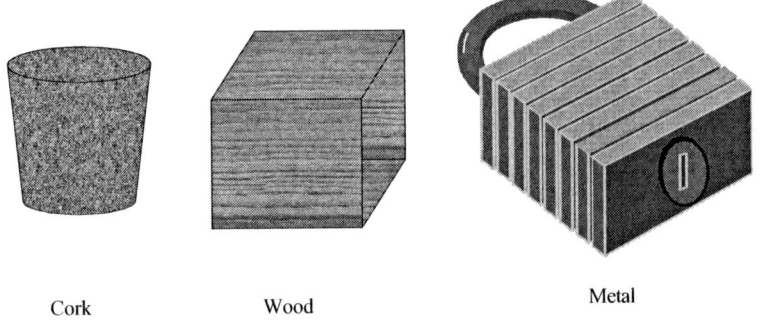

Cork   Wood   Metal

FIGURE 5.9 Density of Objects

TABLE 5.2 Densities of Some Common Substances at Standard Temperature and Pressure

Substance	Physical State	Density (g/cm³)
Hydrogen	Gas	0.000089
Oxygen	Gas	0.0014
Water	Liquid	1.0
Table salt	Solid	2.16
Silicon	Solid	2.33
Aluminum	Solid	2.70
Gold	Solid	19.3

A term related to density is specific gravity. *Specific gravity* (*SG*) refers to the density of liquids and gases at 4°C and is the ratio of the density of a substance to the density of water.

Surface Tension ■ When a liquid is on a flat surface, it has a surface area of contact (see Figure 5.10). The *surface tension* of the liquid is the energy required to increase the surface area of contact. For the surface area to increase, liquid molecules must break intermolecular attraction forces and move away from the interior of the liquid and toward its surface. This movement requires energy. The concept of surface tension is used in semiconductor manufacturing to measure the ability of liquid coatings to uniformly adhere to the wafer surface.

FIGURE 5.10 Surface Tension of a Liquid on a Wafer

Thermal Expansion ■ When an object is heated, it expands due to the increased vibrations of the atoms. Due to this thermal expansion, the dimensions of a heated object will actually increase (see Figure 5.11). When the same object is cooled, its dimensions will decrease.

Some materials will expand thermally more than others. The amount a material expands due to heating is known as its *coefficient of thermal expansion*, or *CTE*. Amorphous materials that have a cubic crystal orientation expand to thermal exposure in all directions. For all other crystals, such as monocrystal silicon, thermal expansion varies with the type of crystal orientation.

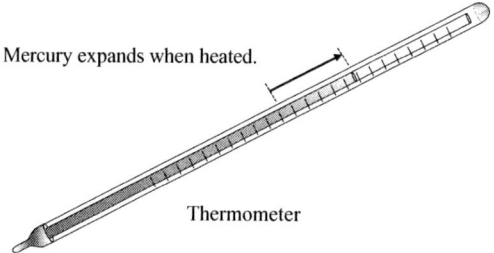

FIGURE 5.11 Thermal Expansion of a Heated Object

Stress ■ *Stress* occurs when an object is exposed to a force. The magnitude of the stress depends on two factors: the amount of force, and the area over which the force is applied. The units of stress are pounds per square inch (psi) or pascal (Pa) in the metric system. The formula for stress is,

$$\text{Stress} = \frac{\text{Force}}{\text{Area}} \text{ (psi)}$$

The force that creates stress in a wafer can come from various sources. Stress can come from work damage on the surface of the wafer; from internal forces due to dislocations, excess vacancies and impurities; and from growth around included foreign material.[2] If two dissimilar materials with significantly different coefficients of thermal expansion (CTE) are bonded together and heated, then a force occurs because the two materials expand at different rates and pull against each other to cause a stress. This stress can create a warped wafer due to the CTE mismatch. Such stress is a concern in semiconductor manufacturing because a microchip is a planar structure with layers of different materials with different CTEs. Deposited films usually have some form of stress that can be compressive or tensile, and the nature of the stress depends on process conditions (see Figure 5.12). The reliability of a chip is improved by ensuring materials have minimal stress.

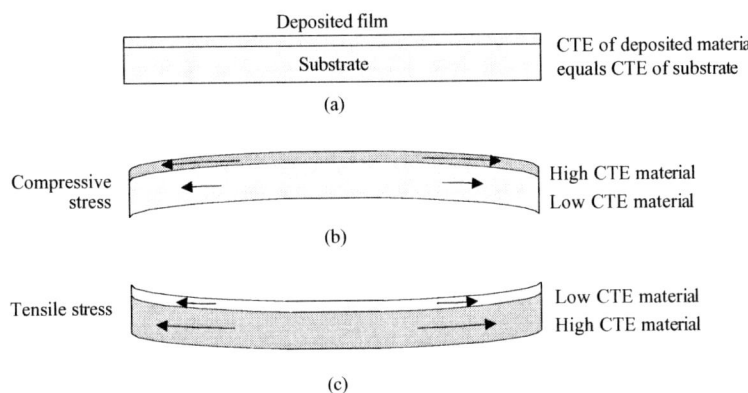

FIGURE 5.12 CTE Mismatch of Two Materials

PROCESS CHEMICALS

Semiconductor manufacturing is a chemically intensive process involving many different types of process chemicals of ultrahigh purity. The process chemicals are used in all three states: liquids, solids, and gases. Chemicals are used in the process of manufacturing semiconductors to:

- Clean or prepare the wafer surface with wet chemical solutions and ultrapure water rinse.
- Dope the wafer with energetic atoms to create p-type and n-type silicon.
- Deposit the different metal conductor layers and the necessary dielectric layers between the conductors.
- Grow the thin silicon dioxide film to be used as the critical MOS gate dielectric.
- Etch thin films with plasma or wet chemicals to selectively remove material and form the required pattern in the film.

There is a vast array of chemicals used in the fab, yet some predominant chemicals are repeatedly used in the different process areas of the fab. Cleaning is one of the most chemically intense steps of semiconductor manufacturing. It is estimated that 30% of all processing steps in the wafer fabrication process are for wafer cleaning or wafer surface preparation.[3]

Liquids

Liquids can be either a pure substance, such as pure water, or a chemical mixture. Automotive gasoline is an example of a chemical mixture of hydrocarbons and additives blended together to produce efficient combustion.

If a chemical mixture is so well mixed that the molecules or ions of the components are the same throughout, then we call this a *chemical solution*. Gasoline is a chemical solution. Another example of a solution is the household first aid antiseptic hydrogen peroxide, which consists mostly of water, with only 5% hydrogen peroxide. The component of the solution present in the larger amount (such as water in the hydrogen peroxide solution) is called the *solvent*. The dissolved substances in the solution are the *solutes*. Solutions in water are called *aqueous solutions*, meaning water is the solvent.

There are many types of liquids used in the wet processes of semiconductor manufacturing. All liquids used in wafer fabs must have extremely high purity, with no contamination from particulates, metallic ions, or unwanted chemicals. Contamination of chemicals is a relative concept. A term that is frequently used to describe small concentrations of an impurity is *parts per million* (*ppm*), which is measured by volume or by mass (weight). An example of ppm is if you consider the amount of an impurity in air. To calculate the ppm of an impurity by mass, you determine the weight of the impurity in a fixed volume of air, divide this number by the weight of the air, and multiply by one million. Chemical purity specifications have become more stringent so that ultrapure chemicals used in the fab often have less than a *parts per billion* (*ppb*) or *parts per trillion* (*ppt*) of an impurity.[4] The amount of impurity measured in a wafer fab process chemical is often limited by the measurement capability of the instrument.

There has been a long-term effort to reduce the quantity of liquid chemicals in the wafer fab. Many liquid chemicals used in semiconductor manufacturing are hazardous and require special handling and disposal procedures. Furthermore, chemical residues can contaminate wafers, as well as produce vapors that diffuse through the air and deposit on wafer surfaces. Although it has not been possible to eliminate wet chemicals from the wafer fab, there has been a significant reduction in their use, such as in reducing the number of wet cleaning steps.[5]

Liquid process chemicals in a wafer fab are grouped in the following manner:

- Acid
- Base
- Solvent

Acid ■ In the classical definition, *acids* are solutions that contain hydrogen and dissociate in water (meaning bonds break down) to yield hydronium ions, H_3O^+. Since an acid contains hydrogen, it contains H in its formula, such as phosphoric acid (H_3PO_4) or hydrochloric acid (HCl).

As an example of how acid dissociates in water, consider the following reaction for hydrochloric acid:

$$HCl \text{ (gas)} + H_2O \text{ (liquid)} \rightarrow Cl^- \text{ (aqueous)} + H_3O^+ \text{ (aqueous)}$$

We can see that when HCl is dissolved in water, it reacts with the water to produce the hydronium ion, H_3O^+. This defines HCl as an acid.

There are many different acids used in semiconductor manufacturing. Table 5.3 lists some common acids and their typical use in a wafer fab.

TABLE 5.3 Common Acids Used in Semiconductor Manufacturing

Acid	Symbol	Example of Use*
Hydrofluoric acid	HF	Etching silicon dioxide (SiO_2) and cleaning quartzware
Hydrochloric acid	HCl	Wet cleaning chemical that is part of the standard clean 2 (SC-2) solution to remove heavy metals from wafer
Sulfuric acid	H_2SO_4	Solution known as "Piranha" [7 parts H_2SO_4 to 3 parts of 30% hydrogen peroxide (H_2O_2)] used to clean wafers
Buffered oxide etch (BOE): Solution of hydrofluoric acid and ammonium fluoride	HF and NH_4F	Etching of silicon dioxide (SiO_2) film
Phosphoric acid	H_3PO_4	Etching of silicon nitride (Si_3N_4)
Nitric acid		HNO_3 Used in mixture of HF and HNO_3 to etch phosphosilicate glass (PSG)

*Explanation of wafer cleaning is in Chapter 6 and etching in Chapter 16.

Acids are further divided into two categories: organic and inorganic. Organic acids such as carboxylic acids contain hydrocarbons, while inorganic acids such as hydrofluoric acid (HF) do not.

Base ■ A *base* is a substance that contains the OH chemical group (e.g., NaOH, sodium hydroxide and KOH, potassium hydroxide) and dissociates in water to yield the hydroxide ion, OH^-. Another name for a base substance is *alkali* or an *alkaline substance*. A base will increase the hydroxide ion in an aqueous solution. For example, sodium hydroxide is an ionic compound containing metal ions and hydroxide ions. It is a base because when it dissolves in water it dissociates to yield Na^+ and OH^- ions:

$$NaOH \rightarrow Na^+ + OH^-$$

Common alkaline substances used in semiconductor manufacturing are shown in Table 5.4.

TABLE 5.4 Common Bases Used in Semiconductor Manufacturing

Base	Symbol	Example of Use
Sodium hydroxide	NaOH	Wet etchant
Ammonium hydroxide	NH_4OH	Cleaning solution
Potassium hydroxide	KOH	Positive photoresist developer
Tetramethyl ammonium hydroxide	TMAH	Positive photoresist developer

pH ■ An acid or a base varies in strength and is classified as strong or weak. The *pH scale* is used to assess how strong or weak a solution is as an acid or base. This scale ranges from 0 to 14, with 7 being the neutral point. Acids have pH values below 7 and bases have pH values above 7. Pure water is the reference substance for the pH scale and is neutral with a pH of 7. Strong acids, such as sulfuric acid (H_2SO_4) have low pH values between 0 and 3. Strong bases, such as sodium hydroxide (NaOH) have pH values greater than 7 and approaching 14. Figure 5.13 shows some common chemicals and their location on the pH scale.

pH	Household Chemicals	
1	Car battery acid (sulfuric acid, H_2SO_4)	Corrosive
2		
3	Lemon juice, vinegar	
4	Soda, wine	
5	Tomato juice, beer	
6	Urine	
7	Tap water, milk, saliva	
8	Blood, saliva	
9	Milk of magnesia	
10	Detergents	
11	Household ammonia	
12		
13	Household drain cleaners	Caustic
14	Nickle-cadmium battery (NaOH base)	

FIGURE 5.13 The pH Scale for Different Chemicals

Solvents ■ A *solvent* is a substance capable of dissolving another substance to form a solution. A good solvent will dissolve (solubilize) a broad range of substances. Most solvents, such as alcohol and acetone, are volatile and flammable. Common solvents used in a wafer fab are listed in Table 5.5 on page 102.

TABLE 5.5 Common Solvents Used in Semiconductor Manufacturing

Solvent	Common Name	Example of Use
Deionized water	DI Water	Widely used to rinse wafers and to dilute cleaning solutions.
Isopropyl alcohol	IPA	General-purpose cleaning solvent.
Trichloroethylene	TCE	Solvent used for wafer and general cleaning.
Acetone	Acetone	General-purpose cleaning solvent (stronger than IPA).
Xylene	Xylene	Strong cleaning solvent; may also be used for photoresist edge bead removal.

Deionized water (DI water) is a widely used solvent in semiconductor manufacturing that has all conductive ions removed. With a pH of 7, which is neutral, DI water is neither an acid nor a base. It has the ability to dissolve other substances, including many ionic and covalent compounds. When H_2O water molecules dissolve an ionic compound, they separate, surround, and disperse the ions into the liquid. Water does this by overcoming the electrostatic force of attraction between the ions.

Chemical Distribution ■ There is a wide range of chemicals used in the semiconductor industry, with many being toxic and hazardous. A review of key chemical safety issues is provided in Appendix 1. Safe, high-purity, uninterrupted delivery of chemicals from storage vessel to process tools is critical. For liquid chemicals, this delivery is often accomplished through bulk chemical distribution (BCD).

A *bulk chemical distribution* system consists of a chemical source, such as a storage vessel, a chemical delivery module, and a piping system (see Figure 5.14).[6] The BCD storage vessel is often housed below the main production floor in the groundlevel sub-fab. The chemical delivery module will filter, blend, and transport the chemicals through the piping system. A piping system delivers the chemical to the individual process stations. Modern BCDs are integrated into a fully computerized and networked system for real-time chemical monitoring and control.

The decision regarding how to store and deliver process chemicals depends on factors such as chemical compatibility, reduction of chemical contamination, and safety. Purity requirements of

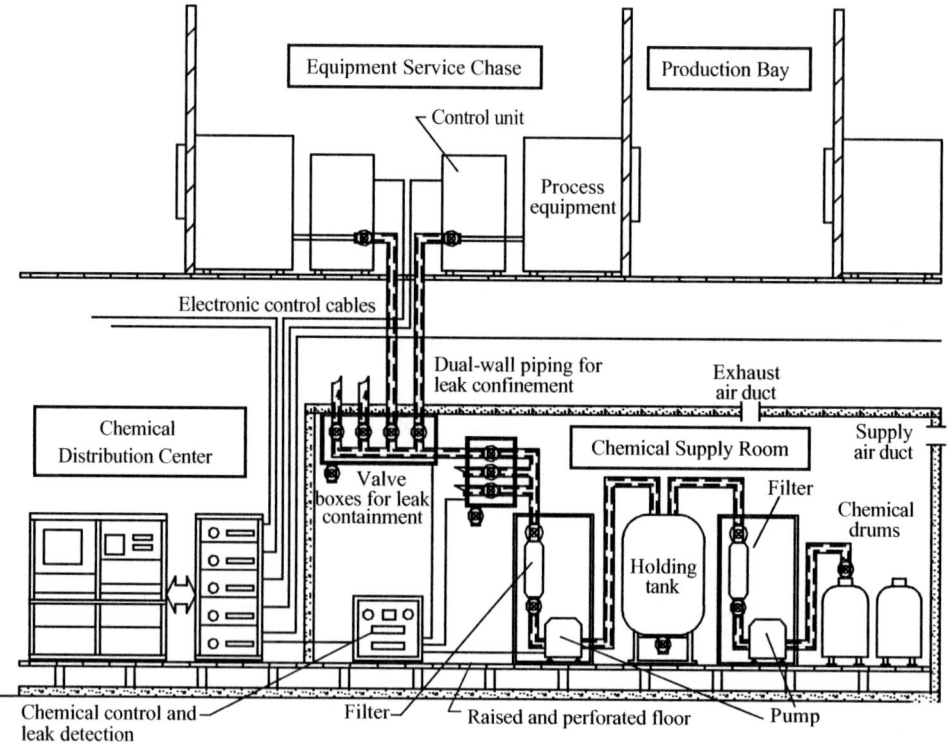

FIGURE 5.14 Bulk Chemical Distribution

Bulk Chemical Distribution
(Photo courtesy of Advanced Micro Devices)

chemicals used in semiconductor manufacturing are described as *ultrahigh purity* (*UHP*), with contaminants being controlled to the parts per billion (ppb) to parts per trillion (ppt) range.

Some chemicals are not suitable for BCD. They may be used in small quantities or have a limited shelf life (how long they can be stored before use). In this case, these chemicals will have special packaging systems for *point-of-use* (*POU*) delivery, which means they are stored and used at the process station. An example of this type of chemical is the photoresist chemical used in photolithography (discussed in Chapter 13).

Gases

There are about 50 different types of gases used throughout the 450 or so process steps in semiconductor manufacturing.[7] The types and quantities of gases are changing because there are new materials being introduced into the fab process, including copper metallurgy and new diffusion barriers (see Chapter 12). Gases are usually categorized as one of two types: bulk gases or specialty gases. *Bulk gases* are traditionally defined as oxygen (O_2), nitrogen (N_2), hydrogen (H_2), helium (He), and argon (Ar). The *specialty gases* are also referred to as process gases, and are the other important gases used to manufacture semiconductor microchips.

Extremely high purity is required of all gases, with bulk gases controlled to seven nines purity (99.99999%) and specialty gases controlled at least to four nines (99.99%). Particulate contamination in gases is controlled to the 0.1-μm range. Other controlled contaminants are oxygen, moisture content, and trace impurities such as metals.[8] Many process gases are toxic, corrosive, reactive, and pyrophoric (they combust or burn when exposed to air). For these reasons, gases are contained in a gas delivery system that delivers the gases in a safe, clean, and accurate manner to the various process stations in the wafer fab.

Bulk Gases ■ Bulk gases are relatively simple gases for the gas suppliers to manufacture and are stored outside of the wafer fab manufacturing area in large storage tanks or 1000-lb bulk tube trailers. These gases are distributed to the workstation through the bulk gas distribution (BGD) system. Benefits from the centralized gas control of BGD are a reliable gas supply, less sources of particulate contamination, and less human involvement in the daily delivery of the gas. Bulk gas is often the lowest cost method for gas delivery with higher gas purity. Furthermore, on-site production of the bulk gases is done at large wafer fabs in order to reduce costs. Bulk gases are divided into inert, reducing, and oxidizing gases (see Table 5.6 on page 104).

TABLE 5.6 Bulk Gases

Type of Gas	Gas	Symbol	Example of Use
Inert	Nitrogen	N_2	Purge gas lines and process chambers of moisture and residual gas. N_2 is sometimes used as a process gas during some deposition processes.
	Argon	Ar	Used in process chambers during wafer processing.
	Helium	He	Process-chamber gas also used to leak check vacuum chambers.
Reducing	Hydrogen	H_2	Carrier gas for epitaxial layer process. Also used in combination with O_2 to form steam during some furnace oxidation processes. Used in many wafer fabrication processes.
Oxidizing	Oxygen	O_2	Process-chamber gas.

In recent years, safety concerns have driven the development of on-site generation for other gases, in particular arsine and phosphine.[9] This arrangement permits these highly toxic gases to be produced close to the process tool.

Specialty Gases ■ Specialty gases are those gases supplied in relatively low volumes. These gases are frequently more hazardous than the bulk gases. The specialty gases are the source of many of the materials needed to manufacture a microchip. The greatest challenge for dealing with specialty gases is that many of them are a chemical hazard. They are corrosive (e.g., HCl and Cl_2), pyrophoric (e.g., silane), toxic (e.g., arsine and phosphine), and extremely reactive (e.g., WF_6). The specialty gases are typically used in process chambers at the workstation tools.

Specialty gases traditionally have been transported to the wafer fab in 100-lb metal containers called cylinders for distribution of the gas to the workstations. The cylinders are placed in a gas cylinder cabinet containing a control panel with regulators to control pressure, a flow control, shut-off valves, and a purge panel to control purge sequences needed between cylinder changes. The cabinet also may contain filters, purifiers, or equipment to monitor gas purity, as well as safety equipment such as fire sensors and leak-detection sensors.

A local gas distribution system in the process area is used to deliver the specialty gas from the cylinder to the work chamber of the process tool (see Figure 5.15). This piping can cover hundreds of feet with many permanent welds, bends and fittings and is connected to the process tool at the tool "drop."

Bulk Gas Distribution System
(Photo courtesy of Air Products and Chemicals, Inc.)

Chemicals in Semiconductor Fabrication 105

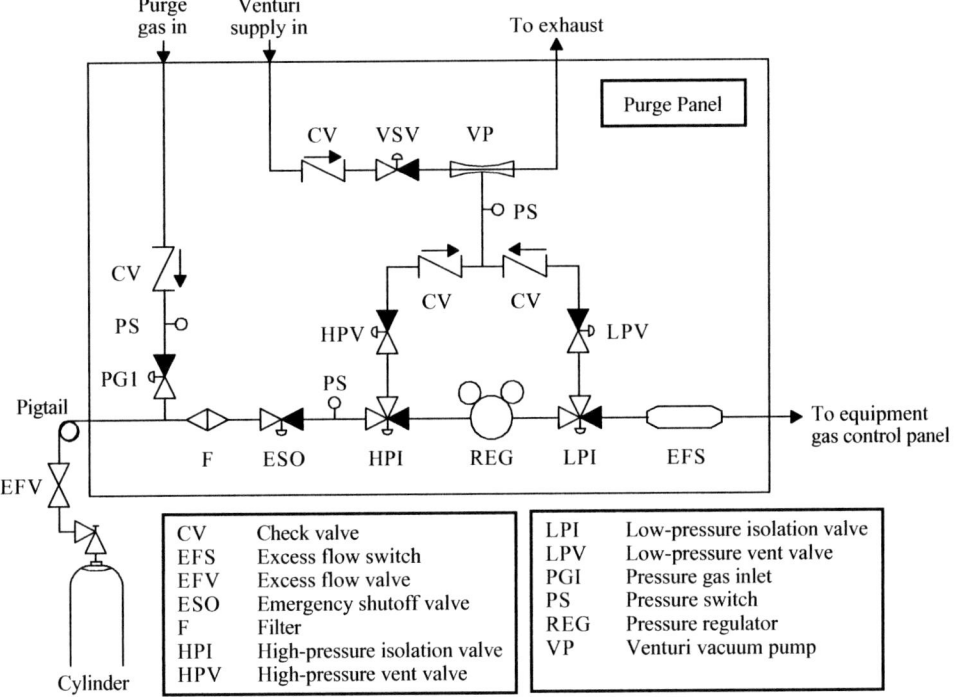

FIGURE 5.15 Typical Specialty Gas System Design
(Schematic used with permission from International SEMATECH)

Gas Purge. A *gas purge* is a method of flushing undesirable residual gases, atmospheric gases, or water vapor from a process chamber and the gas delivery system. Purging can eliminate stagnation points and contamination in gas lines. It is done with an inert gas such as nitrogen and involves replacing the undesirable gas with the purge gas, either through displacement or by pulling the gas out of the system by vacuum flow. Purging of the system on automated equipment is done automatically through software control of gas line valves before and after situations such as a gas cylinder change or opening a process chamber.

Gas Piping. Gas delivery piping is constructed with electropolished 316L stainless steel tubes to transport the gas (316L is a specific type of stainless steel). There are no plastic parts in a gas piping system, except for some membrane gas filters. Double-walled tubing is often used for hazardous gases (see Figure 5.16). The inner walls of the double-walled tubing are electropolished to minimize contamination. *Electropolishing* is a chemical process done to remove about 30 microns of the pipe's inner surface, creating a clean, smooth surface that minimizes the possibility of chemical reactions that produce contaminants. An electropolished surface finish brings a thin layer of chromium to the steel's surface, which emits very few particles.[10]

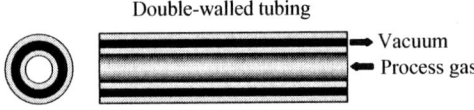

FIGURE 5.16 Double-Walled Tubing
(Used with permission from International SEMATECH.)

Gas Line Connections. For safety reasons, gas lines will have a 360-degree bend, referred to as a pigtail, to make the line more flexible. The gas line terminates in a CGA (Compressed Gas Association) connector for connection to the gas cylinder valve (see Figure 5.17 on page 106). A recently developed DISS (Diameter Index Safety System) cylinder valve has been introduced with specific connector diameters for different gases, which creates a better seal with less abrasion to reduce the potential for contamination. DISS valves are generally used with high-purity products, toxics, and corrosives. Gas lines need a continuous flow through the piping system with no dead spots (known as a deadleg) to trap gas and cause undesirable chemical degradation due to stagnation.

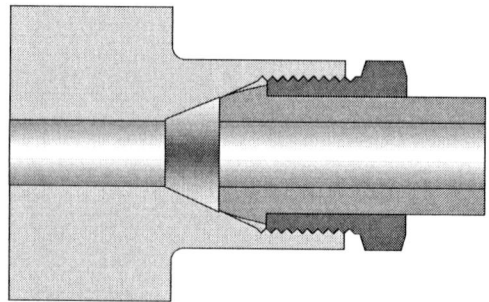

FIGURE 5.17 CGA Gas Line Connector
(Used with permission from International SEMATECH)

Gas Stick. The gas line from the local gas distribution system connects into another gas panel at the process tool. The tool panel is made up of a series of gas "sticks," each controlling one type of incoming gas. Each stick will typically have an on/off valve, flow controller, pressure regulator, and filters (see Figure 5.18). The number of incoming gas lines depends on the process, with a typical number being from six to 30 or more lines for multi-chamber process tools.

Cylinder Change-Out. Cylinders for specialty gases require technicians to change them when they are empty. This is referred to as a *cylinder change-out* and is sometimes done as often as several times per week. Because specialty gases are often hazardous, gas cylinder change-outs require care to avoid safety and product problems.[11] One safety concern during a cylinder change is incorrectly purging gas lines, which causes residual gas to escape, leading to a flame or vapor cloud. Another safety concern is an improperly supported cylinder that may fall over, causing a leak. With regard to the product performance, an incorrect cylinder change can lead to contamination in the gas line and affect product yield. Furthermore, a cylinder change may require the process chamber to be requalified to ensure the gas is acceptable, which interrupts manufacturing production.

Due to safety hazards, contamination concerns, and high maintenance costs, there has been an effort to convert some specialty gases to bulk distribution. For example, a tube trailer of silane is estimated to eliminate the change-out of 300 to 400 cylinders per year, while the tube trailer is changed only once per year.[12] This is a substantial reduction in risk for a problem during change-outs. Eventually more bulk specialty gas distribution systems similar to those used for wet chemicals and bulk gases may replace the cylinders commonly used to deliver specialty gases.[13] This

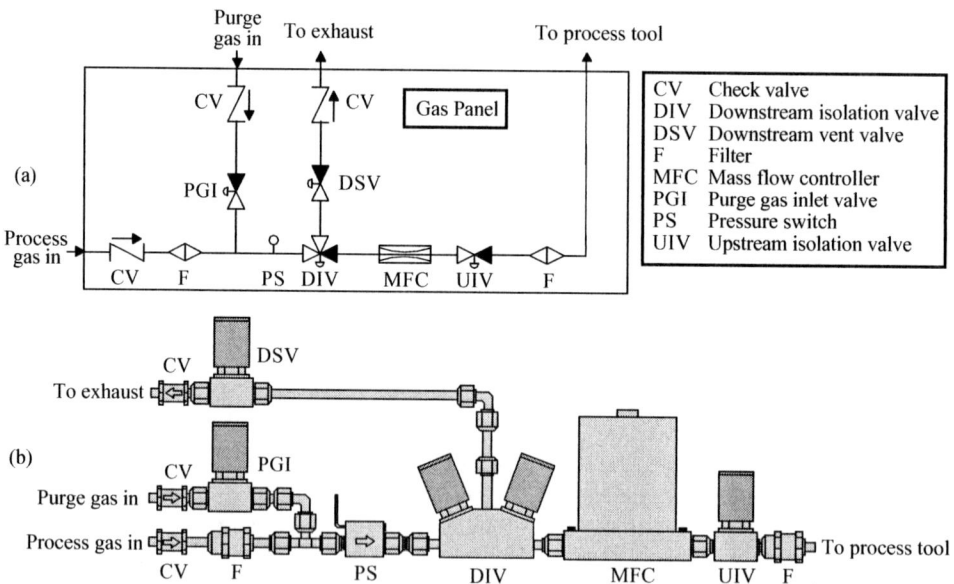

FIGURE 5.18 Gas Stick at Process Tool.
(Schematic (a) used with permission from International SEMATECH)
Component diagram (b) redrawn from Swagelok components, *Swagelok Co. Catalog* provided by Arthur Valve & Fitting Co., Austin, TX.

trend initially started with silane because it is pyrophoric (it burns upon exposure to air) and now encompasses other specialty gases used in high volume, such as nitric oxide (NO). The goal is to reduce the human involvement in the gas delivery system and the potential for error.

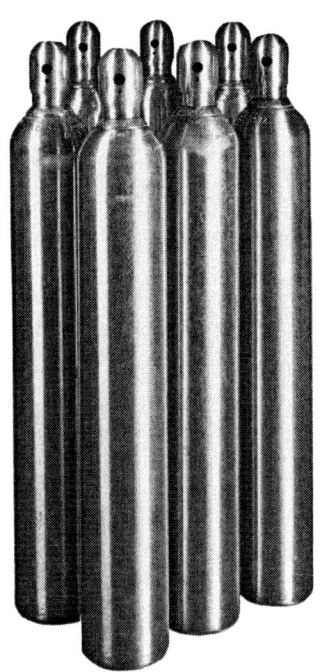

Specialty Gas Cylinder
(Photo courtesy of Praxair)

Classifying Specialty Gases ■ Specialty gases may be classified as hydrides (or similar gases), fluorinated compounds, or acid gases. Hydride gases contain hydrogen, while fluorinated compounds contain fluorine. Some common specialty gases are listed in Table 5.7 on page 108, with an example of their use.

TABLE 5.7 Some Common Specialty Gases for Semiconductor Manufacturing

Class of Gas	Gas	Symbol	Example of Use
Hydrides	Silane	SiH_4	Source of silicon in deposition processes (Chapter 11)
	Arsine	AsH_3	Source of arsenic for n-type doping of silicon wafers (Chapter 17)
	Phosphine	PH_3	Source of phosphorous for n-type doping of silicon wafers (Chapter 17)
	Diborane	B_2H_6	Source of boron for p-type doping of silicon wafers (Chapter 17)
	Tetraethyl orthosilicate (TEOS)	$Si(OC_2H_5)$	Source of silicon dioxide used in deposition processes (Chapter 11)
	Silicon tetrachloride (also tetrachlorosilane)	$SiCl_4$	Source of silicon used in deposition processes (Chapter 11)
	Dichlorosilane (DCS)	SiH_2Cl_2	Source of silicon used in deposition processes (Chapter 11)
Fluorinated Compounds	Nitrogen trifluoride	NF_3	Source of fluoride ions used in plasma etch processes (Chapter 16)
	Tungsten hexafluoride	WF_6	Source of tungsten used in metal deposition processes (Chapter 12)
	Tetrafluoromethane	C_2F_4	Source of fluoride ions used in plasma etch processes (Chapter 16)
	Carbon tetrafluoride	CF_4	Source of fluoride ions used in plasma etch processes (Chapter 16)
	Silicon tetrafluoride	SiF_4	Source of silicon and fluorine ions for use in deposition, implant, and etch processes (Chapters 11, 16, 17)
	Chlorine trifluoride	ClF_3	Process chamber cleaning gas
Acid Gases	Boron trifluoride	BF_3	Source of boron for p-type doping of silicon wafers (Chapter 17)
	Chlorine	Cl_2	Source of chlorine used in metal etching (Chapter 16)
	Boron trichloride	BCl_3	Source of boron for p-type doping of silicon and a source of chlorine for metal etching (Chapters 16, 17)
	Hydrogen chloride	HCl	Process chamber cleaning gas and contamination getterer
Other Gases	Ammonia	NH_3	Process gas used with SiH_2Cl_2 for deposition of SiN_3 (Chapter 11)
	Nitrous oxide	N_2O	Source of oxygen for reaction with silicon to from silicon dioxide (Chapter 11)
	Carbon monoxide	CO	Used in etch processes (Chapter 16)

SUMMARY

There are four states of matter: solid, liquid, gas, and plasma. Important chemical properties for wafer fabrication are: temperature, pressure and vacuum, condensation, vapor pressure, sublimation and deposition, density, surface tension, thermal expansion, and stress. Process chemicals are widely used to manufacture microchips. Liquids are grouped as acid, base, or solvent. An acid or a base varies in strength and is measured on the pH scale. Chemicals are delivered to the workstation through bulk chemical distribution or point-of-use delivery. Gases used in the wafer fab are classified as either bulk or specialty. Bulk gases are the relatively simple gases used to manufacture chips, such as nitrogen and oxygen. Specialty gases are often a chemical hazard and are transported and stored in the fab in metal cylinders. There are special procedures and delivery systems to ensure that the specialty gases are cleanly and safely used at the workstations. Specialty gases are generally classified as either hydrides, fluorinated compounds, or acid gases.

KEY TERMS

process chemicals
solid
liquid
gas
inert gas
plasma
properties
physical properties
chemical properties
chemical reaction
temperature
heat (thermal energy)
Celsius
Kelvin
pressure
vacuum
condensation
vaporization
absorption
adsorption
vapor pressure
volatile
sublimation
deposition
density
specific gravity
surface tension
coefficient of thermal expansion (CTE)
stress
chemical solution
solute
aqueous solutions
parts per million (ppm)
parts per billion (ppb)
parts per trillion (ppt)
acids
base
pH scale
solvent
bulk chemical distribution (BCD)
ultrahigh purity (UHP)
bulk gases
specialty gases
gas purge
electropolishing
cylinder change-out

REVIEW QUESTIONS

1. What are the four states of matter? Describe each state.
2. What is a material property?
3. Describe the two types of properties of materials.
4. What happens in a chemical reaction? Give an example of a chemical reaction.
5. What is temperature? How is temperature related to heat?
6. State the three temperature scales. Which scale is most common for scientific work?
7. What is pressure? The pressure of a gas is dependent on what conditions?
8. Define vacuum. What is the most common vacuum unit and how is it defined?
9. Define condensation and vaporization. What is the difference between absorption and adsorption?
10. What is vapor pressure? Describe a volatile material.
11. Define sublimation and deposition.
12. Define and describe density. What is the specific gravity of a material?
13. What is surface tension?
14. Define the coefficient of thermal expansion (CTE) of a material.
15. Describe stress and state its formula.
16. How are stress and coefficient of thermal expansion related?
17. What is a chemical solution and what are its components?
18. What is an aqueous solution?
19. What do the abbreviations *ppm*, *ppb*, and *ppt* represent?
20. What is an acid? List three common acids used in a wafer fab.
21. What is a base? List three common bases used in a wafer fab.
22. Explain the pH scale and how it measures acids and bases. Identify the regions in the pH scale attributed to weak solutions of a strong acid, a strong base, and water.
23. What is a solvent? List three common solvents used in a wafer fab.
24. Describe deionized (DI) water for the wafer fab.
25. Describe the bulk chemical distribution system.
26. What is the point-of-use chemical delivery?
27. What are the two types of gas categories?
28. Describe the purity required for gases in semiconductor manufacturing.
29. What five gases are classified as bulk? Give three benefits of bulk gas distribution.
30. What is a specialty gas?
31. What is the greatest challenge for dealing with specialty gases?
32. How are specialty gases usually transported and stored in the wafer fab?
33. What is a gas purge? What type of gas is this done with?
34. How is gas piping constructed? What is electropolishing, and why is this process beneficial?
35. What is CGA? What is DISS and why is it used?
36. Describe a gas stick.
37. Describe a specialty gas cylinder change-out. What are the safety concerns associated with this procedure?
38. State the three classes of specialty gases. Give an example of a common gas for each class.

CHEMICAL SUPPLIERS' WEB SITES

AERONIX Inc.	http://www.aeronex.com
Air Products and Chemicals	http://www.airproducts.com/
Ashland Specialty Co.	http://www.ashchem.com/
ATMI Inc.	http://www.atmi.com/
BOC Edwards	http://www.boc.com/edwards/
Dow Chemical/Filmtec	http://www.dow.com/liquidseps/
Dow Corning	http://www.dowcorning.com/
DuPont	http://www.dupont.com/semiconductor
Eastman Chemical Co.	http://www.eastman.com/
EKC Technology Inc.	http://www.ekctech.com/
J. T. Baker	http://www.jtbaker.com/
Leybold Inficon Inc.	http://www.leyboldinficon.com/
Linde	http://www.linde.de/english/Home.htm
Matheson Gas Products	http://www.mathesongas.com/
Millipore Corp.	http://www.millipore.com/
MKS Instruments Inc.	http://www.mksinst.com/
Parker Hannifin Corp.	http://www.parker.com/
Praxair	http://www.praxair.com/

PTI Advanced Filtration Inc.	http://www.pti-afi.com/
Scott Specialty Gases	http://www.scottgas.com/
Solkatronic Chemicals	http://www.solkatronic.com/
Swagelok Company	http://www.swagelok.com/
Union Carbide	http://www.unioncarbide.com/
Voltaix Inc.	http://www.voltaix.com/

REFERENCES

1. M. Silberberg, *Chemistry: The Molecular Nature of Matter and Change,* (St. Louis: Mosby, 1996), p. 23.
2. W. Runyan and T. Shaffner, *Semiconductor Measurements and Instrumentation,* 2nd ed., (New York, NY, McGraw-Hill, 1997), p. 206.
3. T. Hattori, "Trends in Wafer Cleaning Technology," *Solid State Technology* (May 1995): p. 7.
4. A. Braun, "PPT – Time for a Reality Check?" *Semiconductor International* (June 1998).
5. J. Sargent, V. Starov, and R. Werner, "Transition in the Post-Etch Wafer-Cleaning Market and Technologies," *Solid State Technology* (May 1997): p. 180.
6. R. DeJule, "Bulk Chemical Distribution Addresses Tightening Specs," *Semiconductor International* (August 1996): p. 75.
7. P. Singer, "Effective Gas Handling: A Balance of Cost and Purity," *Semiconductor International* (September 1994): p. 64.
8. P. Singer, "Trends in Gas Management and Use," *Semiconductor International* (April 1998): p. 112.
9. Ibid.
10. A. Braun, "Cleanroom Technologies Continue to Keep Contamination at Bay," *Semiconductor International* (March 1998): p. 59.
11. L. Laurin, "Bulk Silane—a Potential Hazard or a Potential Hazard Reducer," *Solid State Technology,* (January 1998): p. 104.
12. P. Singer, "Trends," p. 112.
13. N. Chowdhury and L. Mostowy, "Developing a Bulk Distribution System for High-Purity Hydrogen Chloride," *Micro Contamination Identification, Analysis and Control* (September 1995): p. 33.

CHAPTER 6
CONTAMINATION CONTROL IN WAFER FABS

Contamination-free wafer fabrication is absolutely essential in order for devices on chips to function properly. As device critical dimensions decrease, contamination requirements become more stringent. We will learn in this chapter about the different types of contamination important for wafer fabrication, their sources, and how to control contamination to effectively manufacture high-performance ICs with minimal contamination-generated defects.

To control unacceptable contamination during manufacturing, the semiconductor industry has developed cleanrooms. A cleanroom is essentially a purified space with ultraclean air that isolates the chip manufacturing from the dirty conditions of the outside world, including chemicals, humans, and ordinary work conditions.

It is important to understand the conditions of a cleanroom because this is where the intricate microchips are manufactured. Numerous details of the work procedures in a wafer fab are defined by how we maintain the integrity of the cleanroom. One of the primary elements for successful cleanrooms is human discipline.

OBJECTIVES

After studying the material in this chapter, you will be able to:

1. State and describe the five different types of cleanroom contamination, and discuss the problems associated with each type of contamination.
2. List seven sources of contamination in a cleanroom, and describe how each one affects wafer cleanliness.
3. Interpret and use the class number for cleanroom air quality.
4. State and discuss seven appropriate actions for workers entering a cleanroom that follow acceptable protocol.
5. Describe the different aspects of an ultraclean cleanroom facility, including air filtering, electrostatic discharge, ultrapure DI water, and process gases.
6. Explain how modern workstation design and a minienvironment contribute to contamination reduction.
7. State the chemistry of the two standard wet-cleaning methods, explain the type of contamination removed by each, and discuss wet-cleaning modifications and alternatives.
8. Describe the different types of wet-cleaning equipment, and state how each cleaning process contributes to wafer cleanliness.

INTRODUCTION

A wafer has multiple microchips on its surface, and each chip has literally millions of devices and interconnection circuitry that are highly sensitive to contamination. As the feature size on a chip shrinks to accommodate higher performance and denser circuitry, the need to control surface contamination becomes more critical (see Figure 6.1 on page 114). To achieve contamination control, all wafer fabrication is done in a cleanroom where contaminants are strictly controlled.

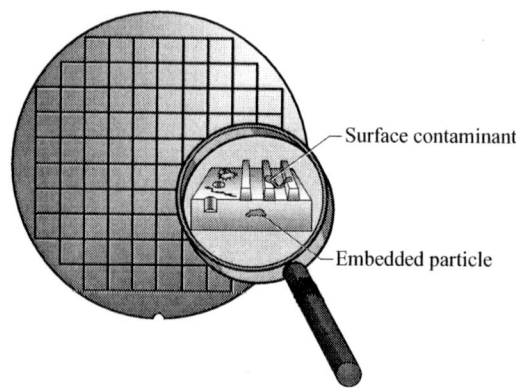

FIGURE 6.1 Wafer Contaminants

Cleanroom Background

When semiconductor manufacturing began nearly half a century ago, the need to control contamination was evident. Early cleanrooms were based around local clean zones, with clean benches used by operators wearing smocks and gloves. The introduction of the high-efficiency particulate air filter (HEPA filter) in the 1960s was a first step toward significant particulate reduction in the wafer fab. The HEPA filters delivered clean air at the workbench to efficiently move particles away from the product.

Modern semiconductor manufacturing is performed in a sophisticated facility known as a *cleanroom*. This is a wafer fabrication facility that is isolated from the outside environment and free from contaminants such as particles, metals, organic molecules, and electrostatic discharge (ESD). When we say free, that means that these contaminants are not detectable at the detection level of the most advanced test instrument. A cleanroom represents the comprehensive set of procedures and practices that are followed to ensure a wafer fabrication facility is contamination-free for semiconductor manufacturing.

Wafer Fab Cleanroom (Photo courtesy of Advanced Micro Devices)

TYPES OF CONTAMINATION

Contamination in semiconductor manufacturing is any undesirable substance introduced to the semiconductor wafer that impacts the production yield or electrical performance of a microchip. Since our interest is wafer manufacturing, we will focus on the various types of surface contamination introduced during the manufacturing process.

Contamination often leads to a defective chip. *Killer defects* are those causes of failure where the chip on the wafer fails during electrical test. It is estimated that 80% of all chip failures at electrical test are due to killer defects from contamination.[1] Failure at electrical test results in a yield loss, causing the defective die on the wafer to be scrapped (thrown away) at a significant cost to the chip manufacturer.

Cleanroom contamination is grouped into five categories:

- ◆ Particles
- ◆ Metallic impurities
- ◆ Organic contamination
- ◆ Native oxides
- ◆ Electrostatic discharge (ESD)

Particles

Particles are small objects that can adhere to the surface of a wafer. Airborne particles suspended in the air are referred to as *aerosols*. The relative size distribution of various particles from pebbles down to atoms is shown in Figure 6.2.[2]

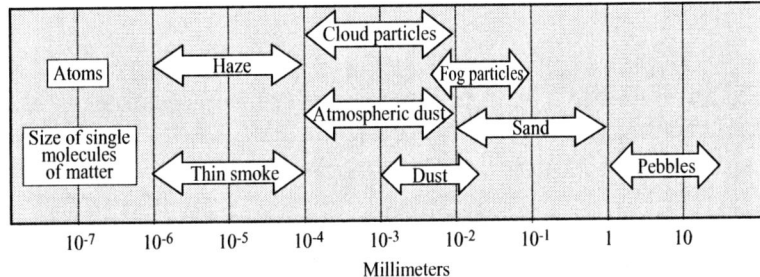

FIGURE 6.2 Relative Size of Particles

Problems From Particles ■ For semiconductor manufacturing, our objective is to control and reduce wafer exposure to particles. Particles can cause defects during the wafer fabrication process leading to open or bridged circuitry. They can create an electrical short between adjacent conductors. Particles also can be sources of other types of contamination, as discussed in the following section.

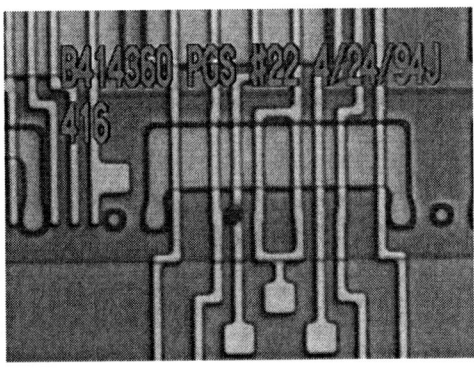

Defect from Particles (Photo courtesy of Advanced Micro Devices)

The rule of thumb for an acceptable particle size in semiconductor manufacturing is that it must be less than one-half the minimum device feature size.[3] Particles larger than this size will cause killer defects. For example, a 0.18-μm feature size cannot be exposed to 0.09 μm and larger particles. To appreciate these dimensions, consider that a human hair is about 90 μm in diameter. A dimension of 0.18 μm would be about 500 times smaller than the human hair (see Figure 6.3).

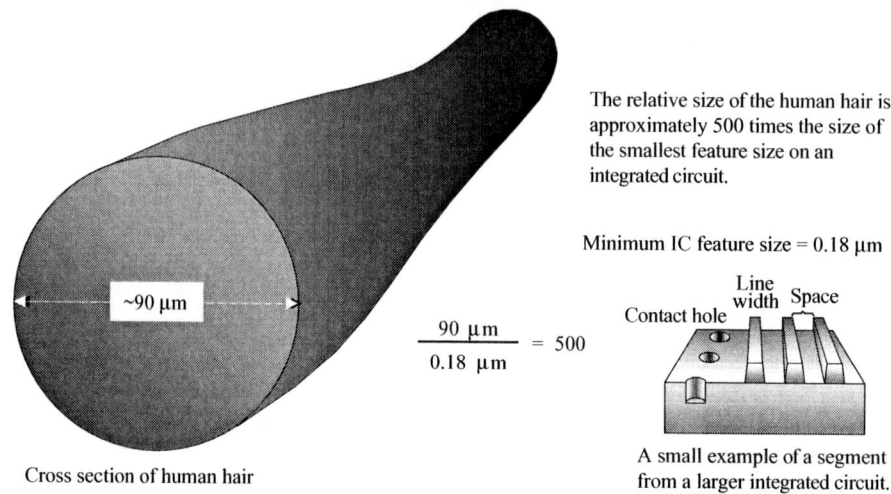

FIGURE 6.3 Relative Size of Human Hair to 0.18 μm Particle Size

The *particle density* on a wafer surface represents the number of particles in a given area. A higher particle density produces a higher chance of a killer defect. The number of particles above a certain critical size that are added to a wafer surface at an operation is termed the *particles per wafer per pass* (*PWP*). In the early days of semiconductor manufacturing, skilled operators visually inspected wafers for particles using simple tools such as a microscope. However, this approach is unacceptable for the VLSI and ULSI eras. Since the mid-1980s, particle detection has been widely done by scanning the wafer surface with a laser beam and detecting the position and intensity of the scattered light caused by particles (see Figure 6.4). The smallest detectable diameter of current particle-detection equipment used in production is roughly 0.1 μm.[4]

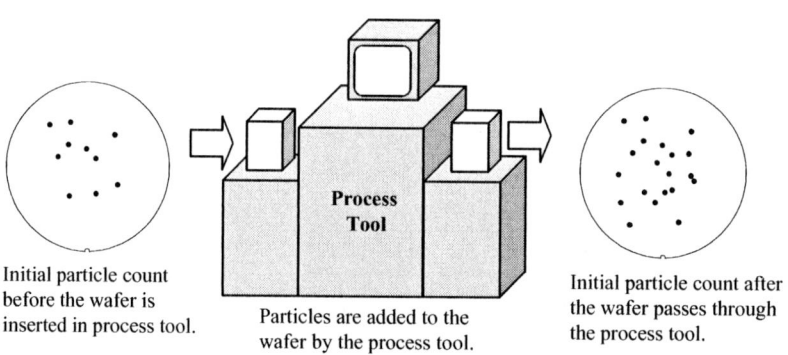

FIGURE 6.4 Particles Per Wafer Per Pass on a Wafer

Semiconductor manufacturing in the ULSI era also requires stringent control of a wide range of airborne molecular particles, including water vapor, acid vapors, hydrocarbons, and gases that add impurities to the wafer surface. The measurement of airborne molecular contamination at the ppb level is now becoming standard for advanced fabrication facilities. Molecular compounds include metals, nonmetals, and organic and inorganic contaminants, all of which are damaging for device performance.

Metallic Impurities

Contamination in a wafer fab also can come from metal compounds. The typical *metal impurities* damaging to semiconductor processing are the alkali metals that are found in many common chemicals and processes. These metals are strictly controlled in all materials used in a wafer fab (see

Table 6.1).[5] The alkali metals are from Group IA of the periodic table and are extremely reactive elements because they give up one valence shell electron, thus forming a cation that reacts with a nonmetal anion to form an ionic compound.

TABLE 6.1 Typical Metal Impurity Elements

Heavy Metals	Alkali Metals
Iron (Fe)	Sodium (Na)
Copper (Cu)	Potassium (K)
Aluminum (Al)	Lithium (Li)
Chromium (Cr)	
Tungsten (W)	
Titanium (Ti)	

Metals come from chemical solutions or different process steps during semiconductor manufacturing. The process of ion implantation (discussed in a later chapter) exhibits the highest metal contamination, on the order of 10^{12} to 10^{13} atom/cm^2.[6] Another source of metallic contamination is through reaction of the chemicals with the transport piping and containers. For instance, carbon monoxide gas, used as an additive gas to improve various wafer processes, can react with the nickel in stainless steel, gaskets, and other components of the gas delivery system.[7] From this reaction gas-phase carbon monoxide forms nickel tetracarbonyl particles that are then distributed on the wafer surface. These particles can be redistributed into the bulk wafer and lead to increased device defects.

There are two basic ways for metallics to deposit on a wafer surface.[8] In the first way, metal binds to the silicon surface by the charge exchange between a metallic ion and a hydrogen atom located on the wafer surface. These types of metallic impurities are difficult to remove. The second way metal is deposited on the wafer surface is when the surface oxidizes and metallic impurities are located in the oxidation layer. For this reason cleanliness is critical in oxidation processes (see Chapter 10). Metallic impurities in an oxide layer are removed only by removing the oxide from the wafer surface.

Metallic ions are highly mobile in semiconductor materials and are referred to as *mobile ionic contaminants* (*MICs*). When introduced into a wafer, MICs move throughout the silicon and seriously damage the device's electrical performance and long-term reliability. Sodium is typically one of the most prevalent MICs in untreated chemicals, with people as its greatest carrier. The human body contains a high percentage of sodium in the form of fluids (e.g., saliva, tears, perspiration, and so on). Sodium contamination is rigorously controlled in wafer fabs.

It is hard to imagine how mobile an MIC material is in silicon. Consider an MIC such as sodium. A single crystal of table salt (NaCl) contains enough sodium to deposit one quadrillion (10^{12}) atoms of sodium per square centimeter on 5,000 wafers (150-mm diameter) and therefore destroy all chips on the wafers.

Problems From Metallic Impurities ■ Metallic impurities lead to reduced device yield in semiconductor manufacturing, including structural defects in the oxide-polysilicon gate structure.[9] Additional problems include increased leakage currents at a pn junction and reduced minority carrier lifetime. MIC contamination can migrate to the oxide-silicon interface in the gate structure and alter the threshold voltage required to turn on a transistor (see Figure 6.5 on page 118). Because they are so mobile, metallic ions can move around a device long after electrical test and shipment and cause the device to fail during usage. A major goal of semiconductor manufacturing is to minimize exposure to metallic impurities and MICs.

Organic Contamination

Organic contaminants are those that contain carbon, nearly always bonded to itself and to hydrogen, and sometimes to other elements as well. Some sources of organic contamination are bacteria, lubricants, vapors, detergents, solvents and moisture. Equipment used in wafer fabs today is designed with components that require no lubricant, for example, oil-free pumps and bearings.

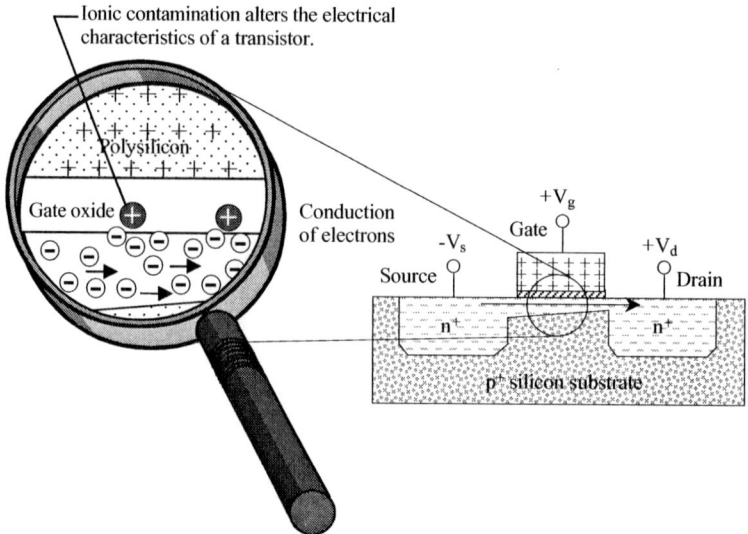

FIGURE 6.5 Mobile Ionic Contaminant Altering Threshold Voltage

Problems From Organic Contaminants ■ Trace organic contaminants degrade the integrity of the gate oxide material under certain processing conditions.[10] Another problem with organic materials on the surface of semiconductor wafers during processing is incomplete cleaning of the surface. This condition permits contaminants such as metal impurities to remain intact on the wafer surface after cleaning.

Native Oxides

The surface of a silicon wafer will oxidize if exposed either to air at room temperature or to DI water that contains dissolved oxygen. This thin oxide layer is termed *native oxide*. The initial native-oxide growth on a silicon wafer occurs in the presence of moisture. When the wafer surface is exposed to air, several tens of molecular layers of moisture adsorb on the wafer within a second. Oxygen from the room is dissolved into the adsorbed moisture layer on the wafer surface and penetrates into the silicon surface. This process causes the silicon surface to oxidize even at room temperature. The thickness of the native oxide increases as exposure time lengthens.

Problems From Native Oxides ■ A silicon surface that is native oxide-free is important for semiconductor performance and reliability. Native oxide interferes with other process steps, such as the growth of single-crystal film on the wafer or the growth of the ultrathin gate oxide.[11] Native oxide also includes some metallic impurities, which can move into the silicon wafer and cause electrical defects.[12]

Another problem created by native-oxide growth occurs in contact regions for metal conductors. Contacts make electrical connections between interconnect wiring and the source and drain region of the semiconductor device. If there is a native-oxide layer, then this will increase the resistance of the contact and reduce and perhaps even prevent current flow (see Figure 6.6).[13]

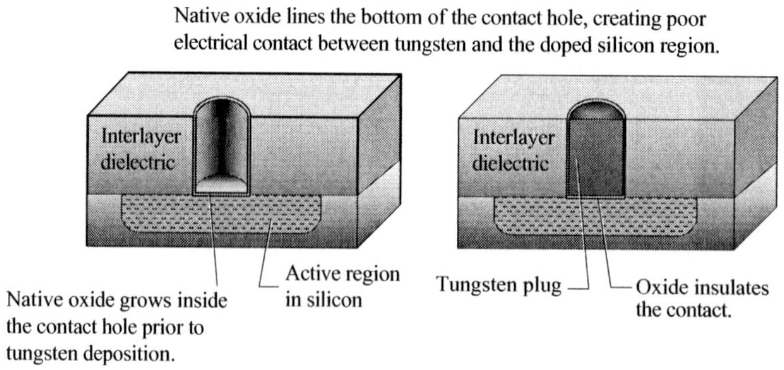

FIGURE 6.6 Native Oxide

Native oxide requires removal through a cleaning step with a mixture of HF acid (see the section on wet cleaning later in this chapter). Another approach to inhibiting native oxide is to integrate multiple process steps into a multichamber tool with an evacuated, high-vacuum chamber so that the wafers are not exposed to ambient atmosphere and moisture.

Electrostatic Discharge

Electrostatic discharge (ESD) is a form of contamination because it is an uncontrolled transfer of static charge from one object to another that potentially damages the microchip. ESD is generated when two materials with differing static-charge potential are touching or rubbing each other (this action is referred to as *triboelectricity*). Atoms with excess negative charges are attracted to adjacent atoms that have an excess of positive charges. This attraction creates a discharge of electricity that can be up to tens of thousands of volts.

Semiconductor manufacturing is especially prone to developing static electrical charges because the wafer fab is maintained at a low humidity, typically 40% ± 10% relative humidity (RH). This condition is conducive to the generation of high levels of static charge.[14] Although increasing the RH would decrease static-charge generation, it would also increase contamination through corrosion, thus making it an impractical approach.

Problems From Electrostatic Discharge ■ Although the amount of static charge transferred during ESD is usually small (on the order of nanocoulombs), the discharge deposits its energy into a very small area of the wafer. An electrostatic discharge that occurs in a few nanoseconds can generate peak currents over 1 amp, literally vaporizing metal conductor lines or punching through oxide layers.[15] This discharge also can be the cause of gate oxide breakdown. Another significant problem with ESD is that once a wafer surface has a charge buildup, its resulting electric field can attract charged particles or polarize and attract neutral particles to its surface (see Figure 6.7). An example of this is how a TV screen attracts dust particles. Furthermore, the smaller the particle, the more effect the electrostatics have on particle attraction.[16] As device critical dimensions are reduced, smaller particles attracted by ESD become more significant and can create killer defects. In order to minimize particle contamination, wafer charging must be controlled.

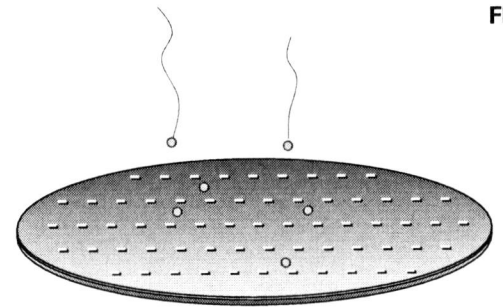

FIGURE 6.7 Particles Attracted to a Charged Wafer

SOURCES AND CONTROL OF CONTAMINATION

The wafer fab cleanroom is stringently controlled for contamination to reduce killer defects that impact microchip performance. Nearly anything that comes in contact with a wafer is a potential source of contamination. Seven sources of contamination in a wafer fabrication facility are:

- Air
- Humans
- Facility
- Water
- Process chemicals
- Process gases
- Production equipment

Air

The most fundamental concept of a cleanroom is the control of particles in the wafer fab air. The air that we normally breathe is not acceptable for semiconductor manufacturing because it contains excessive airborne contaminants. These small airborne aerosol particles float and remain in the air for a long period of time, depositing on wafers as contamination and creating killer defects.

The *class number* designates the air quality inside a cleanroom by defining the particle size and density in the cleanroom air. This figure represents how well particles are controlled to reduce particulate contamination. Class numbers originated from Federal Standard 209, first released in 1963 and revised several times until the most recent version of 209E (see Appendix B). Table 6.2 shows the number of acceptable particles per cubic foot for the different cleanroom class numbers and particle size.

TABLE 6.2 Definition of Airborne Particulate Cleanliness Classes Per Federal Standard 209E

Class	Particles/ft^3				
	0.1 μm	0.2 μm	0.3 μm	0.5 μm	5 μm
1	3.50×10	7.70	3.00	1.00	
10	3.50×10^2	7.50×10	3.00×10	1.00×10^1	
100		7.50×10^2	3.00×10^2	1.00×10^2	
1,000				1.00×10^3	7.00
10,000				1.00×10^4	7.00×10
100,000				1.00×10^5	7.00×10^2

If the cleanroom class is stated only by the number of particles, such as a cleanroom of class 1, then assume a particle size of 0.5 μm. This means there is a maximum of one particle of size 0.5 μm or larger per cubic foot. For particle sizes different from 0.5 μm, the cleanroom class should be expressed as the class number at a specific particle size. Examples are: class 10 at 0.2 μm (read from Table 6.2 as a maximum of 75 particles per cubic foot at 0.2 μm or larger) and class 10 at 0.1 μm (a maximum of 350 particles per cubic foot at 0.1 μm or larger).

Ultrafine Particles ■ There recently has been usage of a class 0.1, with particle sizes down to 0.02 to 0.03 μm. The latest clean air standard also has a provision for the number of *ultrafine particles* in a cubic meter of air, which is called the "U" descriptor. The U descriptor specifies ultrafine particles as those smaller than 0.1 μm in diameter, down to the smallest diameter detectable with a discrete particle counter. Without referring to a specific particle size, the U descriptor defines cleanliness as U(x), where (x) is the maximum allowable number of ultrafine particles per cubic meter of air.

Humans

A human is a particle generator. People continually enter and leave cleanrooms and are the greatest sources of contamination in a cleanroom.[17] Particles come from hair and hair products (hair spray, gel), lint from clothes, flakes of dead skin, and so on. On the average, a person sheds over one ounce of particles per day, which can amount to an astonishing 10,000,000 particles per minute of 0.3 μm size and larger (see Table 6.3).

TABLE 6.3 Particles Emitted by Human Activities

Source of Particles	Average Number of Particles per Minute > 0.3 μm
Motionless (sitting or standing)	100,000
Moving hands, arms, trunk, neck, and head	500,000
Walking at 2 miles per hour	5,000,000
Walking at 3.5 miles per hour	7,500,000
Cleanest skin (per square foot)	10,000,000

Simple activities in the wafer fab such as opening and closing doors or excessive movements around process tools create particulate contamination. Normal human activities such as talking, coughing, and sneezing are damaging to semiconductors.

Cleanroom Garments ■ To attain ultraclean conditions in a cleanroom, humans must follow certain procedures, known as *cleanroom protocol,* and be covered with special cleanroom garments (also referred to as a "bunny suit"). The garment is made up of a hood with a facemask, coveralls, boots, and gloves, and will completely cover the body. The goal of the cleanroom garment system is to meet these functional criteria:

◆ Total containment of body-generated particles and aerosols.
◆ Zero particle release from the garment system.
◆ Zero electrical-charge buildup for ESD.
◆ No release of chemical or biological residues.

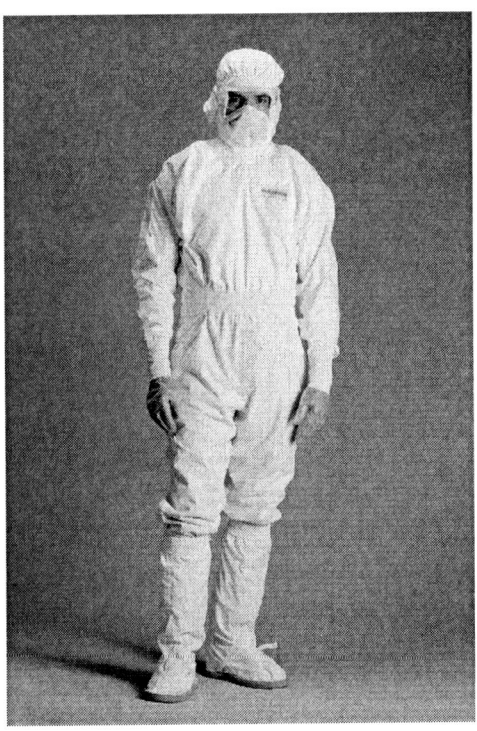

Technician in Cleanroom Garment. (Photo courtesy of International SEMATECH) James Minor, photographer © 1990

The modern cleanroom garment is a high-technology membrane fabric or densely woven polyester fabric. Advanced materials will have a 99.999% efficiency rating (stops 99.999% of all particles from passing) for 0.1 micron particles and greater. The extent to which the cleanroom garment system covers the human body can vary. Some fabs require inner clothing layers, such as polyester underwear, in addition to the outer bunny suit. People may be required to take a shower and use lotion to prevent skin flaking prior to final dressing and entering fabs. Some wafer fabs also require fab workers to put on a bubble helmet with a breathing recirculator and exhaust blower that pumps the user's breath through a filter pack strapped to the waist. This helmet prevents particles from human saliva from contaminating the fab work area.

Cleanroom Protocol ■ Each semiconductor company has a strict procedure for cleanroom protocol to minimize contamination in the cleanroom. Some standard cleanroom protocols are listed in Table 6.4.[18]

TABLE 6.4 Proper Cleanroom Protocol

Should Do	Should Not Do	Why?
Only authorized personnel are allowed within the cleanroom.	No people allowed that have not been properly trained in what the cleanroom expects of them. The cleanroom supervisor has the last word.	Authorized personnel are familiar with the many strict and demanding restrictions of cleanroom operations.
Take only what is necessary into the cleanroom.	No cosmetics, tobacco products, handkerchiefs, tissues, food, drink, candy, wooden/mechanical pencils or pens, perfumes, colognes, watches, jewelry, cassette players, phones, beepers, video cameras, audiocassette recorders, gum, combs, hair brushes, cardboard or noncleanroom-approved paper. No blueprints, operations manuals or instruction sheets.	To prevent entry of unwanted contamination sources which create defects in semiconductor devices.
Gown in the prescribed manner according to your company training.	No uncovered street clothes allowed within the cleanroom. Do not touch garments with bare skin.	To keep cleanroom apparel free of contaminants that could be carried into the cleanroom.
Always make sure that all head and facial hair is covered.	Do not expose any facial or head hair.	Hair is a source of contamination.
Follow procedures for entering the cleanroom, such as an air shower and shoe cleaner (if required).	Do not open any door into the cleanroom until all procedures are complete.	All showers may assist in removing contaminants; many firms have stopped using this procedure due to problems with airborne contamination.
Keep cleanroom garment closed at all times while in the cleanroom.	Do not expose any street clothing while in the cleanroom. Do not allow any part of your skin to touch the outer parts of the cleanroom garment.	Sources of unwanted contaminants.
Move slowly.	Do not congregate or move quickly.	This disrupts the airflow pattern.

Facility

For semiconductor manufacturing to function as an ultraclean environment, a systems approach is necessary to control all inputs and outputs to the cleanroom area. There are three basic strategies for eliminating particles from cleanrooms:[19]

1. Start out with a cleanroom that is free of particles.
2. Minimize the introduction of particles into the cleanroom through equipment, tools, personnel, and cleanroom supplies.
3. Continuously monitor the cleanroom for particles for timely response to cleaning maintenance.

Cleanroom Layout ■ In the early 1970s, the LSI manufacturing environment had an overall cleanliness class of 10,000 in the manufacturing space and a local class of 100 at the individual workbenches. The industry developed a *ballroom layout* approach, with one large fabrication room with a class 10,000 rating and laminar flow benches that provided the class 100 work environment (see Figure 6.8).

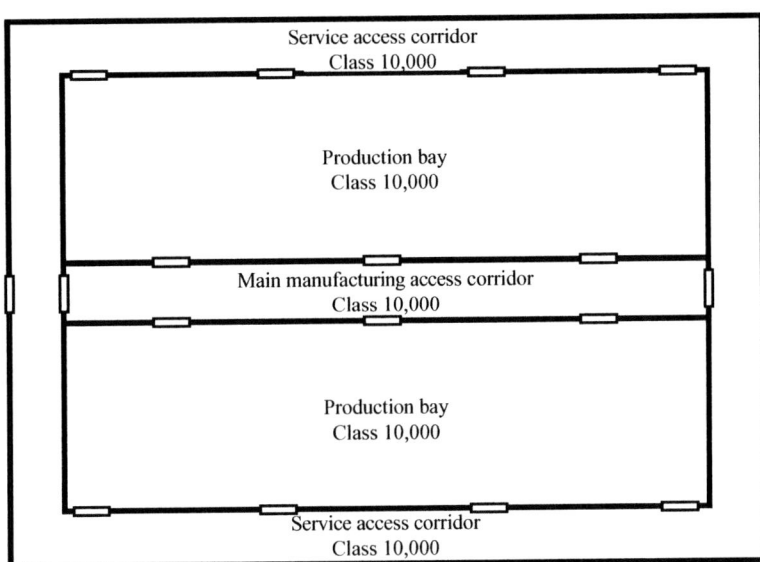

FIGURE 6.8 Early Ballroom Cleanroom Layout

With the submicron technologies of the 1980s came the introduction of the *bay and chase layout* approach to cleanrooms. In this cleanroom layout, a common corridor separates the production area (referred to as production bay or process bay) from the service area (referred to as service chase, equipment chase, or gray area), as shown in Figure 6.9 on page 124. Inside the production bay where wafers are processed, the cleanliness class is typically class 1. Most of the equipment maintenance occurs in the class 1,000 service chase.

Modern wafer fabs are built based on either type of cleanroom design. The ballroom approach is promoted as an advanced cleanroom design that encompasses automated wafer handling and localized contamination control at the process tools.[20] For both types of cleanrooms, there is typically a *sub-fab* area below the cleanroom that contains much of the facilities equipment (e.g., pumps, piping, ductwork, cabling, and so on).

Airflow Principles ■ To achieve ultraclean conditions in a cleanroom, the nature of the airflow is critical. For a cleanroom class of 100 and below, *laminar airflow* is necessary.[21] Laminar airflow means that the airflow is smooth with no turbulence in the airflow pattern (see Figure 6.10 on page 124). The vertical laminar airflow has a slight positive pressure relative to the outside pressure and acts as a curtain to minimize cross-contamination from equipment or personnel to any exposed product.[22]

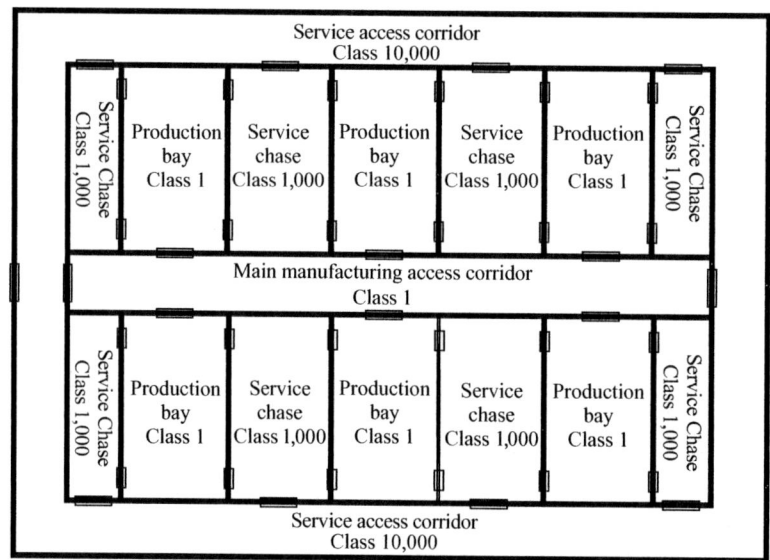

FIGURE 6.9 Bay and Chase Concept for Cleanroom

Air Filtering ■ A simplified diagram of the air handling system in a wafer fab cleanroom is shown in Figure 6.11. Air enters the cleanroom through special particulate filters in the ceiling and passes with laminar flow to the floor and into the recirculation air system to return back to the air filtering system with makeup air. In a modern fab, air may turn over every six seconds to improve recovery of ultraclean conditions during disturbances such as shift changeover. An exhaust system is used to remove undesirable heat and chemicals from the process tools and work areas.

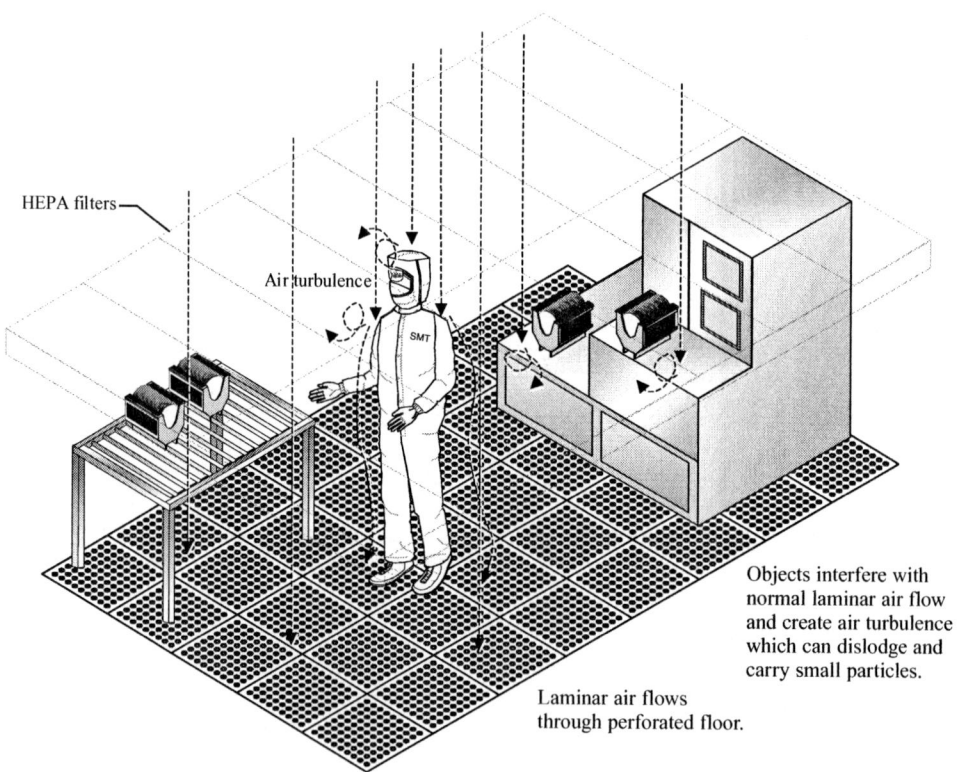

FIGURE 6.10 Laminar Airflow

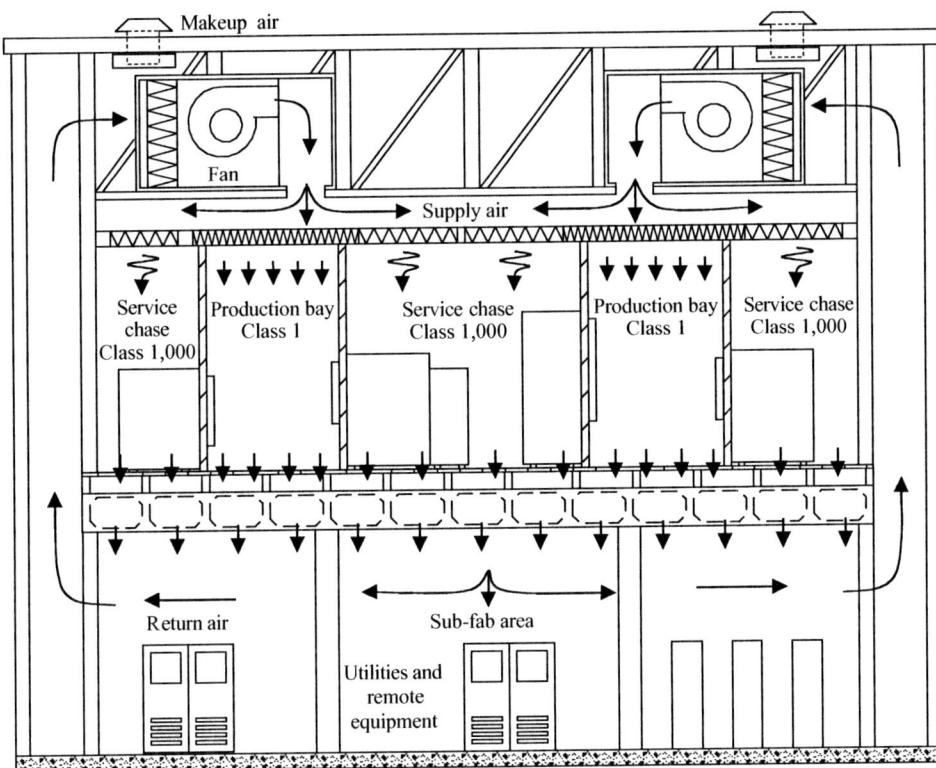

FIGURE 6.11 Wafer Fab Air Handling System

The special particulate filters located in the ceiling are either *high-efficiency particulate air* (*HEPA*) or *ultra-low penetration air* (*ULPA*). HEPA fibrous filters are made with fiberglass fibers and constructed to create laminar airflow (see Figure 6.12). In general, ULPA represents filters that have an efficiency of 99.9995% or better for particulate diameters greater than 0.12 microns.[23]

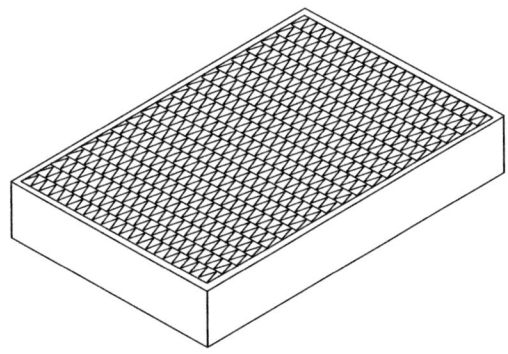

FIGURE 6.12 HEPA Filter

Temperature and Humidity ■ Temperature and humidity settings are specified for wafer fab facilities. An example of temperature control in a class 1 at 0.3 μm cleanroom is 68 ± 0.5°F.[24] Relative humidity (RH) is important because of its contribution toward corrosion (higher RH promotes more corrosion, just as moisture causes steel to rust). A typical RH setting may be 40% +/- 10%.

Electrostatic Discharge ■ Most electrostatic discharge (ESD) problems are controlled through the proper use of equipment and procedures. The principal ESD control methods are:

- Static-dissipative cleanroom materials
- ESD grounding
- Air ionization

Cleanroom materials such as carts, cassettes, equipment, and so on must be *static dissipative*. This term means that the resistivity of the material is lowered through the use of conductive additives, permitting mobile electrostatic charges to flow through the material. At the same time, people and objects in the cleanroom must be continuously connected to ground. In this manner, electrostatic discharge is conducted through the human body and all cleanroom materials that the wafer comes in contact with and flows harmlessly away from the product through ground. This setup avoids potentially discharging through a device on a wafer and causing irreparable damage to the chip.

Air Ionization. The wafer has various insulating materials placed on it during processing, such as an oxide film. These materials are easily charged and will hold this charge for a long period. Since this insulation material is in intimate contact with the product, it requires a method to neutralize the charge buildup. The most common way to neutralize insulating materials on a wafer is with *air ionization*. Special ionizer emitters located on the ceiling in the cleanroom produce a high electric field that ionize the air molecules and make it conductive by gaining or losing an electron. When this conductive air contacts a charged surface, such as an insulator material on a wafer, the surface attracts ions from the opposite polarity and neutralizes the electrostatic surface charge (see Figure 6.13).

Air ionization by emitters can be limited because many ions are eliminated before reaching the wafer surface due to recombination. A recent development is air neutralization using soft-X-ray radiation.[25] Exposing the ambient air surrounding the charged wafer to soft X-rays will generate large ion pairs. This process effectively neutralizes the charge on the wafer surface to 0 volts after roughly two seconds.

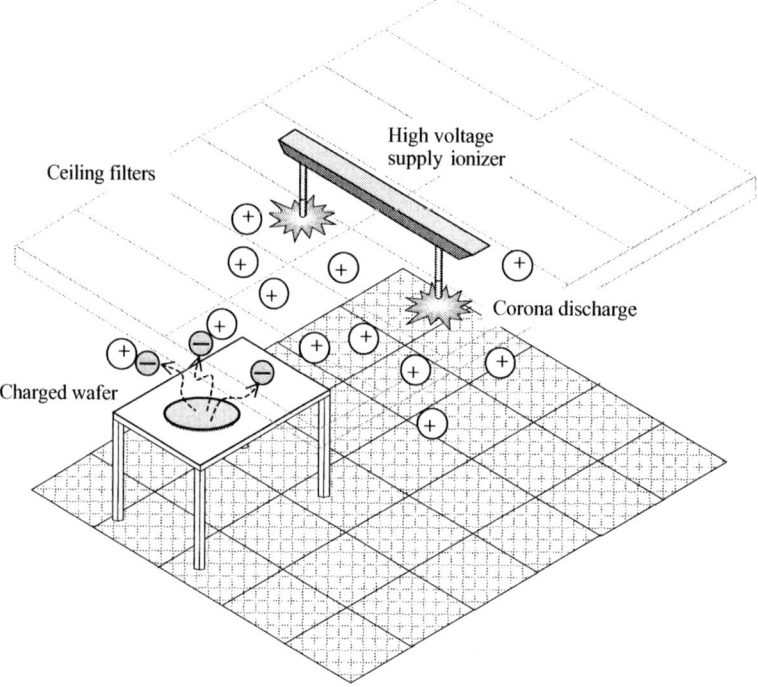

FIGURE 6.13 Neutralizing Static Charge on a Surface with Air Ionization

Water

To manufacture semiconductors, a high quality, *ultrapure deionized* (*DI*) water (sometimes referred to as UPW) is required in large quantities. City water has too many contaminants to be acceptable for the production of wafers. DI water is the most heavily used chemical in semiconductor manufacturing, primarily in the chemical wafer cleaning solutions and as a postclean rinse. It is estimated that current DI water consumption in a wafer fab runs up to 2,000 gallons of ultrapure deionized water for each wafer produced in a modern 200-mm process line.[26]

Unacceptable contaminants in ultrapure DI water are:

- Dissolved ions
- Organic materials
- Particulates
- Bacteria
- Silica
- Dissolved oxygen

Figure 6.14 shows different water particles and their sizes.[27]

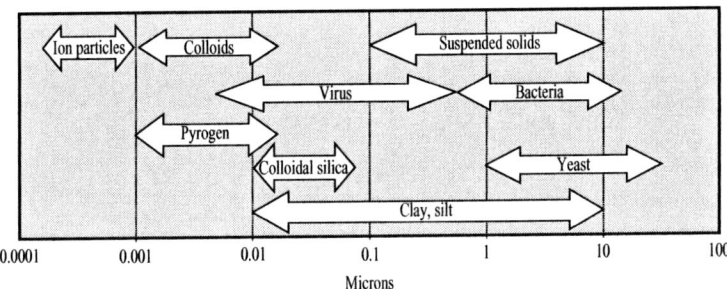

FIGURE 6.14 Size of Particles in Water

Dissolved ions in water are from minerals such as sodium and potassium that easily form ions. An example is salt (NaCl) that breaks up into Na^+ and Cl^- ions. These mobile ionic contaminants (MICs) are undesirable and create performance problems in semiconductor devices.

Organic materials, referred to as *total organic carbon* (*TOC*), are the sum of all carbon-containing compounds dissolved in the water. These organic contaminants have a damaging effect on the ability to grow oxide films.

Bacteria are live, reproducing organisms in the water. Fragments shed by bacteria are known as *pyrogens*. Bacteria found in water lead to defects in oxidation, polysilicon, and metal-conducting layers. Some bacteria contain phosphorus that could cause uncontrolled doping.

Silica is found in city water as finely suspended particles. These particles range in size from ten angstroms to ten microns. A high silica content fouls filtration equipment in the water purification equipment and also decreases the reliability of thermally grown oxides.

Another type of contaminant in water is dissolved oxygen. It creates a problem because it leads to native-oxide formation on the wafer surface. Dissolved oxygen in water also creates problems if the water is pressurized as it is in some semiconductor processes. When the water is depressurized, dissolved gases may come out of solution forming bubbles that can cause incomplete water wetting on the wafer surface.[28]

DI-Water Installation ■ A DI-water installation has two major parts for purifying water, referred to as the makeup loop and the polishing loop (see Figure 6.15 on page 128).[29] The *makeup loop* removes particles, total organic carbons (TOCs), bacteria, microorganisms, ionic impurities, and total dissolved minerals from the raw water. It has a prefilter that removes particles above one micron and a purifier section for removal of ionic impurities, bacteria, and dissolved gases. The *polishing loop* is the final part of the water purification system that removes the remaining contaminants.

Deionization ■ *Water deionization* to make DI water is the process of removing the electrically active salt ions using specially manufactured ion-exchange resins. This process changes the water from a conductive medium to a resistive medium with a resistivity of 18 megohm-cm (18,000,000 ohm-cm) at 25°C. DI water used in the wafer fab is referred to as 18-megohm (MΩ) water. Ultrapure DI water is passed through two deionizers, once in the makeup loop and another time in the polishing loop.

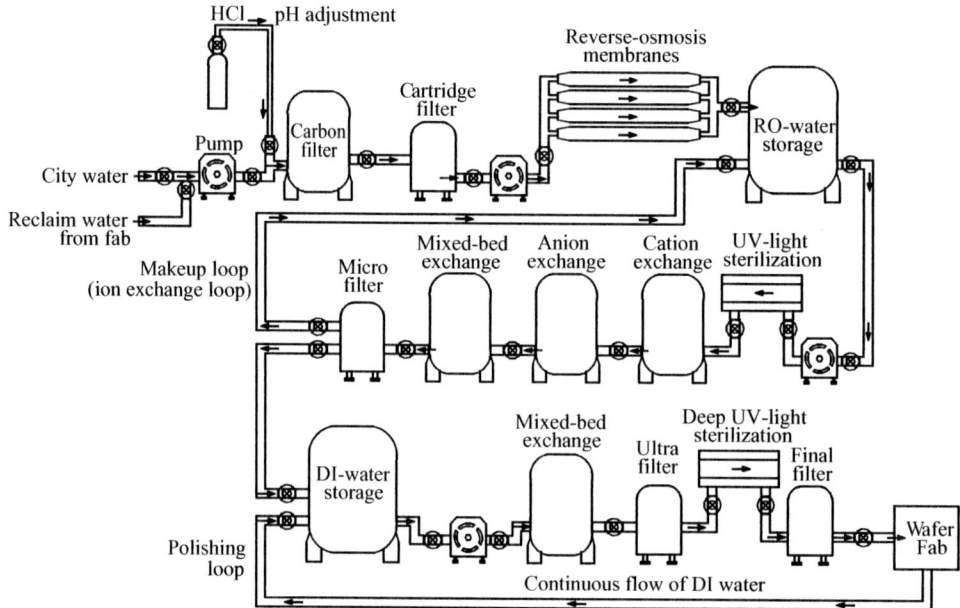

FIGURE 6.15 DI-Water Makeup and Polishing Loops

DI-Water Filtration ■ The DI-water makeup loop employs various filters with the intent of letting water pass through while trapping the particulates and colloids (extremely fine particles) in the filter media. A common filtering technique for ultrapure DI water is *reverse osmosis* (*RO*) to remove smaller particles and metallic ions. The operating principle for RO filters is to flow the water under pressure across a membrane filter to separate ionized salts, colloids, and organic materials down to 150 molecular weight (see Figure 6.16). RO can separate impurities as small as 0.005 microns and is also referred to as *hyperfiltration*.

Ultrafiltration is used at point-of-use (POU), meaning its filtering location is at the equipment, to remove submicron-size particles. This filter uses pressure and flow through a membrane with pore sizes ranging from 10 angstroms to 0.2 microns.

Other parts of an ultrapure DI-water system include a degasifier unit used to remove dissolved gases in DI water, such as oxygen, to reduce contamination such as native-oxide growth on the wafer surface. An improved technique for removing dissolved gases (primarily oxygen) to less than 10 ppb is a *membrane contactor* (see Figure 6.17).[30] Membrane contactor filters consist of

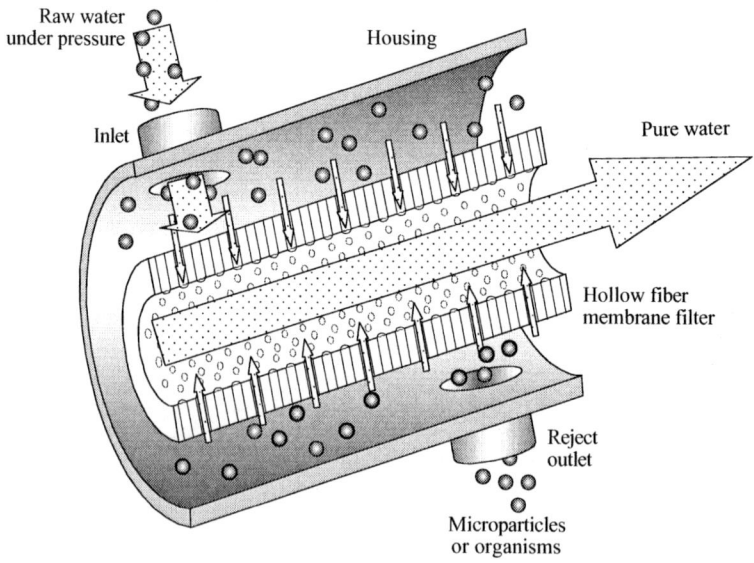

FIGURE 6.16 Principle of Reverse Osmosis Filtration

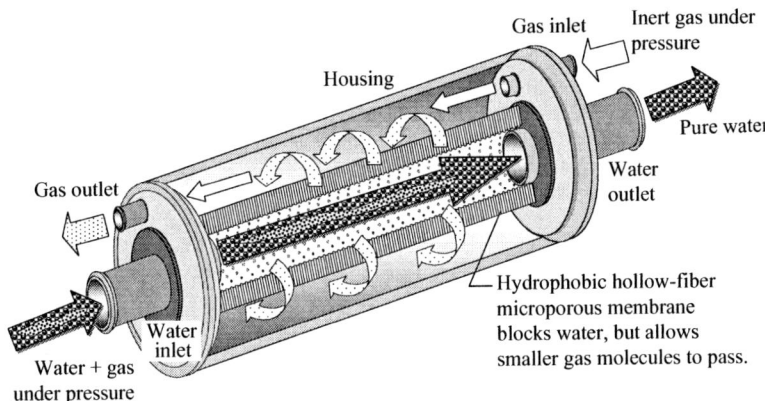

FIGURE 6.17 Membrane Contactor Filter

hydrophobic polypropylene microporous hollow fibers. The fibers have small pores in their walls to allow dissolved gases to pass through. However, the fiber's hydrophobic nature does not permit liquid water to pass through the pores. The membrane removes oxygen and all free gases.

Zeta Potential ■ The *zeta potential* represents a positive or negative electrical charge that can build up in colloids (very fine suspended particles in a liquid). Particles in water generally have a zeta potential with a negative charge, such as colloidal silica, bacteria, and pyrogens. These colloids can be filtered out of the water using a positively charged filter that traps particles that are smaller than the pore diameter of the filter.[31] This form of particle removal uses electrostatics to attract particles with opposite zeta potential rather than relying on the pore size of the membrane (see Figure 6.18).

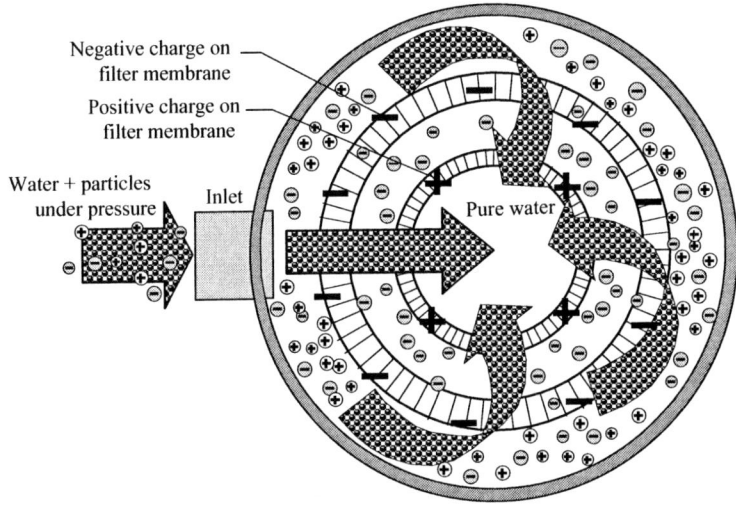

FIGURE 6.18 Electrostatic Filtration Using the Zeta Potential

Bacteria Control. Ultrapure water systems use *ultraviolet (UV) lamps* for bacterial sterilization. Water is exposed at a controlled rate to UV light waves, which deactivates certain molecules to reduce bacteria. UV systems are simple and reliable and can reduce bacteria to less than 1%.

Biological control of an ultrapure electronic-grade water system is also done by ozonating the water. Ozone (O_3) is created by discharging an electric current through dry air. The resulting ozone mixture is purified and then injected into the ultrapure water to kill the bacteria. All ozone is removed from the water by exposing the ozonated water to ultraviolet light, which breaks down the ozone to oxygen. Since most bacteria have a physical diameter in excess of 0.2 microns, point-of-use microfiltration can also be effective in removing bacteria.

Process Chemicals

Liquid chemicals used in semiconductor processing must be free of contamination for successful device yield and performance. Chemical purity is identified by an *assay number*, which describes the percentage of a particular chemical in the container (without reference to other substances present). For example, an assay of 99.99% for a bottle of hydrofluoric acid means the bottle contains 99.99% hydrofluoric acid and 0.01% other substances.

Filters are used to prevent chemical degradation during delivery or to maintain chemical purity during a recirculation step. Filters should be located at appropriate locations (e.g., near input to the gas controller) and as close to the process chamber as possible through point-of-use filtration. The different filter classes are:

> Particle filtration: Depth-type filtration (see Figure 6.19) for particles from approximately 1.5 microns and larger.
>
> Microfiltration: Membrane filtration of a liquid that removes particles in range of 0.1 to 1.5 microns.
>
> Ultrafiltration: Pressure-driven membrane process that rejects large molecules from approximately 0.005 to 0.1 microns.
>
> Reverse osmosis: Also called hyperfiltration. A pressure-driven solution process which transports liquid through a semipermeable membrane with the exclusion of particles and metal ions as small as approximately 0.005 microns.

A *membrane filter* uses a thin membrane of polymer or ceramic with small penetrating pores as the filter medium (see Figure 6.20). The size and distribution of the pores are controlled throughout the membrane. The membrane serves as a barrier that permits the passage of materials only up to a certain size or shape as defined by the pores. It is also called a surface filter because it removes particles from the air stream through interception at its surface. Membrane filters are used in reverse osmosis, microfiltration, and ultrafiltration. A membrane filter is often used as a point-of-use filter and is placed just before the process tool to provide final filtration.

A good filter will not create a significant pressure drop of the required flow, will not introduce secondary contamination, and will be compatible with the chemical. The *filter efficiency* is the percentage of particles of a specific size and above that are stopped in the filter. For liquid filters in a ULSI process, a typical efficiency rating for a membrane filter is 99.9999999% of 0.02 micron particles and above (referred to as *nine nines efficiency*).

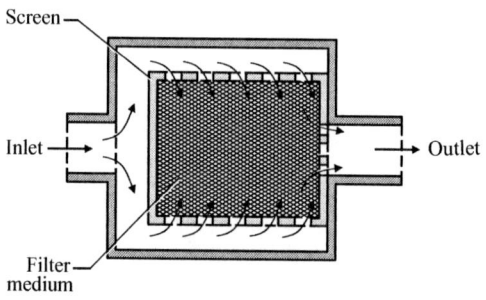

FIGURE 6.19 Depth-Type Filter (Used with permission from International SEMATECH)

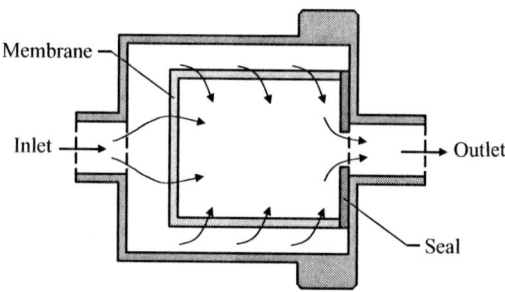

FIGURE 6.20 Membrane Filter (Used with permission from International SEMATECH)

Process Gases ■ The delivery and use of ultraclean gas is critical for semiconductor manufacturing in the ULSI era. However, the handling and delivery system can introduce impurities that adversely affect the yield of semiconductor devices. Gases pass through purifiers and gas filters to remove impurities and particles. Some gas filters are all-metal (e.g., nickel) that do not shed particles or outgas organic contaminants. These filters have a nickel membrane that is able to withstand the corrosive gases with a proven efficiency for particles as small as 0.003 microns. Other gas filters are made of a Teflon polymer.

Production Equipment

The production equipment used to manufacture the semiconductor wafer is the most significant source of particles in a wafer fab.[32] During the wafer fabrication process, silicon wafers are repeatedly loaded from cassettes into tools, processed through multiple equipment operations, unloaded back into cassettes, and then transported to the next workstation. This sequence occurs over and over for the 450 or more process steps necessary to fabricate a wafer, exposing the wafer to a multitude of mechanical and chemical operations at the different tools.

Many wafer fabrication operations occur in a vacuum, which requires special design considerations to avoid contamination. Examples of different sources of particle contamination from process tools are:

- Flaking of by-products built up on chamber walls.
- Automated wafer handling and transportation.
- Mechanical operations such as shaft rotation and valve opening or closing.
- Pumping and venting in vacuum environments.
- Cleaning and maintenance procedures.

Since equipment automation means fewer humans interacting with the product, there is less concern about particles from humans and more emphasis on reducing particles from equipment.[33] The number of particles on a wafer surface will increase during the fabrication process as the wafer is exposed to more tool operations. This condition is shown in Figure 6.21.[34]

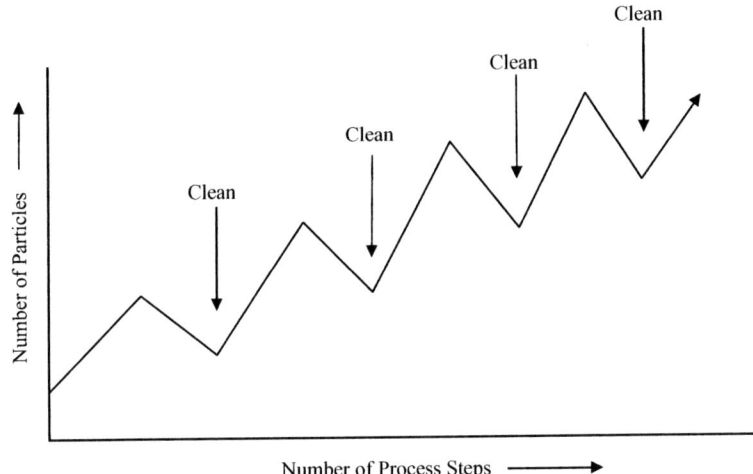

FIGURE 6.21 Wafer Surface Particles as a Function of Process Steps

Workstation Design

The right materials for *workstation design* of process tools and general workstations are necessary for achieving an ultraclean cleanroom. All materials emit some particles, but the goal is to reduce the emission to an acceptable level. Smooth, highly polished surfaces are best to reduce particulate contamination. Stainless steel has become a widely used material for work surfaces and equipment in the cleanroom. When properly processed, stainless steel has a relatively low particle emission rate. Electropolishing is a crucial finishing step.

Bulkhead Installation ■ Some fabs choose to install their equipment with a *bulkhead equipment layout*. In this approach, the major portion of the equipment is located behind the production bay fab wall in the service chase (see Figure 6.22 on page 132). Only the user-interface operator's controls and the wafer cassettes are located in the fab. This configuration isolates the equipment and its servicing in the chase, which is typically at a lower-class level of contamination.

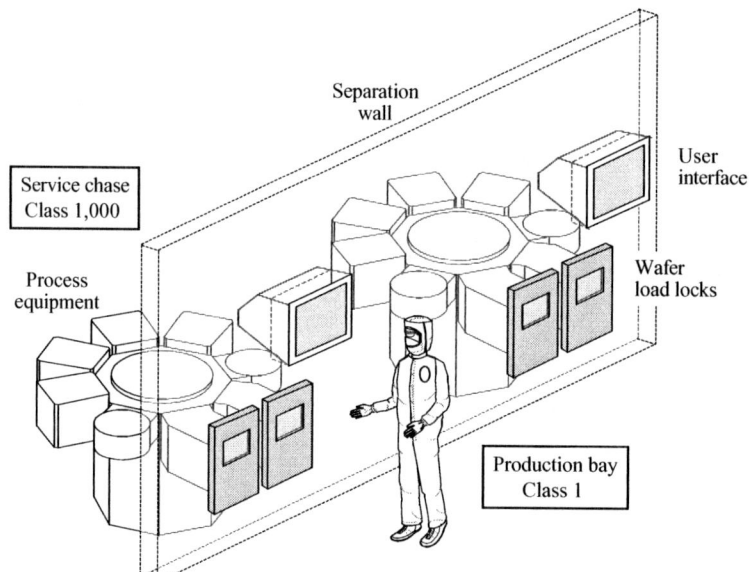

FIGURE 6.22 Bulkhead Installation

Handling ■ From the early days of semiconductor manufacturing until the 1970s, wafers were handled either manually with tweezers and vacuum wands (wafer holders that use vacuum to hold the wafer). Manual handling created particulate contamination and caused killer defects as device geometries decreased. Eventually manufacturers used *wafer cassettes* to transport wafers between tools (typically 25 wafers per cassette), with conveyor systems and elevators as a way to pick up and transfer wafers into and out of the tool (see Figure 6.23). Cassettes are designed to minimize particle generation, be electrostatic dissipative, and have minimal chemical outgassing.

Currently wafer handling within tool operations is largely done by robotic technology. *Robotic wafer handlers* do most cleanroom wafer handling to load and unload wafers from cassettes to the work area of the process tools and to manipulate wafers during the defined sequence of operations in a tool. This development has significantly reduced wafer particulate contamination over manual handling.

Once wafers arrive at a process chamber, they are placed on a chuck that holds the wafer during processing. In the early days of semiconductor manufacturing, this chuck was a mechanical clamp. It changed to a vacuum chuck to reduce particles on the topside of the wafer. However, a vacuum chuck tends to distort the shape of the wafer, which is undesirable during processing. In

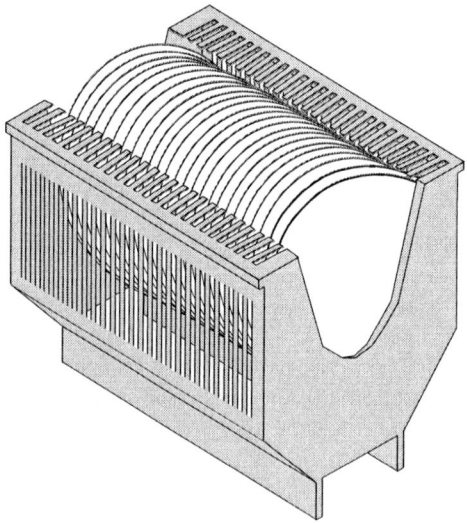

FIGURE 6.23 Wafer Cassette

order to improve process uniformity across the wafer surface, an *electrostatic chuck* (*ESC*) is commonly used today. It generates minimum particles and holds the wafer flat during processing (see Figure 6.24). An electrostatic chuck works by applying a voltage to an electrode on the chuck and generating a static charge. This electrode is isolated from the rear surface of the wafer by means of a dielectric material. An opposite charge is induced into the bottom side of the wafer, which pulls the wafer toward the chuck.

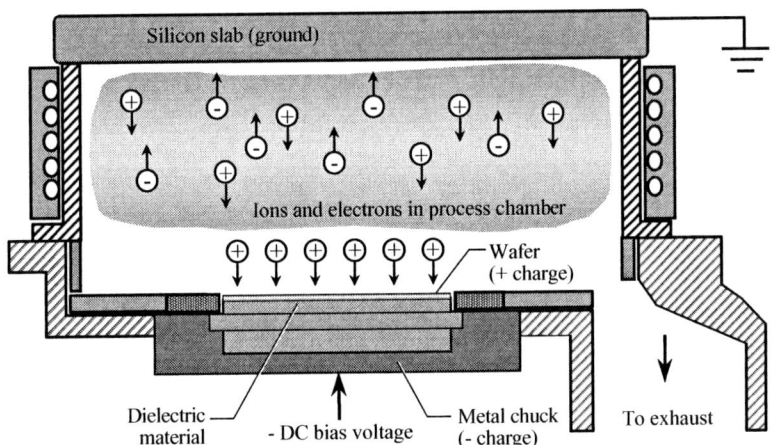

FIGURE 6.24 Electrostatic Wafer Chuck
(Used with permission from Applied Materials, Inc.)

Minienvironment ■ The concept of a cleanroom is continually being reassessed, primarily due to the need for more stringent control of contaminants and the high cost of constructing cleanrooms. There has been increased interest in controlling contamination at the specific workstation location where the wafer is processed through the use of minienvironments.

A *minienvironment* is a localized environment created by an enclosure that isolates wafers from the cleanroom environment while they are not in a process chamber (see Figure 6.25). This concept is also referred to as *wafer isolation technology*. A minienvironment clean zone can include the wafer cassettes used to hold wafers, wafer processing stations, loading ports, and storage locations.

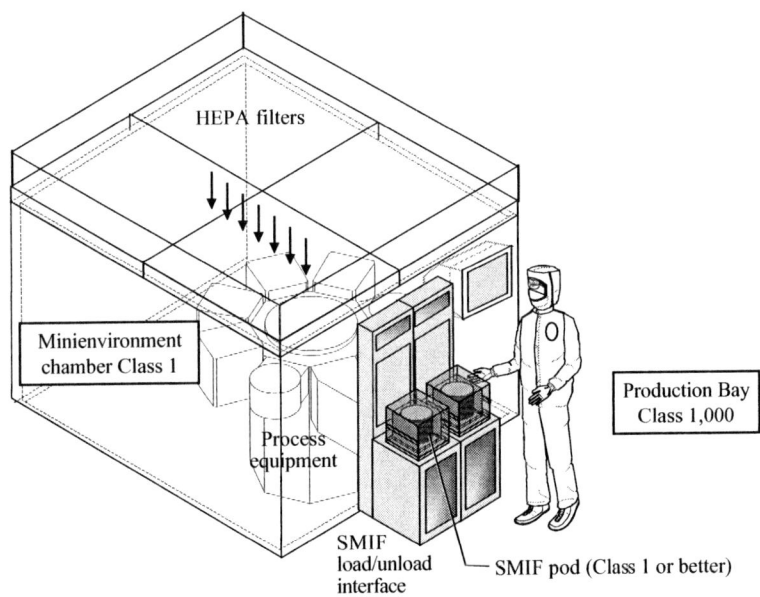

FIGURE 6.25 Minienvironment Concept

Minienvironments are controlled to an extremely pure cleanliness class (e.g., class 0.1 at 0.1 micron), while the cleanroom itself is at a higher class, such as a class 10. This condition makes it much simpler to achieve an ultraclean environment around the wafer during load/unload and processing at a workstation.

To transfer wafers between tools within a minienvironment, a standardized pod is used to enclose and transport cassettes of wafers. The pod has a *standard mechanical interface* (*SMIF*) to interface with the various tools, which was originally developed and patented by Hewlett-Packard. When a pod is presented to a tool, a robot on the tool automatically opens a door on the pod, removes the cassette, and presents it to the tool robot. SMIF systems can be added to existing tools or integrated into the tool enclosure.

For fabs processing 300-mm wafers, the cassette and pod will become one, permitting wafers to be handled directly from the pod. This new pod is referred to as a *front-opening unified pod* (*FOUP*). Handling will be completely automated with no manual lifting of wafer pods. The SMIF interface is fully integrated into the tool.

An advantage of a minienvironment is the possibility to control molecular contamination.[35] Wafers are isolated from molecular contamination through inert-gas purging of the minienvironment with nitrogen during loading, unloading, and transport. This control would be much more difficult to achieve in a large cleanroom. Minienvironments also reduce wafer exposure to water vapor during the nitrogen purge, which serves to inhibit native-oxide growth.

SMIF Pod on Bulkhead Installation
(Photo courtesy of Applied Materials, Inc.)

WAFER WET CLEANING

Due to the continual reduction in critical dimensions on a wafer, the wafer surface must be clean prior to undergoing processing. The most important way to control contamination on a wafer is to prevent contaminating the wafer. However, once a wafer surface is contaminated, then the contaminant must be removed by cleaning.

The goal for wafer cleaning is to remove all surface contaminants: particles, organics, metallics, and native oxides. Every wafer process step is a potential source of contamination to the devices on the wafer. Throughout the entire ULSI fabrication process, it is estimated that the surface of an individual wafer is wet cleaned up to 100 times.[36]

Wet-Cleaning Overview

The predominant wafer surface cleaning process is with *wet chemistry*. In the early 1980s, the consensus was that wet cleaning would be replaced by dry cleaning methods by the turn of the century. There has been a substantial effort to achieve this, but no completely successful replacement for wafer wet cleaning has been found. Wet cleaning of wafers is thriving and is being improved to attain more effective surface cleaning. The typical chemicals used in wafer wet cleaning and the contaminants they remove are shown in Table 6.5.[37]

TABLE 6.5 Wafer Wet-Cleaning Chemicals

Contaminant	Name	Chemical Mixture Description (all cleans are followed by a DI water rinse)	Chemicals
Particles	Piranha (SPM)	Sulfuric acid/hydrogen peroxide/DI water	$H_2SO_4/H_2O_2/H_2O$
	SC-1 (APM)	Ammonium hydroxide/hydrogen peroxide/DI water	$NH_4OH/H_2O_2/H_2O$
Organics	SC-1 (APM)	Ammonium hydroxide/hydrogen peroxide/DI water	$NH_4OH/H_2O_2/H_2O$
Metallics (not Cu)	SC-2 (HPM)	Hydrochloric acid/hydrogen peroxide/DI water	$HCl/H_2O_2/H_2O$
	Piranha (SPM)	Sulfuric acid/hydrogen peroxide/DI water	$H_2SO_4/H_2O_2/H_2O$
	DHF	Hydrofluoric acid/water solution (will not remove copper)	HF/H_2O
Native Oxides	DHF	Hydrofluoric acid/water solution (will not remove copper)	HF/H_2O
	BHF	Buffered hydrofluoric acid	$NH_4F/HF/H_2O$

One of the most critical surface cleaning process steps during wafer fabrication occurs prior to growing a thermal oxidation layer on the wafer. Ultrathin oxide layers must start with a completely clean wafer surface.

RCA Clean ■ The industry standard wet-clean process is referred to as the *RCA clean* process, developed by W. Kern and D. Puotinen at RCA in the 1960s and first published in 1970. RCA wet clean consists of sequential immersion in two different chemical baths, Standard Clean 1 (SC-1) and Standard Clean 2 (SC-2).

The chemical mixture of *Standard Clean 1 (SC-1)* is $NH_4OH/H_2O_2/H_2O$ (ammonium hydroxide/hydrogen peroxide/DI water). The three chemicals are mixed with a ratio range of 1:1:5 to 1:2:7. *Standard Clean 2 (SC-2)* is a composition of $HCl/H_2O_2/H_2O$ (hydrochloric acid/hydrogen peroxide/DI water), and is mixed in a ratio range of 1:1:6 to 1:2:8.[38] Both of these chemical mixtures are based on hydrogen peroxide (H_2O_2) and are traditionally used at a temperature of 75 to 85°C, with a 10 to 20 minute exposure time.

Standard Clean 1 (SC-1). As seen in Table 6.5, the SC-1 clean is an alkaline solution capable of removing particles and organic materials. For particles, the SC-1 wet-chemical process works primarily through oxidation of the particle or by electric repulsion.[39] To understand the oxidation mechanism, realize that hydrogen peroxide is a powerful oxidizing agent that oxidizes the wafer surface and the particle. The oxidation layer on the particle can provide a liftoff mechanism that degrades and dissolves the particle, which breaks the adhesion forces between the particle and the surface. The particle then becomes soluble in the SC-1 solution and leaves the surface. This action is shown in Figure 6.26. This oxidizing action from the hydrogen peroxide also forms a protective layer on the silicon surface that keeps the particle from reattaching to the wafer surface.

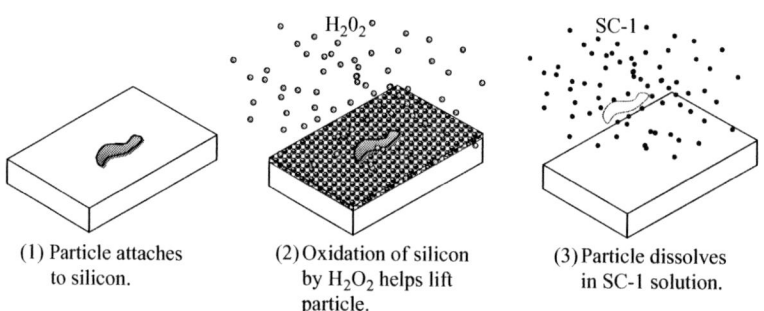

FIGURE 6.26 Oxidation and Solubility of Particle in SC-1

The SC-1 particle removal mechanism actually achieves an electrical repulsion of a particle. The hydroxide ion (OH$^-$) from ammonium hydroxide (NH$_4$OH) slightly etches the wafer surface and undercuts beneath the particle. The hydroxide ion also builds up a negative charge on the silicon surface and the particle. This negative charge on the particle and surface serves to repulse the particle from the surface and move it into the SC-1 solution (see Figure 6.27). Another benefit of the negative surface charge is that it prevents particles from redepositing.

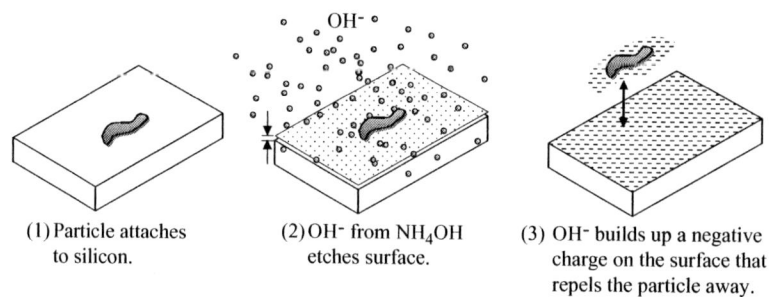

FIGURE 6.27 Particle Removal Through Negative-Charge Repulsion

Because the SC-1 step removes particles through an etching process on silicon, there is some microroughening of exposed silicon. This microroughening is a concern because it makes it difficult to grow very thin oxidation layers required in ULSI technology. Another concern regarding the use of SC-1 is a documented correlation between metal contamination on the wafer and the decomposition of the hydrogen peroxide in the mixture.[40] This contamination requires the bath to be replaced frequently, which is undesirable because it increases the use of chemicals in the fab.

It should be noted that the buildup of a charge potential on the wafer surface and the particle could be described by their zeta potential.[41] For particle removal, the zeta potential of the wafer and particle should be in the same polarity direction in the solution in order to have repulsive electrical forces between the wafer and particle.

Standard Clean 2 (SC-2). The SC-2 wet-clean process is used to remove metals from the surface of the wafer. To remove metallic (and some organic) contaminants from the silicon surface, it is necessary to have a solution with a high oxidation potential and low pH. In this manner, metals become ionized and dissolve in the acid solution that has a strong oxidizing action. Now the cleaning solution can capture electrons from the metallic and organic contaminants and oxidize them. Metals are ionized to be dissolved in the solution while organic impurities are decomposed.

Modifications to RCA Clean ■ There have been modifications to the RCA clean process, mainly because this process uses chemicals and ultrapure water in high volume and at high temperature.[42] Very few chip manufacturers still use RCA clean in the same ratio as the original solution. Companies now will use mixtures that are up to 100 times more dilute with ultrapure H_2O, referred to as *dilute cleaning chemistries,* which achieves equal to or better cleaning effectiveness as the original solution.[43] For instance, a dilute SC-1 chemical ratio may be 1:4:50 for $NH_4OH:H_2O_2:H_2O$ instead of the traditional 1:1:5 ratio. Dilute chemistries are an improvement for safety and health, plus they have the cost benefits of reduced chemical usage and disposal.

An important reason for the continued success of the RCA wet-clean is the availability of ultrapure water and chemicals. New cleaning approaches such as point-of-use chemical generation provide even higher levels of purity than ever before, which lead to more effective cleaning action. RCA clean does generate a large amount of chemical vapor, increasing the load on the cleanroom exhaust system to keep the chemical vapors from getting into the cleanroom. Another problem with bath evaporation is its effect on changing the bath composition over time.

Piranha Mixture. Piranha is a strong cleaning solution that combines sulfuric acid (H_2SO_4) and hydrogen peroxide (H_2O_2) to remove organic and metallic impurities on the wafer surface. Piranha is used at different steps in the process, sometimes before the SC-1 and SC-2 cleaning steps. The most common mixture is seven parts of concentrated H_2SO_4 to three parts of 30% (by volume) of H_2O_2. The usual cleaning method is to immerse the wafers in the piranha at 125°C for about 19 minutes followed by a thorough DI-water rinse. A variation of piranha is *Caro's acid,* which is prepared by mixing 380 parts of concentrated H_2SO_4 with 17 parts of 30% H_2O_2 and 1 part of ultrapure water.

HF Last Step. Many cleaning steps expose the wafer surface to a last step of hydrofluoric acid (HF) to remove native oxides on the wafer surface. Native oxide-free silicon is critical for producing a high-purity epitaxy film and ultrathin (50 angstroms and less) oxide layers for the gate region of MOS circuits. After the HF exposure, the wafer surface is completely terminated with hydrogen atoms and has a high stability against reoxidation in air.[44] A hydrogen-terminated silicon surface is maintained in the same condition as if it were bulk silicon crystal.[45]

Chemical Vapor Cleaning. Another method used in a few fabs is a chemical vapor to remove residual oxide and metallic contamination from a single wafer in a process chamber. The wafer is exposed to a fine mist spray of dilute $HF:H_2O$, followed by a DI-water rinse and an IPA (isopropyl alcohol) vapor-dry step. This method was developed to minimize HF chemical use, but it is not widely used because the increase in cleaning performance is minimal.

Wafer Cleaning Steps ■ A typical wafer cleaning sequence is shown in Table 6.6[46] on page 138. There are variations where some of the HF/H_2O steps are omitted.

TABLE 6.6 Typical Wafer Wet-Clean Sequence

Cleaning Step	Purpose
H_2SO_4/H_2O_2 (Piranha)	**Organics and Metals**
UPW rinse (ultrapure water)	Rinse
HF/H_2O (dilute HF)	Native oxides
UPW rinse	Rinse
$NH_4OH/H_2O_2/H_2O$ (SC-1)	**Particles**
UPW rinse	Rinse
HF/H_2O	Native oxides
UPW rinse	Rinse
$HCl/H_2O_2/H_2O$ (SC-2)	**Metals**
UPW rinse	Rinse
HF/H_2O	Native oxides
UPW rinse	Rinse
Drying	Dry

Wet-Clean Equipment

Because of the extensive use of wet cleaning in semiconductor manufacturing, the type of equipment used is a factor in reducing the chemical concentration and use of chemicals during wet cleaning. Traditional wet-clean processing has been done in *wet sinks,* consisting of a series of acid and rinse tanks housed in fume hoods. Self-contained cleaning equipment with microprocessor controls, robotic handling, and auto-dispensing of chemicals is common. The trend for wet cleaning and rinsing is for robotic handling of wafers with cassetteless operation within a minienvironment. Without the cassette, there is less obstruction of chemical flow to the wafer surface. This condition can improve the cleaning efficiency, reduce the amount of chemical usage, and lead to shorter rinse times.

Megasonics ■ One of the most widely used technologies with SC-1 for wet cleaning is megasonic cleaning. *Megasonic cleaning* uses ultrasonic energy with frequencies near 1 MHz during the clean process (see Figure 6.28). This process achieves much more effective particle removal at lower bath temperatures (30°C versus the original 80°C). This fact is important because of the difficulty in removing smaller particles, simply because of the difficulty in delivering the necessary force to such minuscule particle dimensions.[47]

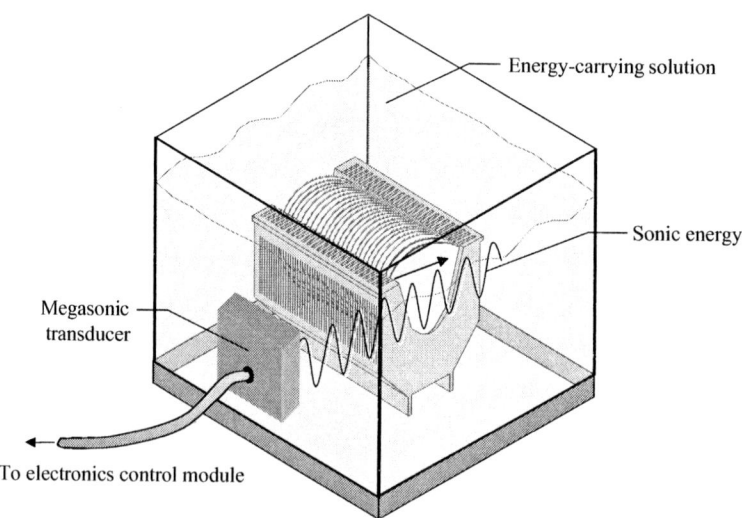

FIGURE 6.28 Megasonic Cleaning

Megasonic cleaning is accomplished when the vibrations of a megasonic transducer creates pressure waves in the liquid of the cleaning tank. The main particle removal mechanisms with megasonic cleaning are cavitation (formation of bubbles) and acoustic streaming.[48] *Cavitation* occurs when the low-pressure portion of the pressure waves create small bubbles that become filled with gas and/or vapor. These cavity bubbles oscillate through the liquid medium and violently collapse (implode) due to the sonic energy. This collapsing action is cavitation, which enhances particle removal without damage to the silicon wafer.[49] *Acoustic streaming* is the steady flow of liquid induced in the megasonic tank from the ultrasonic energy. Flowing liquid has more cleaning action than stationary water because it transports the particles away from the surface.

When the vibration frequencies are less than 100 kHz, then the process is referred to as ultrasonic. However, cavitation-induced pitting of silicon wafers occurs at ultrasonic frequencies that have not been found in the megasonic frequency range (800-1200 kHz).[50] Another reason for megasonics becoming more widespread in chemical cleaning and subsequent DI-water rinse operations is because it reduces the volume of chemicals required.

Spray Cleaning ■ With *spray cleaning* technology, the wet-cleaning chemicals are sprayed onto wafers, placed in a cassette and rotated inside a sealed chamber (see Figure 6.29). A DI-water rinse is sprayed on the wafers after each cleaning step and the DI water's resistivity is monitored to determine when all the chemicals are removed. The spray chamber is sealed during the process to isolate the chemicals and their vapors. After completion of the cleaning and rinsing cycles, the chamber is purged with hot nitrogen and the spin rate is accelerated to dry the wafers.

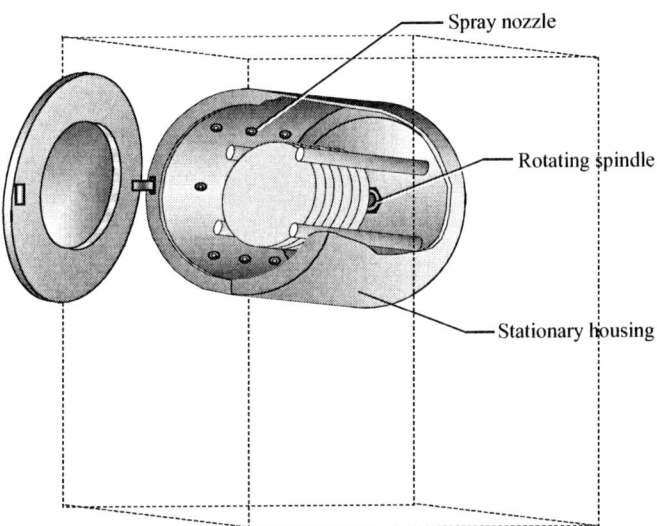

FIGURE 6.29 Spray Tool Designs for Wafer Cleaning

An advantage of spray tools is a continuous supply of premixed, freshly blended chemicals sprayed on the wafer. There is a physical force created during cleaning from the many small droplets of the spray hitting the wafer with a high turnover of chemicals and rinse water. The use of spray and centrifugal force from the rotating wafer ensures effective rinsing at a reduced chemical and water usage rate. There are also improvements in metallic and organic contamination removal when using spin-spray technology. However, spray cleaning does not provide uniformity of cleaning and rinsing because the center of the wafer is not turning at the high velocity of the outer wafer edge. This problem becomes worse with larger diameter wafers.

Scrubbers ■ Wafer *brush scrubbing* is an effective method for removing particles from the wafer surface. Brush scrubbing is widely used following wafer chemical mechanical planarization (known as CMP, which will be discussed in Chapter 18) due to the extensive particles generated during CMP. Brush scrubbing is able to remove particles one micron in diameter and smaller.

Early versions of wafer scrubbers made with nylon brushes damaged the wafer surface because of the stiffer nylon in conjunction with high pressure water sprays. Modern brushes are made with polyvinyl alcohol (PVA), which is a soft, compressible, spongelike material (see Figure 6.30). PVA brushes are effective at removing particles without wafer damage.[51] Brush scrubbers are available in a double-sided version that brushes both sides of the wafer at once. Brush scrubbers are often used with room temperature solutions of nontoxic chemicals or DI water that is sprayed on the wafer while brushing is underway.

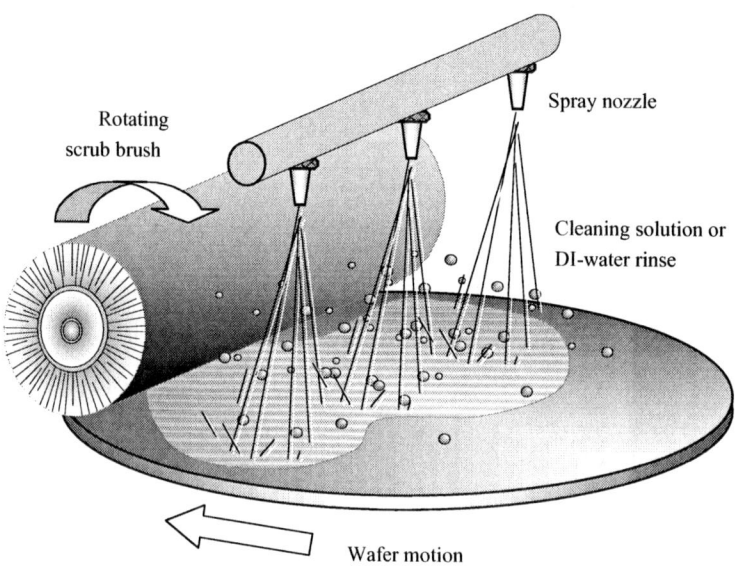

FIGURE 6.30 Wafer Brush Scrubber

Wafer Rinse ■ Chemical residues must be removed from the wafer after cleaning. Every wet-clean process step is followed by an ultrahigh purity *DI-water rinse*. This rinsing step requires a continuous supply of clean water to the wafer surface. The water rinse can also serve to end oxide-etching action from an HF cleaning step.

Overflow Rinsers. Traditionally the most common type of DI-water rinse has been the *overflow rinser* (see Figure 6.31). DI water is brought into the rinse system to flow through and around the wafers, sometimes with a nitrogen bubbler to aid in the mixing of chemicals at the wafer surface. The fluid motion in overflow rinsers serves to sweep away contaminants that have diffused into the flow stream from the surface of the wafers. High flow rates with no dead spots are goals for rinsing.

The overflow rinse process has also been applied to a cascade rinse system. In this case, the DI water cascades between two or three overflow rinsers connected to each other. The wafers start

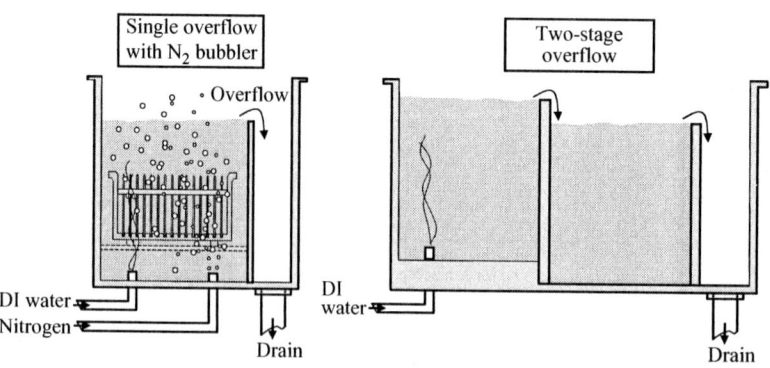

FIGURE 6.31 Overflow Rinser

the rinse process in the downstream rinser and are moved sequentially through the rinsers to the first one with the direct DI-water supply.

Overflow rinsers are the conventional system, but they consume a large amount of DI water. Given the large amount of water used in semiconductor facilities, there has been extensive investigation into alternative rinse processes that consume less water.

Dump Rinse. A common rinse method is the *dump rinser*. DI water is sprayed on the wafers while filling the rinse tank. At a certain water level, a drain in the bottom of the tank quickly opens and the water instantly dumps out (see Figure 6.32). The drain then closes and the cycle repeats for a set number of cycles. An inert gas, such as nitrogen, is often bubbled through the water to aid in removing contaminants such as particles by creating a scrubbing action.

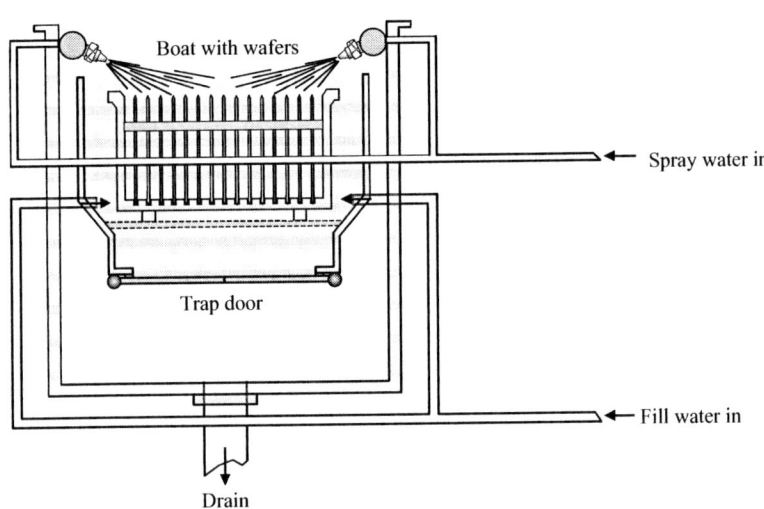

FIGURE 6.32 Dump Rinse

Spray Rinse. A *spray rinser* is typically used in conjunction with wafer cleaning and drying. It uses the physical force of the flowing water to dislodge residual chemicals on the wafer surface.

Hot DI-Water Rinsing. *Hot DI-water rinsing* (70 to 80°C) has become widely used for rinsing the wafer. The benefits of hot DI-water rinsing are that it aids in the removal of residual chemicals (especially if some type of HF solution has been used) and it improves the performance of the dry wafer. However, there has been some research that shows that the use of DI-water rinse at elevated temperatures creates a problem with etching the silicon surface, leading to surface microroughness.[52] Surface microroughness is undesirable for the extremely thin films required in ULSI technology.

Wafer Drying ■ Given the amount of aqueous rinse in a wafer process, it is important to dry the wafer, ideally with no drying spots. How a wafer responds to water refers to its wettability. Water will either adhere to a clean hydrophilic wafer or bead up on a hydrophobic surface. A *hydrophilic surface* has an affinity toward water, allowing water to spread across it in large puddles. Oxide-coated and RCA-cleaned wafer surfaces are hydrophilic. A *hydrophobic surface* has an aversion to water and does not support large pools of water. The water generally beads up, such as water on a newly waxed car, and is called *dewetted*. Oxide-free surfaces that have been hydrofluoric acid (HF) etched are hydrophobic due to the hydrogen-terminated surface. As the wafer undergoes processing, semiconductors require a hydrophobic surface to promote adhesion during subsequent layering processes.[53] Because HF-cleaned wafer surfaces are susceptible to contamination, there must be an adequate effort to properly dry the wafer surface.

Spin Dryers. The *spin dryer* has been widely used in the past. A cassette of wafers is placed in the spin equipment and high-speed rotation removes moisture while the wafers are being sprayed with heated nitrogen. However, there are problems, such as the difficulty of removing moisture in holes and the generation of particulate contamination from the mechanism. In addition, the high-speed rotation of the wafer causes a charge buildup, which attracts particles. Adding static eliminators inside the equipment controls this static electricity buildup.

IPA Vapor Dry. In this process, wafer drying occurs through the displacement of water by heated solvent vapors of isopropyl alcohol (IPA). This *IPA vapor dry* method has a lower probability of contamination from particles, but the purity level of the IPA must be controlled. The IPA is heated in a tank with the wafers suspended in the vapor above the liquid level. When the wafers are removed from the vapor dryer, the solvent vapors evaporate, leaving the wafer dry.

Alternatives to RCA Clean

There are several alternative cleaning technologies to RCA clean that have been researched and used in varying degrees.

Dry Cleaning ■ Wet-cleaning methods are still meeting most cleaning requirements in wafer fabs. Dry-surface cleaning techniques are implemented primarily in integrated cluster tools, rather than stand-alone tools. Ultimately, the cost benefits through increased yield or reduced cost of ownership will decide the frequency of dry cleaning in wafer fabrication.

Plasma-Based Dry Cleaning. *Plasma-based dry cleaning* must be considered as a serious alternative to wet clean. Today, dry-plasma technology is used for removing organic photoresist from wafers (known as *ashing*), or as an integrated preconditioning step in process tools to remove native oxides. In the plasma process, gases and plasma energy are used to cause the chemical reaction that removes the contaminant (plasma processes will be discussed in detail in later chapters).

An example of plasma cleaning is a microwave downstream-plasma process.[54] This plasma can remove film residues containing organic material plus metallic and other inorganic components. The plasma can remove the sidewall film while minimizing removal of the underlying dielectric layer. This plasma clean is followed by a DI-water rinse.

There will be further development and use of dry-plasma cleaning processes for semiconductor manufacturing since the advanced processes required for 0.18-μm critical dimensions and below are using new materials with different forms of residues (within the same classes we have defined). In many cases, the chemical reactions of wet cleaning do not possess enough activation energy (energy required to cause a reaction) to remove all the residues. This is why the higher-activation energies associated with plasma techniques and nonplasma methods are needed for wafer cleaning.

Chelating Agents ■ A *chelating agent* binds and removes metallic ions. If this agent, such as ethylenediamine-tetra-acetic acid (EDTA) is added to a cleaning solution, it can reduce the redeposition of metals in the solution. It does this by altering the reduction-oxidation potential of the metallic species. An example would be adding a chelate to the ammonium hydroxide of SC-1 as a means of preventing metals from any chemical impurities from adhering to the wafer surface.

Ozone ■ *Ozone* (O_3) injected into ultrapure water has been identified as a possible replacement for some RCA wet cleaning. It has been shown that a combination of ozonized ultrapure water followed by an SC-2 cleaning step is effective in removing metals such as copper (Cu) and silver (Ag), as well as organic contaminants.[55] Research has indicated that ozone injected into DI water could replace a piranha step used for light organic cleaning.

Cryogenic Aerosol Cleaning ■ The principle of *cryogenic aerosol cleaning* is to sufficiently cool a gas (argon) to form solid ice particles that are injected onto a wafer surface to remove particulate contamination. The cooling of the argon is done by expansion cooling (lowering of pressure) in a vacuum chamber as a mixture of argon and nitrogen flows through a jet nozzle array. The nitrogen is used to dilute the argon and control the diameter of the solid argon particles. Research data has shown that cryogenic aerosol cleaning can be superior to wet cleaning. A benefit of this process is that it is environmentally benign. This process is not currently in wide use.

SUMMARY

Wafer fabrication is done in ultraclean cleanrooms. There are five types of contamination controlled in cleanrooms: particles, metallic impurities, organic contamination, native oxides, and electrostatic discharge (ESD), all of which can affect device performance. Particles must remain less than one-half the critical dimension or else they are known as a killer defect. Air is controlled by filtering and has a class number to designate the particle size and density. Humans must follow strict cleanroom protocol to minimize contamination. Facilities have special floor layouts to minimize the introduction of contaminants, with laminar airflow and HEPA filters used to attain ultraclean air. Air ionization is done to control ESD. Ultrapure DI water controls many types of contaminants through reverse osmosis, ultrafiltration, and bacteria control. Process chemicals and gases have different levels of filtering, transportation, and handling procedures to achieve high purity.

Cleanroom equipment has special workstation designs to minimize contaminants and is becoming more controlled through use of the minienvironment.

The predominant method of wafer cleaning is a wet process using the RCA cleans of SC-1 and SC-2. Particles and organics are removed by SC-1 and metallics are removed by SC-2. Additional wet cleans are the piranha mixture and HF last step. Two common cleaning methods with RCA clean are megasonics and spray cleaning. Brush scrubbers are often used at the chemical mechanical planarization (CMP) operation to remove particles. Different types of DI-water rinse procedures are overflow rinsers, dump rinsers, spray rinsers, and hot DI water. Wafers are dried by spin dryers or IPA vapor dry. Alternatives to RCA wet clean are dry cleaning with plasma and the use of chelating agents, ozone, and cryogenic aerosol cleaning.

KEY TERMS

cleanroom

contamination

killer defects

particles

aerosols

particle density

particles per wafer per pass (PWP)

metal impurities

mobile ionic contaminants (MICs)

organic contaminants

native oxide

electrostatic discharge (ESD)

class number

ultrafine particles

cleanroom protocol

ballroom layout

bay and chase layout

sub-fab

laminar airflow

high-efficiency particulate air (HEPA)

ultra-low penetration air (ULPA)

air ionization

ultrapure DI-water

total organic carbon (TOC)

pyrogens

silica

makeup loop

polishing loop

water deionization

reverse osmosis (RO)

hyperfiltration

ultrafiltration

membrane contactor

zeta potential

ultraviolet (UV) lamps

assay number

filters

membrane filter

filter efficiency

workstation design

bulkhead equipment layout

wafer cassettes

robotic wafer handlers

electrostatic chuck (ESC)

minienvironment

standard mechanical interface (SMIF)

front-opening unified pod (FOUP)

wet chemistry

RCA clean

Standard clean 1 (SC-1)

Standard clean 2 (SC-2)

dilute cleaning chemistries

piranha
Caro's acid
wet sinks
megasonic cleaning
cavitation
acoustic streaming
spray cleaning
brush scrubbing
DI-water rinse
overflow rinser
dump rinser

spray rinser
hot DI-water rinsing
hydrophilic surface
hydrophobic surface
spin dryer
IPA vapor dry
plasma-based dry cleaning
chelating agent
ozone (O_3)
cryogenic aerosol cleaning

REVIEW QUESTIONS

1. Give a general description of a cleanroom.
2. What is contamination in semiconductor manufacturing?
3. Define a killer defect.
4. State the five categories of cleanroom contamination.
5. What is a particle? What is an aerosol? Why are particles a problem for semiconductor manufacturing?
6. Explain the rule of thumb for acceptable particle size in semiconductor manufacturing.
7. Describe the particle density on a wafer surface. What is PWP?
8. Give an example of a typical metal impurity. Give the two basic ways for metallics to deposit on a wafer surface.
9. Describe what an MIC is.
10. Identify a problem that results from metallic impurities in semiconductor manufacturing.
11. What is an organic contaminant? Provide two possible sources of organic contamination in wafer fabrication.
12. Identify two problems that result from organic contamination in semiconductor manufacturing.
13. Explain native oxide. Identify three problems that result from native oxide.
14. Explain electrostatic discharge (ESD).
15. Give three different problems in wafer fabrication resulting from ESD.
16. List seven sources of contamination in a wafer fabrication facility.
17. Explain the class number for air quality.
18. Interpret the following: (a) class 10 at 0.3 μm and (b) class 1 at 0.5 μm.
19. Describe an ultrafine particle.
20. Explain how a human can generate particles.
21. List four criteria that cleanroom garments should meet. Describe the modern cleanroom garment.
22. List seven proper cleanroom protocols that should be followed by cleanroom personnel.
23. Give three strategies for eliminating particles from cleanrooms.
24. Describe the ballroom layout for cleanrooms.
25. What is the bay and chase approach to cleanrooms?
26. What is laminar airflow, and at what class number does laminar airflow become critical?
27. How often may air turnover in a modern wafer fab?
28. What are HEPA and ULPA filters? What is the efficiency of an ULPA filter?
29. What are typical temperature and relative humidity (RH) settings in a wafer fab? Why is RH control important?
30. List and explain three ESD control methods.
31. Briefly describe each of the six unacceptable contaminants in ultrapure DI water.
32. What are the two major parts of a DI-water installation?
33. Explain the deionization of water. At what resistivity level is water considered deionized?
34. Describe reverse osmosis (RO) filtration. What is ultrafiltration?
35. Explain how a membrane contactor filter works.
36. Explain zeta potential.
37. How is bacteria controlled in ultrapure DI water?
38. What is the assay number for chemical purity?
39. List and discuss four classes of filters.
40. What is a membrane filter?
41. Describe the filter efficiency. What is the typical filter efficiency of a membrane filter?
42. Describe the filtration of a process gas.
43. What is the most significant source of particles in a wafer fab?
44. Give four examples of particle contamination from process tools.
45. What is a bulkhead equipment layout?
46. Describe how wafer cassettes reduced contamination over manual handling.
47. How is most cleanroom wafer handling done?
48. Explain the function and purpose of an electrostatic chuck.
49. Describe a minienvironment, and explain why this situation improves contamination control in a cleanroom.
50. What is a SMIF? What is a FOUP? How do they benefit wafer fabrication?

51. What is the goal for wafer cleaning?
52. What is the predominant wafer surface cleaning process?
53. Describe the RCA cleaning process.
54. What chemical mixture is used in SC-1? What contaminants are removed by SC-1?
55. Describe how the SC-1 wet-clean process removes wafer surface particles.
56. State two concerns that arise from cleaning with the SC-1 wet-clean process.
57. What chemical mixture is used in SC-2? What contaminants are removed by SC-2?
58. Explain what are dilute cleaning chemistries.
59. What is a piranha mixture, and what contaminants are removed from the wafer?
60. Discuss the HF last cleaning step and why it is used.
61. List a typical wafer wet-clean sequence. What is a wet sink?
62. Describe megasonics cleaning and why it is used, including cavitation and acoustic streaming.
63. Discuss spray-cleaning technology. What is an advantage from this cleaning method?
64. Describe wafer brush scrubbing. What contaminant does it remove, and at what process step is this method often used?
65. When is a DI-water rinse done? Describe three different methods for wafer rinsing.
66. Explain the difference between a hydrophilic and a hydrophobic surface.
67. Describe two different methods for wafer drying.
68. Discuss plasma cleaning. Why will the usage of dry plasma cleaning increase?
69. What is a chelating agent, and how is it used in cleaning?
70. How could ozone be used for wafer surface cleaning?
71. Describe cryogenic aerosol cleaning.

CHEMICAL AND EQUIPMENT SUPPLIERS' WEB SITES

Adept Technology Inc.	http://www.adept.com
AERONEX Inc.	http://www.aeronex.com
Air Kontrol Inc.	http://www.airkontrol.com/
Air Products and Chemicals	http://www.airproducts.com
Amerimade Technology	http://www.amerimade.com/
Apex Industries	http://www.apexind.com/
Applied Science and Technology	http://www.astex.com/
Aquionics Inc.	http://www.aquionix.com/
Asahi/America Inc.	http://www.asahi-america.com/
Ashland Specialty Co.	http://www.ashchem.com/
ASI Technologies	http://www.asidoors.com/
AST Products	http://www.astp.com
BOC Edwards	http://www.boc.com/edwards/
Clean Air Products	http://www.cleanairproducts.com/
Contamination Control Products	http://www.ccpcleanroom.com/
Dow Chemical/Filmtec	http://www.dow.com/liquidseps/
Dow Corning	http://www.dowcorning.com/
Dryden Engineering Co.	http://www.drydeneng.com/
Eastman Chemical Co.	http://www.eastman.com/
EKC Technology Inc.	http://www.ekctech.com/
Entegris Inc.	http://www.entegris.com/
Environflex Inc.	http://www.enviroflex.com/
Filtration Technology Inc.	http://www.filtrationtechnology.com/
FSI International	http://www.fsi-intl.com/
General Chemical Corp.	http://www.genchem.com/
IN USA Inc.	http://www.inusaozone.com
Integrated Designs LP	http://www.pumpless.com/
Ion Systems	http://www.ion.com/
J. T. Baker	http://www.jtbaker.com/
Kappler Protective Apparel	http://www.kappler.com/
Koch Microelectronics Service Co.	http://www.kochmicroelectronic.com/
Meissner Filtration Products Inc.	http://www.meissner.com/

MicroChem Corp.	http://www.microchem.com/
Micro Magazine	http://www.micromagazine.com/
Millipore Corp.	http://www.millipore.com/
Modutek Inc.	http://www.modutek.com
NetMotion Inc.	http://www.netmotion.com/
Pall Corp.	http://www.pall.com/
Parker Hannifin Corp.	http://www.parker.com/
Pope Scientific Inc.	http://www.popeinc.com/
PTI Advanced Filtration Inc.	http://www.pti-afi.com/
PURAC America Inc.	http://www.purac.com/
Schumacher	http://www.schumacher.com
SCP Global Technologies	http://www.scpglobal.com/
Sage Technologies Corp.	http://www.sagetech.net/
Semitool	http://www.semitool.com/
Simco Static Control	http://www.simco-static.com/
TEL, Tokyo Electron Ltd.	http://www.teainet.com
The Texwipe Co. LLC	http://www.texwipe.com/
Ultrapure & Industrial Services	http://www.ultrapure.com/
US Filter/Filterlite	http:///www2.usfilter.com/
Verteq Inc.	http://www.verteq.com/

REFERENCES

1. T. Hattori, "Particle Reduction in VLSI Manufacturing," *Contamination Control and Defect Reduction in Semiconductor Manufacturing III* (Pennington, NJ: The Electrochemical Society, 1994) p. 3.
2. A. French, *Newtonian Mechanics,* (New York: W. W. Norton, 1971), p. 32.
3. C. Gross et al., "Assessing Future Technology Requirements for Rapid Isolation and Sourcing of Faults," *Micromagazine* online edition (July/August 1998).
4. T. Hattori, "Detection and Analysis of Particles in Production Lines," *Ultraclean Surface Processing of Silicon Wafers: Secrets of VSLI Manufacturing,* ed. T. Hattori, (Berlin: Springer, 1998), p. 245.
5. W. Kern, *Handbook of Semiconductor Wafer Cleaning Technology,* (Park Ridge, NJ: Noyes Publications, 1993), p. 9.
6. C. Chang and T. Chao, "Wafer-Cleaning Technology," *USLI Technology,* ed. C. Chang and S. Sze (New York: McGraw-Hill, 1996), p. 67.
7. G. Cooper, "The Effect of CO Cylinder Materials on Wafer Contamination," *Semiconductor International* (July 1997): p. 301.
8. C. Chang and T. Chao, "Wafer-Cleaning Technology," p. 67.
9. S. De Gendt et al., "Silicon Surface Metal Contamination Measurements Using Grazing-Emission XRF Spectrometry," *Science and Technology of Semiconductor Surface Preparation, Symposium Proceedings* 477 (Warrendale, PA: Materials Research Society, 1997) p. 397.
10. K. Saga, "Influence of Surface Organic Contamination on the Incubation Time in Low-Pressure Chemical Vapor Deposition of Silicon Nitride on Silicon Substrates," *Science and Technology of Semiconductor Surface Preparation Symposium Proceedings* 477 (Warrendale, PA: Materials Research Society, 1997): p. 379.
11. T. Ohmi, "Total Room Temperature Wet Cleaning for Si Substrate Surface," *Journal of the Electrochemical Society* (September 1996): p. 2957.
12. C. Chang and T. Chao, "Wafer-Cleaning Technology," p. 80.
13. M. Suzuki et al., "Etching Characteristics During Cleaning of Silicon Surfaces by NF3-added Hydrogen and Water-Vapor Plasma Downstream Treatment," *Science and Technology of Semiconductor Surface Preparation, Symposium Proceedings* 477 (Warrendale, PA: Materials Research Society, 1997): p. 167.
14. D. Tolliver, *Handbook of Contamination Control in Microelectronics: Principles, Applications and Technology*, (Park Ridge, NJ: Noyes Publications, 1988), p. 175.
15. L. Levit and J. Menear, "Measuring and Quantifying Static Charge in Cleanrooms and Process Tools," *Solid State Technology* (February 1998): p. 85.
16. A. Braun, "Cleanroom Technologies Continue to Keep Contamination at Bay," *Semiconductor International* (March 1998): p. 58.
17. R. Kraft, "Proper Cleanroom Protocol," *Semiconductor International* (March 1998): p. 73.
18. Summarized from R. Kraft, "Proper Cleanroom Protocol," *Semiconductor International* (March 1998): p. 73.
19. R. Jarvis and L. Armentrout, "Full-Fab Surface Particle Detection Improves Yields," *Semiconductor International* (June 1997): p. 199.
20. J. Schroeder, "Automation-Centric Processing Bay Layout," *Semiconductor International* (June 1997): p. 209.

21. H. Tseng and R. Jansen, "Cleanroom Technology," *ULSI Technology*, ed. C. Chang and S. Sze (New York: McGraw-Hill, 1996), p. 14.
22. S. Middleman and A. Hochberg, *Process Engineering Analysis in Semiconductor Device Fabrication*, (New York: McGraw-Hill, 1993), p. 125.
23. H. Tseng and R. Jansen, "Cleanroom Technology," p. 28.
24. D. Mahoney, "The Construction of a Class 1 Cleanroom: A Case Study," *Semiconductor International* (October 1997): p. 141.
25. H. Inaba, "Effect of Electrostatic Charge on Particle Adhesion on Wafer Surfaces," *Ultraclean Surface Processing of Silicon Wafers: Secrets of VLSI Manufacturing*, ed. T. Hattori (Berlin, Springer, 1998): p. 143.
26. T. Roche, T. Peterson, and E. Hanson, "Water Use Efficiency in Immersion Wafer Rinsing," *Science and Technology of Semiconductor Surface Preparation, Symposium Proceedings* 477, (Warrendale, PA: Materials Research Society, 1997): p. 527.
27. R. Mohindra and W. Kern, "New Process for Producing Particle-Free Deionized Water," *Semiconductor International* (July 1997): p. 191.
28. M. Dax, "Membrane Contactor Technology Gives PPB Dissolved Oxygen in Water," *Semiconductor International* (December 1996): p. 36.
29. H. Tseng and R. Jansen, "Cleanroom Technology," *ULSI Technology*, ed. C. Chang and S. Sze (New York: McGraw-Hill, 1996), p. 39.
30. M. Dax, "Membrane Contactor," p. 36.
31. R. Mohindra and W. Kern, "New Process," p. 194.
32. T. Hattori, "Detection and Analysis of Particles in Production Lines," *Ultraclean Surface Processing of Silicon Wafers: Secrets of VLSI Manufacturing*, ed. T. Hattori (Berlin: Springer, 1998), p. 245.
33. A. Braun, "Cleanroom Technologies Continue to Keep Contamination at Bay," *Semiconductor International* (March 1998): p. 58.
34. T. Hattori, "Trends in Wafer Cleaning Technology," *Ultraclean Surface Processing of Silicon Wafers: Secrets of VLSI Manufacturing*, ed. T. Hattori (Berlin: Springer, 1998), p. 438.
35. R. McIllvaine, "Cleanroom Demands for the Next Generation," *Semiconductor International*, (March 1998): p. 68.
36. T. Ohmi, "Revolution of Silicon Substrate Surface Cleaning," *ULSI Science and Technology Proceedings* 97-3, (Pennington, NJ: The Electrochemical Society): p. 197.
37. T. Hattori, "Trends in Wafer Cleaning Technology," Supplement to *Solid State Technology* (May 1995): p. S8.
38. C. Chang and T. Chao, "Wafer-Cleaning Technology," p. 61.
39. Ibid., p. 64.
40. M. Meuris et al., "The IMEC Clean: A New Concept for Particle and Metal Removal on Si Surfaces," *Solid State Technology* (July 1995): p. 109.
41. T. Ohmi, "Total Room Temperature," p. 2957.
42. Ibid., p. 2958.
43. M. Heyns et al., "Advanced Wet and Dry Cleaning Coming Together for the Next Generation," *Solid State Technology* (March 1999): p. 40.
44. J. Rosamilia et al., "Hot Water Etching of Silicon Surfaces: New Insights of Mechanistic Understanding and Implications to Device Fabrication," *Science and Technology of Semiconductor Surface Preparation, Symposium Proceedings* 477 (Warrendale, PA: Materials Research Society, 1997): p. 181.
45. T. Ohmi, "Total Room Temperature," p. 2958.
46. T. Ohmi, "Total Room Temperature," p. 2961.
47. R. DeJule, "Trends in Wafer Cleaning," *Semiconductor International* (August 1998): p. 64.
48. J. Liu et al., "Si_3N_4 Particle Removal Efficiency Study," *Science and Technology of Semiconductor Surface Preparation, Symposium Proceedings* 477 (Warrendale, PA: Materials Research Society, 1997), p. 27.
49. G. Gale, A. Dai, and I. Kashkoush, "How to Accomplish Effective Megasonic Particle Removal," *Semiconductor International* (August 1996): p. 133.
50. Ibid.
51. R. DeJule, "Trends in Wafer Cleaning," p. 65.
52. J. Rosamilia, "Hot Water Etching," p. 181.
53. J. Park et al., "The Formation of Water Marks on Both Hydrophylic and Hydrophobic Wafers," *Science and Technology of Semiconductor Surface Preparation, Symposium Proceedings* 477 (Warrendale, PA: Materials Research Society, 1997): p. 513.
54. K. Lao and W. Wu, "Microwave Downstream Plasma Removes Metal Etch Residue," *Semiconductor International* (July 1997): p. 231.
55. T. Ohmi, "Total Room Temperature," p. 2959.

CHAPTER 7
METROLOGY AND DEFECT INSPECTION

Inspection has been done since the beginning of wafer fabrication. Semiconductor manufacturing veterans can recall when simply observing the color of the oxide film on the wafer surface was acceptable to estimate film thickness. Whichever color tint the oxide film appeared to be was compared to a color chart made up of pieces of wafers with different film thickness associated with each color tint (see Appendix D).

Wafer process inspection technology has undergone major changes. Feature sizes continue to shrink below the quarter micron era and chip density on wafers is continually increasing. Each process step has critical issues that define success or failure: contamination, junction depth, and film quality, to name a few. Furthermore, the introduction of new materials and processes will also bring new ways chips can fail. Measurements are critical to characterize the wafer materials and verify their acceptability.

To maintain good process yield and improve device performance, wafer fabs have improved control over process parameters and reduced sources of defects during fabrication. Some of these improvements have come from areas such as equipment automation, robotic handling, contamination reduction, and more consistent flow of wafers through the fab to avoid long waiting periods. Other improvements could not have occurred without the ability to measure wafers and evaluate the process performance. This evaluation is done using highly accurate measurement equipment that provides real-time data about the wafer fabrication performance, giving key information to the engineer and technician for the decision-making process.

OBJECTIVES

After studying the material in this chapter, you will be able to:

1. Explain why IC metrology is performed, and discuss equipment, yield and data collection issues associated with metrology.
2. Identify twelve different quality measures used in wafer fabrication, and identify the fabrication processes where each is used.
3. Describe the various metrology methods and equipment associated with the different quality measures.
4. List and discuss the purpose of seven different types of analytical equipment used to support IC fabrication.

INTRODUCTION

Metrology is the science of measurement to determine dimensions, quantity, or capacity. For IC fabrication, *metrology* refers to the techniques and procedures for determining physical and electrical properties of the wafer during the fabrication process. Metrology used during fabrication employs measurement equipment and sensors to collect and analyze data about wafer parameters and defects. *Defects* are the characteristics of the wafer or the results of the wafer fabrication process that cause nonconformance to the specified wafer requirement. The wafer *defect density* is the number of defects per unit area of wafer surface, usually cm^2. Wafer defects are specified by type and dimension. Metrology measurements are used by manufacturing personnel (e.g., technicians, engineers, and managers) to ensure product performance and to make meaningful decisions regarding changes to improve processperformance.

It is essential that accurate evaluation of wafers be performed throughout the fabrication process to verify that the product meets the specified requirements. To achieve this, every process step in wafer fabrication has stringent quality measures that define the requirements needed at each step in order for the die to pass electrical test and meet reliability specifications during usage. Quality measures require extensive data collection on either test wafers or production wafers to demonstrate that the chip product's process requirements have been met.

IC METROLOGY

IC metrology is a necessary means for measuring the performance of fabrication processes to ensure the specific quality standards are attained. To accomplish this requires a sample wafer, a measurement tool, and the means to analyze the data. Traditionally, most in-process data has been collected on *monitor wafers* (also referred to as *test wafers*) which are blank (or unpatterned) wafers included during process runs specifically to characterize the process. These blank wafers are recirculated for repeated usage (after undergoing appropriate refurbishing steps, such as surface stripping and cleaning). For instance, monitor wafers are included during thermal oxidation in a vertical diffusion furnace to measure oxide thickness and to check for particles inside the process chamber. Depending on the process step, blank wafers may include a predeposited film layer on the wafer surface.

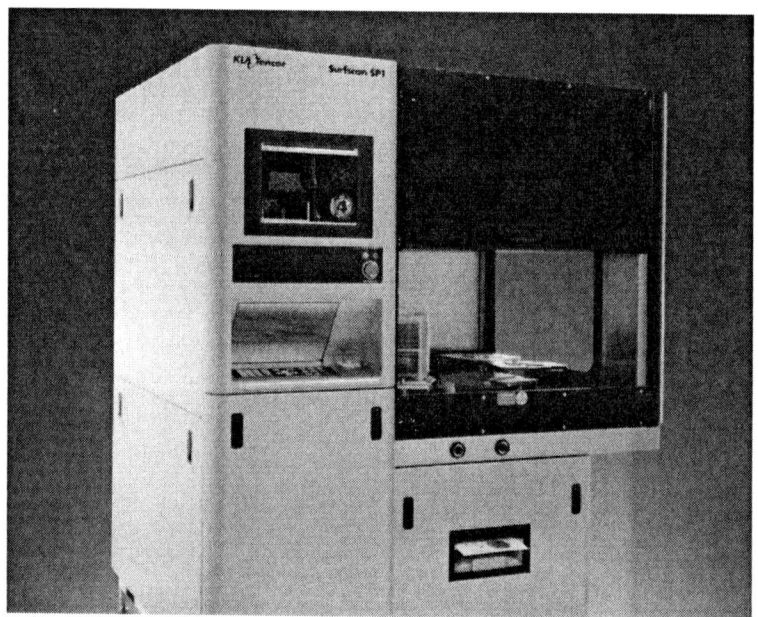

Unpatterned Surface Inspection System
(Photo courtesy of KLA-Tencor Corporation)

Many semiconductor manufacturers have begun using production wafers, sometimes patterned wafers, for in-line tool monitoring (see Figure 7.1). Using actual production wafers more closely simulates what is happening during the process run, providing the manufacturing team members better information for making decisions.

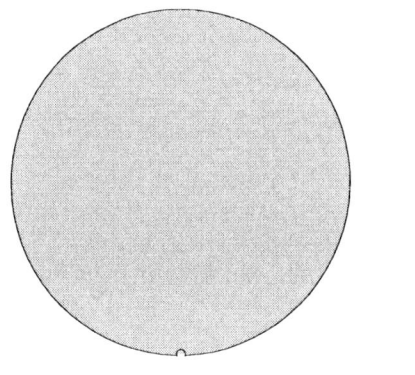
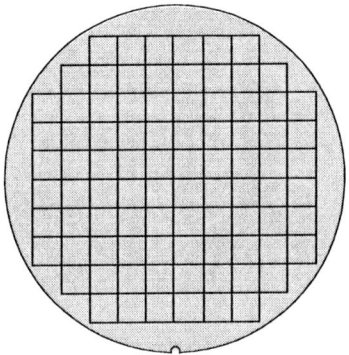

Monitor wafer   Patterned wafer

FIGURE 7.1   Monitor Wafer vs. Patterned Wafer

Measurement Equipment

There are different types of metrology equipment used to perform measurements in wafer fabrication. A major way to distinguish these tools is by how the measurement is done—with a stand-alone tool independent of the process or with metrology equipment integrated with the processing equipment. The two dominant categories for measurement equipment are shown in Table 7.1.[1] A *stand-alone measurement tool* functions independent of the process to perform metrology measurements. An *integrated measurement tool* has sensors that permit it to function as a part of the process and deliver in situ (real-time) measurement data.

Yield

Yield is an important indicator of a wafer fab's ability to produce high quality die. *Yield* is defined as the percent of good parts produced out of the total group of parts started. For example, if there are 200 die on a wafer and 190 die are acceptable, then the wafer yield is:

$$\text{Yield (\%)} = \frac{190}{200} \times 100 = 95\%$$

Different methods are used to calculate yield. One method is to measure yield relative to the type of part produced for a period of time (e.g., the yield of all wafers produced within a week). Yield can also be measured with respect to a particular location in the process, such as the die yield at the etch process. An important yield measurement for semiconductor manufacturing is wafer sort yield, which indicates the percentage of die that are acceptable after functional test (see Chapter 19).

Yield is widely used in semiconductor manufacturing to describe the health of the fabrication process. A high yield means that the process is producing good parts and performing as intended. A low yield is a strong indicator that there are quality problems in the product design or production process that must be resolved through improvement.

TABLE 7.1   Measurement Tool Classification

	Stand-Alone Tools
Off-line	Available only outside fab (usually destructive or contaminating).
At-line	Available in the fab (measures monitor wafers that are destructive, contaminated or unpatterned).
In-line	Used during production (can measure patterned wafers).
	Integrated Tools
On-line	Available at the process workstation to measure patterned wafers but not able to measure during wafer processing.
In situ	Measures wafer, process, or equipment during processing (real-time measurement).

In situ measurements are a recent development that were not very common even in the mid-1990s. As equipment efficiency improvements occur, there will be more integration of in situ metrology with fabrication equipment.

Data Management

Detecting defects is a challenge in IC fabrication because of shrinking feature sizes. The semiconductor manufacturer must have a method of categorizing the defects, sorting between false and real defects, and determining the root cause of each defect for determining the appropriate corrective action. Defect analysis should differentiate between random or nonrandom causes and correlate defect data with electrical and other tests to track how different defects affect wafer yield.

To a large degree, defect analysis is supported by an integrated software system. Large software programs connect major measurement tools and provide the ability to set up sample plans, analyze defect data, and detect trends amongst defects. Defects must be categorized in order to conduct failure analysis. The most advanced metrology software has *automatic defect classification* (*ADC*) to identify and classify defects based on software recognition patterns. Advanced defect management software also catalogs wafers and images based on user-defined criteria and includes statistical process control (SPC) capability for process monitoring and control.

QUALITY MEASURES

The breadth of metrology for IC fabrication is demonstrated by the many quality measures used throughout wafer processing. Semiconductor *quality measures* define the requirements for specific aspects of wafer fabrication to ensure acceptable device performance as measured by electrical tests and device reliability. Major quality measures used in the wafer fab are shown in Table 7.2, including the process areas where each measure is applicable. Semiconductor manufacturers specify their particular wafer quality measures by stipulating the exact requirements for their products at each process step. The manufacturer may specify other quality requirements beyond those listed in Table 7.2.

TABLE 7.2 Quality Measures in Wafer Fabrication by Production Areas

	Quality Measure	Implant	Diffusion*	Thin Films		Polish	Etch	Photo
				Metals	Dielectric			
1	Film Thickness		√	√	√	√	√	√
2	Sheet Resistance	√	√	√				
3	Film Stress		√	√	√			
4	Refractive Index		√		√			
5	Dopant Concentration	√	√					
6	Unpatterned Surface Defects	√	√	√	√	√	√	
7	Patterned Surface Defects						√	√
8	Critical Dimensions (CDs)						√	√
9	Step Coverage					√	√	
10	Overlay Registration							√
11	Capacitance-Voltage		√					
12	Contact Angle							√

*The diffusion bay processes include: oxidation, deposition, diffusion, anneal, and alloy.

Film Thickness

Since wafer fabrication is a layering process, there are many different types of films found on the wafer surface throughout the fabrication process. Some of these different types of films are metal, dielectric, photoresist, and polysilicon. The quality of these films is essential to a high-yield fabrication process to produce reliable chips.

A critical quality parameter for films are their thickness. *Film-thickness metrologies* can be divided into two general types: whether they measure opaque (i.e., light-blocking, such as metal) films or transparent films. In some instances, such as for the gate oxide dielectric, the film thickness must be measured with an accuracy of an angstrom (Å) or less. Additional film quality parameters are surface roughness, reflectivity, density, and lack of pinholes and voids.

Resistivity and Sheet Resistance ■ One of the most practical methods for evaluating thickness for conductive films is to measure the *sheet resistance*, R_s. To discuss R_s, we need to understand how resistivity is related to resistance in a thin film layer. Consider current flow through a square sheet of conductive material, with the thickness, length, and width shown in Figure 7.2. Recall from Chapter 2 that resistance, R, of a conductor is given as:

$$R = \frac{\rho(l)}{a} \quad \text{(ohms)}$$

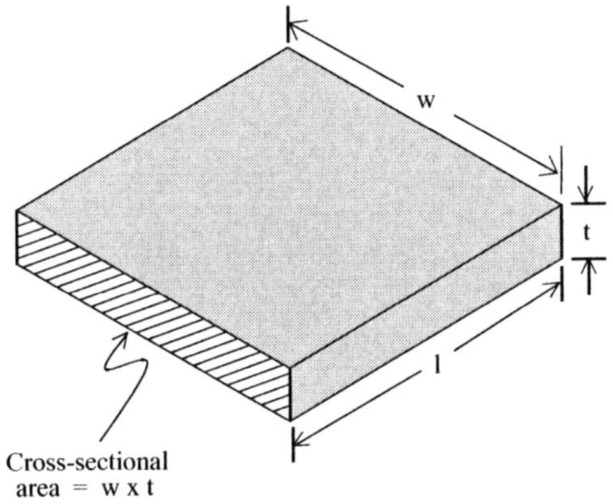

FIGURE 7.2 Illustration of Square Thin Film

The cross-sectional area is given by the product of width and thickness (w × t) and can be substituted in the equation for the area. Based on the assumption that the length (l) is the same dimension as width (w), then *l* can be substituted for *w*. Making these substitutions yields:

$$R_s = \frac{\rho(l)}{w \times t}$$

$$R_s = \frac{\rho(l)}{l \times t}$$

$$R_s = \frac{\rho}{t} \quad \text{(ohms/square, or } \Omega/\square\text{)}$$

Where,
R_s = Sheet resistance in ohms/square
ρ = Film resistivity in ohm-cm
t = Thickness of film

Sheet resistance, R_s, can be interpreted as the end-to-end resistance of a square sample of a thin film on a wafer. It depends on the film resistivity and thickness. Sheet resistance is independent of the size of the square film sheet. Measuring the sheet resistance between two points at the same distance apart on any size square of the same thickness will produce the same resistance. For this reason, the units for R_s are given as ohms/square ($\Omega/\square$). Dimensionally the units for R_s are the same as ohms, but ohms/square is typically used as a reminder that this is actually the resistance of a sheet of film. If the thickness and sheet resistance are known, then the *sheet resistivity*, ρ_s, can be calculated:

$$\rho_s = R_s (t) \quad \text{(ohms-cm)}$$

The terms sheet resistance and sheet resistivity are sometimes used interchangeably since they only differ by the thickness of the film. A tool used for measuring sheet resistance is the four-point probe.

Four-Point Probe. Resistance cannot be practically measured on thin films with two simple probes, as usually done with multimeters commonly used by electronic technicians. This is because of excessive contact resistance at the contact interface between the probes and the wafer material. In the semiconductor industry, a widely used method to measure sheet resistance is the *four-point probe*. This method has four in-line probes equally spaced that touch the wafer surface in a single file (see Figure 7.3). A known value of current (I) is passed between the two outer probes and the potential difference (v) developed across the two inner probes is measured. This approach avoids dealing with the effect of contact resistance.

The spacing between probes, s, should be less than the wafer diameter and less than the film thickness. The resistivity of the film sheet is related to the four-point probe current and voltage by:

$$\rho_s = \frac{V}{I} \times 2\pi s \quad \text{(ohms-cm)}$$

Where,
ρ_s = Sheet resistivity in ohm-cm
V = DC voltage across the voltage probes (in volts)
I = Constant DC current passing through the current probes (in amperes)
S = Spacing between the probes

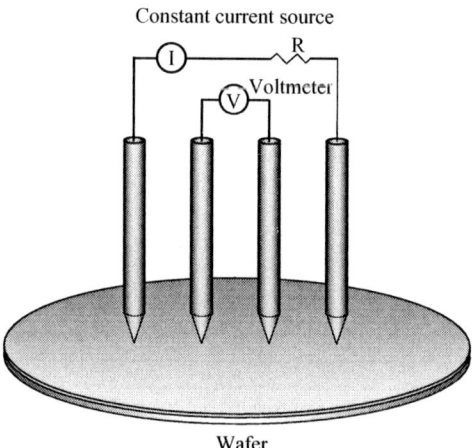

FIGURE 7.3 Four-Point Probe

Resistivity measurements with a four-point probe can be reproduced to within ± 2% if care is taken in selecting instrumentation, and in determining probe pressure and current levels.[2] The four-point probe is a nondestructive technique (e.g., the wafer does not have to undergo permanent damage to perform the test). However, due to the potential for damage to the wafer from the probe contacts, the semiconductor industry has moved toward noncontacting probes.[3] One contactless method is the use of eddy-current probes. A high-frequency ac current flowing through a coil induces eddy currents in conducting films placed under the coil. The eddy currents are a loss in energy that can be attributed to the loading effects produced by the resistance of the conducting film. The resulting change in electrical energy can be measured and used in computing the value of the sheet resistance of the film being measured.

Sheet Resistance (Opaque Films) ■ Sheet resistance is used to indirectly measure the thickness of opaque conductive films (e.g., metal, silicide, or semiconductor films) deposited on an insulator substrate such as a silicon wafer. As long as the film layer is large and the probe spacing is small, the sheet resistance, R_s, is given by:[4]

$$R_s = 4.53 \frac{V}{I} \quad \text{(ohms/square, or } \Omega/\square\text{)}$$

This formula is actually derived from the equation for sheet resistivity listed on page 154. The constant 4.53 results from a correction factor based on the assumption of an infinitely large sheet of film with small probe spacing. The correction factor can be modified to a different value if the film sheet is not infinitely large.

Van der Pauw. A modification of the four-point probe is the *Van der Pauw Method* that measures sheet resistivity with four probes on the periphery of an arbitrarily shaped sample (see Figure 7.4). The concept is based on measuring electrical current on two corners of a square pattern and measuring voltage on the other two. The results are the same as those achieved with the four-point probe.

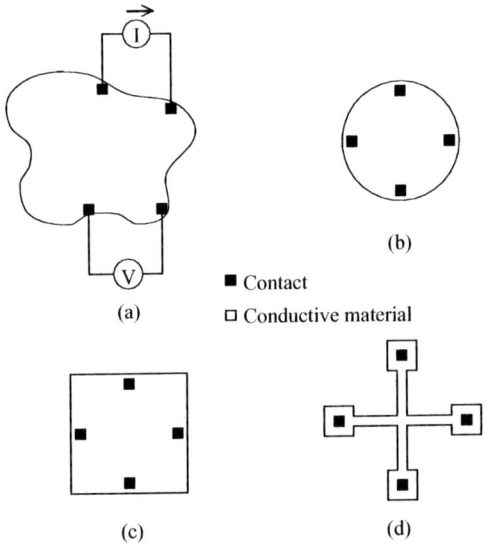

FIGURE 7.4 Van der Pauw Sheet Resistivity

Contour Maps. A practical benefit of sheet resistance measurements with the four-point probe is that the technique can be done at many sites across a test wafer to provide *contour maps* (see Figure 7.5). A contour map will typically present the data with a contour line showing the nominal $\Omega/\square$ and then deviations above and below this nominal value. The sheet resistance data results from a predetermined number of measurement sites across the wafer.

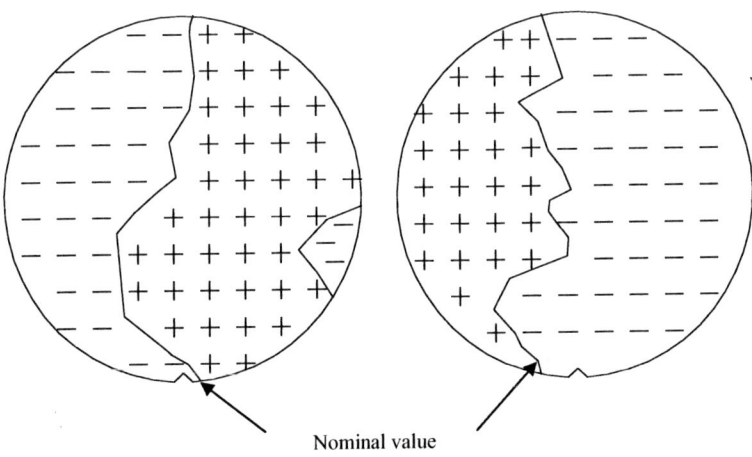

FIGURE 7.5 Sheet Resistance Contour Map

Ellipsometry (Transparent Films) ■ *Ellipsometry* is a nondestructive, noncontact optical film-thickness measurement technique primarily used to measure thin, transparent films. It is the leading method for film thickness metrology in the wafer fab.[5] The basic principle of ellipsometry is to use a linearly polarized laser light source that, when reflected from the sample, becomes elliptically polarized (see Figure 7.6). Polarized light consists of all light rays traveling in one plane. Ellipsometry measures the shape of this reflected ellipse, and, based on known inputs such as the angle of reflection, it accurately determines the film thickness. It is common in ellipsometry tools for the angle of the incident light to be varied to provide more measurement samples that are optimized for the film material. This practice is referred to as *variable-angle spectroscopic ellipsometry* (*VASE*). VASE has improved ellipsometry for measuring multilayer stack structures, which is common in ULSI (see Chapter 12 for a description of multilayer stacks).[6]

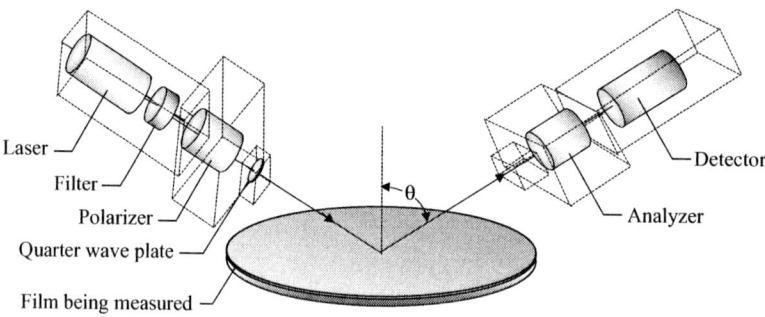

FIGURE 7.6 Basic Principle of Ellipsometry

Spectroscopic ellipsometry analyzes the polarization change of light over a wide spectral range (spectroscopy uses the characteristic light wavelengths emitted by different materials). Information about how the different wavelengths vary relative to one another after being reflected from the film is very useful in a production environment because it makes ellipsometry sensitive to different physical properties of the thin film. Ellipsometry is capable of measuring different types of very thin films on the order of tens of angstroms thick.[7] The types of films include dielectrics, metals, and polymer coatings. The main requirement is that the film be transparent or semitransparent. Fully automated thin film ellipsometry tools routinely measure gate oxide films < 40 Å thick with a repeatability better than 0.1%. Another example of ellipsometry measurements is sub-100 Å stacked oxide/nitride/oxide (ONO) film structures used in MOS capacitors with measurement repeatability in the 1% range. Thin metal layers (< 500 Å) are considered semitransparent and can be measured with ellipsometry. An example of a semitransparent metal layer is the thin copper seed layer used in copper interconnects. Metal layers (> 1000 Å) are generally considered opaque and cannot be measured with ellipsometry.

Ellipsometry measurement-tool features such as small measurement spots, pattern-recognition software, and high-precision wafer positioning hardware have made it possible to do many transparent film process-control measurements directly on production wafers. This reduction in unpatterned monitor wafers has reduced process-control costs and increased ellipsometry efficiency.[8] Ellipsometer tools are also being directly integrated into process tools for in situ (real time) measurement applications in areas such as etch and planarization. Real-time thickness measurement is desirable because it permits the process to define a precise endpoint of the film thickness during runs for better accuracy and repeatability in film thickness.

Reflection Spectroscopy ■ When light reflects off a surface, *reflection spectroscopy* (also referred to as *reflectometry*) is one of the three general types of optical measurement techniques, with the other two being optical microscopy (discussed later in this chapter) and ellipsometry (see above). Reflection of a structure is often used to characterize layer thickness for light-absorbing dielectric layers on a nonabsorbing wafer substrate (see Figure 7.7). Based on the relationship of how light reflects off the top and bottom surface of the film layer, reflectometry can be used to compute the film thickness. An advanced reflection-spectroscopy technique utilizes a dual-beam spectrometer in order to give a clearer light signal for imaging. One light beam is used for reflection measurements from the film layer, while the second light source provides a reference source to correct for real-time lamp deviation or noise.

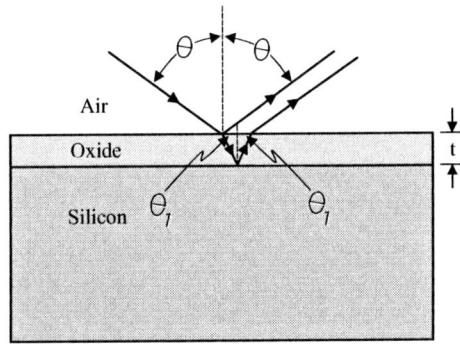

FIGURE 7.7 Light Reflection from a Thin Layer

X-Ray Film Thickness ■ X-ray beams can be focused on a surface and used for making film thickness measurements through a lesser-used technique known as *X-ray fluorescence* (*XRF*). When X-rays strike a film, the absorption of radiation excites electrons in the film. The excited electrons drop into a lower energy state, emitting X-ray photons (known as fluorescence) whose energy represents the identity of the film atom. By measuring these X-ray photons, film thickness can be determined (see Figure 7.8). A modification is to use *total-reflection XRF* (*TRXRF*), which uses a small angle and reduces the amount of X-ray scattering to improve the measurement sensitivity.[9]

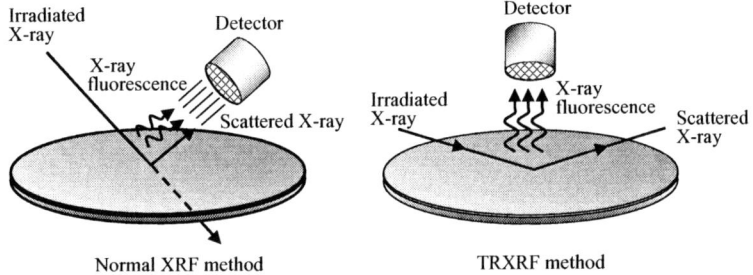

FIGURE 7.8 Film Thickness Measured with X-Ray Fluorescence (XRF)

A thin film's X-ray reflectivity will vary depending on the angle of incidence and wavelength. The reflectivity can be used to calculate the thickness and density of the film layer. It is difficult to use X-rays to measure films with mixed compositions or film stacks with two or more layers. X-ray fluorescence applies primarily to single-layer films.

Photoacoustic Technology ■ A recent development in metal thin film measurement is *photoacoustics*. This noncontact technology is based on light-induced sound pulses that generate an acoustic pulse that is directed toward the film stack. When the acoustic pulse strikes the surface and underlying film interface, an echo is created that bounces back toward the surface. This echo causes a slight change in reflectivity that is detected at the wafer surface (see Figure 7.9 on page 158). The time it takes for the pulse echoes to bounce back is used to calculate the film thickness.

This technique has a spot size of < 8 μm, which because of its small size, makes it capable of probing structures on patterned wafers. It can measure film stacks with an individual layer thickness down to < 20 Å, which is critical as device scaling requires smaller structures for increased performance.[10]

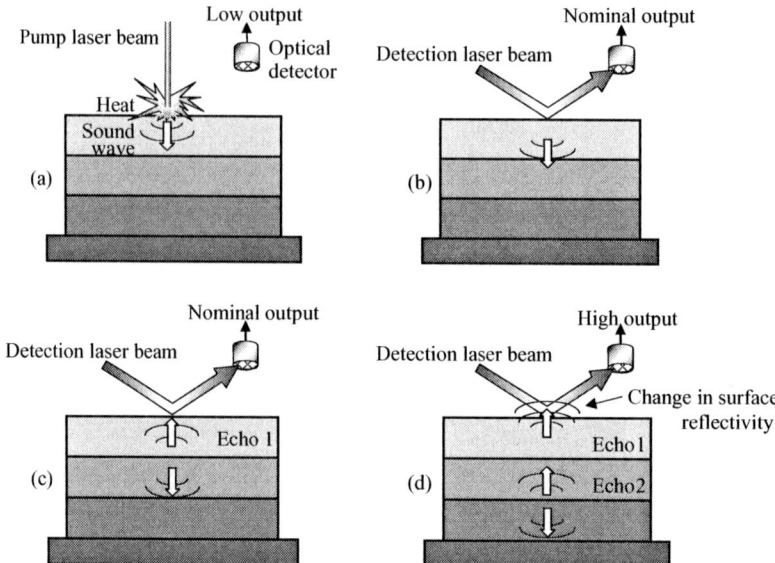

FIGURE 7.9 Photoacoustic Film-Thickness Measurement
Redrawn from *Solid State Technology,* (June 1997), p. 86.

Film Stress

Highly localized *film stress* can be introduced during routine manufacturing processes into thin films that may deform the substrate and create reliability concerns. The amount of this deformation is measured with thin film stress measurement tools. Stress measurement is done by analyzing the changes in the radius of curvature of the substrate resulting from the film deposition and applies to all types of standard thin films, including metals, dielectrics, and polymers. The wafer radius is measured before and after film deposition using either a scanning laser-beam technique or split-beam laser technique to create a stress profile map across the wafer (see Figure 7.10). Automated stress measurement tools have SMIF handling capability.

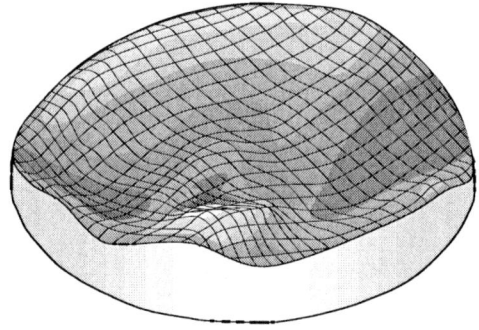

FIGURE 7.10 Detailed Stress Map of Wafer

Refractive Index

Refraction is the property of a transparent substance that addresses how much a light beam bends as it travels through it (see Figure 7.11). Variations in the refractive index can indicate contamination in the film and lead to incorrect thickness measurements. The index of refraction for pure oxide is 1.46. The index of refraction for thin films is measured by interference and ellipsometry techniques with the same ellipsometer measurement tool used to determine thin film thickness.

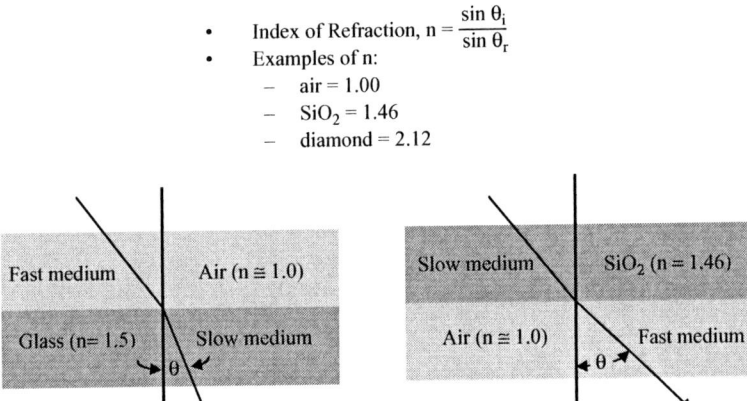

FIGURE 7.11 Index of Refraction

Dopant Concentration

The performance of semiconductor devices is directly affected by distribution of dopant atoms in the silicon in areas such as the formation of pn junctions, the epilayer, and the doping of poly-silicon (see Figure 7.12). Modern processes employ dopant concentrations ranging from about 10^{10} atoms/cm2 to about 10^{18} atoms/cm^2 (see Chapter 17).[11] There are several techniques for measuring the dopant concentration, or dose, of atoms in silicon. A common in-line method is the four-point probe measurement, typically used for high-dopant concentration. The thermal-wave system is also used in-line and is acceptable for low-dose readings. For off-line measurements, secondary-ion mass spectrometry (SIMS, described later in this chapter) with whole-wafer positioning has recently been used as an alternative method for doping-concentration process control.[12] The capacitance-voltage test (described later in this chapter) can also be used to characterize dopant concentration.

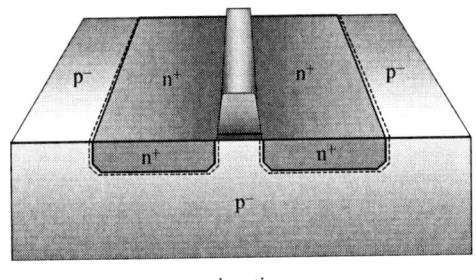

--- pn junctions

FIGURE 7.12 pn Junction

The four-point probe method has been previously discussed in this chapter, and is used in the same manner for dopant concentration as for sheet resistance (since sheet resistivity of the implanted silicon layer is related to the dopant concentration). The contour mapping technique possible with the four-point probe method is useful for routinely monitoring the implantation dose concentration.

Thermal-Wave System ■ A widely used method for monitoring ion-implant dose concentration is the *thermal-wave system*. This method measures the lattice damage in the implanted wafer due to ion implantation.[13] This is done by measuring changes in the wafer surface reflectivity from two lasers focused on the same localized spot on the wafer (see Figure 7.13 on page 160). One laser heats the wafer with a modulated Ar laser to create waves of heat (thus the term thermal wave). The thermal waves cause a change in the reflectance of the other HeNe probe laser that is proportional to the number of crystal defect sites in the wafer. A thermal-wave signal detector with calibration data is used to characterize the implant process by correlating the amount of crystal damage to the dopant concentration and other implant parameters.

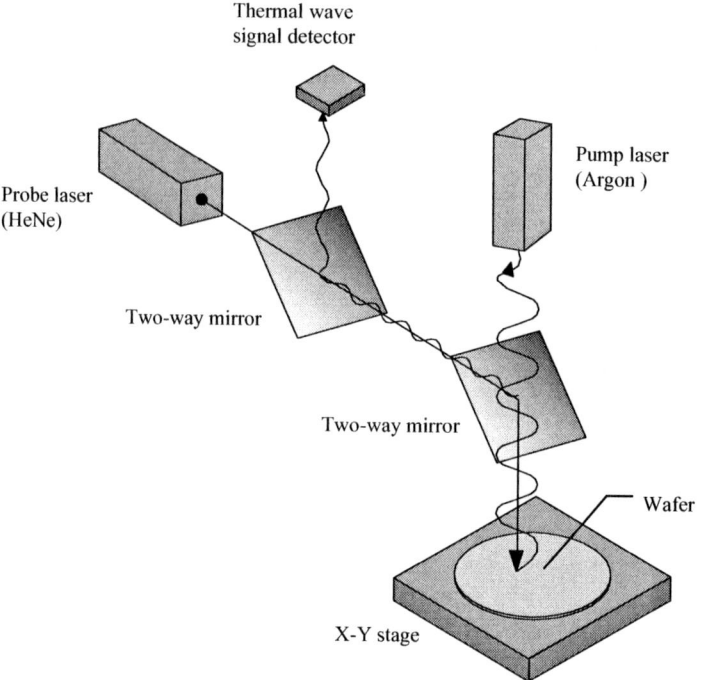

FIGURE 7.13 Thermal-Wave System for Measuring Dopant Concentration

The thermal-wave system has become common for ion-implant monitoring because of its ease of use and ability to be used on both patterned and unpatterned wafers. Its main disadvantage is that it measures damage and therefore needs calibration curves to indirectly assess dopant concentration.

Spreading Resistance Probe ■ The *spreading resistance probe* (SRP) has been a metrology tool in wafer fabrication since the 1960s and is used to measure both dopant concentration depth profiles and resistivity.[14] It is capable of profiling very shallow pn junction depths. The spreading resistance probe has two carefully aligned probes that are moved in steps along a beveled wafer surface, with the resistance between the probes measured at each step (see Figure 7.14). As the probes pass through the junction, the probes sense the change in conductivity type (n or p). The sample must be carefully prepared with a bevel angle, usually < 1°, which makes the SRP a destructive test. As the probes step through the wafer bevel, the spreading resistance, R_{sp}, of a flat circular contact of radius r on a planar surface of semiinfinite material of resistivity ρ is given by,

$$R_{sp} = \frac{\rho}{4r} \quad (\text{ohms})$$

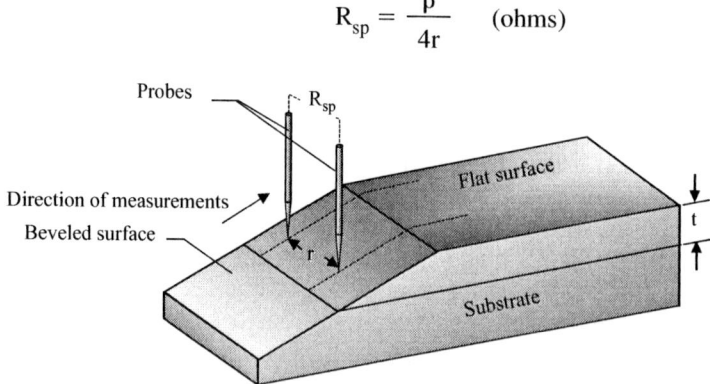

FIGURE 7.14 Spreading Resistance Probe (SRP)

The SRP employs calibration curves specific for a manufacturer's instrument to overcome a number of error sources associated with this measurement (e.g., silicon surface finish, contact resistance, tip deformation, bevel surface preparation, and so on). Computer algorithms are used to correlate the depth and resistance values to the dopant concentration at each level. The weak points of SRP are the need for skilled operators, the sample preparation, and the destructive nature of the test. Its strong point is its ability to provide accurate dopant concentration measurements with no junction-depth limitation.

Unpatterned Surface Defects

Unpatterned wafers are bare silicon wafers or wafers with various blanket films used as test wafers and incorporated in process runs to provide a source of characterization information about process conditions. The unpatterned wafer may be polished to a mirror finish or have a film that has a rougher surface. After use in a process run, the unpatterned wafer is typically cleaned and reused, which adds cost to the manufacturing process and reduces a company's profit margin. Typical defects inspected for on unpatterned wafers used for process monitoring include particles, scratches, slip lines, and other material defects.

Defect-detection equipment for wafer-surface defects falls into two general categories: darkfield and brightfield optical detection.[15] *Brightfield detection* is the traditional light source for microscope equipment—it examines the wafer surface for defects with directly reflected visible light. With brightfield detection, horizontal surfaces reflect most of the light while slanted or vertical surfaces reflect less. *Darkfield detection* examines light scattered off defects located on the wafer surface. This is done by directing light to the wafer surface at a shallow angle through the outside of the optic's objective body (see Figure 7.15). This light impinges on the wafer surface and passes back up through the center of the optics. This action renders all flat surfaces black, while irregularities appear as bright lines. This fact makes darkfield detection useful for bringing out small defects on the wafer surface that might be difficult to see with brightfield detection. An example of darkfield detection is seeing dust particles in a ray of sunshine in a dark room. Both systems usually use some form of signal or image processing to locate defects based on the light signal received from the wafer surface.

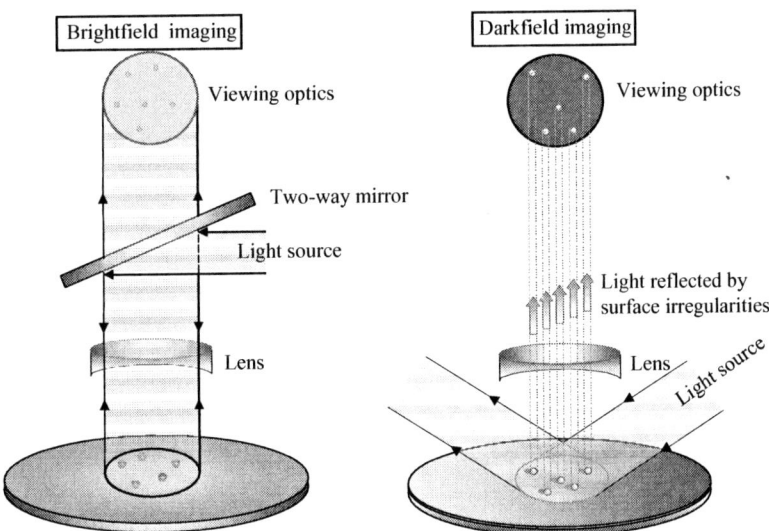

FIGURE 7.15 Darkfield and Brightfield Detection

Optical Microscopy ■ The *optical microscope,* or *light microscope,* has been used in science since the seventeenth century. For semiconductor manufacturing, it has traditionally been one of the most widely used methods of inspecting the wafer surface for defects such as particles and scratches. The optical microscope provides low-magnification views of the wafer, typically less than 1,000X magnification. Previously, optical microscopy was not capable of performing in the submicron regime. With shrinking device geometries, the size of killer defects to be detected has decreased, which has required the optical microscope to also improve over the years. Depending on the type of optical system used, particulate defects down to a size of 0.1 μm are currently able to be detected.[16] Not all layers on a chip are critical in terms of dimensions, and in many instances it is necessary to detect only gross defects. In this case, the optical microscope is quick and cost-effective.

Wafer Inspection System
(Photo courtesy of Inspex)

Optical System. An optical microscope uses light reflection to detect surface defects. The modern optical microscope is integrated into a wafer inspection station that includes robotic wafer handling and software interface with video and defect classification. A typical optical system is shown in Figure 7.16. Microscope manufacturers have achieved optical systems capable of detecting smaller objects by improving the optics and using new light sources with shorter wavelengths (the ability to detect smaller objects improves with shorter light wavelengths). The technology of optics and light are discussed in detail in Chapter 14.

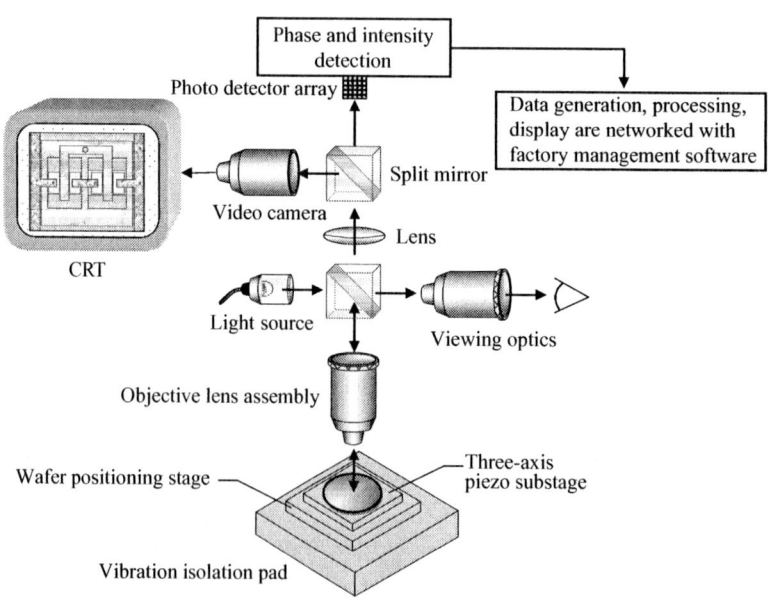

FIGURE 7.16 Schematic of Optical System

An important parameter for optical microscopy is *contrast,* which is the ability to distinguish between parts of an object. Techniques used in modern optical microscopes to enhance contrast are brightfield and darkfield detection (described on page 161), confocal contrast, and color interference contrast. A *confocal contrast microscope* uses a scanning technique to view a single point of the object at a time, thus providing better image contrast and therefore better visualization of the object (see Figure 7.17). A confocal microscope uses either visible light or laser scanning. *Color interference contrast* splits a beam of light into a direct and a reference beam. The direct beam is altered by the sample and then recombined with the reference beam, creating an image based on interference.

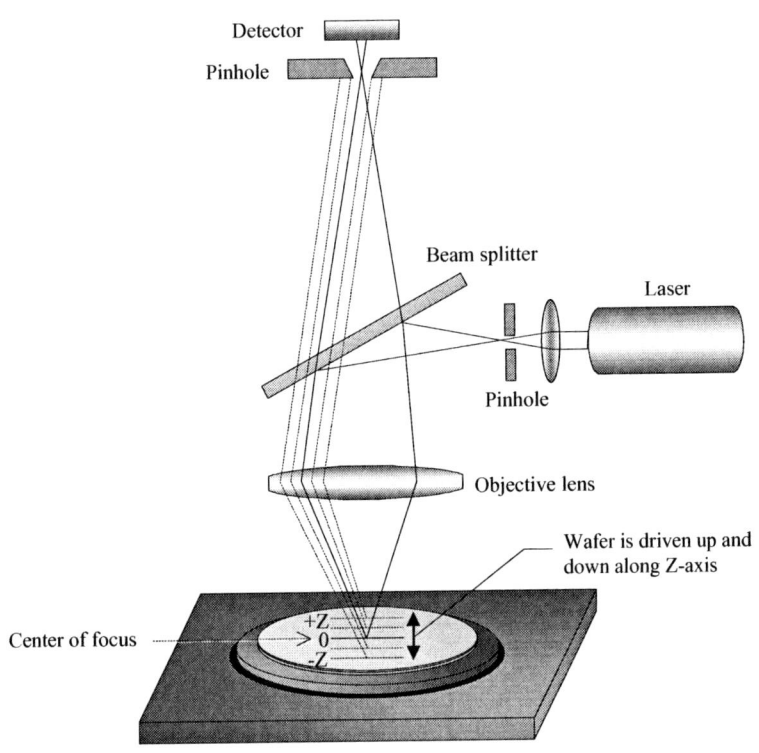

FIGURE 7.17 Principle of Confocal Microscopy

Light Scattering Defect Detection ■ In the beginning of semiconductor manufacturing, particles and surface defects were inspected by an operator using a light source and a microscope. This method was unreliable, leaving much to the subjectivity of the operator. A technique for identifying particles based on *light scattering* (also referred to as *laser scattering* or *scatterometry*) came into widespread use in the mid-1980s. This darkfield-detection method identifies surface particles and other defects by illuminating the surface with laser light and then using optical imaging to detect any light scattered by particles (see Figure 7.18 on page 164). The strength of the scattered light signal received from an individual particle depends on factors such as size, shape, composition, surrounding wafer surface (e.g., roughness and haze), and the type of equipment used. Particle diameters down to about 0.1 μm are currently capable of being detected with light-scattering techniques.[17]

Initially, when light-scattering particle detection became widespread in the mid-1980s, it could only detect particles on polished wafer surfaces. As the technology has advanced, it is now possible to detect particles on patterned product wafers and wafers with a thin film surface. As device geometries decrease, the critical particle size that creates killer defects also decreases. It was thought that the need to detect smaller particles would cause light-scattering technology to disappear, mainly because the particle size detection limit of about 1-μm was on the order of the light wavelength. When the light wavelength is the same as the particle size, it is difficult to detect particles because the scattered light intensity is not clearly defined.[18] Improvements in light scattering have been attained as sensitivity factors such as the choice of wavelength of the incident light beam are reduced. In this manner, light-scattering surface measurement has become a dominant

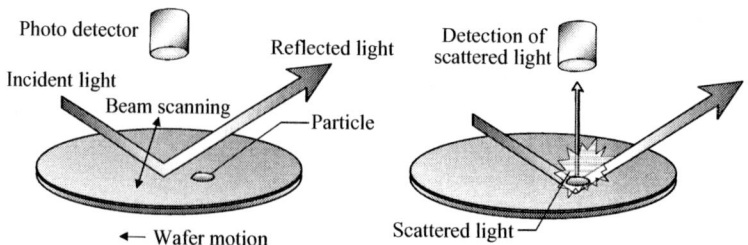

FIGURE 7.18 Particle Detection by Light Scattering

technology for tool monitoring of surface defects with both unpatterned and patterned wafers, primarily because its high-speed measurement provides rapid results during manufacturing process runs.[19]

Particles Per Wafer Per Pass. A practical aspect of measurement tools that detect wafer surface defects is the ability to create particle maps to count the particles on the wafer, showing where they occur as well as the distribution of particle diameters. A major effort in the fab is to perform *particles per wafer per pass* (PWP) measurements with unpatterned wafers to check the defect counts from process runs (see Figure 7.19). Examples of processes analyzed for PWP are photoresist spin tracks, dielectric deposition chambers, and chemical mechanical planarization (CMP) equipment. To perform a PWP procedure, a technician uses a wafer inspection tool to count the number of defects on a test wafer, runs the test wafer through the production equipment, and then recounts the number of defects. Corrective action is taken to reduce the number of defects and thereby increase the number of functional die on a wafer. It is often difficult to identify the smaller particles. In this case, the light-scattering detection equipment may work in conjunction with the scanning electron microscope (SEM) to identify particle composition for corrective action.

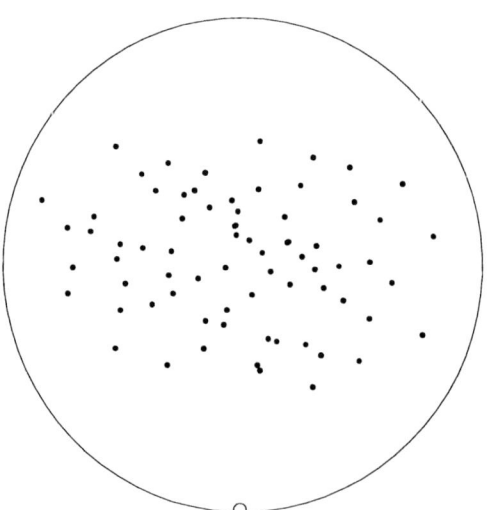

FIGURE 7.19 Particle Map

Patterned Surface Defects

Patterned wafers for fabrication metrology are regular production wafers used for in-line monitoring of surface defects. Production wafers common in IC fabrication may consist of multilayer metals with up to six or more layers that have been grown or deposited in furnaces and cluster tools. These multilayers make it challenging to distinguish between surface defects and the circuit pattern on the wafer. The majority of defects on patterned wafers are detected using light-scattering techniques. Optical microscopy (see previous section) is commonly used to detect surface defects on patterned wafers. There is also a less-common method of digital comparison, which correlates the wafer surface to a defect-free reference wafer.

Light Scattering on Patterned Wafers ■ Patterned wafers are becoming common for tool monitoring during process runs. The major defects on product wafers are particles, scratches, and pattern defects. With reduced geometries and increased process complexity, pattern defects occur more frequently. This condition makes defect detection on patterned wafers all the more critical.

Defect detection on patterned wafers with light scattering is similar to unpatterned wafers. However, the light-scattering process must be modified so that the measurement equipment can distinguish between scattered light from a particle and scattered light from a pattern edge. One technique is to use the regularity of the repeated circuit patterns to identify this particular scattered light and block it using a filter. Other patterned-wafer measurement equipment has a function that electrically cancels the cyclical signal pulses from the edge of the repeated patterns.

Critical Dimension (CD)

An important reason for critical dimension (CD) measurements is to achieve precise control over all line widths of the product being built. In CMOS technology, the transistor's gate structure is very critical. The gate width determines the channel length and the length of the channel affects speed. CD variation often indicates some instability in a critical part of the semiconductor manufacturing process. It is projected that the critical dimension (CD) for the 16-Gbit DRAM will be 0.1 μm in the year 2006.[20] To achieve this CD control, the metrology tool will need a precision and accuracy of better than 2 nm—the size of four silicon atoms side by side.[21] The measurement tool that is capable of attaining this level of metrology measurement is the scanning electron microscope (SEM).

Scanning Electron Microscope (SEM) ■ The *scanning electron microscope* (*SEM*) has been the dominant measurement tool to verify acceptable CD control in all submicron generations since the early 1990s. The first commercial SEMs were produced in the 1960s. The SEM can achieve magnifications up to 100,000 to 300,000X, which is significantly higher than optical microscopes. The resolution (smallest feature detectable) for SEMs is on the order of 40 to 50 Å. Cross sections of wafers viewed with the SEM can provide defect information. Because of the concern for controlling submicron linewidths, SEM technology was developed specifically for CD measurement in the 1980s, and is sometimes referred to as CD-SEM. The SEM measurement tool is often coupled with other analytical techniques like EDX or FIB (see the following section).

SEM Basics. The SEM is a sophisticated microscope that functions by creating a highly focused beam of electrons that scans an object while detectors measure the resulting scattered electrons.[22] It is a nondestructive and noncontact metrology tool. The SEM has an electron gun, focusing elements for shaping the electrons into a beam, and a final electrostatic-magnetic focusing system that makes the electrons strike the sample within a small, 2 to 6 nm spot (see Figure 7.20 on page 166). Since electrons have a very short wavelength (for instance, 1.22 Å for 100 eV electrons, versus about 5,500 Å for visible light), atomic-scale objects can be viewed with SEM electron imaging.

The electron gun produces a beam of electrons in a vacuum chamber of about 10^{-6} torr. It is desirable to have a high current of stable electrons with a narrow energy spread. Near the wafer, the electrons are focused into a narrow beam with a cylindrical magnetic objective lens often combined with electrostatic focusing elements to create a high-energy beam. The beam undergoes x-y deflection with an electrostatic deflector to scan the wafer. When the incident beam strikes the wafer, secondary and backscattered electrons, along with other electrons, X-rays, and photons, are emitted or transmitted due to interaction between the beam and the sample surface. The Everhart-Thornly (E-T) detector is used to collect the secondary electrons and create an electronic image that represents the sample surface. Backscattered electrons are also collected and offer superior compositional contrast between different materials.

The energy of the electron beam is directly related to the image needed. A low-energy beam (<2 keV) has a low accelerating voltage needed for nondestructive, in-line CD measurement. A high-energy electron beam (100 – 200 keV) is used to image underlying or deep structures such as contact holes. High-energy beams make it possible to nondestructively image below the surface of the wafer (say ~ 20 μm for a high-energy beam). Nevertheless, the problems with SEMs are most severe when imaging dense or deep structures, such as very fine resist lines or deep and narrow contact holes on insulator layers. This problem is mainly due to the difficulty of getting a satisfactory signal of

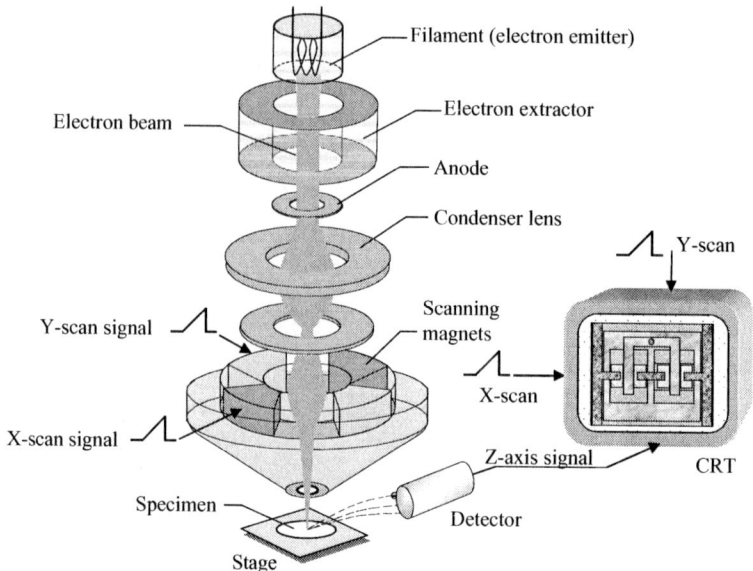

FIGURE 7.20 Simple Schematic of CD-SEM

secondary electrons from such structures. The limitations of the SEM are that it requires a high vacuum and it is necessary to coat insulating samples with conductive films prior to imaging.

CD-SEM. The *CD-SEM* has high resolution imaging with automated equipment control that rapidly evacuates the chamber to the desired vacuum, automatically positions the wafer, and uses a preprogrammed process recipe to select certain measurement sites. It is ideal for the CD-SEM to have a high wafer throughput to support production (up to 70 wafers/hour at five measured sites per wafer). The instrument can examine and measure wafers from all orientations and tilt angles of up to 60° (critical for measuring fine process patterns, sidewalls, and holes). A CD-SEM also can be used to perform defect review and analysis. Additional benefits of SEM imaging include the integration of other metrology tools, such as X-ray compositional analysis, and focused ion beam milling (see the following section).

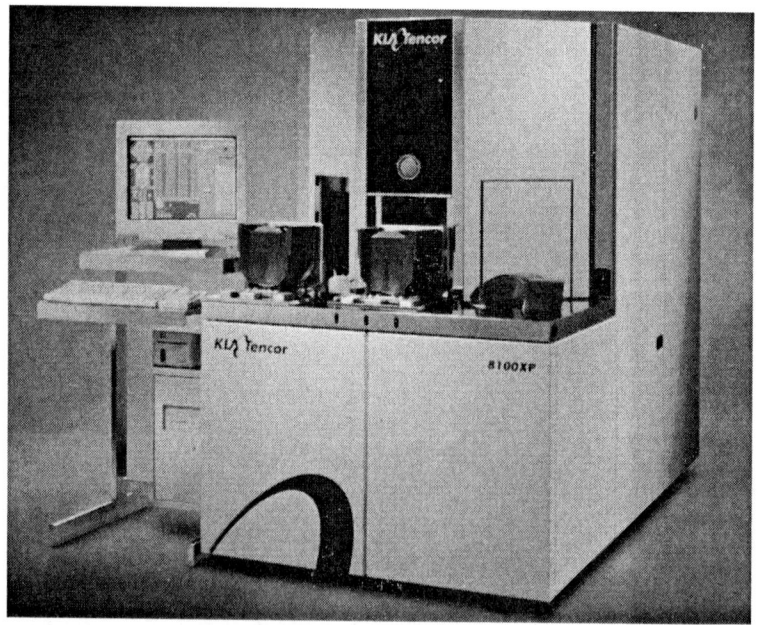

CD-SEM
(Photo courtesy of KLA-Tencor)

Step Coverage

Because of surface topography during the fabrication of wafers, the ability to achieve conformal step coverage is a desirable material property (see Figure 7.21). *Conformal step coverage* has uniform material thickness in all regions of the step, including the sidewalls and corners.

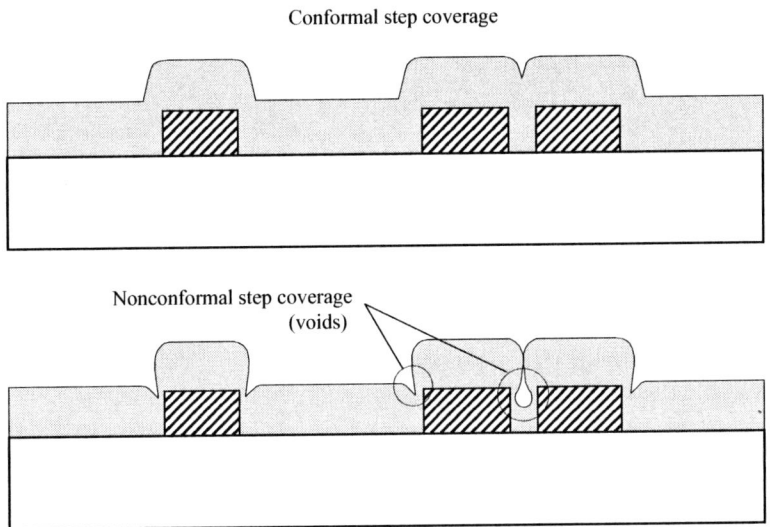

FIGURE 7.21 Step Coverage

A high resolution, nondestructive *surface profiler* with a stylus tip is often used to measure step coverage and other features on the wafer surface. This automated surface metrology tool uses a stylus that comes in contact with the wafer surface with a force as low as 0.05 mg, capable of profiling soft production-wafer films without damage to the wafer surface (see Figure 7.22). The stylus usually has a diamond tip with a radius of 0.1 μm, with older stylus having radii up to 12.5 μm.[23] Current profilers can measure wafer features as small as 0.1 μm with a 7.5 Å step height repeatability.[24]

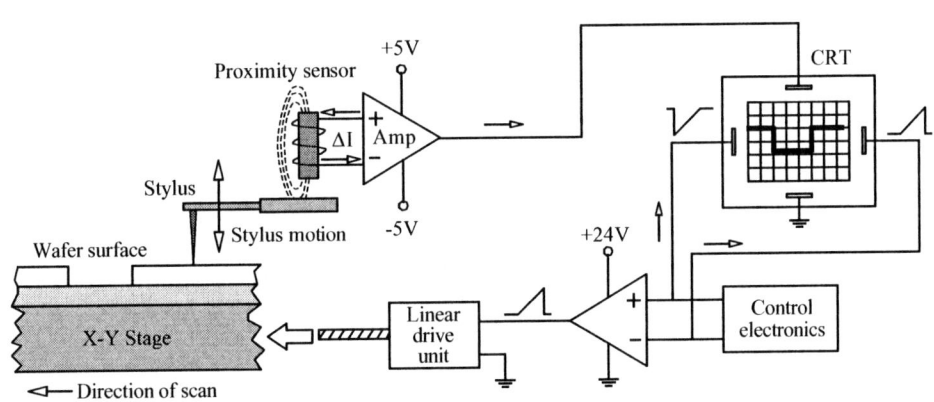

FIGURE 7.22 Surface Profiler

Overlay Registration

Overlay registration is used following the photolithography process to measure the ability of the tools and processes to accurately print photoresist patterns over a previously-etched pattern on a wafer. Due to shrinking feature sizes, the registration of the mask pattern to the wafer is a challenge because of reduced tolerances available for overlay registration. Furthermore, the increased use of chemical mechanical planarization (CMP) creates very low-contrast images on the wafer that are difficult to distinguish. This condition makes alignment of the wafer to the mask more complicated.

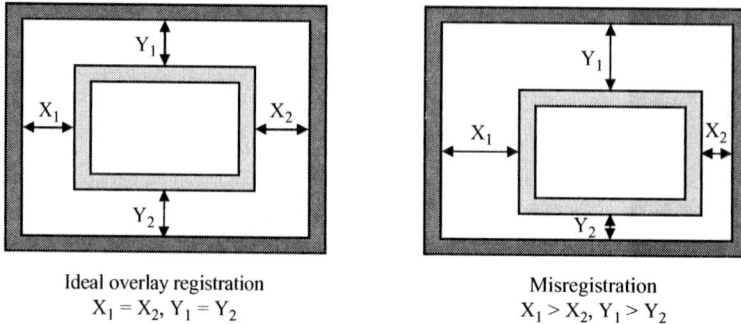

FIGURE 7.23 Overlay Registration Inspection Patterns

An automated-overlay metrology tool is used in photolithography to compare the registration of the special overlay mask pattern transferred to the photoresist with the etched overlay features on the wafer. It is important for this tool to be capable of measuring alignment targets on process layers that have low-contrast and grainy patterns in the wafer. Using brightfield reflected light is not sufficient to focus and measure images of low-contrast alignment targets on the wafer surface.

The primary method used today to measure registration of overlay targets is *coherence probe microscopy* (*CPM*). It is also referred to as a *correlation microscope*. Coherent light has waves with a definite phase relationship to one another, meaning that the individual light waves do not move relative to one another. Using coherent light permits CPM to image not only the amount of light scattered off a surface but also its phase relationship. This method is able to capture wafer surface information in the Z axis along with the wafer surface plane, which improves focusing of the wafer target and enhances the low-contrast image of a polished overlay target. Out-of-focus and diffracted signals are ignored.[25]

Capacitance-Voltage (C-V) Test

MOS device reliability is highly dependent on the thin layer of high-quality oxide in the gate structure. Contamination in the gate oxide region can lead to a shift in the threshold voltage requirement, causing device failure. Mobile ionic contaminants (MICs) and other undesirable charge conditions are detected at the SiO_2 layer after oxidation steps with a test known as *capacitance-voltage test,* or *C-V test*. The C-V test is commonly done to detect ionic contamination after oxidation steps. In addition, the C-V test provides information about gate oxide integrity (GOI), including the dielectric thickness, the dielectric constant, *k*, the resistivity of the silicon between the electrodes (to characterize majority carrier concentration), and the flatband voltage (a voltage level where there is no potential difference across the oxide structure).

The appropriate model for understanding gate oxide performance is the parallel-plate capacitor, which was explained in Chapter 2. For a MOS device, there are two capacitors in series that are functioning when the threshold voltage is applied. The first is the gate oxide, sandwiched between the doped polysilicon gate and the channel region below the gate structure. The second capacitor is formed in the silicon substrate material. This results from the attraction of charges into the gate region to form the conducting channel (known as inversion) during application of the threshold voltage. The oxide and silicon substrate are modeled as series capacitors during the C-V test (see Figure 7.24).

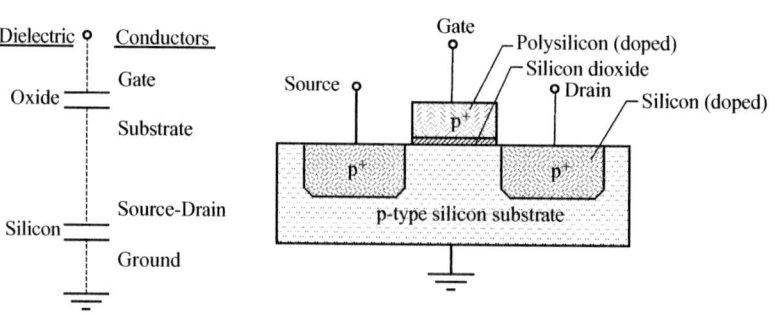

FIGURE 7.24 MOS Model of Two Capacitors at Gate Region

Since two capacitors in series have a combined lower capacitance than any individual capacitor, there is a drop in the capacitance of the gate structure when the threshold voltage is applied. This expected drop in capacitance is used in the C-V test to verify that no undesirable charges (e.g., mobile ionic contaminants) exist.

Steps for C-V Test ■ During a C-V contamination test, the two series capacitors in the gate region are modeled by using a special wafer. A variable voltage is applied between a metallized area on an oxide layer and the lightly-doped silicon beneath the oxide (see Figure 7.25).

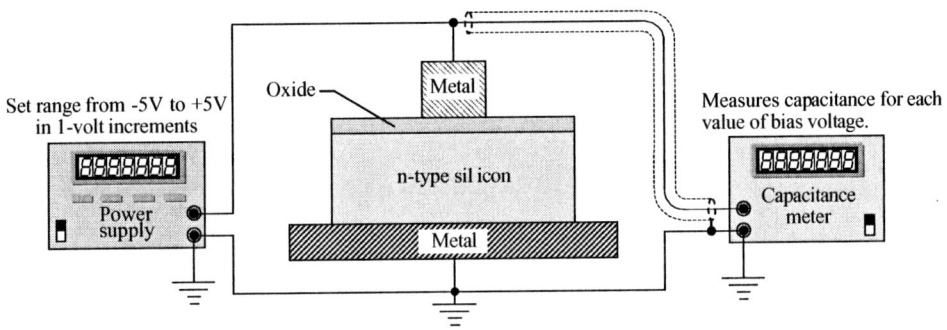

FIGURE 7.25 C-V Test Setup and Plotting

The first step for a C-V test is to apply a variable-voltage bias to both the metal contact of the oxide film under test and the lightly-doped silicon beneath the oxide. The bias is varied from positive to negative for a p-type silicon and negative to positive for n-type silicon. The objective of this first step is to deplete the silicon directly beneath the metallized area of majority carriers. In this manner, the silicon functions as a dielectric, which serves to reduce the total capacitance of the test structure because the oxide and silicon substrate are capacitors in series. The capacitance versus voltage during this test is plotted (see Figure 7.26).

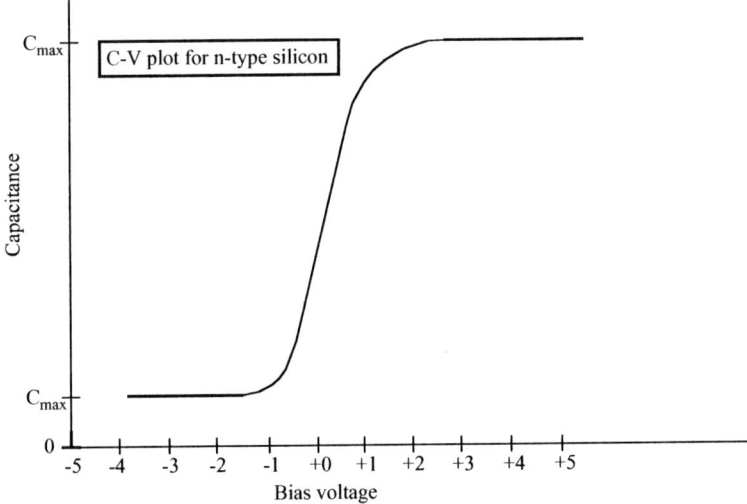

FIGURE 7.26 Capacitance versus Voltage for n-Type Silicon (First Step of C-V Test)

The second step is to apply a constant positive DC voltage to the metallized area (its magnitude depends on the oxide thickness) while heating the wafer to 300°C for five minutes followed by a cooldown prior to removing the bias (see Figure 7.27 on page 170). The elevated temperature serves to increase the mobility of the ionic contaminants. The positive voltage bias repels the positive ionic contaminants and drives them toward the oxide-silicon interface.

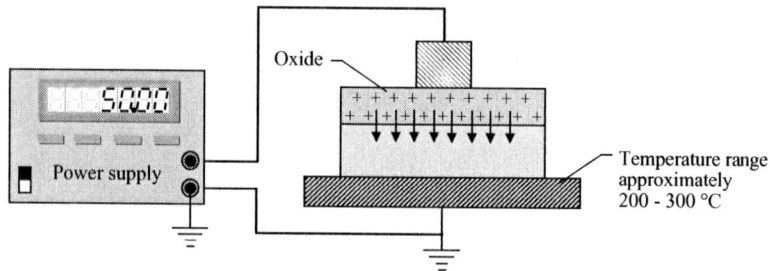

FIGURE 7.27 Ionic Charge Collection in C-V Test

The third step is to repeat the first C-V plot. However, now if there are positive ionic contaminants collected at the Si/SiO$_2$ interface, then a more negative voltage is required to get an equivalent charge in capacitance. This is a voltage shift, Vs, that is a measure of the amount of contamination in the oxide (see Figure 7.28). The magnitude of the voltage shift is proportional to the ionic contamination in the oxide, the oxide thickness, and the wafer doping. The actual amount of ionic contamination can be calculated from this plot.

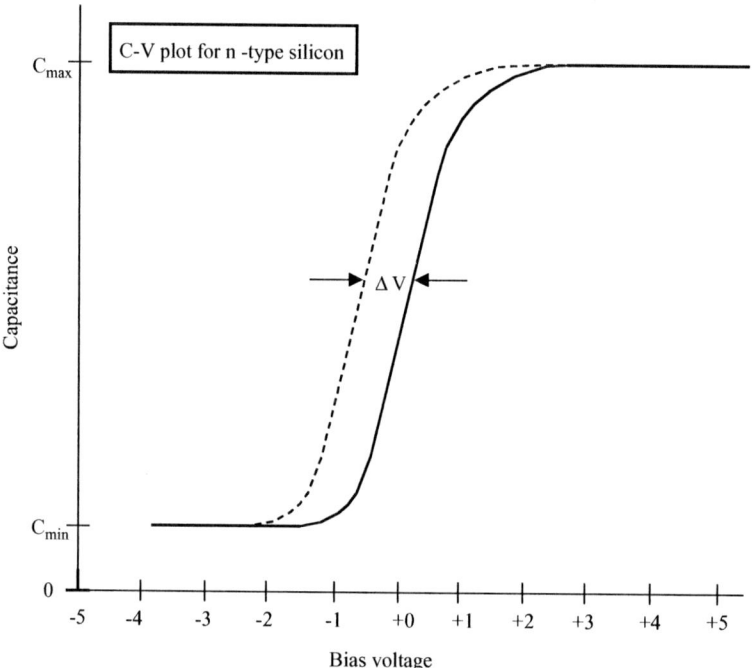

FIGURE 7.28 Voltage Shift in n-Type Silicon

The final step is to verify that this shift is due to contamination and not oxide charging. A negative voltage is applied while heating the substrate to move the ionic contaminants away from the Si/SiO$_2$ interface. Performing the C-V test again should produce the original plot, verifying that it was actually the ionic contaminants that produced the voltage shift.

Benefits of C-V Test ■ The C-V analysis is used to monitor the wafer to ensure that the cleaning process removes ionic contaminants. This analysis is done either on fabricated test wafers or production wafers with a special test structure (test structures are described in Chapter 19). The analysis cannot tell you where the contamination comes from, such as the wafer surface, cleaning step, equipment maintenance, or furnace process. The C-V test is usually a part of any evaluation for process changes. In this case, wafers are tested in two groups. One group undergoes normal processing prior to C-V testing while the other group follows the new process and then is C-V tested. The results are compared to verify that the proposed process is acceptable. There is a small amount of voltage shift that is acceptable, depending on the oxide thickness and the sensitivity of the measurement equipment.

Metrology and Defect Inspection

Contact Angle

Contact-angle meters are used to measure adhesion of liquids to the wafer surface and to calculate surface energies or adhesion tension. This measurement characterizes wafer surface parameters such as wettability, cleanliness, finish, and adhesion (see Figure 7.29). The contact (tangent) angle formed between a liquid drop and its supporting surface is relative to the forces at the liquid/solid or liquid/liquid interface and can be used as a wafer-test specification or as a quality characteristic. Both direct-angle measurements and indirect dimensional measurements methods are employed to obtain highly accurate and repeatable contact angle measurements.

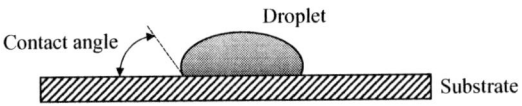

FIGURE 7.29 Contact Angle

ANALYTICAL EQUIPMENT

This section provides an overview of the major analytical equipment used to support wafer fabrication. These analytical tools provide highly accurate wafer measurements and are typically located in an off-line lab to support production problems. Figure 7.30 shows when some of these instruments were first used and how important each one is or is projected to become in either process development or manufacturing. The analytical equipment reviewed includes:

- Secondary-ion mass spectrometry (SIMS)
- Time of flight secondary-ion mass spectrometry (TOF-SIMS)
- Atomic force microscope (AFM)
- Auger electron spectroscopy (AES)
- X-ray photoelectron spectroscopy (XPS)
- Transmission electron microscope (TEM)
- Energy- and wavelength-dispersive spectrometer (EDX and WDX)
- Focused ion beam (FIB)

	Year							Importance to Manufacturing
	1950s	1960s	1970s	1980s	1990s	2000s	2010s	
AES			Development Manufacturing					Useful
AFM					Research, Development			Useful
FIB					Research, Development, Manufacturing			Critical
SEM		Research, Development, Manufacturing						Critical
SIMS		Research, Development, Manufacturing						Critical
TEM		Research, Development, Manufacturing						Critical
TOF SIMS					Research, Development			Useful
XPS			Development					Useful

FIGURE 7.30 Relative Importance of Analytical Equipment

Secondary-Ion Mass Spectrometry (SIMS)

Secondary-ion mass spectrometry (*SIMS*) is a method of eroding a wafer surface with ions accelerated in a magnetic field to analyze the surface material composition. These ions strike the surface of the wafer and dislodge, or sputter, other ions, some of which are known as secondary ions (see Figure 7.31 on page 172). Secondary ions contain the wafer material and any impurities such as dopants. They are collected and analyzed in a vacuum chamber with a mass spectrometer that can

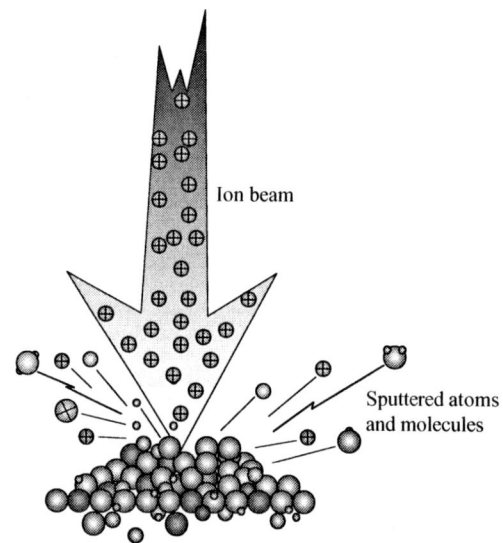

FIGURE 7.31 Ion Beam Sputtering of Surface Material

identify the type of dopant and its concentration in the silicon.[26] SIMS is essentially a destructive measurement technique and requires an ultrahigh vacuum environment (around 10^{-10} to 10^{-12} torr).

Recall that ions are charged particles, able to be accelerated or deflected in a magnetic field. When accelerated ions strike a wafer surface, this ion-beam sputtering process involves violent collisions at the atomic level. This action could be compared to a sandblaster, which propels sand with such energy that it removes surface pieces but leaves behind craters in the base material. However, with an ion beam, there are also chemical reactions that take place during the collision, creating many complex molecules in the sputtered material. The particular incident ion could be nearly any element, but is typically chosen as Cs^+, O_2^-, O_2^+ or Ar^+ and is selected based on the collision chemistry that will best produce secondary ions.

SIMS Tool Description ■ A common method to produce ions used in SIMS is the duoplasmatron (see Figure 7.32). A gas, commonly O_2 or Ar, is converted into a plasma by a low-pressure arc formed between a hot filament and an anode. An extraction electrode is used to draw charged atoms from the plasma.

Cs^+ ions are produced by a different method. A pellet of cesium chromate is vaporized in a heated reservoir. The vapor feeds into a hot (~1100°C) and porous tungsten plug in an ionization chamber. The vapor diffuses through the pores of the plug and becomes ionized, followed by an extractor that collects the ions and forms them into an ion current.

After the ions are produced, they go through an analyzer magnet to select the ions of interest (e.g., Cs^+ or O^-). These ions are then either focused into a small spot with a magnet, known as an ion microprobe, or used to flood the surface of the sample. The latter is referred to as an ion microscope and is the most common technique today. With an ion microscope, the sputtered secondary ions are collected simultaneously from many points on the surface and then identified by analyzing their mass-to-charge ratio in the mass spectrometer. A common mass spectrometer, the quadrupole, is based on an oscillating electric field that separates ions into defined oscillations based on their mass, permitting certain ions to pass through an aperture and be identified.

The rate of material removal using SIMS can range from a high sputter rate, known as *dynamic SIMS* and often used for ion implanter characterization, to a slow sputter rate of a single monolayer of atoms every few hours. The latter is referred to as *static SIMS,* and is frequently used to analyze contamination in thin oxide and nitride films. The SIMS measurement tool is sensitive enough to measure ppba (parts per billion–atomic) impurities in a junction or contact that is only 1 to 10 μm across. It has become the primary measurement tool used to verify the performance of ion implanters because it can characterize the dose and depth of the junction plus reveal any undesirable metal impurities at the junction.[27]

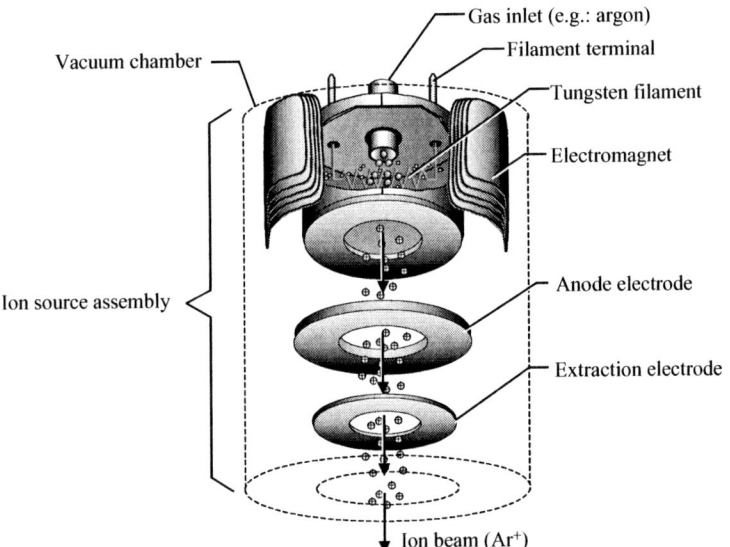

FIGURE 7.32 Ion Production in a Duoplasmatron

Time of Flight SIMS (TOF-SIMS) ■ A limiting factor for SIMS is that the mass spectrometer only detects around 0.001% of all the ions emitted from the sample. Another approach, referred to as *time of flight SIMS* (*TOF-SIMS*), can detect as many as 10 to 50% of the ions emitted by the sample. The TOF-SIMS identifies ions of different mass by measuring the time it takes for an ion to travel the length of a fixed path, since the charged particle velocity is a function of mass (see Figure 7.33). This fact makes it possible to reduce the incident beam current by a factor of up to 10^5 relative to a SIMS with a quadrupole spectrometer. The rate of material removal is so low that only a fraction of a single-surface monolayer is removed in an hour. For this reason TOF-SIMS is essentially nondestructive and is ideal for very thin films on the wafer surface.[28]

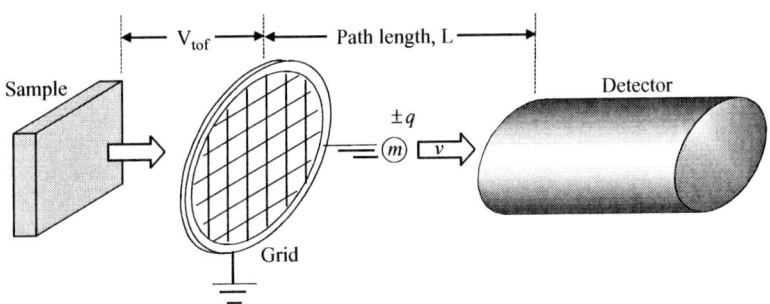

FIGURE 7.33 TOF-SIMS Mass Spectrometer Principle

Atomic Force Microscope (AFM)

The *atomic force microscope* (*AFM*) is a surface profiler that scans a small, counterbalanced tip probe over the wafer to create a 3-D surface map.[29] It was first demonstrated in 1986 and uses optical technology, direct tip contact, and a laser to sense the position of the tip relative to the surface (see Figure 7.34 on page 174). Probe and surface separation is so small (on the order of 2 Å) that atomic forces affect the probe between the surface and the tip. The tip geometry is critical and must be characterized to provide accurate measurements. AFM measurements are exceedingly slow, which is a problem for in-line measurements in a production environment.

There are several variations of AFM technology.[30] In the simplest system, a laser beam reflects off the top surface of the probe tip and is directed to a photodiode. When the surface topography moves the probe, it changes the position of the laser in the photodiode, creating an electronic map of the surface. A more advanced method has the probe oscillate very near the surface. Atomic van der Waals forces interact with the vibrating tip, inducing phase shifts in the vibration that are detected by an imaging system. The benefit of this approach is that it can measure vertical

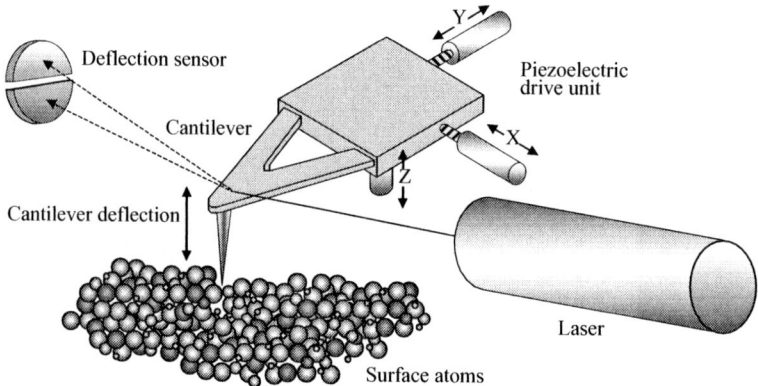

FIGURE 7.34 Schematic of Atomic Force Microscope

photoresist profiles. AFM is a recent measurement technology and is considered a promising alternative to SEMs.

Auger Electron Spectroscopy (AES)

Auger electron spectroscopy (*AES*) measures the energy of auger electrons emitted by the surface of a sample when struck by an incident electron beam. It is very sensitive to the surface, with a depth of only 10 to 50 Å. Auger (pronounced ōzhā´) electrons were discovered in 1923 by Pierre Auger in France, and make up a very small percent (< 0.1%) of the total electrons produced in a sample (secondary electrons are the dominant species). The energy associated with Auger electrons provides a distinct link back to the parent atom that is used for identification of the sample elements.

Because Auger electrons are easily absorbed by the sample, only those Auger electrons on the outer monolayers of the surface escape and are detected in AES. This makes Auger technology especially suited for analyzing the surface of materials, usually to a depth of about 2 nm. Oxides of metals, silicides, and wafer surfaces are readily sensed by Auger spectroscopy. Furthermore, the Auger spectrometer uses a highly focused electron beam, with the smallest beam spots on the order of 12 nm, which is beneficial in microcircuit analysis. AES requires an ultrahigh vacuum environment in order to reduce contaminant formation on the sample.

X-Ray Photoelectron Spectroscopy (XPS)

X-ray photoelectron spectroscopy (*XPS*) is primarily used to identify chemical species on a sample surface, analyzing to a sample depth of about 2 nm (equivalent to AES). All elements are detectable except hydrogen and helium (their detection requires a good spectrometer). In XPS, X-ray photons are directed toward the wafer surface and interact with certain core-level electrons referred to as XPS electrons (see Figure 7.35). If the X-ray energies exceed the XPS electrons' binding energy, then the electrons are emitted from the sample. Because the binding energy of an electron is influenced by its chemical surroundings, XPS identifies both the element and its chemistry.[31]

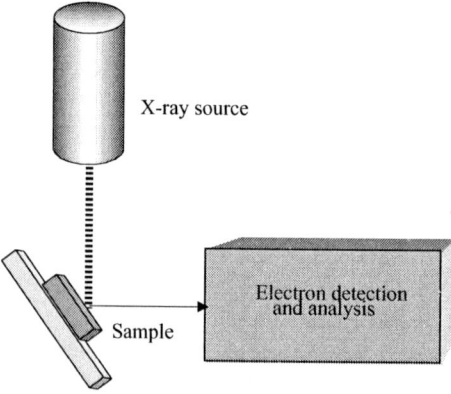

FIGURE 7.35 Schematic of XPS Measurement

Transmission Electron Microscope (TEM)

The principle of operation for the *transmission electron microscope* (*TEM*) is similar to an SEM, with the major difference that the beam of electrons is transmitted through an ultrathin slice of the sample (on the order of 10 to 100 nm thick). Based on factors such as the electron wavelength, accelerating voltage and specimen thickness, an image is formed and magnified on a screen with a resolution of about 2 Å (see Figure 7.36).

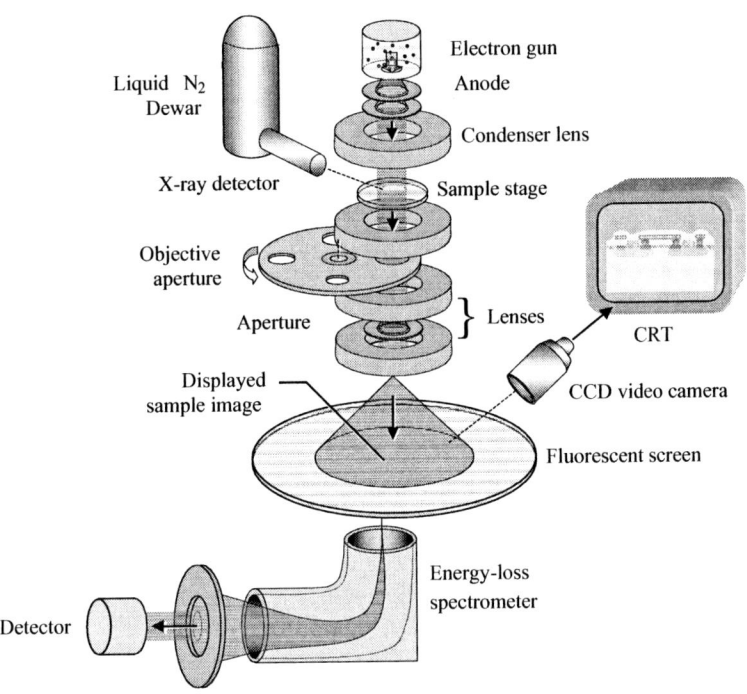

FIGURE 7.36 Schematic of TEM

TEM is the only metrology tool capable of quantifying some very small features on a wafer. For instance, silicon crystal point defects, such as single dislocations introduced into an active junction by ion implant (which leads to junction leakage), can be imaged on an atomic scale by TEM. Table 7.3 outlines some process areas of wafer fabrication that benefit from TEM analysis.[32] The most difficult aspect of TEM is sample preparation, with a wide array of techniques used, such as mechanical polishing, chemical etching, and ion-beam milling.

TABLE 7.3 Sample TEM Applications in Semiconductor Manufacturing

	TEM Measurement
Silicon Material	Dislocation and stacking fault densities in Si shallow junctions
Patterning	Precise sidewall profile in polysilicon and metal structures
Metallization	Characterization of metal silicides and alloys
Implantation	Surface and buried implant damage
Contamination	Thin organic and oxide films at submicron contact interfaces

Energy- and Wavelength-Dispersive Spectrometer (EDX and WDX)

A useful signal generated by the incident electron beam of an SEM is the characteristic X-ray produced by the energetic primary electrons emitted by the sample material. These X-rays carry information related to the atom species present in the sample, permitting the identification of a variety of thin films, particles, and defects created in semiconductor manufacturing. X-ray detectors are found on nearly all electron microscopes today.

The *energy-dispersive spectrometer* (*EDX*) is the most widely used X-ray detection method for identifying elements and is complementary to SEM. This is primarily because the EDX can quickly detect X-rays of all energies simultaneously, which means it is a relatively fast measurement. However, because it penetrates well beyond the sample surface, it should not be considered a surface analysis. EDX operation is based on a large diode made from a high-quality doped Si crystal and isolated from the SEM vacuum chamber by a thin (~25 μm) beryllium window (see Figure 7.37). An X-ray passing through the window produces a series of electron-hole pairs, which are electronically detected and identified back to the energy level of the X-ray. EDX can acquire all energy peaks and make a spectrum plot within a few minutes.[33]

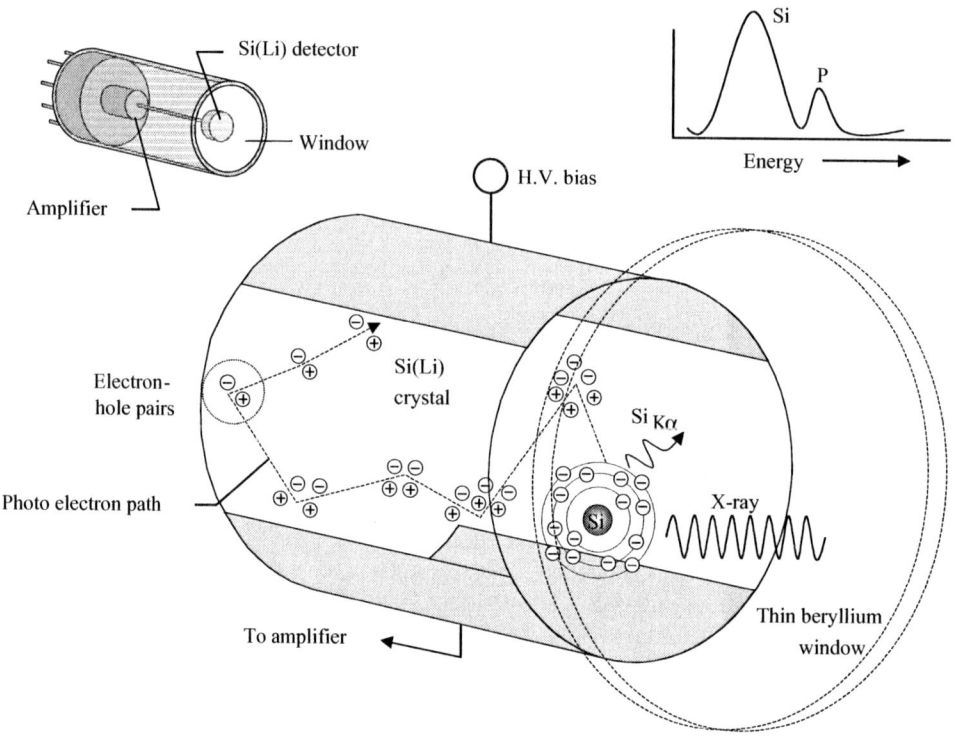

FIGURE 7.37 Energy-Dispersive Spectrometer (EDX)

The *wavelength-dispersive spectrometer* (*WDX*) operates based on a diffracting crystal and photo counter. The crystal separates and disperses the incoming X-rays by wavelength, which are then collected by a photo counter detector. The appropriate crystal is selected and placed in the path of the X-rays. Critical parameters are X-ray wavelength, lattice spacing of the crystal, incident angle, and order of reflection. WDX is a slow measurement but yields superior measurements with high accuracy.

Focused Ion Beam (FIB)

Traditionally, and often still today, a wafer is analyzed by removing it from the cleanroom to a lab where it is cross-sectioned and examined on an SEM. This process started changing when *focused ion beam* (*FIB*) systems became prevalent in the early 1990s because of their convenient cross-sectioning capabilities. The FIB is a destructive technique similar to an SEM in design and operation, except that the primary beam is made of Ga⁺ ions instead of electrons. These ions are focused through a set of lenses into a small spot. At the location where they impact the wafer, atoms are ejected into the vacuum, creating a small void of precisely controlled shape and depth in the sample material. This precision gives the FIB the possibility of making cross sections in specific locations of a wafer. The FIB technique is, however, a time-consuming process.

Focused ion beam (FIB) milling is capable of carving a 10 to 100-nm thick cross section from any area on a wafer inside the cleanroom (see Figure 7.38). It can cut through metal, polysilicon, oxide, and nitride layers with little or no damage to adjacent structures. Typically a high-current beam is made for the initial cut and is followed by a low-current beam with a tighter focus for final sample polishing. FIB can make the thin samples necessary for TEM. For this reason, equipment suppliers make custom FIB designs specifically for this application.

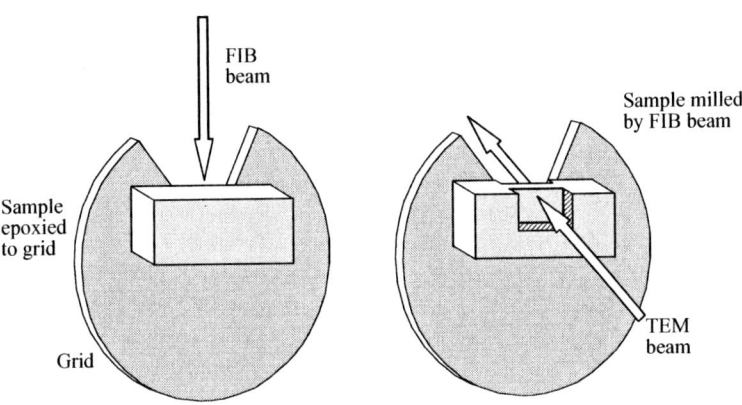

FIGURE 7.38 Focused Ion Beam (FIB) Milling

SUMMARY

IC metrology measures the performance of fabrication processes, typically with monitor or production wafers. Metrology tools are either stand-alone or integrated into the process tool, and they often use advanced defect-analysis software. Specified quality measures are used at different process locations to verify acceptable performance. Film thickness is most often measured with the four-point probe sheet resistance (opaque films), ellipsometry (transparent films), and reflection spectroscopy. Lesser-used methods are X-ray and photoacoustic. Film stress on a wafer is determined by measuring the wafer curvature. The refractive index of a transparent substance is measured using ellipsometry. The dopant concentration is measured using four-point probe, infrared interference, and spreading resistance probe. Surface defects are measured with unpatterned or patterned wafers, with the latter becoming more cost-effective. The optical microscope is used, with advanced features to improve wafer-surface contrast. Light scattering is often used to detect particles on the wafer surface, with contour maps generated to highlight the defects. The recent trend is for more surface defect analysis to be done with patterned wafers. Critical dimensions are measured using the scanning electron microscope (SEM). Enhanced techniques are used to measure overlay registration. The capacitance-voltage test provides information on the quality of the gate oxide structure, primarily to ensure the oxide is contamination-free. The contact angle verifies acceptable wafer surface quality. Sophisticated analytical tools are used to characterize the wafer and verify acceptable process performance.

KEY TERMS

metrology
defects
defect density
monitor wafers (test wafers)
yield
stand-alone measurement tool
integrated measurement tool
automatic defect classification (ADC)
quality measures
film-thickness metrologies
sheet resistance, R_s
sheet resistivity, ρ_s
four-point probe
Van der Pauw method
contour maps
ellipsometry
variable-angle spectroscopic ellipsometry (VASE)
spectroscopic ellipsometry
reflection spectroscopy (reflectometry)
X-ray fluorescence (XRF)
Total-reflection XRF (TRXRF)
photoacoustics
film stress
refraction
dopant concentration (dose)
thermal-wave system
spreading resistance probe
unpatterned wafers
brightfield detection
darkfield detection
optical microscope (light microscope)
contrast
confocal contrast microscope
color interference contrast
light scattering (laser scattering or scatterometry)
particles per wafer per pass (PWP)
patterned wafers
scanning electron microscope (SEM)
CD-SEM
conformal step coverage
surface profiler
overlay registration
coherence probe microscopy (CPM)
correlation microscope
capacitance-voltage test (C-V test)
contact angle meters
secondary ion mass spectrometry (SIMS)
dynamic SIMS
static SIMS
time of flight SIMS (TOF-SIMS)
atomic force microscope (AFM)
Auger electron spectroscopy (AES)
X-ray photoelectron spectroscopy (XPS)
transmission electron microscope (TEM)
energy-dispersive spectrometer (EDX)
wavelength-dispersive spectrometer (WDX)
focused ion beam (FIB)

REVIEW QUESTIONS

1. What is metrology? What is the purpose of metrology in IC fabrication?
2. Define a defect. What is meant by wafer defect density?
3. Discuss the difference between monitor wafers and production wafers for metrology measurements.
4. What is the difference between a stand-alone measurement tool and an integrated measurement tool?
5. Describe automatic defect classification (ADC), and explain how this analysis can improve metrology.
6. Define semiconductor quality measures. List twelve different quality measures used in IC fabrication. State the process areas where the different quality measures are used.
7. List the two general areas of film-thickness metrology.
8. State the formula for sheet resistance for a thin film. Explain how sheet resistance is interpreted for a square sample. What are the units for sheet resistance?
9. What is the formula for the sheet resistivity of a thin film?
10. Describe the four-point probe and give its benefits for measuring sheet resistance.
11. What is the Van der Pauw method?
12. Explain what a contour map is.
13. Explain the basic principles of ellipsometry. What are the benefits for using ellipsometry to measure thin film thickness?
14. Explain what VASE is. What applications does VASE offer improved ellipsometry?

15. Describe spectroscopic ellipsometry.
16. Describe reflection spectroscopy. What is a dual-beam spectrometer?
17. How is film thickness measured using X-rays? What does XRF stand for? What is total reflection XRF?
18. Describe how photoacoustic technology is used for film-thickness measurement.
19. How is thin film stress measured on a wafer?
20. How is refraction measured on a wafer? What does a variation in the refractive index indicate?
21. Can the four-point probe be used to characterize dopant concentration in a film?
22. Explain the principle of thermal wave for measuring dopant concentration.
23. Describe the spreading resistance probe method for dopant concentration.
24. What is brightfield detection? What is darkfield detection?
25. Describe the general state of optical microscopy for wafer-surface defect detection.
26. What is contrast? Discuss two methods to improve inspection contrast on the wafer surface.
27. Explain how light scattering can be used to detect surface defects.
28. Explain particles per wafer per pass (PWP).
29. What is the dominant measurement tool for wafer critical dimension?
30. Explain the basic operation of an SEM.
31. What is a CD-SEM?
32. How is step coverage often measured?
33. What is overlay registration? State and discuss the primary technique for measuring overlay registration.
34. Why is the capacitance-voltage test done? Describe four steps necessary to perform this test.
35. State the purpose of the contact-angle metrology measurement.
36. Describe secondary-ion mass spectrometry (SIMS).
37. What is TOF-SIMS and in what condition is it desirable to use?
38. Explain atomic force microscopy (AFM).
39. Describe the technique for using the Auger electron spectroscope.
40. What is X-ray photoelectron spectroscopy?
41. Explain transmission electron microscopy.
42. Explain the difference between EDX and WDX.
43. Describe focused ion beam (FIB) milling and explain its benefits.

METROLOGY EQUIPMENT SUPPLIERS' WEB SITES

Applied Materials	http://www.appliedmaterials.com/products/
Carl Zeiss Microelectronics	http://www.zeiss.de/
Cerprobe Corp.	http://www.cerprobe.com
FEI Company	http://www.feic.com
Gaertner Scientific Corp.	http://www.gaertnerscientific.com/
Hitachi	http://www.hitachi.com/semiequipment/products.html
Inspex	http://www.inspex.com/
International SEMATECH	http://www.sematech.org/public/index.htm
JA Woollam Co. Inc.	http://www.jawoollam.com/
JEOL	http://www.jeol.com/
Kaman Instrumentation	http://www.kamaninstrumentation.com/
Keithley Instruments	http://www.keithley.com/
Kernco Instruments Co.	http://www.kerncoinstr.com/cam.htm
Kevex Spectrace	http://www.kevexspectrace.com/
KLA-Tencor	http://www.kla-tencor.com/splash.html
Leica	http://www.leica.com/
Leybold Inficon Inc.	http://www.leyboldinficon.com/
The Micromanipulator Co. Inc.	http://www.micromanipulator.com/
Nanometrics	http://www.nanometrics.com/
National Institute of Standards	http://www.nist.gov/
Nicolet Instruments	http://www.nicolet.com/
Nikon	http://www.nikonusa.com/
Olympus America Inc.	http://www.olympus.com/
Perkin-Elmer	http://www.perkinelmer.com/
Rudolph	http://www.rudolphtech.com/home/
Schlumberger	http://www.1.slb.com/ate/diagsys
SEMI	http://www.semi.org/

Sonoscan Inc.   http://www.sonoscan.com/
Therma-Wave   http://www.thermawave.com/index.htm
Veeco Instruments   http://www.veeco.com/

REFERENCES

1. S. Butler, "Process Control Through Integrated Metrology," *Solid State Technology* (January 1999): p. 34.
2. C. W. Pearce, "Crystal Growth and Wafer Preparation," *VLSI Technology,* 2nd ed., ed. S. Sze, (Boston: McGraw Hill, 1988), p. 300.
3. W. Runyan and T. Shaffner, *Semiconductor Measurements and Instrumentation,* 2nd ed., (New York: McGraw-Hill, 1998), p. 117.
4. J. C. C. Tsai, "Diffusion," *VLSI Technology,* 2nd ed., ed. S. Sze, (Boston: McGraw-Hill, 1988), p. 300.
5. Semiconductor Industry Association, *The National Technology Roadmap for Semiconductors: Technology Needs,* (San Jose, CA: SIA 1997), p. 80.
6. P. Burggraaf, "Thin Film Metrology: Headed for a New Plateau," *Semiconductor International* (March 1994): p. 58.
7. R. DeJule, "Advances in Thin Film Measurements," *Semiconductor International* (May 1998): p. 64.
8. C. Morath et al., "Ultrasonic Multilayer Metal Film Metrology," *Solid State Technology* (June 1997), p. 85.
9. M. Dax, "X-Ray Film Thickness Measurements," *Semiconductor International* (August 1996), p. 98.
10. R. DeJule, "Advances in Thin Film Measurements," *Semiconductor International* (May 1998): p. 56.
11. B. El-Kareh, *Fundamentals of Semiconductor Processing Technologies,* (Boston: Kluwer Academic, 1995), p. 371.
12. Semiconductor Industry Association, *National Technology Roadmap,* p. 80.
13. E. Rimini, *Ion Implantation: Basics to Device Fabrication* (Boston: Kluwer Academic, 1995), p. 70.
14. D. Schroder, *Semiconductor Material and Device Characterization,* 2nd ed., (New York: Wiley, 1998), pp. 31–35.
15. J. Baliga, "Defect Detection on Patterned Wafers," *Semiconductor International* (May 1997): p. 64.
16. A. Braun, "Defect Detection Overcomes Limitations," *Semiconductor International* (February 1999): p. 52.
17. A. Braun, "Defect Detection and Review Enter New Era," *Semiconductor International* (May 1998): p. 61.
18. W. Runyan and T. Shaffner, *Semiconductor Measurements and Instrumentation,* 2nd ed., p. 227.
19. A. Braun, "Defect Detection and Review Enter New Era," *Semiconductor International* (May 1998): p. 66.
20. Semiconductor Industry Association, *National Technology Roadmap,* p. 14.
21. M. Davidson and A. Vladar, "The Physics of Metrology Instruments," *Solid State Technology* (June 1998): p. 135.
22. Material for description of CD-SEM taken from M. Davidson and A. Vladar, "Physics of Metrology Instruments," pp. 136-142.
23. A. Braun, "Analytical Techniques for Process Problem Solving," *Semiconductor International* (October 1997): p. 112.
24. KLA-Tencor, Product literature for *High Resolution Profiler, KLA-Tencor HRP-220,* (August 1998).
25. KLA-Tencor, Product literature for *Automated Overlay Metrology Tool, KLA-Tencor 5200 XP,* (May 1998).
26. W. Runyan and T. Shaffner, *Semiconductor Measurements and Instrumentation,* 2nd ed., p. 88.
27. Ibid., p. 402.
28. Ibid., p. 418.
29. K. Wilder, B. Singh, and W. Arnold, "Novel In-Line Applications of Atomic Force Microscopy," *Solid State Technology* (May 1996), p. 109.
30. M. Davidson and A. Vladar, "Physics of Metrology Instruments," p. 144.
31. D. Schroder, *Semiconductor Material and Device Characterization,* 2nd ed., p. 701.
32. W. Runyan and T. Shaffner, *Semiconductor Measurements and Instrumentation,* 2nd ed., p. 332.
33. Ibid.

CHAPTER 8
GAS CONTROL IN PROCESS CHAMBERS

Semiconductor manufacturing is a cyclic repetition of several major process steps. Many fabrication processes involve chemical reactions that take place in process chambers. The driving force for these chemical reaction processes is to optimize the required chemical reactions by introducing the correct chemicals in the proper environment (e.g., vacuum) while providing energy to drive the reaction. At the same time, detrimental aspects of the reaction are minimized, such as exposure to moisture, the ambient environment, and contaminants. This optimum condition is reached by carefully introducing the necessary mix of precursor chemicals into the process chamber, often as a gas, and then monitoring the chemical reaction to achieve the desired conditions on the wafer surface.

OBJECTIVES

After studying the material in this chapter, you will be able to:

1. Explain why process chambers are used in semiconductor manufacturing.
2. Describe the benefits of a vacuum, the vacuum ranges, and appropriate pumps.
3. Explain the need for gas flow in process chambers, and describe how it is controlled.
4. Explain what an RGA is and why it is beneficial in process chambers.
5. Describe what plasma is and how it is obtained.
6. Discuss the effects of contamination in process chambers and explain how to minimize it.

INTRODUCTION

During the beginning of the semiconductor industry, only two wafer fabrication steps required a vacuum chamber for processing. A vacuum was used for the evaporation of aluminum for the one and only metal layer and for the evaporation of gold on the back of the silicon wafer so that the circuit die could be mounted in its transistor package.[1] Vacuum processes in those early days occurred in a chamber known as a bell jar (see Figure 8.1 on page 182).

Present-day wafer processing often requires chemical reactions that take place in process chambers. The *process chamber* is a controlled vacuum environment where intended chemical reactions occur under controlled conditions. For this reason, a process chamber for chemical reactions is sometimes referred to as a *reactor*. Process chambers have many functions:

- Controlling how gas chemicals flow into and react in the chamber in close proximity to the wafer.
- Maintaining a prescribed pressure inside the vacuum environment.
- Removing undesirable moisture, air, and reaction by-products.
- Creating an environment for chemical reactions such as plasma to occur.
- Controlling the heating and cooling of the wafer.

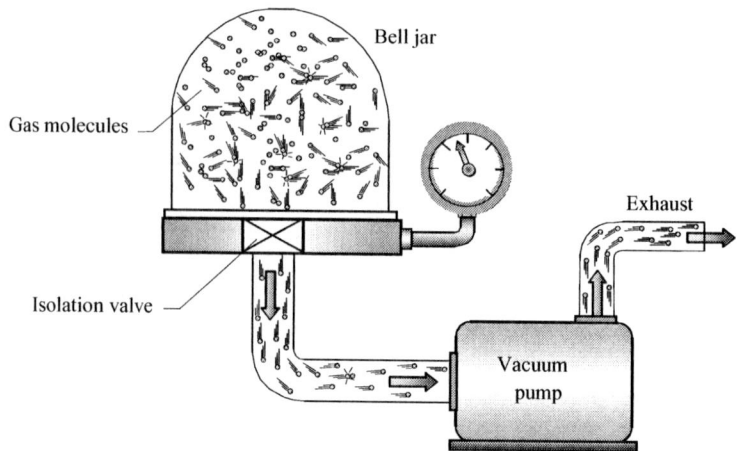

FIGURE 8.1 Early 1960s Vacuum Bell Jar

Gas is typically the material state used to transport the various chemicals needed in the process chambers. Gas flows into the process chamber as a result of the pressure differential between the supply system and the vacuum conditions in the chamber. Sometimes solid materials are used in the process chamber. An example is the solid metal target used as a source of material during sputtering (see Chapter 12).

Since the late 1980s, process chambers have been configured in the form of a *cluster tool*. The multiple process chambers are clustered around a central transfer chamber with a wafer transport system that is typically a robot arm (see Figure 8.2). This equipment design permits integration of multiple process steps. Wafers are transported from process chamber to process chamber under vacuum, eliminating native-oxide formation and reducing contamination on the wafer. Cluster tools also improve wafer manufacturing throughput (defined as the number of wafers processed per unit time) because there is no need to vent the chamber during wafer transfer steps.

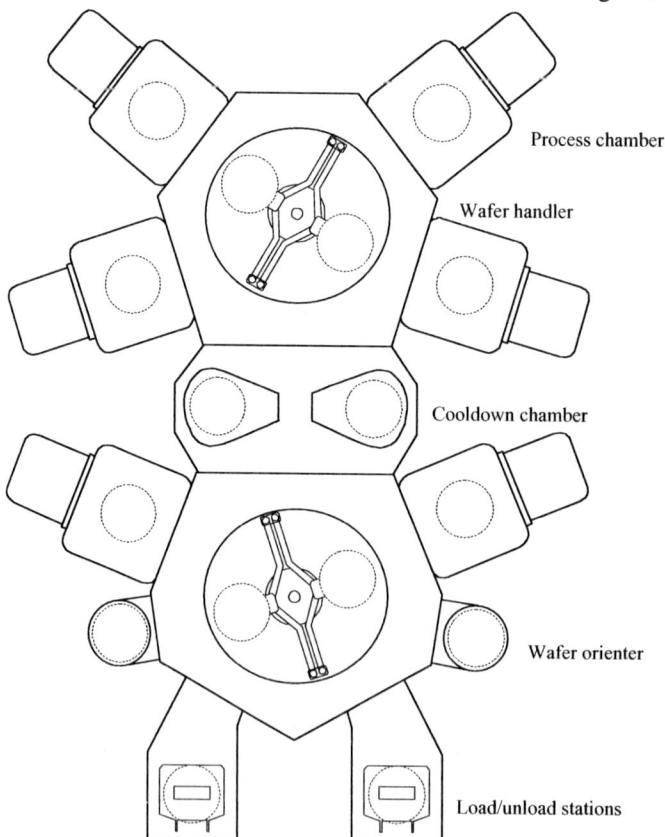

FIGURE 8.2 Integrated Cluster Tool
(Used with permission from Applied Materials, Inc.)

Cluster Tool with Integrated Process Chambers
(Photo courtesy of Applied Materials, Inc.)

VACUUM

Many chemical reactions in wafer fabrication are performed in a vacuum environment. A *vacuum* exists when there is less pressure in an enclosed volume than in the surrounding atmospheric pressure. The benefits of a vacuum in semiconductor manufacturing are shown in Table 8.1.[2]

TABLE 8.1 Benefits of Vacuum in Semiconductor Manufacturing

	Vacuum Condition	Benefit
1.	Create clean environment	Removes particles, unwanted gases, moisture, and contaminants.
2.	Low molecular density	Reduces the number of molecules in the system to reduce contamination and to move a gas out of the way (lower the molecular interference).
3.	Extend distance between collisions of molecules (Mean Free Path)	Provides the necessary condition for creating the plasma needed in semiconductor processes such as sputtering and etch.
4.	Accelerate reactions	Helps accelerate processes by lowering the vapor pressure of materials so they can react faster with other chemicals.
5.	Create a force	Creates a force, such as a vacuum pickup on a robot arm for wafer handling.

Vacuum Ranges

The following terms describe the different vacuum ranges: low vacuum, medium vacuum, high vacuum (also called high vac), and ultrahigh vacuum (UHV). These ranges are shown in Table 8.2 on page 184.[3] As a reference, the vacuum of deep space is about 10^{-16} torr.

Low vacuum (also referred to as *rough vacuum*) has two important characteristics:[4] gas flow is primarily by collisions between molecules (also known as *viscous flow*) and the pressure is high enough that true mechanical pressure gauges can be used. Low vacuum is commonly used in fabrication processes that depend on gas-phase chemical reactions, momentum transfer between molecules and/or a high rate of interactions between the gas and surfaces. *Medium vacuum* is from about 1 torr to 10^{-3} torr and is a transition between low and high vacuum. *High vacuum* is characterized by having few collisions between gas molecules (molecular flow). This condition results in very clean wafer surfaces. *Ultrahigh vacuum* is a continuation of high vacuum with stringent control of the vacuum chamber design and materials to minimize undesirable gas contaminants.

TABLE 8.2 Vacuum Ranges

Wafer Fab Processes	Vacuum Ranges in Torr				Chapter in book
	Rough 759 - 10^0	Medium 10^0 - 10^{-3}	High 10^{-3} - 10^{-6}	Ultrahigh 10^{-6} - 10^{-9}	
Oxidation	Wafer handlers / Atmospheric tools				10
Photo	Vacuum chucks / Wafer handlers				13 - 15
Polish	Wafer handlers / Slurry removal				18
Etch	Plasma resist strippers / Plasma etchers				16
Deposition	Batch deposition tools / Single wafer deposition tools				11
Metallization	Metal evaporators / Metal sputtering tools				12
Ion Implant	Batch ion implanters / Single wafer ion implanters				17
Metrology	Wafer inspection tools for quality checks and diagnostics / Analytical tools for research and failure analysis				7

Mean Free Path

The average distance a gas molecule moves before it strikes another molecule is known as the *mean free path (MFP)*. When the pressure is lowered in a vacuum, the space between the gas molecules increases, which is an important factor for how gases flow through the system and for creating a plasma in the process chamber. The mean free path for air at different pressure regimes and standard temperature is given in Table 8.3.

TABLE 8.3 Mean Free Path and Molecular Density versus Pressure

	760 Torr (atmosphere)	1×10^{-3} Torr	1×10^{-9} Torr
# of molecules/cm^3	3×10^{19} (30 million trillion)	4×10^{13} (40 trillion)	4×10^7 (40 million)
Mean Free Path	5×10^{-6} cm	5 cm	48 km

VACUUM PUMPS

There are many different vacuum pumps used in semiconductor manufacturing. For our purposes, they can be categorized into two general types: roughing pumps and high vacuum pumps. *Roughing pumps* have several purposes: to achieve a rough to medium vacuum (pressure down to 10^{-3} torr) in a chamber, to evacuate the entry area for wafers into a cluster tool (known as the loadlock), and to exhaust a high vacuum pump (see Figure 8.3). *High vacuum pumps* are used for achieving high and ultrahigh vacuum from 10^{-3} torr to 10^{-9} torr. Modern vacuum pumps for new wafer fabs are dry, meaning they contain no oils or lubricants that can backstream into the process chamber and contaminate wafers as well as the process chamber. This chapter addresses only dry pumps as the intent is to focus on current technologies needed to support sub-quarter micron processes.

Gas Control in Process Chambers **185**

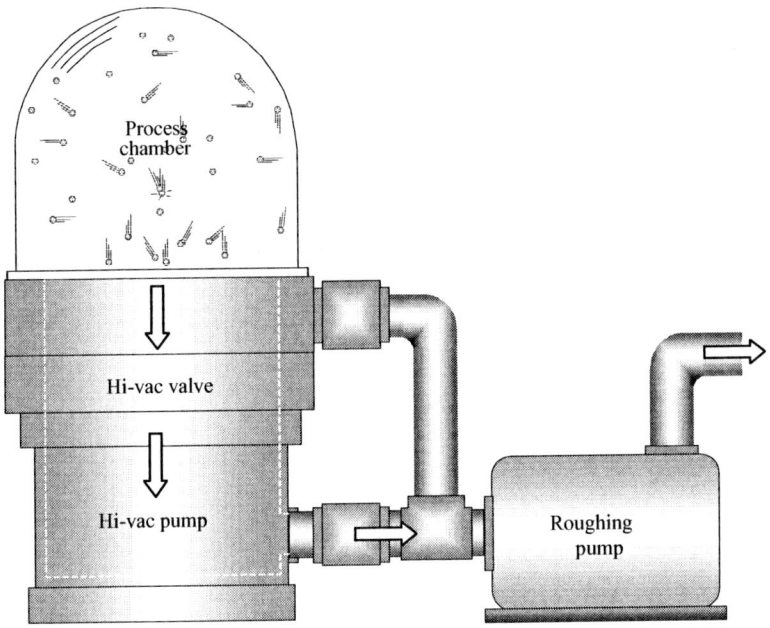

FIGURE 8.3 Roughing Pump Exhausting a High Vacuum Pump

Roughing Pump

When chamber pumpdown starts, roughing pumps remove more than 99.99% of the initial air or contaminants in the chamber. Removing this material with the roughing pump permits the high-vacuum pump to be optimized for removal of the remaining water vapor and gas molecules attached to the walls of the chamber. There are various types of roughing pumps, each with characteristics that work for a specific application. Two types of roughing pumps are:

- Dry mechanical pump
- Blower/booster pump

Dry Mechanical Pump ■ A *dry mechanical pump* employs mechanical devices to remove gases, such as the rotary claw dry pump (see Figure 8.4). The pump principle is typically based on increasing the chamber volume, thus lowering the pressure (i.e., Boyle's law). A mechanical pump often uses nonmetallic materials on the moving surfaces to avoid the use of sealing or lubricating fluids.

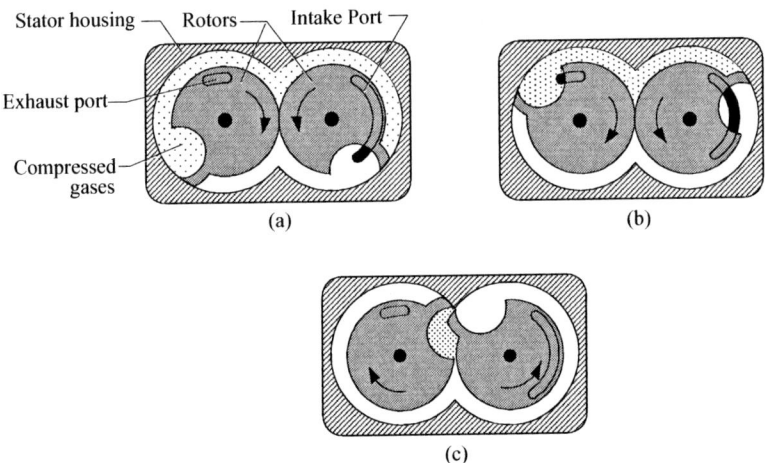

FIGURE 8.4 Rotary Claw Dry Mechanical Pump
(Used with permission from International SEMATECH)

Blower/Booster Pump ■ The *blower* or *booster pump* (also referred to as a *lobe pump*) is a widely used mechanical pump because it provides for high throughput of gas and requires no lubricants. It is desirable for systems where a high volume of gas must be pumped at a rough vacuum range. A blower is also referred to as a *Roots blower* or a *Roots-type blower*. The principle for how a blower works with the mating lobes is shown in Figure 8.5. A blower is often exhausted into a roughing pump because it will not pump at atmosphere under viscous flow. Newer blower pumps are being designed to exhaust directly to the atmosphere without the use of a roughing pump.

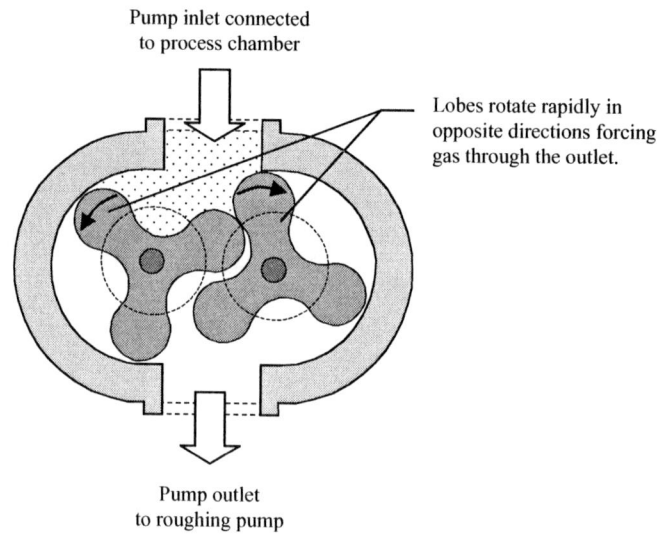

FIGURE 8.5   Roots Blower Pump

High Vacuum Pump

Two common high vacuum pumps are:

- Turbomolecular pump
- Cryopump

Turbomolecular Pump ■ The *turbomolecular pump* (usually called a *turbo pump*), is a versatile, reliable, and clean pump that is widely used in wafer fab equipment. The turbo pump is capable of reaching a pressure of 10^{-10} torr if a bakeout is done to drive out moisture in the process chamber. This pump also gives a fast start-up to full pumping action once turned on.

A turbo pump works on the principle of mechanical compression. There are a number of high-speed rotating blades (resembling blades in a jet engine) positioned between fixed blades that impart momentum and direction to the gas molecules (see Figure 8.6). Each set of rotating and fixed blades is a compression stage—a pump may have from ten to forty of these stages. Turbo pump blades will rotate at speeds up to 90,000 RPM. A turbo pump exhausts into a roughing pump because it cannot pump with viscous flow at atmospheric pressure.

Turbo pumps are specially designed for semiconductor applications. Magnetically levitated bearings (referred to as mag-lev) are common to eliminate the need for lubrication and special antivibration designs. The turbine blades can be designed to produce high gas throughput and high compression ratio for fast pumpdown of light gases.

The most common causes for failure of a turbo pump are exposing the pump suddenly to atmospheric pressure (called dumping the pump), particulate entering the pump, or physical shock. When a turbo pump is suddenly dumped to atmosphere, the turbine blades flex and come in contact with each other, causing catastrophic failure. Because the rotors on a turbo pump are precisely balanced, the pump should never be moved or bumped while it is in use. Extensive maintenance is not required on turbo pumps; most fabs run pumps until they crash and then change them.[5]

Gas Control in Process Chambers **187**

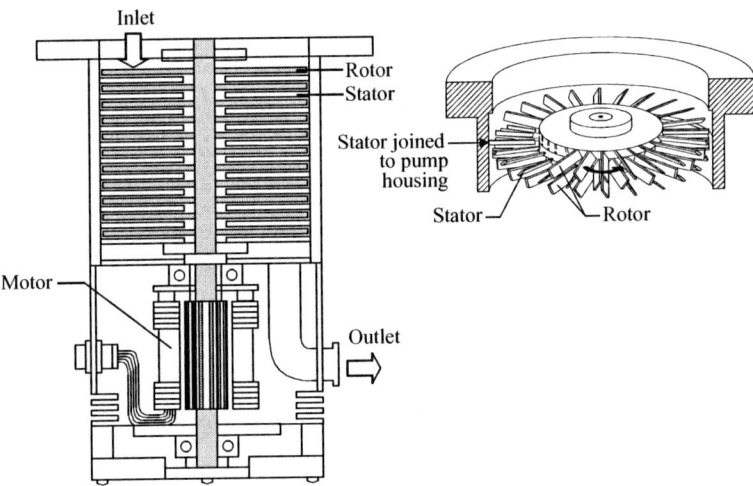

FIGURE 8.6 Turbo Pump Blades
(Used with permission from Varian Vacuum Systems)

Cryopump ■ A *cryopump* is a capture pump that removes gases from the process chamber by making them so cold that they are frozen and captured in the pump, thus providing the pumping action. The cryopump has become the industry standard in semiconductor equipment for pumping to the high and ultrahigh vacuum range. It has high gas throughput while pumping, plus has a very high water vapor pumping speed. This feature is useful when pumping down from atmosphere to remove the moisture in the chamber. The cryopump is extremely clean, with no oils or moving parts exposed to the vacuum. This attribute is desirable for wafer fabrication, which is a reason why this pump is common on new production equipment. There are two main components to a cryopump: a gaseous helium compressor and a pump module with a cold head, baffle, and body (see Figure 8.7).

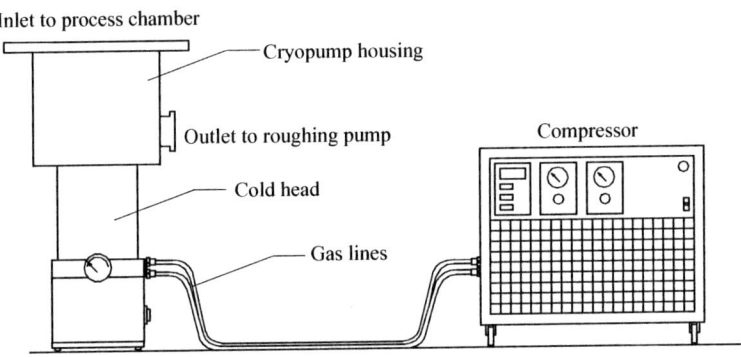

FIGURE 8.7 Cryopump Compressor and Pump Module
(Used with permission from Varian Vacuum Systems)

The helium compressor works similar to a regular refrigeration compressor (e.g., in the air conditioning system at home), except that it supplies high-pressure, high-purity, room-temperature helium to the expander module in the pump. When the gas expands in the pump module from a high-pressure to a low-pressure condition, the helium takes in heat and refrigeration occurs. This process produces cryogenic temperatures from 80 K down to 50 K. The pump has a second stage that expands the helium again and reaches approximately from 20 K to 10 K. Once the helium expands, it comes in contact with and cools multiple surfaces called *cryoarrays* (see Figure 8.8 on page 188). It is on these surfaces that the gases from the vacuum chamber are cooled and condensed or adsorbed. The condensed gases are immediately frozen on the cold cryoarray surfaces and trapped, which is essentially the pumping action.

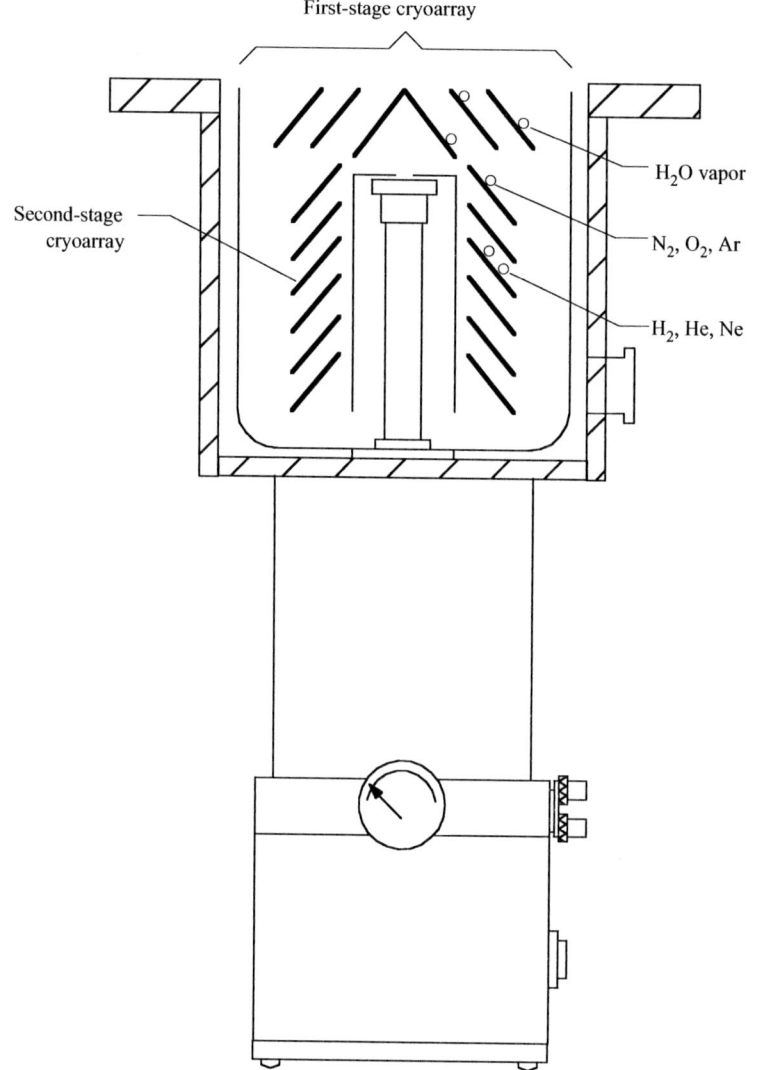

FIGURE 8.8 Cryoarray Surfaces in Pump Module
(Used with permission from Varian Vacuum Systems)

Since cryopumps capture gases by freezing the molecules rather than compressing and expelling them, the gases actually accumulate as frozen solids on the cryoarray surfaces. This process is similar to ice (frozen water) that forms on the inside of a freezer. These captured gases are removed periodically through a process called *regeneration,* where the pump is warmed to room temperature or above and the gases are vented through an appropriate vent line. Cryopumps require a roughing pump to remove the air from the pump and vacuum system.

Vacuum in Integrated Tools

The vacuum environments of an integrated cluster tool depend on the requirements of each individual process chamber (see Figure 8.9).[6] Chambers are isolated from one another, with a progressively better level of vacuum from the loadlock to the process chamber. The *loadlock* is where wafers enter the cluster tool, isolating the inner regions of the tool from the workplace environment. The system is designed to provide a well-controlled, low-contamination environment for wafer preparation and processing.

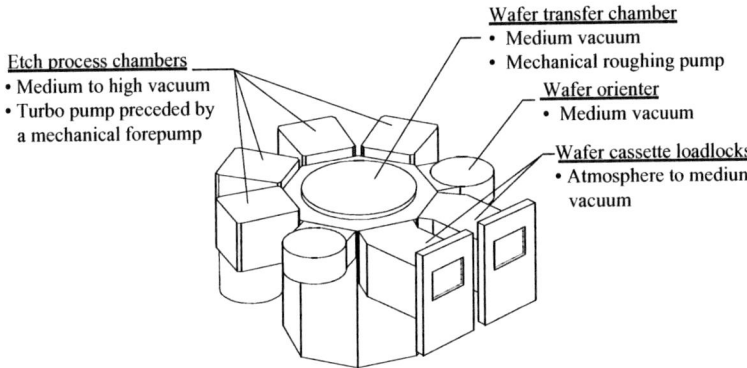

FIGURE 8.9 Cluster Tool Layout with Vacuum Environment

PROCESS CHAMBER GAS FLOW

The flow of gases into the process chamber is critical to attain the desired chemical reaction. The basic process chamber requirements for gas flow are:[7]

- Ability to handle a wide variety of bulk and specialty gases, many of which are corrosive and toxic.
- The control of gas flow into the process chamber is accurate and repeatable.
- The gas mix proportions are able to be controlled during the process run.
- Materials used in the chamber are not affected by the process gases and do not introduce contaminants into the gas stream.

When discussing gas delivery systems or vacuum, the mass quantity of gas flow at standard conditions is called *throughput* (Q). Throughput is the net number of gas molecules that pass through a point of the vacuum system in a specified period of time. Throughput defines the volumetric gas flow in a system at standard conditions. The most common units for throughput are torr-liters per second, standard cubic centimeters per minute (sccm), or standard liters per minute (slm).

Pumps are specified for a certain *pump speed*, which indicates how effectively the pump can remove gases. Pump speed is typically expressed in units of volume per unit time (such as liters/second or cubic feet per minute).

Throughput and pump speed are important in many semiconductor process steps. The chemical reactions that take place at the surface of the wafer may require high gas flow rates (which means high throughput). Factors such as the type of pump, its pumping speed, its location in the system, and the gas throughput are important variables that can determine whether an acceptable chemical reaction occurs at the wafer surface.

Mass Flow Controllers

Chemical reactions involve physical processes where molecular quantities are important for proper control of the reactions. From the ideal gas law we know that the number of gas molecules in a given volume changes in proportion to the absolute pressure and temperature. Thus, controlling gas flow into a chamber only by volume will not always yield the same number of gas molecules, which is undesirable for controlling chemical reactions.

To overcome this problem, gas flow into process chambers is controlled by use of a *mass flow controller* (*MFC*), shown in Figure 8.10 on page 190. MFCs use the heat-transfer property of the gas to directly measure the mass flow rate into the chamber. It employs a thermal sensor to detect changes in the mass flow of the gas. Integrated tools will typically use many MFCs to control the flow of the various gases into the process chambers. A pressure regulator is required ahead of the MFC to ensure that a constant specified pressure is delivered to the MFC.

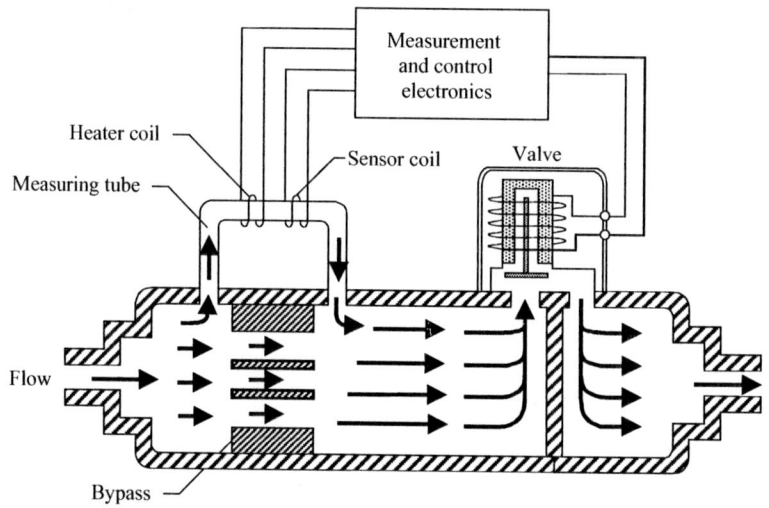

FIGURE 8.10 Thermal Mass Flow Controller
(Used with permission from International SEMATECH)

RESIDUAL GAS ANALYZER (RGA)

The *residual gas analyzer* (*RGA*) is an important process chamber instrument to identify the types of gas molecules remaining in an evacuated system. In this manner, it can be used for leak detection, and analysis of contamination in process chambers, and as a troubleshooting tool to solve vacuum-based problems in chambers.[8] Its most common applications are leak detection and process troubleshooting.

RGA Basics

The principle of the RGA is to separate, identify, and measure the quantity of all gas molecules in the chamber. An RGA measures the partial pressure contribution of each gas present in the vacuum system, as well as the total pressure from all gas molecules. It typically operates only in the high and ultrahigh vacuum ranges, but it is acceptable to use an RGA for vacuum conditions up to 10 millitorr.

Mass Flow Controller
(Photo courtesy of MKS Instruments, Inc.)

There are four basic parts to an RGA: an ionizer, an aperture, an analyzer, and a detector (see Figure 8.11). These are essentially the parts of a mass spectrometer, but an RGA is much smaller and is designed for attaching to process tools. These four parts are located in the sensing head of the RGA behind a process-specific valve inlet attached directly to the process chamber.

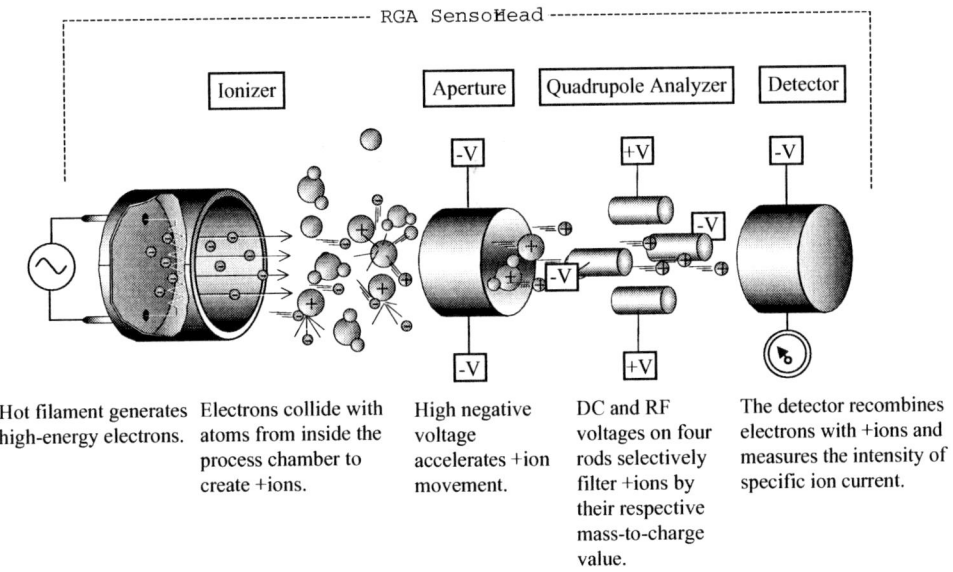

FIGURE 8.11 Basic Parts of Residual Gas Analyzer (RGA)

The ionizer creates ions from the gas molecules present in the system by bombarding the molecules in the chamber with electrons generated in the ionizer. The bombardment removes an electron from the gas molecules and creates positive ions. These ions are directed toward the analyzer by applying a voltage potential or magnetic field in the aperture. The analyzer is where the ions are separated by mass. There are several different types of analyzers; a common type is the *quadrupole mass analyzer* (*QMA*). It consists of four cylindrical rods that have both a constant DC potential and a high-frequency RF component (see Figure 8.12). For a given voltage level applied to the cylinders, only ions of a given atomic mass-to-charge pass through the filter. All other ions are grounded on the cylinder rods. The detector on modern RGAs will have better than 1 atomic mass unit (amu) resolution to differentiate between the various ions. As voltages are progressively changed on the filter, different ions are passed through the cylinders. In this manner, various gases of different masses are separated and identified.

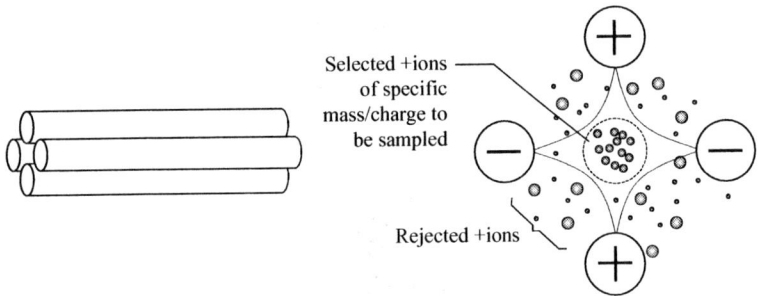

FIGURE 8.12 Quadrupole Mass Filter

RGA as Real-Time Monitor

RGA data can be an important part of verifying that a process chamber is ready for processing, especially for large-diameter wafers (see Figure 8.13 on page 192). This is because RGA data can provide real-time information about the cleanliness and stability of the process chamber during pumpdown.

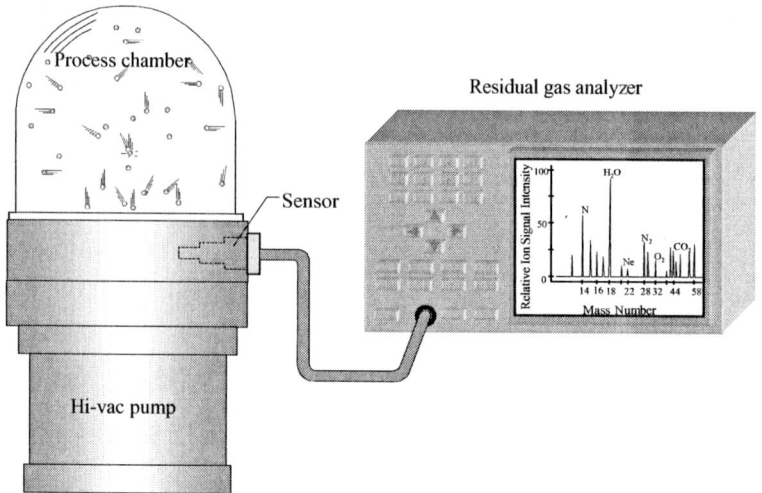

FIGURE 8.13 RGA Data Output

Using the RGA as an in situ (real-time) process monitor is a recent development. While the wafer is undergoing processing, the RGA will directly monitor the chemical constituents in the chamber to verify that the process is proceeding correctly.[9] The RGA can be used to replay conditions in the event of a wafer quality problem. This replay ability allows for more rapid diagnosis of process problems.

There is a growing need for RGAs to monitor chemical processes, such as plasma etching and plasma-enhanced chemical vapor deposition (PECVD).[10] For plasma applications, an RGA can follow the different process chemicals as the reaction occurs in the chamber and give insight into plasma behavior and how the chemical species change over time.

PLASMA

Plasma is a neutral, highly energized, ionized gas consisting of neutral atoms or molecules, positive ions, and free electrons. Positive ions and free electrons are formed when a valence electron is removed from a neutral atom. For instance, fluorine is neutral when there are an equal number of protons and electrons in its atomic structure. Fluorine is ionized when an electron is separated from its host atom (see Figure 8.14). Ionization of gas atoms in a confined process chamber can occur by subjecting the gas to strong DC or AC electromagnetic fields or by bombarding the gas atoms with some sort of electron source. These methods will be covered later in greater detail.

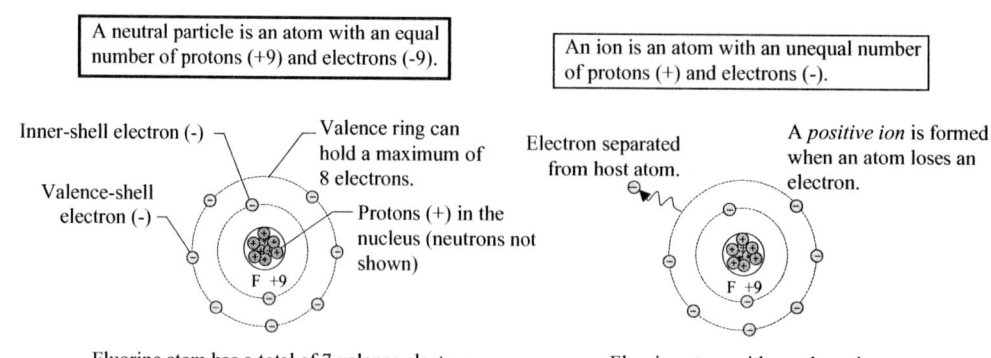

FIGURE 8.14 Creation of an Ion

Plasma is used at various process steps during wafer fabrication because it supplies much of the energy needed to support a gas reaction near the wafer surface in a process chamber. For example, plasma is used in lieu of thermal energy to ionize and excite a source gas to deposit thin films in high-density plasma chemical vapor deposition (HDP-CVD), discussed in Chapter 11. Another plasma application is to selectively remove metal through plasma etching (see Chapter 16). The most common indication that a plasma exists in a process chamber is the characteristic observable light referred to as a glow discharge (see Figure 8.15).

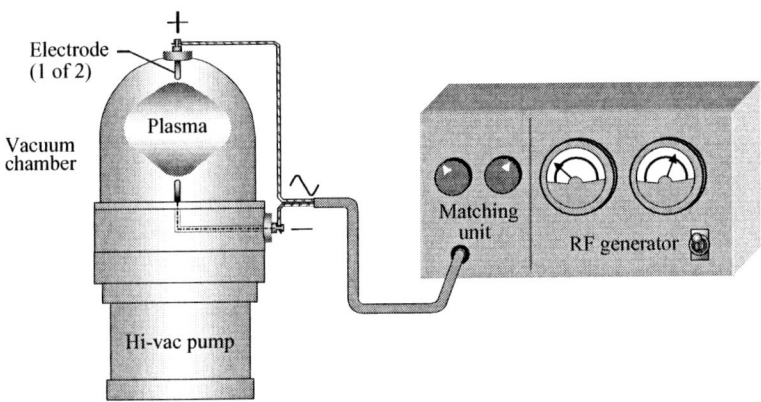

FIGURE 8.15 Plasma Glow Discharge

Glow Discharge

Although a glow discharge across a gas mixture can be created by applying DC power, the most common methods use AC power in the radio frequency (RF) range. When the electrical power is first applied, free electrons in the gas mixture are greatly influenced by the presence of electric fields. Electrons are accelerated through the gas mixture and collide with atoms and molecules, releasing additional electrons during the collisions.

In a weakly ionized plasma, or glow discharge, such as that commonly used in wafer fabrication, the high-energy electrons collide with neutral atoms or molecules and excite them. These excited atoms or molecules are short-lived, with lifetimes measured in nanoseconds. When an excited atom or molecule returns to its lowest energy state, energy is released during this relaxation in the form of a radiated photon (or light). This release causes the characteristic glow in the glow discharge (see Figure 8.16), with different glow colors for different species (e.g., oxygen, nitrogen, fluorine, and so on). The high-energy electrons in a plasma transfer energy to the neutral atoms and molecules by impact and initiate a reaction which could not have occurred in a high pressure environment. Typical parameters sustaining a glow discharge include the RF power and frequency, pressure, gas mixture and flow rate, vacuum pumping speed, and surface temperature.

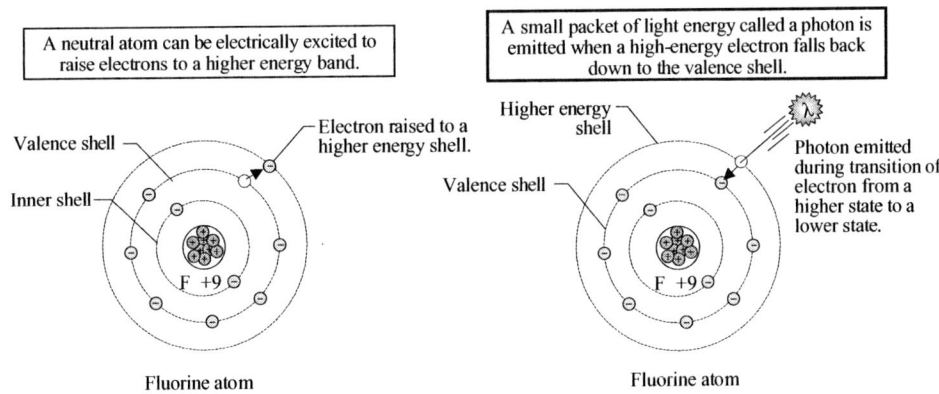

FIGURE 8.16 Electrically Exciting and Relaxing an Atom

Radicals ■ Radicals are highly reactive chemical species. As electrons collide with atoms or molecules, different kinds of products are created. The dissociation (or fragmentation) of a molecule occurs when a large molecule is broken into smaller particles (see Figure 8.17). Recall that the number of electrons in the valence shell are related to the chemical properties of an atom. A *radical* is created when a neutral molecule is bombarded by an energetic electron. This action cleaves a bond without adding or removing an electron, creating a highly reactive species. This radical is an uncharged atom or molecular fragment that has incomplete bonding (or unpaired electrons) in the valence shell.

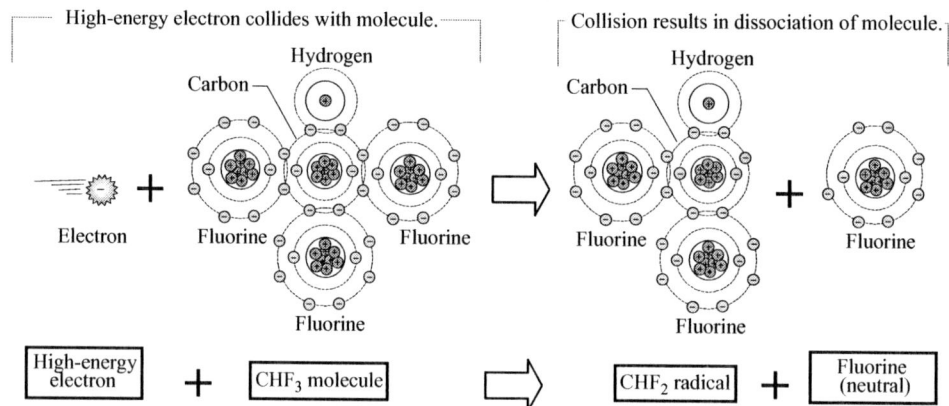

FIGURE 8.17 Dissociation of a Molecule

RF Energy ■ The energy of plasma is sustained through the absorption of RF radiation in an applied alternating current (AC) electric field of several hundred volts rms. The use of RF power, typically at 13.56 MHz (chosen for industrial purposes by the FCC, or Federal Communication Commission), creates a high-efficiency plasma. In recent years, different frequencies have started being used for plasma generation, such as 400 KHz, 2 MHz, 4 MHz, and 2.45 GHz. The frequency has a direct influence on the ion mobility, thus affecting process uniformity and the rate of the process (e.g., the metal removal rate for etching). The RF field is applied between a negative cathode and a positive anode, referred to as electrodes. The wafer typically will be held against the grounded electrode, with the RF power applied to a parallel electrode (see Figure 8.18).

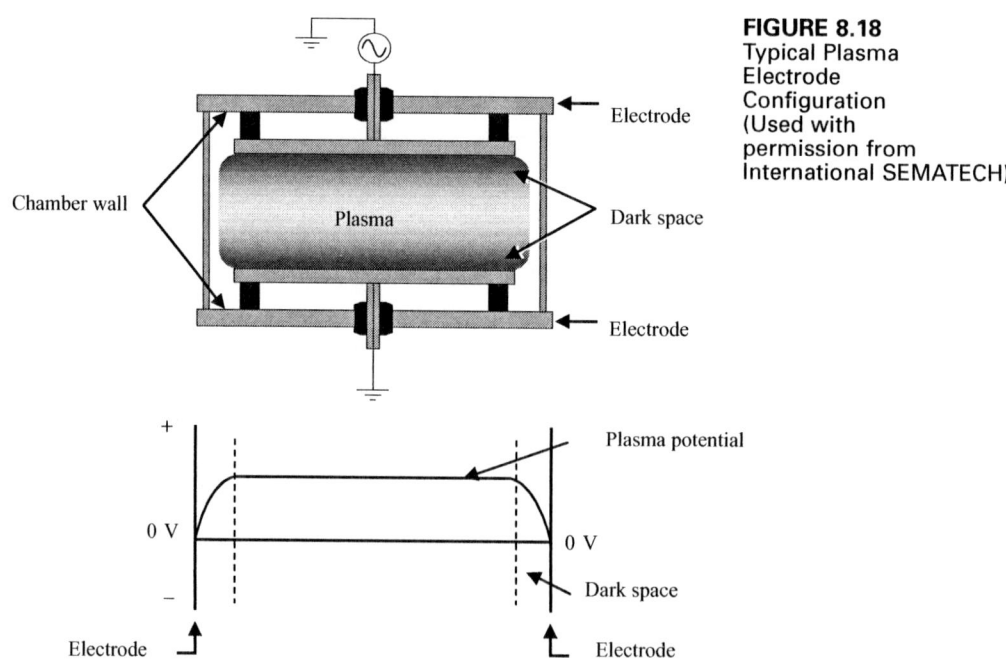

FIGURE 8.18
Typical Plasma Electrode Configuration (Used with permission from International SEMATECH)

The electrons, ions, and other species generated in a glow discharge will move toward the electrodes. Electrons in the glow discharge can move faster toward the positive electrode when compared to the slow moving massive positive ions. This movement produces a positive potential in the glow discharge region. The positive ions move toward the cathode and cross through a *dark space,* or ion sheath, adjacent to the cathode. The dark space results from the lack of electrons in this region. With respect to the glow discharge region, the dark space has a large voltage drop and strong electric field. This condition accelerates the positive ions toward the cathode, which then ejects secondary electrons. These secondary electrons are repelled from the cathode and cross back through the dark space where they maintain the glow discharge.

PROCESS CHAMBER CONTAMINATION

Wafer fabrication equipment with vacuum can be viewed as two general types: tools that process large batches of wafers (e.g., 150 or 200 wafers) or single-wafer, multichamber tools with loadlocks and several integrated process chambers. In general, chamber contamination on wafers is lower in single-wafer reactors than in large batch tools. The improvement in single-wafer cluster tools is attributed mostly to maintaining the processing chambers under relatively constant temperature and pressure conditions, which minimizes factors that contribute to particle generation.

Trace amounts of water are probably the most significant source of contamination in process chambers with vacuum.[11] Water is a problem in vacuum systems since it is reduced to ions as well as H_2 or O_2 gas molecules because of the reactor chemistry and vacuum. It sticks tenaciously to surfaces, outgasses slowly, and is a chemical poison in reactors. For example, the tenacity of water sticking to chamber walls is so high that it is estimated that there are 10,000 times more water molecules on the chamber walls than in the vacuum space of one cubic meter of processing chamber. It is also believed that water creates particle contamination during pumpdown by the formation of particles.[12]

In order to minimize water adsorption contamination in process chambers, it is desirable to reduce the need to open or disassemble process reactors for cleaning purposes. This goal is often accomplished by using in situ cleaning techniques (cleaning while the wafers are processed). Nevertheless, process equipment still requires maintenance actions, which necessitate equipment shutdown and repair. The specific steps followed by the maintenance technician during servicing have a significant effect on the control of contamination in the equipment. Recommendations for minimizing contamination during the servicing of wafer fabrication equipment are listed in Table 8.4.[13]

TABLE 8.4 Recommendations to Minimize Contamination During Equipment Servicing

Recommendations
1. Maintain good temperature and humidity control in the cleanroom environment where the equipment is located.
2. Control the equipment's pump and vent cycles to minimize turbulence and prevent particle generation when processing wafers.
3. Avoid abrasive cleaning materials.
4. Use exact replacement parts and materials to avoid subtle sources of equipment contamination and leaks.
5. Use low particle-generating gas-handling components, such as regulators and automatic valves that have a tendency to generate particles.

SUMMARY

Process chambers have many functions in wafer fabrication, such as controlling gas flow. A vacuum in a process chamber creates a clean environment with low molecular density and high mean free path. Vacuum ranges are low, medium, high, and ultrahigh. The most common vacuum pumps in the wafer fab generally can be categorized as roughing (dry mechanical and blower/booster) and high vacuum (turbo pump and cryopump). The mechanical pump is a low vacuum pump. The blower is a low to medium vacuum that has high throughput. The turbo pump is a common high vacuum pump that uses turbine blades, while the cryopump is a capture pump that removes gases by freezing them. Gas flow in the process chamber is controlled using a mass flow controller (MFC). The residual gas analyzer (RGA) is used to detect leaks or to analyze residual gas in a chamber after vacuum pumpdown. It also can do real-time monitoring of the reaction in a chamber during a process run. Plasma is a highly energized gas that is commonly used in process chambers to excite source gases during a reaction. It involves a glow discharge that results from collisions between energetic electrons that produce radicals. Plasma energy is sustained by the absorption of RF radiation. Cleanliness in the process chamber is critical, with water moisture being the most serious contaminant.

KEY TERMS

process chamber
reactor
cluster tool
vacuum
torr
low vacuum (rough vacuum)
medium vacuum
high vacuum
ultrahigh vacuum
mean free path (MFP)
roughing pump
high vacuum pump
dry mechanical pump
blower (booster pump or lobe pump)

turbomolecular pump (turbo pump)
cryopump
cryoarrays
regeneration
loadlock
throughput
pump speed
mass flow controller (MFC)
residual gas analyzer (RGA)
quadrupole mass analyzer (QMA)
plasma
glow discharge
radical
dark space

REVIEW QUESTIONS

1. What is a process chamber? What are its five functions?
2. Describe a cluster tool, and explain why it is beneficial in IC fabrication.
3. What is a vacuum?
4. What are the benefits of vacuum in semiconductor manufacturing?
5. State the most common vacuum unit, and describe how to interpret it.
6. List and describe four vacuum ranges.
7. What is mean free path? Why is this condition important?
8. Give reasons for using a roughing pump and a high vacuum pump.
9. Describe two types of roughing pumps. Which one has the higher gas throughput?
10. Describe two types of high vacuum pumps.
11. What are the most common reasons for failure of a turbo pump?
12. Describe the purpose of the cryoarray, and explain how it creates pumping action.
13. What is regeneration?

14. Draw a picture of a cluster tool and show chambers with at least three different vacuum levels.
15. What is the purpose of the loadlock in a cluster tool?
16. State the four basic process chamber requirements for gas flow.
17. Describe gas throughput.
18. What is pump speed?
19. What is the purpose of a mass flow controller?
20. What does a residual gas analyzer (RGA) do?
21. List and describe the three basic parts to an RGA.
22. How does the quadrupole mass analyzer work on an RGA?
23. What is plasma? How is plasma beneficial in a process chamber?
24. Describe the plasma glow discharge region.
25. What is a radical in a plasma?
26. Why is RF energy used in plasma?
27. Why is moisture a problem in process chambers?
28. List the recommended steps to minimize contamination during equipment servicing.

VACUUM EQUIPMENT SUPPLIERS' WEB SITES

Alberta University Vacuum Page	http://nyquist.ee.ualberta.ca/~schmaus/vacf/
Alcatel Vacuum Products	http://www.alcatel.com/
Apiezon Products	http://www.apiezon.com/
AVS, American Vacuum Society	http://www.vacuum.org/
BOC Edwards	http://www.boc.com/edwards/
CTI Cryogenics	http://www.ctivacuum.com/
Ebara Technologies	http://www.ebaratech.com/
Granville-Phillips	http://www.helixtechnology.com/
Inficon Inc.	http://www.leyboldinficon.com/
Leybold Vacuum	http://www.leyboldvac.de/
Millipore Corp.	http://www.millipore.com/
MKS Instruments	http://www.mksinst.com/
Omega Engineering Inc.	http://www.omega.com/
Osaka Vacuum Ltd.	http://www.osakavacuum.com/
Parker Hanniflin Corp.	http://www.veriflo.com
Pfeiffer Vacuum Tech., Inc.	http://www.pfeiffer-vacuum.com/
SEMI	http://www.semi.org/
Unit Instruments	http://www.unit.com/
Varian Inc.	http://www.varianinc.com/
Varian Vacuum Technologies	http://www.varianinc.com/vacuum/
VAT Valve	http://www.vatvalve.com/
Veeco Instruments Inc.	http://www.veeco.com/

REFERENCES

1. R. Waits, "Semiconductor Manufacturing and Vacuum Technology: A Memoir," *Solid State Technology* (May 1997): p. 105.
2. H. Tompkins, *The Fundamentals of Vacuum Technology*, (New York: American Vacuum Society, 1997), p. 6.
3. C. Tilford and J. P. Looney, "Vacuum Measurement: The Basics," *Semiconductor International* (May 1994): p. 73.
4. Ibid.
5. J. Baliga, "Vacuum Pump Designs Adjust to Harsher Conditions," *Semiconductor International* (October 1997): p. 88.
6. S. Hansen, *Introduction to the Creation and Control of the Vacuum Process Environment*, (Andover: MKS Instruments, 1995), p. 118.

7. Ibid., p. 110.
8. L. Peters, "Residual Gas Analysis: A Technology at a Crossroads," *Semiconductor International* (October 1997): p. 95.
9. T. Banks, G. Diamond, and S. Ruck, "Integrating Mass Spectrometry Data into the Fab Environment," *Semiconductor International* (June 1997): p. 138.
10. Ibid., p. 98.
11. A. Rapa and A. Bross, "Contamination Control in Multilevel Interconnection Manufacturing," *Handbook of Semiconductor Interconnection Technology,* ed. G. Schwartz, K. Srikrishnan, and A. Bross, (New York: Marcel Dekker, 1998), p. 552.
12. Ibid.
13. Ibid., p. 553.

CHAPTER 9
IC FABRICATION PROCESS OVERVIEW

The typical IC wafer fab process may take as many as six to eight weeks and involve 350 or more process steps to complete the full manufacturing flow. The complexity of this process can be overwhelming.

Recall that most semiconductor processing occurs in the top few microns of a silicon wafer. This activity corresponds to the first part (front end) of the manufacturing process flow. All materials on top of the silicon are part of the layering strategy needed to interconnect the many devices on the chip. To add multiple metal and insulating layers, the process flow requires the wafer to cycle through the different process steps. Once you understand the process, you will realize that only a few process areas are used many times over to fabricate a high-performance microchip.

OBJECTIVES

After studying the material in this chapter, you will be able to:

1. Draw a diagram showing how a typical wafer flows in a sub-micron CMOS IC fab.
2. Give an overview of the six major process areas and the sort/test area in the wafer fab.
3. Describe the primary purpose of each of the 14 CMOS manufacturing steps.
4. Discuss the key process and equipment used in each CMOS manufacturing step.

INTRODUCTION

This chapter describes a simple manufacturing process for producing 0.18-μm CMOS integrated circuits on a silicon wafer. This is presented to help the reader gain a better understanding and appreciation for semiconductor manufacturing. Specific details about each process step is provided later in this text in the process chapters.

An overview of the entire wafer fab process is first presented, with a model that depicts how wafers repeatedly cycle through a few major process areas. This explanation simplifies the concept of keeping wafer flow to a manageable level in a fab. Recognize that when changes are made to the process, such as parameter or tool changes, the results of the change may not be known until many days or weeks later during wafer test at the end of the process. This fabrication complexity makes it important for each process area to flawlessly perform its tasks by meeting the defined quality measures at each step.

CMOS PROCESS FLOW

IC manufacturing is a complicated sequence of chemical and physical operations that are performed on a silicon wafer. Simply stated, the operations fall into four basic categories: layering, patterning, etching, and doping. Figure 9.1 on page 200 illustrates how complex the process can be even for manufacturing a single MOS transistor.

Since CMOS technology is the most popular of the process families, we have chosen it as an example to describe wafer process flow. The specific example is that of a 0.18 μm CMOS integrated circuit process flow. And, since this is an overview of IC manufacturing, you will be introduced to a variety of terms and concepts, which will be further explained in subsequent chapters. As you study this chapter, keep in mind the manufacturing area where specific process operations are being performed. Note the purpose of each operation, the type of equipment and materials that are used, and the quality measures that are followed to determine the integrity of the process at each step.

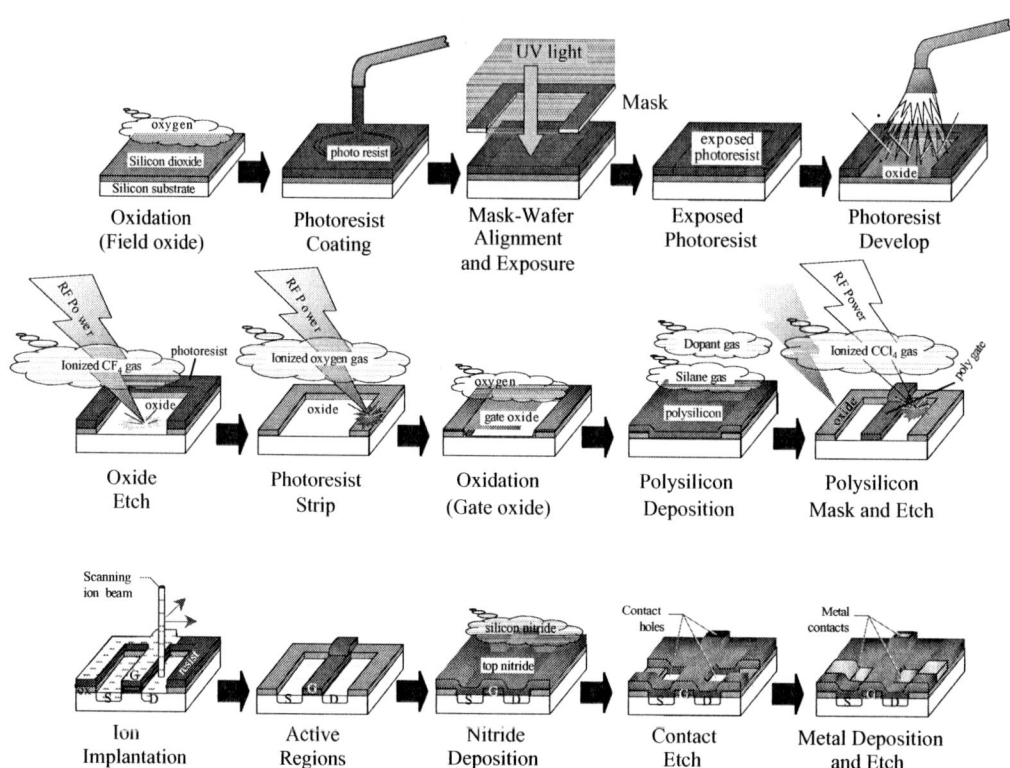

FIGURE 9.1 Major Fabrication Steps in MOS Process Flow (Used with permission from Advanced Micro Devices)

Overview of Areas in a Wafer Fab

ICs are manufactured in a wafer fab. As shown in Figure 9.2, wafer fabs are generally divided into six distinct production areas: diffusion (which includes oxidation, film deposition, and doping processes), photolithography, etch, thin films, ion implant, and polish. These six major production areas and their related process and metrology tools are all housed in the ultraclean cleanrooms of the wafer fab. The polish area is a new addition to semiconductor manufacturing for high-performance ICs and is gaining popularity in the industry. Although located near the wafer fab, the test/sort area for testing individual die on the wafer is not housed in the same cleanroom environment as the other areas of the fab. Assembly and packaging plants are generally located in other facilities, perhaps even in other countries.

Diffusion ■ The *diffusion* bay is recognized as the area where high-temperature processing and film depositions are performed. The primary tools in the diffusion area are a *high-temperature diffusion furnace* and a *wet cleaning station*. High-temperature diffusion furnaces (see Figure 9.3) can operate at temperatures near 1200°C and are configured to run a variety of processes, including oxidation, diffusion, deposition, anneals, and alloys. These processes will be covered in detail in later chapters. Wet cleaning stations are the secondary tools used in the diffusion area. Wafers must be cleaned thoroughly to remove contamination and native-grown oxide on the surface before inserting the wafers into the furnaces.

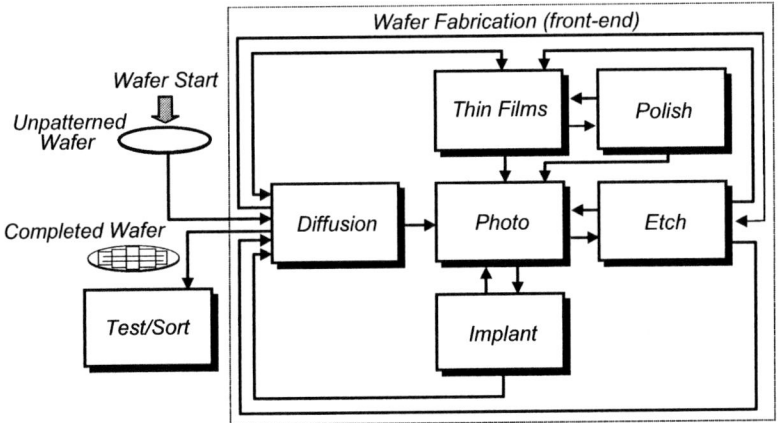

FIGURE 9.2 Model of Typical Wafer Flow in a Sub-Micron CMOS IC Fab
(Used with permission from Advanced Micro Devices)

Photolithography ■ The *photolithography* bay is recognizably different than the other areas in the fab due to the yellow fluorescent tubes that light up this production bay. The purpose of photolithography is to photograph the image of a circuit pattern onto the photoresist that coats the wafer surface. Photoresist is a light-sensitive chemical that captures the image of a mask pattern resulting from exposure to ultraviolet (UV) light. Photoresist is sensitive to certain wavelengths of light, such as UV and white light, but the wavelengths of yellow light do not affect it.

The *coater/developer track* is a cluster tool used to perform many of the operations in photolithography. This tool primes the wafer, coats it with photoresist, spins the wafer to smooth out the photoresist, bakes the wafer, and uses robotics to transfer the resist-coated wafer to the alignment and exposure tool. The purpose of the *stepper* is to align the wafer to an array of die patterns etched on a chrome-coated quartz reticle. When properly aligned and focused, the stepper exposes a small area of the wafer, then steps to the next field and repeats the process until the entire wafer surface has been exposed to the die patterns on the reticle (see Figure 9.4 on page 202). When completed, the wafer returns to the coater/developer tracks where the resist is developed, then the wafer is rinsed and baked again.

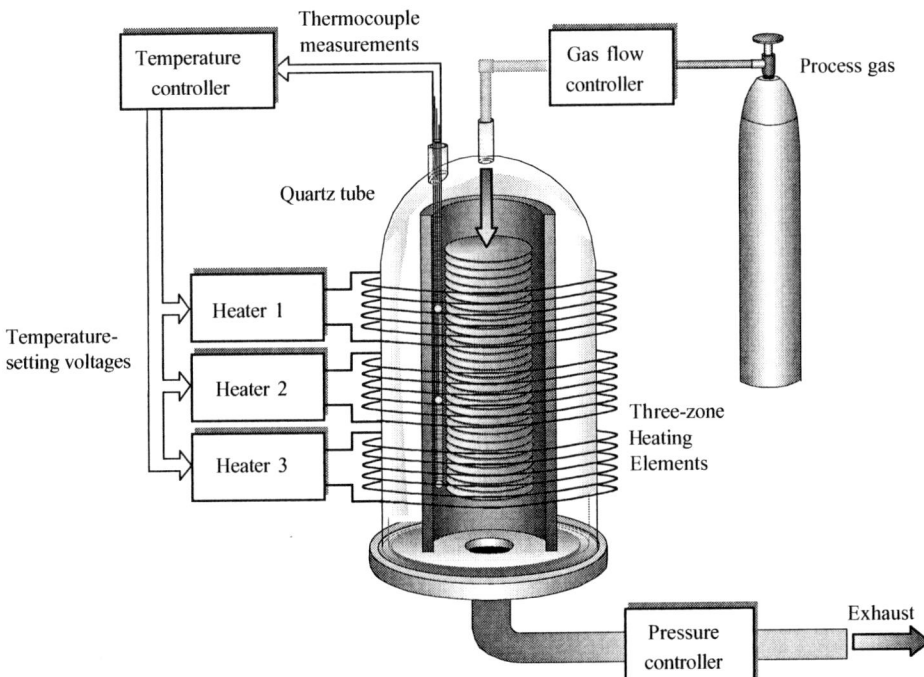

FIGURE 9.3 Simplified Schematic of a High-Temperature Furnace

Photolithography Bay in a Sub-Micron Fab
(Photo courtesy of Advanced Micro Devices)

Referring once again to Figure 9.2 on page 201, photolithography is shown to be at the center of the wafer fab. This position is due to the fact that wafers flow into photolithography from all other areas of the fab. Contamination control is especially important here because particles or defects can become imbedded in or on the resist films during the photographic process. A defect in the photographic mask or reticle or a particle on the stepper can be photographed onto all wafers that are processed with these tools.

To reduce contamination, open containers of chemicals are prohibited from being used in this area. Thus, cleaning stations and photoresist strippers are usually located in other areas of the fab other than photolithography. Note that in Figure 9.2 on page 201, wafers flow from photolithography into only two other areas: etch and ion implant. Consequently, there are only three production areas where photoresist-coated wafers can be found—photo, etch, and ion implant.

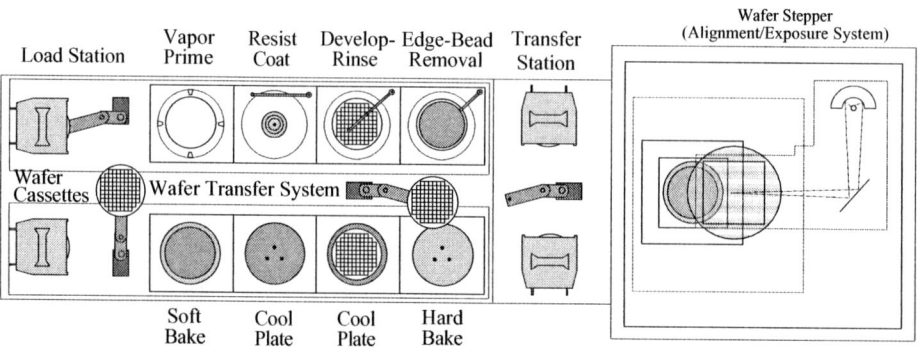

FIGURE 9.4 Simplified Schematic of a Photolithography Processing Module

Etch ■ The *etch* process creates a permanent pattern on the wafer in areas not protected by the photoresist pattern. The most common tools in the etch bay are the *plasma etcher*, the *plasma resist stripper*, and the wet cleaning station. Today most etching steps are done with dry plasma etchers (see Figure 9.5), although some wet etch processes are still in use. The plasma etchers are tools that use radio frequency (RF) energy to ionize gas molecules inside a vacuum chamber. The plasma is the glow given off by the electrically-excited gases. The gases react with the top layer of material on the wafer. After the etch process, another plasma system, called the plasma stripper, uses ionized oxygen to remove the photoresist from the wafer. This step is followed by thoroughly cleaning the wafers with a combination of chemicals.

IC Fabrication Process Overview **203**

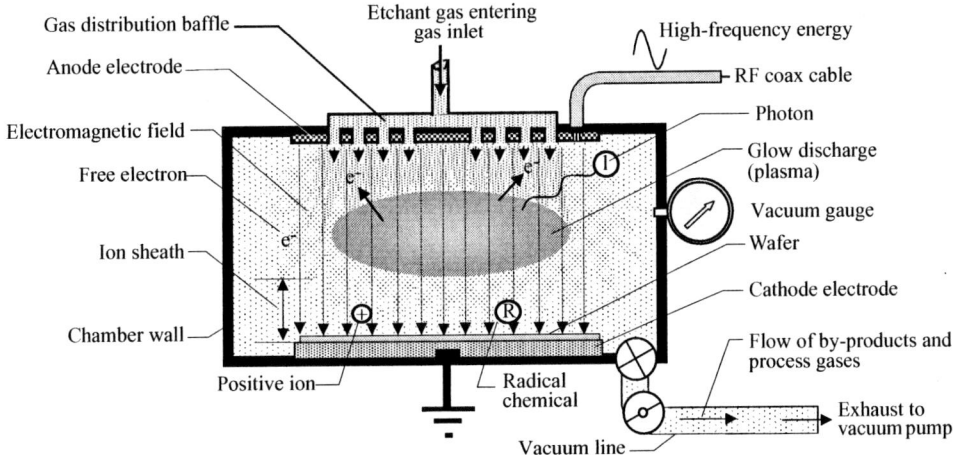

FIGURE 9.5 Simplified Schematic of a Dry Plasma Etcher

Ion Implant ■ The *ion implanter* is the most popular tool for doping wafers in a sub-micron process. Gases carrying the desired dopant, e.g., arsenic (As), phosphorus (P), and boron (B), are ionized inside the implanter (see Figure 9.6). High voltages and magnetic fields are used to control and accelerate the ions. The high energy of the dopants penetrates through the surface of the resist-coated wafer. After implantation is completed, the photoresist is stripped off and the wafer is thoroughly cleaned.

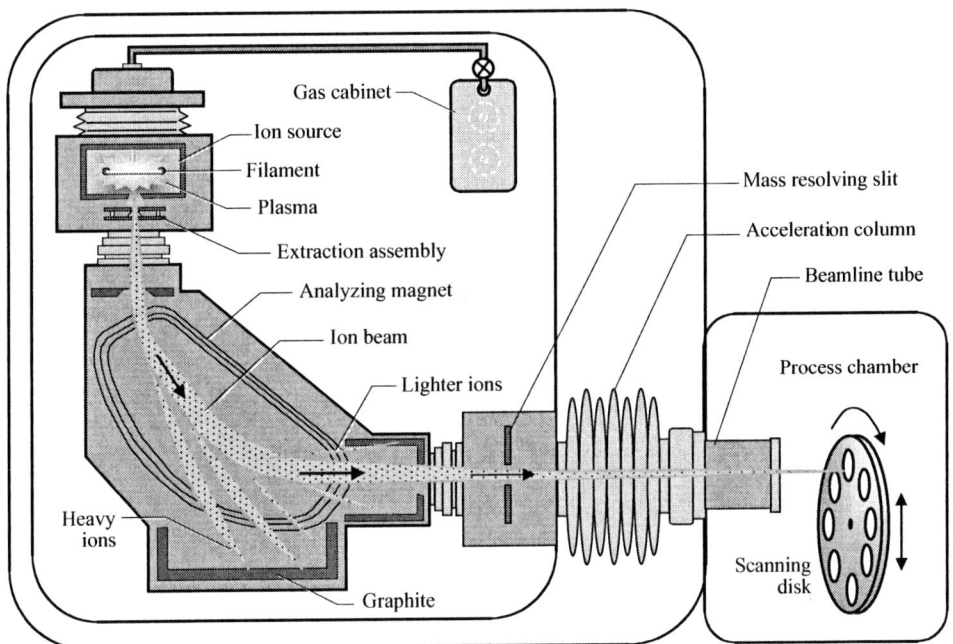

FIGURE 9.6 Simplified Schematic of an Ion Implanter

Thin Films ■ The *thin films* bay is primarily responsible for depositing dielectric and metal layers during different steps of the manufacturing process. The processes in thin films operate at lower temperatures than the furnaces in the diffusion area. There are many diverse tools in this area. All thin film deposition equipment operates under low to medium vacuum conditions (see Figure 9.7 on page 204), including chemical vapor deposition (CVD) and metal sputtering tools (PVD, or physical vapor deposition). Other tools used in this area may include the spin-on-glass (SOG) system,

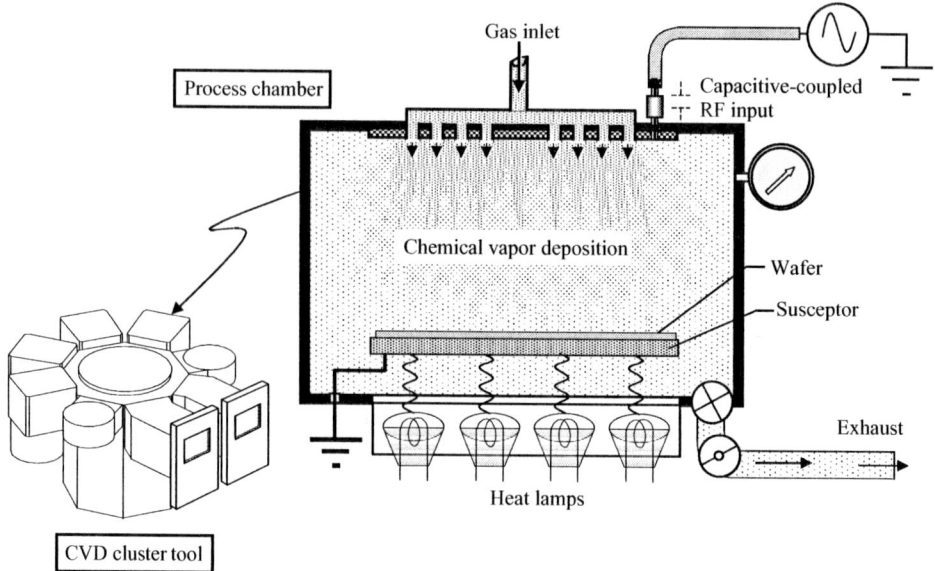

FIGURE 9.7 Simplified Schematics of a CVD Cluster Tool and Process Chamber

rapid thermal processor (RTP) system, and wet cleaning station. The SOG is used to fill in low areas of the wafer for the purpose of planarizing (smoothing) the surface of the wafer. RTAs are used to anneal ion implant damage of the silicon substrate and for metal alloy processing steps.

Polish ■ The purpose of the CMP (*chemical mechanical planarization*) process is to planarize the top surface of the wafer by lowering the high topography to be level with the lower surface areas of the wafer. CMP minimizes uneven wafer surfaces that make further processing difficult. The polisher is the primary tool in the CMP area, and this process is also referred to as *polish*. CMP uses a combination of chemical etching and mechanical abrading to remove a desired amount of the upper layer of the wafer. Other tools that support CMP include *wafer scrubbers,* cleaning stations, and metrology tools.

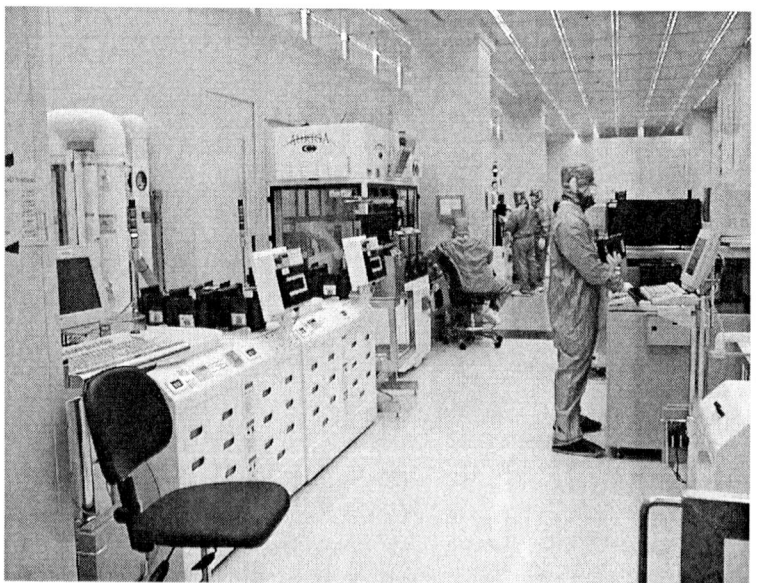

Polish Area in a Sub-Micron Wafer Fab
(Photo courtesy of Advanced Micro Devices)

Thin Film Metallization Work Bay
(Photo courtesy of Advanced Micro Devices)

CMOS MANUFACTURING STEPS

The remainder of this chapter focuses on the manufacturing steps that a wafer follows in a typical CMOS process. Although there are some operations in the process flow that can be performed in batches of wafers at one time, this process flow describes the manufacturing of a single wafer. For further simplification, this process describes the production of only a microscopic area of a single CMOS inverter consisting of two transistors—nMOS and pMOS. Cross-sectional views of the product will be provided for each major operation.

The CMOS manufacturing steps to be described are:

1. Twin Well Process
2. Shallow Trench Isolation Process
3. Poly Gate Structure Process
4. Lightly Doped Drain (LDD) Implant Processes
5. Sidewall Spacer Formation
6. Source/Drain (S/D) Implant Processes
7. Contact Formation
8. Local Interconnect Processes
9. Via-1 and Plug-1 Formation
10. Metal-1 Interconnect Formation
11. Via-2 and Plug-2 Formation
12. Metal-2 Interconnect Formation
13. Metal-3 to Pad Etch and Alloy
14. Parametric Testing

1. Twin Well Process

The first step in some CMOS wafer fabrication processes is to define the active regions of the MOSFETs. Many of today's sub-quarter micron processes use a *twin-well* (also referred to as *twin-tub*) approach to define the active regions of the nMOS and pMOS transistors. A twin-well consists of a p-well and an n-well, with each well requiring at least three to five major processing steps to fabricate. A *retrograde implant* technique is used to obtain optimum electrical characteristics for the FETs. This technique starts with a high-energy, high-dose implant that penetrates approximately 1 μm into the epilayer. Subsequent well implants are done at the same active site with a progressive decrease in energy, junction depth, and dose of dopant material such as phosphorus or boron (measured in atoms/cm^2). The results help set the threshold operating voltage for the FETs as well as prevent common CMOS problems, such as *latchup* and other reliability issues. Figure 9.8 and Figure 9.9 illustrate, respectively, the major steps required in the formation of the n-well and p-well.

n-well Formation ■ The five major steps in n-well formation are described below in the following process and illustrated in Figure 9.8.

Description of Five Major Steps in n-well Formation

	Process Step	**Description**
1.	Epitaxial Growth	The silicon wafer, having already a thin layer of *epitaxial* silicon (epilayer), arrives in the diffusion bay. The epilayer has the exact crystal structure as the substrate, except with improved purity and fewer crystal defects. The epilayer has already been lightly doped with p-type dopant (boron).
2.	Initial Oxide Growth	In diffusion the wafer is cleaned in a series of chemical baths to remove particles, organic and inorganic contamination, and *native oxide* growth (natural silicon dioxide growth) from the wafers. After the wafer is rinsed and dried, the wafer is placed in a high temperature (~1000 °C) furnace process chamber. Oxygen is flowed into the process chamber to react with silicon to grow approximately 150 Å of oxide. This initial oxide serves several functions: (1) it protects the top surface of the epi silicon from contamination, (2) it prevents excessive damage to the silicon during implantation, and (3) as a screen oxide, it helps control the depth of the dopants during implantation.
3.	1st Mask, n-well Implant	In photolithography the wafer undergoes a series of process steps in the coater/developer track tool. The "tracks," as commonly referred to, prime the top surface of the wafer, coat the wafer with photoresist (liquid photographic film), spin the wafer, and bake the wafer. An internal automatic wafer-handling system transfers the wafer between process stations within the tracks. Another wafer handler removes resist-coated wafers, one at a time, and transfers them to the *alignment-and-exposure system* (an extremely complex and precise camera). The alignment-and-exposure system photographs the image of a specific mask layer directly onto the resist-coated wafer. In this case, the mask layer defines the areas of the product that are to be implanted in order to form the n-wells for the pMOS transistors. The exposed wafer is then transferred back to the tracks, where the image first appears when the wafer is sprayed with the chemical developer. The developed wafer undergoes another bake step and is inspected before being transferred to the ion implant bay. The printed pattern is inspected for proper *linewidths,* called *critical dimensions* (CD). In the event of a major defect, the photoresist can be stripped off and the wafer can then be reworked. Photolithography is the only area in the fab where a wafer can be easily reworked.
4.	n-well Implant (High energy)	The patterned wafer arrives in the ion implant bay. The photoresist pattern covers specific areas of the wafer that are to be protected from ion implantation. Windows, or openings, in the photoresist allow high-energy positive dopant ions to penetrate into the upper surface of the epilayer (~1 μm *junction depth*). In this case, phosphorus is the desired dopant. The *ion implanter* is the main process tool in this bay. Its purpose is to ionize dopant atoms, accelerate them with high voltage (~200 KeV), select the most appropriate dopant species to implant, focus the ions into a narrow beam, and, finally, scan the wafer to provide uniform doping across all unprotected areas of the wafer. At this point the dopant ions penetrate the crystal lattice of the silicon causing damage to the covalent atomic structure. This damage will be repaired later in a diffusion and annealing step.
		Note: After each ion implant operation, an oxygen-plasma reactor tool strips the photoresist off each implanted wafer. The wafer is then cleaned by a wet chemical process to remove residual photoresist and polymers created by the plasma process. This note applies to all subsequent implant operations but will be listed only this one time.
5.	Anneal	The implanted wafer is transferred to diffusion where the wafer is cleaned before being inserted into an *anneal* furnace. Four things occur as a result of this anneal process. (1) A new barrier oxide layer is grown over the bare silicon. (2) The higher temperatures cause the movement of dopants further into the silicon (called diffusion). (3) The implant damage is repaired. (4) The bonds between the dopant atoms and the silicon atoms are activated, making the dopant atoms a part of the crystal lattice structure (electrical activation).

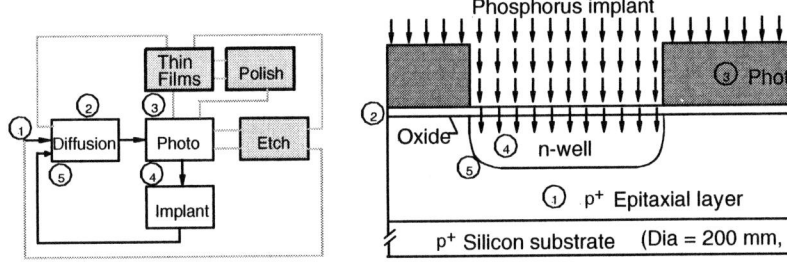

FIGURE 9.8 n-well Formation

p-well Formation ■ The three major steps in p-well formation are described in the following process and illustrated in Figure 9.9.

Description of Three Major Steps in p-well Formation

Process Step	Description
1. 2nd Mask, p-well Implant	The photolithography steps for the p-well implant mask are exactly the same as for the first mask. The only difference is the mask is the direct opposite of the n-well implant mask. Compare Figure 9.8 and Figure 9.9 to see the differences.
2. p-well Implant (High energy)	The retrograde p-well implant energy levels are considerably lower than the n-well. This condition is due to the difference in the masses of the elements being implanted. Comparing the atomic mass unit (amu) of boron (~11) to phosphorus (~31), boron is about one-third the mass of phosphorus. Thus, the energy needed to implant boron should only require about one-third of that used to implant phosphorus to the same junction depth.
3. Anneal	This anneal step is basically the same as the first annealing step.

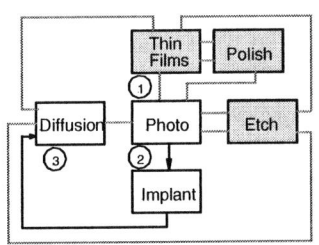 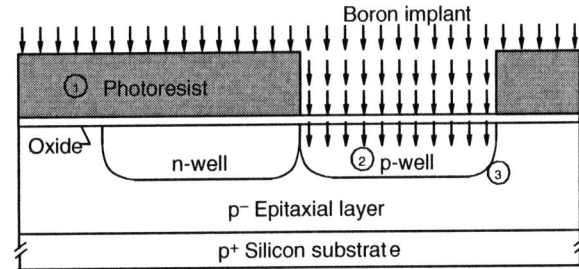

FIGURE 9.9 P-well Formation

2. Shallow Trench Isolation Process

Shallow trench isolation (STI) is an alternative method for creating isolation regions between active transistor areas on a substrate.[1] This method is especially useful in the manufacturing of quarter-micron devices. The preferred isolation technique up until the late 1990s had been the LOCOS (local oxidation of silicon) technique that was instituted in the early 1970s.[2] Despite its complexity, STI is gaining in popularity with ULSI chip manufacturing. STI formation is explained as follows in three major steps: trench etch (see Figure 9.10 on page 208), oxide fill (see Figure 9.11 on page 209), and oxide polish (see Figure 9.12 on page 209).

STI Trench Etch ■ The four major steps in STI trench etch are described in the following process and illustrated in Figure 9.10.

Description of Four Major Steps in STI Trench Etch

	Process Step	Description
1.	Barrier Oxide	The wafer arrives in the diffusion bay where the wafer is again cleaned to remove contamination and oxide. After rinsing and drying, the wafer is inserted into a high-temperature oxidation furnace. Approximately 150 Å of new oxide is grown on the wafer. This layer will form a barrier to protect active regions from chemical contamination during nitride strip.
2.	Nitride Deposition	The wafer is inserted into a high-temperature (~750 °C) *LPCVD* (*low-pressure chemical vapor deposition*) furnace. Inside the process chamber ammonia and dichlorosilane gases react to form a thin layer of silicon nitride (nitride, Si_3N_4) on the wafer. The nitride will serve two functions throughout the STI formation process steps. (1) Nitride is a durable masking material, which protects active regions during the STI oxide deposition process and (2) nitride serves as a polish-stop material during the chemical mechanical planarization (CMP) step.
3.	3rd Mask, STI	The wafer arrives in photolithography from diffusion. This photolithography process step is similar to the previous masking step, except a different mask is used. This masking step is much more critical than the first mask due to the smaller dimensions. Inspection of the wafer includes CD measurements as well as checking for *defect inspection* (*DI*), and optical *visual inspection* (*VI*). These measurements require checking the accuracy of the alignment-exposure system relative to the written specifications.
4.	STI Trench Etch	The photoresist pattern is designed to protect areas of the silicon that are not to be etched. The windows in the photoresist allow ions and highly reactive *radical chemicals* to etch nitride, oxide, and silicon in unprotected areas. The preferred tool for etching these deep trenches is the *dry plasma etcher*. The etcher uses high-powered radio frequency (RF) energy to ionize a fluorine- or chlorine-based gas inside a vacuum process chamber. The RF energy dissociates molecules and ionizes atoms to create a chamber filled with a variety of plasma components. These plasma components provide the chemical and physical etching that results in the removal of silicon in locations designated as *isolation regions*. The slanted profile and the rounded bottom of the trenches improve the filling process and the electrical characteristics of the isolation structure. *Note:* Following completion of each etch operation, the wafer is stripped of photoresist and wet cleaned in a series of chemical baths. Key inspection procedures include measurements that verify proper step height (Å), etch rate (Å/min), CD, and DI. This note applies to all subsequent etch operations but will be listed only this one time.

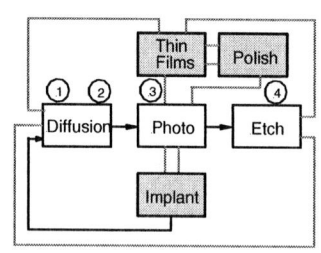

FIGURE 9.10 STI Trench Etch

STI Oxide Fill ■ The basic steps in STI oxide fill are described in the following process and illustrated in Figure 9.11.

Description of Basic Steps in STI Oxide Fill

	Process Step	Description
1.	Trench Liner Oxide	The wafer is again cleaned in the diffusion bay to remove contamination and oxide. After rinsing and drying, the wafer is inserted into a high-temperature oxidation furnace. Approximately 150 Å of oxide grows in the exposed walls of the isolation trenches. The nitride mask prevents oxygen diffusion into the active regions. The purpose of the liner oxide is to improve the interface between the silicon and the trench CVD oxide.
2.	Trench Fill with CVD Oxide	This deposition process is performed either in diffusion using an LPCVD furnace or in thin films using a variety of oxide CVD systems. Higher throughput and deposition rates are obtained with the furnace.

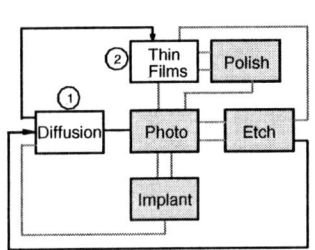
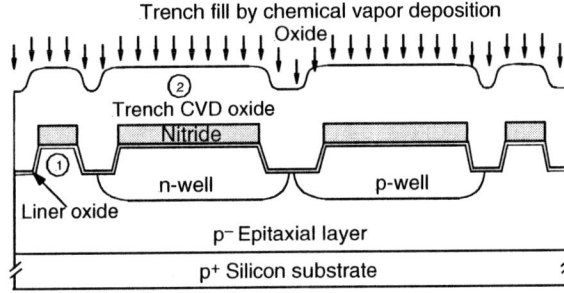

FIGURE 9.11 STI Oxide Fill

STI Oxide Polish-Nitride Strip ■ Planarization of the wafer surface is accomplished by several methods. In the past, filling the gap to planarize the wafer could be done using *spin-on-glass* (*SOG*), consisting of 80% solvent and 20% silicon dioxide. Baking the SOG after the deposition evaporated the solvent, leaving oxide inside the gaps. A complete surface *etchback* also could be performed to reduce the overall thickness. However, no method has been as effective a planarization technique as CMP, chemical mechanical planarization (also referred to as *polish*). The following table and Figure 9.12 on page 210 describe and illustrate the process.

Description of Basic Steps in STI Oxide Polish and Nitride Strip

	Process Step	Description
1.	Trench Oxide Polish (CMP)	In the CMP area the wafer is turned upside down and held by vacuum to a wafer carrier. The wafer rotates in the same direction as a polishing pad attached to a rotating polishing table. A chemical slurry keeps the pad and wafer wet during the polishing process and provides reactants to help abrade the oxide surface. The nitride, being a harder material than the oxide, serves as the polish-stop material to prevent over-polishing the isolation structures. *Note:* The following statement applies to all polishing operations and will not be repeated again. The polishing process creates excessive particles and chemical contaminants that must be thoroughly removed before transferring the wafer to other areas in the wafer fab. Special scrub stations clean the top and bottom of the wafer and ensure it is rinsed and dried. Inspections for film thickness, particles, and defects are also conducted before wafer is transferred.
2.	Nitride Strip	The diffusion bay is traditionally equipped with hot phosphoric acid stations for stripping the nitride from the wafer. In the diffusion bay the wafer is stripped, cleaned, rinsed, and dried, then inspected to check the thickness of the isolation oxide.

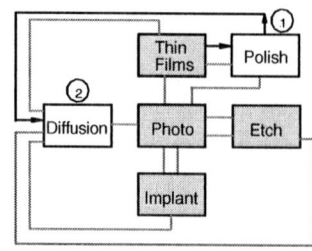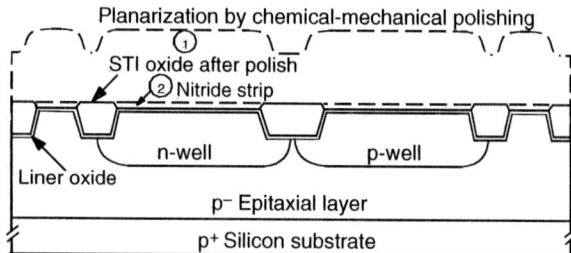

FIGURE 9.12 STI Formation

3. Poly Gate Structure Process

Forming the structure of the transistor gate is a critical step in the process because it includes thermally growing the thinnest *gate oxide* and patterning and etching the *polysilicon gate,* one of the smallest physical structures in the entire IC process (see Figure 9.13). The width of the polysilicon gate is often the most critical CD linewidth of the entire wafer.

Description of Basic Steps in Forming the Poly Gate Structure

	Process Step	Description
1.	Gate Oxide Growth	The wafer is cleaned to remove contamination and oxide. There is only a matter of a few hours before the wafer must be inserted in the oxidation furnace. Oxide will form as long as the silicon is exposed to oxygen in the atmosphere. The wafer is inserted in the oxidation furnace and a thin oxide of approximately 20 to 50 Å is grown.
2.	Polysilicon Deposition	The wafer is immediately loaded into an LPCVD furnace where silane (SiH_4) is introduced into the process chamber. The silane dissociates and polysilicon is deposited on the wafer. Approximately 5000 Å of polysilicon (also referred to as poly) is deposited. Some processes may call for doping the polysilicon immediately following the deposition step. This action can be done in the same process chamber or in a separate furnace.
3.	4th Mask, Poly Gate	In photolithography, deep UV lithography (assuming sub-0.25 μm technology) techniques are used to pattern the fine structures of the polysilicon gates. An *antireflective coating* (ARC) is commonly applied between the poly and the photoresist to reduce undesirable reflections. The resist linewidths for the gates are the narrowest structures on the IC; therefore, various quality measurements are required, such as CD and overlay registration (OL), and defect inspection (DI).
4.	Poly Gate Etch	By far one of the most critical etch steps in the IC process requires the use of the best *anisotropic plasma etchers* available in the fab. These single-direction etchers etch the polysilicon to provide the vertical profile as shown in Figure 9.13.

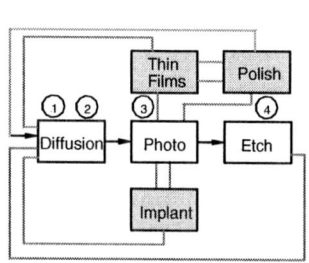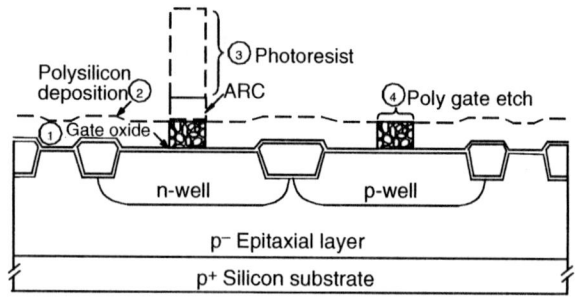

FIGURE 9.13 Poly Gate Structural Process

4. Lightly Doped Drain (LDD) Implant Process

As gate widths get smaller, the *channel length* below the gate structure (the silicon region between the source and drain) also decreases. This reduction increases the potential for charges to *punchthrough* the transistor source and drain and cause undesirable channel leakage current. There are techniques that are used to reduce the occurrence of this leakage in the channel.

The next series of ion implant steps begins to define the source and drain regions of transistors. Each pMOS and nMOS transistor will be implanted twice—once with a shallow implant called the *lightly doped drain (LDD) implant*, then followed by a medium or high dose *source/drain implant*. Lightly doped drain (LDD) implants (see Figure 9.14 below and Figure 9.15 on page 212) use the larger mass of arsenic and BF_2 dopant materials to create an *amorphous* (disorganized with no long-range order or crystal structure) upper layer of silicon. The combination of large mass and amorphous surface conditions helps maintain a shallow junction, which also helps reduce channel current leakage effects between the source and drain.

n⁻ LDD Implant ■ The steps for creating n⁻ lightly doped drain implants are described in the following process and illustrated in Figure 9.14.

Description of n- Lightly Doped Drain Implant

	Process Step	Description
1.	5th Mask, n⁻ LDD Implant	The purpose of this masking step is to pattern the wafer in a manner that opens windows in the photoresist where n-channel transistors can be implanted. All other areas are protected by the remaining photoresist.
2.	n⁻ LDD Implant (Low energy, shallow junction depth)	Arsenic ions are selectively implanted through the windows in the patterned resist. The energy, dose and depth are significantly lower than in the previous n-well implant steps. Arsenic is preferred over phosphorus because its larger mass amorphizes the silicon surface to create a more uniform dopant depth during implant.

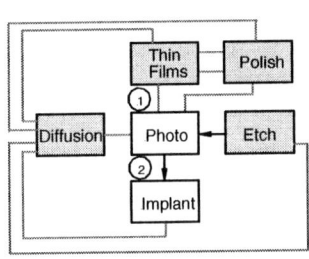
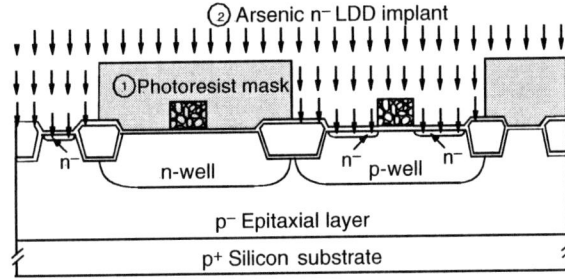

FIGURE 9.14 n⁻ LDD Implant

p⁻ LDD Implant ■ The steps for creating p⁻ lightly doped drain implants are described in the following process and illustrated in Figure 9.15.

Description of p- Lightly Doped Drain Implant

	Process Step	Description
1.	6th Mask, p⁻ LDD Implant	The purpose of this masking step is to pattern the wafer in a manner that opens windows in the photoresist where p-channel transistors can be implanted. All other areas are protected by the remaining photoresist.
2.	p⁻ LDD Implant (Low energy, shallow junction depth)	Boron difluoride, BF_2, is preferred over boron for this implant step. BF_2 is a heavier substance than boron. The heavier BF_2 helps amorphize the silicon surface.

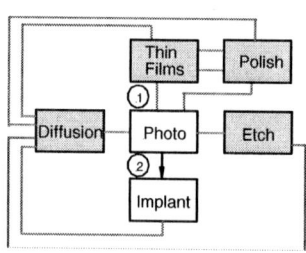
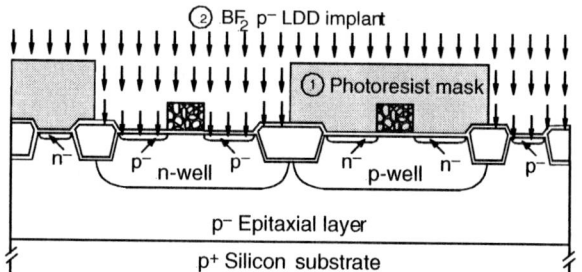

FIGURE 9.15 p⁻ LDD Implant

5. Sidewall Spacer Formation

Sidewall spacers will be used alongside the poly gates to prevent the higher source/drain (S/D) implant from penetrating too close to the channel where S/D punchthrough could occur. Sidewall spacer formation requires two major process steps (see Figure 9.16). First, an oxide layer is deposited across the surface of the wafer, then the oxide is etched back using a dry etch process. There is no requirement for a masking step because the anisotropic (single-direction) etch tool uses ions to sputter away most of the oxide. The etch process ends when the polysilicon is exposed. However, not all of the oxide is removed. Some oxide remains on the sidewalls of the polysilicon gates.

Description of Two Major Steps in Sidewall Spacer Formation

	Process Step	Description
1.	Spacer Oxide Deposition	This procedure is performed in the thin film bay. Approximately 1000 Å of oxide is deposited with a chemical vapor deposition (CVD) process. This layer will be used to form spacers on the sides of the polysilicon gates.
2.	Spacer Oxide Etchback	The dry plasma etcher removes most of the CVD oxide leaving behind the thicker oxide on the sidewalls of the polysilicon gates.

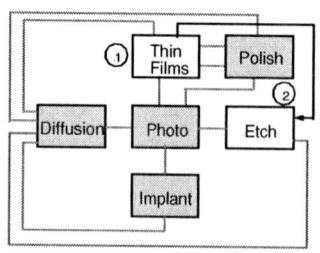
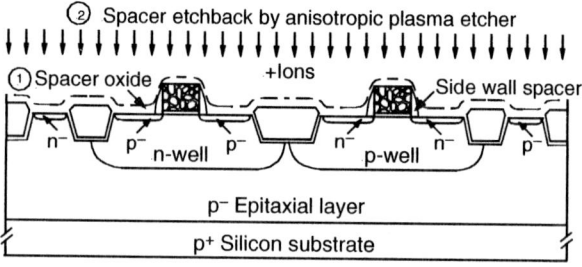

FIGURE 9.16 Sidewall Spacer Formation

6. Source/Drain (S/D) Implant Processes

To complete the retrograde implant technique, medium-dose implants are made to penetrate the silicon slightly beyond the LDD junction depth, but not as deep as the original twin-well implants (see Figure 9.17 and Figure 9.18). The spacer oxide from the previous step will protect the channel from the dopant atoms during the implant process.

n^+ S/D Implant ■ The steps for creating n^+ source/drain implants are described in the following process and illustrated in Figure 9.17.

Description of n+ Source/Drain Implant

	Process Step	Description
1.	7th Mask, n^+ S/D Implant	This masking step defines areas of the nMOS transistors that are to be implanted.
2.	n^+ S/D Implant (Medium energy)	This is a medium-energy implant step that penetrates the silicon deeper than the LDD junction depth. Spacer oxide prevents the arsenic dopant from encroaching into the narrow channel.

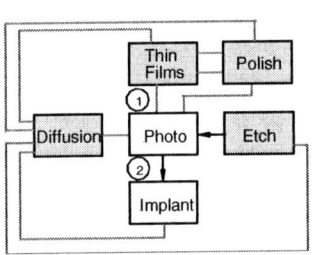 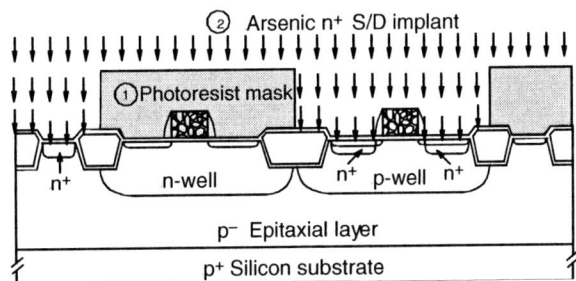

FIGURE 9.17 n^+ S/D Implant

p^+ S/D Implant ■ The steps for creating the p^+ source/drain regions are similar to n^+ S/D formation. The steps for p^+ S/D are described in the following process and illustrated in Figure 9.18.

Description of p+ Source/Drain Implant

	Process Step	Description
1.	8th Mask, p^+ S/D Implant	This masking step defines areas of the pMOS transistors that are to be implanted.
2.	p^+ S/D Implant (Medium energy)	This implant step penetrates the silicon slightly beyond the LDD junction depth. Spacer oxide prevents boron dopant from encroaching into the narrow channel.
3.	Anneal	The implanted wafer is annealed in a *rapid thermal process* (*RTP*) tool. RTP tools have the ability to quickly reach temperatures of ~1000 °C and maintain the setpoint for several seconds. This condition is important to prevent structures from spreading and to control the diffusion of dopants in the S/D regions.

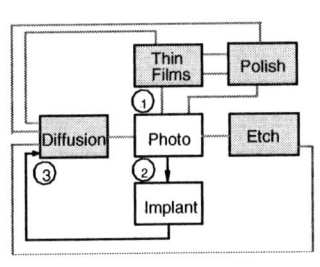 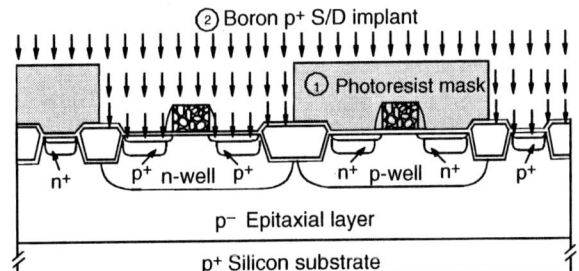

FIGURE 9.18 p^+ S/D Implant

7. Contact Formation

The purpose of the contact formation process steps is to form *metal contacts* on all active areas of silicon. These metal contacts serve to promote adhesion between the silicon and the metal conductor material that will be deposited later (see Figure 9.19). Titanium is a good choice for the metal contact material, although cobalt can also be used. Titanium has low resistivity characteristics and reacts very well with silicon. When raised to a temperature >700 °C, titanium bonds with silicon to form a *titanium silicide* ($TiSi_2$) compound, or *tisilicide*. Titanium and silicon dioxide do not react; therefore, no chemical bonding or physical connection is made between these two materials. Thus, titanium can be easily etched off the oxide without the need for a masking step. Tisilicide remains at all locations where active silicon exists (e.g. source, drain, and gate).

Description of Major Steps in Titanium Contact Formation

	Process Step	Description
1.	Titanium Deposition	The wafer is cleaned thoroughly to remove contaminants and oxides from the silicon. Titanium (Ti) is deposited on wafers using a sputtering process. *Sputtering* is a *physical vapor deposition* (*PVD*) process that is performed in a plasma process chamber where energetic argon ions bombard a metal target to release metal atoms, which deposit onto a wafer.
2.	Anneal	The wafer is inserted in the RTP tool. The high temperature triggers a chemical reaction between titanium and silicon that forms $TiSi_2$ (tisilicide).
3.	Titanium Etch	Chemicals etch away the unreacted titanium, leaving behind tisilicide over the active silicon areas.

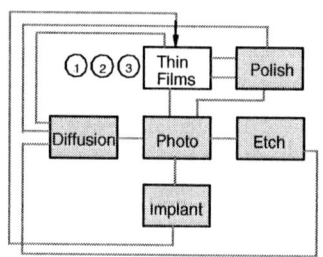 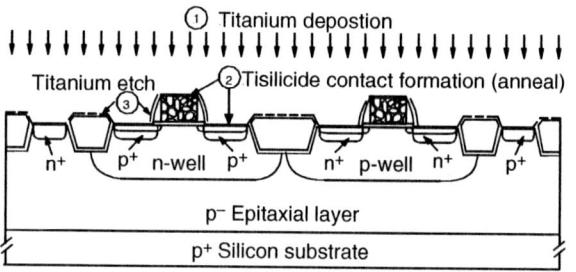

FIGURE 9.19 Contact Formation

8. Local Interconnect (LI) Process

The next step in the IC manufacturing process is to form metal connecting lines between transistors and other tisilicide contacts. The method used in this process flow is referred to as *local interconnect* (*LI*). The steps leading up to the formation of LI are as complicated as the formation of the STI. The process begins with the deposition of dielectric films, followed by CMP, patterning, etch, and tungsten metal deposition and finished off with a metal polish (see Figure 9.20 and Figure 9.21). This process is referred to as *damascene*—a name adopted from a practice that began thousands of years ago by artisans in Damascus, Syria.[3] The result of these steps produces a top surface that resembles intricate inlaid jewelry or artwork. The graphic in Figure 9.22 illustrates how the metal lines are formed within the walls of the etched oxide.

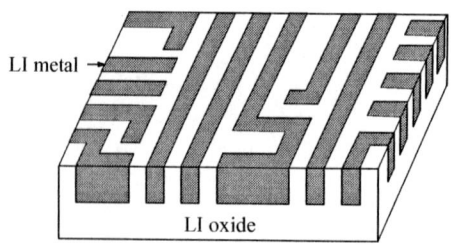

FIGURE 9.20 LI Oxide as a Dielectric for Inlaid LI Metal

Description of LI Oxide Dielectric Formation

	Process Step	Description
1.	Nitride (Si_3N_4) CVD	A barrier layer of silicon nitride is deposited using a CVD process. The nitride provides protection for the active regions from the dopants in the next film layer.
2.	Doped Oxide CVD	The LI dielectric component of the local interconnect structure is formed from CVD oxide (SiO_2). This oxide is lightly doped with phosphorus or boron. The addition of dopants to the oxide improves the dielectric qualities of the glass. An additional RTP step allows the glass to flow and smooth the surface.
3.	Oxide Polish (CMP)	CMP is used to planarize the LI oxide. The resulting thickness of the oxide after polish is ~8000 Å.
4.	9th Mask, LI Etch	The wafer is patterned in photolithography and then etched in the etch bay. Narrow trenches are created in the LI oxide that will serve as the forms for defining the paths for the LI metal.

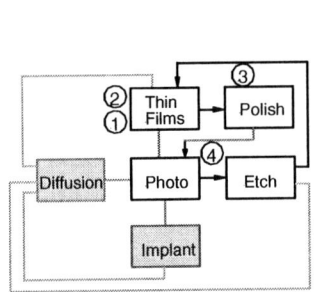 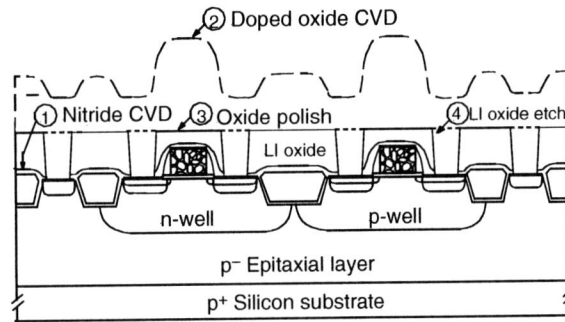

FIGURE 9.21 LI Oxide Dielectric Formation

Description of LI Metal Formation

	Process Step	Description
1.	Titanium Deposition (PVD process)	A thin barrier layer of titanium (Ti) lines the bottom and the inside of the LI walls. The Ti serves as a "double-adhesive tape" to hold the tungsten (W) to the SiO_2.
2.	Titanium Nitride Deposition	Titanium nitride (TiN) is deposited immediately over the Ti to serve as a diffusion barrier for the tungsten metal.
3.	Tungsten Deposition (CVD process polish)	Tungsten fills the LI trenches and coats the entire wafer. Tungsten is preferred over aluminum for LI metal due to its ability to fill holes without leaving voids in the formed metal plug. Another reason is for its good polishing characteristics.
4.	Tungsten polish	The tungsten is polished down to the upper surface of the LI oxide.

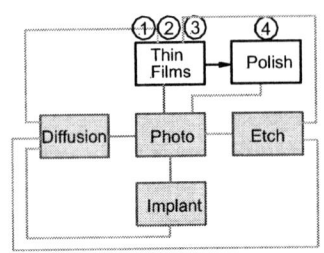 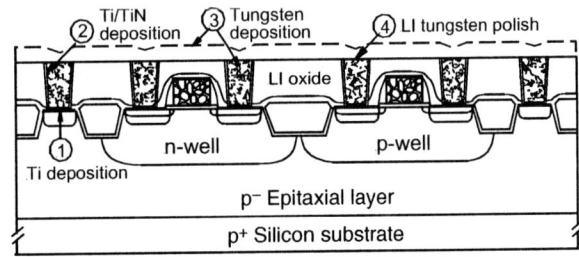

FIGURE 9.22 LI Metal Formation

9. Via-1 and Plug-1 Formation

The *interlayer dielectric* (*ILD*) serves as an insulator material between each metal layer or between the first metal layer and silicon. The ILD will have many small *vias,* which are small openings in the ILD that provide an electrical pathway from one metal layer to adjacent metal layers. Vias are filled with a conductive metal (usually tungsten, and referred to as a *tungsten plug*) and are placed at the appropriate locations to form the electrical circuit between metal layers (see Figure 9.23 and Figure 9.24). ILD-1 is the first of a series of interconnection process to be described.

Description of Major Steps in Via-1 Formation

	Process Step	Description
1.	ILD-1 Oxide Deposition (CVD)	Oxide is deposited across the surface of the wafer with a CVD tool in thin films. This layer (ILD-1) will serve as the dielectric material from which vias are to be formed.
2.	Oxide Polish	CMP is used to planarize the ILD-1 oxide. The resulting thickness of the oxide after polish is approximately 8000 Å. The wafer is scrubbed and cleaned to remove particles generated during the polish step.
3.	10th Mask, ILD-1 Etch	The wafer is patterned in photolithography and then etched in the etch bay. Small via holes, less than a quarter micron in diameter, are etched into the ILD-1 oxide. This step is critical in terms of CD, OL, and defects.

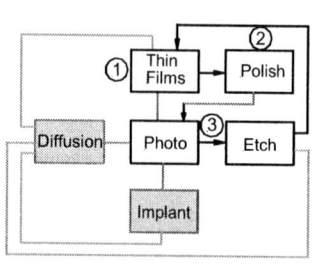
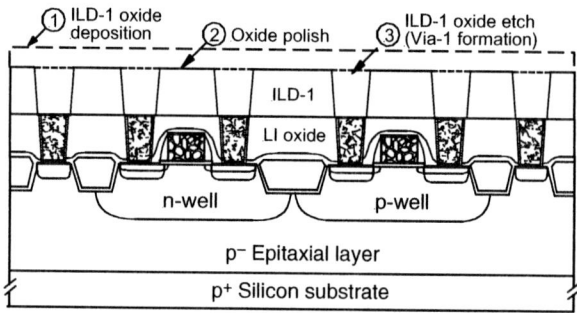

FIGURE 9.23 Via-1 Formation

Description of Major Steps in Plug-1 Formation

	Process Step	Description
1.	Titanium Barrier Metal Deposition (PVD)	A thin layer of titanium is deposited across the surface of the wafer with a PVD tool in thin films. Titanium lines the bottom and the walls of the via holes. The Ti serves as the glue that holds the tungsten plug inside the hole.
2.	Titanium Nitride Deposition (CVD)	TiN is thinly deposited on top of the Ti layer. The TiN serves as a diffusion barrier for the tungsten in the next deposition step.
3.	Tungsten Deposition (CVD)	Another CVD tool is used to deposit tungsten on the wafer. Tungsten fills the small openings to form the plug.
4.	Tungsten Polish	The tungsten-coated wafer is polished down to the upper surface of the ILD-1 oxide.

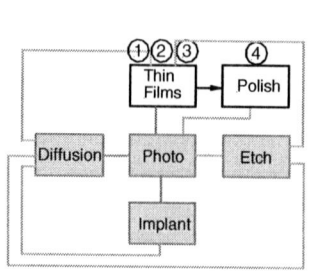
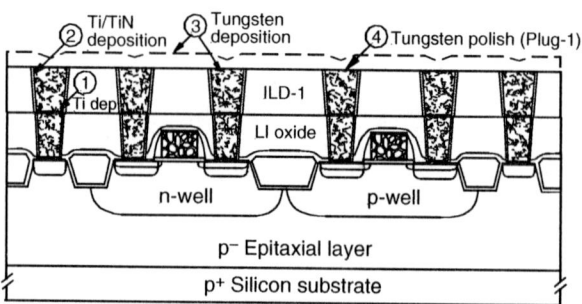

FIGURE 9.24 Plug-1 Formation

IC Fabrication Process Overview **217**

SEM Micrograph of Polysilicon, Tungsten LI and Tungsten Plugs (Polysilicon is represented by the thin lines between the taller LI lines). (Micrograph courtesy of Integrated Circuit Engineering)

10. Metal-1 Interconnect Formation

The next series of operations has to do with depositing a three-layer metal film, referred to as a *metal stack* or *sandwich,* on the wafer, followed by a masking and etching step (see Figure 9.25). The multilayer metal stack consists of different refractory metals—titanium, aluminum/ copper, and titanium nitride. When finished, the wafer will have the first of five metal stacks to be constructed on the device. The number of metal layers on a device varies depending on die complexity, with advanced die currently having around eight metal layers. Including LI metal, the device covered in this process flow has a total of six metal layers.

Description of Steps in Metal-1 Interconnect Formation

	Process Step	Description
1.	Titanium Barrier Metal Deposition (PVD)	As with other metal processes, titanium is the first metal to be deposited on the entire wafer. It provides a good bond between the tungsten plugs and the next metal, aluminum. It also bonds well to the interlayer dielectric material and improves the reliability of the metal stack.
2.	Aluminum-Copper Deposition (PVD)	The aluminum-copper (99% Al, 1% Cu) metal is sputtered onto the titanium-coated wafer by a PVD tool in thin films. The 1% Cu is added to the aluminum to improve the reliability of the aluminum.
3.	Titanium Nitride Deposition (PVD)	TiN is thinly deposited on top of the aluminum-copper layer to serve as an antireflective coating for the next photomasking step.
4.	11th Mask, Metal Etch	The wafer is patterned with photoresist, then the three-layer metal stack is etched using a plasma etcher.

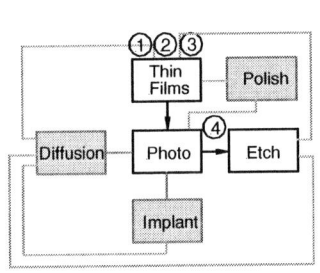
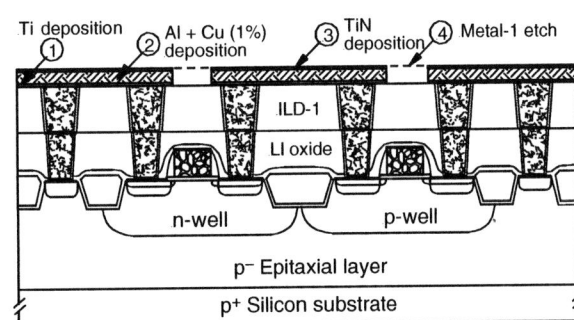

FIGURE 9.25 Metal-1 Interconnect Formation

11. Via-2 and Plug-2 Formation

The next four operations will result in the formation of the second interlayer dielectric (ILD-2) and the via openings (see Figure 9.26 and Figure 9.27). ILD-2 is similar to ILD-1 except for the very first step of filling small and large gaps that were etched into Metal-1. *Gap fill* is the filling of these gaps with a dielectric material that can be placed into the narrow spaces without creating voids or other defects that could affect electrical performance. There are two common methods for filling gaps—spin-on-glass (SOG) with etchback and *high-density plasma chemical vapor deposition* (HDPCVD). HDPCVD is the preferred method for sub-quarter micron processes. Once the gaps are filled, a plasma-enhanced CVD (PECVD) system finishes depositing the remainder of the ILD-2 oxide. Following the deposition of the ILD-2, the oxide is polished, patterned, and then etched to open the vias for the tungsten plugs to be formed.

Description of Major Steps in Via-2 Formation

	Process Step	Description
1.	ILD-2 Gap Fill	The latest gap-fill method for ULSI devices utilizes a high-density plasma process to alternately deposit and etch the interlayer oxide as its being deposited. The result is a dense oxide with few or no voids in the metal gaps.
2.	ILD-2 Oxide Deposition	A PECVD system is used to deposit the remainder of the ILD-2 oxide layer.
3.	ILD-2 Oxide Polish	The wafer is polished to planarize the surface prior to the next patterning step.
4.	12th Mask, ILD-2 Etch	The wafer is patterned with photoresist, then the ILD-2 oxide is etched using a plasma etcher.

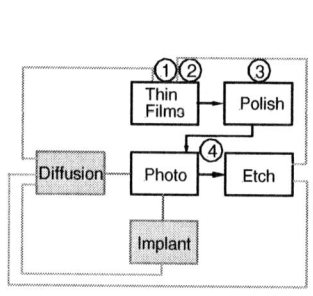
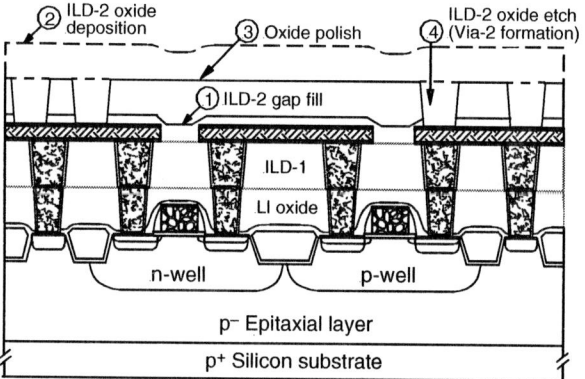

FIGURE 9.26 Via-2 Formation

Description of Major Steps in Plug-2 Formation

	Process Step	Description
1.	Titanium Barrier Metal Deposition (PVD)	As with other metal processes, titanium is the first metal to be deposited on the wafer. It provides a good bond between the tungsten plugs and the next metal, aluminum. It also bonds well to the interlayer dielectric material and improves the reliability of the metal stack.
2.	Titanium Nitride Deposition (CVD)	TiN is thinly deposited on top of the Ti layer. The TiN serves as a diffusion barrier for the tungsten in the deposition step.
3.	Tungsten Deposition (CVD)	Another CVD tool is used to deposit tungsten on the wafer. Tungsten fills the small via openings to form the plug.
4.	Tungsten Polish	The tungsten-coated wafer is polished down to the upper surface of the ILD-2 oxide, leaving the tungsten plugs in the vias.

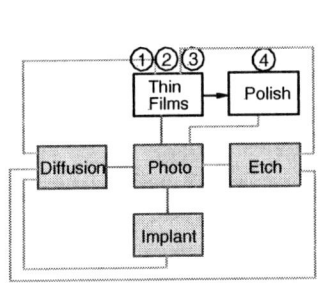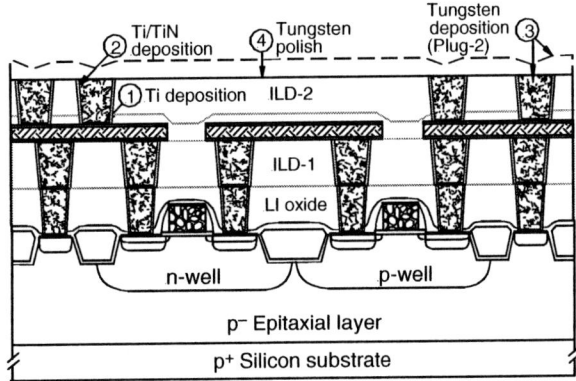

FIGURE 9.27 Plug-2 Formation

SEM Micrograph of First Metal Layer over First Set of Tungsten Vias (Note TiN Cap over Aluminum)
(Micrograph courtesy of Integrated Circuit Engineering)

12. Metal-2 Interconnect Formation

The next series of operations describes the process of forming the interconnects between layers (see Figure 9.28 on page 220). The process is repeated in the same manner for all remaining metal stacks.

Description of Major Steps in Metal-2 Interconnect Formation

	Process Step	Description
1.	Metal-2 Deposition to Etch	The second metal stack is deposited exactly as the first metal stack. The stack is a composite of three layers—Ti, Al/Cu, and TiN. A plasma etcher is used to etch the Metal-2 lines through windows in the patterned resist.
2.	ILD-3 Gap Fill	Following the Metal-2 etch, HDPCVD is used to fill the metal gaps with dense oxide.
3.	ILD-3 Oxide Deposition to Polish	ILD-3 oxide is deposited using a PECVD method. This action is followed by polishing the oxide to attain surface planarity.
4.	Via-3 Etch, Ti/TiN Deposition, Tungsten Deposition, Polish	Via-3 openings are etched, then these are filled with barrier metal (Ti/TiN). Tungsten is deposited across the wafer surface. The tungsten is polished off the wafer until the ILD-3 oxide layer is reached. Tungsten plugs are left in the vias to provide the interconnect between Metal-2 and Metal-3.

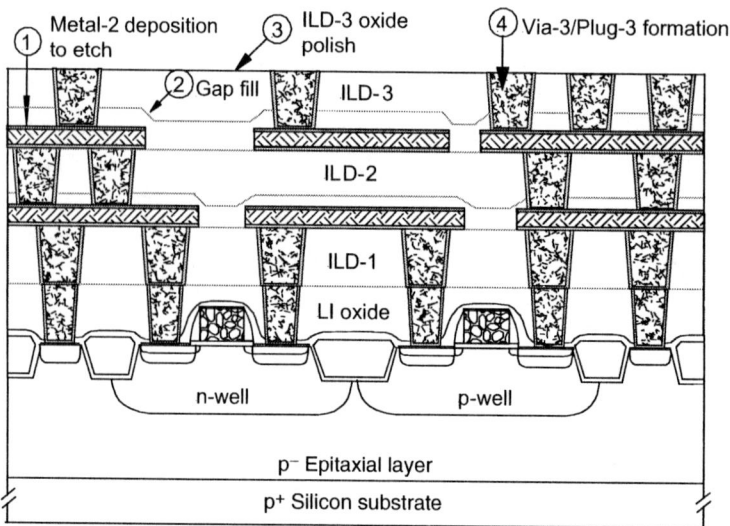

FIGURE 9.28 Metal-2 Interconnect Formation

13. Metal-3 to Pad Etch and Alloy

After repeating the layering process for layers 3 and 4 and at the completion of the Metal-4 etch, ILD-5 oxide is deposited using the thin film process (Figure 9.29). CMP is not required for this dielectric layer because the structures to be patterned are much larger than the quarter micron dimensions formed earlier in the process. The ILD-5 layer is etched to allow metal filling by the Metal-5 deposition. The Metal-5 layer is deposited thicker than previous metal stacks. This metal layer is etched as necessary to form bonding pads and to remove metal from areas where it is not needed.

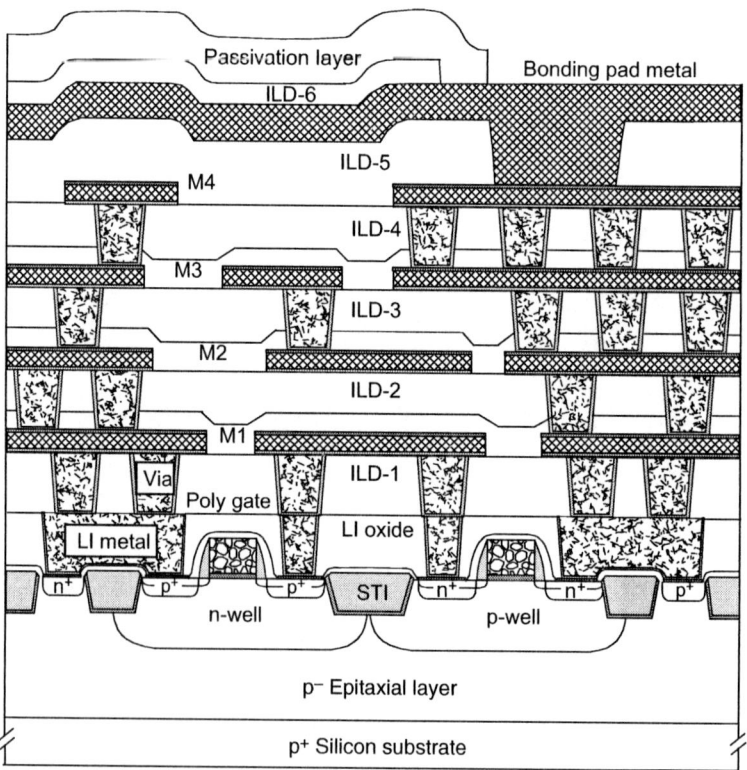

FIGURE 9.29 Full 0.18 μm CMOS Cross Section

This product incorporates 0.18 μm processing technology, including twin-tub technology, retrograde implants, shallow trench isolation, shallow S/D diffused regions, sidewall spacers, titanium silicide contacts, titanium barrier metal, tungsten local interconnect, tungsten plugs, three-layer metal stacks, HDP oxide gap fill, and PECVD oxide as the interlayer dielectric.

The final steps of the process include one more oxide layer (ILD-6) followed by a top layer of silicon nitride (~2000 Å). This layer is referred to as the *passivation layer*. Its purpose is to protect the product from moisture, scratches, and contamination. A final low-temperature alloy step in a diffusion furnace is performed. This heat treatment helps improve the metallurgical bonds between metal connections, thus improving electrical performance and reliability. Care must be taken with this alloy step to prevent overheating the product, which can cause permanent structural defects.

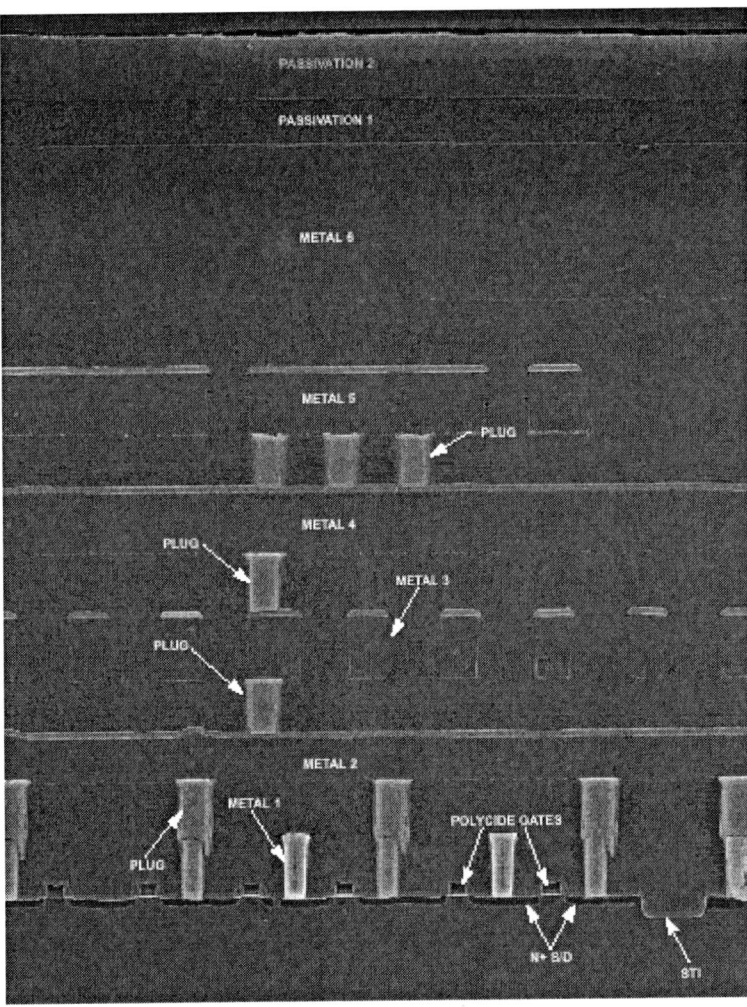

SEM Micrograph of Cross-section of Microprocessor (Micrograph courtesy of Integrated Circuit Engineering)

14. Parametric Testing

The wafer is tested twice to determine its product-worthiness—once when the wafer completes the first metal etch and again after the completion of the last wafer fab process step. Following metal etch, special microprobes connected to electronic test instruments are used to measure certain electrical parameters of specific device test structures on the wafer. This in-line parametric testing procedure is also referred to as *wafer electrical test,* or WET.

Wafer Electrical Test Using a Micromanipulator Prober (Photo courtesy of Advanced Micro Devices)

The last process step for the wafer is at wafer test/sort just outside the wafer fab. The wafer is probed and tested automatically in the electrical test and sort department (also referred to as *E-sort*). Each die on the wafer is tested for electrical functionality. Those die that fail are tracked through software using x-y position coordinates (previously marked with an ink spot) so that the locations of good and bad die are known for later processing. This data is used to determine die yield; that is, the percentage of good die versus bad die on a wafer. The wafer is now ready to be shipped to the assembly department. In some companies, the assembly plants are located in other parts of the world.

If the wafer passes the wafer sort, the wafer is sent to the backgrind department. Here the backgrinder grinds the backside of the wafer. This process makes the wafer thinner and easier to slice for eventual separation into individual die.

SUMMARY

An overview of the wafer fab process defines six major process areas: diffusion, photo, etch, ion implant, thin films, and polish, plus wafer test/sort at the end of the process. Diffusion is for high-temperature processing; photolithography patterns the wafer with photoresist; etch replicates the photoresist pattern in the wafer; ion implant dopes the wafer; thin films deposit dielectric and metal layers; and polish planarizes the top surface of the wafer.

A simplified CMOS process is defined with 14 manufacturing steps. (1) Twin-well implants create the n-well and p-well in the silicon. (2) Shallow-trench isolation isolates the silicon-active areas for sub-0.25 μm technology. (3) The gate structure is formed through gate oxide growth, polysilicon deposition, and patterning. (4) Lightly doped drain (LDD) implant forms a shallow implant in the source and drain. (5) Sidewall spacer formation protects the channel from the subsequent source/drain implant. (6) Source/drain implant is a medium-energy implant that forms a junction depth deeper than the LDD implant. (7) Contact formation forms a silicide contact for tungsten metal to alloy with the silicon. (8) Local interconnect (LI) forms the first metal connecting lines between transistors and contacts. (9) First interlayer dielectric (ILD) to Via-1 deposits the dielectric and creates vias between LI and first metal layer. (10) First metal layer to first metal etch deposits the metal sandwich and patterns the metal. (11) Second ILD to Via-2 deposits the second interlayer dielectric and creates the via openings. (12) Second metal layer to Via-3 deposits the second metal stack and deposits and etches ILD-3. (13) Metal-3 to pad etch and alloy repeats the layering process until the metal-5 bonding pad is deposited, followed by ILD-6 and a passivation layer. (14) The final process step is parametric testing to verify acceptability of each die on the wafer.

KEY TERMS

diffusion
high-temperature diffusion furace
wet cleaning station
photolithography
coater/developer track
stepper
etch
plasma etcher
plasma resist stripper
ion implanter
thin films
CMP (chemical mechanical planarization)
wafer scrubbers
twin-well (twin-tub)
retrograde implant
latchup
epitaxial silicon
native oxide
alignment-and-exposure system
linewidths
critical dimension (CD)
junction depth
ion implanter
anneal
shallow trench isolation (STI)
LPCVD (low-pressure chemical vapor deposition)
sidewall liner
overlay registration (OL)
defect inspection (DI)
visual inspection (VI)
radical chemicals

dry plasma etcher
isolation regions
spin-on-glass (SOG)
etchback
polish
gate oxide
polysilicon gate
anti-reflective coating (ARC)
silane
anisotropic plasma etcher
channel length
punchthrough
lightly doped drain (LDD) implant
source/drain implant
amorphous
sidewall spacers
rapid thermal process (RTP)
metal contacts
titanium silicide (tisilicide)
local interconnect (LI)
damascene
interlayer dielectric (ILD)
vias
tungsten plug
metal stack
sandwich
gap fill
high-density plasma chemical vapor deposition (HDPCVD)
plasma enhanced CVD (PECVD)
passivation layer
wafer electrical test (WET)

REVIEW QUESTIONS

1. List the six distinct production areas in a wafer fab and give a short description of each area.
2. What activity is performed in the diffusion area?
3. List five processes that are done in a high-temperature furnace.
4. What is the purpose of photolithography?
5. What is a coater/developer track used for? List five operations performed using this tool.
6. What is the major concern for particulate contamination in photolithography?
7. Identify the three production areas where photoresist-coated wafers can be found.
8. What is the purpose of the etch process? Name the most common tools used in this area?
9. What is the ion implanter used for?
10. What is the purpose of the thin films bay?
11. List four different tools used or processes done in the thin films area.
12. What does CMP stand for, and what is its purpose? What is another name for CMP?
13. List the 14 manufacturing steps outlined for a typical CMOS process.
14. Describe the epitaxial layer on the silicon wafer.
15. What is a twin-well?

16. Explain the retrograde implant technique. What problem does it help resolve?
17. What are the reasons for the thermal anneal process after ion implantation?
18. Why are the retrograde p-well implant energy levels lower than the n-well implant energy levels?
19. What is shallow trench isolation (STI)? What process did it replace?
20. What tool is used to etch silicon for STI, and why?
21. What tools are used to deposit the oxide in STI oxide fill?
22. Describe the most effective wafer planarization technique.
23. Why is the formation of the transistor gate structure a critical process step?
24. What problem can occur due to smaller gate widths?
25. How do lightly doped drain (LDD) implants reduce the channel-current leakage effects?
26. Explain the purpose of sidewall spacers.
27. What are medium-dose implants after the LDDs used for?
28. What is the purpose of the metal contact?
29. What is a local interconnect (LI)?
30. What is the purpose of the ILD?
31. What is a via? What is a tungsten plug?
32. Describe the materials used in a metal stack.
33. What is gap fill?
34. Why is CMP not required for the final ILD layer?
35. What is the purpose of the passivation layer?
36. Describe what happens in the wafer test/sort area.

REFERENCES

1. S. Ghandhi, Chap. 11 in *VLSI Fabrication Principles: Silicon and Gallium Arsenide,* 2nd ed., (New York: Wiley, 1994), pp. 704–800.
2. S. Wolf, *Process Integration,* vol. 2, *Silicon Processing for the VLSI Era,* (Sunset Beach: Lattice Press, 1990), pp. 17–19.
3. P. Singer, "Making the Move to Dual-Damascene Processing," *Semiconductor International* (January 1999): pp. 70–82.

CHAPTER 10
OXIDATION

A cornerstone of silicon IC wafer fabrication is the ability to thermally grow an oxide layer on the surface of a silicon wafer. A major development from the 1950s was *oxide masking,* which patterned and etched openings in a thermally grown oxide layer to diffuse dopants into the silicon substrate. This development was a key factor in developing a process to fabricate large quantities of transistors.[1] In this way, oxidation played a major role in the development of silicon planar technology and remains viable today, which explains the widespread usage of silicon in wafer fabrication.

With the proper manufacturing control, oxide layers have high quality, stability, and desirable dielectric properties.

Because of these qualities, oxidation is critical, especially for the thin oxide that is an essential part of the gate structure for MOS technology. Thermal oxide is used as a dielectric material, as well as for device isolation, oxide screens for implants, stress-relief oxides (pad oxides), and reoxidizing nitride and polysilicon surfaces for photoresist adhesion and stress reduction.

Oxides can be either deposited or grown. The focus in this chapter will be on thermally grown oxides. We will review the nature of oxide films and how they are grown, including detailed information on the growth mechanism and the high-temperature chamber where growth occurs.

OBJECTIVES

After studying the material in this chapter, you will be able to:

1. Describe an oxide film for semiconductor manufacturing, including its atomic structure, its various uses, and its benefits.
2. State the chemical reaction for oxidation, and describe how oxide grows on silicon.
3. Explain selective oxidation and give two examples.
4. Identify the three types of thermal processing equipment, describe the five parts of a vertical furnace, and discuss the attributes of a fast ramp vertical furnace.
5. Explain what is a rapid thermal processor, its usage and design.
6. Describe the critical aspects of the oxidation process, including its quality measures and some common troubleshooting problems.

INTRODUCTION

Oxide on a silicon wafer is created by either the grown or deposited method. A *grown oxide layer* occurs on a wafer by providing externally supplied high-purity oxygen in an elevated-temperature environment to react with the silicon substrate. The high-temperature oxidation process occurs in the diffusion area of the wafer fab. This is the first process area when a silicon wafer enters the fabrication process (see Figure 10.1 on page 226). A *deposited oxide layer* is generated by using an external silicon source and O_2 and reacting these materials in a chamber to form a thin film on the wafer surface. This chapter will cover the basic technology of oxidation and explain the high-temperature process of thermal oxidation. Chapter 11 will cover the deposition of oxide and other materials.

Since a silicon wafer is a flat surface (planar), oxide is grown or deposited essentially as a layer. Because of this, it is referred to as a *thin film.* Once a layer is on the wafer surface, it is then modified during subsequent processing to attain the necessary three-dimensional shape used in the formation of circuit components, such as a trench capacitor or gate oxide. This modification produces three-dimensional shapes on the wafer surface referred to as *topography* or *surface topology.*

Oxide growth is a natural phenomenon that occurs by exposing the silicon wafer to oxygen at an elevated temperature. The ability to grow an oxide film on silicon is often listed as a significant reason for the use of silicon as the most common semiconductor substrate material (another major reason is silicon's relatively high melting temperature). We use the term "grow" to indicate that temperature is used to cause the oxide to grow out of the silicon semiconductor material, actually consuming silicon in the process.

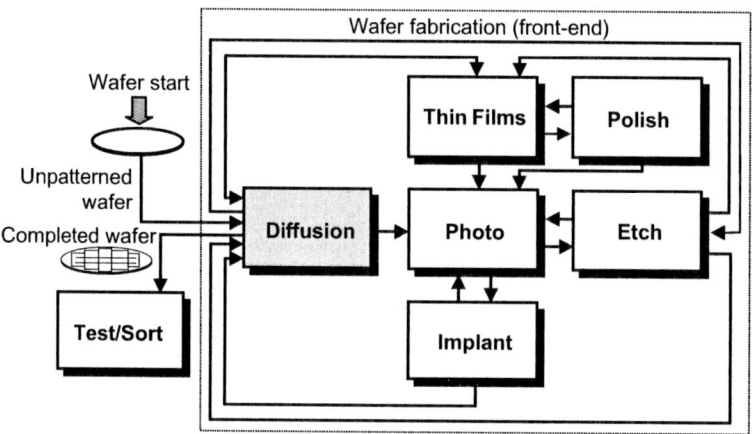

FIGURE 10.1 Diffusion Area of Wafer Fabrication (Used with permission from Advanced Micro Devices)

The amount of thermal exposure (i.e., temperature multiplied by time) for a wafer during processing is referred to as the *thermal budget*. The thermal budget requirement for wafer fabrication is rapidly dropping.[2] A goal throughout semiconductor processing is to minimize the amount of thermal (heat) exposure to the wafer. As the critical dimension on device structures decreases to 0.18 μm and beyond, scaling requirements dictate shallower junction depths. To minimize the unacceptable diffusion of dopants out of a shallow junction region, the thermal budget must decrease accordingly. Another problem caused by excessive thermal budget is the increase in ohmic contact resistance of metal-layer interconnects (see Chapter 12), which increases the overall resistance of the conductive paths. A factor in determining the process conditions for many wafer fabrication process steps is the ability to minimize the thermal budget, either by reduced temperature or by minimizing the time at temperature.

OXIDE FILM

An oxide layer is grown on silicon typically at thermal oxidation temperatures between 750°C to 1100°C. An oxide layer grown on silicon is referred to as *thermal oxide* or *thermal silicon dioxide* (SiO_2). Since silicon dioxide is an oxide material, the two terms are used interchangeably. Another term for silicon dioxide is *glass*. Silicon dioxide is a dielectric material and will not conduct electricity.

Nature of Oxide Film

When a silicon surface is exposed to oxygen, an amorphous silicon dioxide film grows immediately. Although undoped silicon is a semiconductor material, SiO_2 is an insulator. The atomic structure of this SiO_2 film consists of a silicon atom surrounded by four oxygen atoms (see Figure 10.2). We can refer to this atomic structure as a silicon dioxide *tetrahedron cell*. Amorphous SiO_2 has no long-range periodic crystal order at the atomic level. Long-range order is absent because the tetrahedra are not arranged in a regular, three-dimensional array as in a crystal.[3]

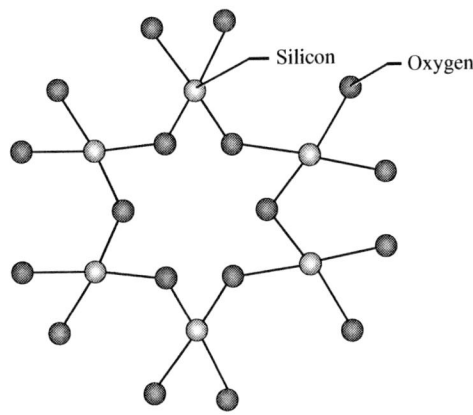

FIGURE 10.2 Atomic Structure of Silicon Dioxide
(Used with permission from International SEMATECH)

Silicon dioxide is a form of intrinsic (pure) glass with a melting temperature of 1732°C.[4] Thermally grown SiO_2 has strong adhesion to silicon and exhibits excellent dielectric properties. A layer of SiO_2 always covers the wafer surface because silicon oxidizes immediately upon exposure to air with a few monolayers of native oxide (a monolayer is one layer of atoms on a surface). Over time, the native oxide layer will thicken to an upper limit of about 40 Å, even at 25°C room temperature.[5] This type of oxide is nonuniform and usually considered a contaminant, with little use in semiconductor fabrication. There are some applications for use of native oxide in wafer fabrication, such as part of a dielectric film stack in memory cells. Native oxide grown from air is contaminated, but native oxide grown in chemical baths using high-purity chemicals is very clean.

Uses of Oxide Film

Oxide is important for silicon semiconductor fabrication because of its ease of formation and its excellent interface with the underlying silicon material. It is the most common layer used in semiconductor fabrication. Different ways in which an oxide layer is used to fabricate a microchip are:

- Device scratch protection and contaminant isolation
- Field isolation to confine charged carriers (termed surface passivation)
- Dielectric material in the gate oxide or memory cell structures
- Impurity-mask barrier during doping
- Dielectric layer between metal conductor layers

Device Protection and Isolation ■ Silicon dioxide grown on the surface of the wafer serves as an effective barrier to isolate and protect sensitive devices in the silicon. SiO_2 physically protects devices because it is a very hard and nonporous (dense) material that effectively insulates active devices in the silicon surface. The hard SiO_2 layer will protect the silicon from scratches and processing damage that might occur during fabrication. Transistors have traditionally been electrically isolated by thermally growing thick SiO_2 in the region between them using the LOCOS process (discussed later in this chapter). However, this process is unacceptable for sub-0.25 μm technology and has been replaced by shallow trench isolation (STI). The STI process uses deposited oxide as the main dielectric (see Chapters 9 and 11).

Surface Passivation ■ A major benefit from thermally grown SiO_2 is the reduction in the surface-state density of silicon that occurs by tying up dangling silicon bonds. This action is referred to as *surface passivation,* which prevents the deterioration of the electronic properties and reduces current-leakage paths of the semiconductor caused by moisture, ions or other external contaminants.[6] The hard SiO_2 layer will protect the silicon from scratches and processing damage that might occur during backend final fabrication. SiO_2 growth on the silicon surface can serve to tie up electrically active contaminants (mobile ionic contaminants) in the oxide layer away from the active surface of the silicon. Passivation is also important for controlling the leakage current of junction devices and growing a stable gate oxide.[7] For the oxide layer to function as a good passivation layer, it has a quality requirement to be a uniform thickness and not have pinholes or voids in the material.

Another factor in using the oxide layer to passify the wafer surface is the oxide thickness. Adequate oxide thickness is necessary to prevent electrical charging in a metal layer caused by a charge buildup in the silicon wafer surface, much like charge storage and breakdown characteristics of ordinary capacitors. This charging would lead to electrical shorting and other undesirable electrical effects. A thick layer of oxide that inhibits charge buildup from metal layers is the *field oxide layer*, which typically is between 2,500 Å to 15,000 Å thick (see Figure 10.3).

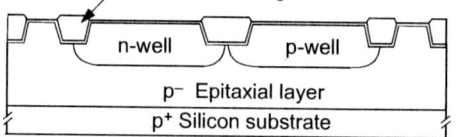

FIGURE 10.3 Field Oxide Layer

Silicon dioxide also has a coefficient of thermal expansion very similar to that of silicon. During high-temperature processing, the wafer will expand and then contract during cooling. The SiO_2 will expand and contract at close to the same rate as silicon, thus minimizing wafer warpage during the thermal excursions. This action also prevents film stress from separating the oxide film from the silicon substrate.

Gate Oxide Dielectric ■ An extremely thin layer of oxide is used as the dielectric material for the important *gate oxide structure* common in MOS technology (see Figure 10.4). The gate oxide is thermally grown because of its high quality and stability with the underlying semiconductor silicon. SiO_2 has a high dielectric strength of 10^7 volts/cm and high resistivity of approximately 10^{17} ohms-cm.[8]

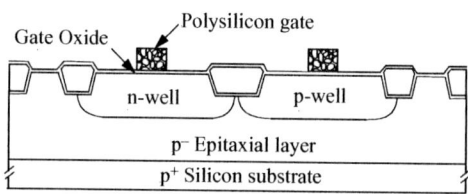

FIGURE 10.4 Gate Oxide Dielectric

The thickness of the gate oxide is chosen specifically for the scaling requirements of the device technology. For 0.18-μm generation technology, a typical gate oxide thickness is 20 ± 1.5 Å. The gate oxide has a specific thickness so that it scales properly with the entire gate structure to permit induction of a charge in the silicon wafer under the oxide. The widespread use of MOS technology in the ULSI era has made the formation of the gate oxide a primary concern in process development. A critical aspect of device reliability is *gate oxide integrity*. The gate structure in a MOS device is the part that controls the flow of electrical current through the device. Because this oxide is fundamental to the proper function of microchips based on field-effect technology, it is essential that it have high quality, with excellent film-thickness uniformity and no contaminants. Any contamination that degrades the proper functioning (integrity) of the gate oxide structure must be scrupulously controlled.

Dopant Barrier ■ Silicon dioxide can be used as an effective barrier to selectively introduce dopants into the silicon surface (see Figure 10.5). Once an oxide layer is grown on the silicon surface, then mask openings are etched into the SiO_2 to create windows where dopant materials can enter the wafer. The oxide protects the surface of silicon from impurity diffusion where there are no openings and therefore allows selective impurity doping. Dopants have a slow rate of movement through SiO_2 when compared to silicon, requiring only a thin oxide layer to block dopants (note that this rate is temperature dependent).

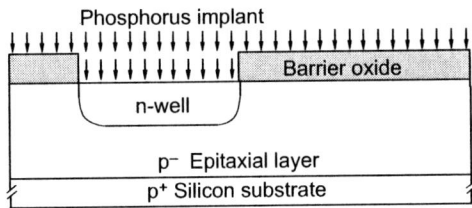

FIGURE 10.5 Oxide Layer Dopant Barrier

A thin oxide layer (e.g., 150 Å) is also grown in areas where ion implant occurs. This oxide screen serves to reduce damage to the silicon wafer and obtain better control over the depth that the dopant is implanted into the silicon by reducing channeling effects (see Chapter 17). Once implantation is done, HF (hydrofluoric acid) can be used to selectively remove the oxide, resulting in a planar silicon surface again.

Dielectric Between Metal Layers ■ Under normal conditions, silicon dioxide will not conduct electricity. For this reason, SiO_2 is an effective insulator between metal layers on the microchip. Silicon dioxide prevents shorting of an upper metal layer to the underlying metal layer, just as the insulator material on an electrical cord prevents an electrical short. The oxide quality is critical with no holes or voids allowed. This oxide is often doped to produce more effective flow properties and to better minimize the diffusion of contaminants (i.e., it acts as a getter). It is typically deposited using chemical vapor deposition (not thermally grown) and will be covered in more detail in Chapter 11.

A summary of different types of oxides and their uses in semiconductor fabrication is shown in Table 10.1.[9]

TABLE 10.1 Oxide Applications

Application	Purpose	Structure	Comments
Native Oxide	This oxide is a contaminant and is generally undesirable. Sometimes used in memory storage or film passivation.	Silicon dioxide (oxide) / p+ Silicon substrate	Growth rate at room temperature is 15 Å per hour up to a maximum thickness of about 40 Å.
Gate Oxide	Serves as a dielectric between the gate and source-drain parts of MOS transistor.	Gate oxide / Gate / Source / Drain / Transistor site / p+ Silicon substrate	Common gate oxide film thickness range from about 20 Å to several hundred Å. Dry thermal oxidation is the preferred growth method.

Application	Purpose	Structure	Comments
Field Oxide	Serves as an isolation barrier between individual transistors to isolate them from each other.	Field oxide / Transistor site / p$^+$ Silicon substrate	Common field oxide film thickness ranges from 2,500 Å to 15,000 Å. Wet oxidation is the preferred growth method.
Barrier Oxide	Protects active devices and silicon from follow-on processing.	Barrier oxide / Metal / Diffused resistors / p$^+$ Silicon substrate	Thermally grown to several hundred angstroms thickness.
Dopant Barrier	Masking material for depositing or implanting dopants into wafer.	Dopant barrier spacer oxide / Ion implantation / Gate / Spacer oxide protects narrow channel from high-energy implant	Dopants diffuse into unmasked areas of silicon by selective diffusion.
Pad Oxide	Provides stress reduction for silicon nitride (Si_3N_4).	Passivation Layer / Bonding pad metal / Nitride / Pad oxide / M-4 ILD-5 / M-3 ILD-4	Thermally grown and very thin.
Implant Screen Oxide	Used to reduce implant channeling and damage.	Screen oxide / Ion implantation / p$^+$ Silicon substrate / High damage to upper Si surface + more channeling / Low damage to upper Si surface + less channeling	Thermally grown.
Insulating Barrier Between Metal Layers	Serves as a protective layer between metal lines.	Bonding pad metal / Interlayer oxide / Passivation layer / M-4 ILD-5 / M- ILD-4	This oxide is not thermally grown, but is deposited.

THERMAL OXIDATION GROWTH

The various applications for thermally grown oxide have different thickness requirements. Silicon dioxide thickness ranges for different requirements are summarized in Table 10.2. The color chart in Appendix D shows silicon dioxide color as a function of the film thickness, which was used in the early days of wafer fabrication to estimate the oxide thickness. Other critical quality parameters for oxide layers in semiconductor fabrication are thickness, uniformity, pinholes and voids.

TABLE 10.2 Oxide Thickness Ranges for Various Requirements

Semiconductor Application	Typical Oxide Thickness, Å
Gate Oxide (0.18 μm generation)	20–60
Capacitor Dielectrics	5–100
Dopant Masking Oxide	400–1,200 (Varies depending on dopant, implant energy, time, and temperature)
STI Barrier Oxide	150
LOCOS Pad Oxide	200–500
Field Oxide	2,500–15,000

Chemical Reaction for Oxidation

Thermal oxide is grown by a chemical reaction between silicon and oxygen. Uniform oxide growth is achieved by exposing silicon to elevated temperature in the presence of high-purity oxygen. If this growth occurs with *dry oxygen,* meaning no moisture, then the following chemical equation describes the reaction:

$$Si\ (solid) + O_2\ (gas) \rightarrow SiO_2\ (solid)$$

The time and quality of this reaction varies and is affected by the purity of oxygen gas supplied to the silicon wafer surface and the reaction temperature. It occurs naturally when silicon is exposed to air at room temperature. The reaction rate is increased with an increase in temperature. The typical temperature for the oxidation of silicon during wafer fabrication is normally between 750°C to 1100°C and can vary for different oxidation process steps. The furnace temperature at any one operation is precisely controlled. The rate of oxide thickness for dry oxidation versus temperature and time is shown in Figure 10.6 on page 232.[10]

Wet Oxidation ■ When water vapor is introduced into the reaction, known as *wet oxidation*, the rate of theoxidation reaction is increased further. The chemical reaction for wet oxidation is:

$$Si\ (solid) + 2H_2O\ (vapor) \rightarrow SiO_2\ (solid) + 2H_2\ (gas)$$

For wet oxidation, oxygen saturated with water vapor is used instead of dry oxygen as the oxidizing gas. The water vapor could also be supplied as steam, referred to as *pyrogenic steam* (see Figure 10.7 on page 232). A wet oxidation reaction produces a silicon dioxide film and hydrogen gas during oxidation growth. The faster growth rate in a wet atmosphere is due to the faster diffusion and higher solubility of water vapor than oxygen in silicon dioxide.[11] However, the hydrogen molecules produced in the reaction are trapped in the solid silicon dioxide layer, making the layer less dense than oxide grown in dry oxygen. This condition can be improved to achieve a similar oxide in structure and properties such as that produced in dry oxidation by heating the oxide in an inert atmosphere.

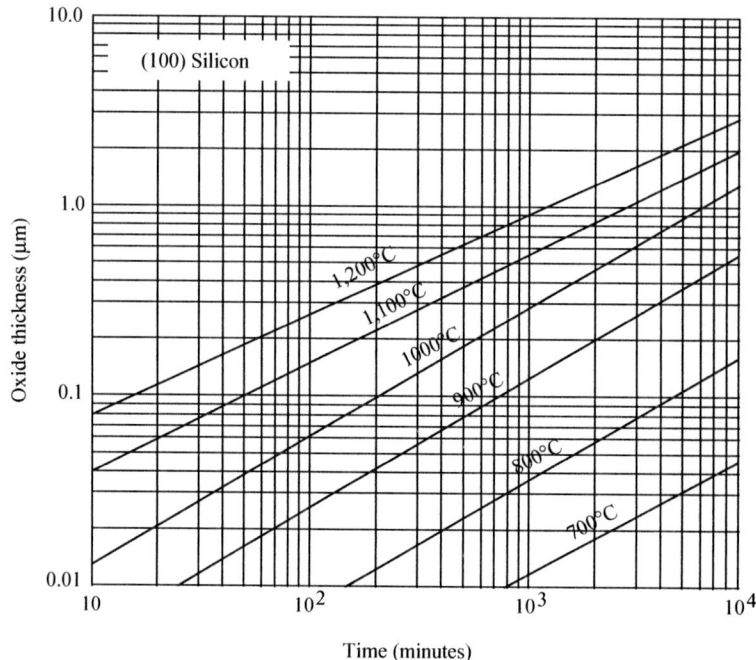

FIGURE 10.6 Dry Oxidation Time (Minutes)

Oxidation Growth Model

Silicon dioxide grows by consuming silicon, which is the case for either the dry or wet oxidation process. This is shown in Figure 10.8. The thickness of the silicon consumed is 0.46 of the total oxide thickness.[12] This means that for every 1,000 Å of oxide thickness, 460 Å of Si is consumed.

The growth of the oxidation layer is controlled and limited by the movement of oxygen through the oxide at the oxide-silicon interface. For the oxide layer to continue growing, the oxygen must come in contact with the silicon. However, the SiO_2 separates the oxygen gas from the silicon wafer. Oxide growth occurs when the oxygen gas molecules move through the existing silicon dioxide layer to the silicon wafer. This movement is referred to as diffusion (more precisely, gas diffusivity through a solid barrier). *Diffusion* is the movement of one material through another.

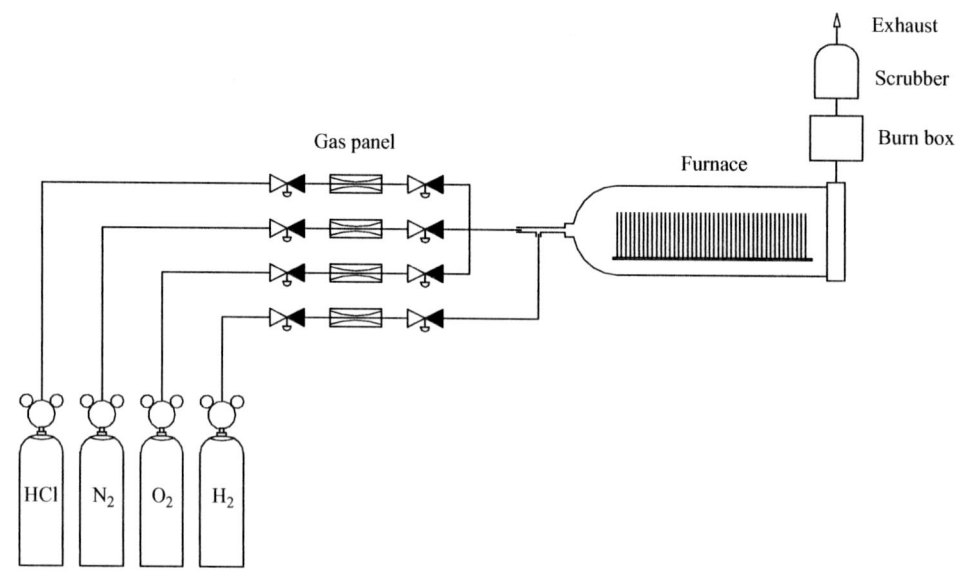

FIGURE 10.7 Wet Oxidation

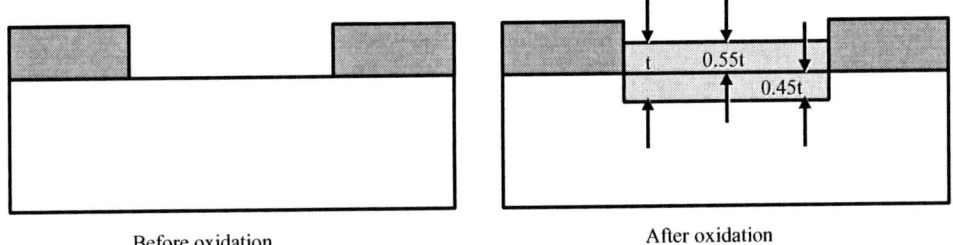

FIGURE 10.8 Consumption of Silicon During Oxidation

Atoms diffuse from regions of high concentration to regions of low concentration for solid, liquid, or gas states, and this diffusion is accelerated by thermal energy. It is a natural physical process that occurs regularly in everyday life. An example of liquid-state diffusion is putting a drop of red food colorant into a glass of water (see Figure 10.9). The colorant is initially concentrated when it enters the water, and then it will diffuse into the water until it is uniformly red. If the water is heated, then the diffusion will spread throughout the water more quickly.

The workbay in the wafer fab where oxidation occurs is still referred to as diffusion or the diffusion bay. In the early days of wafer fabrication, diffusion was an important process for creating pn junctions. Dopants were supplied to the silicon by chemical sources and then diffused to the desired junction depth by subjecting the wafers to elevated temperatures. Diffusion of dopants to create pn junctions is no longer done in wafer fabrication, although the use of the term diffusion bay is still common today. It has been replaced by ion implantation (see Chapter 17). Nevertheless, diffusion of materials (in this case oxygen) occurs in processes such as oxidation. Oxidation is well defined and governed by certain laws of diffusion that are based on a set of mathematical relationships referred to as Fick's laws. *Fick's laws* describe the rate of movement of diffusing materials based on temperature, concentration, and the energy necessary to drive diffusion. The detailed study of Fick's laws is not covered in this text.

Oxide-Silicon Interface ■ There is an abrupt transition at the oxide-silicon (Si/SiO_2) interface between single crystal silicon to amorphous SiO_2. Recall that for the SiO_2 molecule, each silicon is bonded to four oxygen atoms and each oxygen atom is bonded to two silicon atoms. At the Si/SiO_2

FIGURE 10.9 Liquid-State Diffusion

interface, some silicon atoms in the structure remain unbonded (see Figure 10.10). This incomplete oxidation of silicon less than 2 nm away from the Si/SiO$_2$ interface is the origin of a positive *fixed oxide charge*. Other charges built up at the interface include an *interface-trapped charge* consisting of positive or negative charges that result from structural defects, oxidation-induced defects, or metal impurities, and a *mobile oxide charge* due to mobile ionic contaminants (MICs). There is also the possibility of an *oxide-trapped charge* that is positive or negative and trapped in the bulk of the oxide away from the interface.[13] This accumulation of charge at the Si/SiO$_2$ interface is undesirable for normal device operation and can cause the threshold voltage of a MOS device to shift to unacceptable values.[14] Some of these undesirable charges can be minimized through a low-temperature (~ 450°C) anneal step in hydrogen or forming gas (hydrogen-nitrogen mixture).[15]

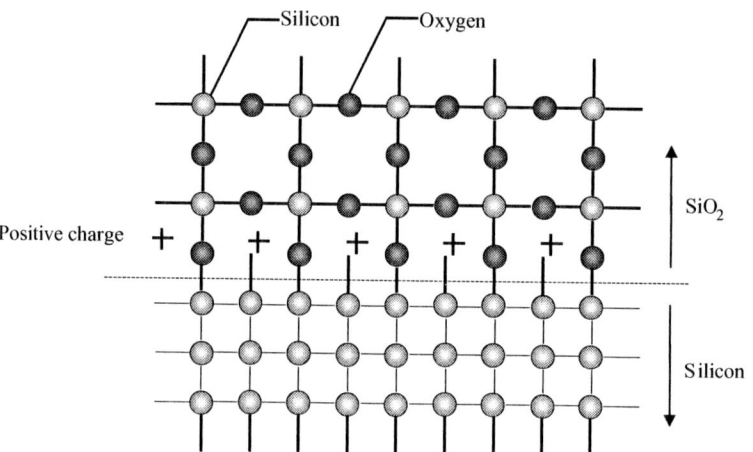

FIGURE 10.10 Charge Buildup at Si/SiO$_2$ Interface
(Used with permission from International SEMATECH)

Use of Chlorinated Agents in Oxidation. Using a chlorine-containing gas during the oxidation process can serve to neutralize the charge accumulation at the interface. The chlorine ions will diffuse to the positively charged layer and form a neutral layer. The chloride concentration is kept below 3% because excessive chloride ions will cause device instability. Another important advantage of chloride ions added to the thermal oxidation process is that they tend to increase the oxidation growth rate by about 10–15%. Furthermore, the presence of chlorine can actually immobilize (referred to as gettering) mobile ionic contaminants (MICs) that come from the furnace equipment, processing materials, and handling.

Early work used chlorine or vapor-phase hydrogen chloride (HCl). Chlorine is extremely toxic and HCl vapor in the presence of water vapor is corrosive. A common chlorine compound often used to transport the chlorine into the furnace is dichloroethylene (DCE) or variants of this compound.[16] Other less-corrosive sources of chlorine are trichloroethylene (TCE) and 1,1,1-trichloroethane (TCA).[17] TCE is carcinogenic and not used anymore, while TCA has fallen out of favor because it is an ozone-depleting chemical. Also, some semiconductor firms have switched back to HCl because of significant improvements in high-purity piping and fittings used to transport this chemical.

Rate of Oxide Growth ■ The *rate of oxide growth* describes how fast the oxide grows on the wafer. It depends on parameters such as temperature, pressure, oxidizing condition (dry or wet), silicon crystal orientation, and doping levels. The growth rate is of interest because the wafer processing time can be reduced if diffusion occurs quickly, which serves to reduce the thermal budget. The model for oxide growth on silicon is referred to as a linear-parabolic model developed by Deal and Grove and accurately represents oxide growth over a wide range of thicknesses (most optimally from 300 to 20,000 Å).[18] Oxide growth is described by two growth stages: the linear stage and the parabolic stage.

The initial growth of silicon dioxide is referred to as the *linear stage* and consumes silicon on the wafer surface as a linear function of time. This means the oxide layer is growing into the wafer at a linear rate over time. The linear stage of oxide growth is valid up to about 150 Å of oxide thickness.[19] It is described by the linear equation:[20]

$$X = \left(\frac{B}{A}\right)t$$

Where, X = the thickness of the growing oxide
(B/A) = the linear rate constant
t = the time it takes to grow the oxide

In the linear stage, oxidation varies linearly with time. Oxidation is *reaction-rate controlled* in the linear region because the limiting factor for oxidation growth is the reaction occurring at the Si/SiO$_2$ interface. You will note that the linear rate constant, B/A, is the slope of this linear relationship, and therefore it controls the rate of the reaction. As temperature increases, the values for B/A increase, which means that the rate of oxidation also increases.

The *parabolic stage* of oxidation growth is the second phase of oxidation growth and starts after about 150 Å of oxide thickness. The equation that describes the parabolic stage is:[21]

$$X = (Bt)^{\frac{1}{2}}$$

Where, X = the thickness of the growing oxide
B = parabolic rate constant
t = the time it takes to grow the oxide

Note that this equation generates the shape of a parabola. Oxide growth in the parabolic stage is much slower than the linear stage. This is because as the oxide layer becomes thicker, the oxygen must diffuse through a larger distance to arrive at the Si/SiO$_2$ interface (see Figure 10.11). The reaction is thus limited by the rate at which the oxygen diffuses through the oxide. For this reason, the parabolic stage of oxide growth is said to be *diffusion controlled*. As the parabolic rate constant increases, the rate of oxide growth will increase. For instance, the parabolic rate constant, B, for wet oxidation is found to be much larger than that for dry oxidation.[22] Thus the rate of oxidation increases with wet oxidation.

A general curve shows both the linear and parabolic stage, as illustrated in Figure 10.12 on page 236.[23] This is a simplified curve from the original curve published by Deal and Grove in 1965 used to describe the thermal oxidation of silicon.[24]

Factors Affecting Oxide Growth ■ There are other factors that can affect the rate of oxide growth besides temperature and the presence of H$_2$O. We will review some of these other factors.

Dopant Effects. Heavily doped silicon oxidizes at a faster rate than lightly doped material. In the parabolic stage, boron doping will oxidize faster than phosphorus. Boron tends to become incorporated in the oxide film, which weakens its bond structure and leads to a subsequent increase in the diffusivity of the oxygen through it.[25] There is little difference in the linear rate constant between boron and phosphorus doping.

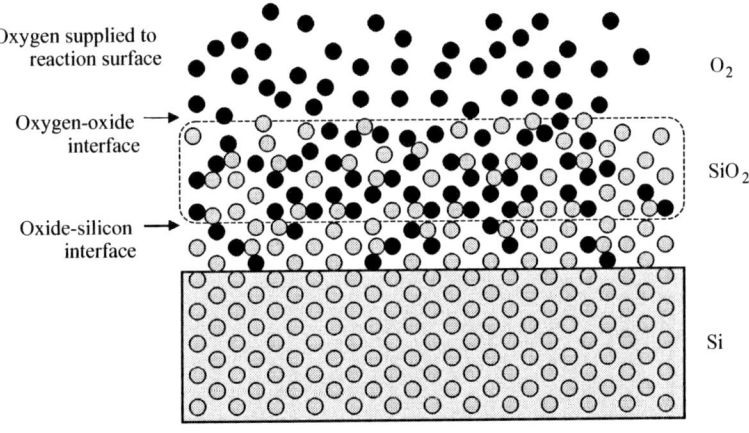

FIGURE 10.11 Diffusion of Oxygen Through Oxide Layer
(Used with permission from International SEMATECH)

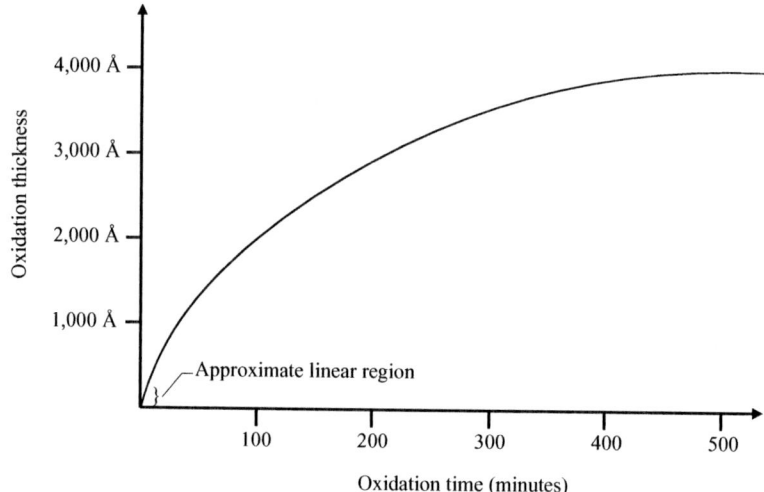

FIGURE 10.12 Linear and Parabolic Stages for Dry Oxidation Growth at 1100°C (Used with permission from International SEMATECH)

Crystal Orientation. The linear oxidation rate is dependent on the orientation of the crystal because the density of silicon atoms on the (111) plane is greater than on the (100) plane. Therefore, (111) silicon crystal will oxidize in the linear stage at a faster rate than (100) silicon crystal, but (111) also has a greater charge buildup.

During the parabolic stage, the parabolic rate constant, B, is independent of the crystal orientation of the silicon substrate. There is no difference in oxidation growth rate for (111) vs. (100) in the parabolic stage. This is logical since the parabolic rate of oxidation growth is based on the diffusivity of the oxygen through the existing SiO_2 and is not affected by the reaction at the Si/SiO_2 interface.[26] However, charge buildup remains high because this factor is a function of atom density at the surface.

Pressure Effect. Since the growth rate of an oxide layer depends on the movement of the oxidizer from the gas phase to the silicon interface, the growth rate will increase with pressure. Higher pressure significantly increases the linear and parabolic rate constants, forcing the oxygen atoms to penetrate through the growing oxide more rapidly. This condition allows the temperature to be reduced to achieve equivalent growth rates or a more rapid oxide growth at the same temperature. A rule of thumb for the increased oxide growth versus temperature reduction is that for each 1-atmosphere increase in pressure, the furnace temperature can be decreased 30°C.[27] This ratio can be used to reduce the thermal budget. A high-pressure oxidation process could be used, for example, to grow a thick field oxide layer.

Plasma Enhancement. Plasma-enhanced oxidation is another method that increases the oxide growth rate at low temperature, thus reducing the thermal budget.[28] The technique is usually carried out in an oxygen-plasma discharge generated by an RF source. The silicon is biased to be at a potential less than the plasma potential, which collects the charged oxygen in the plasma on the wafer. This activity results in rapid oxidation of the silicon and allows oxides to be grown at temperatures less than 600°C. The drawbacks to this technique are problems associated with particle generation, higher film stress that is different from thermally grown oxides, and inferior film quality compared to thermally grown oxides.[29] For these reasons, this method is not widely used in wafer fabrication.

Initial Growth Phase ■ The Deal and Grove linear-parabolic model accurately predicts oxidation growth for thickness above 300 Å. However, below this thickness it is found that dry oxidation is faster than predicted. This is an important oxidation growth range because the gate oxide thickness for subquarter micron MOS technology is currently around 20 to 60 Å. The manufacturing process must be capable of producing these oxide layers with high yield and long-term reliability.

Because the gate oxide is becoming so thin, there is not one model that accurately predicts the rate of oxidation. One particular model has detected the presence of pores, about 10 Å in diameter, in very thin oxides grown in dry oxygen. These pores allow the oxidant to remain in direct contact with the silicon in the early phases of growth, thus causing the rapid initial growth of oxide.[30] The investigation into thin gate oxides is an area of ongoing research.

Selective Oxidation ■ The oxidation of selective areas on a wafer uses SiO_2 to electrically isolate adjacent devices on the silicon surface. The traditional method for isolating devices with greater than 0.25 μm feature sizes has been *local oxidation of silicon* (*LOCOS*). A deposited layer of silicon nitride (Si_3N_4) serves as an oxidation barrier. It is etched to allow selective thermal oxide growth, since oxide will not grow where the nitride covers the silicon (see Figure 10.13). After thermal oxidation, the nitride and any underlying barrier oxide are removed to expose bare silicon surface regions ready for device formation.

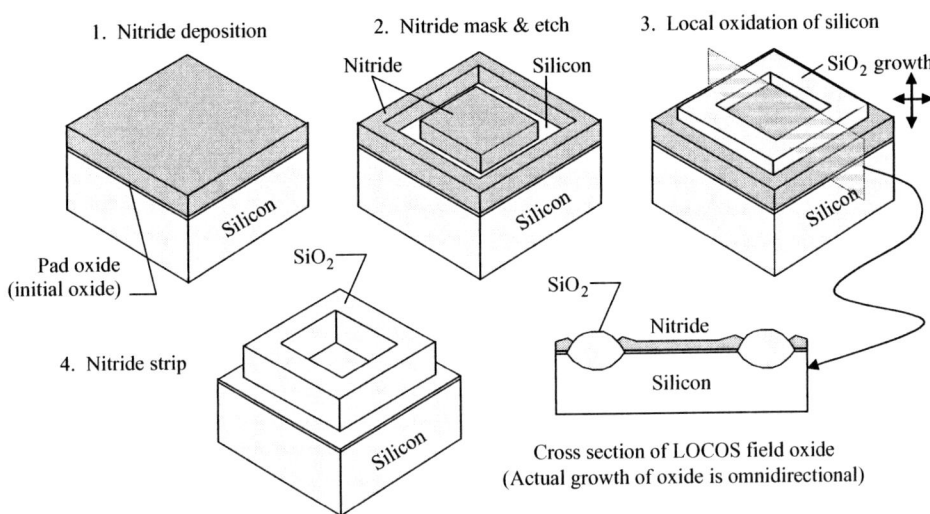

FIGURE 10.13 LOCOS Process

As the oxygen diffuses through the grown oxide, it moves in all directions. Some of the oxygen moves down into the silicon and other oxygen atoms move sideways. This means there is a slight lateral growth of the oxide under the nitride mask. Since the oxide is thicker than any silicon consumed, growth under the nitride mask serves to push up the nitride edges. This action is referred to as the *bird's beak effect*. This phenomenon is an undesirable by-product of LOCOS type oxidation processes (see Figure 10.14). The bird's beak effect is most pronounced when the oxide is relatively thick. To reduce stress between the nitride mask and silicon, a thin barrier layer of thermal oxide is used between them, which is termed a *pad oxide*.

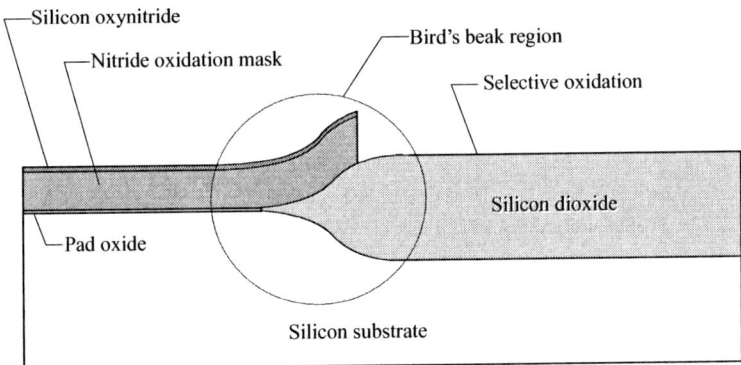

FIGURE 10.14 Selective Oxidation and Bird's Beak Effect
(Used with permission from International SEMATECH)

Shallow Trench Isolation (STI). The primary form of selective oxidation used for sub-0.25-μm technology is *shallow trench isolation* (*STI*). The main dielectric material in STI is a deposited oxide (see Chapter 11). Selective oxidation is accomplished by creating a mask, usually of silicon nitride (Si_3N_4). The mask is deposited and patterned, followed by etching of the silicon to form a trench. The etching of silicon trenches is followed by thermal oxidation of a 150–200 Å thick oxide layer in areas exposed through the mask (see Figure 10.15). This thermally grown oxide passivates the silicon surface and serves as a barrier between the silicon and deposited trench-fill oxide. It is also an effective barrier that prevents sidewall current leakage in finished devices.

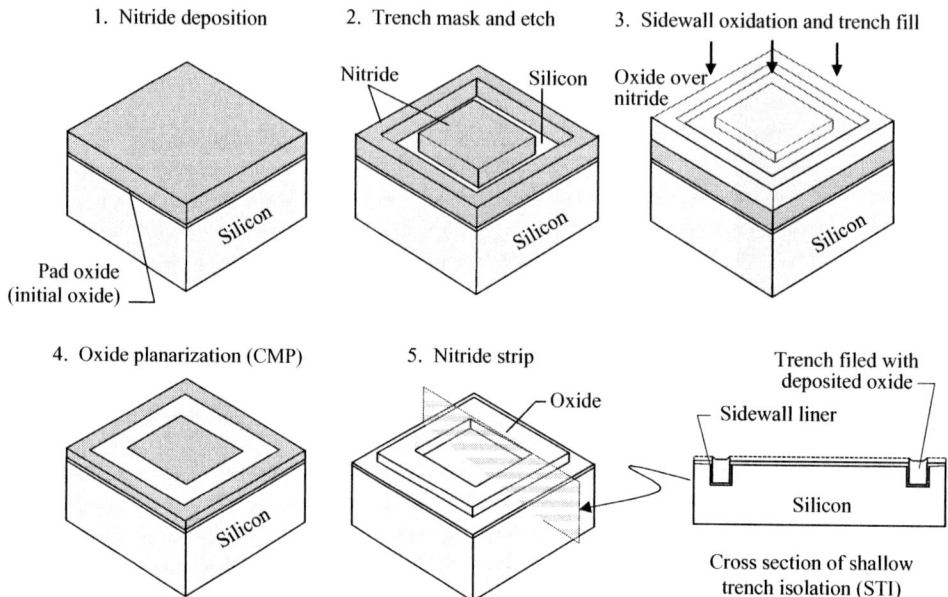

FIGURE 10.15 STI Oxide Liner

Silicon nitride is usually deposited by chemical vapor deposition (see Chapter 11), followed by etching to open up the selective areas for thermal oxidation. The silicon surfaces on the wafer covered by the silicon nitride layer do not oxidize as long as the nitride mask is sufficiently thick (the nitride mask covering the silicon will oxidize slightly). Silicon nitride can be grown on silicon by reacting nitrogen with surface silicon atoms at elevated temperatures, usually above 900°C. However, nitride grown using this method can only be grown to a very thin layer (about 5 nm), which limits its use to only special applications that take advantage of its high dielectric constant.[31]

Stress in Silicon Dioxide ■ Silicon dioxide stress is undesirable because it can contribute to wafer warpage and possibly to defect generation in the form of a slip in the silicon wafer. Measured stress in a thermally grown SiO_2 film is found to be compressive and have a relatively small magnitude. The stress in an oxide film results from the difference between the coefficients of thermal expansion of Si and SiO_2.[32] Stress from an oxide film layer can cause the wafer to bow so that the oxidized surface becomes convex. Optical measurements of the curvature can be used to quantify stress.

Oxidation-Induced Stacking Faults ■ Wet or dry thermal oxidation can cause *oxidation-induced stacking faults* (*OISFs*) at the interface between Si and SiO_2. Recall from Chapter 4 that stacking faults are a form of unit-cell dislocation due to layer stacking errors. It is believed that OISFs are caused by incomplete oxidation at the Si/SiO_2 interface, which leads to excess interstitial silicon in this region.[33] Stacking faults can cause increased leakage current if the fault is in the vicinity of a pn junction. OISF formation is greatly reduced by performing thermal oxidation with chlorine because chlorine promotes vacancy formation at the silicon surface, providing a means to remove the excess interstitial silicon atoms.

Horizontal Diffusion Furnace
(Photo courtesy of International SEMATECH)

FURNACE EQUIPMENT

In this section we will review basic furnace equipment technology. Furnace equipment serves a variety of purposes during the wafer fabrication process. Thermal oxide growth, including the formation of the critical gate oxide, is a major reason for the use of furnaces. Other applications include thermal anneal of the wafer surface after ion implantation (Chapter 17); the deposition of a variety of films, such as doped and undoped polysilicon, silicon nitride, and silicon dioxide (Chapter 11); reflow of glass (Chapter 11); and the formation of silicide films (Chapter 12). We will focus on the equipment used for thermal oxidation processes.

There are three basic types of thermal processing equipment used:

- Horizontal furnace
- Vertical furnace
- Rapid thermal processor (RTP): single-wafer

The *horizontal furnace* was the workhorse in thermal wafer processing since the early days of the semiconductor industry. Its name derives from the horizontal position of the quartz tube where wafers are located and heated. This furnace was largely replaced in the early 1990s by the

Vertical Diffusion Furnace
(Photo courtesy of International SEMATECH)

vertical furnace, mainly because the vertical furnace is easier to automate, improves operator safety, and reduces particulate contamination.[34] The vertical furnace, also called a *vertical diffusion furnace,* or VDF, has better temperature control and uniformity than a horizontal furnace. Both the horizontal and vertical furnaces are considered conventional *hot wall* batch furnaces because both the wafer and the furnace walls are heated, and they generally process large quantities or batches of wafers (100 to 200 wafers). Conventional furnaces ramp up and down the temperature of the wafers at about 20°C/minute or less.

The *rapid thermal processor* (*RTP*) is a small, fast-heating system that typically processes a single wafer at a time with a radiant heat source and cooling source. The RTP is also referred to as a rapid thermal anneal (RTA) system when used to anneal the silicon substrate (see Chapter 17). Due to its extremely fast localized heating times, the RTP only heats the wafer (not the wall of the furnace). The typical RTP is able to achieve up and down ramp rates in the tens of degrees per second range, with models available that use dual-sided wafer heating to attain ramp-up rates up to 250°C/second.[35] The RTP has been in use since the late 1980s and is used in areas such as barrier-layer formation (see Chapter 12) and oxide reflow.

HORIZONTAL VERSUS VERTICAL FURNACES

A comparison of performance factors for the conventional horizontal and vertical furnaces is given in Table 10.3.[36]

TABLE 10.3 Performance Comparison of Horizontal and Vertical Furnace Systems

Performance Factor	Performance Objective	Horizontal Furnace	Vertical Furnace
Typical wafer loading size	Small, for process flexibility	200 wafers/batch	100 wafers/batch
Cleanroom footprint	Small, to use less space	Larger, but has four process tubes	Smaller (single process tube)
Parallel processing	Ideal for process flexibility	Not capable	Capable of loading/unloading wafers during process, which increases throughput
Gas flow dynamics (GFD)	Optimize for uniformity	Worse due to paddle and boat hardware. Buoyancy and gravity effects cause nonuniform radial gas distribution.	Superior GFD and symmetric/uniform gas distribution
Boat rotation for improved film uniformity	Ideal condition	Impossible to design	Easy to include
Temperature gradient across wafer	Ideally small	Large, due to radiant shadow of paddle	Small
Particle control during loading/unloading	Minimum particles	Relatively poor	Improved particle control from top-down loading scheme
Quartz change	Easily done in short time	More involved and slow	Easier and quicker, leading to reduced downtime
Wafer-loading technique	Ideally automated	Difficult to automate in a successful fashion	Easily automated with robotics
Pre- and post-process control of furnace ambient	Control is desirable	Relatively difficult to control	Excellent control, with options of either vacuum or neutral ambient

Horizontal furnaces are still in use and have undergone advances in technology, leading to some renewed interest in their use in fabs. Their low cost relative to vertical furnaces makes them appealing to wafers with geometries larger than 0.5 μm. This capacity permits a mix-and-match approach, with horizontal furnaces used for certain less-demanding applications and vertical furnaces used for critical applications.

Both horizontal and vertical furnaces have been configured for atmospheric oxidation and diffusions, as well as low-pressure chemical vapor deposition (LPCVD) applications. Examples of materials LPCVD furnaces are used to deposit as thin layers are SiO_2, Si_3N_4, and polycrystalline silicon (poly) on silicon wafers. The topics of LPCVD and diffusion will be discussed in Chapter 11.

Vertical Furnace

The vertical furnace was first introduced in the early 1990s. The change from horizontal furnaces was driven primarily by the reduced cleanroom footprint achieved by vertical furnaces and the improvement in automated handling. To understand basic furnace technology, we will analyze a conventional vertical furnace. There are five major control systems to a vertical furnace system (these same five systems also apply to a horizontal furnace):

- Process chamber
- Wafer transfer system
- Gas distribution system
- Exhaust system
- Temperature control system

A block diagram of a vertical furnace system showing the five major systems is shown in Figure 10.16. Note that these are all under the control of a single microcontroller.

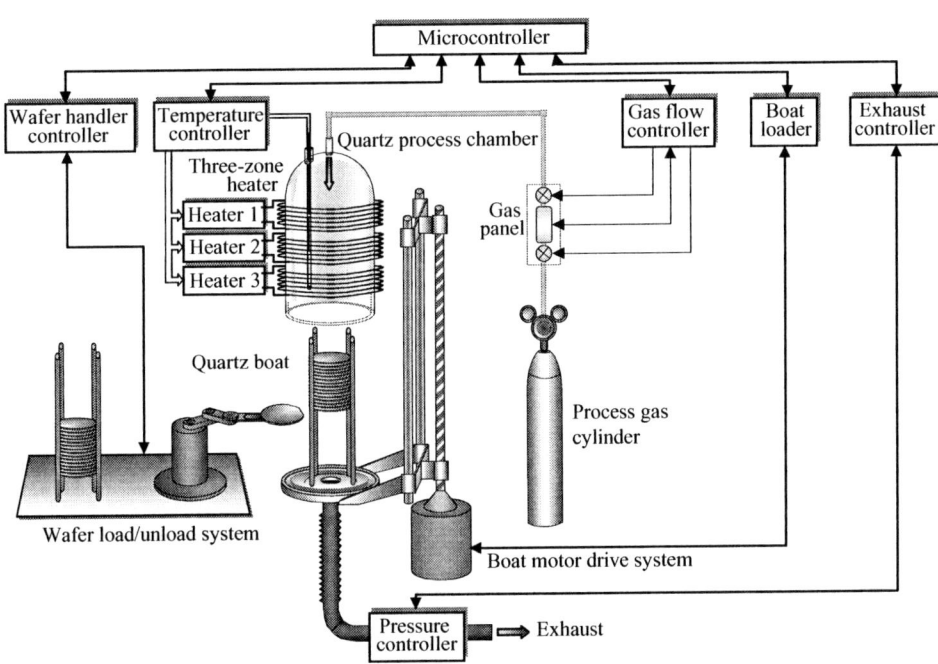

FIGURE 10.16 Block Diagram of Vertical Furnace System

Process Chamber ■ The *process chamber*, or *furnace tube*, is where wafers are heated in the furnace. A vertical furnace consists of a vertical quartz bell-jar style furnace tube that surrounds the wafers for thermal processing in the furnace, a heating element with multiple heat zones, and a heating jacket (see Figure 10.17 on page 242). Furnace tube is also a term used to describe the heated chamber on horizontal furnaces. The furnace tube must be easy to remove so that it can be cleaned when required.

Furnace Tube Materials. The wafers are placed in a vertical *wafer boat* that holds the wafers horizontal in the furnace tube. This boat and the other high-temperature components for the

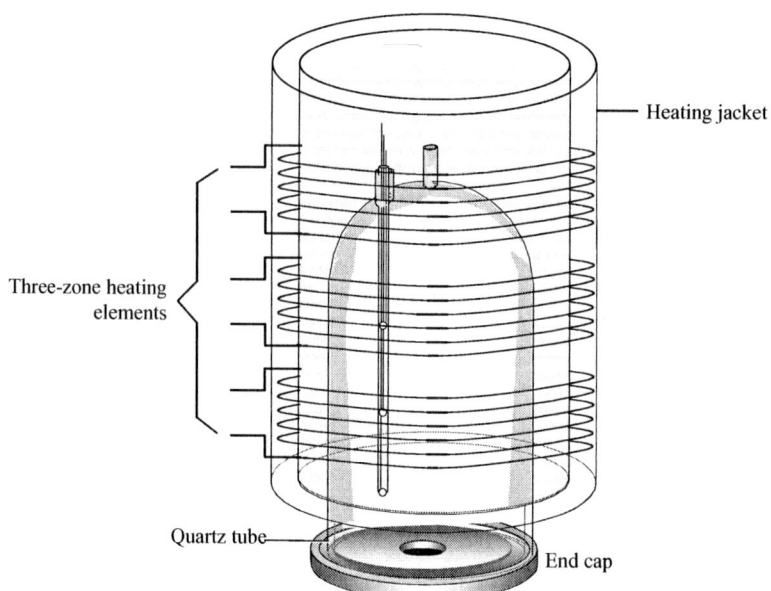

FIGURE 10.17 Vertical Furnace Process Tube

furnace tube are made of amorphous quartz that is resistant to high temperature. Quartz is single crystal SiO_2, while amorphous quartz used for high-temperature applications is fused silica. Furnace containers are commonly called *quartzware,* and the supporting structure that holds wafers in the furnace tube is called a *quartz carrier.*

Films are deposited on the vertical wafer boat and also the inside wall of the furnace tube during processing, especially when the furnace is used for CVD deposition (see Chapter 11). The deposited films will crack and flake, sometimes after only a few cycles. The particles can become airborne and settle on the wafers causing defects and yield loss. This action requires furnace components such as the process tube to be removed and cleaned to reduce particles. Another option to reduce flaking of films is to manufacture the furnace components with silicon carbide (SiC), since the adhesion of the deposited thin film improves with SiC.[37] The drawback to SiC material for furnace components is that it is substantially more expensive than quartz.

Heat Zones. Each quartz process tube is surrounded by a heating element that can be controlled to produce multiple heating zones. It is common to have anywhere from three to seven zones, with 300-mm furnaces having up to nine zones.[38] The number of zones is important because it facilitates control of the furnace tube to attain a *flat zone* near the middle of the tube where thermal processing takes place. The heating zones on each side of the flat zone serve to optimize the wafer ramp up and ramp down to process temperature. This action allows the temperature in the flat zone to be controlled to less than 0.25°C, even at temperatures of over 1000°C.[39]

Wafers are loaded horizontally into a quartz holder called a tower or boat, which is positioned on a quartz mount referred to as the pedestal. The pedestal and furnace tube rest on a water-cooled base plate. Some furnaces rotate the tower during processing for improved heating uniformity.

An important feature of an advanced vertical furnace is its control over the atmosphere around the wafers in the furnace tube. This atmosphere is referred to as the *ambient.* Some furnaces use load locks to keep the ambient in the furnace tube from being exposed to atmospheric gases, while others have nitrogen purge to remove any remaining gases after processing a batch of wafers. Some designs also include an extra quartz enclosure called a liner or inner tube around the wafers to improve the ambient temperature control.[40]

Heating Element. The heating element in a vertical furnace is a metal resistance wire wrapped around the outside of the process tube to provide uniform heat throughout the zone it is heating. Heating elements for three zones are shown in Figure 10.18. The heater elements are switched on and off in response to signals from the temperature controller using a switching system consisting of silicon-controlled rectifiers (SCRs). This system permits control of the amount of power delivered to the heaters (e.g., 50% power or 100% power).

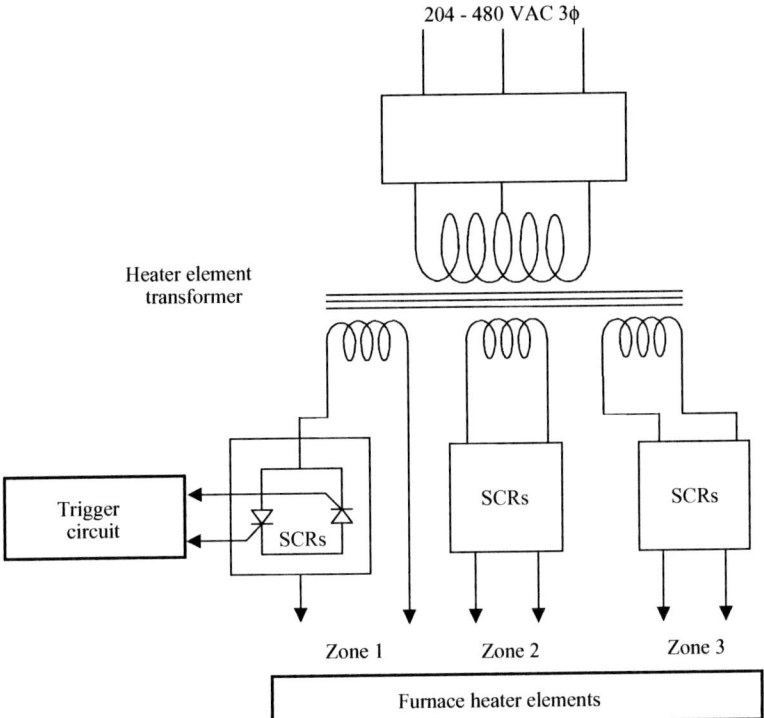

FIGURE 10.18 Heater Element Power Distribution
(Used with permission from International SEMATECH)

Temperature Control. The ability to precisely control temperature in the furnace tube is critical in a furnace. An important part of this temperature control is the sensor known as a *thermocouple* (*TC*) that detects the temperature and provides a corresponding millivolt signal to the furnace controller. Thermocouples are often used because they are rugged, accurate, inexpensive, and work over a wide range of temperatures.

There are multiple thermocouples for each zone of the process tube (see Figure 10.19). The *profile thermocouples* are positioned inside the process chamber in close proximity to the wafer stack, with one in each temperature zone. The profile thermocouples approximate the wafer surface temperature. There are *control thermocouples* (also referred to as spike thermocouples) positioned outside the process tube near the heater element windings in each temperature control zone. These

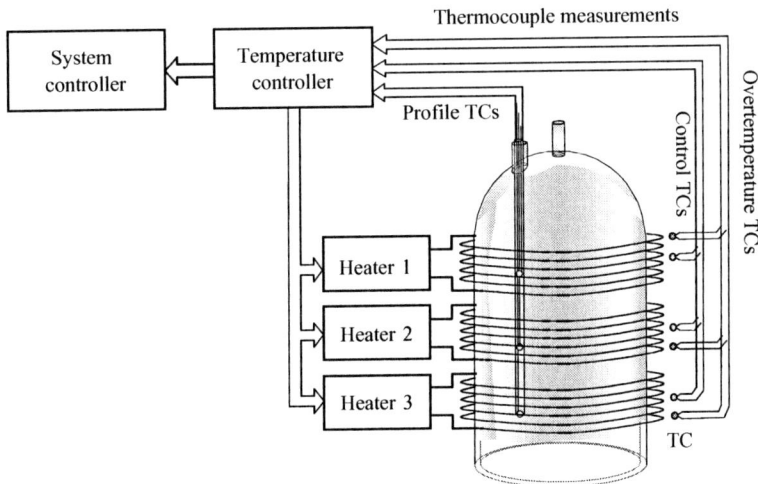

FIGURE 10.19 Locations of Thermocouples in the Furnace Chamber

thermocouples measure the heater element temperature.[41] In addition, there are typically *over-temperature thermocouples* near the control thermocouples to monitor the maximum heater temperature to ensure the furnace never operates at an excessive temperature.

Wafer Transfer System ■ The primary function of the *wafer transfer system* in a vertical furnace is to load and unload wafers from the process chamber. All wafer loading and unloading during the wafer transfer is done using a wafer transfer robot. The robot moves wafers between four stations: cassette station, furnace station, load station, and cooling station. The load station during wafer transfer is maintained at a class 10 or better.

Gas Distribution System ■ The gas distribution system maintains the ambient inside the furnace by delivering the correct flow of gases to the process tube. Depending on the process, there are different bulk and specialty gases that are delivered to the furnace through the distribution system. Some of these typical furnace gases used for oxidation and other furnace processes are listed in Table 10.4.[42]

TABLE 10.4 Common Gases Used in Furnace Processes

Gases	Classifications	Examples
Bulk	Inert gas	Argon (Ar), Nitrogen (N_2)
	Reducing gas	Hydrogen (H_2)
	Oxidizing gas	Oxygen (O_2)
Specialty	Silicon-precursor gas	Silane (SiH_4), dichlorosilane (DCS) or (H_2SiCl_2)
	Dopant gas	Arsine (AsH_3), phosphine (PH_3) Diborane (B_2H_6)
	Reactant gas	Ammonia (NH_3), hydrogen chloride (HCl)
	Atmospheric/purge gas	Nitrogen (N_2), helium (He)
	Other specialty gas	Tungsten hexafluoride (WF_6)

It is important to properly remove gases and their by-products. In a vertical furnace, this removal is done through a port located at one end of the vertical process tube. The gases move into an exhaust manifold to control the direction of flow for each gas. Flammable gases, such as silane (SiH_4), phosphine, and hydrogen use a chamber known as a *burn box* to actually combust the gas in the presence of air far downstream of the process chamber, reducing the flammable gas to less harmful by-products (see Figure 10.20). Special particle filters are then used to remove the solids. The gases pass through a plant *scrubber* where the toxic gases are absorbed. Most furnace exhausts use wet scrubbers that use water solutions to absorb the gases.

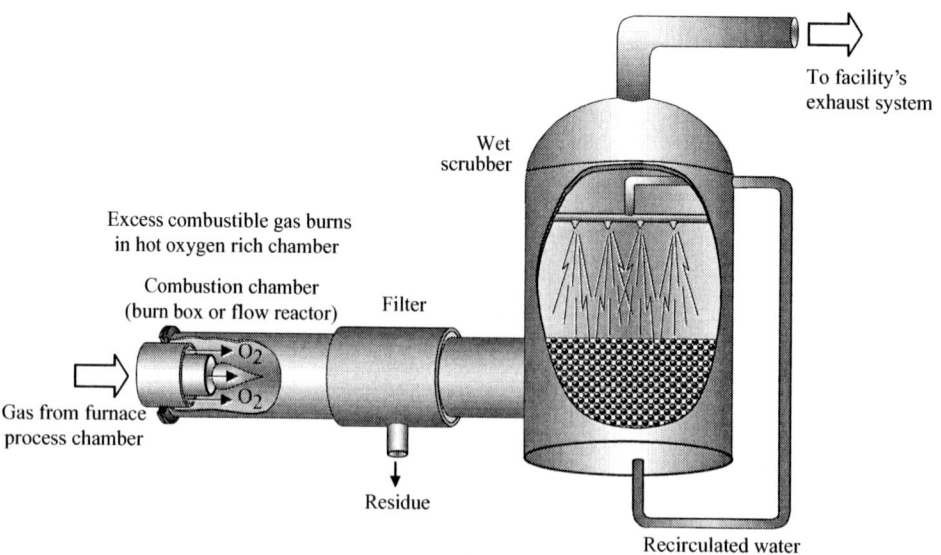

FIGURE 10.20 Burn Box to Combust Exhaust
(Used with permission from International SEMATECH)

Furnace Control System ■ A furnace microcontroller regulates all operations of the furnace, such as process time and temperature control, sequence of process steps, type of gas, the gas flow rates, temperature ramp rates, and wafer loading and unloading. The *temperature ramp rates* are the rate of temperature change that the wafers are exposed to during heating and cooling and are measured in °C/minute. A typical ramp rate for a vertical furnace is 10°C/minute while maintaining the flat zone temperature to within 0.6°C.[43] Other functions such as diagnostics and data collection are also performed by the microcontroller.

Each furnace microcontroller is typically interfaced into a host computer. The host computer is capable of downloading a specific wafer process recipe, which contains all the necessary data for the microcontroller. The host computer will also perform functions such as wafer lot tracking, recipe programming, and automatic scheduling of lots.

Fast Ramp Vertical Furnace

Key issues in furnace performance (and the resulting wafer throughput) are the heating and cooling times required by the furnace. New vertical diffusion furnaces, called *fast ramp furnaces,* are able to quickly raise the temperature of a batch of wafers to the processing temperature (ramp-up time), reducing the time needed to stabilize at the process temperature, followed by a rapid cooldown after processing (ramp-down time). The development of fast ramp furnaces makes it possible to process a wafer batch size of 100 wafers with a ramp-up rate of 100°C/minute and a cooling ramp-down of 60°C/minute.[44] This compares with a ramp-up rate of a few degrees per minute for a conventional vertical furnace.

This fast ramp performance becomes critical with large-diameter wafers due to the larger wafer mass required for heating and cooling. Figure 10.21 shows a typical thermal profile of a fast ramp vertical furnace versus a conventional vertical furnace. Another important aspect of fast ramp furnaces is control of the ambient atmosphere in the furnace. For instance, nitrogen ambient control during loading will inhibit uncontrolled oxide growth and produce more uniform thickness for oxide films.

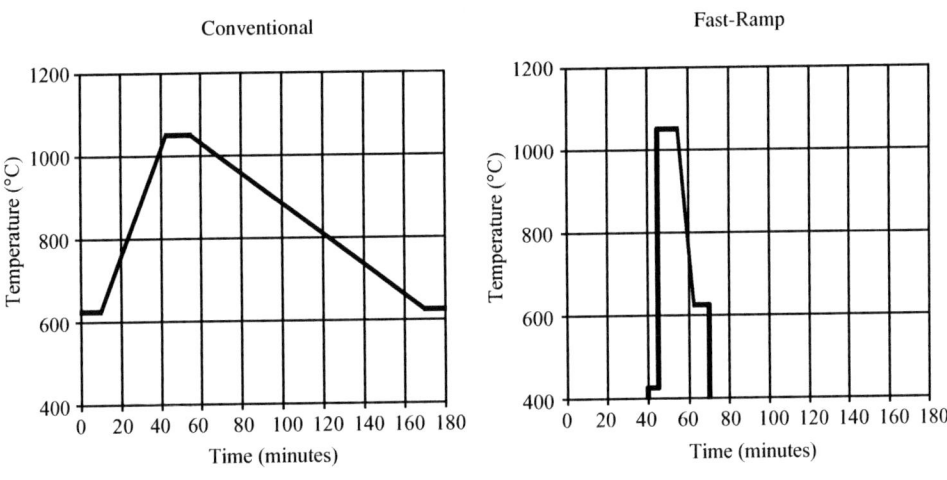

FIGURE 10.21 Thermal Profile of Conventional Versus Fast-Ramp Vertical Furnace
Printed from the June 1996 edition of *Solid State Technology,* copyright 1996 by PennWell Publishing Company

Advanced Temperature Control ■ Improved temperature control and thermal uniformity across the wafer during thermal processes are critical factors for reducing the thermal budget during wafer fabrication. A typical fast-ramp furnace can control the temperature uniformity to within ± 0.1°C at process temperature, while maintaining ± 5°C on all wafers during an 80°C/minute ramp cycle.[45] The key differences in a fast-ramp furnace are fast heating elements, special wafer carriers that increase the wafer-to-wafer gap for more uniform heating and cooling across the wafer, forced air cooling, and properly tuned temperature controllers.

Temperature in conventional furnaces is measured with thermocouples (see previous section), with a thermocouple controlling each furnace zone. A new technique of temperature control is called *model-based temperature control,* which allows individual wafers to be controlled for heating and cooling rather than merely controlling the ambient of the furnace. This is done by optimizing the control software for each individual furnace with the size of the wafer load. The furnace controller is able to monitor each zone relative to one another to optimize the temperature at the wafer.

Another factor for fast ramp furnaces is the size of the load. There is a trade-off between a large batch size of 150 to 200 wafers and the ramp rate. Increasing the ramp rate for a large batch of wafers fast will create stress on the wafers because the edge circumference of the wafers will heat up as quickly as the furnace does, while the temperature at the center of the wafer will lag by several hundred degrees. Fast ramp furnaces will typically use a smaller batch size of 50–100 wafers to assist in increasing the ramp rate. The smaller batch size also improves the flow of parts through the furnace process because there are fewer wafers to batch together for a run.

Rapid Thermal Processor

The rapid thermal processor (RTP) is a method of heating a single wafer to a temperature range of 400 to 1300°C in a very short time (often fractions of a second). The main advantages for an RTP over a conventional vertical furnace are:

- Reduced thermal budget
- Minimized dopant movement in the silicon
- Ease of clustering multiple tools
- Reduced contamination due to cold wall heating
- Cleaner ambient because of the smaller chamber volume
- Shorter time to process a wafer (referred to as cycle time)

The comparison of a conventional vertical furnace to a rapid thermal processor is outlined in Table 10.5.[46]

TABLE 10.5 Comparison of Conventional Vertical Furnace and RTP

Vertical Furnace	RTP
Batch	Single-wafer
Hot wall	Cold wall
Long time to heat and cool batch	Short time to heat and cool wafer
Small thermal gradient across wafer	Large thermal gradient across wafer
Long cycle time	Short cycle time
Ambient temperature measurement	Wafer temperature measurement
Issues:	Issues:
Large thermal budget	Temperature uniformity
Particles	Minimize dopant movement
Ambient control	Repeatability from wafer to wafer
	Throughput
	Wafer stress due to rapid heating
	Absolute temperature measurement

RTP Design ■ A schematic of a rapid thermal processor (RTP) is shown in Figure 10.22. A single wafer is rapidly heated in a chamber, commonly called a *reactor,* under atmospheric conditions or at low pressure. The RTP has a gas-handling system and a computer that controls system operation.

Most RTPs use multiple tungsten halogen lamps organized into zones as the heat source. The lamps are usually located on the top and bottom of the wafer and can range from 25 lamps to over 150 lamps. They are configured in multiple zones, such as from 4 to 14 zones to permit temperature contouring on the wafer. This contouring can compensate for nonuniform heating and cooling that may occur during ramp up and ramp down in cold wall systems. The silicon wafers are heated by selectively absorbing radiation from the lamps, which produce short-wavelength radiation. In this manner, RTP transfers energy between a radiant heat source and a wafer and does not heat the

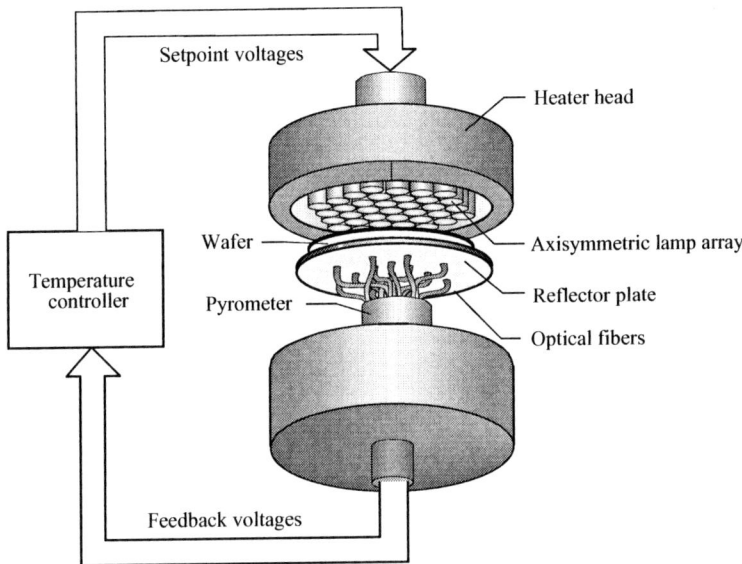

FIGURE 10.22 Rapid Thermal Processor (RTP)

surrounding reactor wall. This is why the term *cold wall* is used. A cold wall reactor is usually polished metal (such as stainless steel to increase reflectivity), with a quartz window to pass the radiation from the tungsten halogen lamp heat sources. Note that there are some systems that use lamps or resistance heaters to heat a susceptor, which then heats the wafer through thermal contact. In this case, the RTP would be classified as a warm or hot wall heater.

Temperature control in an RTP is done with thermocouple or optical pyrometers. Thermocouples are used for physical contact with the wafers to establish the actual wafer temperature. While relatively reliable, thermocouples have a slow response time and suffer from a shortened life span at higher temperatures.

An *optical pyrometer* can measure temperature from a distance with a fast response time. This measurement is done by sensing the emitted infrared radiation of the wafer while it is being heated. However, the emissivity (amount of emitted radiation) of the wafer surface can vary depending on the type of film on the top or bottom surface of the wafer. Recent advances in optical pyrometry

Rapid Thermal Processor
(Photo courtesy of Advanced Micro Devices)

have made emissivity less of a factor by utilizing solid-state pyrometers for better reliability and temperature measurement control.

The RTP has been hampered in the past with poor temperature uniformity across the wafer and nonequilibrium heating.[47] These problems lead to crystal slip and wafer warpage. There have been significant advances in RTP systems in the past few years, resulting in improved temperature control.

Equipment Integration ■ An RTP introduces flexibility into thermal processing because it can be integrated with other process steps into a multichamber cluster tool. This is beneficial because all processing steps occur in a vacuum environment to eliminate concerns with native oxide and contamination. The movement of single wafers through a cluster tool also reduces the time it takes to process wafers by reducing the waiting times associated with large wafer batches. Single wafers can also be processed in parallel with one another. These benefits occur in conjunction with the important action of reducing the thermal budget of the wafer.

RTP Applications ■ The RTP has become widely used in many processes throughout wafer fabrication. Advances in RTP chamber design and temperature uniformity are making this equipment acceptable with better temperature uniformity across wafers. Examples of operations that often use the RTP in wafer fabrication are:

- Anneal of implants to remove defects and activate and diffuse dopants
- Densification of deposited films, such as deposited oxide layers
- Borophosphosilicate glass (BPSG) reflow
- Anneal of barrier layers, such as titanium nitride (TiN)
- Silicide formation, such as titanium silicide ($TiSi_2$)
- Contact alloying

The first widely used application for RTP processing was annealing after ion implant. The advantage of using an RTP over a conventional diffusion furnace is that the wafer spends a much shorter time at temperature, thus reducing the thermal budget. However, the uniform temperature control of conventional furnaces makes this approach competitive.

OXIDATION PROCESS

The goal of thermal oxidation is the growth of a defect-free, uniform layer of SiO_2 at the thickness required. The type of oxidation process conditions used for a particular wafer fabrication step depends on the thickness of the oxidation layer and the properties required. Thin oxides, such as gate oxides, are usually grown in dry oxygen. Because sodium ion contamination is a concern, HCl is added to the O_2 supply during the oxidation of high-quality oxides. Where thick oxides are required, such as the field oxidation layer, then steam is used (HCl is not used in steam oxidation processes). Higher pressure during the growth process allows thick oxide growth to be achieved at a reduced temperature in reasonable time periods. The flow illustrated in Figure 10.23 shows the typical steps for thermal oxidation.

Pre Oxidation Cleaning

Cleanliness of the wafer is critical to achieve high quality oxidation (see Chapter 6). Contaminants such as particulates and mobile ionic contaminants (MICs) will have serious impact on device performance and yields. For example, the presence of MICs in the thermally grown gate oxide structure can result in long-term changes in the device threshold voltage as the MICs drift from the gate oxide to the Si/SiO_2 interface.[48] This condition is detrimental to device electrical performance. For a thermal oxidation process, the ability to keep MICs and particulate matter out of the system depends on maintaining the system to a high state of cleanliness control. The following areas are critical for minimizing contamination:

- Maintenance of the furnace and associated equipment (especially quartz components) for cleanliness
- Purity of processing chemicals
- Purity of oxidizing ambient (the source of oxygen in the furnace)
- Wafer cleaning and handling practices

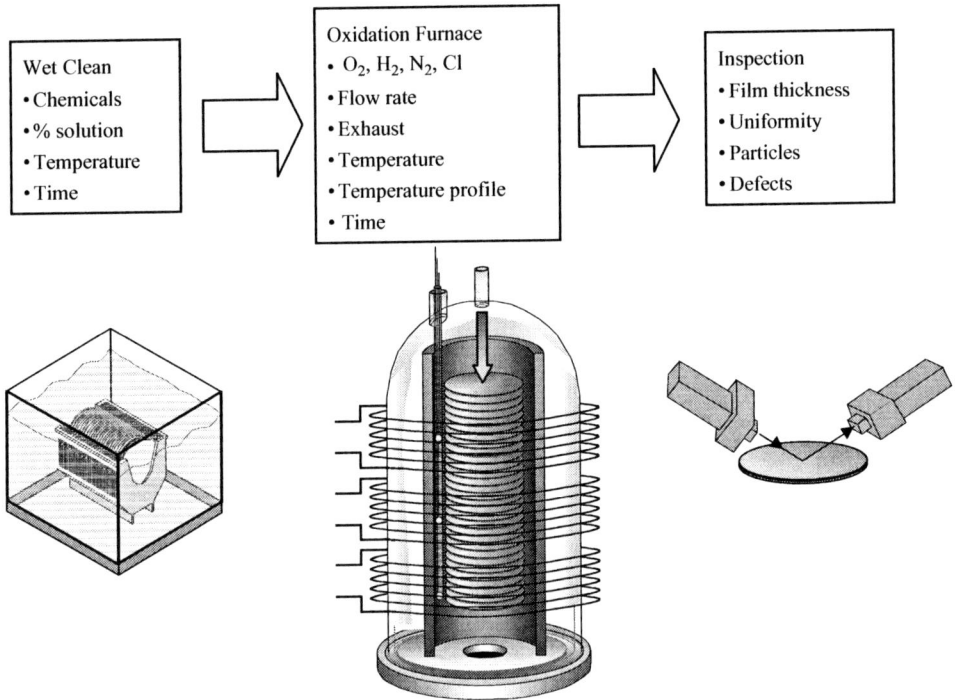

FIGURE 10.23 Thermal Oxidation Process Flow Chart

The basic wet-cleaning chemistries that have withstood the test of time are the RCA SC-1 and SC-2 clean systems and the piranha clean (sulfuric acid, hydrogen peroxide, and water mixture). These wafer cleaning methods and their modifications were discussed in Chapter 6. A wide selection of cleaning equipment is found in the oxidation area of the fab, including manual and automated wet sinks, ultrasonic systems (megasonics), acid spray processors, and rinser/dryer systems.

Oxidation Process Recipe

For thermal oxidation to occur, certain process conditions are followed in a specific format in the furnace equipment. This is known as a *process recipe*. For wafer fabrication, it is common that the parameters of a process recipe are stored in a software database and downloaded into the tool software for a particular wafer lot. We will review a sample process recipe for dry oxidation in a batch furnace.

Dry Thermal Oxidation ■ The most common reason for dry thermal oxidation is to grow the very thin layer of oxide used as the gate oxide.[49] As the devices become smaller in order to increase the component density on a chip (following Moore's law from Chapter 1), then the device dimensions and electrical parameters must also be scaled down proportionally. Gate oxide layers are considered the most critical oxidation application in wafer fabrication, with a gate oxide thickness on the order of 20–40 Å for 0.18-μm devices and 20–30 Å for 0.15-μm devices.[50] Dry thermal oxidation produces a high-quality oxide that is uniformly dense, free of pinholes, and reproducible. It is important for gate oxides to be produced in a clean, sodium-free system that is truly dry, since as little as 25 ppm of water will alter the oxide growth rate and properties of the oxide.[51] There is ongoing research to evaluate other replacement materials that could function as a dielectric in the gate structure, but there is no suitable alternative in production.

Process Recipe ■ Table 10.6 shows a sample process recipe for dry oxidation of a gate oxide.[52] During idle conditions, the process recipe shows there is a nitrogen (N_2) purge gas introduced into the chamber. It flows continuously until the process recipe is started and then the purge is turned off and N_2 process gas is turned on. Once the furnace tube is loaded, the temperature is raised from idle at 850°C to 1,000°C at a ramp rate of 20°C per minute. The wafers have a five-minute stabilization period. Since this is dry oxidation, oxygen at a flow rate of 2.5 slm is introduced into the process chamber. HCl flows in at 67 sccm and is used to reduce the interface charges and to getter MICs. During the anneal step, which is done to minimize charge buildup, the O_2 and HCl are turned off and N_2 is turned on. Oxygen still diffuses through the oxide even though the oxide flow has been cut off, which is taken into account to produce stoichiometric (uniform composition) silicon dioxide. A five-minute period is allowed to unload the furnace tube.

TABLE 10.6 Process Recipe for Dry Oxidation Process

Step	Time (min)	Temp (°C)	N_2 Purge Gas (slm)	Process Gas N_2 (slm)	Process Gas O_2 (slm)	Process Gas HCl (sccm)	Comments
0		850	8.0	0	0	0	Idle condition
1	5	850		8.0	0	0	Load furnace tube
2	7.5	Ramp 20°C/min		8.0	0	0	Ramp temperature up
3	5	1000		8.0	0	0	Temperature stabilization
4	30	1000		0	2.5	67	Dry oxidation
5	30	1000		8.0	0	0	Anneal
6	30	Ramp −5°C/min		8.0	0	0	Ramp temperature down
7	5	850		8.0	0	0	Unload furnace tube
8		850	8.0	0	0	0	Idle

Note: Gas flow units are slm (standard liters per minute) and sccm (standard cubic centimeters per minute).

QUALITY MEASUREMENTS

Representative quality measures for thermal oxide growth are listed in Table 10.7.[53]

TABLE 10.7 Quality Measures for Thermal Oxide Layer

Quality Parameter	Types of Defects	Remarks
1. Oxide thickness.	A. Thickness of gate oxide outside of specification (representative gate oxide specification is 20 ± 1.5 Å).	Possible reasons for this problem are: • Incorrect process gas flow (e.g., MFC calibrated improperly). Since HCl enhances oxidation rate, verify HCl:O_2 ratio is correct. • Verify O_2 leak integrity of the furnace with a bare silicon test wafer. • Check the metrology equipment by measuring the oxide thickness against a standard thickness wafer. • Excess native oxide growth due to overexposure of wafer to air either before or after normal furnace oxidation.

Quality Parameter	Types of Defects	Remarks
2. Gate oxide integrity (GOI).	A. Gate oxide breakdown. B. Mobile ionic contamination (MIC) in film.	Gate oxide defects are often related to processing conditions: • Perform a C-V test to demonstrate gate oxide integrity using an unpatterned test wafer. • Perform an oxide-charge analysis on a test wafer with a surface-charge analyzer. • Review the preoxidation clean steps to assess sources of contamination (e.g., particles or MICs). • Verify no contamination from an incoming gas line or defective filter.
3. Particles in the oxide film.	A. Contaminated quartzware. B. Wafer broken inside furnace. C. Contaminated carrier. D. Contaminated gas filter or line.	Actions to correct these sources of particles added during the oxidation process are: • Check cleanliness of quartzware and carrier. • Verify the alignment of the robotic handling systems. • Check incoming gas filters.
4. Particles under the oxide film.	A. Contaminated preoxidation clean.	Source of particles prior to the oxidation step are: • Verify the preoxidation clean steps are properly set up and performed. • Check cleanliness of quartzware and carrier.

When a batch of wafers is loaded into a furnace tube for oxidation, a number of test wafers with bare surfaces (also referred to as unpatterned wafers, as described in Chapter 19) are placed at strategic locations in the tube (see Figure 10.24). These test wafers are used to perform the various evaluations after the oxidation step to ensure the oxide has acceptable quality.

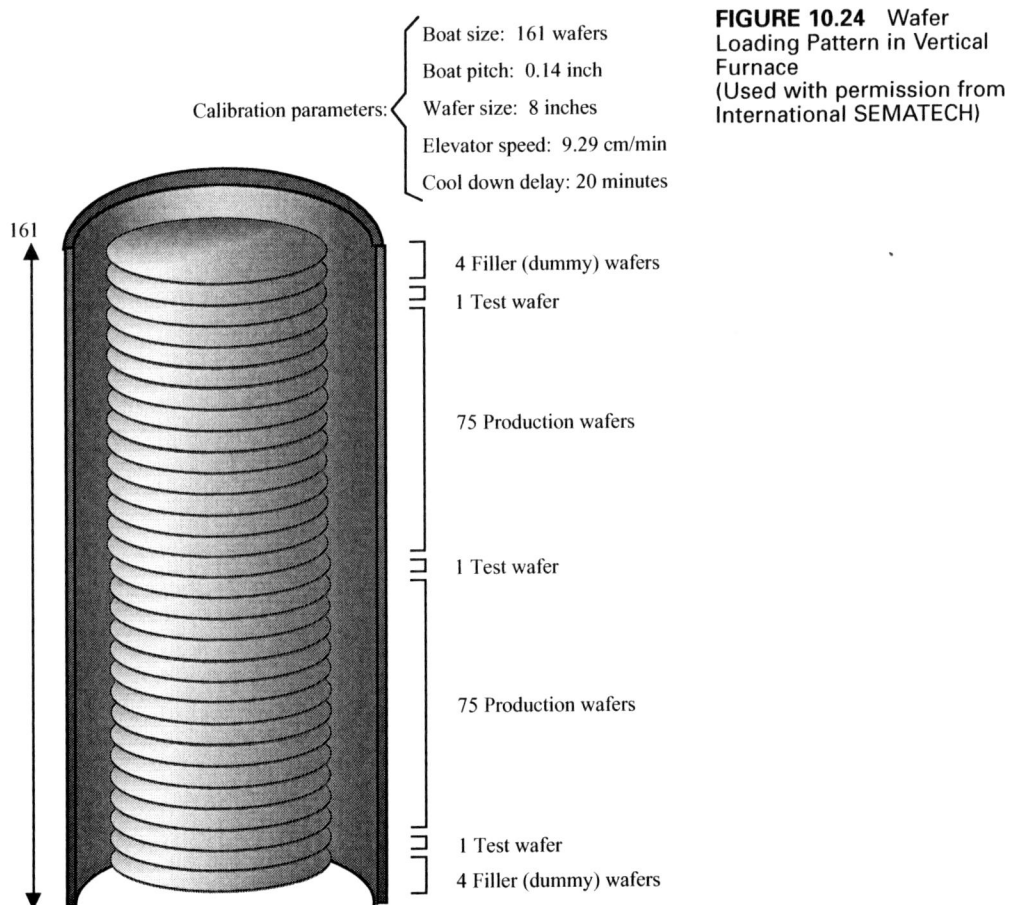

FIGURE 10.24 Wafer Loading Pattern in Vertical Furnace (Used with permission from International SEMATECH)

OXIDATION TROUBLESHOOTING

Common problems and corrective actions for thermal oxidation are described in Table 10.8.

TABLE 10.8 Common Oxidation Troubleshooting Problems

Problem	Probable Causes	Corrective Actions
1. Incorrect gas flow into furnace tube.	A. Incorrect process recipe. B. Malfunctioning MFC. C. Incorrect $H_2:O_2$ ratio for steam process (O_2 starved).	Possible corrective actions for this problem are: • Verify that correct process recipe is downloaded into furnace software for for the wafer being processed. • Check process gas MFCs (O_2, N_2, H_2, Cl) to verify calibration was performed and operating properly. • Verify gas valves are functioning properly (e.g., no leaks, and so on). • Check that no room air is leaking into the furnace tube from outside the quartz or furnace door seal.
2. Incorrect temperature uniformity in vertical furnace chamber.	A. Wrong process recipe for product. B. Incorrect operation of thermocouples.	Possible corrective actions are: • Verify that the correct process recipe is loaded for the wafer being processed. • Assess whether the temperature nonuniformity is during the ramp up, flat zone, and ramp down for corrective action. • Check all thermocouples for proper operation. Verify no degradation due to excessive heat or corrosion. Replace defective thermocouples. Verify no drift in thermocouple reference temperature.
3. Inadequate temperature uniformity in rapid thermal processor (RTP).	A. Malfunctioning heating system (e.g., tungsten halogen lamps). B. Verify correct operation of of temperature measurement sensor.	Possible corrective actions are: • Verify heating lamps are operating and have the correct lamp intensity. • Verify correct calibration and temperature measurement of the optical pyrometer temperature sensor. Check there is no variation in wafer emissivity by performing reflectivity measurement of the wafer surface.

SUMMARY

Silicon dioxide (or oxide) can be thermally grown or deposited on a wafer. SiO_2 has a tetrahedron cell atomic structure of a silicon atom surrounded by four oxygen atoms and is amorphous. Five uses of oxide film in wafer fabrication are (1) protection and isolation, (2) passivation, (3) dielectric, (4) dopant mask and (5) deposited insulator between metal layers. The thickness of an oxide layer depends on its applications. The chemical reaction for thermally grown dry oxide requires silicon and an oxidizer gas, consumes silicon during the reaction, and is speeded up with the use of wet oxidation. There is an undesirable oxide charge at the Si/SiO_2 interface that can be neutralized by the use of HCl. The growth of thermal oxide follows a linear growth stage up to about 150 Å, followed by a slower parabolic growth stage. Factors affecting the oxide growth rate are dopants, crystal orientation, pressure, and temperature. SiO_2 can be selectively grown or deposited.

The three basic types of thermal processing equipment are a horizontal furnace, vertical furnace, and rapid thermal processor (RTP). The vertical furnace has become the primary batch furnace in fabrication. It consists of a process chamber, wafer transfer system, gas distribution system, exhaust system, and temperature control system. Chamber materials are made of quartz or silicon carbide. The temperature is precisely controlled with the use of multiple thermocouples. The goal in thermal processing is to minimize the thermal budget. Fast-ramp vertical furnaces quickly ramp the temperature up or down with a smaller batch of wafers by using advanced temperature control. The rapid thermal processor (RTP) can reduce the thermal budget by heating wafers with ramp rates of up to hundreds of degrees per second and is often integrated with other process steps. Preoxidation cleaning is critical for attaining a high-quality oxide.

KEY TERMS

grown oxide layer
deposited oxide layer
topography (surface topology)
thermal budget
thermal oxide (thermal silicon dioxide [SiO_2])
tetrahedron cell
surface passivation
field oxide layer
gate oxide structure
gate oxide integrity
dry oxygen
wet oxidation
pyrogenic steam
diffusion
Fick's laws
fixed oxide charge
interface-trapped charge
mobile oxide charge
oxide-trapped charge
rate of oxide growth
linear stage
reaction-rate controlled
parabolic stage
diffusion controlled
local oxidation of silicon (LOCOS)
bird's beak effect
pad oxide

shallow trench isolation (STI)
oxidation-induced stacking faults (OISFs)
horizontal furnace
vertical furnace
hot wall batch furnaces
rapid thermal processor (RTP)
process chamber or furnace tube
wafer boat
quartzware
quartz carrier
flat zone
ambient
thermocouple (TC)
profile thermocouples
spike thermocouples
over-temperature thermocouples
wafer transfer system
burn box
scrubber
temperature ramp rates
fast ramp furnaces
model-based temperature control
reactor
cold wall
optical pyrometer
process recipe

REVIEW QUESTIONS

1. What is the difference between a grown and a deposited oxide layer?
2. Describe what topography is on the wafer surface.
3. Define thermal budget and explain why it is undesirable.
4. State what thermal silicon dioxide is and give another name for it.
5. Describe the atomic structure for silicon dioxide.
6. What is surface passivation and why is it beneficial?
7. Describe the field oxide layer, and state its range of thickness.
8. Why is the gate oxide thermally grown?
9. Explain gate oxide integrity.
10. Describe how SiO_2 can be used as a dopant barrier.
11. List six applications for thermal oxides in wafer fabrication and give a purpose for each application.
12. State the chemical reaction for dry oxidation. At what temperature range does this reaction usually take place?
13. State the chemical reaction for wet oxidation. Is this faster or slower than dry oxidation? Why?
14. If an oxide layer is thermally grown to be 2,000 Å thick, how much silicon is consumed?
15. What is diffusion? How does this process occur in thermal oxidation?
16. List the four types of oxide charges that can occur at the Si/SiO_2 interface. Are oxide charges desirable? Why or why not?
17. Give two advantages to using chlorinated agents during oxidation.
18. Describe rate of oxide growth. What parameters influence this rate?
19. Describe the linear growth stage for thermal oxidation. What is the thickness range where it occurs? State the equation that describes this growth.
20. What does it mean to be reaction-rate controlled for thermal oxidation growth?

21. Describe the parabolic growth stage for thermal oxidation. What is the thickness range where it occurs? State the equation that describes this growth.
22. What does it mean to be diffusion controlled for thermal oxidation growth?
23. What effect does doping have on oxide growth?
24. Explain the effect from crystal orientation on oxide growth.
25. Pressure has what effect on oxide growth?
26. What is the effect of plasma on oxide growth?
27. What is LOCOS and how does this process use thermal oxidation? What is a bird's beak in forming an oxidation layer, and why is this condition undesirable?
28. Explain shallow trench isolation (STI).
29. What causes stress in an oxide layer?
30. What are oxidation-induced stacking faults (OISFs)?
31. List the three basic types of thermal processing equipment.
32. What is a hot wall furnace?
33. List five performance factors for horizontal and vertical furnaces, and stipulate which type of furnace is the optimum.
34. What are the five components in a vertical furnace system?
35. Describe the process chamber (or furnace tube).
36. What two common materials are used for process tube materials?
37. How many zones are there typically in a furnace, and why is this number important?
38. What is a wafer boat?
39. What is meant by the term *ambient* when used in a high-temperature furnace?
40. What are the three locations for thermocouples in a vertical furnace? Describe the purpose of each thermocouple.
41. Explain the purpose of the wafer transfer system in a vertical furnace.
42. A gas delivery system in a vertical furnace serves what purpose?
43. Explain what a burn box is.
44. Describe the purpose of the furnace microcontroller.
45. Explain what a fast ramp vertical furnace is. What ramp-up and ramp-down rates can this furnace achieve?
46. Describe the advanced way to control temperature in a fast ramp vertical furnace.
47. What is a rapid thermal processor (RTP)? What are six advantages it has over the conventional furnace?
48. Describe how an RTP heats a wafer. Is an RTP typically a hot wall or cold wall heating system?
49. Describe the purpose of an optical pyrometer in an RTP and how it functions.
50. Why is preoxidation cleaning important?

FURNACE AND RTP EQUIPMENT SUPPLIERS' WEB SITES

Amtech Systems Inc.	http:/www.amtechsystems.com/
Applied Materials	http:/www.appliedmaterials.com/products/
Asahi Glass Electronic Materials	http:/www.agem-usa.com/
ASM	http:/www.asm.com/
Axcelis (formerly Eaton)	http:/www.axcelis.com/
CVD Equipment Corporation	http:/www.cvdequipment.com/
GaSonics International	http:/www.gasonics.com/
Eaton Corporation	http:/www.semiconductor.eaton.com/
Heraeus Amersil Inc.	http:/www.heraeus-amersil.com/
Kokusai Semiconductor Equipment	http:/www.ksec.com/
MRL Industries	http:/www.mrlind.com/
Omega Engineering Inc.	http:/www.omega.com/
Semitool Inc.	http:/www.semitool.com/
Silicon Valley Group	http:/www.svg.com/
TEL, Tokyo Electron Ltd.	http:/www.teainet.com
Tystar Corporation	http:/www.tystar.com/

REFERENCES

1. G. Moore, "The Role of Fairchild in Silicon Technology in the Early Days of 'Silicon Valley,'" *Proceedings of the IEEE* 86–1 (January 1998): p. 62.
2. S. Nag et al., "Low-Temperature Pre-Metal Dielectrics for Future ICs," *Solid State Technology* (September 1998): p. 70.
3. G. Anner, *Planar Processing Primer,* (New York: Van Nostrand Reinhold, 1990), p. 146.
4. S. Ghandhi, *VLSI Fabrication Principles: Silicon and Gallium Arsenide,* 2nd ed., (New York: Wiley, 1994), p. 452.
5. Ibid.
6. SEMATECH, *Semiconductor Dictionary,* (Austin, TX: SEMATECH, 1995).
7. S. Ghandhi, *VLSI Fabrication,* p. 452.
8. B. El-Kareh, *Fundamentals of Semiconductor Processing Technologies,* (Boston: Kluwer Academic, 1995), p. 40.
9. SEMATECH, "Furnace Equipment Overview," Module 3 in *Furnace Processes and Related Topics,* (Austin, TX: SEMATECH), p. 3–2.
10. Ibid.
11. B. El-Kareh, *Fundamentals,* p. 47.
12. B. El-Kareh, *Fundamentals,* p. 45.
13. D. Schroder, *Semiconductor Material and Device Characterization,* 2nd. ed., (New York: Wiley, 1998), pp. 337–339.
14. J. Mayer and S. Lau, *Electronic Materials Science: For Integrated Circuits in Si and GaAs,* (New York: Macmillan, 1990), p. 265.
15. D. Schroder, *Semiconductor Material,* pp. 337–339.
16. SEMATECH, "Oxidation Processes," module in *Furnace Processes and Related Topics,* p. 8.
17. G. Anner, *Planar Processing Primer,* p. 176.
18. S. Wolf and R. Tauber, *Process Technology,* vol. 1, *Silicon Processing for the VLSI Era,* 2nd ed. (Sunset Beach: Lattice Press, 2000), p. 269.
19. SEMATECH, "Oxidation Processes," module in *Furnace Processes and Related Topics,* p. 6.
20. S. Ghandhi, *VLSI Fabrication,* p. 460.
21. Ibid.
22. Ibid.
23. SEMATECH, "Oxidation Processes," module in *Furnace Processes and Related Topics,* p. 7.
24. B. Deal and A. Grove, "General Relationship for the Thermal Oxidation in Silicon," *Journal of Applied Physics* (1965): pp. 3770–78.
25. S. Ghandhi, *VLSI Fabrication,* p. 463.
26. S. Wolf and R. Tauber, *Process Technology,* p. 278.
27. S. Wolf and R. Tauber, *Process Technology,* p. 283.
28. B. El-Kareh, *Fundamentals,* p. 60.
29. S. Wolf and R. Tauber, *Process Technology,* p. 218.
30. S. Ghandhi, *VLSI Fabrication,* p. 462.
31. B. El-Kareh, *Fundamentals,* p. 69.
32. S. Wolf and R. Tauber, *Process Technology,* p. 296.
33. S. Ghandhi, *VLSI Fabrication,* p. 473.
34. P. Singer, "Furnaces Evolving to Meet Diverse Thermal Processing Needs," *Semiconductor International,* (March 1997), p. 85.
35. Industry Watch, "Thermal Processing: Meeting the Challenges of 300 mm," *Semiconductor International,* (April 1998), p. 19.
36. SEMATECH, "Furnace Equipment Overview," module 3 in *Furnace Processes and Related Topics,* pp. 3–6.
37. J. Tomanovich, "LPCVD Components Trend Toward SiC," *Solid State Technology,* (June 1997): p. 135.
38. P. Singer, "Furnaces Evolving," p. 86.
39. A. Helms et al., "Status and Future of Batch, Hot-Wall Furnaces," *Solid State Technology,* (November 1999): p. 83.
40. P. Singer, "Furnaces Evolving," p. 86.
41. C. Porter et al., "Improving Furnaces with Model-Based Temperature Control," *Solid State Technology,* (November 1996): p. 120.
42. SEMATECH, "Gas Delivery Systems," module 8 in *Furnace Processes and Related Topics,* (Austin, TX: SEMATECH, 1994), pp. 8–3.
43. A. Helms et al., "Status and Future," p. 83.
44. L. Peters, "Thermal Processing's Tool of Choice: Single-Wafer RTP or Fast Ramp Batch?" *Semiconductor International,* (April 1998): p. 84.
45. Ibid.
46. R. Fair, "Conventional and Rapid Thermal Processes," *ULSI Technology,* ed. C. Change and S. Sze, (New York: McGraw-Hill, 1996), p. 151.
47. K. Reid and A. Sitaram, "Rapid Thermal Processing for ULSI Applications: An Overview," *Solid State Technology,* (February 1996): p. 64.
48. S. Wolf and R. Tauber, *Process Technology,* p. 225.
49. SEMATECH, "Oxidation Processes," module in *Furnace Processes and Related Topics,* p. 11.
50. L. Peters, "Thermal Processing's Tool," p. 86.
51. S. Ghandhi, *VLSI Fabrication,* p. 466.
52. Adapted from SEMATECH, "Oxidation Processes," module in *Furnace Processes and Related Topics,* p. 13.
53. SEMATECH, "Troubleshooting Process and Equipment Problems," module in *Furnace Processes and Related Topics,* (Austin, TX: SEMATECH, 1994), pp. 1–16.

CHAPTER 11
DEPOSITION

Microchip fabrication is a planar process involving many different film layers on the wafer surface. Deposition is a process that places the film layers on the wafer. Conductor and insulator films are essential to the successful fabrication of semiconductor devices on silicon substrate wafers. Film layering technology is used to manufacture the circuits and to connect the different IC devices, primarily with metal conducting layers sandwiched between insulating dielectric layers.

Many different types of films are deposited on wafers during the fabrication process. In some cases, these deposited films become an integral part of the device structure, while other sacrificial films are used for particular processing steps and then removed. Films deposited during microchip fabrication are often referred to as thin films because they are so thin that their electrical and mechanical properties differ from a thicker bulk film of the same material.

This chapter will discuss the processes and equipment used to deposit thin films, with a focus on dielectric films such as silicon dioxide and silicon nitride and the deposition of polysilicon. The deposition of metal and metal compound thin films is discussed in Chapter 12.

OBJECTIVES

After studying the material in this chapter, you will be able to:

1. Describe multilayer metallization. Discuss the acceptable characteristics of a thin film. State and explain the three stages of film growth.
2. Provide an overview of the different film deposition techniques.
3. List and describe the eight basic steps in a chemical vapor deposition (CVD) reaction, including the different types of chemical reactions.
4. Describe how CVD reactions are limited, and explain reaction dynamics and the effect of dopant addition to CVD films.
5. Describe the different types of CVD deposition systems, explain how the equipment functions, and discuss the benefits/limitations of a particular tool for film applications.
6. Explain the importance of dielectric materials for chip technology, giving examples of applications.
7. Discuss epitaxy and three different epilayer deposition methods.
8. Explain spin-on-dielectrics.

INTRODUCTION

The design and fabrication of early semiconductor wafers from the MSI and LSI eras was relatively straightforward. This process consisted of fabricating the semiconductor devices in silicon and interconnecting the devices to one metal conducting layer sandwiched between silicon dioxide as the dielectric material. Figure 11.1 on page 258 shows the deposited layers needed to make an early nMOS transistor. This technology was an extension of the first planar transistors made in the SSI era. The critical dimension was well over one micron. The wafer layers were not flat structures due

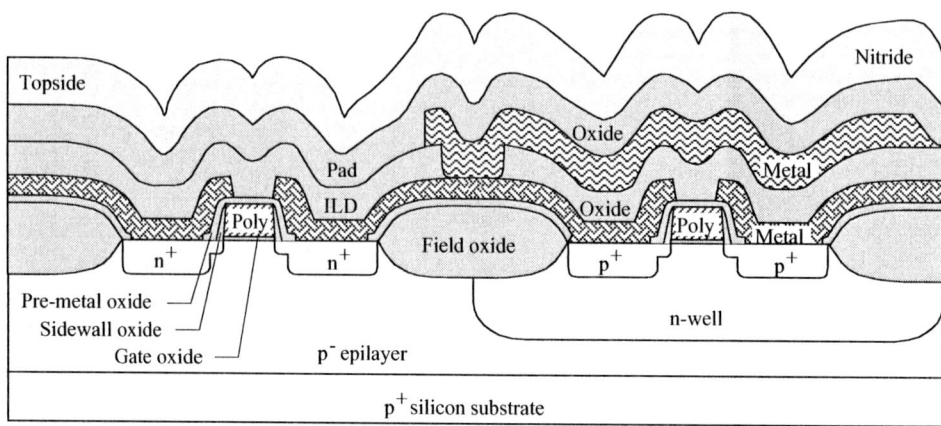

FIGURE 11.1 Film Layers for an MSI Era nMOS Transistor

to varying feature heights, which becomes a limiting factor for high-density chips with multiple metal layers needed in the ULSI era.

As wafers progress to higher density chips with shrinking geometries of 0.18 μm and below, the materials and processes used in wafer fabrication are undergoing dramatic changes. There is a concurrent scaling of all device features to maintain electrical performance. To make electrical connections in today's advanced microchips, six or more conducting metal layers are required. New metal conductor materials are needed to maintain electrical performance. Advanced dielectric materials are deposited between the metal films to provide adequate insulation protection. At the same time, each chip has literally tens of billions of electrical connections between the various metal layers and silicon devices. The ability to deposit reliable thin-film materials is a critical operation in a wafer fab. Thin films is a major process step in wafer fabrication, as shown in the process flow model in Figure 11.2.

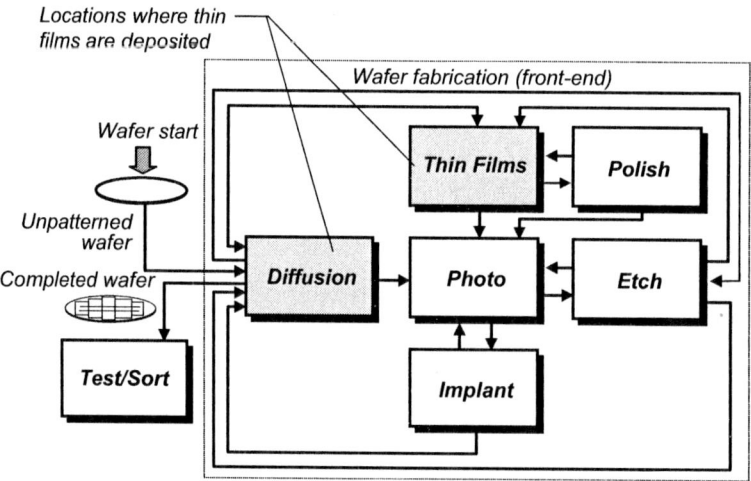

FIGURE 11.2 Process Flow in a Wafer Fab
(Used with permission of Advanced Micro Devices)

Film Layering Terminology

Multilevel metallization refers to the metal and dielectric layers needed to interconnect the densely packed devices on the wafer. An example of this process is shown in Figure 11.3. Without the dielectric insulating layers, electrical shorts would occur (similar to electrical wiring without its insulating cover). The metal layers are connected by openings in the dielectric film referred to as *vias*.

Adding metal levels is costly to manufacture. It is estimated that the addition of a single metal level in a CMOS process adds approximately 15% to the total wafer fabrication cost.[1] The

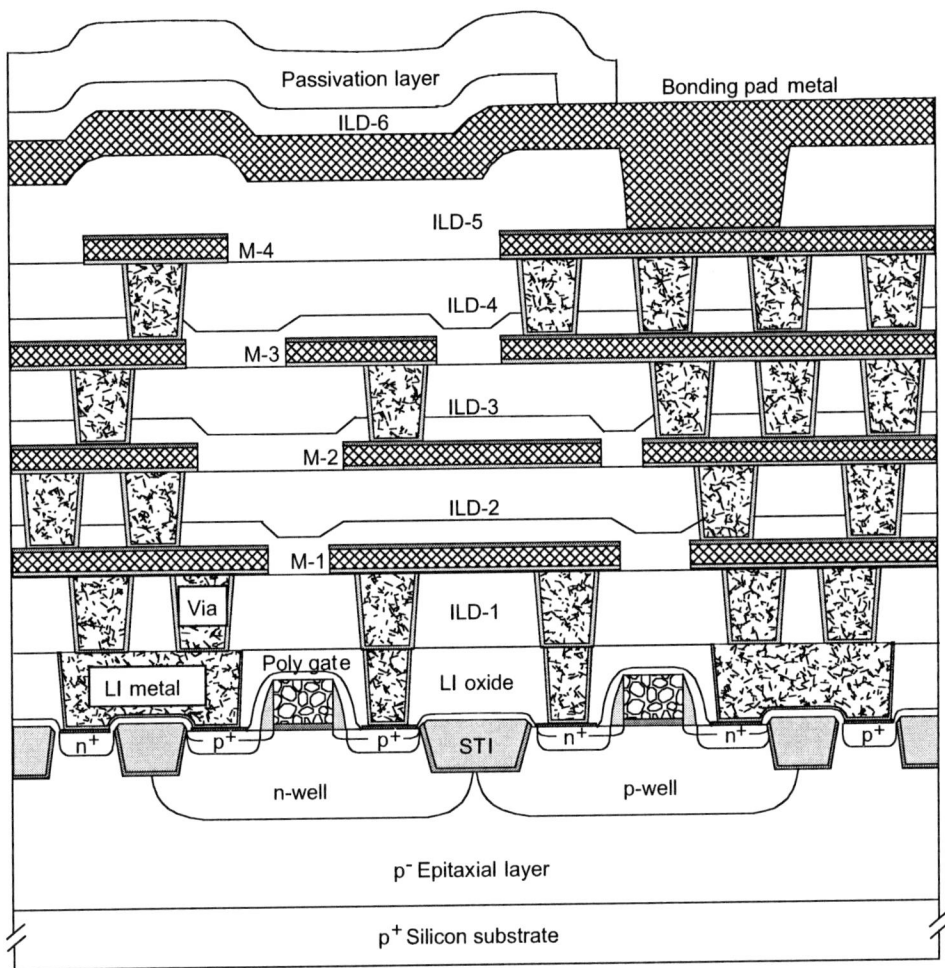

FIGURE 11.3 Multilevel Metallization on a ULSI Wafer

complexity added to wafer manufacturing by additional metal levels makes it important to reduce defects to ensure the chip yield is not impacted. The chip designer considers the tradeoffs among cost, complexity, and performance across the different wafer fab process areas for implementing additional metal levels.

Metal Layers ■ *Aluminum metallization* is the use of aluminum alloy for interconnect wiring. Aluminum alloy has been used since the beginning of semiconductor manufacturing. The Al metal is deposited on the entire surface of the wafer in a solid thin film and then etched to define the width and spacing of the interconnect lines. The industry is undergoing a transition to copper metallization to achieve increased chip speed with fewer process steps (copper metallization is reviewed in detail in Chapter 12). Each metal layer can be referred to as Metal-1, Metal-2, and so on. *Critical layers* are those metal layers with linewidths etched to the critical dimension of the device (e.g., CD of 0.15 μm). For ULSI, the CD typically occurs at the polysilicon features of the gate, the oxide structure, and the metal layers closest to the silicon surface. Critical layers are sensitive to particulate contamination (killer defects) and reliability issues such as electromigration are more pronounced in fine geometry linewidths. *Noncritical layers,* usually the upper metal levels, have much wider line widths, often 0.5 μm and larger, and are less sensitive to particulate contamination. However, factors such as long conductor lengths in the noncritical upper levels can affect chip speed and power consumption.[2]

Dielectric Layers ■ The dielectric layer between the active devices in silicon and the first metal layer is termed the *first interlayer dielectric (ILD-1).* This layer is also called the *premetal dielectric (PMD).* ILD-1 is typically a doped silicon dioxide, or glass (explained later in this chapter). The

Metal Layers in a Chip
(Micrograph courtesy of Integrated Circuit Engineering)

important function of the ILD-1 layer is to isolate transistor devices in two ways: electrically from the metal interconnect layers, and physically from contamination sources such as mobile ions. The ILD-1 has a restricted thermal budget in high-performance logic devices in order to avoid degrading transistor characteristics.[3]

The *interlayer dielectric* (*ILD*) is used between different metal layers in the device. The ILD serves as an insulating film between two conducting metals or adjacent metal lines. The ILD has traditionally been silicon dioxide (SiO_2) with a dielectric constant of around 3.9 to 4.0.[4] The dielectric constant is an important property of deposited insulating films because it directly affects circuit speed performance (see the following section).

FILM DEPOSITION

A *thin film* is a thin, solid layer of a material created on a substrate. If a solid material has three dimensions (thickness, width, and length), then a thin solid film has one of its dimensions (usually thickness) much smaller than the other two (see Figure 11.4). The thin film is bonded to a wafer substrate, which has a much thicker dimension than the film. The thin-film surface is so close to the substrate that it has a profound influence on the physical, mechanical, chemical, and electrical properties of the thin film material.[5] The most widely used unit to describe the thickness of a thin film in wafer fabrication is the angstrom (Å).

Thin film deposition in semiconductor fabrication is any process that physically deposits a film on the wafer substrate. The film can be either a conducting, insulating, or semiconducting material. Examples of deposited films are polysilicon, silicon dioxide (SiO_2), silicon nitride (Si_3N_4), polysilicon (silicon in a polycrystalline structure), and metals such as copper and refractory metals (e.g., tungsten).

Thin-Film Characteristics

For a thin film to be acceptable in wafer fabrication, it must have desirable film characteristics. General characteristics of an acceptable thin film for device performance are:[6]

- Good step coverage
- Ability to fill high aspect ratio gaps (conformality)
- Good thickness uniformity
- High purity and density

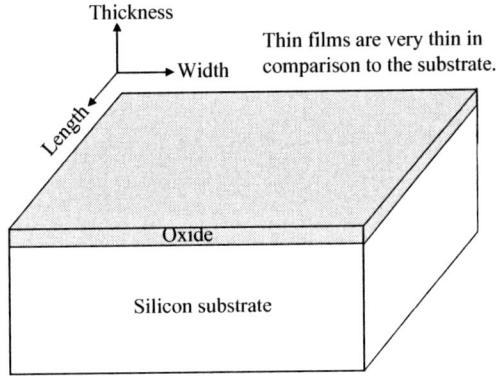

FIGURE 11.4 Solid Thin Film

- Controlled stoichiometries
- High degree of structural perfection with low film stress
- Good electrical properties
- Excellent adhesion to the substrate material and subsequent films

Film-Step Coverage ■ It is desirable for thin films to maintain a uniform thickness over surface features (see Figure 11.5). Steps on the wafer surface occur due to three-dimensional shapes from patterned features that create surface topography. If the film thins excessively at a step, this can cause high stress, electrical shorts, or undesirable induced charges in the device. Films with minimum stress are important, since stress can lead to substrate deformation in a convex or concave shape. All deposition techniques result in the formation of some stress in films.

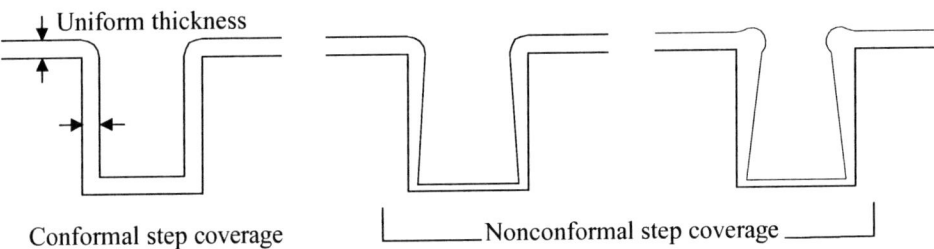

FIGURE 11.5 Film Coverage Over Steps

High Aspect Ratio Gaps ■ A small gap (trench or hole) is characterized by its *aspect ratio,* which is defined as the ratio of its depth to width (see Figure 11.6 on page 262). Aspect ratio is expressed as a ratio, such as 2:1, which in this case means the gap depth is two times the width. A gap can also be specified by its opening width, such as 0.25 μm, which may be different than the CD of other structures on the wafer. The ability to fill very small gaps and holes on the surface of the wafer has become one of the most important film characteristics for devices fabricated with sub-0.25 micron geometries. Examples of gaps that require effective gap-fill capability are vias passing through the interlayer dielectric (ILD) and trenches for shallow trench isolation (STI).[7] A high aspect ratio is typically above 3:1, with some applications having aspect ratios of 5:1 or larger. High aspect ratio gaps make it difficult to uniformly deposit the film, leading to pinch-off and voiding. Deposition processes that can produce uniform, void-free films in high aspect ratio openings are critical in the sub-0.25 μm era of high-density IC circuits with shrinking geometries.

Thickness Uniformity ■ Acceptable thin films are conformal with good thickness uniformity, which means the film follows the outline of the base material in all areas. The resistance of a material will change with varying thickness, which would be undesirable. Thinner layers of film also

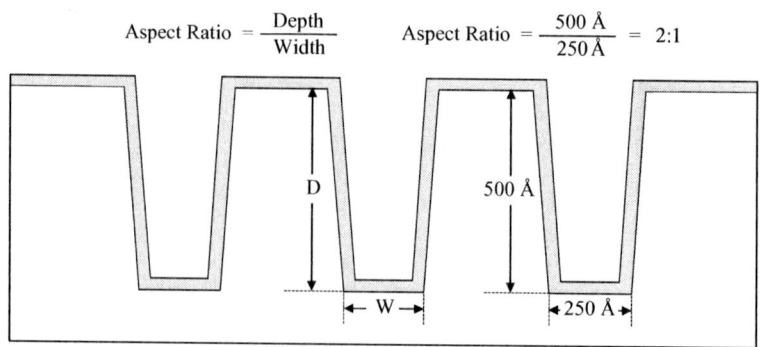

FIGURE 11.6 Aspect Ratio for Film Deposition

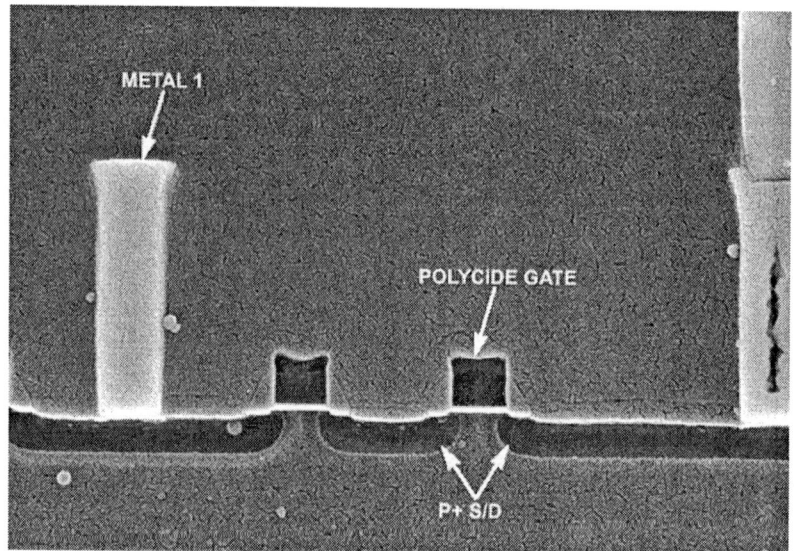

High Aspect Ratio Gap
(Micrograph Courtesy of Integrated Circuit Engineering)

tend to have more defects such as pinholes, which leads to less mechanical strength. It is desirable for thin films to have good surface flatness to minimize steps and cracking.

Film Purity and Density ■ High film purity means that the film does not have any unwanted chemical elements or molecules, which is important if the film is to function properly. This includes the avoidance of contaminants such as mobile ionic contaminants (MICs) and particulates. The occurrence of film contaminants such as hydrogen can degrade film properties. The density of the film is an important indicator of film quality because it means the film is free of pinholes and voids. A porous film has lower density and in some instances a lower refractive index than a nonporous film.

Stoichiometry ■ A desirable film will have uniform composition. As chemicals undergo change in a reaction, stoichiometry describes how the quantities of reactants and products are produced once the reaction ends or stabilizes. *Stoichiometry* is the ratio of one component amount to another component amount in a compound or molecule (e.g., the H_2O stoichiometric ratio is 2:1). If you know what is in a reaction, then stoichiometry tells you how much of the ingredients was needed. The chemical reactions that occur in deposition are complex and can lead to films that have a different composition than intended. A goal in deposition is to have the correct quantity of molecules in the reaction so that the deposited film approaches the ratio of the elements in the nominal chemical formula of the incoming gas.

Film Structure ■ The film structure is critical, especially related to the grain size. Materials tend to collect and grow as grains during the deposition process. If the grain size varies in a film, then it will have varying electrical and mechanical properties that can affect the long-term reliability of the film, especially concerning electromigration (discussed in Chapter 12).[8] The growth of a film can cause undesirable stress in the film layer that deforms the wafer substrate, leading to film cracking, delamination or void formation. An example of this condition is hydrogen contamination in nitride film deposition, which leads to a compressive stress.[9] The objective during film deposition is always to minimize the amount of stress.

Film Adhesion ■ Film adhesion to the substrate material is important for thin films to avoid delamination and cracking. A cracked film can cause surface roughness and allows contamination to pass through the film. For insulating films, cracks can lead to electrical shorts or current-leakage paths. Adhesion of a film to a surface is determined by the cleanliness of the surface and also by the type of material to which the film can alloy. Metals such as chromium, titanium, and cobalt are often useful for their adhesion properties (see Chapter 12). Good adhesion of films in multilevel metallization is critical to maintaining the overall electrical and mechanical integrity of the structure. This is true in the as-deposited condition and after subsequent processing.

Film Growth

A deposited film grows in three distinct stages (see Figure 11.7).[10] The first stage is *nucleation* where clusters of stable nuclei are formed. This stage occurs when the first few atoms or molecules of the reactants combine to form isolated patches of film that attach to the wafer surface. Nucleation occurs directly on the wafer and is essential to further thin film growth. The second stage is the *nuclei coalescence* into clusters, also referred to as *island growth*. These randomly oriented island clusters grow based on surface mobility and the density of the cluster. The island clusters continue to grow and eventually develop into the third stage, a *continuous film* where they meet and form a solid sheet that spreads across the substrate surface.

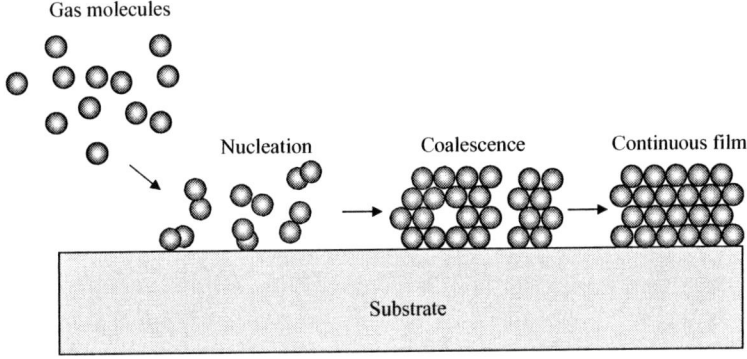

FIGURE 11.7 Stages of Film Growth

The size of the individual cluster, before meeting a neighboring cluster, is related to the surface mobility of the reacting species and the density of the nuclei. A high surface mobility and/or low nucleation rate promote the formation of relatively large clusters. This condition generally leads to polycrystalline films with short-range crystal order. On the other hand, low surface mobility combined with a high nucleation rate leads to amorphous film growth with no short-range order.[11] Low deposition temperatures usually favor amorphous films because the low thermal energy slows the surface mobility of the reacting materials.

Deposited films can be amorphous, polycrystalline or single crystalline.[12] Semiconductor films are used in all three forms. Insulator or metal films are usually amorphous or polycrystalline. Silicon deposited on an oxide layer is polycrystalline, as in the polysilicon used in the gate structure. Epitaxial single-crystal films are necessary for reliable semiconductor properties. To obtain a single-crystal deposited film, it is deposited on a single-crystal wafer substrate.

Film Deposition Techniques

Deposition of materials onto the wafer surface places a continuous thin film on the wafer. The film material derives from an external source, which may be a gas source that produces the material to be deposited through a chemical reaction or a solid target source that requires the physical removal of the material. With the increased complexity of semiconductor processing in the ULSI era, the number of deposited films on a wafer is high. As a result, the deposition of thin films is one of the most common processes in the semiconductor industry.

There are different techniques used in wafer fabrication to deposit films on a substrate. The major deposition methods shown in Table 11.1 are generally grouped into chemical or physical processes.

TABLE 11.1 Techniques of Film Deposition

Chemical Processes		Physical Processes		
Chemical Vapor Deposition (CVD)	**Plating**	**Physical Vapor Deposition (PVD or Sputtering)**	**Evaporation**	**Spin-On Methods**
Atmospheric Pressure CVD (APCVD) or Sub-Atmospheric CVD (SACVD)	Electrochemical deposition (ECD), commonly referred to as electroplating	DC Diode	Filament and Electron Beam	Spin-on-glass (SOG)
Low Pressure CVD (LPCVD)	Electroless Plating	Radio Frequency (RF)	Molecular Beam Epitaxy (MBE)	Spin-on dielectric (SOD)
Plasma Assisted CVD: • Plasma Enhanced CVD (PECVD) • High Density Plasma CVD (HDPCVD)		DC Magnetron		
Vapor Phase Epitaxy (VPE) and Metal-organic CVD (MOCVD)		Ionized metal plasma (IMP)		

Adapted from F. Barlow III, A. Elshabini-Riad, and R. Brown, "Film Deposition Techniques and Processes," *Thin Film Technology Handbook,* eds. A. Elshabini-Riad and F. Barlow III (New York: McGraw-Hill), pp. 1–2.

This chapter reviews the dielectric thin films deposited with chemical vapor deposition (CVD), epitaxy, and spin-on-dielectric (SOD) methods. All metallization is discussed in Chapter 12, including chemical vapor deposition (CVD) for metals, sputtering, electroplating, and evaporation.

CVD is commonly used for depositing dielectric and metal films. Spin-on-dielectric (SOD) is used for applying liquid dielectric films followed by a high-temperature curing process. Sputtering, or PVD, is the most common method in wafer fabrication for applying metals such as aluminum. Electroplating, widely used in thin film deposition for magnetic recording heads in disk drives, has not been used in silicon wafer fabrication. It has recently become the most promising process for depositing copper metallization. Evaporation is the traditional method for applying a metal layer, but its poor gap-fill property led to its demise at the start of the VLSI era, when it was replaced by sputtering.

CHEMICAL VAPOR DEPOSITION

Chemical vapor deposition (*CVD*) is the process of depositing a solid film on the wafer surface through a chemical reaction of a gas mixture. The wafer surface or its vicinity is heated in order to provide additional energy to the system to drive the reactions. The essential aspects of CVD are:

1. Chemical action is involved, either through chemical reaction or by thermal decomposition (referred to as *pyrolysis*).
2. All material for the thin film is supplied by an external source.
3. The reactants in a CVD process must start out in the vapor phase (as a gas).

CVD deposition by use of a chemical reaction occurs when chemical compounds are mixed and then reacted in a deposition chamber, referred to as a *reactor*. The atoms or molecules deposit on the wafer surface and form the film.

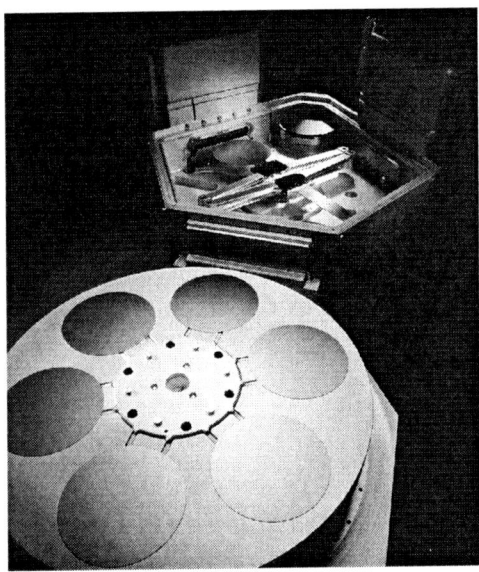

Chemical Vapor Deposition Tool
(Photo courtesy of Novellus Systems, Inc.)

CVD Chemical Processes

There are five basic chemical reactions that can be used in CVD:[13]

1. **Pyrolosis:** a compound dissociates (breaks bonds, or decomposes) with the application of heat, usually without oxygen.
2. **Photolysis:** a compound dissociates with the application of radiant energy that breaks bonds.
3. **Reduction:** a chemical reaction occurs by reacting a molecule with hydrogen.
4. **Oxidation:** a chemical reaction of an atom or molecule with oxygen.
5. **Reduction-oxidation (redox):** a combination of reactions 3 and 4 with the formation of two new compounds.

Among these five general types of reactions, there are many specific CVD reactions used to deposit films on a wafer substrate. The choice of a particular reaction is usually defined by parameters such as deposition temperature (which must be acceptable for the wafer materials), the film properties, and manufacturing issues. An example of a specific CVD reaction is the deposition of a silicon dioxide film by an oxidation reaction of silane with oxygen. This reaction produces silicon dioxide that is deposited on the heated wafer surface, with a by-product of hydrogen.

$$\underset{silane}{SiH_4} + \underset{oxygen}{O_2} \xrightarrow{(heat)} \underset{silicon\ dioxide}{SiO_2} + \underset{hydrogen}{2H_2}$$

CVD Reaction

The CVD chemical reaction that deposits the solid film takes place on (or very close to) the wafer surface. This is a *heterogeneous reaction* (also referred to as *surface catalyzed*). Some reactions also occur in the gas phase above the wafer surface, which is termed a *homogeneous reaction*. Homogeneous reactions are undesirable because they form gas-phase clusters of the depositing material, which results in poorly adhering, low-density films with higher defects.[14] In CVD chemical reactions, heterogeneous reactions at the wafer surface are favored for the production of high-quality films.

CVD Reaction Steps ■ The fundamental CVD reaction has eight major steps that explain the reaction mechanism. The steps are summarized in the following list and are shown in Figure 11.8:[15]

1. **Gas transport to deposition zone:** Mass transport of gas in the main gas flow region from the reactor inlet to the deposition zone of the wafer.
2. **Formation of film precursors:** Gas-phase reactions leading to the formation of the film precursors (initial atoms and molecules that will constitute the film) and by-products.
3. **Film precursors at wafer:** Mass transport of the film precursors to the wafer growth surface.
4. **Precursor adsorption:** Adsorption (binding) of film precursors to the surface.
5. **Precursor diffusion:** Surface diffusion of film precursors to the film growth sites.
6. **Surface reactions:** Surface chemical reactions leading to film deposition and by-products.
7. **By-product removal from surface:** Desorption (removal) of the by-products of the surface reactions.
8. **By-product removal from reactor:** Mass transport of the by-products in the bulk gas-flow region away from the deposition zone and towards the reactor exit.

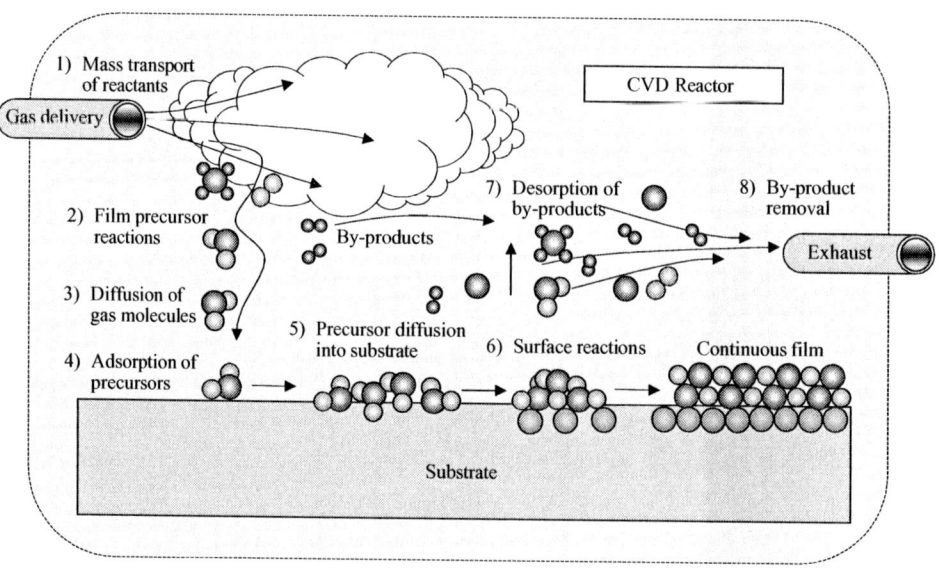

FIGURE 11.8 Schematic of CVD Transport and Reaction Steps

Adsorption is the chemical binding that occurs during deposition, causing the gaseous atoms and molecules to chemically attach to the solid wafer surface. *Desorption* is the removal or release of by-products from the wafer surface reactions. In chemical reactions, *species* describes a particular chemical substance that is an atom, ion, or molecule. In gas-phase reactions, there are often intermediate reactions called *precursors* that form a gas species that does not contain the original gas components. In CVD, gas precursors are transported to the wafer surface for adsorption and reaction. For example, for the following three reactions, the first gas-phase reaction shows that when silane decomposes, a silicon hydride precursor is produced. This precursor then reacts with silane to form disilane. During the intermediate CVD reaction, the silicon hydride is adsorbed onto the wafer surface along with silane. Disilane then decomposes to form the final silicon solid-film product.

1) SiH_4 (gas) $\rightarrow$ SiH_2 (gas) + H_2 (gas)   (pyrolysis)
   silane   *silicon hydride*   *hydrogen*

2) SiH_4 (gas) + SiH_2 (gas) $\rightarrow$ Si_2H_6 (gas)   (Reactive intermediate formation)
   silane   *silicon hydride*   *disilane*

3) Si_2H_6 (gas) $\rightarrow$ 2Si (solid) + $3H_2$ (gas)   (Final product formation)
   disilane   *solid film product*   *by-product*

Rate Limiting Step ■ The length of time it takes for the CVD reaction to occur is important for productivity in manufacturing. The surface reaction rate increases with increasing temperature. Since CVD reaction steps are sequential, the slowest step defines the bottleneck in the process. In other words, the step that occurs at the slowest rate will determine the rate of deposition.[16]

The rate of a CVD reaction cannot proceed more rapidly than the mass-transport rate at which reactant gases are supplied from the main gas stream to the wafer substrate. This is true no matter how high the temperature is increased. This situation is referred to as a *mass-transport limited* deposition process. A mass-transport limited process is only weakly dependent on temperature.[17] This means that regardless of the temperature, there is insufficient reactant gas being supplied to the wafer surface to speed up the reaction. Because of this, a high-temperature, higher pressure CVD process that wants to proceed rapidly is usually mass-transport limited—it cannot supply enough reactant gas to the substrate because of diffusion or physical limitations.

At lower reaction temperatures and pressure, the surface reaction rate is reduced because there is less energy available to drive the surface reaction. Eventually the arrival rate of reactants will exceed the rate at which they are consumed by the chemical reaction process at the wafer surface. In this case, the deposition rate is *reaction-rate limited*.[18] This condition is similar to pouring too much liquid through a funnel—the liquid will not flow faster than the funnel permits. Even though more reactant gas is supplied, the reaction will not speed up because the low temperature provides insufficient energy. A reaction-rate limited CVD process is also referred to as *kinetically controlled*, meaning the diffusivity of the reactants is not as critical as the kinetic energy of the reactants at the surface. For reaction-rate limited processes, it is important to maintain a uniform temperature in order to achieve a uniform deposition rate across the wafer surface.

CVD Gas-Flow Dynamics ■ The gas-flow dynamics for CVD are important for uniform film deposition. Gas flow refers to the physical transport of reactant gases to the CVD sites on the wafer (see Figure 11.9 on page 268). At the molecular level, this requires having a sufficient number of molecules at the right place and the right time. This condition is necessary for a stoichiometric reaction to occur between the reactants and products.

The principal factors for CVD gas flow are the mass transport of the reactant gases from the main gas stream to the wafer surface and the chemical reaction rate at the surface. It is assumed that the dominant transfer mechanism from the gas phase to the surface is diffusion.[19] Considering the gas flow right at the wafer surface during a CVD reaction process, there is zero or near-zero gas flow. This condition creates a *boundary layer* of gas flow that increases from the zero gas-flow velocity at the wafer surface (due to frictional forces) to some typical gas flow farther from the surface.[20] These dynamics are shown in Figure 11.10 on page 268. The gas flow farther from the surface reaches some average gas-flow velocity that represents the average velocity of the main gas stream in the CVD reactor. If the boundary layer is narrow, then it can be treated as not moving at all near the wafer surface. In this case, it is referred to as a *stagnant layer*. The size of the stagnant layer will influence the design of the different types of CVD reactors considered later in this chapter.

Pressure in CVD ■ If a CVD reaction occurs at low pressure, then the diffusivity of the gas species increases significantly through the boundary layer to reach the surface. This diffusivity

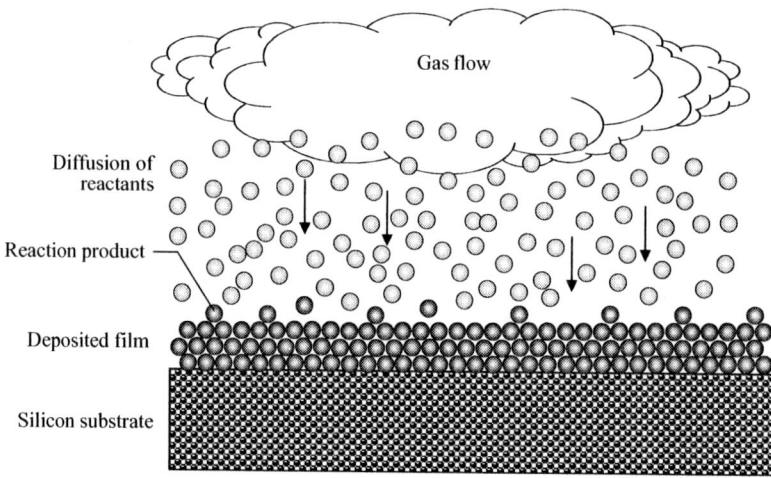

FIGURE 11.9 Gas Flow in CVD

increases the transport of the reactants to the substrate surface (and also increases the removal of by-products away from the substrate). The net effect of lower pressure in a CVD reactor is that the reactants reach the surface more quickly. In this case, the rate-limiting step becomes the surface reaction. At lower pressures, the CVD process is reaction-rate limited. This means that wafers can be stacked vertically and at very close spacing in the reactor since the transport of the reactants from the gas stream to the wafer is not limiting the process.

Doping During CVD ■ The introduction of dopants into SiO_2 during CVD deposition is desirable because it leads to important benefits for wafer fabrication. For instance, the introduction of phosphine (PH_3) to the gas stream during silicon dioxide deposition will lead to the formation of *phosphosilicate glass* (*PSG*). The chemical reaction is:

$$\underset{silane}{SiH_4\ (gas)} + \underset{phosphine}{2PH_3\ (gas)} + \underset{oxygen}{O_2\ (gas)} \rightarrow \underset{silicon\ dioxide}{SiO_2\ (solid)} + \underset{phosphorus}{2P\ (solid)} + \underset{hydrogen}{5H_2\ (gas)}$$

Within the glass, the phosphorous is in the form of a phosphorus pentoxide (P_2O_5), making the glass a dual compound as it codeposits with silicon dioxide. The P_2O_5 content in films is limited to about 4% by weight for films that are left permanently on the wafer because PSG becomes increasingly hygroscopic (water-absorbing).

PSG can be deposited at 600 to 650°C using high-density plasma CVD (HDPCVD) and has recently become popular for the first interlayer dielectric (ILD-1) layer because of its low deposition temperature, relatively planar surface, and excellent gap-fill characteristics. Incorporation of P_2O_5 into silicon dioxide films leads to a reduction in the film stress, which improves the film integrity. The dopant increases the moisture-barrier property of glass. Another benefit is that PSG layers are effective in serving as a getter to immobilize ionic contaminants. The ions become attached to the phosphorus and are thus restrained from diffusing through the film to the wafer surface.

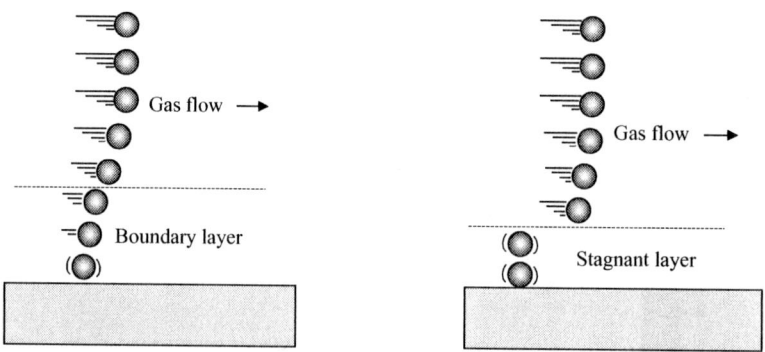

FIGURE 11.10 Gas-Flow Dynamics at the Wafer Surface

Borosilicate Glass. If diborane (B_2H_6) is used instead of phosphine (PH_3), then the result is *borosilicate glass* (*BSG*). BSG requires reflow at a high temperature (e.g., 1000°C) to smooth out sharp steps in the wafer surface and densify the film. However, a high-temperature reflow is undesirable for the wafer thermal budget. BSG also has the disadvantage of not being a good barrier to mobile ionic contaminants (MICs).[21]

Borophosphosilicate Glass. Another method for doping oxide is to incorporate about 2 to 6 % weight each of boric oxide (B_2O_3) and phosphorus pentoxide (P_2O_5) into the silicon dioxide deposition to form *borophosphosilicate glass* (*BPSG*). To achieve a dense oxide with good step coverage after deposition, the oxide is reflowed at a high temperature until it softens and flows. For BPSG, the reflow temperature is 800 to 1000°C for one hour. Reflow also improves its gettering characteristics to block mobile ionic contaminants (MICs). BPSG was commonly deposited on the wafer as the ILD-1 but has been recently replaced by PSG deposited by HDPCVD to reduce the thermal budget.[22]

High temperatures adversely affect the wafer thermal budget in sub-0.25 μm devices. For example, refractory metals (e.g., titaniun and cobalt) used to reduce contact resistance at junctions undergo material transformations at about 600°C that increase their contact resistance to unacceptable levels. In addition, the oxide planarity achieved by a high-temperature reflow is not sufficient for reduced depth of field requirements in ULSI photolithography. For this reason, reflow of doped oxide for planarity and gap fill has been largely replaced by high-density plasma CVD to achieve a stable film with good gap fill, followed by chemical mechanical planarization (CMP) to attain a flat surface (see Chapter 18).

Fluorosilicate Glass. *Fluorosilicate glass,* or *FSG,* is a fluorinated oxide that is being investigated as a first generation low-*k* dielectric for ILD deposition on 0.18 μm generation devices. By adding fluorine to silicon dioxide, there is a decrease in the dielectric constant of the material from about 3.9 for SiO_2 to about 3.5 for FSG.

To form an FSG film, silicon tetrafluoride (SiF_4) is added to the silane (SiH_4) and O_2 gases during the CVD reaction. A potential problem with using FSG is that the fluorine bonds are not always stable and can lead to corrosion, limiting the fluorine content to about 6%. If fluorine ions encounter moisture, hydrofluoric acid (HF), which etches oxide, is created. A method to avoid this problem is to remove weakly bonded fluorine atoms through the introduction of hydrogen during the deposition reaction.[23]

Oxide Versus Silicon Doping. Note that the doping of silicon dioxide is not the same as doping of silicon. When silicon is doped, there is a donation or acceptance of electrons between the dopant and silicon, within a single crystal lattice structure. Deposited oxide is an amorphous crystal structure and the dopants do not act as donors or acceptors of electrons. Doping is done to modify a physical characteristic of SiO_2, such as improving its ability to getter mobile ions.

CVD DEPOSITION SYSTEMS

There are different types of production systems for the deposition of dielectric and metal film layers. Some systems use large batch processing to perform CVD deposition on a large quantity of wafers at one time. However, the most advanced deposition systems currently in use for wafer manufacturing employ single-wafer processing in a cluster tool approach.

Table A.3 in Appendix A summarizes different gases used for CVD reactions. An example of a CVD gas is pure silane, which is used to deposit Si or SiO_2 in CVD systems. Silane reacts with air to form solid particles, causing particle contamination with the silane gas lines and leading to contaminated pipes or combustion. The gas line piping must be carefully installed to avoid any leaks. Many of the gases used in CVD film deposition are toxic. These gases are classified in Table A.3 based on four different safety hazards: pyrophoric (ignites spontaneously in air at or below 54.5°C), poisonous, corrosive, and dangerous combinations of gases, such as silane with hydrogen.

CVD Equipment Design

CVD has a range of equipment reactor designs, with each producing slightly different types of film quality. CVD reactors are broadly categorized based on the reaction chamber pressure regime used during operation: atmospheric-pressure CVD (APCVD) reactors and reduced-pressure reactors. The reduced-pressure CVD reactors have two general types. First, there are low-pressure CVD (LPCVD) reactors where the energy input is thermal. Second, there are plasma-assisted CVD reactors, either plasma-enhanced CVD (PECVD) or high-density plasma CVD (HDPCVD), where the energy is partially supplied by a plasma as well as thermally. Figure 11.11 depicts the different CVD reactor types.

CVD Reactor Types	Atmospheric	Low-pressure	Batch	Single-wafer
Hot-wall	✓	✓	✓	
Cold-wall	✓	✓	✓	✓
Continuous motion	✓		✓	
Epitaxial	✓		✓	
Plenum	✓		✓	
Nozzle	✓		✓	
Barrel	✓		✓	
Cold-wall planar		✓	✓	✓
Plasma-assisted		✓	✓	✓
Vertical-flow Isothermal		✓	✓	✓

FIGURE 11.11 CVD Reactor Types

CVD Reactor Heating ■ A major distinction in CVD reactors is whether they are hot-wall or cold-wall reactors. A *hot-wall reactor* employs a heating method that heats not only the wafer but also the wafer holder (referred to as a *susceptor*) and the walls of the reactor. Hot-wall reactors have film-forming reactions on both the reactor chamber walls and substrate, requiring frequent cleaning or in situ cleaning to minimize particle contamination. A resistance heater surrounding the reactor tube is an example of a hot-wall reactor. A *cold-wall reactor* heats only the wafers and susceptor, which is where film is deposited. The walls of the reactor are cold and therefore do not have sufficient energy to allow a deposition reaction to occur. RF induction heating or infrared lamps mounted in the reactor are examples of this type of reactor. Localized heating at the wafer minimizes particle formation in the reactor.

CVD Reactor Configuration ■ The geometry of the reactor design depends closely on the pressure attained during the deposition process. Atmospheric-pressure reactors operate in the mass-transport limited region, so they must be designed so that an equal amount of reactant gases are delivered to each wafer. To achieve this condition, wafers are usually laid flat on a horizontal surface. A drawback to this approach is that wafers are susceptible to contamination from falling particles.

Low-pressure reactors (LPCVD) are reaction-rate limited. This means that they can be vertically stacked at close spacing with a large number of wafers per run, since adequate reactant gas is always supplied to the wafer surface. Quartz wafer holders (boats) hold up to 200 wafers. At the same time, LPCVD reactors must have precise temperature control because of their limitation from the reaction rate.

CVD Reactor Summary ■ The different types of CVD processes and their principal characteristics are shown in Table 11.2.[24]

TABLE 11.2 Types of CVD Reactors and Principal Characteristics

Process	Advantages	Disadvantages	Applications
APCVD (Atmospheric Pressure CVD)	Simple reactor, fast deposition, and low temperature.	Poor step coverage, particle contamination, and low throughput.	Low-temperature oxides (both doped and undoped).
LPCVD (Low Pressure CVD)	Excellent purity and uniformity, conformal step coverage, and large wafer capacity.	High temperature, low deposition rate, more maintenance intensive, and requires vacuum system.	High-temperature oxides (both doped and undoped), silicon nitride, polysilicon, W, and WSi_2.
Plasma-Assisted CVD: • Plasma-Enhanced CVD (PECVD) • High-Density Plasma CVD (HDPCVD)	Low temperature, fast deposition, good step coverage, and good gap fill.	Requires RF system, higher cost, stress is much higher with a tensile component, and chemical (e.g., H_2) and particle contamination.	High aspect ratio gap fill, low-temperature oxides over metals, ILD-1, ILD, copper seed layer for dual damascene, and passivation (nitride).

APCVD (Atmospheric Pressure CVD)

The first type of chemical vapor deposition used in the semiconductor industry is *atmospheric pressure CVD (APCVD)*.[25] As previously discussed, APCVD generally operates in the mass-transport limited regime. At any given time, there may not be sufficient gas molecules present at the wafer surface for a reaction to occur. Therefore, the reactor must be designed to have optimum reactant gas flow to every wafer in the system. The basic system design should never permit the reaction to slow down because insufficient reactant gas is available. Since the reactor is at atmospheric pressure, the reactor design can be simple and allows for high deposition rates.

Two different types of continuous-processing APCVD systems are shown in Figure 11.12 on page 272. These equipment designs use a belt or conveyor to carry the wafer samples through the reactor gases, which flow through the center of the reactor.

Continuous-processing APCVD systems have high equipment throughput, good uniformity, and the capability to process large-diameter wafers. APCVD systems do have problems with high gas consumption and often need frequent reactor cleaning. Since film deposition also takes place on the conveyor as well as the wafer, the belt transport system must be cleaned (sometimes it is in cleaned in situ, or during use). APCVD deposited films often exhibit poor step coverage.

Silicon Dioxide ■ The most common use of APCVD is in the deposition of silicon dioxide (SiO_2) and doped oxides (e.g., PSG, BPSG, and FSG, as discussed earlier in this chapter). These films have been used traditionally as an interlayer dielectric (ILD), as a protective overcoat, or to planarize (smoothen) a nonuniform surface.

SiO_2 Deposition With Silane. SiO_2 is deposited by oxidizing *silane* (SiH_4) with oxygen. Pure silane is a highly pyrophoric, unstable gas that burns on exposure to air. For this reason, it is commonly supplied in a low dilution (typically 2 to 10% by volume) in argon or nitrogen in order to make it safer to handle. This reaction can be done at a low temperature of 450 to 500°C, which is an advantage for depositing SiO_2 on the aluminum metal lines as an ILD.[26] However, due to the small mean free path and poor surface migration through the boundary layer, this method has poor step coverage and poor gap-fill characteristics. This makes it unacceptable for critical ULSI applications.

SiO_2 Deposition With TEOS-Ozone. A common APCVD application is to deposit oxide by reacting *TEOS*, which is tetraethylorthosilicate ($Si(C_2H_5O)_4$) or tetraethoxysilane, with ozone (O_3). TEOS is an organic liquid precursor. It usually uses a carrier gas, typically nitrogen, bubbled through it for delivery of the TEOS gas mixture to the reactor. There are also flow controllers for delivering the TEOS source in liquid state to the reactor. Ozone is a triatomic oxygen molecule that is much more reactive than oxygen; therefore, the process can be done at low temperatures (e.g., 400°C) without plasma. Because the decomposition of TEOS is induced by O_3 without a plasma, the reaction takes place at atmospheric pressure (APCVD) of 760 torr or subatmospheric pressure (SACVD) at about 600 torr. There appears to be no significant difference between APCVD and

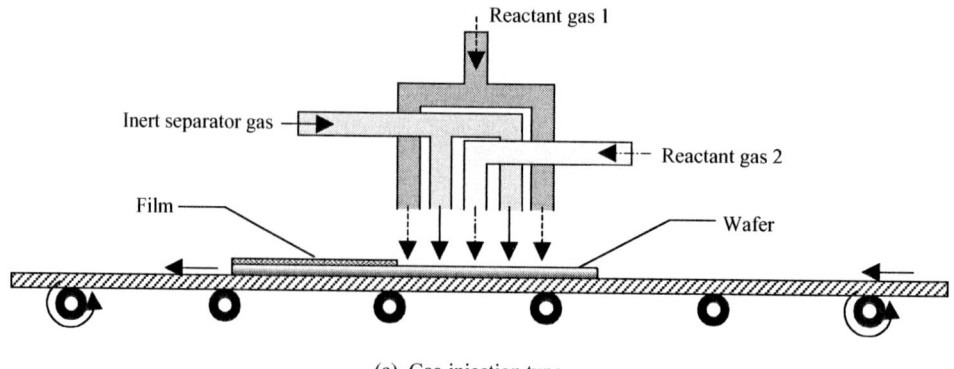

(a) Gas-injection type

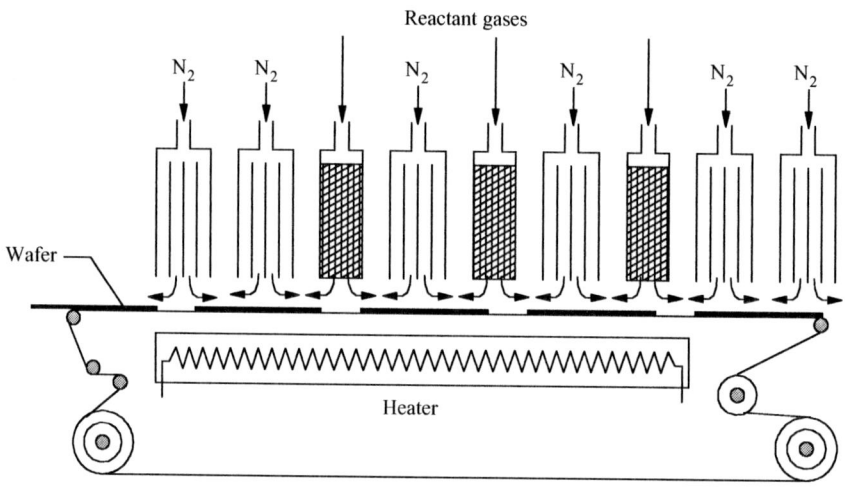

(b) Plenum type

FIGURE 11.12 Continuous-Processing APCVD Reactors

SACVD.[27] The deposited oxide film has improved profiles over steps, is conformal, and has excellent electrical characteristics as a dielectric.[28] The reaction equation between TEOS and O_3 to obtain an oxide is:

$$\underset{TEOS}{Si(C_2H_5O)_4} + \underset{ozone}{8O_3} \rightarrow \underset{silicon\ dioxide}{SiO_2} + \underset{water}{10H_2O} + \underset{carbon\ dioxide}{8CO_2}$$

An APCVD TEOS-O_3 as-deposited SiO_2 film is porous and often requires a reflow step to remove moisture as well as densify the film. This reflow adds a process step and reduces the thermal budget.[29] Doping uniformity is also a concern, as this can lead to problems for nonuniform removal rates during chemical mechanical planarization. The primary advantage of APCVD TEOS-O_3 is the improved step coverage for high-aspect ratio gaps for applications such as shallow trench isolation (see Figure 11.13). Another advantage to APCVD TEOS-O_3 is that it is just a basic thermal CVD process to deposit oxide, which could avoid wafer surface damage and wafer corner damage at steps. This type of damage is possible from high-density plasma CVD (discussed later in this chapter). An aspect of TEOS-O_3 films is that they are usually deposited in combination with other oxide film layers (e.g., PECVD oxide) for reasons such as reducing the TEOS-O_3 tensile stress on a thick film or to reduce sensitivity with the underlying material.[30]

Doped Oxides. APCVD oxides are usually doped with chemicals such as phosphorus and boron (discussed earlier in this chapter). If SiO_2 is undoped, it is referred to as *undoped silicate glass* (*USG*) or *undoped oxide* (*UDOX*). Phosphorus doped oxide is phosphosilicate glass (PSG). In the traditional deposition process, a high-temperature reflow (as high as 950°C for 15 to 30 minutes) heats and softens the doped oxide, causing it to planarize the surface (see Figure 11.14).

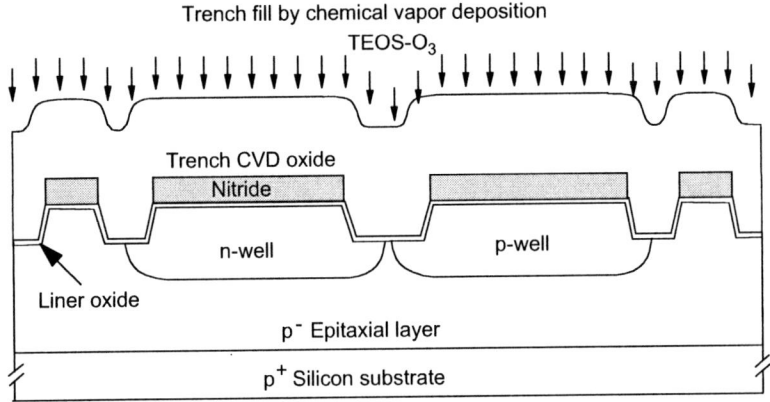

FIGURE 11.13 Improved Step Coverage of APCVD TEOS-O_3

Silane-based borophosphosilicate, or BPSG films, are deposited at low temperatures (400 to 450°C) and then immediately densified and stabilized at about 800°C for one hour. A smoother surface makes it easier to deposit and pattern the next film. Note that for the ULSI device technology, reflow has been replaced with chemical mechanical planarization (CMP) to planarize the surface. CMP is discussed in Chapter 18.

LPCVD (Low Pressure CVD)

Low pressure CVD (LPCVD) systems are more common now than APCVD because of their lower cost, higher production throughput, and superior film properties. LPCVD operates at a medium vacuum (about 0.1 to 5 torr), and employs temperatures between 300 and 900°C.[31] Conventional oxidation type furnaces (horizontal or vertical) and multichamber cluster tools can be used for LPCVD processing.

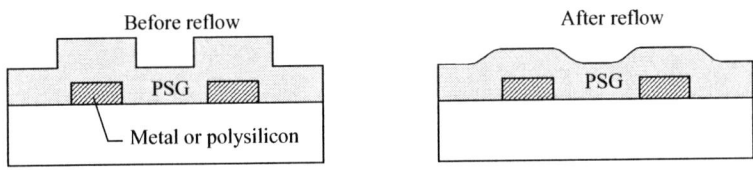

FIGURE 11.14 Planarized Surface after Reflow of PSG

LPCVD reactors typically operate in the reaction-rate limited regime. In this reduced-pressure regime, the diffusivity of the reactant gas molecules increases so that the mass-transfer of the gas to the wafer no longer limits the rate of the reaction. Because of this transfer state, the gas-flow conditions inside the reactor are not important, permitting the reactor design to be optimized for high wafer capacity (e.g., wafers can be closely spaced). Films are uniformly deposited on a large number of wafer surfaces as long as the temperature is tightly controlled.

The boundary layer in a LPCVD reactor is different from APCVD because it extends farther from the wafer surface due to the low-pressure condition (see Figure 11.15 on page 274). The boundary layer has a low molecular density, which permits incoming gas molecules to easily diffuse through this layer, exposing the wafer surface to more molecules than it can possibly consume. This condition explains why an LPCVD reactor operates in the reaction-rate limited regime. It is the reaction rate that is the limiting factor in the deposition rate, not the reactant supply. Furthermore, there are many collisions during LPCVD, so that material strikes the wafer in an undirected manner. This action is beneficial for conformal film coverage on high-aspect ratio steps and trenches. As a result, step coverage is usually good with LPCVD.[32]

The LPCVD reactor design favors the hot-wall reactor type so that uniform temperature control is achievable over a large operating length. A representative reaction chamber is shown in Figure 11.16 on page 274.

Since LPCVD reactors are often hot wall, particles deposit on the reactor wall. These deposits are minimized by reducing the partial pressure of the gas-phase reactants, which leads to fewer

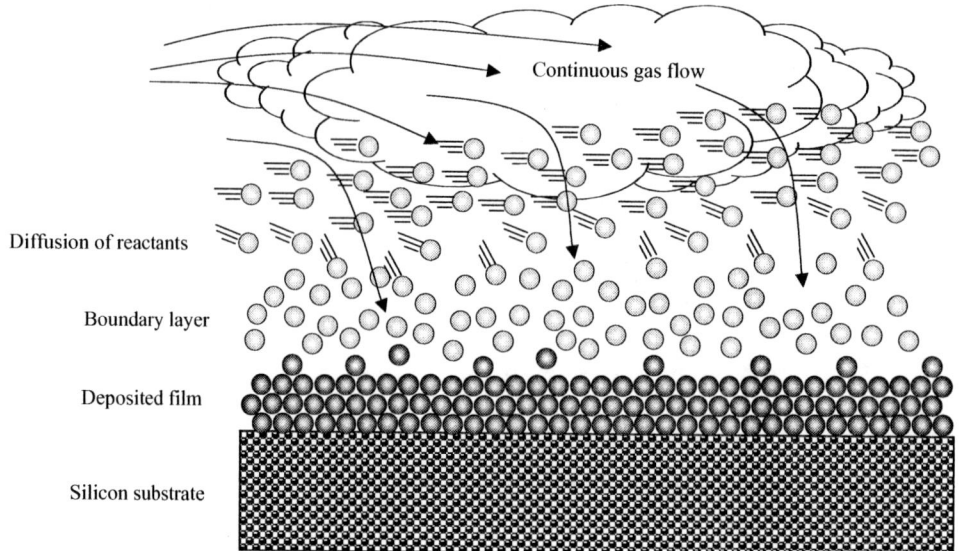

FIGURE 11.15 Boundary Layer at Wafer Surface

gas-phase reactions. For this reason, LPCVD reactors are usually operated at lower growth rates than APCVD systems.[33] Hot-wall reactors require periodic reaction chamber cleanup of particles by routine maintenance.

The traditional method of tube cleaning was to remove a dirty quartz tube, install a previously cleaned tube, and clean the dirty tube for future use. Cleaning involved manually rolling the tubes half-submerged in an acid bath such as aqueous HF. For production and safety reasons, an in situ clean is desirable. One in situ method is the use of plasma-generated fluorine gases that react with the solid residue in the reactor, forming volatile reaction products that are pumped from the system. Another in situ approach is to use chlorine trifluoride (ClF_3) with elevated temperature to activate a thermal clean. In situ tube cleaning for LPCVD reduces equipment downtime, lowers particle counts, and minimizes personnel exposure to chemicals.[34]

LPCVD reactors with a large number of wafers (e.g., 150 to 200 wafers) experience reactant depletion as gases are transported along the reaction chamber. This depletion leads to a reduction in growth rate. Adjusting the reactor temperature so that it increases slightly from inlet to outlet (typically by 25 to 50°C) compensates for this temperature nonuniformity.

Silicon Dioxide ■ There are numerous applications for LPCVD oxides (doped and undoped) in ULSI multilevel metallization. LPCVD oxides are used for interlayer dielectric (ILD), shallow trench isolation oxide fill, and sidewall spacers.

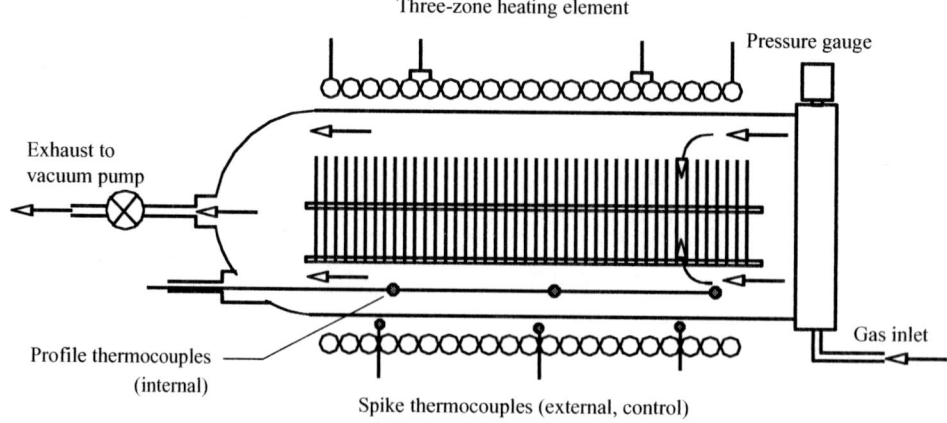

FIGURE 11.16 LPCVD Reaction Chamber

SiO₂ With TEOS. A common LPCVD method for depositing silicon dioxide is the pyrolysis (decomposition) of TEOS, with or without oxygen, at low pressure and at a temperature between 650 to 750°C. This method is sometimes referred to as LPTEOS, for low-pressure TEOS. LPTEOS yields very good oxide conformality due to rapid surface diffusion of gas molecules. A liquid TEOS source is used by bubbling a carrier gas through it (e.g., N_2, O_2, or He). The liquid source is heated by its own independent temperature source. The concentration of the liquid source vapor entering the reactor is controlled by the carrier gas-flow rate and liquid source temperature. A schematic of a typical TEOS deposition system is shown in Figure 11.17. The film growth rate for LPCVD oxides is significantly slower (100-150 Å/minute) than for APCVD films.[35]

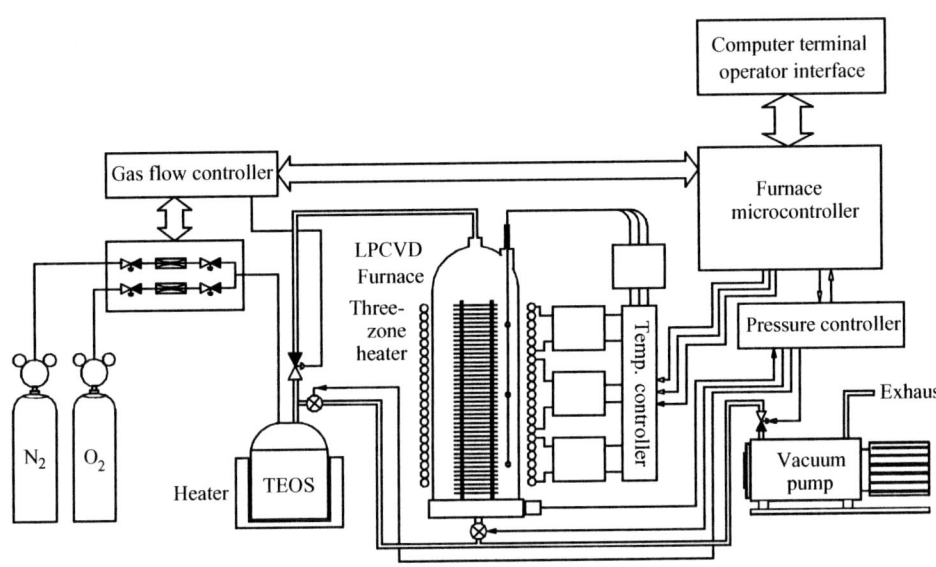

FIGURE 11.17 Oxide Deposition with TEOS LPCVD

SiO₂ With SiH₄. SiO_2 can also be deposited by LPCVD by oxidizing silane (SiH_4) at a relatively low temperature of about 450°C. Similar to APCVD oxide with SiH_4, this process has poor step coverage. A higher temperature deposition is done with dichlorosilane (SiH_2Cl_2 or DCS) and nitrous oxide (N_2O) at about 900°C to produce a good quality SiO_2 film. However, the high temperature is undesirable for the wafer thermal budget.

Silicon Nitride ■ *Silicon nitride* (Si_3N_4) is a film often used as the final wafer passivation layer because it provides good protection against the diffusion of impurities and moisture. It achieves excellent step coverage with high conformal coverage with LPCVD deposition. Silicon nitride is also used as a mask material (referred to as a hard mask), such as in the shallow trench isolation (STI) process. Silicon nitride has a high dielectric constant (i.e., a *k* value of 6.9, versus about 3.9 for CVD SiO_2), which makes it undesirable for the ILD dielectric due to the resultant high capacitance between conductor layers.

Reacting dichlorosilane ($SiCl_2H_2$) and ammonia (NH_3) in an LPCVD reactor produces silicon nitride at a reduced pressure and temperature between 700 and 800°C. The chemical reaction is:

$$3SiCl_2H_2 \text{ (gas)} + 4NH_3 \text{(gas)} \rightarrow Si_3N_4 \text{(solid)} + 6HCl \text{ (gas)} + 6H_2 \text{(gas)}$$
$$\text{dichlorosilane} \qquad \text{ammonia} \qquad \text{silicon nitride} \qquad \text{hydrogen chloride} \qquad \text{hydrogen}$$

The important variables that affect the properties of silicon nitride in an LPCVD process are total pressure, reactant concentrations, deposition temperature, and temperature gradients. For example, increasing the total pressure and partial pressure of dichlorosilane in the reaction chamber will increase the deposition rate.

Silicon nitride can also be produced in an APCVD process by reacting silane and ammonia, but better film uniformity and higher wafer throughput is achieved in the low-pressure process.[36] Deposited silicon nitride is an amorphous film that often contains large amounts of hydrogen. A high ammonia ratio and low deposition temperature will increase the hydrogen content.

Polysilicon ■ Polysilicon film (also referred to as *poly-si*, or *poly*) is commonly deposited using LPCVD. Recall that the term polysilicon means that the silicon is a polycrystal, with many small single-crystal regions separated by grain boundaries. The stress in thin polysilicon films is compressive.

Doped polysilicon serves as the gate electrode in MOS devices (see Figure 11.18). Key reasons for the use of doped polysilicon in the gate structure are:[37]

1. Ability to be doped to a specific resistivity.
2. Excellent interface characteristics with silicon dioxide.
3. Compatibility with subsequent high-temperature processing.
4. Higher reliability than possible metal electrodes (e.g., aluminum).
5. Ability to be deposited conformally over steep topography.
6. Allows for self-aligned gate process (see Chapter 12).

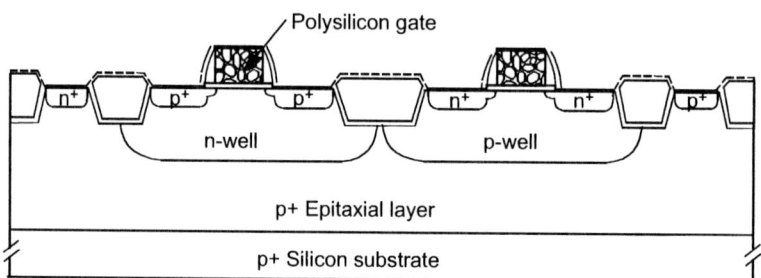

FIGURE 11.18 Doped Polysilicon as a Gate Electrode

Polysilicon is deposited in LPCVD by the pyrolysis (thermal decomposition) of silane at a temperature of 575 to 650°C. Low pressure reactors use either pure silane or 20 to 30% silane diluted with nitrogen, which is fed into the system at a pressure of 0.2 to 1.0 torr to fabricate polysilicon films. A practical deposition rate is about 100 to 200 Å/minute. The addition of diborane can enhance the polysilicon deposition rate because the diborane forms borane radicals, BH_3, which catalyze gas-phase reactions. The chemical reaction is:

$$SiH_4 \text{ (gas)} \rightarrow Si \text{ (solid)} + 2H_2 \text{ (gas)}$$
$$\textit{silane} \qquad \textit{silicon} \qquad \textit{hydrogen}$$

The polysilicon film is usually doped in situ during the deposition process by adding arsine (AH_3), phosphine (PH_3), or diborane (B_2H_6) to the gas mixture. An inert gas is often accompanied to improve film uniformity. Doping can be done after deposition by ion implantation (see Chapter 17). The resistivity of the polysilicon after doping is highly dependent on parameters such as deposition temperature, dopant concentration, and annealing temperature effects on grain size.

A refractory metal is deposited over the polysilicon gate to form a *polycide* or *silicide* structure (see Chapter 12). This action is taken to reduce electrical resistivity at the polysilicon interface with the other material (e.g., tungsten) that forms the electrical connection to other device components. Lightly doped polysilicon also serves as a resistor in memory cells, capacitors, and thin-film transistors.

Silicon Oxynitride ■ Silicon nitride films that contain oxygen are referred to as *silicon oxynitride* (SiO_xN_y) and combine the advantages of oxides and nitrides. Oxynitride films also have improved thermal stability, cracking resistance, and reduced film stress when compared to silicon nitride. In general, increased oxygen in oxynitride film also decreases the film's refractive index, which makes it less reflective and a useful antireflective layer in photolithography (see Chapter 14).[38] Another benefit of a silicon oxynitride film is the nitrogen in the film accumulates at the silicon interface to reduce the concentration of strained Si-O bonds, potentially reducing the creation of hot electrons (undesirable charge carriers) by as much as three orders of magnitude. An oxynitride layer at the Si/SiO_2 interface of thin gate oxides has been found to be beneficial for improving the device electrical performance.

Oxynitride films are formed by different techniques, including the oxidation of silicon nitride, the nitridation of SiO_2 with ammonia (NH_3), and the direct growth of SiO_xN_y using nitrous oxide as the oxidant species. One way of forming oxynitride films is by reacting silane (SiH_4) with nitrogen dioxide (N_2O) and NH_3. Nitride or oxynitride films are used frequently as an "etch stop" for selective etching of dielectric films.[39]

Plasma-Assisted CVD

The third major type of CVD equipment relies on plasma energy, in addition to thermal energy, to initiate and sustain the chemical reactions necessary for CVD. The advantages of using plasma during CVD are:

1. Lower processing temperature (250 to 450°C).
2. Excellent gap-fill for high aspect ratio gaps (with high-density plasma).
3. Good film adhesion to the wafer.
4. High deposition rates.
5. High film density due to low pinholes and voids.
6. Wide range of applications due to lower processing temperature.

Film Formation ■ The plasma-assisted CVD reaction necessary to form a film occurs when RF power is used to break up gas molecules in a vacuum. The frequency of the RF power depends on the application, with typical frequencies found at 40 kHz, 400 kHz, 13.56 MHz, and 2.45 GHz (microwave power). The molecular fragments (radicals) are chemically reactive species and readily bond to other atoms to form a film at the wafer surface (see Figure 11.19). Gaseous by-products are removed by the vacuum pumping system. The wafer is usually heated in order to assist the surface reactions and reduce the level of undesirable contaminants, such as hydrogen.

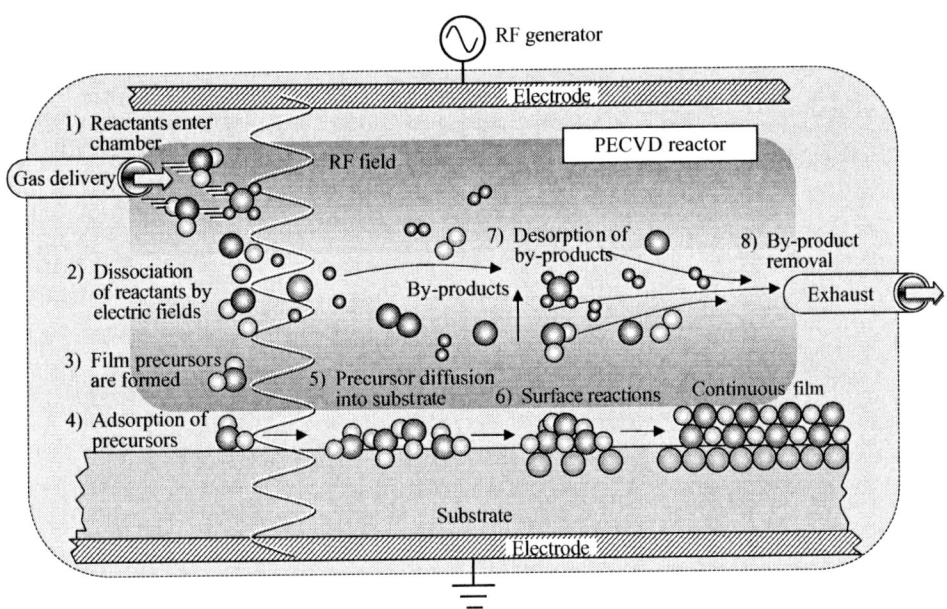

FIGURE 11.19 Film Formation During Plasma-Assisted CVD

The plasma-assisted CVD reactions at the wafer surface are very complex. The properties of plasma-deposited films depend on many variables, such as electrode configuration and separation, power level and frequency, gas composition, pressure and flow rate, and substrate temperature.[40] The essential aspect of these variables for production is embodied in the chamber design. The specific details of the intermediate reactions that occur in plasma glow discharge are not well understood. For these reasons, deposited films are often found to be not very stoichiometric. This means that the deposited film will not necessarily have the same ratio of elements found in its nominal chemical formula.

There are two types of plasma processes used in CVD:

- Plasma-enhanced CVD (PECVD)
- High-density plasma CVD (HDPCVD)

Plasma-Enhanced CVD (PECVD) ■ The development of *plasma-enhanced CVD (PECVD)* uses plasma energy to create and sustain the CVD reaction. It is a natural follow-on to LPCVD since the system pressure for both types of CVD processes is comparable. The important difference is the much lower PECVD deposition temperature. For instance, silicon nitride (Si_3N_4) is deposited using LPCVD at 800 to 900°C, yet cannot be deposited over aluminum metallization because Al melts at 660°C. On the other hand, silicon nitride deposited with PECVD at a temperature of 350°C is suitable for this application.

PECVD is performed in a vacuum chamber between parallel conducting plates positioned several inches apart, typically with a variable gap for process optimization. A modern reactor is a multichamber cluster tool. The wafer (or wafers) is placed on the grounded bottom plate and RF power is applied to the top electrode. A plasma develops when the source gas is introduced through the gas manifold and the center of the electrodes. Exhaust gases are pumped out from the periphery of the bottom electrode. Sometimes the reactant gases are introduced at the periphery and pumped out at the center of the bottom electrode. A schematic of an overall PECVD system is shown in Figure 11.20.

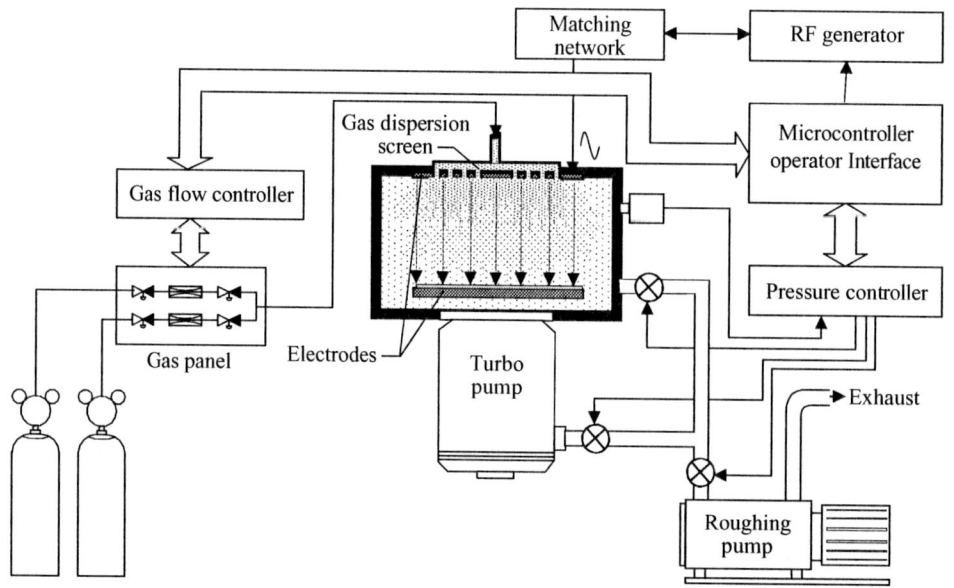

FIGURE 11.20 General Schematic of PECVD

A PECVD reactor is typically a cold-wall plasma reactor, with the wafer heated in its chuck while the remaining parts of the reactor are unheated. Deposition parameters must be controlled to ensure the temperature gradient does not affect the film thickness uniformity. Cold-wall reactors create fewer particles and require less downtime for cleaning. Deposition chambers often have in situ cleaning to reduce particles. With parallel-plate processing, there is the possibility that particles formed in the gas phase will fall on the wafer surface.

Silicon Dioxide PECVD. Oxide films with PECVD are usually formed by reacting silane (SiH_4) with either oxygen (O_2), nitrous oxide (N_2O), or carbon dioxide (CO_2) in a plasma. The processing temperature is about 350°C.[41] The oxide can be doped with boron or phosphorus to form BSG or PSG, respectively, or with both to deposit BPSG. PECVD PSG tends to be more crack resistant, conformal, and free of pinholes than APCVD PSG.

A plasma gas mixture of silane and oxygen is not normally used because oxygen reacts readily in the gas phase, which promotes particle generation and poor film quality (e.g., pinholes). The silane and nitrous oxide gas mixture produces more uniform films than through the use of an oxygen reactant. The reaction of silane with nitrous oxide to form silicon dioxide is:

$$\underset{silane}{SiH_4\ (gas)} + \underset{nitrous\ oxide}{2N_2O\ (gas)} \rightarrow \underset{silicon\ dioxide}{SiO_2\ (solid)} + \underset{nitrogen}{2N_2\ (gas)} + \underset{hydrogen}{2H_2\ (gas)}$$

The silicon dioxide is nearly stoichiometric, although it does contain some hydrogen, with a small amount of nitrogen present in the film. Hydrogen can exist as a hydride (Si-H), silanol (Si-O-H), or as water (H-O-H). The presence of O-H groups is undesirable for MOS transistor electrical characteristics and must be minimized.[42]

Parallel-plate PECVD can also deposit SiO_2 using TEOS, and is referred to as *PETEOS*. However, PETEOS is not acceptable for filling narrow-spaced metal lines due to voids. PETEOS is combined with APCVD TEOS or high-density plasma CVD (HDPCVD) to attain conformal gap fill, followed by chemical mechanical planarization. PETEOS has a relatively high SiO_2 deposition rate, which makes it desirable for wafer throughput in integrated tools.

Silicon Nitride PECVD. Silicon nitride PECVD is a film used as a final passivation layer on the chip as scratch protection, and a moisture barrier and as a barrier to prevent sodium diffusion. It is also widely used as a mask material in shallow trench isolation (STI) and in self-aligned contact technology (see Chapter 12). The Si_3N_4 nitride film deposited by PECVD is not very stoichiometric, and is sometimes written as $Si_xN_yH_z$. This equation highlights its nonstoichiometric composition and identifies that the film contains hydrogen (usually 9 to 30%). Table 11.3 compares the properties of silicon nitride for LPCVD vs. PECVD.

TABLE 11.3 Properties of Silicon Nitride for LPCVD vs. PECVD

Property	LPCVD	PECVD
Deposition temperature (°C)	700 to 800	300 to 400
Composition	Si_3N_4	$Si_xN_yH_z$
Step coverage	Fair	Conformal
Stress at 23°C on silicon (dynes/cm^2)	$1.2 - 1.8 \times 10^{10}$ (tensile)	$1 - 8 \times 10^9$ (tensile or compressive)

PECVD nitride films are commonly produced from silane (SiH_4) and reacted with ammonia or nitrogen. The reaction equations are:

$$\underset{silane}{SiH_4\ (gas)} + \underset{ammonia}{NH_3\ (gas)} \rightarrow \underset{silicon\ nitride}{Si_xN_yH_z\ (solid)} + \underset{hydrogen}{H_2\ (gas)}$$

$$\underset{silane}{SiH_4\ (gas)} + \underset{nitrogen}{N_2\ (gas)} \rightarrow \underset{silicon\ nitride}{Si_xN_yH_z\ (solid)} + \underset{hydrogen}{H_2\ (gas)}$$

PECVD silicon nitride can have increased compressive film stress caused by ion bombardment during deposition that ruptures the Si-N or Si-H bonds. High compressive stress in a nitride layer can cause voids and cracks in the underlying aluminum metallization. There is usually substantial hydrogen content in the film. Hydrogen can reduce film stress, but it also degrades film properties.[43] When nitrogen is substituted for ammonia, the hydrogen content is reduced. However, ionization of nitrogen is difficult for plasma formation.

Silicon Oxynitride. Films of silicon oxynitride are deposited using PECVD by reacting Si_3N_4 with nitrous oxide (N_2O) at a temperature of 200 to 250°C.[44] Silicon oxynitride combines some properties from nitrides and oxides. It has good resistance to moisture and sodium penetration, with excellent mechanical, chemical, and electrical properties. These properties make silicon oxynitride suitable for topside passivation.

Gap Fill. A major limitation for PECVD in fine-geometry devices is gap fill. For gaps below 0.5 μm spacing (actual gap width), PECVD cannot fill the high aspect ratio gap without pinching off the top and leaving a void. A void in a dielectric material that is filling a gap is undesirable because of its effect on electrical performance and long-term reliability. For devices at the

High Density Plasma Deposition Chamber
(Photo courtesy of Applied Materials, Inc.)

0.25 μm technology node and below, HDPCVD has replaced PECVD because of its superior gap-fill properties.

It should be pointed out that other CVD methods can achieve fine-geometry gap-fill but have other limitations. For example, APCVD using TEOS oxide has good conformality and produces excellent gap fill. However, TEOS has different deposition characteristics depending on the underlying surface. One problem is that TEOS layers have dangling bonds that absorb moisture. To inhibit this condition, PECVD undercoat and overcoat layers are required. These layers add cost and also risk being removed during the subsequent chemical mechanical planarization (CMP).[45]

High-Density Plasma CVD (HDPCVD) ■ A recent development in plasma assisted CVD is *high-density plasma CVD (HDPCVD)*. This deposition method became widely accepted in advanced wafer fabs in the mid-1990s. As its name implies, the plasma in HDPCVD is a high-density mixture of gases at low pressure that is directed toward the wafer surface in the reaction chamber. Its main benefit is that it can deposit films to fill high aspect ratio gaps with a deposition temperature range of 300 to 400°C. HDPCVD was initially developed for interlayer dielectric (ILD) applications, but is also being used for deposition in ILD-1, shallow trench isolation, etch-stop layers, and deposition of low-k dielectrics.[46]

The HDPCVD reaction involves a chemical reaction between two or more gas precursors. For the deposition of oxide ILD, oxygen (or ozone) is often used with a silicon containing gas such as silane or TEOS, along with argon. To form the high-density plasma, a source excites the gas mixture with RF or microwave power (2.45 GHz) and directs the plasma ions into a dense region above the wafer surface. There are different high-density plasma sources, such as electron cyclotron resonance (ECR), inductively coupled plasma (ICP), and Helicon. These plasma sources are explained in detail in Chapter 16.

Wafer Bias and Heat Load. An RF bias (often between 1,500 to 3,000 kW of RF power) is applied to the wafer, pulling energetic ions out of the plasma and directing them toward the wafer surface. Biasing the wafer gives directionality to the energetic ions of the plasma. The high-density plasma ion density is about 10^{11} to 10^{12} ions/cm^3 at low pressures (between 2 to 10 mtorr). This high plasma density, in conjunction with directionality from wafer biasing, is an important reason for HDPCVD's ability to deposit films into narrow gaps with high-aspect ratio geometries of 3:1 to 4:1 and higher.

Much of the challenge for the use of high-density plasma is related not only to the performance of the plasma source but also to the details of the chamber design so that the technology works in high-volume wafer fabrication.[47] A particular problem is that the high-density plasma will increase the thermal load to the wafer, since 2000W of RF bias can result in power density application of approximately 6 watts/cm^2 on the wafer surface. This condition leads to high wafer temperatures. However, the ILD must have process temperatures below 400°C to avoid harming the aluminum metallization; in addition, high heat loads cause thermal stress to the wafer.[48] This temperature limit requires cooling the wafer by applying a backside blanket of helium gas to the wafer through access ports in the electrostatic chuck (ESC). This action creates a thermal conductivity path between the wafer and the ESC, thus cooling the wafer and the chuck.

Simultaneous Deposition and Etching. An HDPCVD process uses a simultaneous deposition and etching action that is the basis for its ability to fill high-aspect ratio gaps with dielectric material without voids (see Figure 11.21). This is referred to as the *dep:etch (D:E) ratio,* which typically has a value of approximately 3:1 for HDPCVD.[49] Interpret this ratio as meaning the deposition rate (i.e., the rate material is deposited) is occurring three times faster than the etch rate. Increasing the ratio will increase deposition and therefore wafer throughput, but if the ratio is too high, then voids occur because the gap is not completely filled.

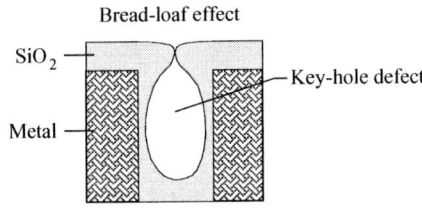

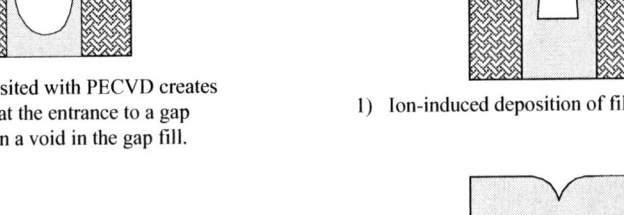

Film deposited with PECVD creates pinch-off at the entrance to a gap resulting in a void in the gap fill.

1) Ion-induced deposition of film precursors

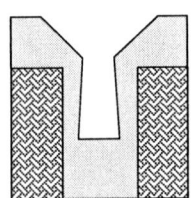

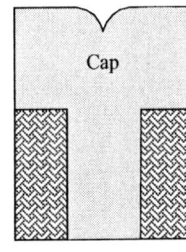

2) Argon ions sputter-etch excess film at gap entrance resulting in a beveled appearance in the film.

3) Etched material is redeposited. The process is repeated resulting in an equal 'bottom-up" profile.

FIGURE 11.21 Dep-Etch-Dep Process

The surface reactions for the simultaneous deposition-etch include five steps, with the first three considered the dominant mechanisms:[50]

1. **Ion-induced deposition:** The ion is pulled from the plasma and deposited to produce the main gap-filling phenomena. It is driven by the kinetic energy of reactant ions breaking surface bonds to form reaction sites.
2. **Sputter etch:** Energetic argon and reactant ions attracted to the surface due to wafer bias will bombard the surface and etch (dislodge) atoms.
3. **Redeposition:** Atoms are dislodged from the bottom of the gap and usually redeposit on sidewalls. This is important for uniform thickness on sidewalls and the bottom of the gap.
4. **Hot-neutral CVD:** This is a minor contribution to the reaction where thermal energy drives some deposition.
5. **Reflection:** Another minor contribution of ions reflecting off sidewalls and then deposited.

Simultaneous deposition and etching is a beneficial by-product of the plasma directionality created in the oxygen and argon gas mixture in HDPCVD. For a silicon dioxide (SiO_2) deposition process, the oxygen reacts with silane (SiH_4) to form SiO_2, while argon acts to remove the deposited material away by sputtering (see Chapter 12 for an explanation of sputtering). Some factors that affect the dep:etch ratio are the ratio of the oxygen and argon gases, the chamber pressure, the ion energy, and the RF bias on the wafer.[51] Low pressure is important to reduce the mean free path, which reduces collisions and maintains good directionality of the plasma. For acceptable deposition rates and wafer throughput in HDPCVD, high gas-flow rates are required. As shown in Figure 11.22, the wafer often will sit directly on the throat of a high-speed turbo pump (e.g., 4,000 to 5,000 liters/sec pump speed for 300-mm diameter wafers).

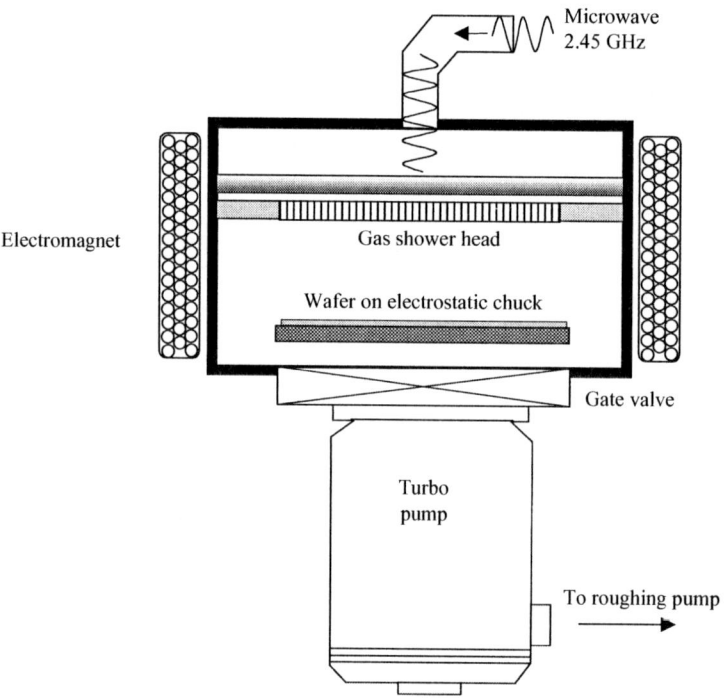

FIGURE 11.22 HDPCVD with Wafer at Throat of Turbo Pump

In practice, HDPCVD is sometimes used as the first step in a three-part process for dielectric gap fill.[52] High aspect ratio gaps (> 3:1) are filled with HDPCVD, followed by a regular density PECVD to deposit a cap film (see Figure 11.23). This capping film is then planarized by chemical mechanical planarization (see Chapter 18). Planarization maintains a smooth and constant oxide thickness on top of the metal conductor lines.

DIELECTRICS AND PERFORMANCE

The dielectric materials used in different film layers directly affect the performance of a microchip. Two important aspects of dielectrics are the dielectric constant and device isolation.

Dielectric Constant

The *dielectric constant* (k) of a nonconductive material represents its effectiveness at storing potential electrical energy under the influence of an electric field. In other words, this is the insulating material's ability to act like a capacitor. The lowest attainable k is 1.0 and represents air. A high k dielectric stores more electrical energy. The dielectric constant of thermal silicon dioxide (SiO_2) has a k value of about 3.9. The k value of a plasma-enhanced CVD (PECVD) oxide is about 4.1 to 4.3.

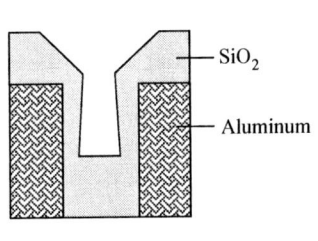
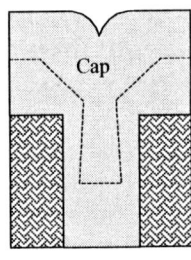
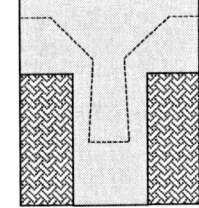

1) HDPCVD gap fill
2) PECVD cap
3) Chemical mechanical planarization

FIGURE 11.23 Three-Part Process for Dielectric Gap Fill

Doped SiO_2 has been traditionally the most common interlayer dielectric (ILD) material. Extensive research is underway to replace doped SiO_2 for the ILD with another dielectric material that has a lower dielectric constant (see Table 11.4).[53] Reducing the k value of the dielectric reduces capacitive losses between adjacent conductors because the dielectric stores less electric field and therefore takes less time to charge, allowing for an increase in speed performance of the metal conductors. A low-k dielectric material for the ILD becomes critical for smaller linewidths with less spacing between metal lines. As the metal linewidths decrease, the capacitive effects of the conductors and insulators increase, and using low-k materials compensates for this.

TABLE 11.4 Potential Low-k Materials for ILD of ULSI Interconnects

Potential Low-k Dielectric	Dielectric Constant (k)	Gap Fill (μm)	Cure Temp. (°C)	Remarks
FSG (silicon oxyfluoride, Si_xOF_y)	3.4 to 4.1	<0.35	No issue	FSG has almost the same k-value as SiO_2 and a reliability concern that fluorine will attack and corrode tantalum barrier metal.
HSQ (hydrogen silsesquioxane)	2.9	<0.10	350 to 450	Silicon-based resin polymer available in solution as FOx (Flowable Oxide) for spin-on coating application. May require surface passivation to reduce moisture absorption. Cure is done in nitrogen.
Nanoporous silica	1.3 to 2.5	<0.25	400	Inorganic material with tunable dielectric constant that relies on pore density. Increased porosity reduces mechanical integrity—porous material must withstand polishing, etching, and heat treatments without degradation.
Poly(arylene) ether (PAE)	2.6 to 2.8	<0.15	375 to 425	Spin-on aromatic polymer with excellent adhesion and ability to be polished with CMP.
a-CF (fluorinated amorphous carbon or FLAC)*	2.8	<0.18	250 to 350	Leading candidate for CVD deposition with high-density plasma CVD (HDPCVD) to produce film with good thermal stability and adhesion.
Parylene AF4 (aliphatic tetrafluorinated poly-p-xylylene)	2.5	<0.18	420 to 450	CVD film that meets adhesion and via resistance requirements with need to maintain gas delivery system at 200°C to control parylene precursor flow rate.

* P. Singer, "Technology News: Wafer Processing," *Semiconductor International* (October 1998): p. 44.

Chip Performance ■ An indicator of chip performance is the speed at which signals are transmitted. Continual die shrinks to achieve sub-0.25 μm technology nodes translate into reduced interconnect linewidths. This reduction leads to increased line resistance (R) for signals. Furthermore, reduced spacing between conductor lines creates more parasitic capacitance (C). The result is an increase in RC signal delay, which slows chip speed and lowers chip performance. This condition is a recent development due to sub-0.25 μm geometries and is often referred to as the *interconnect delay* (see Figure 11.24). In essence, reducing the interconnect dimensions results in more signal delay because of parasitic resistance and capacitance effects. This is the opposite of what occurs in transistors, where reducing gate length serves to reduce gate delay and increase transistor speed.[54]

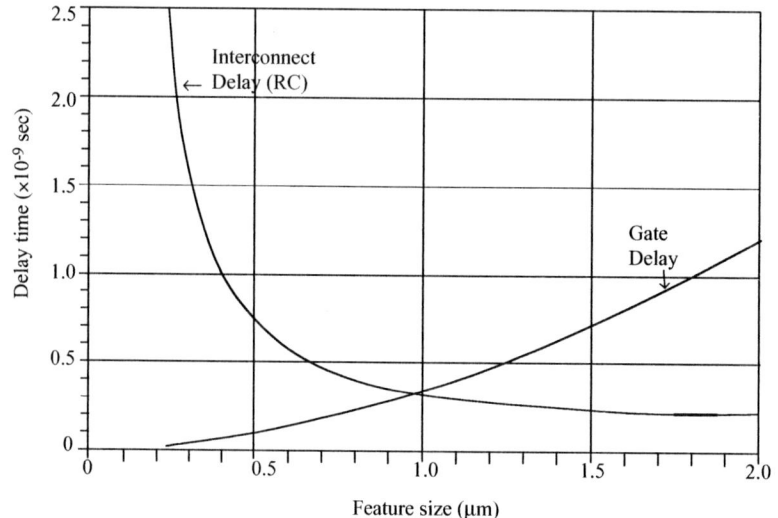

FIGURE 11.24 Interconnect Delay (RC) Versus Feature Size (μm)
(S. Murarka, "Low Dielectric Constant Materials for Interlayer Dielectric Applications, *Solid State Technology* (March 1996): p. 83.)

The line capacitance, C, is directly proportional to the *k*-value of the dielectric material. A low-*k* dielectric reduces the total interconnect capacitance of the chip (see Figure 11.25), reduces the RC signal delay, and improves chip performance. Lowering the total capacitance also decreases power consumption.[55] The use of a low-*k* dielectric material in conjunction with a low-resistance metal line provides the interconnect system with optimum performance for ULSI technology.

Low-*k* Dielectric Requirements ■ Table 11.5 outlines the typical requirements for a low-*k* dielectric constant film for introduction into wafer fabrication. It is anticipated in the semiconductor industry that the introduction of a low-*k* ILD material will follow the introduction of low-resistance metal lines made of copper (see Chapter 12 for a discussion of copper metallization). The investigation of alternative low-*k* dielectrics is being researched at this time.

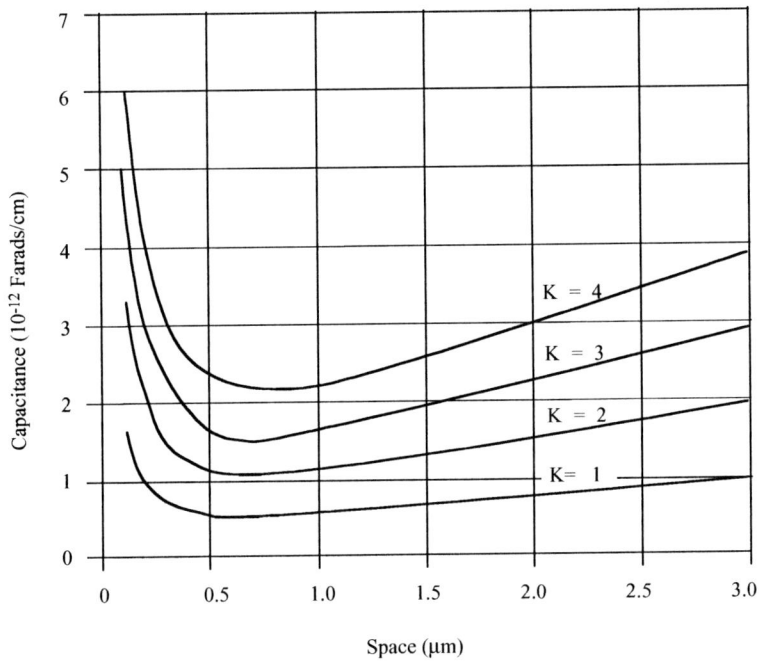

FIGURE 11.25 Total Interconnect Wiring Capacitance
(Used with permission from *Semiconductor International* (September 1998): p. 66.)

TABLE 11.5 Low-k Dielectric Film Requirements

Electrical	Mechanical	Thermal	Chemical	Processing	Metallization
Low dielectric constant	Good adhesion	Thermal stability	Resistant: acids and bases	Patternability	Low contact resistance
Low dielectric loss	Low shrinkage	Low coefficient of thermal expansion	Etch selectivity	Good gap fill	Low electromigration (corrosion)
Low leakage	Crack resistant	High conductivity	Low impurities	Planarization	Low stress voiding
High reliability	Low stress		No corrosion	Low pinhole	Hillock (smooth surface)
	Good hardness		Low moisture uptake	Low particulate	Compatible with barrier metals (Ta, TaN, TiN, and so on.)
			Acceptable storage life		

Adapted from P. Singer, "The Future of Dielectric CVD: High-Density Plasmas?" *Semiconductor International* (July 1997): p. 134.

High-k Dielectric Constant ■ There is ongoing investigation in the industry for new high-k dielectric materials, primarily for DRAM storage capacitors and as an eventual replacement for the extremely thin gate oxide (about 20 Å for 0.18 μm devices). DRAM technology has undergone a 4X increase in storage density every three years for the past 25 years due to processing and design improvements.[56] To attain the required charge storage with a SiO_2 and/or SiN_x dielectric, DRAM storage cell design has evolved into complex stacked capacitor structures (see Figure 11.26 on page 286). If the conventional SiO_2/SiN_x dielectric were replaced with a high-k dielectric, a simpler stack structure could be used for lower fabrication costs. A potential high-k material is tantalum pentoxide (Ta_2O_5), which has a value for k of 20 to 30 and can be easily integrated into the existing process. However, DRAM memory cells are sensitive to leakage and breakdown voltage of storage materials and interfaces. To compensate, the DRAM memory cell that requires a thicker Ta_2O_5

reduces its benefit. Another potential high-*k* material is barium strontium titanate ($(BaSr)TiO_3$, or BST), which gives a significant improvement in capacitance per unit area over conventional dielectrics.

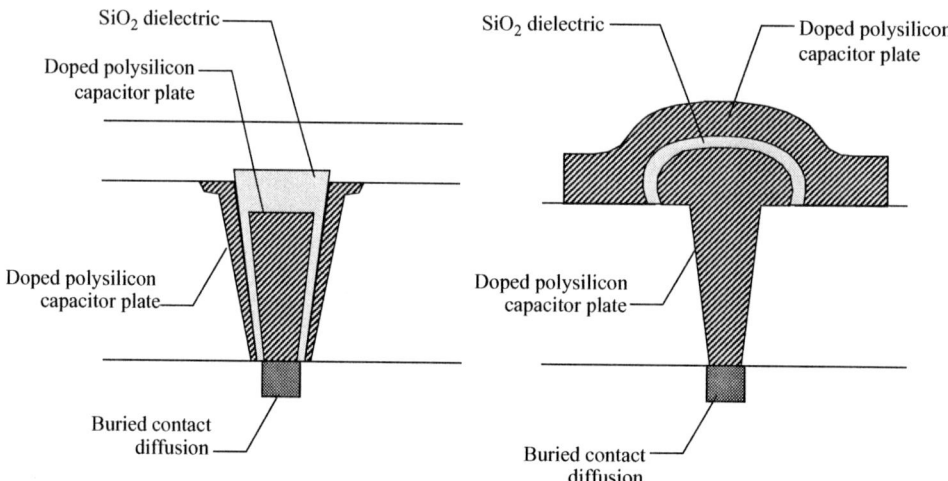

FIGURE 11.26 General Diagram of DRAM Stacked Capacitor

A new high-*k* gate dielectric is needed for the gate oxide because device scaling is forcing the existing SiO_2 to become extremely thin, with the future oxide thickness expected to be <10 Å for the 50 nm device generation. In a MOS transistor, the gate dielectric material must withstand a significant voltage difference between the gate electrode and the underlying silicon substrate. Thin gate oxides are affected by *tunneling currents,* which becomes a problem when the SiO_2 dielectric layer is below about 15 to 20 Å.[57] With the small gate dimensions in ULSI circuitry, electrons can tunnel through the gate material when the transistor is switched on and off. This leads to threshold voltage drift and eventual circuit failure because the device cannot switch states.

Device Isolation

Device isolation in MOS wafer fabrication provides for electrical isolation between devices in the wafer. Isolation techniques serve to reduce or eliminate parasitic field transistors common in MOS planar fabrication (see Chapter 3). Isolation must accommodate scaling between the different device technologies (e.g., junction depth, gate oxide thickness, and so on.). This means that the space allocated for device isolation is being reduced for high-performance ICs. The two basic techniques for device isolation on MOS technologies are local field isolation through local oxidation of silicon (LOCOS), and shallow trench isolation (STI).[58]

Local Oxidation ■ The *local oxidation of silicon* (*LOCOS*) has been the traditional isolation technique for wafer fabrication of devices with critical dimensions of 0.35 μm and larger (see Chapter 10). This technique uses patterned islands of silicon nitride (Si_3N_4) to define the regions for oxidation growth. LOCOS isolation structures are too large for isolation in devices with deep submicron scaling.[59] The major factor limiting LOCOS for 0.25-μm geometries and below is the lateral growth of oxide during silicon oxidation, which imposes natural limits on the minimum area and topography achievable. LOCOS is not acceptable for high-density ULSI technology.

Shallow Trench Isolation ■ *Shallow trench isolation* (*STI*) is the preferred isolation process for wafers fabricated at the 0.25 μm and below technology nodes. The reasons why STI replaced the LOCOS process are:[60]

1. The need for more robust device isolation, especially in DRAM devices.
2. A significant reduction in the surface area for transistor isolation.
3. Superior latchup protection.
4. No channel encroachment.
5. Compatibility with chemical mechanical planarization (CMP).

In the basic STI process (see Chapter 10), trenches about 0.3 to 0.8 μm deep (up to several microns deep for DRAM trenches) are dry etched into the wafer substrate with nearly straight sidewalls and rounded corners (see Chapter 16 for an explanation of etching). The aspect ratio for STI trenches can vary from 2:1 to 5:1, with higher aspect ratios needed in DRAM devices due to their sensitivity to leakage. The trenches have a liner oxide grown on their surface, filled with CVD oxide and planarized using chemical mechanical planarization (CMP). STI technology is a more expensive process than LOCOS technology, primarily because it involves more complex process steps. Nevertheless, the benefits of STI outweigh its higher cost. The process steps for STI, including barrier layers and the liner oxide, are outlined in the CMOS process flow model in Chapter 9.

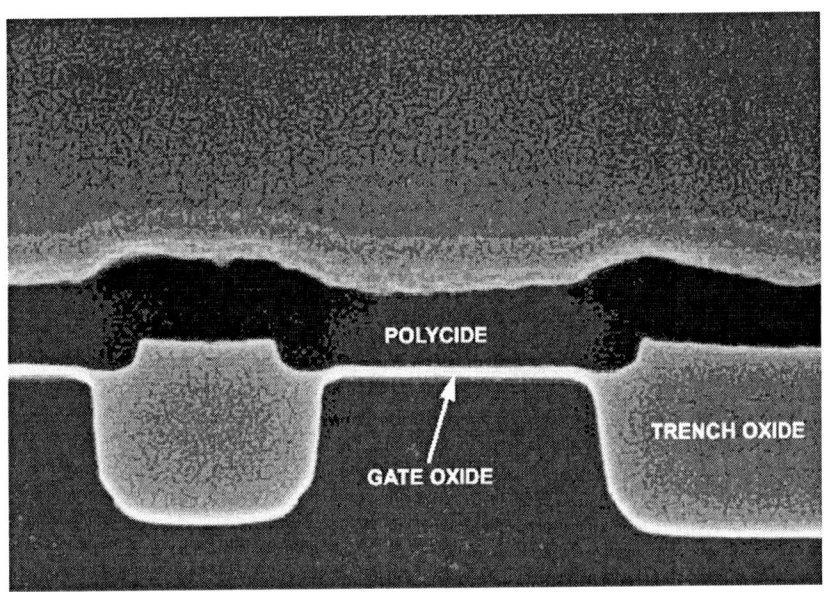

Shallow Trench Isolation
(Micrograph courtesy of Integrated Circuit Engineering)

SPIN-ON-DIELECTRICS

There are a wide range of low-k dielectrics that are designed to use wafer spin as an application method, referred to as spin-on-dielectric (SOD). The process for applying these materials is similar to a familiar process of spin-on-glass (SOG) used in low-cost IC fabrication, which we will review first.

Spin-On-Glass (SOG)

Spin-on-glass (SOG) was frequently used for gap fill and planarization of the ILD before the wide acceptance of chemical mechanical planarization (CMP) in the 1990s. SOG materials usually consist of two basic types: organic and inorganic.[61] The organic is based on siloxane, while the inorganic is based on a silicate. The organic siloxane SOG has significant water absorption after curing, is thermally unstable, and does not tolerate plasma exposure. After curing, the silicate SOG behaves like SiO_2, does not absorb excessive moisture, and is thermally stable. However, it shrinks significantly during curing, which leads to stress buildup and cracking if the layer is too thick. For these reasons, the siloxane SOG is usually used as a planarizing layer that undergoes an etchback process with a deposited oxide layer to render it smooth (see Chapter 18). The silicate SOG is used mainly for gap fill applications. There are many different commercial modifications of these two types.

Before applying the SOG, the wafer usually has a predeposited oxide layer, such as a PECVD oxide.[62] The SOG is a liquid and is applied by spinning the wafer at a predetermined speed in either a closed or open bowl. The spin speed and parameters such as ramp-up rate define the SOG thickness. The film is cured, typically at 400°C, and then etched back to planarize the surface. A capping oxide film is applied to seal and protect the SOG, which minimizes water absorption (see Figure 11.27). Without the cap oxide, there is substantial risk of reliability problems from absorbed moisture, such as through poisoning where the via between the metal layers is corroded and has high resistance.

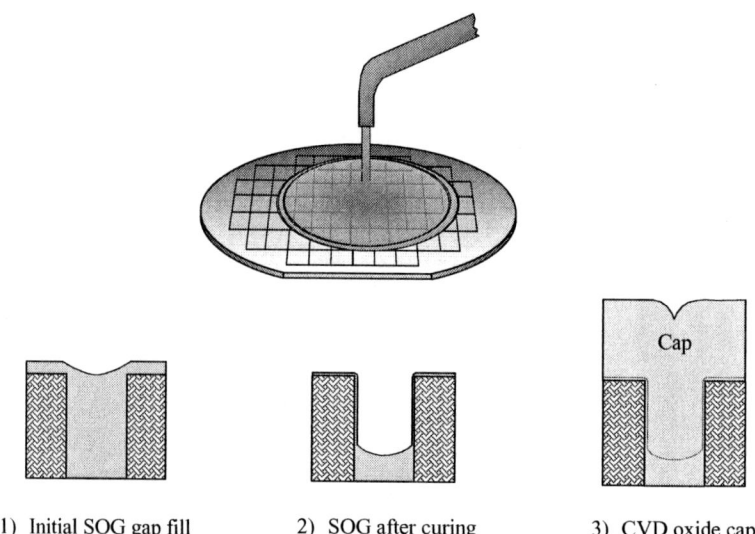

1) Initial SOG gap fill 2) SOG after curing 3) CVD oxide cap

FIGURE 11.27 Gap-Fill with Spin-On-Glass (SOG)

Spin-On-Dielectric (SOD)

Low-k dielectric films are currently being investigated as *spin-on-dielectrics* (*SODs*). An example of an SOD low-k film is hydrogen silsesquioxane (HSQ), presented earlier in this chapter in Table 11.4 on page 283. The SOD technology is being considered as a cost-effective alternative to CVD processes for depositing low-k films. Most SOD applications have used standard spin coaters (used in photolithography which is discussed in Chapter 13) with batch curing in generic furnace tubes. There has been investigation into a single-wafer cluster tool approach to spin coat and cure the film, using a high-temperature hotplate at 350 to 475°C.[63] A proposed spin coat and cure process for HSQ is outlined in Table 11.6.

TABLE 11.6 Proposed HSQ Low-k Dielectric Processing Parameters

Major Operation	Process Step	Parameter
Spin Coating	Apply bowl speed	50 rpm
	Maximum bowl speed	800 to 1500 rpm
	Backside rinse	800 rpm, 5 sec
	Topside edge bead removal	1000 rpm, 10 sec
	Spin dry	1000 rpm, 5 sec
Cure	Initial soft-bake cure	200°C, 60 sec, N_2 purge
	In-line cure	475°C, 60 sec, N_2 ambient

In some cases, the nature of the SOD material is an extension of an existing SOG material. Other SOD materials are spin-on polymers that are substantially different than SOG materials. New low-k materials have less tendency to absorb moisture and have superior crack resistance. Because of these features, single layer deposition is usually acceptable with no need for a cap oxide layer. In some cases, however, an adhesion promoter is required.[64]

EPITAXY

Epitaxy is the deposition of a thin layer of single-crystal material upon the surface of a single-crystal substrate (see Figure 11.28). This epitaxial layer is often referred to as an epilayer. Epitaxy provides flexibility for the device designer to optimize performance by controlling epilayer doping thickness, concentration, and profile, independent of the wafer. This control is accomplished by the intentional addition of dopants during the epitaxial growth process. An epilayer also reduces the occurrence of latchup in CMOS devices (see Chapter 3). The most common epitaxial reactor in IC production is a high-temperature CVD system.[65]

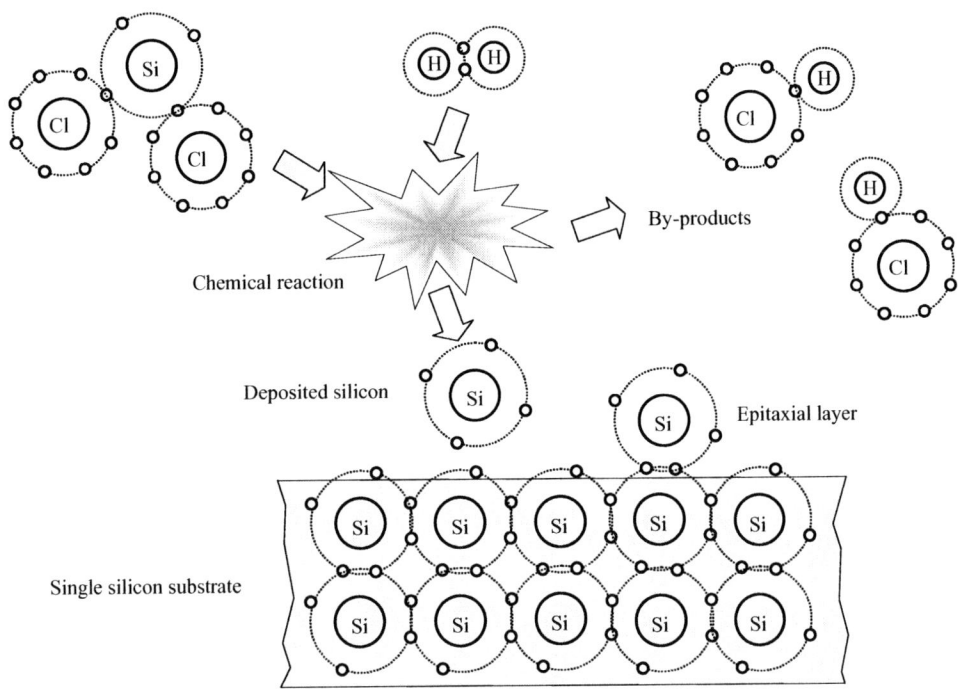

FIGURE 11.28 Silicon Epitaxial Growth on a Silicon Wafer

There are applications under investigation where epitaxy could contribute to advancing future IC performance. An example is *raised source/drain* (*S/D*) structures to achieve low contact resistance.[66] The raised S/Ds are formed by depositing epitaxial silicon on the source, drain, and gate areas of the device. This action effectively increases the surface areas of the S/Ds and therefore lowers the sheet resistance (similar to a larger diameter wire having lower resistance than the same wire with a smaller diameter). These deposits can decrease the contact resistance from 5 Ω/square on conventional devices to 1-2 Ω/square. The raised S/D structure is under investigation for possible implementation at the 0.15-μm technology node.

Undesirable conditions of nonuniform doping can occur during epitaxial deposition.[67] Since lightly doped epilayers are often grown on more heavily doped substrates, there is a condition of *autodoping* into the epitaxial layer. This condition occurs when dopant impurities evaporate from the wafer or are liberated due to chlorine etching of the wafer surface during deposition. These impurities move into the gas stream and cause unintentional doping of the epilayer. As the epilayer grows, less dopant is available from the wafer, and the impurities in the gas stream reach a constant level. Another form of irregular doping is the substrate acting as a source of dopant impurities that diffuse into the epilayer. This action is referred to as *out-diffusion*. Both autodoping and out-diffusion can affect the dopant transition between the substrate and epilayer, causing the transition to be less abrupt than desired.

If the film and substrate are of the same material (e.g., Si film on an Si substrate), then the film growth is *homoepitaxy*. A less common type of epitaxy is heteroepitaxy, which occurs when the film and substrate are of two different materials (e.g., Si on Al_2O_3).

Epitaxy Growth Methods

Epitaxial silicon is commonly deposited using CVD systems. Before epitaxy growth, the wafer must be cleaned of native oxide and any residual organic or metal impurities with a goal of achieving a perfect silicon surface (see Chapter 6). During the epitaxial deposition process, the atoms produced by the gas reaction strike the wafer surface and then move around until they find the correct location to bond to the surface atoms. This action forms the epitaxial layer with the same crystallographic arrangement as the substrate.

Possible gas sources for the epitaxy reaction are the hydrogen reduction of silicon tetrachloride ($SiCl_4$), silane (SiH_4), dichlorosilane (SiH_2Cl_2, or DCS) or trichlorosilane (TCS). Deposition temperatures are from 1050 to 1250°C. Nearly all silicon epitaxy for wafer fabrication is done by the reduction of chlorosilanes (SiH_xCl_{4-x}, where x = 0, 1, 2 or 3) that are diluted in hydrogen.[68] The temperature of the reaction can be lowered if there are fewer chlorine atoms in this precursor. Epitaxial silicon is not commonly grown with silane (SiH_4) because there is excessive particle formation when the silicon deposits on the warm surfaces of the reactor.

Different methods have been used to grow single-crystal layers on silicon wafers, including solid-phase, liquid-phase, vapor-phase, and molecular beam. Three methods used to grow epitaxial layers on silicon wafers for IC production are:

- Vapor-phase epitaxy (VPE)
- Metalorganic CVD (MOCVD)
- Molecular-beam epitaxy (MBE)

Vapor-Phase Epitaxy (VPE) ■ The most common method used for silicon epitaxial growth in wafer fabrication is *vapor phase epitaxy* (*VPE*), which is a subset of CVD.[69] Silicon VPE is achieved by passing gas compounds of the desired chemicals over single-crystal silicon wafers that are heated from 800 to 1150°C. The heat from the high temperature provides the energy necessary to drive the chemical reactions, which take place on the surface of the wafer. This process is shown in Figure 11.29.

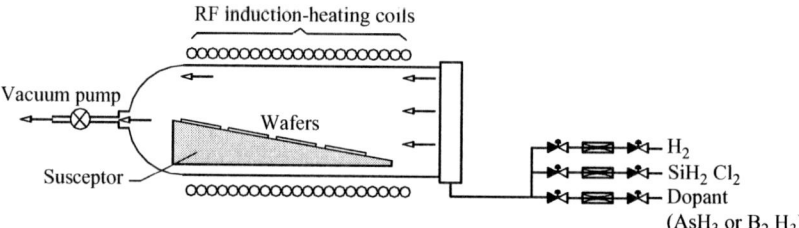

FIGURE 11.29 Illustration of Vapor-Phase Epitaxy

Before processing wafers in a VPE reactor, the system is purged with nitrogen or hydrogen, followed by a vapor HCl etching. The reactant gases, such as chlorosilane, along with dopant gases, are then introduced to the growth chamber where the wafer is heated to the desired temperature. Once the reactant and dopant gases are in the growth reactor, the chemical species undergo the necessary chemical and physical reactions to deposit the doped epilayer.

Typical equipment design for an epitaxial reactor consists of the gas distribution system, a reactor tube, a susceptor to hold and heat the wafer, a control system, and a gas exhaust system. Mass flow controllers and pneumatic valves are used in the gas distribution system to attain tight control over gas flows into the reactor chamber. The susceptor is typically made of graphite or a polysilicon that is coated with silicon carbide or silicon nitride. It must be strong and nonreactive to the reactants and their by-products.[70] Heating the susceptor is accomplished through inductive heating or by radiation from filament lamps. The horizontal and vertical reactors shown in Figure 11.30 are the most common. The susceptor in the horizontal reactor is tilted a few degrees to improve uniformity.

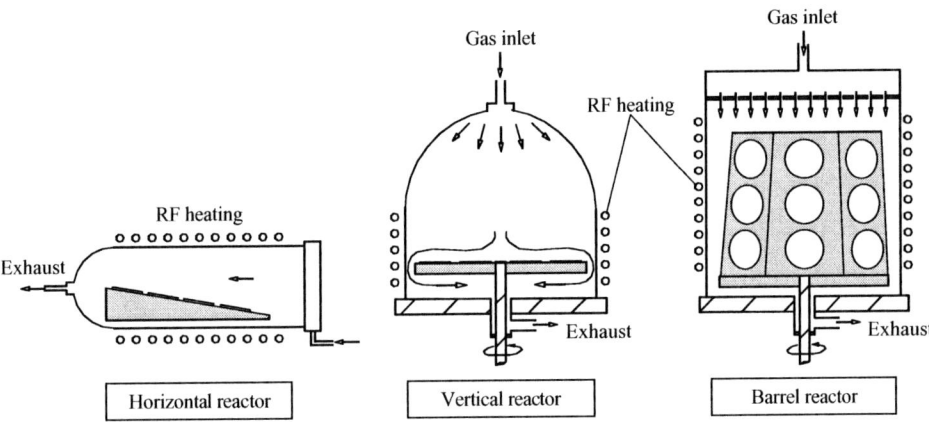

FIGURE 11.30 Silicon Vapor-Phase Epitaxy Reactors

Metalorganic CVD (MOCVD) ■ Another epilayer growth method is commonly referred to as *metalorganic CVD* (*MOCVD*), but this is actually a broader term that is equally applicable to the deposition of polycrystalline and amorphous films of metals and oxides.[71] MOCVD is a form of VPE and is not typically used for silicon epitaxy because no suitable sources are readily available. It has been used to deposit compound semiconductor epitaxial layers, such as the III-V compound of gallium arsenide (GaAs) from organometallic sources at low temperature. As with the VPE process, compounds of the desired materials are transported to a heated substrate where the complex chemical reaction takes place on the wafer surface. MOCVD is important for the controlled deposition of ultra-thin, doped, or undoped semiconductor heterolayers (such as GaAs/Si), primarily used for lasers, light-emitting diodes (LED), and optoelectronic integrated circuits.[72] MOCVD is also used in development activity for depositing some organic low-k dielectrics for future IC fabrication.

Molecular-Beam Epitaxy (MBE) ■ *Molecular-beam epitaxy* (*MBE*) is one of the main methods used to deposit gallium arsenide (GaAs) heteroepitaxy with atomic resolution of thickness. It can also be used for depositing silicon onto a wafer substrate with very tight controls on epilayer thickness and doping uniformity. MBE takes place under conditions of high vacuum, usually at a base pressure of 10^{-10} to 10^{-11} torr or higher, typically with the use of a high vacuum cryopump. The reaction temperature range is between 500 to 900°C.

Most silicon MBE systems produce silicon atoms for the epitaxial reaction through the evaporation of silicon with an electromagnetically focused electron beam source. This process is similar to vacuum evaporation for deposition. The beam of silicon atoms leaving the silicon source travels through the evacuated chamber without collisions and deposits on the single-crystal wafer surface. Another more recent method is to deliver silicon atoms via a gas source at very low flow rates. The silicon growth rate of an MBE system is determined by measuring the number of atoms that leave the source and the fraction that actually strike the wafer surface and stick.

CVD QUALITY MEASURES

Important quality measures for CVD are listed in Table 11.7.

TABLE 11.7 Key Quality Measures for CVD

Quality Parameter	Types of Defects	Remarks
1. Voids during deposition of high-aspect ratio gap (>3:1) with PECVD SiO_2.	A. Keyhole voids are formed during deposition of high aspect ratio gaps. After chemical mechanical planarization (CMP) and removal of top surface, some voids become trenches (see Figure 11.31 on page 294). This trench can lead to an open circuit after metal deposition.	• Gap fill is critical as CDs are reduced and gaps have a high aspect ratio. • Voids are highly stressed regions and trap moisture or solvents that cause corrosion or outgas at high vacuum. • The root cause of this problem is trying to fill a high aspect ratio gap with a PECVD that is limited to deposition system. This application requires HDPCVD.
2. Film stress.	A. High film stress leads to cracking and delamination. B. Film stress can propagate silicon defects in the substrate. C. Stress may cause current leakage.	The presence of dopant in a glass film can reduce stress. Deposition parameters that affect film stress are: • RF power: adjust power to improve stress (e.g., reduce power by 10 watts to lower stress). • Pressure: adjust pressure to affect stress (e.g. increased pressure leads to higher film stress).
3. Film thickness.	A. Thickness of film exceeds requirements.	Deposition parameters that affect thickness are: • Time: reduce deposition time to reduce thickness (e.g., 1 sec reduction for each 100 Å of thickness). • Gas-flow rate: reduce gas flow to reduce thickness (e.g., reduce SiH_4 by 2 sccm for >5% thickness over nominal).
4. Refractive index (R/I).	A. R/I is a good monitor to assess quality of the film. B. R/I is highly dependent on the composition of the deposited film (stoichiometry).	R/I is an optical property of a material. Compared to thermal oxides, CVD oxides are inferior in quality and integrity. CVD oxides exhibit more particles and pinholes. The R/I of CVD SiO_2 can be compared to the R/I of SiO_2 (1.46) to assess relative quality: • High R/I means excessive Si in film. • Low R/I indicates a porous film, which tends to absorb moisture.

CVD TROUBLESHOOTING

Troubleshooting techniques for common CVD problems are described in Table 11.8. Plasma deposition systems have two areas of safety: chemical and electrical. Many plasmas use toxic gases (see Appendix A, Table A.3). There are high voltages and RF energy in the RF power supplies. Use caution when troubleshooting CVD systems.

TABLE 11.8 Common CVD Troubleshooting Problems

Problem	Probable Cause	Corrective Action
1. Particle contamination associated with film.	Source of particles is isolated by determining whether particles are found on top of film, in film, or below film: A. On film: particles formed after deposition. Check for particles on sidewalls of hot-wall reactors and belt-driven reactors. B. In film: Gas-phase nucleation particles from gas too rich in silane or silicon. Gas supply contamination causes particles. C. Under film: Particles from silicon carbide, quartz, or reactor walls fall on wafer before deposition.	• Particles on film: Particles on sidewalls indicate need for more frequent wet cleaning of quartzware and chamber surfaces for preventive maintenance. Inspect in situ dry cleaning process. Also verify the correct procedures for cleaning (manually or in situ). • Particles in film: Improper gas flows from MFC calibration or problem in process recipe of software program. Check for leaks in gas supply system or O-rings. Verify point-of-use filters are acceptable. • Particles under film: Check wafer cleaning process before deposition.
2. Film thickness.	Thickness is related to both equipment and process problems. Variables affecting thickness are: A. Incorrect temperature control. B. System pressure is too high or low. C. System power needs adjustment. D. Improper gas flow.	• Temperature controller may require calibration. A common problem is a defective thermocouple. • System pressure is controlled by process recipe. Check vacuum system for leaks. • Adjust RF power to optimize film thickness. • Check calibration of MFC to ensure proper gas flow. • If checking thickness with test wafer, verify the test wafer is clean and does not have a thickness variation before the test.
3. Cracking of dielectric material on top of electrostatic chuck (ESC).	High thermal load at the ESC causes the dielectric material to erode or crack. This condition leads to plasma arcing or wafer chucking/dechucking problems.	• Investigate the ESC backside cooling system to ensure it is functioning properly. • Inspect ESC materials to verify there is no breakdown due to high power exposure, temperature, or plasma clean conditions.

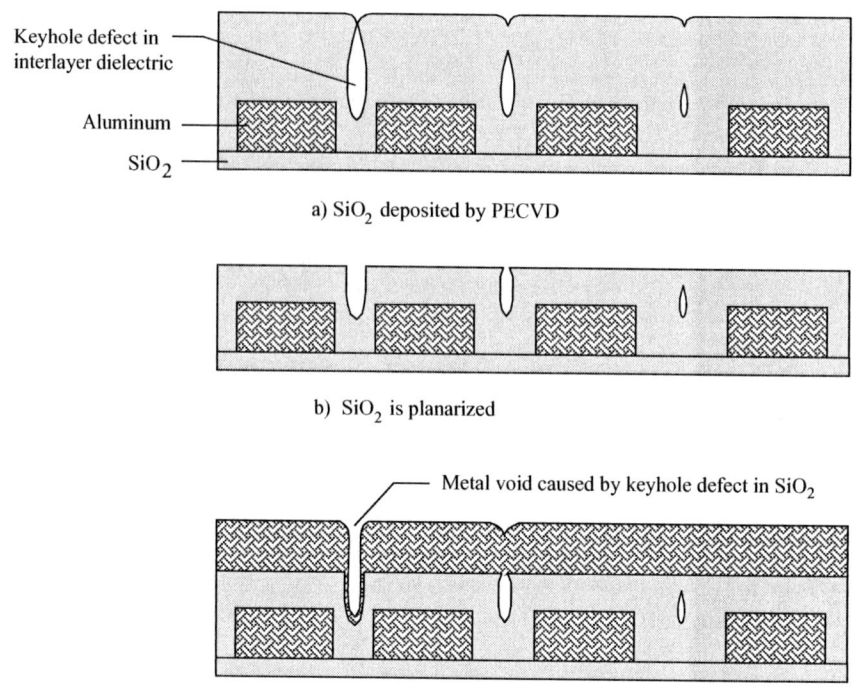

FIGURE 11.31 Effects of Keyholes in ILD on Metal Step Coverage

SUMMARY

Multilevel metallization with dielectric and metal layers is necessary for advanced ICs. Thin films have special characteristics to meet the requirements for wafer fabrication, including high aspect ratio gap fill. A film grows in stages from a cluster of nuclei to a continuous film. There is a wide range of tools for depositing film layers, with the focus on dielectric chemical vapor deposition (CVD). CVD deposits a thin film on the wafer surface through a chemical reaction that is assisted by thermal or plasma energy. CVD surface reactions include pyrolosis, oxidation, and reduction. The CVD chemical reaction follows defined steps and can be limited by the supply of chemicals or by the nature of the surface reaction. The thin film is sometimes doped to obtain improved performance. Reactors for CVD are categorized by atmospheric pressure CVD (APCVD), low pressure CVD (LPCVD), or plasma-assisted CVD. APCVD is simple in design and is primarily used to deposit conformal oxide with a TEOS-O_3 process.

LPCVD deposits various films, including oxide, silicon nitride, and polysilicon. The LPCVD process is reaction-rate limited (gas flow is not critical) and often occurs at lower temperatures. Plasma-assisted CVD is divided into plasma-enhanced CVD (PECVD) and high-density plasma CVD (HDPCVD). PECVD was the first plasma-assisted CVD but had limited use for filling gaps less than 0.5 μm. HDPCVD is used for advanced ICs because of its excellent gap-fill capability for high-aspect ratios at relatively low temperatures. The deposited dielectric material increases chip performance by reducing the RC interconnect delay with a low-k dielectric. Dielectric material is also important for isolation technologies. Dielectric materials have been applied in liquid form by spinning, which may be an application method for future low-k dielectrics. Epitaxy is grown primarily by CVD methods, including vapor-phase epitaxy (VPE), metalorganic CVD (MOCVD), and molecular-beam epitaxy.

KEY TERMS

multilevel metallization
aluminum metallization
critical layers
noncritical layers
first interlayer dielectric (ILD-1)
premetal dielectric (PMD)
interlayer dielectric (ILD)
thin film
stoichiometry
aspect ratio
nucleation
nuclei coalescence
continuous film
chemical vapor deposition (CVD)
reactor
heterogeneous reaction (also surface catalyzed)
homogeneous reaction
adsorption
desorption
species
precursors
mass-transport limited
reaction-rate limited
kinetically controlled
boundary layer
stagnant layer
phosphosilicate glass (PSG)
borosilicate glass (BSG)
borophosphosilicate glass (BPSG)
fluorosilicate glass (FSG)

cluster tools
hot-wall reactor
cold-wall reactor
atmospheric pressure CVD (APCVD)
silane
TEOS oxide
low pressure CVD (LPCVD)
silicon nitride (Si_3N_4)
silicon oxynitride (SiO_xN_y)
plasma-enhanced CVD (PECVD)
PETEOS
silicon nitride PECVD
high-density plasma CVD (HDPCVD)
dep:etch ratio
dielectric constant (k)
device isolation
local oxidation of silicon (LOCOS)
shallow trench isolation (STI)
spin-on-glass (SOG)
spin-on-dielectric (SOD)
epitaxial growth
raised source/drain structures
autodoping
out-diffusion
homoepitaxy
heteroepitaxy
vapor-phase epitaxy (VPE)
metalorganic CVD (MOCVD)
molecular-beam epitaxy

REVIEW QUESTIONS

1. What is multilevel metallization? Why is it necessary for chip fabrication?
2. What is aluminum metallization? Describe a chip critical layer and a noncritical layer.
3. Explain the purpose of an ILD layer. Where is the ILD-1 layer located on a chip?
4. What is a thin film? List and describe eight characteristics of an acceptable thin film.
5. Define the aspect ratio. Why is a high-aspect ratio important for ULSI devices?
6. List and describe the three stages of thin film growth.
7. List the five major techniques for deposition.
8. Define chemical vapor deposition (CVD).
9. What are the five basic CVD chemical reactions? Give a short description of each reaction.
10. In a CVD reaction, is a heterogeneous or homogeneous reaction preferred? Why?
11. Identify and describe the eight steps in a CVD reaction.
12. Explain the difference between a mass-transport limited and a reaction-rate limited CVD process. Which process is temperature dependent?
13. What is a boundary layer in CVD? Under what conditions does it become a stagnant layer?
14. What is the beneficial effect of low pressure in a CVD reaction?
15. What do the abbreviations PSG, BPSG, and FSG stand for? What are three benefits of adding dopants to oxide?

16. Describe the different types of CVD reactors and their principal advantages.
17. Explain the difference between a hot-wall reactor and cold-wall reactor for CVD.
18. Explain APCVD. What is the principal disadvantage with APCVD SiO_2, using silane as a source?
19. What is TEOS? Discuss the primary advantages to using APCVD TEOS-O_3.
20. Why is LPCVD more common than APCVD? Describe LPCVD processing.
21. Why does LPCVD operate in the reaction-rate regime?
22. State what film is deposited using LPCVD TEOS, and explain the advantage of this film.
23. Why is silicon nitride often used as a passivation layer?
24. What CVD tool is used to deposit the polysilicon gate material? List six reasons why polysilicon is used as a gate electrode.
25. What are the benefits to using a silicon oxynitride film?
26. State six advantages to using plasma during CVD.
27. What is PECVD? What is the main difference between PECVD and LPCVD?
28. Describe the main differences between silicon nitride PECVD and LPCVD.
29. Discuss a major limitation for using PECVD in ULSI microchips.
30. Explain what HDPCVD is. What are its main benefits in advanced ICs?
31. Describe the effect wafer biasing has on HDPCVD directionality.
32. Explain simultaneous deposition and etching for HDPCVD. What is the value for the typical dep-etch ratio?
33. List and describe the five steps used for simultaneous deposition-etching in CVD.
34. Discuss the importance of the dielectric constant of the ILD.
35. Explain the interconnect delay and why it is beneficial to lower the ILD constant dielectric.
36. What is an application for a high-k dielectric material and why is it needed?
37. What are LOCOS and STI? Why has STI replaced LOCOS for advanced ICs?
38. List the processing steps for STI.
39. What is spin-on-glass? Explain spin-on-dielectric and what it could be used for in future applications.
40. Describe what epitaxy is. Define the terms *autodoping* and *out-diffusion*.
41. List and discuss three methods for epitaxial growth.

DEPOSITION EQUIPMENT SUPPLIERS' WEBSITES

Amtech Systems Inc.	http://www.amtechsystems.com/
Applied Materials	http://www.appliedmaterials.com/products/
ASM	http://www.asm.com/
CVC Incorporated	http://www.cvc.com/
CVD Equipment Corporation	http://www.cvdequipment.com/
Genus Incorporated	http://www.genus.com/
Kokusai Semiconductor Equipment	http://www.ksec.com/
Novellus Systems Inc.	http://www.novellus.com/index.htm
TEL, Tokyo Electron Ltd.	http://www.teainet.com
Tystar Corporation	http://www.tystar.com/

REFERENCES

1. R. Liu, "Metallization," *USLI Technology,* ed. C. Change and S. Sze (New York: McGraw-Hill, 1996), p. 379.
2. R. Jackson et al., "Processing and Integration of Copper Interconnects," *Solid State Technology* (March 1998): p. 49.
3. S. Nag et al., "Low-Temperature Pre-Metal Dielectrics for Future ICs," *Solid State Technology* (September 1998): p. 69.
4. P. Singer, "The Future of Dielectric CVD: High-Density Plasmas?" *Semiconductor International* (July 1997): p. 134.
5. S. Sivaram, *Chemical Vapor Deposition: Thermal and Plasma Deposition of Electronic Materials* (New York: Van Nostrand Reinhold, 1995), p. 2.
6. Adapted from S. Wolf and R. Tauber, *Silicon Processing for the VLSI Era*, vol. 1 *Process Technology,* 2nd ed., (Sunset Beach, CA: Lattice Press, 2000) pp. 108–114.
7. J. Baliga, "Options for CVD of Dielectrics Include Low-k Materials," *Semiconductor International* (June 1998): p. 139.
8. C. Rye et al., "Barriers for Copper Interconnections," *Solid State Technology* (April 1999): p. 53.
9. P. Singer, "Film Stress and How to Measure It," *Semiconductor International* (October 1992): p. 54.
10. S. Ghandhi, *VLSI Fabrication Principles: Silicon and Gallium Arsenide,* 2nd ed., (New York: John Wiley & Sons, 1994), p. 522.
11. Ibid.

12. G. Anner, *Planar Processing Primer* (New York: Van Nostrand Reinhold, 1990), p. 361.
13. G. Anner, *Planar Processing Primer*, p. 361.
14. S. Wolf and R. Tauber, *Silicon Processing for the VLSI Era*, vol. 1, *Process Technology* (Sunset Beach, CA: Lattice Press, 1986), p. 151.
15. A. Jones and P. O'Brien, *CVD of Compound Semiconductors: Precursor Synthesis, Development and Applications* (Weinheim, Germany: VCH, 1997), p. 31.
16. S. Wolf and R. Tauber, *Silicon Processing for the VLSI Era*, vol. 1, *Process Technology*, p. 151.
17. Ibid., p. 154.
18. Ibid., p. 154.
19. G. Anner, *Planar Processing Primer*, p. 365.
20. SEMATECH, "Desposition Processes," *Furnace Processes and Related Topics* (Austin, TX: SEMATECH, 1994), p. 6.
21. S. Ghandhi, *VLSI Fabrication Principles: Silicon and Gallium Arsenide*, 2nd ed., p. 532.
22. S. Nag et al., "Low-Temperature Pre-Metal Dielectrics," p. 69.
23. J. Baliga, "Options for CVD of Dielectrics," p. 140.
24. H. Cheng, "Dielectric and Polysilicon Film Desposition," *ULSI Technology*, ed. C. Chang and S. Sze (New York: McGraw-Hill, 1996), p. 211.
25. M. Hammond, "Introduction to Chemical Vapor Deposition," *Solid State Technology* (December 1979): p. 61.
26. S. Ghandhi, *VLSI Fabrication Principles: Silicon and Gallium Arsenide*, 2nd ed., p. 529.
27. G. Schwartz and K. Srikrishnan, "Interlevel Dielectrics," *Handbook of Semiconductor Interconnection Technology*, ed. G. Schwartz, K. Srikrishnan, and A. Bross (New York: Marcel Dekker, 1998), p. 231.
28. B. El-Kareh, *Fundamentals of Semiconductor Processing Technologies* (Boston: Kluwer Academic Publishers, 1995), p. 118.
29. J. Baliga, "Options for CVD of Dielectrics," p. 140.
30. G. Schwartz and K. Srikrishnan, "Interlevel Dielectrics," p. 233.
31. H. Cheng, "Dielectric and Polysilicon Film Deposition," *ULSI Technology*, p. 211.
32. S. Ghandhi, *VLSI Fabrication Principles: Silicon and Gallium Arsenide*, 2nd ed., p. 524.
33. Ibid., p. 520.
34. C. Gugliemini and A. Johnson, "Properties and Reactivity of Chlorine Trifluoride," *Semiconductor International* (June 1999): p. 162.
35. S. Ghandhi, *VLSI Fabrication Principles: Silicon and Gallium Arsenide*, 2nd ed., p. 529.
36. H. Cheng, "Dielectric and Polysilicon," p. 225.
37. S. Wolf and R. Tauber, *Silicon Processing for the VLSI Era*, Vol. 1, *Process Technology*, 2nd ed., p. 175.
38. Y. Trouiller et al., "Inorganic Bottom ARC SiO_xN_y for Interconnection Levels on 0.18 μm Technology," *SPIE* 3508, (1998), p. 122.
39. L. Peters, "Pursuing the Perfect Low-k Dielectric," *Semiconductor International* (September 1998), p .122.
40. B. El-Kareh, *Fundamentals of Semiconductor Processing Technologies*, p. 125.
41. Ibid.
42. SEMATECH, "Desposition Processes," *Furnace Processes and Related Topics*, p. 35.
43. C. Apblett et al., "Silicon Nitride Growth in a High-Density Plasma System," *Solid State Technology* (November 1995): p. 73.
44. S. Ghandhi, *VLSI Fabrication Principles: Silicon and Gallium Arsenide*, 2nd ed., p. 535.
45. E. Korczynski, "HDP-CVD: Trying to Lasso Lightning," *Solid State Technology* (April 1996): p. 63.
46. P. Singer, "Future of Dielectric CVD," p. 128.
47. P. Burggraaf, "Advanced Plasma Sources: What's Working?" *Semiconductor International* (May 1994): p. 57.
48. J. Bondur et al., "Impact of Electrostatic Chuck Performance on HDP CVD SiO_2 Films," 52nd Symposium on Semiconductors and Integrated Circuits Technology, Osaka, Japan.
49. P. Singer, "Future of Dielectric CVD," p. 127.
50. E. Korczynski, "HDP-CVD," p. 64.
51. P. Singer, "Future of Dielectric CVD," p. 127.
52. Ibid.
53. L. Peters, "Pursuing the Perfect Low-k Dielectric," p. 68.
54. M. Bohr, "Interconnect Scaling—The Real Limiter to High Performance ULSI," *Solid State Technology* (September 1996): p. 105.
55. S. Murarka, "Low Dielectric Constant Materials for Interlayer Dielectric Applications," *Solid State Technology* (March 1996): p. 83.
56. D. Kotecki, "High-k Dielectric Materials for DRAM Capacitors," *Semiconductor International* (January 1998): p. 38.
57. P. Singer, "Wafer Processing: New Gate Dielectric Material Needed," *Semiconductor International* (January 1998): p. 38.
58. P. Van Cleemput et al., "HDPCVD Films Enabling Shallow Trench Isolation," *Semiconductor International* (July 1997): p. 180.
59. Ibid., p. 182.
60. L. Peters, "Choices and Challenges for Shallow Trench Isolation," *Semiconductor International* (April 1999): p. 69.
61. R. Liu, "Metallization," p. 419.
62. B. El-Kareh, *Fundamentals of Semiconductor Processing Technologies*, p. 569.
63. T. Batchelder et al., "In-Line Cure of SOD Low-k Films," *Solid State Technology* (March 1999): p. 29.
64. L. Peters, "Pursuing the Perfect Low-k Dielectric," p. 66.
65. S. Campbell, *The Science and Engineering of Microelectronic Fabrication* (New York: Oxford University Press, 1996), p. 340.
66. L. Peters, "Is the 0.18 μm Node Just a Roadside Attraction?" *Semiconductor International* (January 1999): p. 52.
67. R. Jaeger, *Introduction to Microelectronic Fabrication* vol. V, (Reading, MA: Addison-Wesley, 1988), p. 125.

68. S. Campbell, *Science and Engineering,* p. 347.
69. S. Ghandhi, *VLSI Fabrication Principles: Silicon and Gallium Arsenide,* 2nd ed., p. 283.
70. P. J. Wang, "Epitaxy," *ULSI Technology,* ed. C. Chang and S. Sze (New York: McGraw-Hill, 1996), p. 124.
71. A. Thompson, R. Stall, and B. Kroll, "Advances in Epitaxial Deposition Technology," *Semiconductor International* (July 1994): p. 173.
72. B. El-Kareh, *Fundamentals of Semiconductor Processing Technologies,* p. 102.

CHAPTER 12
METALLIZATION

Metallization in wafer fabrication is the process of depositing metal film over a dielectric film and later patterning it to form the interconnecting metal lines and plugs of integrated circuits. This is similar to using insulated copper wire in an automobile to interconnect all the electrical components to make a fully-functional electrical system. The metal lines are sandwiched between dielectric layers for electrical integrity. High-performance microprocessors use metal lines to interconnect tens of millions of devices on one chip. Transistor densities are projected to reach 1 billion transistors per chip by the year 2010, with a corresponding increase in interconnect complexity.

Microchip interconnect technology has become a critical challenge for future IC performance due to the need to decrease signal propagation delay. Because the density of ULSI circuit elements is increasing, interconnect resistance and parasitic capacitance increase, thereby slowing the signal propagation.

A significant change currently underway in wafer fabrication technology is the reduction in interconnect metal resistivity, ρ. This reduction is achieved by the replacement of aluminum alloy with copper as the primary conducting metal. With deep submicron linewidths, there is also a need for a low-k interlayer dielectric (ILD). Lowering the dielectric constant will reduce parasitic capacitance, which contributes to signal delay.

Traditionally, the methods used to deposit metals on a wafer have been physical processes. On the other hand, processes used to deposit insulating and semiconducting layers usually involved the CVD chemical reactions studied in Chapter 11. This separation into physical and chemical processes is less defined with the introduction of new IC metallization technology.

OBJECTIVES

After studying the material in this chapter, you will be able to:

1. Explain the terminology for metallization.
2. List and describe the six categories of metals used in wafer fabrication. Discuss the performance requirements and give applications for each metal category.
3. Explain the benefits for using copper metallization in wafer fabrication. Describe the challenges for implementing copper.
4. State the advantages and disadvantages to sputtering.
5. Describe the physics of sputtering and discuss different sputtering tools and applications.
6. Describe the benefits and applications for metal CVD.
7. Explain the fundamentals of copper electroplating.
8. Describe a process flow for dual-damascene processing.

INTRODUCTION

Wafer metallization is the deposition of a thin film of conductive metal onto a wafer by use of a chemical or physical process. This process is closely linked to dielectric deposition. Metal lines conduct the signal through the IC circuit while the dielectric layers ensure signals are not influenced by adjacent lines. Both metal and dielectrics are thin film processes. In some cases, metal and dielectrics are deposited by the same type of equipment.

Metallization has unique terminology for the different metal connections. The term *interconnect* describes the conductor materials, such as aluminum, polysilicon, or copper, that create the metal wiring that carries electrical signals to different parts of the chip. Interconnect is also used as a general term for the wiring between devices on a die and the overall package. A *contact* is the electrical connection at the silicon surface between the devices in the silicon wafer and the first metal layer. *Vias* are openings that pass through the various dielectric layers to form an electrical pathway from one metal layer to the adjacent metal layer. A metal *plug* fills the vias to form the electrical connection (interconnect) between the two metal layers. These connections are illustrated in Figure 12.1.

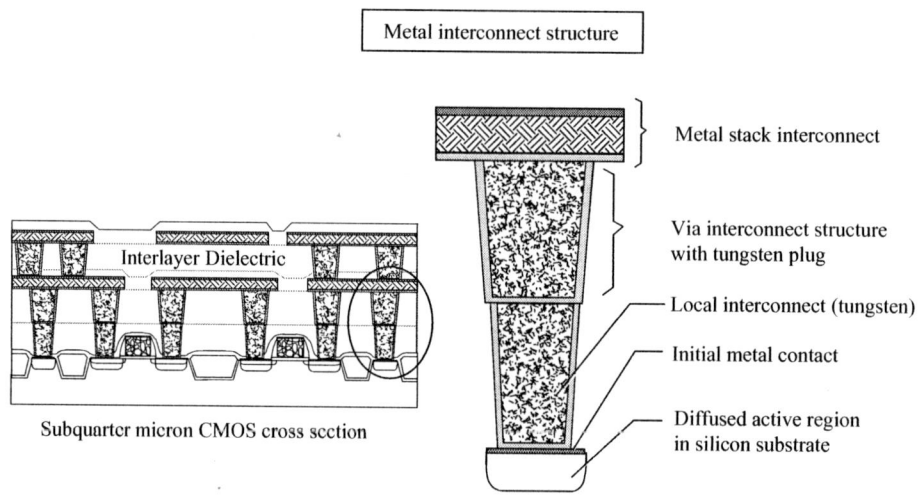

FIGURE 12.1 Multilevel Metallization

The interlayer dielectric (ILD) is an insulating material that electrically separates the metal levels. Once deposited, the ILD is patterned and etched to form via pathways for the various metal layers and the silicon. The vias are filled with a metal, traditionally tungsten (W), to form the via plug. Appreciate that there are many vias on a wafer, up to 10^{11} vias on each individual layer of a 300-mm product wafer.[1] This process of creating vias in the ILD is repeated for every ILD layer on the die. In traditional metallization, a blanket layer of aluminum alloy metal is deposited on the dielectric layer and then patterned and etched to form metal lines. Metal etch is an important technology for traditional metallization.

Metallization is in a transition period and is undergoing rapid change with the introduction of copper metallurgy to replace aluminum alloy. This change is due to the difficulty of etching copper. To overcome this problem, copper metallurgy uses the dual damascene process to form vias and copper interconnects (explained later in this chapter). This metallization process is literally the opposite approach of traditional metallization (see Figure 12.2). Damascene starts with deposition of a blanket dielectric, planarizes the dielectric, patterns and etches holes or trenches in the dielectric for vias and connecting metal lines, deposits blanket metal into the trenches, and then planarizes the metal down to the dielectric to define the metal interconnects.

Metallization technology is critical for increasing performance in advanced ICs, as explained in Chapter 11. Reducing chip performance signal delay caused by interconnect lines was not a significant concern for older IC technologies. The dominant signal delay has traditionally been caused by the device. However, for newer ULSI products manufactured with denser wiring, signal delay due to interconnect has become a larger portion of the clock cycle time and has more effect on limiting the IC performance.[2]

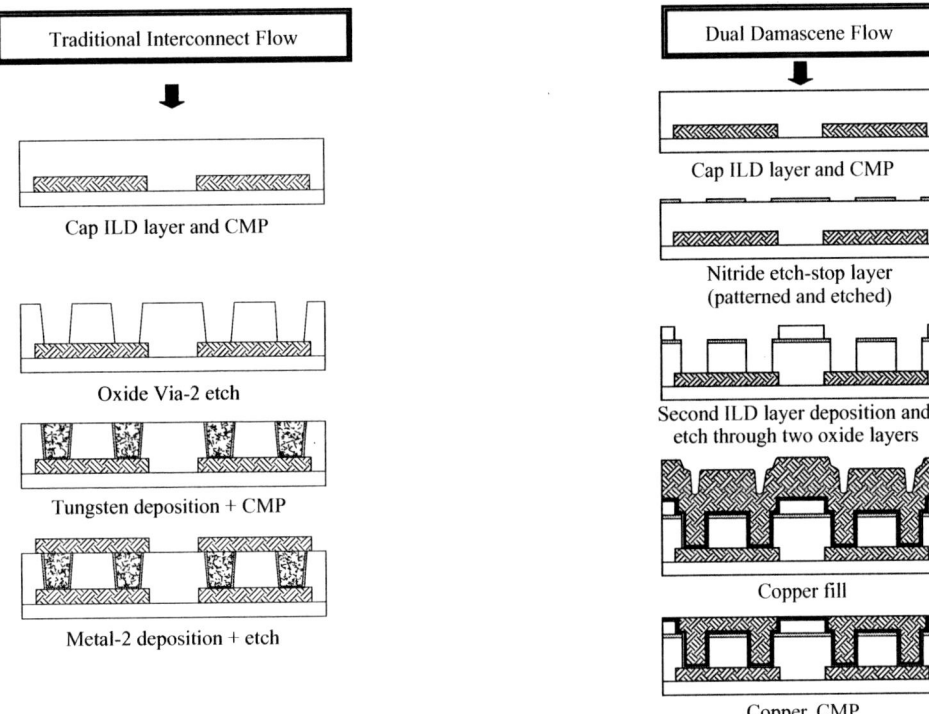

FIGURE 12.2 Traditional Versus Damascene Metallization

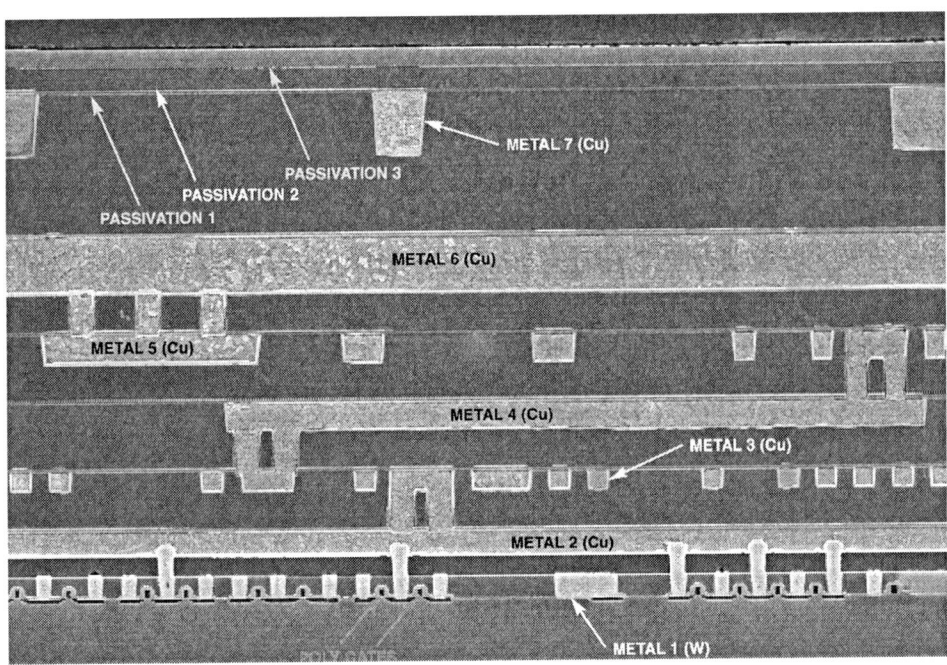

Copper Metallization
(Micrograph courtesy of Integrated Circuit Engineering)

TYPES OF METALS

The types of metals and metal alloys used for chip interconnects are evolving based on the performance requirements. Requirements for a successful metal material are:[3, 4]

1. **Conductivity:** Must be highly conductive and capable of handling high current densities while maintaining electrical integrity.
2. **Adhesion:** Ability to adhere to the underlying substrate and easily connect to external connections. Low contact resistance to semiconductor and metal interfaces.
3. **Deposition:** Readily deposited with a uniform structure and composition (for alloys) by a relatively low-temperature process. Deposition into high-aspect ratio gaps for damascene metallization technique.
4. **Patterning/Planarization:** High-resolution patterning for traditional Al-based metallization without etching the underlying dielectric. Ease of planarization for damascene metallization.
5. **Reliability:** The metal is relatively soft and ductile in order to withstand cyclic temperature variations during processing and service.
6. **Corrosion:** High resistance to corrosion with minimal chemical interactions with the adjacent layers and underlying device regions.
7. **Stress:** Resistance to mechanical stress to reduce wafer distortion and material failures for reasons such as cracking, void formation, and stressed-induced corrosion.

The melting temperature and resistivities of different metals found in wafer fabrication and silicon are shown in Table 12.1.[5]

TABLE 12.1 Silicon and Select Wafer Fab Metals (at 20°C)

Material	Melting Temperature (°C)	Resistivity (μΩ-cm)
Silicon (Si)	1412	$\approx 10^9$
Doped Polysilicon (Doped Poly)	1412	$\approx$ 500 to 525
Aluminum (Al)	660	2.65
Copper (Cu)	1083	1.678
Tungsten (W)	3417	8
Titanium (Ti)	1670	60
Tantalum (Ta)	2996	13 to 16
Molybdenum (Mo)	2620	5
Platinum (Pt)	1772	10

The various metals and metal alloys used in wafer fabrication can be grouped into the following categories:

- Aluminum
- Aluminum-copper alloys
- Copper
- Barrier metals
- Silicides
- Metal plugs

Aluminum

The earliest interconnect metal in semiconductor fabrication was aluminum, and it is still the most common interconnect metal in wafer fabrication. Aluminum is projected to continue as the dominant fab interconnect metal for several more years.[6] Copper interconnect metal is expected to replace aluminum for high-performance IC fabrication during the early 2000s. Nevertheless, it is beneficial to review the background into the selection of aluminum for metallization because many of the same basic technical challenges exist today.

Selection of Aluminum ■ Aluminum has been one of the major materials used in wafer fabrication, along with silicon and silicon dioxide. It is used during wafer fabrication in thin film form to interconnect the different devices in the silicon wafer (see Figure 12.3). At the same time, aluminum is one of the thickest films deposited on the wafer, with the first metal layer deposition about 5,000 Å thick. The upper noncritical layers on a wafer (e.g., metal layer with bonding pads) can range up to 20,000 Å thick.

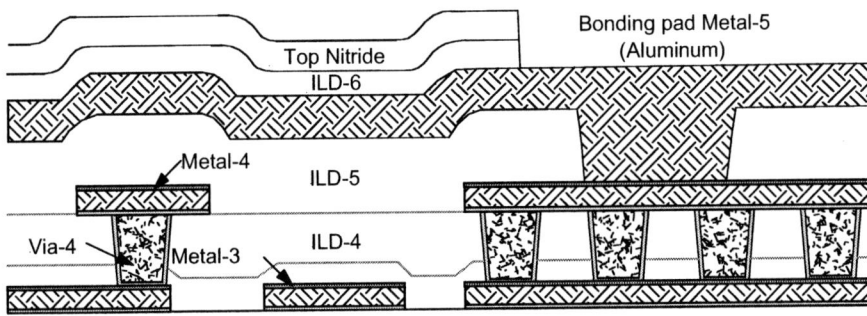

FIGURE 12.3 Aluminum Interconnect

Aluminum's low resistivity of 2.65 µΩ-cm at 20°C has similar but slightly higher resistivity than copper, gold, and silver. However, silver and copper are both prone to corrosion and have high diffusivity in Si and SiO_2, a fact that keeps them from being used for semiconductor fabrication. Gold and silver are much more expensive than aluminum and do not adhere well to oxide films. Gold was sparingly used in the beginning of wafer fabrication, but its high contact resistance with silicon required a platinum barrier metal. On the other hand, aluminum reacts readily with silicon dioxide (SiO_2) when heated to form a thin layer of aluminum oxide (Al_2O_3), which promotes adhesion between the silicon dioxide and aluminum. Aluminum is easily deposited on the wafer and etches in solutions that do not attack underlying films. For these reasons, aluminum was selected as the preferred metal for metallization.

Overall, aluminum is compatible with the major processes in silicon IC fabrication and is relatively inexpensive, making it the choice for metallization since the early days of IC fabrication. However, because of increased circuit density, increased number of metal layers on a wafer, and the reduction in linewidth due to scaling, the metallization technology has evolved from the simple one-level metal layer to a scheme of multiple metal layers. With its lower resistivity, copper is desirable to replace aluminum as the primary interconnect material.

Ohmic Contact ■ Before VLSI circuitry, pure aluminum was used for metallization. Silicon melts at a temperature of 1412°C and pure aluminum melts at 660°C. However, aluminum and silicon alloyed together actually have a lower melting temperature depending on their composition. For an alloy of 88.7% aluminum and 11.3% silicon, the alloy melts at 577°C, which is referred to as the *eutectic temperature*.[7] The eutectic temperature is the lowest temperature that an alloy melts at a particular eutectic composition.

To form a contact between aluminum and silicon, it is necessary to heat the interface, usually in an inert or reducing H_2 atmosphere at a temperature of 450 to 500°C. This thermal bake is also referred to as a low temperature *anneal* or *sinter*. Baking the aluminum on silicon forms a desirable electrical interface referred to as an *ohmic contact*. An ohmic contact has very low resistance (the voltage-current characteristics of the contact interface behave according to Ohm's law). However, there is a small resistance associated with the ohmic contact and it is inversely proportional to the area of the contact. That is, smaller contacts have higher resistance. In modern chip design, ohmic contacts use a special refractory metal (e.g., titanium in the form of a silicide) as a contact at the silicon interface to reduce electrical resistance and improve adhesion (see Figure 12.4 on page 304). Ohmic contacts are fabricated with a salicide (self-aligned silicide) process to achieve optimal positioning above the source/drain and in close proximity to the gate structure. The salicide process is discussed later in this chapter.

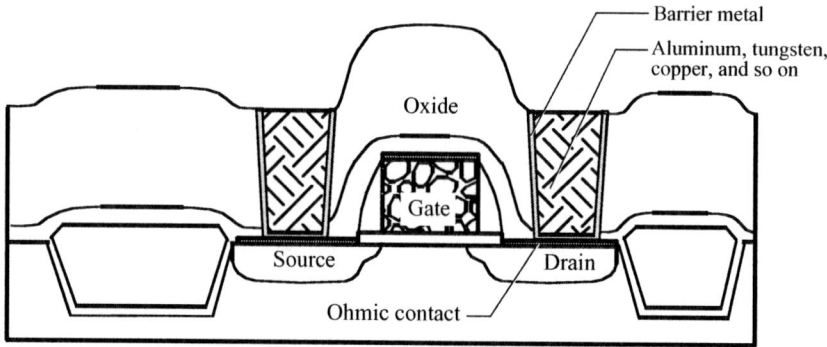

FIGURE 12.4 Ohmic Contact Structure

Given the hundreds of millions of contacts that can exist on a particular die, a reliable contact interface with low resistance and excellent adhesion is extremely important for optimum electrical performance. A failure in any individual contact can cause the entire chip to fail during test or use.

A problem encountered during the earliest work to develop ohmic contacts was undesirable interaction between the aluminum and silicon during the thermal step. This condition led to the microalloying of the contact metal to the silicon, and was referred to as *junction spiking*. Spiking occurs when the interface of pure aluminum and silicon is heated (see Figure 12.5), causing the diffusion of silicon into the aluminum. The amount of silicon dissolved into the aluminum is not uniform and will depend on the time and temperature involved in the heating process. If pure aluminum is heated to 450°C and a source of silicon is present, silicon will begin to dissolve in the aluminum until it reaches a concentration of about 0.5%.[8] The problem is that the source of this silicon is the wafer. As the silicon is dissolved in the aluminum, it leaves behind voids in the wafer, permitting spikes of aluminum to form and penetrate into the silicon contact region. If the aluminum forms an ohmic contact to a shallow junction, the spike may cause a junction short.

The problem of junction spiking was addressed by two methods: the addition of silicon to the aluminum and barrier metallization. The first approach was to use an alloy of aluminum and silicon instead of pure aluminum. If there is already silicon in the aluminum, this will slow down the dissolution of additional silicon from the substrate. However, silicon alloying in aluminum is limited and can lead to silicon nodule formation (small regions of high silicon concentration) in the aluminum due to silicon condensation. Nodule formation can significantly increase the contact resistance and pose a serious reliability concern from local heating at the nodules. The main method to resolve junction spiking was the introduction of barrier metallization to inhibit diffusion (see the following section).

It is important that the oxide at the contact interface be as thin as possible. Wafers with exposed contact regions are often dipped in a dilute hydrofluoric (HF) acid solution immediately before placing the wafers in the deposition chamber. This serves to remove the native oxide layer prior to contact formation.

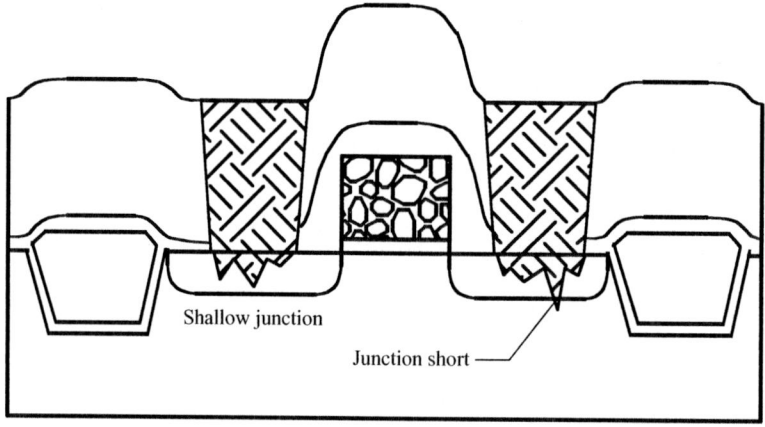

FIGURE 12.5 Junction Spiking

Aluminum-Copper Alloys

Aluminum metal was selected as the primary interconnect material for ICs because of its low resistivity and its compatibility with silicon and the wafer fabrication process. However, aluminum suffers from a reliability problem known as *electromigration*. This is the movement of aluminum atoms in the conductor due to momentum transfer from the electrons carrying the current.[9] Under high current-density conditions, electrons collide with aluminum atoms, causing the atoms to gradually move. This movement of atoms leads to a depletion of atoms at the negative end of the conductor. In conductor regions where depletion occurs, this action leads to voids, thinning of lines, and a potential open circuit. In other regions of the conductor that have an accumulation of metal atoms, metal atoms pile up and form hillocks (see Figure 12.6).[10] *Hillocks* are protrusions on the surface of metal films due to electromigration. If excessive or large hillocks form, adjacent lines or lines on two levels can short together. With advanced circuitry designs in ULSI technology, there is increased current density and chip temperature, both of which make the aluminum chip metallization more prone to electromigration.

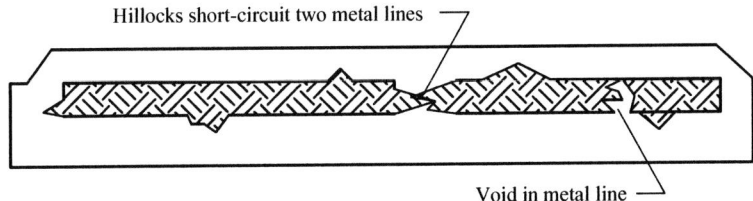

FIGURE 12.6 Hillock on a Metal Line

Electromigration in interconnects is controlled by alloying the aluminum with between 0.5 to 4% copper. This alloying essentially increases the conductor's current-carrying capacity by reducing the grain boundary diffusion effects in the aluminum. At the same time, it has been shown that values of copper in aluminum greater than 8% will actually increase electromigration.[11] The exact reason why the addition of copper reduces the chance of electromigration failure in Al-Cu alloys is not well understood.[12] If copper is alloyed with aluminum, more care must be given to plasma etching this alloyed aluminum (etching is discussed in Chapter 16). Copper is difficult to etch, and any residual copper remaining after etching the aluminum alloy interconnect can promote corrosion.[13] It is also possible to have electromigration in contacts and vias. Electromigration failure in contacts is resolved by the use of barrier metallization.

Electromigration is probably one of the most widely studied failure mechanisms of integrated circuits. Studies of this failure began in the 1950s with the development of the semiconductor and intensified in the late 1960s with the discovery of crack formation in the aluminum conductor leading to failures.[14] The significance of electromigration is that it usually occurs after the chip has been in service for an amount of time, which means that failure occurs catastrophically during customer usage.

Copper

A major transformation is underway in IC design and fabrication by means of the introduction of *copper interconnect* technology. Copper interconnects will replace aluminum metallization with copper to achieve significant benefits in chip performance. A legitimate question that someone may pose is why introduce copper if aluminum has performed so admirably over the years.

Need for Copper ■ The benefits for introducing copper for IC interconnect metallization are:[15]

1. **Reduction in resistivity.** The interconnect wiring resistivity reduces from 2.65 μΩ-cm for aluminum to 1.678 μΩ-cm at 20°C for copper, reducing RC signal delay and increasing chip speed.
2. **Reduction in power consumption.** Narrower lines consume less power.

3. **Tighter packing density.** Narrower lines permit tighter circuitry packing, which means that fewer levels of metal are needed.
4. **Superior resistance to electromigration.** Copper does not have a concern for electromigration.
5. **Fewer process steps.** Potential for 20 to 30% fewer processing steps with damascene processing of copper.

As wafer fabrication design rules reduce to 0.15-μm linewidths and below, the increased packing density of devices on the chip permits more electrical signal speed from device to device (the transistors are closer; therefore, the signal has less distance to travel). This density leads to improved chip performance. However, this improved chip performance is only possible if the interconnect system between the devices is optimized. Narrower linewidths lead to increased line resistance. Tightly spaced conductor lines with a dielectric material between them act as capacitors, leading to a degradation in performance from an increase in resistance (R) and capacitance (C). This condition is the signal delay or interconnect delay discussed in Chapter 11. If either or both of these two parameters are reduced, then the signal delay reduces, leading to increased chip performance.

One possible method reducing interconnect resistance is to increase the conductor cross section. However, this contradicts the goal of increased packing density since wider conductors will take substantially more space. Smaller IC feature sizes are not achievable with larger linewidths. This is why the semiconductor industry is placing emphasis on copper—to lower the interconnect resistance, which lowers R and therefore the overall signal delay. The optimum improvement to RC signal delay is gained when resistance (R) is reduced and the capacitance (C) is lowered by using a low-k dielectric along with thinner barrier metals (see Table 12.2).[16]

TABLE 12.2 Change in Interconnect Delay Compared to 0.25-μm Device Generation

Technology	0.25 μm	0.18 μm	0.13 μm
Conventional Interconnect Technology: • Al/Cu interconnect alloy and TiN barrier metal	0	+21%	+93%
New Technology Introduced by Generation: • Reduced barrier thickness • Low-k (3.0) dielectric • Dual damascene Cu interconnect and plugs	−10%	−27%	−16%

Another benefit from the implementation of copper is that smaller linewidths can carry the same amount of current, permitting a tighter packing density on each metal level. This condition produces fewer overall levels of metal on a chip, leading to significantly reduced manufacturing costs.[17]

Copper is a relatively soft metal. It has superior resistance to electromigration, a common reliability problem with aluminum.[18] This means that chips fabricated with copper can handle higher electrical power densities, which permits the development of new product applications. Table 12.3 compares Cu and Al for different properties and processes common in wafer fabrication.[19]

TABLE 12.3 Comparison of Properties/Processes Between Al and Cu

Property/Process	Al	Cu
Resistivity (μΩ-cm)	2.65 (3.2 for Al—0.5% Cu)	1.678
Electromigration resistance	Low	High
Corrosion resistance (in air)	High	Low
Etch processing	Yes	No
CMP (chemical mechanical planarization) processing	Yes	Yes

Copper Challenges ■ There are three major challenges involved with using copper as semiconductor interconnects that distinguish it from traditional aluminum interconnects. These three challenges differ significantly from aluminum technology and must be resolved before the implementation of copper for IC interconnects:[20]

1. Copper diffuses quickly into oxides and silicon. This is a concern if copper diffuses into the active region of the silicon (i.e., source/drain/gate region of the transistor) because it will damage the device by creating junction or oxide leakage.
2. Copper cannot be easily patterned using regular plasma etching techniques (see Chapter 16 for a discussion of etching). Copper dry etching does not produce a volatile by-product during the chemical reaction that is necessary for economical dry etching.
3. Copper oxidizes quickly in air at low temperatures (<200°C) and does not form a protective layer to stop further oxidation.

These challenges are addressed by converting to dual damascene processing with special barrier metals optimized for copper. Damascene processing eliminates the need to etch copper. Furthermore, tungsten plugs are expected to be used as the first level of metal to contact the source, drain, and gate regions.[21] This use of tungsten overcomes the concern of copper contaminating the silicon (referred to as copper poisoning). The tungsten may even be patterned for wiring of the local interconnect (LI). All other metal lines and vias for the multiple metal layers are expected to be copper.

Another major hindrance to converting to copper in semiconductor manufacturing is the natural reluctance to introducing new materials into production. New materials means there are new sources of contamination, new equipment and procedures, unpredictable results, and so on. Change always brings unforeseen problems that pose increased risks to the semiconductor manufacturer. However, these risks are balanced by the benefits from changing the interconnect material in order to improve IC performance.

Barrier Metals

A more effective way to make reliable ohmic contacts to shallow junctions without material diffusion or problems such as junction spiking is to use barrier metallization. A *barrier metal* is a thin layer of deposited metal or metals that is designed to prevent intermixing of the materials above and below the barrier (see Figure 12.7). The thickness of barrier metals has typically been around 100 nm for features sizes in the 0.25-μm generation, which is substantially thinner than the 400 to 600 nm thickness required for 0.35-μm generation of devices. Thickness of barrier metals is projected to decrease to 23 nm or less for the 0.18-μm technology node and beyond.[22]

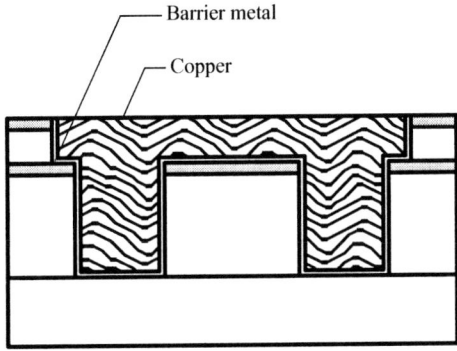

FIGURE 12.7 Barrier Layer for Copper Interconnect Structure

Barrier metals are widely used in semiconductor manufacturing. For contacts between the tungsten (W) plug that connects the aluminum interconnect wiring and the silicon source/drains, the barrier metal prevents the silicon and W from coming into contact with each other. This barrier prevents diffusion between the W and silicon and any junction spiking.

The essential properties of an acceptable barrier layer metal are:

1. Good diffusion barrier properties so that the diffusivity of the two interface materials (e.g., tungsten and silicon) is low at the sintering temperature (*sintering* refers to joining the materials by thermal means).
2. High electrical conductivity with low ohmic contact resistance.
3. Good adhesion between the semiconductor and the metal.

4. Resistance to electromigration.
5. Stability when thin and at high temperature.
6. Resistance to corrosion and oxidation.

Metals commonly used for barrier metals are a class of high melting point metals known as *refractory metals*. Common refractory metals used in multilevel metallization for wafer fabrication are titanium (Ti), tungsten (W), tantalum (Ta), molybdenum (Mo), cobalt (Co), and platinum (Pt). Refractory metals have been used in wafer fabrication since the 1960s, such as in the formation of Schottky barrier diodes in bipolar processes. The benefits of using titanium as a barrier metal for aluminum alloy wiring are improved adhesion, reduced contact resistance, reduced stress, and controlled electromigration. To achieve good barrier properties, wafers undergo a clean step (referred to as *sputter etch* and explained in a later section) before deposition in the processing chamber to remove native oxides and oxide residues on the wafer.

Titanium tungsten (TiW) and titanium nitride (TiN) are two common barrier metal layers that inhibit diffusion between the silicon substrate and aluminum. TiN is widely used in ULSI manufacturing for its superior barrier performance in aluminum alloy interconnect processing. TiN functions as a barrier metal for both tungsten and aluminum. TiN is also widely used as an antireflective coating on aluminum to improve the photolithography patterning process (see Chapter 14). However, TiN does not produce good contact resistance to silicon. To resolve this problem, a thin layer of titanium (e.g., several hundred angstroms or less) is typically deposited before the TiN so that it can react with the underlying material as a silicide (see the following section for an explanation of silicide) and lower the contact resistance. The Ti and TiN are typically deposited in a cluster tool to keep oxides from forming between the two layers.

Copper Barrier Metals ■ Barrier metals are critical for copper metallurgy. Cu has high diffusivity in silicon and silicon dioxide, which will destroy device performance. Traditional barrier metals are not sufficient barriers for copper. Cu requires complete encapsulation by a thin-film barrier layer that functions as an adhesion promoter and effective diffusion barrier.[23] There is a trade-off between these two requirements since good adhesion requires some reaction with Cu, whereas a good barrier metal should not react with Cu. The special barrier metal requirements for copper are:[24]

1. Prevent copper diffusion.
2. Low film resistivity.
3. Good adhesion to both dielectric material and copper.
4. Compatible with chemical mechanical planarization (CMP).
5. Metal layer is continuous and conformal with good step coverage and deposition in high aspect ratio gaps.
6. Minimal thickness to allow the copper to occupy the maximum cross-sectional area.

For Cu interconnect metallurgy, tantalum (Ta), tantalum nitride (TaN), and tantalum silicon nitride (TaSiN) are candidate materials for barrier metals (see Figure 12.8). The diffusion barrier must remain thin (about 75 Å) so that it does not affect the resistivity of the high-aspect ratio plug while still acting as a barrier metal. This condition is difficult to maintain as geometries continue to shrink and metals are deposited into high-aspect ratio vias. Research has shown Ta has good barrier and adhesion properties for copper, whereas TiN, a traditional barrier metal for Al/SiO_2 interconnects, has good barrier properties but poorer adhesion.[25] If TaN is used, it is obtained by small amounts of nitrogen doping or by deposition with a tantalum nitride compound. Investigations have also shown tungsten nitride (WN) functions as an effective barrier for Cu metal.[26] There is ongoing research in the development of barrier metals for Cu metallurgy.

Copper barrier layers can be deposited by using high-density plasma CVD (HDPCVD) or ionized metal plasma physical vapor deposition. Ionized metal plasma PVD for tantalum achieves good step coverage. For deposition with high aspect ratio gaps, then HDPCVD barrier deposition is often the choice.[27]

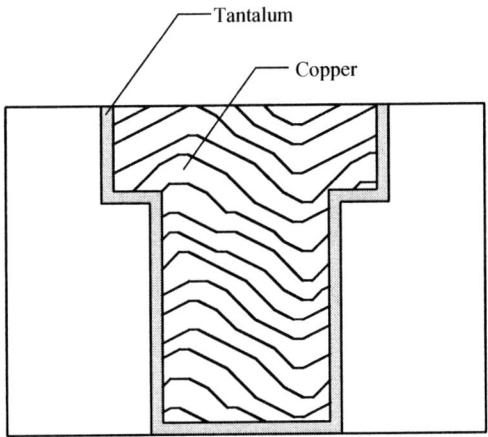

FIGURE 12.8 Ta for Copper Barrier Metal

Silicides

Refractory metals react with silicon when alloyed together to form a silicide. A *silicide* is a metal compound that is thermally stable and provides for low electrical resistivity at the silicon/refractory metal interface (see Figure 12.9). Refractory metal silicides are important in wafer fabrication because of the need to reduce the electrical resistance of the many silicon contacts in the source/drain and gate regions for chip performance. Titanium and cobalt are common refractory metal used for contacts in aluminum interconnect technology.

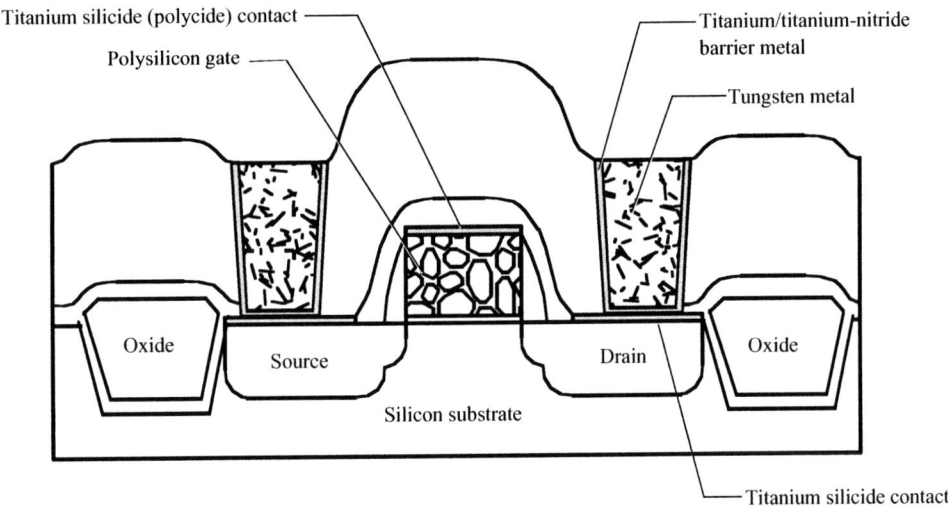

FIGURE 12.9 Refractory Metal Silicide at a Silicon Contact

If the refractory metal is reacted with polysilicon, it is called a *polycide* (see Figure 12.10 on page 310). Doped polysilicon is used as the gate electrode and has a relatively high resistivity (about 500 $\mu\Omega$-cm), which leads to an undesirable RC signal delay. A polycide is beneficial for reducing the series resistance of an interconnection to polysilicon. At the same time, it maintains polysilicon's good interfacial characteristics to oxides.

Silicides improve contact resistance by reducing residual oxides on the silicon surface during silicide formation. Silicides form extremely good metallurgic contact to the silicon, serving as a critical adhesion layer between the contact metal and the silicon junction regions. Many silicides are stable at temperatures exceeding 1000°C with a relatively high eutectic temperature. Table 12.4 on page 310 lists some properties of common silicides used in wafer fabrication.[28] Recall that the eutectic temperature is the lowest temperature where the alloy melts. It is undesirable for the silicide to melt because the liquid alloy can extend into the substrate material and cause junction spiking.

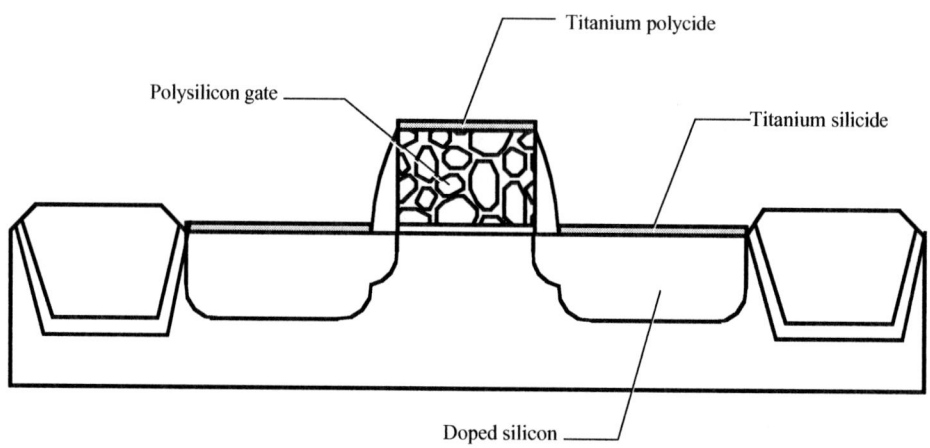

FIGURE 12.10 Polycide on Polysilicon

TABLE 12.4 Some Properties of Select Silicides

Silicide	Lowest Eutectic Temperature (°C)	Typical Forming Temperature* (°C)	Resistivity (μΩ-cm)
Cobalt/silicon (CoSi$_2$)	900	550–700	13–19
Molybdenum/silicon (MoSi$_2$)	1410	900–1100	40–70
Platinum/silicon (PtSi)	830	700–800	28–35
Tantalum/silicon (TaSi$_2$)	1385	900–1100	35–55
Titanium/silicon (TiSi$_2$)	1330	600–800	13–17
Tungsten/silicon (WSi$_2$)	1440	900–1100	31

* B. El-Kareh, *Fundamentals of Semiconductor Processing Technologies*, (Boston: Kluwer Academic Publishing 1995), p. 537.

Silicide formation typically requires the refractory metal be deposited on the silicon wafer, followed by a high temperature thermal anneal process to produce the silicide material. In areas where there is silicon, the metal reacts to form a silicide. In other areas on the wafer surface, such as where there is SiO$_2$, there is little or no silicide formation. Often this thermal anneal step uses a rapid thermal anneal (RTA) process in a multichamber cluster tool.

TiSi$_2$ has traditionally been the most common contact silicide for wafer fabrication, serving as the contact between the transistor active regions of silicon and the tungsten plug. It is often referred to as the glue that holds the tungsten to the silicon. Its beneficial properties are high temperature stability, compatibility with self-aligned contact processing (discussed in the following section), low resistivity when compared to other silicides, and its compatibility with TiN barrier metals.[29] TiSi$_2$ forms two different grain phases (a phase is a physically homogeneous material state) as it is annealed, a low-temperature C49 phase and a high-temperature C54 phase (see Figure 12.11). Both of these phases are TiSi$_2$.

The C49 phase of TiSi$_2$ forms at an anneal temperature of 625 to 675°C with a resistivity of 60 to 65 μΩ-cm. The C54 phase is formed in a second anneal following the C49 phase formation and requires a temperature of 800°C or more. It has a much lower resistivity of only 10 to 15 μΩ-cm, which is desirable to lower the overall contact resistance.[30]

However, at the time of this writing, TiSi$_2$ appears limited in its future use as a contact silicide in sub-0.25 μm technologies. Contacts for ultrashallow source/drain junctions are becoming thin. TiSi$_2$ is undesirable for thin contacts because the silicide resistivity increases.[31] It is also difficult to form this contact during the second anneal because of critical time at temperature requirements.

The silicide that appears promising for the 0.18 μm and below technology nodes is cobalt silicide (CoSi$_2$). This silicide maintains a reduced contact resistance of 13 to 19 μΩ-cm after its

	Sintering Temperature	Resistivity
TiSi$_2$ – C49	625 – 675°C	60 – 65 µΩ-cm
TiSi$_2$ – C54	800°C	10 – 15 µΩ-cm

FIGURE 12.11 Anneal Phases of TiSi$_2$

anneal operation, even with deep submicron geometries of 0.18 µm and less. This reduced resistivity occurs because the grain size of CoSi$_2$ is about ten times smaller than the grain size of TiSi$_2$. Therefore, the low resistance phase is completely nucleated and grown during the thermal anneal operation. A CoSi$_2$ contact is also easier to form because of the smaller grain size.

Note that a silicide is not a barrier metal. In some silicides it is found that silicon diffuses rapidly through them. This diffusion occurs during heat treatment of the metal-silicide-silicon system where silicon diffuses through the silicide to the metal, which then reduces the integrity of the system. The solution to this problem is to place a metal barrier layer between the silicide and the metallization layer (discussed in the preceding section). A common silicide barrier film is titanium nitride (TiN), which is effective for both tungsten and aluminum. A tantalum-based barrier layer is anticipated for copper metallurgy.

Salicide ■ Because of the need to scale device sizes in VLSI and ULSI to optimize performance, the cross-sectional area of the source/drain (S/D) contact between the silicon and the first metal layer is small. This small size leads to increased resistance. A technique to decrease contact resistance at the source and drain areas by providing stable contact structures is referred to as a salicide. A *salicide* approach (taken from the expression *self-aligned silicide*) is used to create silicides that are properly aligned with the exposed silicon of the source, drain, and polysilicon gate. There are numerous chip performance issues that depend on salicide formation (see Figure 12.12).

The basic salicide steps are shown in Figure 12.13 on page 312. This process flow corresponds to Step 6 of the CMOS process flow described in Chapter 9. To form a salicide, oxide has been previously deposited and etched back with a dry plasma etch to leave oxide sidewall spacers

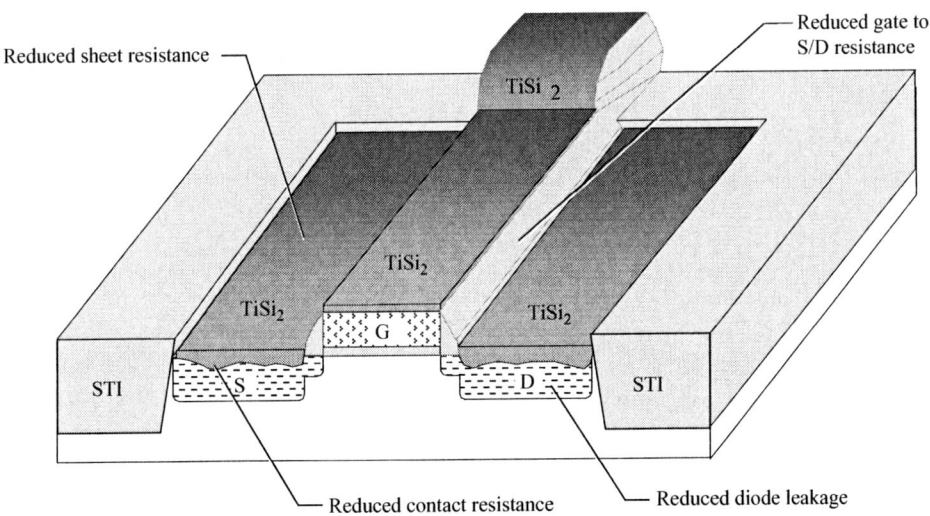

FIGURE 12.12 Chip Performance Issues Related to a Salicide Structure

on the side of the polysilicon gate (Step 5 of the CMOS process flow described in Chapter 9). With the sidewall spacers, only the top poly surface of the gate is exposed. After an HF-dip clean step to remove native oxides, a 250 to 350 Å thin layer of titanium metal is deposited on the wafer. The refractory metal undergoes an RTP rapid anneal at 600 to 800°C to create the high-resistivity C49 titanium silicide phase wherever the refractory metal contacts silicon. The sidewall spacer keeps the S/D $TiSi_2$ from shorting out to the sides of the gate poly, while the oxide of the shallow trench isolation separates the devices. After the first RTP anneal, all unreacted titanium is removed by a wet chemical etch in ammonium hydroxide (NH_4OH) and hydrogen peroxide (H_2O_2). $TiSi_2$ remains and covers the S/D areas and top surface of the poly gate. A second RTP silicide anneal at 800 to 900°C creates the low-resistivity C54 metallic silicide. A major benefit of the salicide process is the avoidance of alignment tolerances that would occur if patterning were required to align the refractory metal on the silicon contact. Care must be taken during the anneal steps to avoid oxygen contamination in the furnace tube. Titanium reacts easily with oxygen to form undesirable titanium oxides. This oxide contamination can promote silicide formation on top of the oxide spacer regions, leading to a short between the polysilicon gate and source or drain.

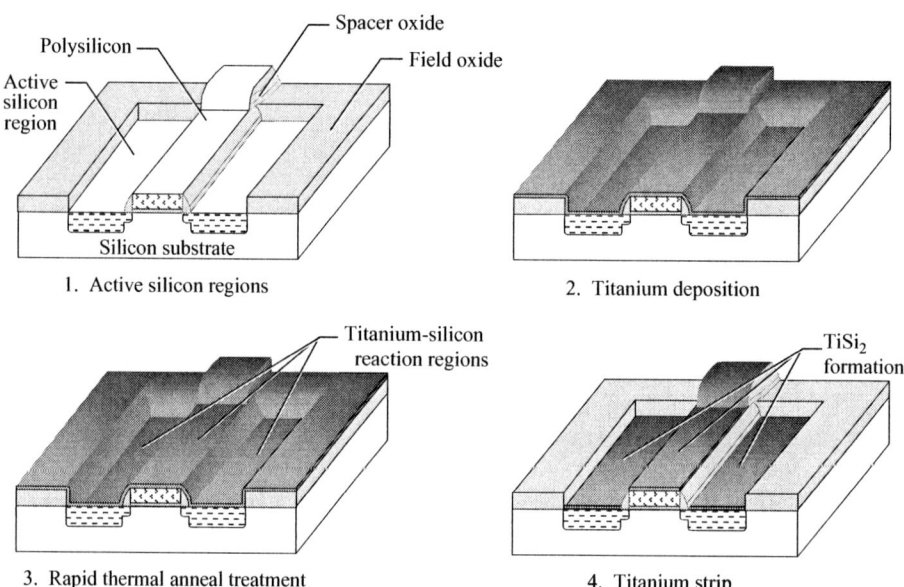

FIGURE 12.13 Formation of Self-Aligned Metal Silicide (Salicide)

Metal Plugs

Multilevel metallization creates the need for billions of vias filled with metal plugs to form electrical pathways between two metal layers. A contact plug is also used to connect the silicon devices in the wafer to the first level of metallization. The most common metal currently used for plugs is tungsten (W), which is why plugs are often referred to as *tungsten plugs* (see Figure 12.14). Tungsten has been traditionally used as a plug material because of its ability to uniformly fill the high-aspect ratio vias when deposited by chemical vapor deposition (CVD) methods. Tungsten is resistant to electromigration failure. It also serves as a barrier to inhibit diffusion and reaction between Si and the first metal layer. Tungsten is a refractory material with a melting point of 3,417°C and a bulk resistivity of 52.8 $\mu\Omega$-cm at 20°C.

Aluminum would be desirable as a plug material because of its lower resistivity (2.65 $\mu\Omega$-cm at 20°C), but sputtered aluminum cannot fill high-aspect ratio vias (see the next section for a description of sputtering). For this reason aluminum is used as the interconnect material and tungsten is limited to the plug. There has been recent interest in reflow aluminum plugs, where aluminum is sputter deposited into the via and then reflowed by a high-temperature anneal with a rapid thermal processor (RTP).[32]

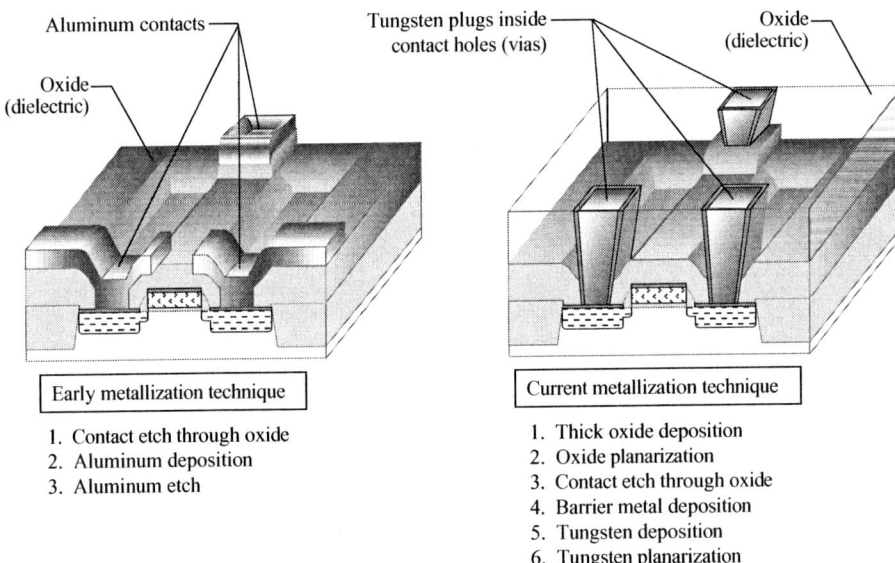

FIGURE 12.14 Tungsten Plug for Multilevel Metal Layers

METAL DEPOSITION SYSTEMS

The traditional metallization techniques used in semiconductor manufacturing fall under a category called *physical vapor deposition* (*PVD*). PVD has been done with filament evaporation, followed by electron beam evaporation and, most recently, sputtering. Chemical vapor deposition (CVD) has recently become a frequently used technique to deposit metal films. Each change in the deposition system has brought improved control of thin film properties and quality.

During the SSI and MSI eras of semiconductor manufacturing, evaporation was the method for metallization. It was replaced by sputtering primarily because evaporation has poor step coverage. Evaporation is still used in research applications and for semiconductors using III–V technologies. It is also used in some specialized areas, such as for C4 bump deposition during packaging (see Chapter 20).

Electroplating technology has been employed in various applications. In recent years it has been used for thin film head metallization in the disk drive industry. However, electroplating is just now being introduced for semiconductor manufacturing as a copper deposition method. It is the early deposition method of choice for creating copper interconnects.

The different metal deposition systems used in traditional and dual damascene metallization are:

◆ Evaporation
◆ Sputtering
◆ Metal CVD
◆ Copper electroplate

Evaporation

In the early days of semiconductor manufacturing, all metal layers were deposited by the PVD method of *evaporation.* Since the late 1970s, sputtering has replaced evaporation in most silicon wafer technologies in order to gain improved step coverage, gap fill, and yield. It is beneficial to review evaporation to understand how it functions and why the silicon industry changed to sputtering.

The process of evaporation consists of placing the material to be deposited in a crucible and heating it inside a vacuum chamber until it vaporizes (see Figure 12.15 on page 314). The most typical method of heating was the use of an electron beam to heat the metal placed in a crucible. By maintaining a high vacuum in the evaporator, the mean free path of the vapor molecules is increased, and the vapor travels in a straight line in the chamber until it strikes a surface and condenses to form a film.

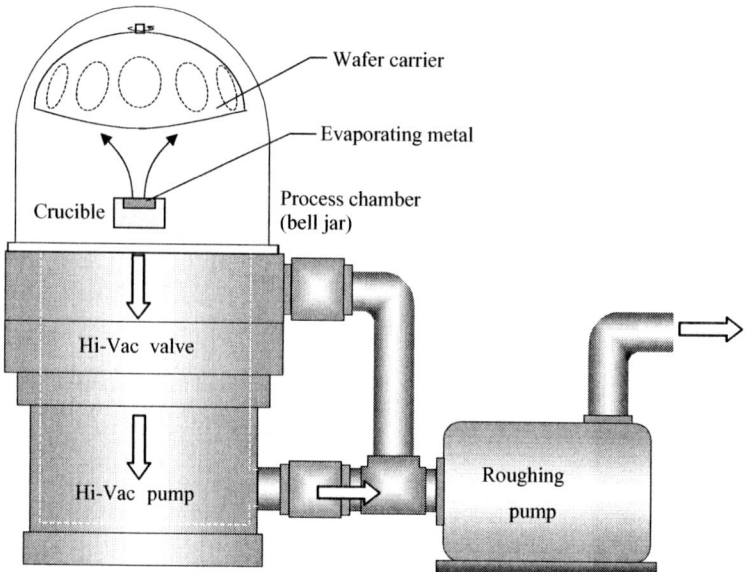

FIGURE 12.15 Simple Evaporator

The biggest drawback to evaporation was the inability to produce uniform step coverage. As the industry progressed to VLSI and ULSI technology, metallization needed to be capable of filling higher aspect ratio holes and producing conformal step coverage. Evaporators made some progress for step coverage by heating the wafers and rotating the wafers in the chamber with a hemispherical cage. However, evaporator technology is not able to form a continuous film over features with an aspect ratio greater than 1.0:1, and is marginal for an aspect ratio of 0.5:1 to 1.0:1. This drawback to evaporation led to its demise in IC production.

Another serious drawback of evaporation is its limitation for depositing alloys. The evaporator requires multiple crucibles for the different materials, which is a problem because of the different vapor pressures of the materials. This limitation makes it difficult to control the ultimate composition of the deposited alloy with any degree of accuracy.

Ultimately, the wafer fabrication industry quickly replaced evaporators with sputtering systems due to their improved step coverage capability. The evaporator is still used in research applications and for some compound semiconductor technologies that can actually use the poor step coverage of evaporators to their advantage during specialized processing. In addition, evaporators are still sometimes used in the chip packaging process to deposit solder bumps on the chip surface (see Chapter 20).

Sputtering

Sputtering, a form of physical vapor deposition (PVD), was discovered in 1852 by Sir William Robert Grove and developed as a thin film deposition technique in the 1920s by Langmuir.[33] As its name implies, sputtering is mainly a physical, rather than chemical, process. In the sputtering process, high-energy particles strike a solid slab of high-purity target material and physically dislodge atoms. These sputtered atoms migrate through a vacuum and eventually deposit on a wafer.

The advantages of sputtering are:[34]

1. Ability to deposit and maintain complex alloys.
2. Capability to deposit high-temperature and refractory metals.
3. Ability to deposit controlled, uniform films on large wafers (200 mm and larger).
4. Capability of multichamber cluster tools to clean the wafer surface for contamination and native oxides before depositing metal (referred to as *in situ sputter etch*).

While sputtering was a big improvement in gap fill over the earlier metallization method of evaporation, its limited capability for step coverage and filling high-aspect ratio gaps makes it insufficient for ULSI applications.[35] Over the years, improvements have been made to the sputtering process to obtain better step coverage, including recent advances with ionized metal plasma PVD. For critical applications, such as the step coverage in tungsten plugs, deposition will usually be done with a CVD metal process. Sputtering continues to be used for depositing critical barrier and seed layers, such as tantalum/tantalum nitride used in copper metallization (when the aspect ratio is reasonable).

Basic Sputtering Steps ■ There are six basic steps to sputtering:[36]

1. Positive argon ions are generated in a plasma in a high vacuum chamber and accelerated toward a target material at a negative potential.
2. During acceleration, the ions gain momentum and strike the target.
3. The ions physically dislodge (sputter) atoms from the target, which has the desired material composition.
4. The dislodged (sputtered) atoms migrate to the wafer surface.
5. The sputtered atoms condense and form a thin film on the wafer surface with essentially the same material composition as the target, following the stages for thin film growth outlined in Chapter 11.
6. Excess material is removed from the chamber by vacuum pump.

To illustrate sputtering, Figure 12.16 shows the basic configuration of a simple, DC diode sputter deposition chamber with parallel plates. It consists of the solid *target* material, a substrate (wafer), and a vacuum enclosure. The target is electrically grounded and referred to as the *cathode*, while the substrate has a positive potential and is termed the *anode*. The anode and cathode are both called *electrodes*. The sputtering target is composed of the necessary material for deposition. Targets are manufactured to achieve a homogeneous composition, fine grain size, and specific crystallographic orientation, all of which contribute to achieving a uniform film deposition rate across the wafer.[37] Target purity levels of 99.999% (5 nines) or better are required for achieving acceptable film purity for sub-0.25 µm geometries.

A high density of positive ions from an argon gas glow discharge is strongly attracted to the negative target material, striking it at high velocity and dislodging the atoms to be deposited. The atoms are sputtered (knocked off) from the target material and scatter in the chamber; eventually some come to rest on the wafer or the chamber walls. This action necessitates cleaning of the chamber in some systems. The atoms on the wafer nucleate and grow a thin film. An important aspect of sputtering is that this process can be used to sputter-deposit alloys, in particular aluminum-silicon

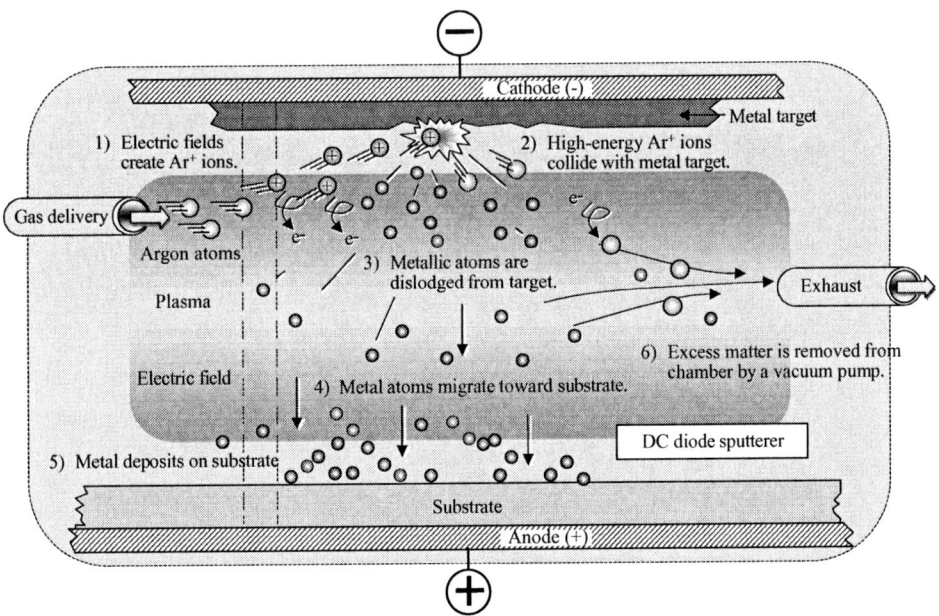

FIGURE 12.16 Simple Parallel Plate DC Diode Sputtering System

and aluminum-copper alloys. For instance, if the target material is aluminum with 1% copper, then the deposited film will be aluminum with 1% copper (at steady-state conditions).

The latest technology of sputtering equipment systems are multichamber cluster tools. The system has a loadlock that moves single wafers from the wafer cassette to its ultimate base vacuum pressure. The loadlock may consist of dual loadlocks to move the wafer through a staged vacuum sequence from ambient atmosphere to base vacuum pressure. A robot transfer system designed specifically for the cluster tool environment moves wafers from the various process chambers and a buffer chamber. It is important that the robot operate without generating particles and be extremely reliable. Cluster tool robots are often magnetically coupled, which permits the robot drive motor to remain on the outside of the process chambers to reduce particles.

Vacuum conditions are important in the sputtering chamber to create the plasma and maintain purity of the deposited films, with base vacuum typically at 10^{-7} torr during initial pump down. The rate that argon gas enters the process chamber is critical because it causes the pressure to rise in the chamber. With the argon and sputtered material in the chamber, the pressure rises to about 10^{-3} torr. The hot, high-vacuum environment of the process chamber promotes the formation of the sputtered atoms into thin films on the wafer surface.

Physics of Sputtering ■ An essential aspect of sputtering is the formation of the ionized argon gas into a plasma (see Chapter 8). Argon is used as the sputtering ion species because it is relatively heavy and is a chemically inert gas, which keeps it from reacting with the growing film or the target. If a high-energy electron strikes the neutral argon, the collision knocks off outer electrons and creates a positively charged argon ion. This energetic particle is used to strike the negatively charged target material to be sputtered.

Sputtering Mechanism. Positively charged argon ions in the plasma are strongly attracted to the negative potential of the cathode target. The charged ions accelerate and acquire kinetic energy (or energy of motion) as they pass through the voltage drop of the glow-discharge dark space (see Chapter 8 for a description of glow discharge). When the Ar ions strike the target surface, the momentum of the Ar ion transfers to the target material to dislodge one or more atoms. This action is referred to as *momentum transfer*. The ejected neutral atom or atoms move through the plasma (with a small chance of being ionized) and land on the wafer. The incident ion energy must be large enough to dislodge target atoms but not so large that the ion penetrates into the target material. Typical sputtering ion energies range from 500 to 5,000 eV.

An analogy for dislodging metal atoms from the surface of the target material during sputtering is hitting billiard balls in a pool game. Even though the cue ball in pool is travelling in one direction, the billiard balls can be knocked in some other direction. This same effect occurs in sputtering, where the argon ion impacts the target and dislodges one or more atoms from the surface of the target (see Figure 12.17).

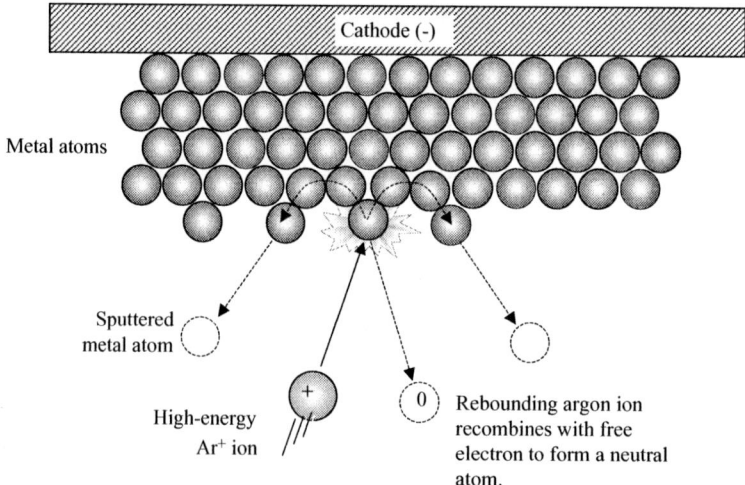

FIGURE 12.17 Dislodging Metal Atoms from Surface of Sputtering Target

The *sputtering yield* is defined as the number of atoms ejected by the target (cathode) per incident ion that strikes it. The yield largely determines the rate of sputter deposition. It varies from about 0.5 to 1.5. A sputter yield of 0.5 means that, on the average, two ions must strike the target for one atom to be ejected. The sputter yield depends on the following conditions:[38]

1. Incident angle of the bombarding ions.
2. Composition and geometry of the target material.
3. Mass of bombarding ions.
4. Energy of the bombarding ions.

A method of increasing the sputtering deposition rate is to confine the plasma to the region between the target and wafers. Due to the ionization process, there are dark spaces just to the front and the sides of the target. Dark space shields are placed on the sides of the target to prevent target material from being sputtered on the sides, since this material will never deposit on the wafers. Shields require periodic changing because sputtered target material builds up on the surface and causes particulate contamination.

Targets undergo erosion from the ion bombardment and are replaced when about 50% or more of the target material is removed.[39] Much of the energy consumed during the sputtering process is dissipated as heat in the target, or by emission of secondary electrons and photons by the target. For this reason, cooling of the target material is done to maintain a low target temperature.

In addition to being struck by the sputtered atoms, the substrate has other species landing on it (see Figure 12.18). These species cause heating of the substrate (up to 350°C) which leads to uneven film deposition. The high temperature can also create an undesirable aluminum oxide during aluminum deposition, which interferes with the sputtering process. The many species impinging on the wafer surface during diode sputtering also increase the possibility of damage due to radiation to sensitive devices.

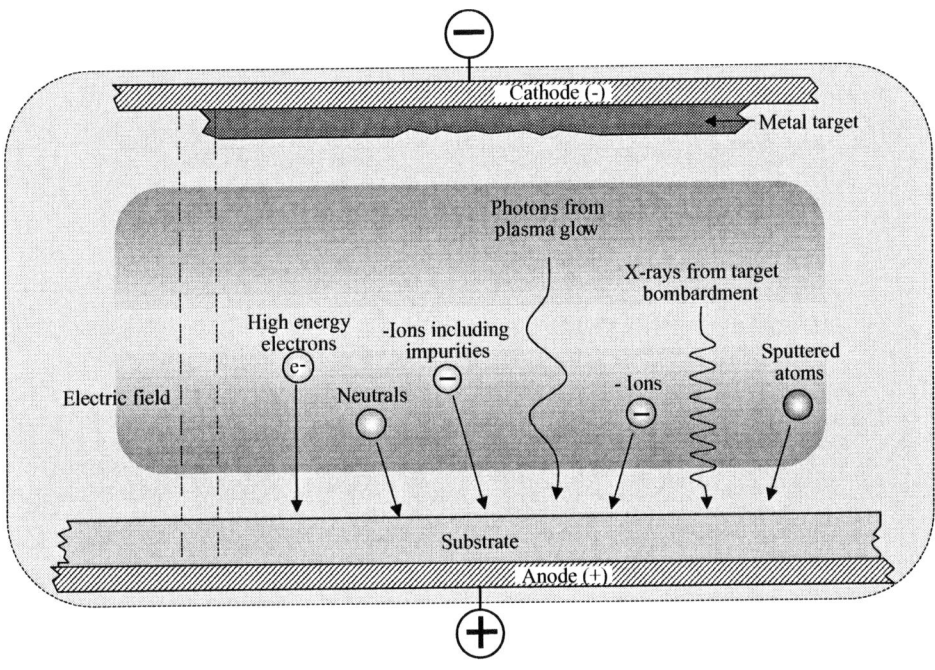

FIGURE 12.18 Different Species Landing on Substrate

Note the presence of the impurity gas atoms during diode sputtering. This is a problem if the impurity atoms are incorporated into the growing film on the substrate. The sources of contaminants are: (1) impure sputtering gas, (2) outgassing from the wafer holder, (3) outgassing from chamber walls, and (4) vacuum leaks.[40] It is necessary to use ultrahigh purity (UHP) sputter gases and to verify there are no chamber leaks.

The spacing between the cathode and anode must be optimized for each sputtering system. The goal is to have as many sputtered atoms as possible be deposited on the wafers and form the desired film.

The sputtering system described thus far is a simple DC diode system that has a DC voltage applied between two electrodes (cathode and anode). The DC diode sputtering system has serious drawbacks that limit its use in manufacturing. A DC diode type sputtering system cannot be used to sputter dielectrics because the electrodes become coated with the dielectric and the glow discharge cannot be sustained. The target rapidly builds up a positive charge, which repels incoming positive ions. DC diode type sputtering is also not capable of performing a sputter etch. *Sputter etch* (or *reverse sputter*) is a preclean step where the sputter process is reversed and argon atoms are used to remove thin native-oxide layers and remaining etch residues that contaminate contacts and vias. In other words, the wafer is sputtered instead of the target. This sputter etch preclean is important in multichamber cluster tools because of the benefit of cleaning followed by deposition without removing the wafer from the vacuum environment.

Three types of sputtering systems are covered in the following pages:

◆ RF (radio frequency)
◆ Magnetron
◆ IMP (ionized metal plasma)

The simple RF sputter system is not used in wafer fabrication because of its inherent inefficiency. The magnetron has traditionally been the most widely used sputter system, and ionized metal plasma (IMP) is becoming more common for sub-0.25 micron technology.

RF Sputtering ■ In an *RF sputtering* system, an RF field is used to create the plasma instead of the DC field described above. The RF frequency has commonly been 13.56 MHz. It is applied to the back surface of the target electrode and capacitively coupled to its front surface (see Figure 12.19). Both the electrons and ions in the plasma are exposed to the RF field, but due to the high frequency, the electrons respond most strongly. The chamber and electrode behave like a diode, creating a high amount of electron flow and resulting in a negative charge on the target electrode. This negative charge (produced by a self-bias) attracts positive argon ions, which sputter material from the insulator or non-insulator target.[41]

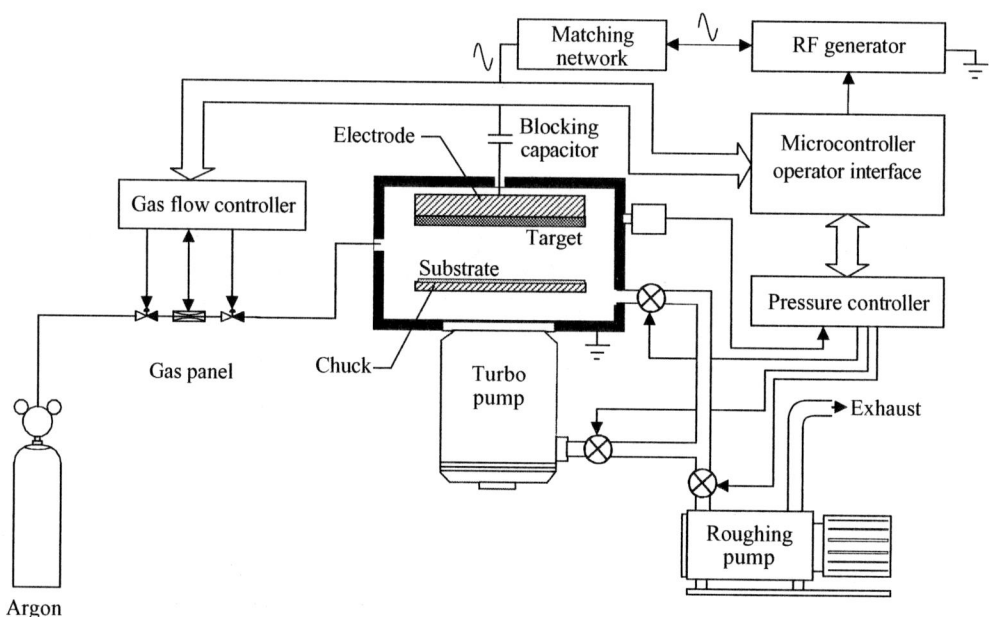

FIGURE 12.19 RF Sputtering System

The wafer can be electrically biased to put it at a different field potential than the argon. The bias applied to the wafer causes the argon atoms to strike the wafer directly. This RF biasing permits the exposed wafer surface to be sputter etched and cleaned.

In practice, an RF sputtering system is limited because it does not have a high sputtering yield, leading to a low deposition rate. Many of the secondary electrons emitted by the target pass through the discharge region without contributing to the creation of plasma. The sputtering rate of the target would be much higher if these secondary electrons were involved in ion collisions, leading to the creation of more ions to bombard the target. To overcome this inefficiency and to achieve high metal deposition rates necessary for wafer fabrication, the concept of magnetron sputtering was developed.

Magnetron Sputtering ■ *Magnetron sputtering* employs magnets configured around and behind the target to capture and restrict the electrons in front of the target (see Figure 12.20). This setup increases the ion bombardment rate on the target in order to produce more secondary electrons, which then increases the ionization rate in the plasma. The end result is that more ions cause more sputtering of the target, which increases the deposition rate of the system. The invention of magnetron sputtering to improve the deposition rate is one of the main reasons why sputtering has become the leading process for single-wafer deposition systems for aluminum and contact alloys used in metallization.

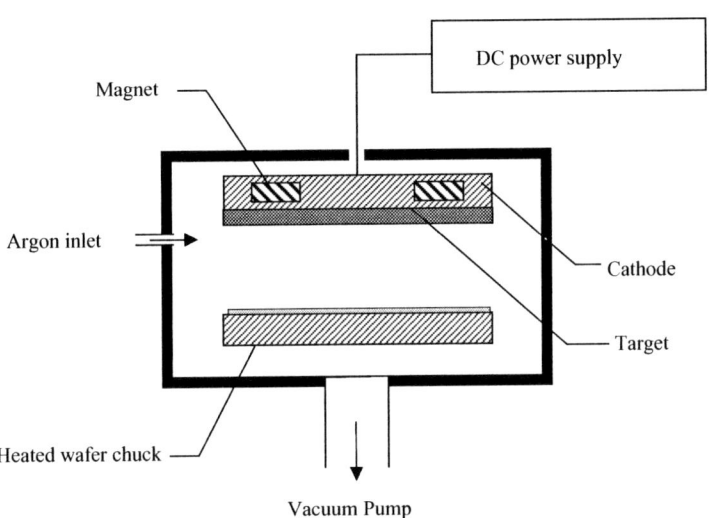

FIGURE 12.20 Magnetron Sputtering

The magnetron sputtering design requires a substantial amount of power supplied to the argon plasma to maximize the sputtering rate (about 3kW to 20 kW of power). Because the sputtering target absorbs most of this power and is in contact with the cathode, cooling of the cathode is required. An important challenge is uniformity of the sputtered deposit on large wafers. To achieve high deposition rates and film uniformity in single-wafer cluster tools, new cathodes have been developed with rotating, rare earth, high-strength permanent magnets.[42]

Step Coverage. Sputtering requires a vacuum environment with high-purity argon to avoid contamination from residual gas. The vacuum during the sputtering process is around 1 mtorr, with a mean free path of several centimeters. This is approximately the distance between the target and wafer. Because of this distance, atoms dislodged from the target essentially pass on to the wafer by following a line-of-sight path.

This line-of-sight path for the ejected atoms exiting at many different angles from the target causes poor coverage on steps and sidewalls in contacts and vias. Due to their geometry, the sidewalls and bottoms of high-aspect ratio contact windows and vias will receive only 10% or less of the metal deposited on the top surface.[43] For this reason PVD is not usually chosen when depositing a material on high aspect ratio steps and trenches.

Collimated Sputtering. To achieve increased coverage on the bottom and side of a contact or via, directional enhancement can be achieved by the use of *collimated sputtering* (see Figure 12.21). The collimator is configured so that it appears as the electrical ground for the plasma. In this manner, any neutral species sputtered from the target at a high angle is intercepted and deposited on the collimator. Other atoms that are ejected straight back from the target will pass by the collimator and deposit on the bottom of the contact hole. At the same time, a collimator reduces sidewall coverage in the contact or via.

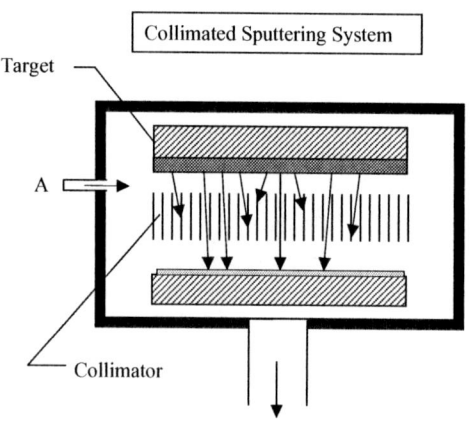
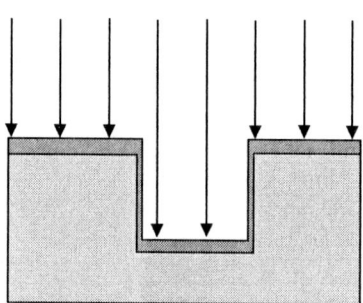

Cross section of via showing coverage of resulting sputtered film.

FIGURE 12.21 Collimated Sputtering

The use of a collimator means that a large portion of the sputtered material will not reach the wafer, since most of the sputtered material ends up on the collimator. This result lowers the sputter yield and increases the cost of deposition, since fewer sputtered atoms reach the wafer. At the same time, maintenance costs increase because the target is less efficient. Given these problems, it is more efficient to use ionized metal plasma (IMP) or CVD deposition over magnetron sputtering if step coverage is a critical factor.

Ionized Metal Plasma (IMP) ■ A problem for sputtering technology in wafer fabrication for high performance ICs is that as feature sizes shrink, the capability to sputter into high aspect ratio vias and narrow trenches becomes limited. A recent development in PVD to overcome this problem is *ionized metal plasma PVD (IMP* or *ionized PVD)*, which was introduced in the mid-1990s. In this approach, the sputtered metal is ionized in an RF plasma at a pressure of 20 to 40 mtorr (see Figure 12.22). The positive metal ions then travel in a highly directional, vertical path toward a wafer configured with a negative voltage bias. This voltage bias can also be used to control the energy of the incoming metal ions, which helps reduce damage to the surface of the wafer.[44] Biasing the wafer enables a higher degree of film conformality in the bottom and corners of high aspect ratio gaps.

Ionized PVD achieves good hole-fill for Ti and TiN on structures with 0.25-micron contacts and vias and 6:1 aspect ratios, with 70% bottom coverage, 10% sidewall coverage, and excellent sidewall integrity. This coverage is critical in the bottom, bottom corner, and lower sidewall areas of holes with a typical 85° sidewall angle.

Ionized PVD is used in production to deposit titanium and titanium nitride barrier films for devices at the 0.25 μm and below generations with high-aspect ratio contact and via structures.[45] It is also capable of depositing tantalum, tantalum nitride, and copper, which are important metals for copper metallization. The development of ionized PVD has further improved sputtering to extend its usefulness in high-aspect ratio gap fill.

Metal CVD

The use of chemical vapor deposition (CVD) for metal deposition is increasing because of its superior conformal step coverage and void-free filling of high-aspect ratio contacts and vias. These factors are critical for wafer fabrication as feature sizes decrease to the 0.15-μm generation and

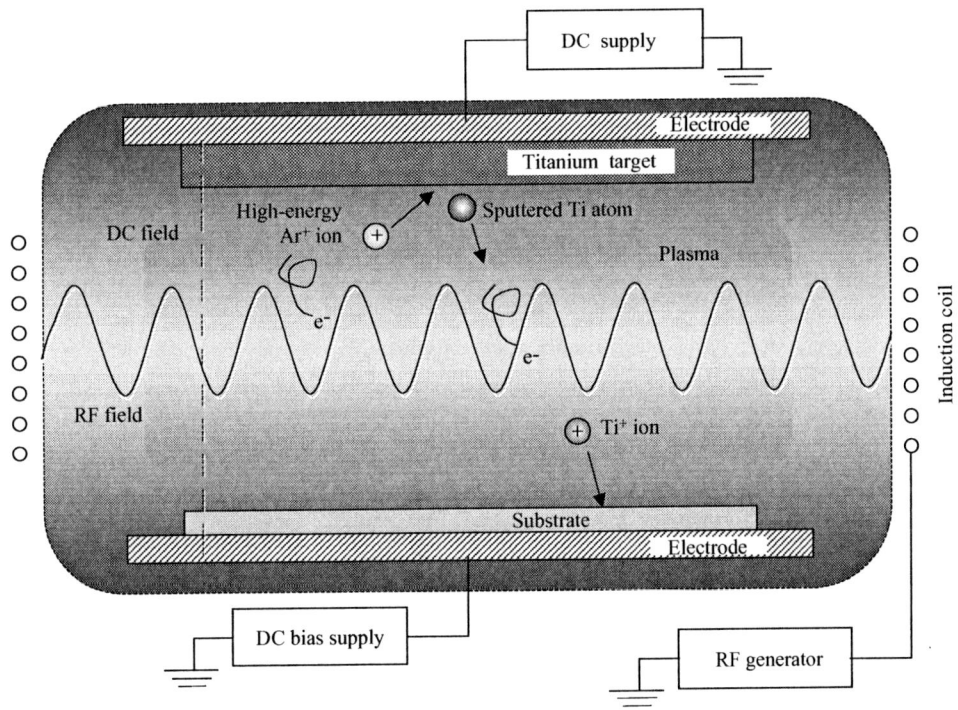

FIGURE 12.22 Concept of Ionized Metal Plasma PVD

beyond. The projected aspect ratio for vias in DRAM memory is projected to be 7:1 for the 0.15-μm generation, and 2.4:1 for logic.[46]

Tungsten CVD ■ *Tungsten (W) CVD* is used in multilevel interconnection technology as the process to deposit the via plug and to act as the contact plug between the first metal layer and the silicide contact. Tungsten CVD is also used as the local interconnect (LI) because of its low resistivity. Tungsten has performed reliably, with billions of W CVD plugs used on one microchip. However, its usefulness as a vertical interconnect may be limited with the introduction of the new damascene copper metallization scheme.

Tungsten CVD is common as a plug material due to:

1. Excellent step coverage and gap-fill, particularly in high-aspect ratio vias.
2. High electromigration resistance.

Low-pressure CVD (LPCVD) is the usual method for depositing W plugs, along with plasma-enhanced CVD (PECVD). Reactors are both hot- and cold-wall and configured as batch or multichamber cluster tools. The most common source gas for W is WF_6.[47] Sputtering costs less to operate than CVD for W deposition but has traditionally poorer directional control. This coverage produces nonuniform tungsten deposition quality in vias, which is why CVD has been the preferred method. This may change with the recent development of ionized PVD.

Tungsten CVD is typically deposited in blanket films. Blanket deposition deposits tungsten nonselectively on the entire wafer surface, including the via pathways, and is the most common approach.

Blanket Tungsten CVD Deposition. Blanket tungsten deposition followed by chemical mechanical planarization (CMP) is the common method used to fill vias to connect between metal layers. As always in CVD, the properties of the film deposited with CVD depend on the chemical reactions that occur at the wafer's surface. For blanket tungsten CVD, the most frequently used reaction is the reduction of tungsten hexafluoride (WF_6) by hydrogen to produce tungsten and hydrogen fluoride:

$$WF_6 \text{ (gas)} + 3H_2 \text{ (gas)} \rightarrow W \text{ (solid)} + 6HF \text{ (gas)}$$

tungsten hexafluoride   *hydrogen*   *tungsten*   *hydrogen fluoride*

PVD Cluster Tool
(Photo courtesy of Applied Materials, Inc.)

Typically the first step in blanket tungsten CVD is to deposit a titanium/titanium nitride (Ti/TiN) barrier layer (see Figure 12.23). The Ti is deposited before TiN to react with the underlying material and lower the contact resistance. If this is a contact in the first interlayer dielectric (ILD-1), then Ti reacts with silicon to form $TiSi_2$ silicide. The TiN functions as a barrier metal and adhesion promoter (glue layer) for tungsten. It needs a minimum bottom thickness of about 50 Å with continuous sidewall coverage to be an effective barrier metal and keep the tungsten from attacking underlying layers.[48] The Ti/TiN barrier layer may be deposited by IMP PVD. The TiN barrier metal may also be deposited either by CVD using ammonia and one of a number of precursor molecules, including the tetrakis dimethylamino titanium (TDMAT) or tetrakis diethylamino titanium (TDEAT). CVD has usually been preferred for TiN because of improved step coverage.

It is necessary to remove the excess blanket tungsten above the dielectric layer. This was previously done by a tungsten etchback process to remove the excess tungsten and leave a planarized plug. In 0.25-μm device generations and below, planarization of tungsten by chemical mechanical planarization (CMP) is the preferred process (CMP and etchback are addressed in Chapter 18).[49] Barrier metals such as TiN and Ti are used to prevent diffusion between the tungsten and the silicon.

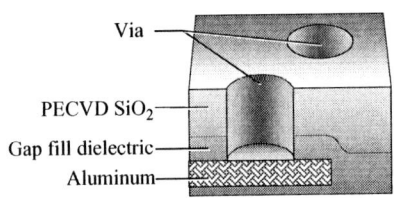
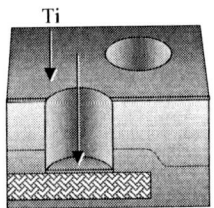
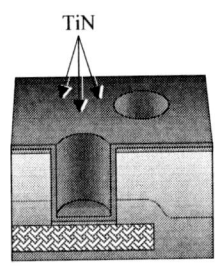

FIGURE 12.23 Blanket Tungsten CVD with Ti/TiN Barrier Metal

Copper CVD ■ *Copper (Cu) CVD* is a potential process for depositing a thin copper seed layer necessary for copper electroplating. The seed layer, or strike layer, is thin (about 500 to 1000 Å) and deposited on top of a diffusion barrier (most likely a Ta-based barrier metal). For successful electroplating, it is critical that this seed layer be continuous and free of pinholes and voids along the feature sidewalls and bottom. If a seed layer is not continuous, this can create a void in the electroplated copper. The excellent conformality of CVD is why this process has been investigated for the seed layer.

Cu Precursors. There are two precursors for CVD Cu, Cu(I), and Cu(II). Note that Cu(I) indicates a copper ion with a +1 positive charge and Cu(II) indicates a +2 positive charge. The most widely used Cu(I) precursor is Cu(hfac)(TMVS), chemical name trimethylvinylsilylhexafluoroacetylacetonate copper (I), chemical formula $C_{10}H_{13}CuF_6O_2Si$, and referred to by its trade name of CurpraSelect.[50] This molecule combines copper in a +1 oxidation state with TMVS and hfac ligands to make a clear yellow, liquid precursor (a ligand is a molecule or ion surrounding a central metal cation). TMVS is flammable, but the compound Cu(hfac)(TMVS) does not ignite easily. Cu(hfac)(TMVS) is compatible with industrial stainless steel delivery vessels. The CVD deposition proceeds by reducing a copper atom to metal while oxidizing another atom to $Cu(hfac)_2$ and releasing free TMVS as by-products. The chemical reaction is:

$$2Cu^{+1}(hfac)TMVS \text{ (gas)} \rightarrow Cu^0 \text{ (solid)} + Cu^{+2}(hfac)_2 \text{ (gas)} + 2TMVS \text{ (gas)}$$

The most common Cu(II) precursor is $Cu(hfac)_2$. For example, the Cu(II) reaction can proceed by reduction in H_2:

$$Cu(hfac)_2 \text{ (gas)} + H_2 \text{ (gas)} \rightarrow Cu^0 \text{ (solid)} + 2H(hfac) \text{ (gas)}$$

Copper Electroplate

The transformation of IC fabrication to copper metallization is underway for all chip manufacturers. Initially, high-performance microprocessors and fast, static RAMs are being converted to copper technology. *Copper electroplate,* also known as *electroplating, electrochemical deposition (ECD),* or *electrofill,* is the first-generation deposition method to be used for copper metallization. Cost and performance will be important issues that affect when copper will replace aluminum as the mainstream interconnect metallization. Electroplating has not traditionally been used in semiconductor manufacturing. However, it has been an important metallization process in other areas of electronic manufacturing, such as thin film heads used in disk storage devices and copper wiring for printed circuit boards. Therefore, its process and equipment requirements are well understood.

Copper Electroplating Tool
(Photo courtesy of Novellus)

Electroplating Fundamentals ■ The basic principle of electroplating copper metal is to immerse a wafer with a conductive surface into a solution of copper sulfate ($Cu(SO_4)$) which contains copper ions to be deposited (see Figure 12.24). The wafer and its seed layer are connected electrically to an external power supply as a negatively charged plate, or cathode. A solid piece of copper is immersed in the solution and configured as the positively charged anode. Electrical current passes through the wafer, into the solution, and through the copper anode.[51] When current flows, the following reduction reaction occurs at the wafer surface to deposit copper metal:

$$Cu^{2+} \quad + \quad 2e^- \quad \rightarrow \quad Cu^0$$

Copper ions   *Electrons*   *Copper metal*

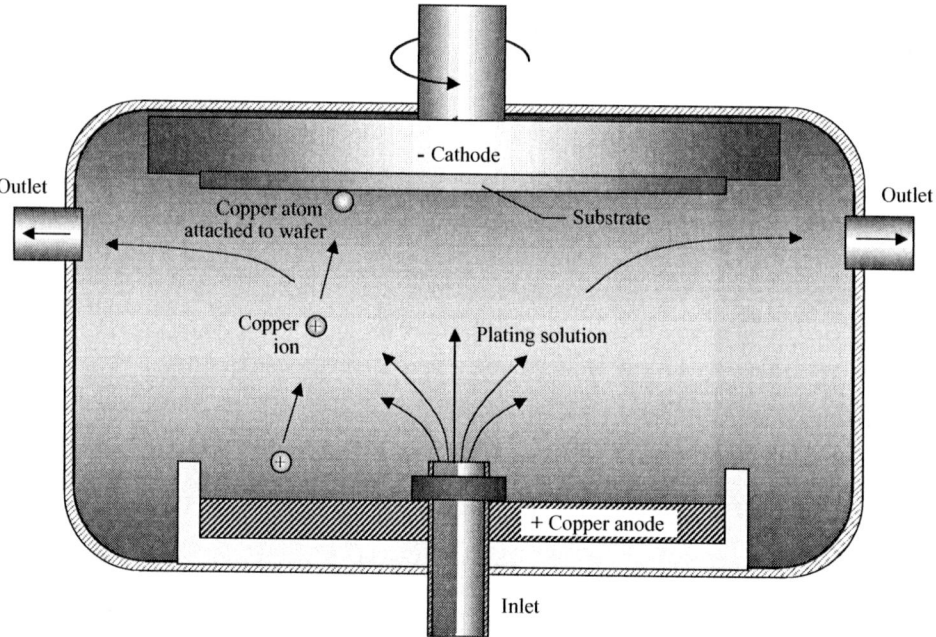

FIGURE 12.24 Copper Electroplating

Metal copper ions on the cathode wafer surface are reduced to metal copper atoms during electroplating. At the same time, an oxidation reaction occurs at the copper anode that balances the current flow in the cathode. This reaction maintains electrical neutrality in the solution.

The quantity of copper deposited is directly proportional to the current delivered to the conductive wafer surface (Faraday's law of electrolysis).[52] With this relationship, the basic parameters to control in electroplating are current and time. With no current flow, there is an equilibrium potential among the anode, cathode, and solution. When an external power source applies a voltage, a current flows between the anode and cathode and metal is deposited on the cathode proportional to the amount of current.

In practice, electroplating is complicated to control, especially the electroplating solution and how the current is applied. Given the high-aspect ratio holes and trenches that must be filled on high-performance ICs, electroplating is challenged to maintain a uniform current density in the holes. This condition is important because the plating rate is a direct function of current density. If there is a high current density at the top of the hole and a lower current density at the bottom of the hole, then copper will plate faster at the top.

Different types of voltage waveforms applied to the cathode/anode system can assist in plating high-aspect ratio holes. For instance, by applying an oscillating electric field and controlling the amplitude of the waveform, a deposit/etch (referred to as dep/etch) sequence is obtained. In this manner, the copper can be slightly removed (etched) in the high-density current regions to balance the copper gap-fill capability.

Because the copper electroplating metallization is a new process for semiconductor manufacturing, there are a host of other issues to address. For instance, it is important that no copper plate on the backside of the wafer. Electroplating also poses concerns for chemical control and disposal. It does simplify the metallization process somewhat by avoiding a high vacuum or complex heating of the wafer.

METALLIZATION SCHEMES

The strategy for metal interconnects for IC fabrication is undergoing dramatic change. After many years of development activities, copper is being implemented as the main chip interconnect conductor for a wide range of ICs.[53] Either aluminum or copper metallization is expected to be used for chip applications during a transition period until copper is fully implemented. The type of metallization scheme used depends on how the chip design is optimized for performance and price.

Traditional Aluminum Structure

The traditional interconnect metal has been an aluminum-copper alloy with an interlayer dielectric of SiO_2 to insulate between the metal layers. The process steps for the traditional aluminum

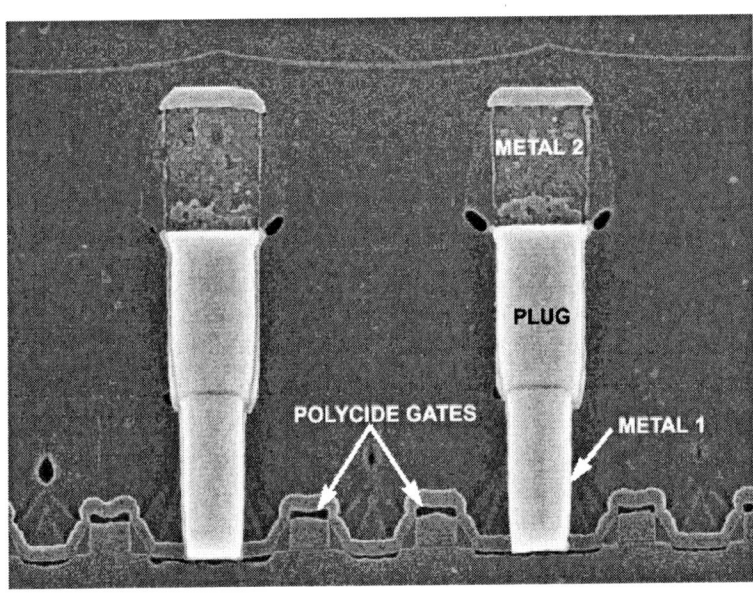

Blanket Aluminum Etch (Micrograph courtesy of Integrated Circuit Engineering)

interconnect technology for an ILD-2 and metal-2 (one complete level) are described in Steps 11 and 12 of the process flow described in Chapter 9. This traditional process can be viewed as an *aluminum subtractive process* because the aluminum is deposited as a blanket film and then etched away (subtracted) to form the circuitry. The critical step that defines the width and spacing of the aluminum interconnect lines is metal etch, while SiO_2 is deposited into the narrow gaps between the lines.

New materials and processes are being rapidly introduced into advanced wafer fabs to support the 0.18-μm technology node with traditional aluminum alloy metallization. This is a transitional period until copper metallization is eventually implemented. As a rule, most fabs will change only one major microchip material at a time (e.g., metal or insulator) to minimize the inherent risk to production. With these new fabrication materials and their stringent processing requirements, the general trend is for increased use of high-density plasma CVD (HDPCVD) for dielectrics, a shift from sputtered metals to metal CVD, and increased use of chemical mechanical planarization (CMP). The goals are to increase product performance and manufacturing productivity.

Copper Damascene Structure

The semiconductor industry is implementing copper as the interconnect material for microchip fabrication. The traditional process flow will not be used for copper metallurgy because copper is not suitable for dry-etching. Copper also has special requirements to minimize its diffusion into silicon. To form copper interconnect wiring, a dual-damascene process is used to avoid copper etching. In the damascene process, it is no longer metal etch that defines the critical line width and spacing, but rather dielectric etch.

Dual Damascene Process ■ Since copper is difficult to etch, early developers of copper technology were forced to consider alternative approaches to creating metal lines. The *dual-damascene process* became the consensus process for copper metallurgy. It creates both the vias and lines for each metal layer by etching holes and trenches in the ILD, depositing copper in the etched features and using chemical mechanical planarization to remove the excess copper (referred to as *copper overburden*).

There are many possible scenarios for a dual-damascene process flow. Table 12.5 explains a basic process flow with the essential technologies that may or may not include all the steps performed by a particular manufacturer.

TABLE 12.5* Copper Metallization using Dual Damascene

Process Step	Description	Structure
1. SiO_2 deposition	ILD oxide deposition with PECVD to desired thickness for via. There is no critical gap fill; therefore, PECVD is acceptable.	
2. SiN etch stop deposition	Thin (250 Å) SiN etch stop is deposited on ILD oxide. The SiN needs to be dense and pinhole-free; therefore, HDPCVD is used.	
3. Via patterning and etch	Photolithography to pattern and dry etch via openings into SiN. Strip photoresist after completion of etch.	

Process Step	Description	Structure
4. Deposit remaining SiO$_2$	PECVD oxide deposition for remaining ILD oxide.	SiO$_2$
5. Interconnect patterning	Photolithography to pattern SiO$_2$ trench with resist. Previously patterned via openings are located in trench.	Photoresist
6. Etch trench for interconnect and hole for via	Dry etch trench in ILD oxide, stopping on the SiN layer. Etch continues to form via opening by passing through opening in SiN.	
7. Deposit barrier metal	Deposit Ta or TaN diffusion layer with ionized PVD on bottom and sidewalls of trench and via.	Barrier metal
8. Deposit Cu seed layer	Deposit continuous Cu seed layer with CVD. The layer must be uniform and free of pinholes.	Cu seed layer
9. Deposit Cu fill	Deposit Cu fill with electrochemical deposition (ECD). Fill both via opening and trench.	Copper layer

Process Step	Description	Structure
10. Remove excess Cu with CMP	Remove excess Cu using chemical mechanical planarization (CMP). This planarizes the surface and prepares for the next level. The resulting surface is a planar structure with metal inlays in the dielectric to form the circuitry.	

* P. Singer, "Making the Move to Dual Damascene Processing," *Semiconductor International* (August 1997): p. 79.

For dual-damascene processing, a two-tiered metal inlay is created that includes a series of holes (for contacts or vias) in addition to the interconnect trench for metal lines. The vias and metal line levels undergo copper fill concurrently, which saves a process step and eliminates the interface between the via and metal line.

There are different approaches to performing dual-damascene processing, each with their own set of challenges. Most of the manufacturing methods integrate the use of etch-stop materials to control how deep etching occurs for the via and trench. Etch stop is typically done by using a barrier metal such as silicon nitride (Si_3N_4 or Si_xN_y) as a hard mask at the bottom of the via or trench. Si_3N_4 etches at a much slower rate than the dielectric material, thus effectively stopping the etch process. Some approaches may even have two separate Si_3N_4 layers, which adds process complexity. In addition, Si_3N_4 has a high-k and increases the interlayer capacitance of the interconnect stack, which is why the Si_3N_4 layer needs to be thin.

The dual-damascene flow shown above has the via etch done in the same etch step as the trench, with the SiN etch stop to define the bottom of the trench. In other variations of dual-damascene, the via is etched first, then the trench is patterned and then etched. Trench structures created over topography from a previous via etch pose a significant challenge for photolithography.

The most important reason for using a damascene approach to copper metallization is to avoid metal etch. A second advantage to damascene processing is that there is no further need for dielectric gap fill between etched metal lines, since the dielectric is applied as a blanket and then etched.

The dual-damascene process in wafer fabrication potentially uses 20 to 30% fewer steps than traditional aluminum metal interconnect. Not only are there fewer manufacturing steps, but dual damascene also eliminates or reduces some of the most difficult steps of traditional aluminum metallization, including aluminum etch and many of the tungsten and dielectric chemical-mechanical polish steps.[54] Reducing process steps in wafer fabrication is important for improving the yield of the process, since fewer process steps translate to fewer sources of error that reduce assembly yield.

METALLIZATION QUALITY MEASURES

Wafer fabrication quality measures for metallization are provided in Table 12.6.

TABLE 12.6 Quality Measures for Metallization

Quality Parameter	Types of Defects	Remarks
1. Sputtered metal adhesion.	A. Metal layer does not adhere to substrate.	Parameters affecting film adhesion are: • Wafer contamination • Stress • Types of materials • Temperature of substrate • Argon pressure
2. Sputtered film stress.	A. Excessive stress leads to: • Cracks in the film surface. • Loss of film adhesion. • Increased resistivity in some materials.	Film stress can be caused by excessive temperature of the wafer substrate. Ways to reduce wafer temperature are: • Lower deposition rate • Increased backside cooling Cracks can cause: • Peeling of the film layer • Contaminant migration • Electrical opens or short. Measure film stress (bow of wafer surface) before and after deposition. Units for stress are dynes.
3. Sputtered film thickness.	A. Metal layer does not meet film thickness specifications (e.g., sheet resistance out of specification).	Parameters that affect film thickners are: • Incorrect recipe • Improper flow rates • Improper substrate temperature • Improper chamber pressure • Improper power supply energy • Wrong time setting
4. Uniformity of electroplated (electrochemical deposition, or ECD) metal film.	A. Nonuniform film thickness indicated by: • Inadequate gap fill and step coverage for high aspect ratio openings (nonuniform metal thickness at top and bottom of high aspect ratio holes). • Nonuniform deposition thickness across the wafer and from wafer to wafer. • Voids in film.	Critical parameters for electroplating film uniformity are: • Depositing a uniform CVD seed layer without voids. • Maintaining proper electroplating bath chemistry (composition and concentration) for organic additives, primarily brighteners and suppressors, to attain void-free deposition that fills bottom and sidewalls of gaps.

METALLIZATION TROUBLESHOOTING

Common troubleshooting problems for metallization are provided in Table 12.7.

TABLE 12.7 Common Metallization Troubleshooting Problems

Problem	Probable Cause	Corrective Actions
1. Degradation in step coverage of metal film.	A. Decrease in substrate temperature. For Al alloy sputter deposition, step coverage is dependent on the wafer temperature during deposition.	Heating the substrate to improve step coverage because surface mobility of deposited metal atoms improves.
	B. Increase in deposition rate.	Decrease deposition rate. Increase in sputter deposition rate may degrade step coverage due to reduced surface mobility because of arrival of more atoms at surface.
2. Vacuum chamber integrity.	A. Chamber cleanliness or moisture in chamber.	• Check for vacuum leaks or chamber outgassing. Residual gas in chamber that can change film reflectivity. • Check for O_2 or N_2 in metal films that can change resistivity and film stress. • Clean chamber and do H_2O bake-out.
	B. Outgassing or system leaks.	• Use helium leak detection to inspect for system leaks. • Assess the chamber conditions before deposition with residual gas analyzer.
3. Contamination of metal film.	A. Particles on surface of film.	Check the following common particle sources: • Dirty input wafers. • Problems in hand-off between robot and chamber mechanism. • Incomplete chamber cleaning between runs. • Dirty cassettes. • Contaminated nitrogen backfill.
4. Voids in copper trench fill after dual-damascene electroplating.	A. Excessive seed layer thickness on the wafer surface (field) that pinches off the via or trench, creating a center void in the film.*	Optimize the CVD Cu seed layer deposition by evaluating the electroplating current from the field to the bottom of vias and trenches. The goal is to deposit adequate seed layer at the bottom of a high-aspect ratio feature without increasing field thickness.
5. Excessive copper dishing after dual-damascene CMP.	A. Cu dishing for Cu CMP is often caused by tantalum diffusion barrier, which must be polished back in the presence of Cu.**	After Cu is polished, the Ta layer must be removed from the level. The Cu pad/slurry field does not effectively remove Ta. Options are: • Optimize the pad/slurry for Ta. • Minimize the Ta field level thickness.

* R. Jackson et al., "Processing and Integration of Copper Interconnects," *Solid State Technology* (March 1998): p. 56.
** Ibid.

SUMMARY

Metallization deposits a thin metal film to form the interconnect wiring and contact or via connections on the chip. There are six metal categories used in wafer fabrication, with specific requirements for optimum performance. Aluminum has been the traditional metal for interconnect lines. An ohmic contact is the low-resistance contact between silicon and the interconnect metal. Aluminum is sometimes alloyed with silicon to reduce junction spiking at the ohmic contact. Aluminum is often alloyed at 0.5 to 4% copper to minimize electromigration reliability problems. New interconnect metallization is based on copper metallurgy to decrease the metal resistivity. Copper combined with low-k dielectric will decrease the chip interconnect delay, but copper does require a new processing approach since it does not etch well. Barrier metals are used when joining metals. Different barrier metals have optimum properties depending on the application. A silicide is a refractory metal alloyed with silicon to improve contact resistance and adhesion. A salicide is a particular silicide that is aligned to the source, drain, and gate structure. Metal layers are connected with vias filled with tungsten plugs. The most widely used metal deposition system is sputtering. Sputtering physically bombards a target to dislodge atoms and deposit the atoms as a thin film on the wafer surface. The three most common types of sputtering are RF, magnetron, and ionized metal plasma. RF sputtering is inefficient, while magnetron has a higher deposition rate but limited gap-fill capability. Ionized metal plasma has improved directionality for high-aspect ratio deposition. CVD metal has the best conformality and is used for tungsten plug fill and copper seed layer. Copper electroplate is the process chosen by major semiconductor manufacturers for copper wiring deposition. Traditional aluminum metallization deposits a blanket aluminum film and then etches line patterns to form interconnect wiring. Copper metallurgy will use a dual-damascene process that etches vias and trenches in dielectric, deposits copper fill, and then removes excess copper through chemical mechanical planarization.

KEY TERMS

interconnect
wiring
contact
vias
plug
eutectic temperature
anneal or sinter
ohmic contact
junction spiking
electromigration
hillocks
copper interconnect
barrier metal
refractory metals
silicide
polycide
salicide
tungsten plugs
evaporation
physical vapor deposition (PVD)
target
momentum transfer
sputtering yield
sputter etch (reverse sputter)
RF sputtering
magnetron sputtering
collimated sputtering
sputtered aluminum
ionized metal plasma PVD (IMP or ionized PVD)
tungsten (W) CVD
copper CVD
electroplate (electroplating, electrochemical deposition (ECD), or electrofill)
evaporation
aluminum subtractive process
dual-damascene process

REVIEW QUESTIONS

1. Define the following terms: interconnect, contact, via, and plug.
2. What metal has traditionally been used for interconnect wiring, and what is its replacement?
3. List and describe the seven performance requirements for metals used in wafer fabrication.
4. List the categories of metals and metal alloys used in semiconductor manufacturing.
5. Explain why aluminum has been the interconnect metal of choice for microchips.
6. What is an ohmic contact and what is its advantage?
7. Describe junction spiking and list the two primary ways it was resolved.
8. Discuss electromigration and how it affects reliability. How is electromigration controlled in aluminum wiring?
9. List and discuss the five benefits for introducing copper metallization.
10. What are the three major challenges for converting to copper for interconnects?
11. Explain what a barrier metal is and the essential properties of a barrier material. What metals are commonly used as a barrier metal?
12. List the special requirements for a copper barrier metal.
13. Define silicide and explain why refractory metal silicides are important in wafer fabrication.
14. What are the benefits of titanium silicide? Explain the C49 and C54 phase structures.
15. What silicide is promising for future wafer technology? Why?
16. What is the salicide process?
17. Describe a tungsten plug, and discuss how it is used in multilayer metallization.
18. Why was evaporation replaced as a metal deposition system?
19. Provide a short explanation about sputtering and how it works.
20. Is sputtering suitable for alloy deposition?
21. What are the advantages to sputter deposition?
22. List and explain the six steps to sputtering.
23. Explain the physics of sputtering.
24. Describe sputter yield.
25. What is the purpose of sputter etch?
26. Describe an RF sputter system. What is its major limitation?
27. Discuss how a magnetron sputtering system improves the deposition rate.
28. Explain why sputtering has poor step coverage. How does collimated sputtering improve step coverage?
29. Describe ionized metal plasma. How does this process improve high-aspect ratio gap fill?
30. What is the typical method for creating a tungsten plug in advanced ICs?
31. Why could CVD be used for the copper seed layer?
32. Explain the fundamental process of copper electroplate.
33. Why is traditional aluminum metallization a subtractive process?
34. What is the primary reason for implementing the dual-damascene process for copper?
35. List ten steps for a dual-damascene metallization process.

METALLIZATION EQUIPMENT AND MATERIALS SUPPLIERS' WEB SITES

Angstrom Sciences Inc. — http://www.angstromsciences.com/

Applied Materials — http://www.appliedmaterials.com/products/

Genus Incorporated — http://www.genus.com/

Materials Research Corporation — http://www.materialsresearch.com/

Nordiko USA Inc. — http://www.nordiko.com/

Novellus Systems Inc. — http://www.novellus.com/index.htm

Process Materials Inc — http://www.processmaterials.com/

TEL, Tokyo Electron Ltd. — http://www.teainet.com/

Veeco-CVC — http://www.veeco.com

REFERENCES

1. C. Weber, D. Jensen, and E. Hirleman, "What Drives Defect Detection Technology?" *Micro* (June 1998): p. 60.
2. M. Bohr, "Interconnect Scaling—The Real Limiter to High Performance ULSI," *Solid State Technology* (September 1996): p. 105.
3. S. Ghandhi, *VLSI Fabrication Principles: Silicon and Gallium Arsenide,* 2nd ed. (New York: Wiley, 1994): p. 548.
4. K. Bachmann, *The Materials Science of Microelectronics,* (New York: VCH Publishers, 1995), p. 484.
5. Compiled from multiple sources, including: S. Wolf and R. Tauber, *Silicon Processing for the VLSI Era,* vol. 1, *Process Technology,* 2nd ed., (Sunset Beach: Lattice Press, 1986), p. 399.
6. L. Peters, "Advanced Aluminum Interconnect Technology," *Semiconductor International* (November 1998): p. 83.
7. R. Jaeger, *Introduction to Microelectronic Fabrication,* (Reading, MA: Addison-Wesley, 1988), p. 136.
8. S. Campbell, *The Science and Engineering of Microelectronic Fabrication,* (New York: Oxford University Press, 1996), p. 406.
9. Ibid., p. 412.
10. E. Amerasekera and F. Najm, *Failure Mechanisms in Semiconductor Devices,* 2nd ed., (New York: Wiley, 1997), p. 103.
11. J. Yue, "Reliability," Sze, *ULSI Technology,* ed. C. Chang and S. Sze (New York: McGraw-Hill, 1996), p. 672.
12. K. Rodbell, "Reliability," *Handbook of Semiconductor Interconnection Technology,* ed. G. Schwartz, K. Srikrishnan, and A. Bross (New York: Marcel Dekker, 1998), p. 480.
13. S. Campbell, *The Science and Engineering of Microelectronic Fabrication,* p. 413.
14. J. Yue, "Reliability," p. 663.
15. P. Singer, "Tantalum, Copper and Damascene: The Future of Interconnects," *Semiconductor International* (June 1998): p. 91.
16. "Technology News," *Solid State Technology,* (February 1998): p. 26.
17. P. Singer, "Tantalum, Copper and Damascene," p. 91.
18. Ibid.
19. A. Sethuraman, J. F. Wang, and L. Cook, "Copper vs. Aluminum: A Planarization Perspective," *Semiconductor International* (June 1996): p. 178.
20. X. Lin and D. Pramanik, "Future Interconnect Technologies and Copper Metallization," *Solid State Technology* (October 1998): p. 63.
21. P. Singer, "Tantalum, Copper and Damascene," p. 94.
22. Semiconductor Industry Association, *National Technology Roadmap for Semiconductors: Technology Needs,* (San Jose: SIA, 1997), p. 101.
23. C. Ryu et al., "Barriers for Copper Interconnections," *Solid State Technology* (April 1999), p. 53.
24. B. Chin et al., "Barrier and Seed Layers for Damascene Copper Metallization," *Solid State Technology* (July 1998), p. 141.
25. C. Ryu, "Barriers for Copper Interconnections," p. 56.
26. A. Braun, "Copper Electroplating Enters Mainstream Processing," *Semiconductor International* (April 1999), p. 64.
27. R. Jackson, et al., "Processing and Integration of Copper Interconnects," *Solid State Technology* (March 1998): p. 56.
28. S. Ghandhi, *VLSI Fabrication Principles,* p. 576.
29. D. Campbell, "Semiconductor Contact Technology," *Handbook of Semiconductor Interconnection Technology,* ed. G. Schwartz, K. Srikrishnan and A. Bross (New York: Marcel Dekker, 1998), p. 176.
30. K. Rodbell, Reliability," p. 480.
31. L. Peters, "Is the 0.18 µm Node Just a Roadside Attraction?" *Semiconductor International* (January 1999), p. 50.
32. R. Liu, "Metallization," *ULSI Technology,* ed. C. Chang and S. Sze (New York: McGraw-Hill, 1996), p. 410.
33. S. Campbell, *Science and Engineering,* p. 292.
34. R. Liu, "Metalization," p. 379.
35. X. Lin and D. Pramanik, "Future Interconnect Technologies," p. 74.
36. S. Wolf and R. Tauber, *Silicon Processing for the VLSI Era,* vol. 1, *Process Technology,* 2nd ed., pp. 443–50.
37. M. Rittner, "Growth Predicted for Sputtering Target and Sputtered Film Markets," *Solid State Technology* (January 2000): p. 26.
38. S. Wolf and R. Tauber, *Silicon Processing for the VLSI Era,* vol. 1, *Process Technology,* p. 445.
39. A. Braun, "Sputtering Targets Adapt to New Materials and Shrinking Architectures," *Semiconductor International* (June 1998): p. 130.
40. S. Wolf and R. Tauber, *Silicon Processing for the VLSI Era,* vol. 1, *Process Technology,* p. 449.
41. Ibid., pp. 348–51.
42. R. Liu, "Metallization," p. 382.
43. Ibid., p. 384.
44. B. Chin et al., "Barrier and Seed Layers," p. 141.
45. A. Braun, "Sputtering Targets," p. 128.
46. Semiconductor Industry Association, *The National Technology Roadmap for Semiconductors: Technology Needs,* (San Jose: SIA, 1997), p. 101.
47. G. Schwartz, "Metallization," *Handbook of Semiconductor Interconnection Technology,* ed. G. Schwartz, K. Srikrishnan and A. Bross, (New York: Marcel Dekker, 1998), p. 320.
48. J. Baliga, "Depositing Diffusion Barriers," *Semiconductor International* (March 1997): p. 77.
49. K. Wijekoon, et al., "Tungsten CMP Process Developed," *Solid State Technology* (April 1998): p. 53.
50. B. Zorich and M. Majors, "Safety and Environmental Concerns of CVD Copper Precursors," *Solid State Technology* (September 1998): p. 101.
51. P. Singer, "Tantalum, Copper and Damascene," p. 94.
52. Ibid.
53. Ibid., p. 91.
54. Ibid.

CHAPTER 13
PHOTOLITHOGRAPHY: VAPOR PRIME TO SOFT BAKE

The essence of photolithography is to print temporary circuit structures on a wafer that will later be used to assist etch and ion implant processes. The structures are first created as patterns on a quartz template that is referred to as a mask. Ultraviolet light passes through the mask to transfer the pattern to a light-sensitive film on the wafer surface. As with common photography work, the photographic film is developed and the image appears on the wafer. Later a chemical etching process is used to image the film pattern on the underlying wafer surface, or the wafer is sent to the ion implant area where selective doping can be performed through the patterns on the wafer. The various patterns transferred to the wafer define the numerous device features, vias, and wiring interconnects necessary for the device layers as well as determine areas of the silicon to be doped.

Photolithography is closely related to the cost and performance of a microchip. The processing cost of a wafer is largely independent of the number of chips on a wafer. That is, a wafer with few chips costs almost as much as the same wafer with many chips since the number of processing steps, amount of materials, wafer handling, and so on, are nearly identical for the two wafers. More chips at the same cost means that the overall cost of each chip is lower, simply by putting more chips on the wafer.

Since the beginning of semiconductor manufacturing, photolithography has been recognized as the driving force behind the integrated circuit fabrication process. This trend continues today as the industry strives to pack more devices and the associated circuitry on a chip. More than any other single fabrication technology, photolithography has contributed to the revolutionary advancement of chip performance, affirming Moore's law over the fifty-plus years of semiconductor manufacturing.

OBJECTIVES

After studying the material in this chapter, you will be able to:

1. Explain the basic concepts for photolithography, including process overview, critical dimension generations, light spectrum, resolution, and process latitude.
2. Discuss the difference between negative and positive lithography.
3. State and describe the eight basic steps to photolithography.
4. Explain how the wafer surface is prepared for photolithography.
5. Describe photoresist and discuss photoresist physical properties.
6. Discuss the chemistry and applications of conventional i-line photoresist.
7. Describe the chemistry and benefits of deep UV (DUV) resists, including chemically amplified resists.
8. Explain how photoresist is applied in wafer manufacturing.
9. Discuss the purpose of soft bake, and explain how it is accomplished in production.

INTRODUCTION

Photolithography produces a three-dimensional pattern on the surface of the wafer using a light-sensitive photoresist material and controlled exposure to light. Other terms for the photolithography process are *photo, lithography, masking,* and *patterning.* In general, lithography refers to any printing process that transfers a pattern to a planar surface. For this reason, photolithography is sometimes referred to as "printing." For semiconductor manufacturing, *microlithography* describes the process used to pattern the ultraminiature features necessary for submicron wafer fabrication. The terms microlithography and photolithography are used interchangeably in this text.

Photolithography Concepts

Photolithography is at the center of the wafer fabrication process, as evidenced by how wafers repeatedly flow into and out of the photolithography operation in the fabrication process (see Figure 13.1). Photolithography is often considered the most critical step in the IC process, requiring superior performance to achieve high yield in other process areas. It is estimated that lithography accounts for nearly one third of the total wafer fabrication cost.[1]

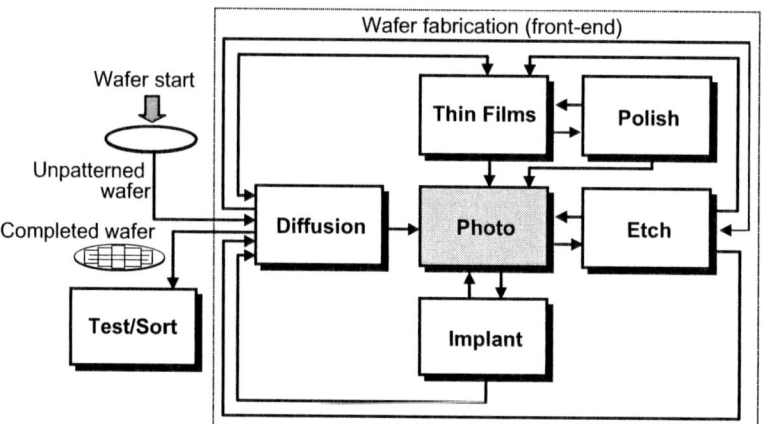

FIGURE 13.1 Wafer Fabrication Process Flow
(Used with permission from Advanced Micro Devices)

The shape of the photolithography pattern transferred to the wafer surface depends entirely on the wafer layer being constructed. Possible patterns are semiconductor devices in the silicon wafer, isolation trenches, contacts, the metal interconnects, and vias to interconnect metal layers. The patterns are transferred into a light-sensitive photoresist material to prepare the substrate for etch (see Chapter 16) or ion implantation (see Chapter 17). The patterned photoresist images are three-dimensional because there are width, length, and height to the pattern in the photoresist (see Figure 13.2). There may be hundreds of identical chips on a wafer, each needing the appropriate pattern to be transferred to each die.

Photolithography technology employs a light-sensitive photoresist, or resist, that is applied on the substrate surface as a polymer solution. The resist is then baked, which drives out the solvent. Next, the resist is exposed to a controlled light. The light passes through a reticle that defines the desired pattern. The photoresist is a temporary material placed on the wafer surface solely to transfer the necessary pattern and is removed once the pattern is etched or implanted onto the wafer surface material.

The *reticle* is a quartz plate that contains the pattern to be reproduced on the wafer, just as the negative of a camera film has the pattern for a photograph. This pattern may contain a single die or several die. A *photomask,* often called a *mask* and often used interchangeably with the term reticle, is a quartz plate that contains the full die array needed to define one process layer for the entire wafer. For this text, reticle refers to the pattern for a die or set of die, whereas mask refers to the entire array or matrix of die necessary for an entire wafer layer. Since light is critical for transferring the pattern to the resist and is controlled through optics, photolithography is sometimes referred to as *optical lithography.*

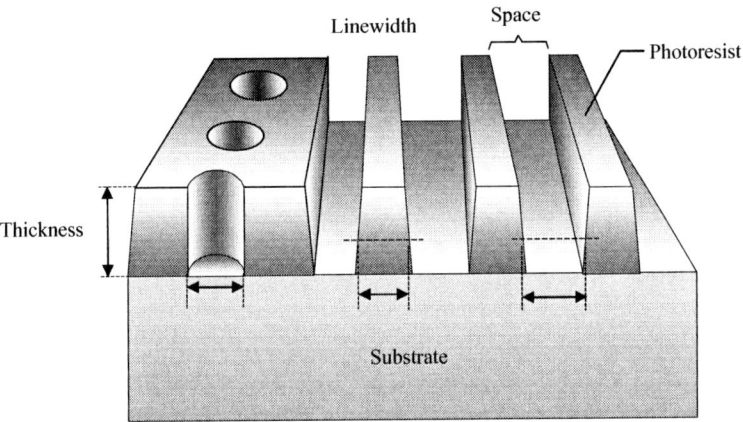

FIGURE 13.2 Three Dimensional Pattern in Photoresist

Advanced CMOS ICs can have up to 30 mask layers needed to pattern the multiple layers on a chip.[2] Each reticle has the unique pattern needed for the particular image, or feature, to be placed on the wafer surface and is stepped across the entire wafer to complete one layer. To fabricate an entire layer, the wafer must be processed into the photolithography operation and then out to undergo subsequent operations (e.g., etch or implantation).

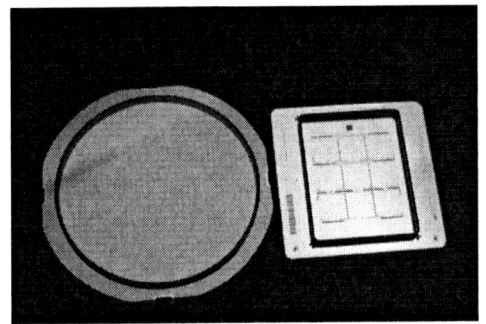

Photomask and Reticle for Microlithography
(Photo courtesy of Advanced Micro Devices)

Critical Dimension Generations ■ Microlithography has traditionally been the technology in semiconductor manufacturing that defines the critical dimension (CD) on a wafer (e.g., the polysilicon gate length on a microprocessor). Since the CD is the most difficult dimension to control during fabrication, it becomes the required dimension for other process areas to achieve. The critical dimension in photolithography is often used to describe device technology nodes or generations. Some sub-0.25 μm technology nodes are 0.18 μm, 0.15 μm, and 0.1 μm. The ability to reduce the critical dimension permits more die to be placed on a single wafer, thus lowering fabrication costs significantly and improving profitability.

Light Spectrum ■ Energy is required to activate the photoresist and transfer the pattern from the reticle. The source of energy is in the form of radiation, typically an ultraviolet (UV) light source. Light-sensitive photoresists are designed to chemically respond to a certain UV wavelength. UV light has been the most common energy source for photolithography patterning and this will probably continue for the near future (including devices fabricated at the 0.10-μm technology node and possibly lower).

The electromagnetic spectrum is used to introduce the UV light spectrum of most interest for photolithography, as shown in Figure 13.3 on page 338. The different UV wavelengths important for photolithography are shown in Table 13.1 on page 338. In general, deep UV (DUV) describes wavelengths less than 300 nm. Additional details about light and the UV spectrum for photolithography will be provided in Chapter 14.

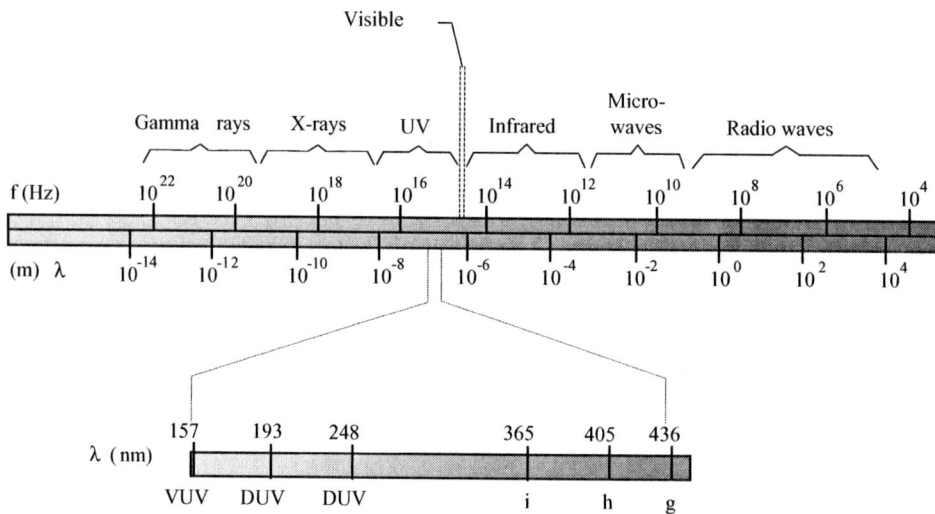

FIGURE 13.3 Section of the Electromagnetic Spectrum

TABLE 13.1 Important UV Wavelengths for Photolithography Exposure

UV Wavelength (nm)	Wavelength Name	UV Emission Source
436	g-line	Mercury arc lamp
405	h-line	Mercury arc lamp
365	i-line	Mercury arc lamp
248	Deep UV (DUV)	Mercury arc lamp or krypton fluoride (KrF) excimer laser
193	Deep UV (DUV)	Argon fluoride (ArF) excimer laser
157	Vacuum UV (VUV)	Fluorine (F_2) excimer laser

Resolution ■ An important performance measurement in photolithography is the resolution of each image. *Resolution* is the ability to differentiate between two closely spaced features on the wafer surface. The actual dimensions of the patterned images on the wafer are the feature sizes. The minimum feature size dimension is the critical dimension (CD), such as 0.18 μm. However, this does not mean that every feature on the wafer has exactly this CD. On a critical layer, there are critical dimensions, such as a contact opening in the first interlayer dielectric or the polysilicon gate length, and other features that are noncritical and larger than the CD. Resolution is important for critical dimensions.

Over the past 30 years, the CD in wafer fabrication has decreased at about 11% per year. The reduction of device feature sizes has been done to achieve improved performance from device scaling, increased circuit density and decreased die size. However, the CD can only be reduced as much as the photolithography process will permit. An important challenge of lithography for achieving the smaller resolution has been the need to reduce the wavelength of the exposing light to be about the same length as the CD (to be discussed in Chapter 14).

Overlay Accuracy ■ As we shall see in Chapter 14, photolithography requires precise alignment between the pattern on the mask and the existing features on the wafer surface. This quality measure is known as *overlay accuracy*. Alignment is critical because the mask pattern must be precisely transferred to the wafer from layer to layer (see Figure 13.4). Since multiple masks are used during patterning, any overlay misalignment contributes to the total placement tolerances between the different features on the wafer surface. This condition is known as the *overlay budget*. A large overlay budget essentially reduces the circuit density, which limits device feature sizes and, therefore, IC performance.

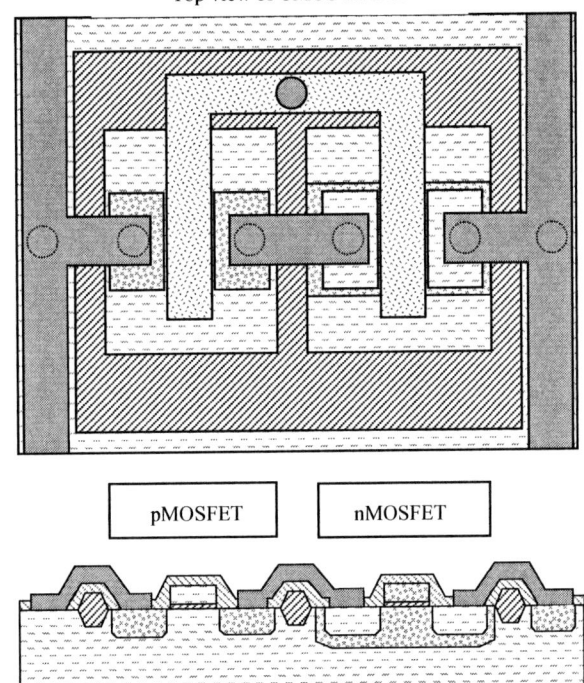

FIGURE 13.4 Importance of Mask Overlay Accuracy

Different types of misalignments affect the overlay budget. Misregistration is caused by poor alignment between the mask and the wafer. *Runout* is a net difference in the distance from die to die.[3] It can be caused by temperature variations. These types of misalignment are controlled in modern alignment systems by global wafer alignment and site-to-site alignment schemes (discussed in Chapter 14).

Process Latitude ■ There are many process variables involved in photolithography. Examples of process variables are equipment settings, types of materials, human performance, machine alignments, and the stability of materials over time. *Process latitude* represents the capability of the photolithography process to consistently produce products that meet the specified requirements. The objective is to achieve the largest process latitude possible to make the process more capable of producing good parts. Process engineers define the settings for the different process variables to obtain the largest process latitude. For photolithography, high process latitude means critical dimensions can be met within the specified limits even with all the process variations encountered during production.

PHOTOLITHOGRAPHY PROCESSES

Photolithography consists of two basic types of processes: negative and positive lithography. *Negative lithography* prints a pattern on the wafer surface that is the opposite of the pattern in the mask. *Positive lithography* prints a pattern on the wafer that is the same pattern as that on the mask. The major difference between these two basic processes is the type of photoresist used. How the photoresist material reacts when exposed to light depends on whether it is a negative or a positive resist material.

Negative Lithography

The primary feature of negative lithography is that the photoresist will become insoluble and harden by crosslinking when exposed to light. Once hardened, the crosslinked resist cannot be washed away in solvents. Resists of this type are known as *negative resists* because the image formed in the resist is the negative of the pattern found on the reticle (see Figure 13.5). Negative resists were the earliest types of photoresist used in semiconductor photolithography.

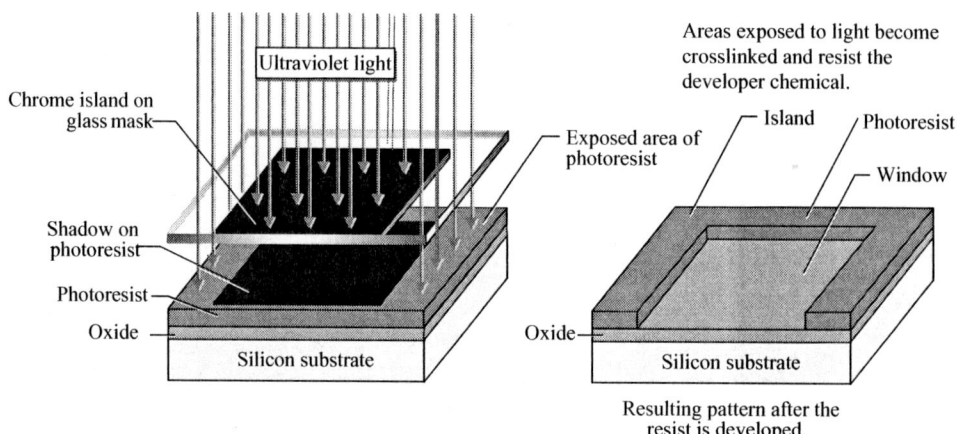

FIGURE 13.5 Negative Lithography

It can be seen in the diagram in Figure 13.5 that the mask for the negative resist is a transparent plate of quartz. The dark portion of the mask is a thin film of chromium deposited and patterned to form the desired mask artwork. The chrome is opaque (not transparent) and does not permit the passage of the ultraviolet light. For a negative resist, the regions underneath the opaque chrome of the mask are not exposed to light and therefore remain unchanged. The resist remains soft and dissolves when exposed to the developer chemical. UV light passes through the clear areas of the mask and hardens the resist so that it does not dissolve in the subsequent developer chemical. In this manner, the negative resist has a pattern that is the opposite of the pattern on the mask.

Positive Lithography

In positive lithography, the pattern printed on the wafer surface has the same image as the pattern on the mask. The areas of the resist exposed to the UV light undergo a photochemical reaction and become soluble and soften in the developer. In this manner, regions of the positive resist exposed to light are removed in a developer solution, whereas resist not exposed to light underneath the opaque mask pattern remains on the wafer (see Figure 13.6). This type of resist is referred to as a *positive resist* because the image formed in the resist is the same pattern formed on the reticle. The resist that remains was already hardened before light exposure and remains on the wafer surface to protect the top layer during subsequent processing, such as etch. The resist is removed when the follow-on processing is done. During the 1970s, positive resist became the dominant resist type for submicron microlithography and remains so today (see the following section for a discussion of the benefits from positive resist).

The pattern transferred into the photoresist can be considered a window or an island. A summary of the results from the two different types of masks and photoresists is shown in Figure 13.7.

Another way to describe masks is by their outward appearance. A mask is referred to as a *dark-field mask* if much of the quartz plate is coated with chrome. A *clear-field mask* is one that has very thin patterns of chrome with large areas of clear quartz. When comparing masks for positive and negative lithography, if a clear-field mask is required for a specific masking layer using a positive resist, then a dark-field mask of the same pattern would be used for negative resist.

Using positive lithography as an example, some common dark-field masks are the masks that are used prior to source/drain implant, LDD implant, and contact etch. A clear-field mask is used for applications such as preceding gate etch and metal interconnect etch. Figure 13.8 illustrates examples of the two types of masks based on positive photoresist lithography.

Photolithography: Vapor Prime to Soft Bake **341**

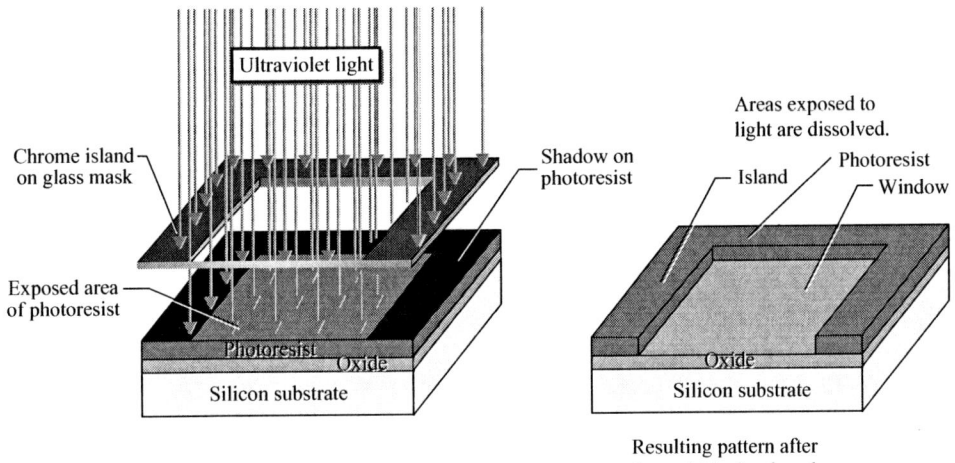

FIGURE 13.6 Positive Lithography

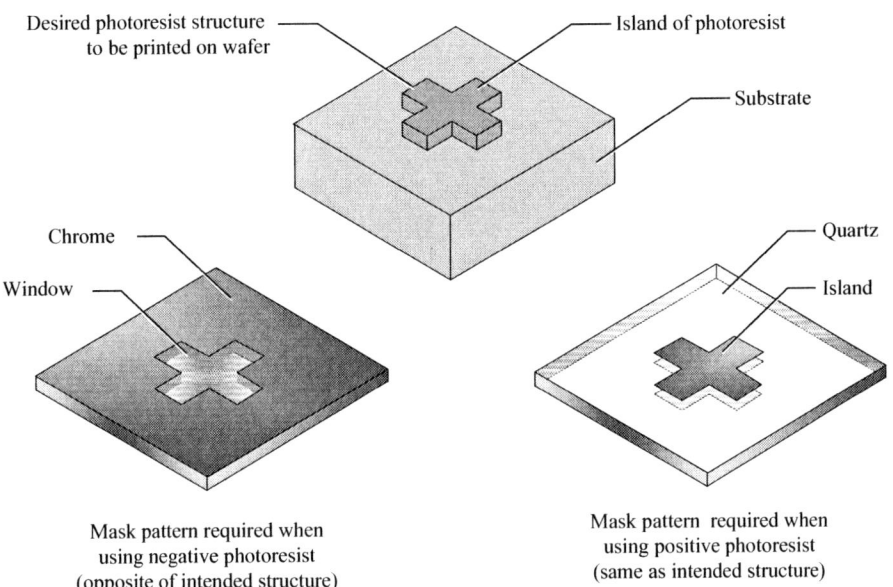

FIGURE 13.7 Relationship Between Mask and Resist

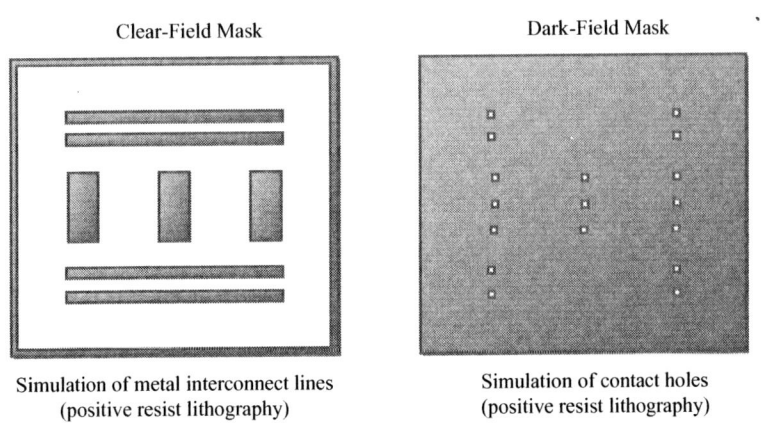

FIGURE 13.8 Clear-Field and Dark-Field Masks

EIGHT BASIC STEPS OF PHOTOLITHOGRAPHY

Photolithography is a complex process with many variables contributing to its process latitude, such as reduced feature size, alignment tolerance, number of masking layers, and cleanliness of the wafer surface. For convenience, we can divide the patterning process of photolithography into an eight-step procedure (see Table 13.2). Figure 13.9 provides a pictorial overview of the eight process steps. In the wafer fab these steps are often referred to as operations. Breaking the large photo process into these eight steps simplifies the various aspects of microlithography. This chapter will first provide a short overview of the eight basic steps, followed by an in-depth analysis of the materials, equipment, and processes used in each photolithography step. The in-depth analyses of the steps are covered at the end of this chapter and in the following two chapters. Table 13.2 indicates the specific steps that are covered in a chapter.

TABLE 13.2 Eight Steps of Photolithography

Step	Chapter
1. **Vapor prime**	13
2. **Spin coat**	13
3. **Soft bake**	13
4. Alignment and exposure	14
5. Post-exposure bake (PEB)	15
6. Develop	15
7. Hard bake	15
8. Develop inspect	15

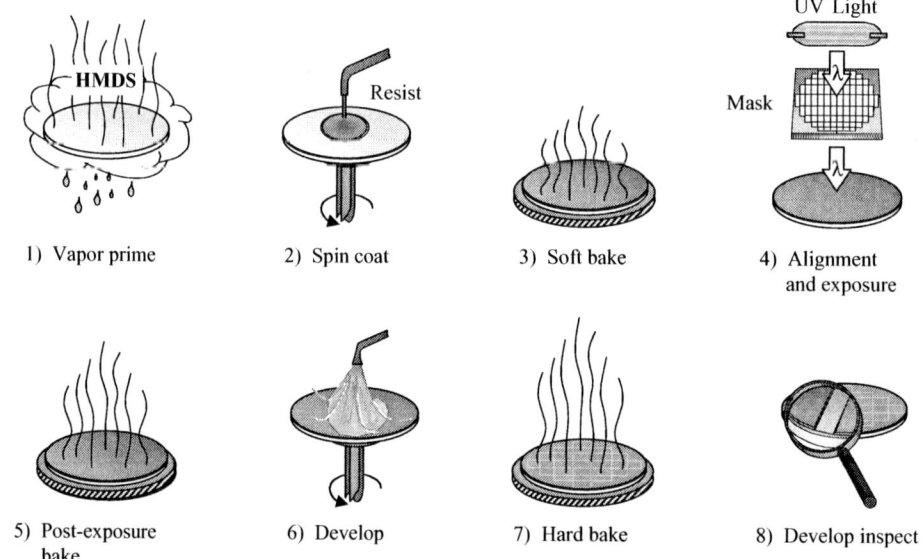

FIGURE 13.9 Eight Steps of Photolithography

The photolithography process has undergone significant equipment integration. Automated equipment, referred to as a *coater/developer track system,* or *tracks,* employs robots, automated material handling, and computers to perform all eight steps without human intervention. Integrated tracks offer many benefits over the previous stand-alone manual tooling for photolithography. This integration improved process control by controlling delays between process steps, processing wafers efficiently, increasing flexibility, reducing contamination due to environmental control and minimal operator handling, and increasing safety due to reduced operator exposure to chemicals.

Photolithography Track System
(Photo courtesy of Advanced Micro Devices)

Step 1: Vapor Prime

The first step in photolithography is to clean, dehydrate, and prime the surface of the wafer. The purpose of these steps is to promote good adhesion between the photoresist and the wafer surface.

Wafer cleaning may involve a wet clean and DI water rinse to remove contaminants. Most wafer cleaning is performed before entering the photolithography workbay. A dehydration dry bake is done in a closed chamber to drive off most of the adsorbed water on the surface of the wafer. The wafer surface must be clean and dry. Immediately after the dehydration bake, the wafer is primed with hexamethyldisilazane (HMDS), which acts as an adhesion promoter.

Step 2: Spin Coat

After priming, the wafer is coated with the liquid photoresist material by a spin coating method. The wafer is mounted on a vacuum chuck, which is a flat metal or teflon disc that has small vacuum holes on its surface to hold the wafer. A precise amount of liquid photoresist is applied to the wafer and then the wafer is spun to obtain a uniform coating of resist on the wafer (see Figure 13.10).

Different resists require different spin coating conditions, such as an initial slow spin (e.g., 500 rpm), followed by a ramp up to a maximum rotational speed of 3,000 rpm or higher. Some important quality measures for photoresist application are time, speed, thickness, uniformity, particulate contamination, and resist defects such as pinholes.

Process Summary:
- Wafer is held onto vacuum chuck
- Dispense ~ 5ml of photoresist
- Slow spin ~ 500 rpm
- Ramp up to ~ 3000 to 5000 rpm
- Quality measures:
  - time
  - speed
  - thickness
  - uniformity
  - particles and defects

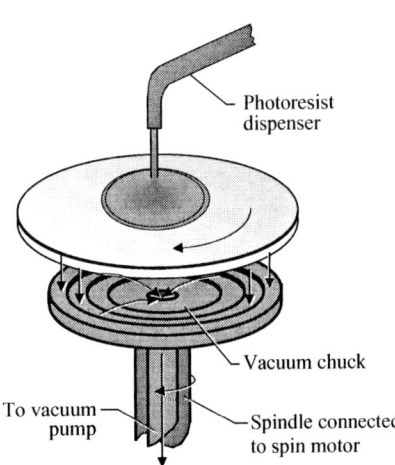

FIGURE 13.10 Spin Coat

Step 3: Soft Bake

After the resist has been applied to the wafer surface, it must undergo a soft bake. The purpose of this bake is to drive off most of the solvent in the resist. The soft bake process improves adhesion, promotes resist uniformity on the wafer, and yields better linewidth control during etching. Typical soft bake temperatures are 90 to 100°C for 30 seconds on a hot plate, followed by a cooling step on a cool plate to achieve wafer temperature control for uniform resist characteristics.

Step 4: Alignment and Exposure

The next step is referred to as alignment and exposure. The mask is aligned to the correct location of the resist-coated silicon wafer. The wafer surface could be bare silicon but usually has an existing pattern previously defined on its surface. Once aligned, the mask and wafer are exposed to controlled UV light to transfer the mask image to the resist-coated wafer (see Figure 13.11). The light energy activates the photosensitive components of the photoresist. Important quality measures for alignment and exposure are linewidth resolution, overlay accuracy, and particles and defects.

Process Summary:
- Transfers the mask image to the resist-coated wafer
- Activates photo-sensitive components of photoresist
- Quality measures:
  – linewidth resolution
  – overlay accuracy
  – particles and defects

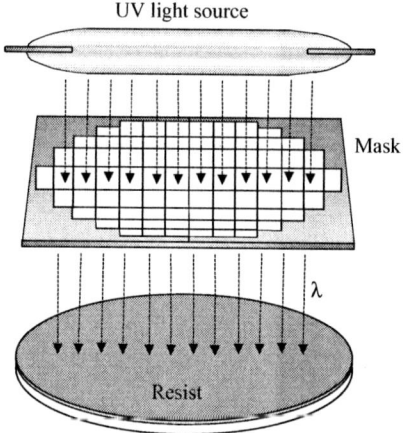

FIGURE 13.11 Alignment and Exposure

Step 5: Post-Exposure Bake (PEB)

There is a post-exposure bake on a hot plate at 100 to 110°C that is necessary for the deep UV (DUV) resists. This bake follows immediately after the photoresist exposure. A few years ago, it was an optional step for non-DUV conventional resists, but it has become a virtual standard even for conventional resists.

Step 6: Develop

Develop is the critical step for creating the pattern in the photoresist on the wafer surface. The soluble areas of the photoresist are dissolved by liquid developer chemicals, leaving visible patterns of islands and windows on the wafer surface. The most common methods for development are spin, spray, and puddle (see Figure 13.12). Following development, the wafers are rinsed in deionized (DI) water and then spin-dried.

Step 7: Hard Bake

A post-development thermal bake, referred to as hard bake, is required to evaporate the remaining photoresist solvent and improve the adhesion of the resist to the wafer surface. This is the critical step for stabilizing the resist for the following etch or implant processing. The hard bake temperature for positive resists is about 120 to 140°C. This is a higher temperature than soft bake, but it cannot be too high or else the resist will flow and deform the pattern.

Process Summary:
- Soluble areas of photoresist are dissolved by developer chemical
- Visible patterns appear on wafer
  – windows
  – islands
- Quality measures:
  – line resolution
  – uniformity
  – particles and defects

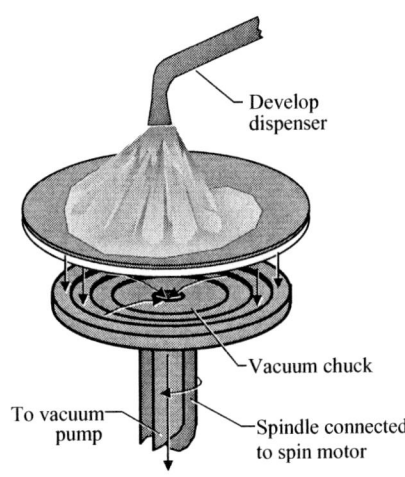

FIGURE 13.12 Photoresist Development

Step 8: Develop Inspect

Once the resist is patterned on the wafer, an inspection is done to verify the quality of the resist pattern. The inspection system is nearly always automated for patterning on highly integrated critical layers. The purpose of this inspection is twofold: to identify wafers that have quality problems with the resist and to characterize the performance of the photoresist process to meet specifications. If the resist is determined to be defective, it can be removed through resist stripping and the wafer can be reprocessed.

The goal of the photolithography process, as in any manufacturing process, is a defect-free product. However, it would be catastrophic to not inspect and leave defects in the resist. Develop inspect is one of the few areas in the wafer fabrication process where errors can be detected and corrected. Once defective wafers are moved to the next patterning step, usually etch, then there is no further chance to correct mistakes. If a wafer is etched incorrectly, it has a fatal defect, is considered scrap, and is of no further value to the company. This is why inspection data to characterize and improve the photoresist process is so important.

VAPOR PRIME

To successfully manufacture integrated circuits, wafers must be meticulously clean at all steps of the wafer fabrication process. Cleaning steps are necessary due to the inevitable contamination that occurs during storage and handling between process steps. Wafer surface preparation is critical to achieving a high-yield photolithography process, since many types of defects can be traced back to contaminated wafers.

Wafer Cleaning

The first step of photolithography is to clean and prepare the surface of the wafer, usually before the wafers arrive at the photolithography workbay. Wafer contaminants and the appropriate cleaning processes were discussed in Chapter 6. As a review, undesirable surface contaminants are particles, metallic impurities, organic contamination, and native oxide. There are many sources of these contaminants, including people, process chemicals, process equipment, packaging and storage, wafer handling methods, and environmental conditions. Thin layers of contaminants on the wafer can consist of ionic (metallic) impurities and atomic and polymeric (organic) layers, which are all difficult to detect. Refer to Chapter 6 for a discussion of types of contamination and cleaning processes.

One of the major effects of contaminants on the wafer surface during photolithography is poor adhesion of the photoresist to the wafer. This condition creates a yield problem of resist lifting during development or the subsequent etch operation. Resist lifting leads to undercutting of the underlying film layer during the etch process (see Figure 13.13 on page 346). Particulate contamination in the resist can lead to uneven resist coating and pinholes in the resist.

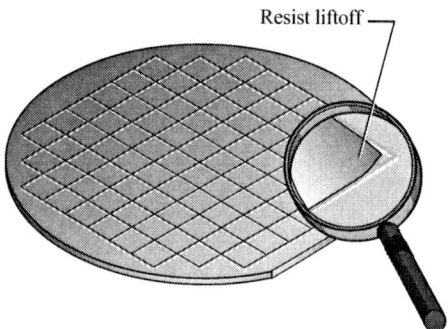

FIGURE 13.13 Effect of Poor Resist Adhesion Due to Surface Contamination

Often wafers entering the photolithography process will have just completed an oxidation or deposition operation and will be in a clean condition. The best situation is to coat these clean wafers with resist as soon as possible. To achieve this, some fabs will place time limits on how long wafers can wait before undergoing photolithography processing to minimize contamination adsorption on the wafer surface.

Dehydration Bake

Silicon wafers readily adsorb moisture on their surface. A wafer surface exposed to moisture is called *hydrophilic* (also called *hydrated*). For resist adhesion, it is important to have a dry, or *hydrophobic*, wafer surface. A hydrophobic wafer is also called *dehydrated* and promotes good resist adhesion. One way to maintain a hydrophobic wafer surface is to coat the wafer with photoresist as quickly as possible after leaving the previous process. Another method is to maintain a controlled room humidity below 50% relative humidity.

Due to the criticality of having a dry wafer surface for resist adhesion, a *dehydration bake* is usually done before priming and resist spin coating. The actual bake temperature varies, with 200 to 250°C commonly used. Dehydration bake temperatures usually do not exceed 400°C, and often cannot attain this temperature due to the temperature sensitivity of the underlying layers (e.g., aluminum). Total wafer surface dehydration occurs at temperatures above 750°C. However, moisture from the atmosphere is quickly reabsorbed back on the wafer surface during cooling; therefore, there is little benefit to heating to this temperature.[4]

Dehydration bake is typically done in a convection oven with a dry inert gas (such as nitrogen) or in a vacuum oven. It can also be done on a hot plate, which is followed by a cold plate to quickly reduce the wafer temperature. Since nearly all wafer fabs use automated wafer track systems, the dehydration bake process is integrated into the wafer handling system.

Wafer Priming

Right after dehydration bake the wafer is primed with *hexamethyldisilazane (HMDS)*, which serves as an adhesion promoter. This process is similar to a paint primer being used to prepare wood for a coat of paint. The HMDS reacts with the silicon surface to tie up molecular water, while also forming a bond with the resist material. It essentially serves as a link between the silicon and the resist so that these materials become chemically compatible.

An important aspect of wafer priming is that the wafer should be coated with resist quickly after the prime operation to minimize moisture problems. It is recommended that coating be performed no later than 60 minutes after completing the priming step.[5] This priming is typically controlled by software on the automated track system.

Priming Techniques ■ HMDS may be applied by puddle, spray, and vapor methods. The method used to prime the wafer is integrated onto the wafer track system.

Puddle Dispense and Spin. The puddle dispense and spin method is used for single wafer processing (see Figure 13.14). The temperature and volume are easily controlled and the system requires a drain and exhaust. This method consumes a large amount of HMDS, which can be a disadvantage.

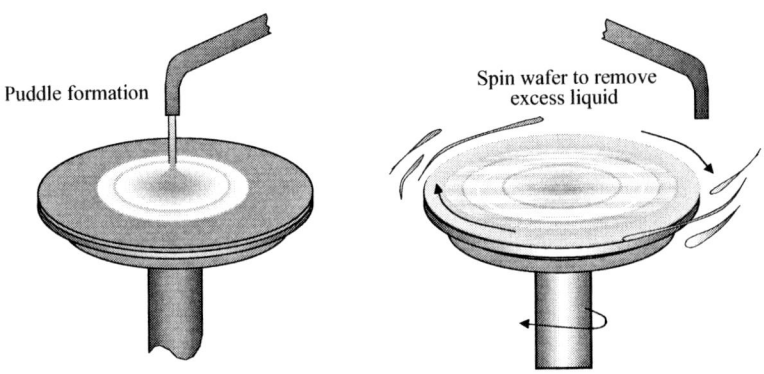

FIGURE 13.14 HMDS Puddle Dispense and Spin

Spray Dispense and Spin. Spray dispense uses a nozzle spray to deposit a fine mist of HMDS on the wafer surface. An advantage of this approach is the spray will assist in particle removal from the wafer surface. A disadvantage is a longer application time and a high consumption of HMDS.

Vapor Prime and Dehydration Bake. The most common method for applying HMDS to the wafer surface is with a *vapor prime coating*. The vapor priming is done at a typical temperature and time of 200 to 250°C for 30 seconds. An advantage of vapor priming is there is no contact with the wafer, which reduces the possibility of particulate contamination from the liquid HMDS. Vapor priming also minimizes consumption of HMDS. Adequate priming of the wafer surface is confirmed with a contact angle meter (see Chapter 7).

One approach is to first perform a dehydration bake followed by a vapor prime of single wafers by thermal conduction heating on a hot plate (see Figure 13.15). The wafer holder is typically made of quartz. Advantages of this approach are inside-out baking of the wafer, low defect density, uniform heating, and repeatability.

Process Summary:

- Dehydration bake in enclosed chamber with exhaust
- Hexamethyldisilazane (HMDS)
- Clean and dry wafer surface (hydrophobic)
- Temp ~ 200 to 250°C
- Time ~ 60 sec.

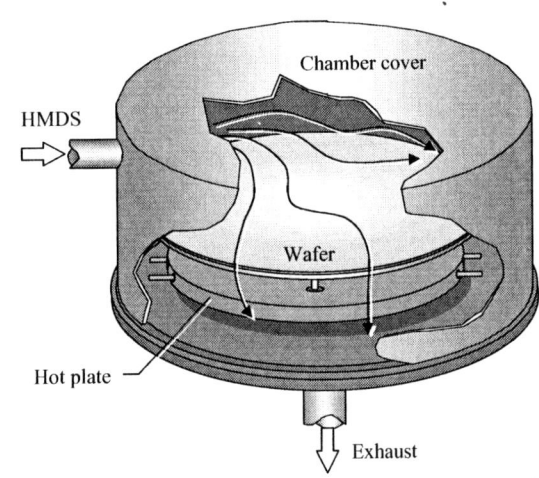

FIGURE 13.15 HMDS Hot Plate Dehydration Bake and Vapor Prime

Another method for dehydration bake in conjunction with vapor priming is to use a vacuum chamber with a nitrogen carrier gas. This is a batch process where the wafers are placed in a quartz holder in the oven chamber. The heated chamber is evacuated and back-filled to a preset pressure with HMDS vapor in the nitrogen carrier gas. At the completion of the pretreatment, the oven is evacuated and back-filled with nitrogen at atmospheric pressure.

SPIN COAT

Once the wafer surface has been cleaned, dehydrated, baked, and primed, it is ready for the application of photoresist. Photoresist is applied by spin coating, where the wafer is spun to form a thin film of resist on the wafer surface. Before considering the photoresist spin process, we will review the different chemistries of photoresists.

Photoresist

Photoresist is an organic compound that experiences a change in solubility in a developer solution when exposed to ultraviolet (UV) light. Photoresists used in wafer fabrication are applied to the wafer surface as a liquid but dried into a film. The purpose of photoresist in wafer fabrication is:

1. To transfer the mask pattern to the resist on the top layer of the wafer surface.
2. To protect the underlying material during subsequent processing (e.g., etch or ion implantation barrier).

As successive generations of device circuit density have reduced critical dimensions, resist technology has undergone improvements in order to transfer the submicron linewidth patterns to the wafer surface. These resist improvements include:

1. Better image definition (resolution).
2. Better adhesion to semiconductor wafer surfaces.
3. Better uniformity characteristics.
4. Increased process latitude (less sensitivity to process variations).

Types of Photoresists ■ The two major types of optical photoresists are negative resist and positive resist. This classification is based on how the resist material responds to UV light. For a negative resist, the UV-exposed regions become crosslinked and hardened. This makes the exposed photoresist less soluble in the developer solution; the photoresist is not removed in the developer liquid. A negative mask image is patterned in the resist. For a positive resist, UV-exposed regions of the resist become more soluble and a positive mask image appears in the resist. A positive resist breaks down during exposure to light and the exposed areas are easily washed away in the developer solution.

We can also group photoresists based on the smallest CD the resist can pattern. A major group is conventional resists capable of patterning linewidth dimensions down to and including 0.35 μm. A new resist technology introduced in the late 1990s was chemically amplified (CA) resist for deep UV (DUV) wavelengths (see the following section). Chemically amplified resist technology can pattern fine-geometry CDs of 0.25 μm and below in high-volume production. This patterning has also been demonstrated in a lab environment for a CD of 0.05 μm. For fabrication of high-performance ICs, a wafer may be patterned with conventional resist for noncritical layers and CA resist for critical layers. We will refer to all optical resists that are not chemically amplified as conventional resists. There are also special resists used for the nonoptical photolithography techniques of e-beam pattern reproduction and X-ray exposure systems (discussed in Chapter 15). Our focus will be on the conventional and chemically amplified optical resists used in wafer fabrication.

Negative Versus Positive Resists ■ As previously outlined, negative photoresist hardens when exposed to light and becomes insoluble. When developed, the image is the opposite of the original mask pattern. Positive photoresist softens when exposed to light and becomes soluble. When developed, the image is the same as the original mask pattern.

The earliest resists in semiconductor microlithography were primarily negative resists until the mid-1970s. Negative resists of this era exhibited good adhesion to the wafer substrate and good resistance to etching. However, due to the swelling and distortion during develop, negative resists typically have a resolution limit of about 2 μm.[6] As long as the circuit linewidth CD remained above this dimension, then negative resists were acceptable. With the introduction of VLSI and ULSI technology and their associated micron and submicron circuit image sizes, the negative resist type was replaced by positive resist. Positive resist had been around for many years, but it had poorer adhesion and its capability for improved resolution was not needed.

This change in the 1970s from negative to positive resist presented a fundamental change in the photolithography process. It required changing the polarity of the photomasks, meaning clear-field regions became dark (opaque) and dark-field regions became transparent. However, this change did not simply involve changing the fields in the mask-making process. Mask dimensions print differently with the two resists. As an example, for a negative resist with a clear-field mask, the dimensions in the resist are smaller than the corresponding photomask dimensions because of light scattering (diffracting) around the image. It is the opposite for a positive resist and a dark-field mask, as diffraction will tend to enlarge the image in the resist. Thus, changing from negative to positive resist required new masks and different adjustments to the mask during construction. Positive resists remain the dominant type of resist today.

Photoresist Physical Properties

In its most elemental form, photoresist is a polymer solution in an organic solvent. The photoresist must perform well under all the different process conditions, including coating, spinning, baking, develop, ion implantation, and etch. There are many different types of photoresists used in wafer applications, each with its own physical properties that are directly related to the photolithography process requirements. A particular resist is selected based on the following physical properties:

- Resolution
- Contrast
- Sensitivity
- Viscosity
- Adhesion
- Etch resistance
- Surface tension
- Storage and handling
- Contaminants and particles

Resolution. Resolution is the ability to differentiate between two or more closely spaced patterned features on the wafer surface. A practical way to interpret resolution is by the smallest feature that can be printed on a wafer and meets specified quality requirements. The smaller the critical dimension produced, the better the resolution capability of the resist and the photolithography system.

Contrast. The *contrast* represents the sharpness of the transition from exposure to non-exposure in photoresists (see Figure 13.16). Contrast represents the ability of the resist to become exposed only in those areas defined by the clear region of the photomask. It is desirable to have high-contrast to produce vertical resist sidewalls.

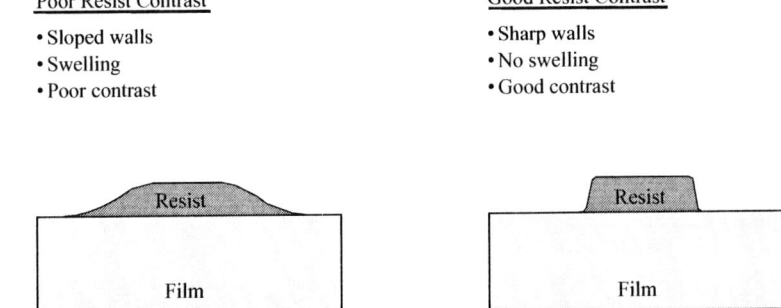

FIGURE 13.16 Resist Contrast

Sensitivity. *Sensitivity* is the minimum amount of light energy (measured in millijoules/cm^2 or mJ/cm^2) at a certain wavelength that is needed to produce a good pattern in the photoresist on the wafer surface. The amount of light energy supplied to a resist is commonly called the *exposure dose.* Sensitivity is important for new resists because of the low energy levels emitted at the shorter UV wavelengths (e.g., deep UV). Resist manufacturers sensitize their resist formulations for maximum absorption in the areas of optimum energy output of the radiation sources available.

Viscosity. The *resist viscosity* represents a quantitative measure of flow characteristics for the liquid resist. Viscosity is time-dependent in that it increases as the solvent in the resist evaporates during usage. Viscosity is important because it affects resist thickness and the uniformity of the resist over the wafer surface topography such as steps and small gaps. As viscosity increases, the resist has less tendency to flow and its thickness on the wafer increases. A thicker resist provides better step coverage and better dry etch resistance but makes the resolution of small openings more difficult. Thick resist also has less probability of having pinholes, which are microscopically small voids that pass completely through the resist to the substrate material. Pinholes are undesirable for subsequent etch processing because the etch chemicals can pass through the small voids and damage the underlying substrate material. Low viscosity resist has more tendency to flow and will lead to thinner coverage of the wafer surface.

Solid content represents the amount of the liquid resist that would remain as a solid if the solvent were evaporated. Specific gravity (SG) is a measure of the resist density and is related to the solid content in the resist. A higher SG means there are more solids in the resist, which indicates a higher viscosity and lower tendency to flow.

Viscosity is measured in the unit of poise. Photoresist viscosity is measured in centipoise (1/100th of a poise), or cps. Viscosity may be reported in centistokes (cs), which is the kinematic viscosity, whereas the centipoise is the absolute viscosity. The kinematic viscosity is found by dividing the absolute viscosity (centipoise) by the density (specific gravity). For instance, a 70-cps resist with an SG of 0.8 would produce a kinematic viscosity of:

$$\text{Kinematic viscosity} = [\text{absolute viscosity (cps)}] / (\text{specific gravity})$$
$$= (70 \text{ cps}) / 0.8 \text{ SG}$$
$$= 87.5 \text{ centistokes (cs)}$$

Adhesion. *Resist adhesion* describes how strongly the resist sticks to the substrate. The resist must adhere to many different types of surfaces, including silicon, polysilicon, silicon dioxide (doped and undoped), silicon nitride, and different metals. Lack of resist adhesion leads to distorted patterns on the wafer surface. The resist adhesion must withstand exposure, development, and subsequent processing (e.g., etch and ion implantation).

Etch Resistance. The resist film must maintain its adhesion and protect the substrate surface from the subsequent wet and dry etch processes (see Chapter 16). This property is known as *etch resistance.* Some dry-etch processes are done at higher temperatures (e.g., 150°C), which require a resist that has thermal stability to maintain its shape.

Surface Tension. *Surface tension* refers to the molecular attraction forces in a liquid that tend to pull surface molecules toward the body of the liquid. An example of high surface tension is water beading up on a waxed automobile surface (see Figure 13.17). Resist has molecular forces that create a relatively high surface tension so that the resist molecules hold together during the various process steps. At the same time, the surface tension of resist must be low enough to provide for good flow and wafer coverage during application.

Storage and Handling. Photoresist chemicals are activated in the presence of energy, either as light or heat, which requires carefully controlled storage and usage conditions. Resists are specified with a shelf life and storage temperature environment. Crosslinking of negative resists and sensitizer-delay of positive resists can occur with extended storage time or elevated temperature. If the solvent on the resist was allowed to evaporate due to an unsealed container, the viscosity would change rapidly and solids could precipitate from the liquid.

Resist manufacturers have developed automated dispense mechanisms to control contamination and evaporation. The resist containers are often a closed system, such as collapsible pouches that are opaque to light and loaded directly into the wafer tracks to avoid opening in the atmosphere.

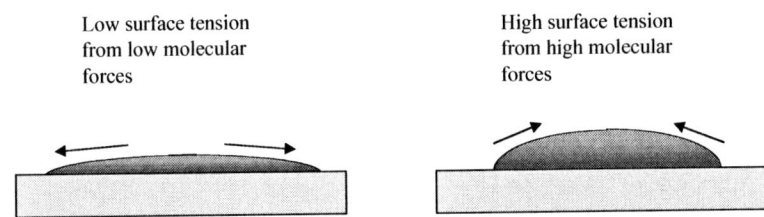

FIGURE 13.17 Surface Tension

Contaminants and Particles. As with any chemical used in wafer fabrication, the purity of the resist material is important. The most critical purity concerns for resist are mobile ionic contaminants (MICs) and particles. The resist is applied to the wafer surface material and can easily introduce detrimental contamination to the wafer. To control contamination and particles, resist suppliers meet extremely tight filtration and packaging procedures. Localized point-of-use resist filtration with membrane filters just before resist application can control contamination in the resist to < 1 ppb.

Conventional I-Line Photoresists

Conventional i-line photoresists for optical microlithography are those suitable for i-line UV wavelengths (365 nm), which corresponds to noncritical layers with a CD down to 0.35-μm. This condition includes both positive and negative resists, although positive resists are the most common. Note that the characteristics of i-line photoresist also represent the same basic properties for resists used at g-line (436 nm) and h-line (405 nm) wavelengths.

I-line photoresists are composed of three basic components,[7] and frequently have a fourth component of special additives (see Figure 13.18):

1. Resin (polymer material)
2. Sensitizer
3. Solvent
4. Additives (optional)

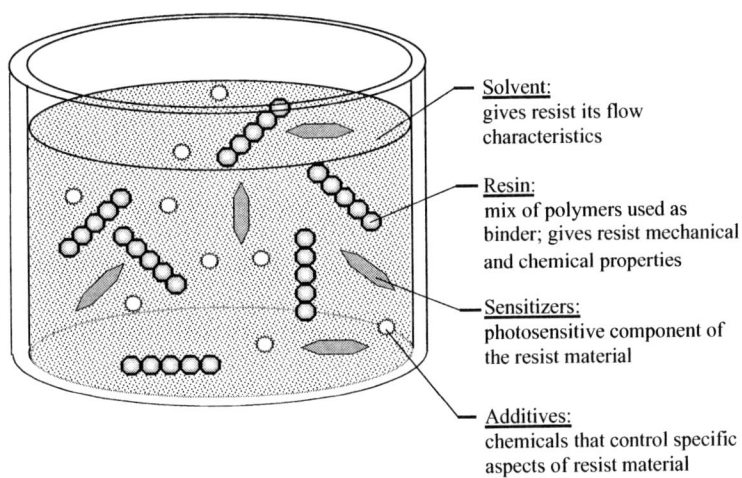

FIGURE 13.18 Components of Photoresist

Resin. The resist resin is an inert matrix of polymers (large organic molecules containing carbon, hydrogen, and oxygen) which is used as a binder to hold together the different materials in the resist. Resin gives the resist its mechanical and chemical properties, such as adhesion, etch resistance film thickness, flexibility, and thermal flow stability. The resin material is typically insensitive to light, meaning it does not undergo chemical change upon exposure to UV light.

Sensitizer. The resist sensitizer is the photosensitive component of the resist material that reacts in response to radiant energy in the form of light, especially in the UV region. The sensitizer reacts photochemically in response to light energy.

Solvent. The resist solvent keeps the resist in its liquid state until it is applied to the wafer substrate. Most solvents are evaporated from the resist before the exposure is done and have little effect on the photochemistry of the resist.

Additives. Resist additives are usually proprietary chemicals, which means the contents are developed by the manufacturer and not disclosed to the public for competitive reasons. Additives are used to control or modify specific chemical or light response aspects of the resist material. It also includes dyes to control the reflectivity of the resist.

Negative I-Line Photoresists ■ The resin for i-line negative resist is typically a chemically inert polyisoprene polymer that is a natural rubber. The polyisoprene is suspended in a solvent such as xylene. The resist sensitizer is a photoactive agent that releases nitrogen gas on exposure to UV light with the proper wavelength, generating free radicals that form crosslinks between the rubber molecules. This crosslinked rubber formed from UV light exposure becomes insoluble in the organic developer solution (see Figure 13.19). At the same time, the unexposed (and therefore unpolymerized) areas of the resist are rinsed away in the developer. These basic steps for crosslinking of negative resist are shown in Table 13.3.

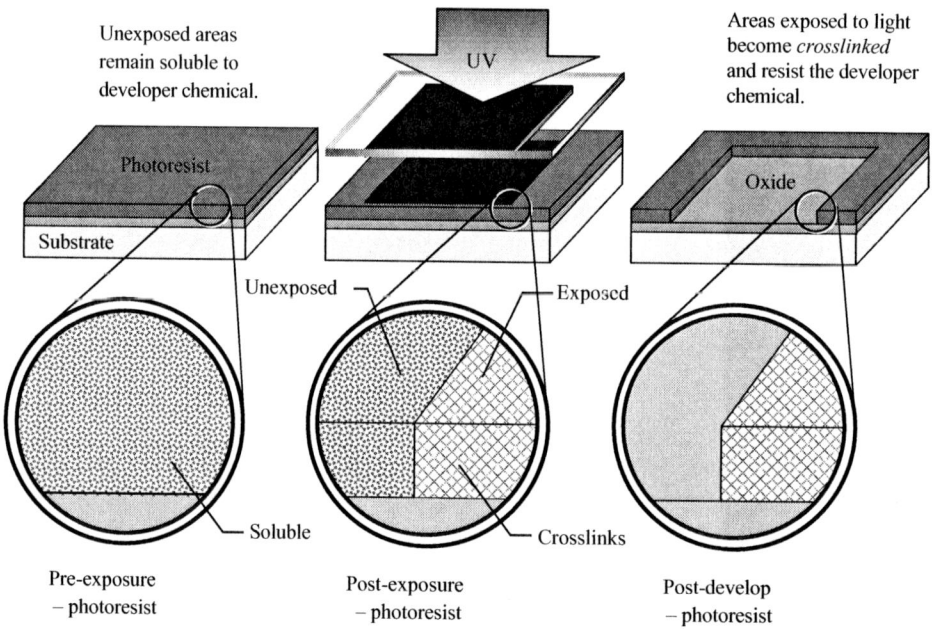

FIGURE 13.19 Negative Resist Crosslinking

TABLE 13.3 Steps to Negative Resist Crosslinking

1. Resist resin is polyisoprene rubber polymer suspended in solvent (soluble in developer solution).
2. Photoactive sensitizer agent releases N_2 during exposure.
3. Release of N_2 generates radicals.
4. Radicals polymerize resist by crosslinking rubber polymer (insoluble in developer solution).

A major disadvantage of conventional negative resists has been their solvent-induced swelling of the exposed resist regions during development. This swelling distorts the resist pattern on the wafer surface and is unacceptable for fine-geometry features with micron and submicron critical dimensions. Another disadvantage is the resist's reactivity with oxygen during UV exposure that can inhibit crosslinking. There is, however, research into new negative resist formulations that do not swell or distort patterns.[8] A negative resist has high exposure speed (which translates into higher throughput of wafers) and good wafer adhesion.

Positive I-Line Photoresists ■ Positive i-line photoresist is the most common resist in IC photolithography. The resin material in positive i-line photoresists is a phenol-formaldehyde polymer called *novolak*. An example of a common use of novolak is the glue that binds the different layers of plywood lumber. Novolak is a long chain polymer that forms the key resist film properties, such as good adhesion and chemical resistance, making it an acceptable barrier to subsequent processes such as etch. Novolak resin dissolves in developer solution when there is no dissolution inhibitor present.

The sensitizer in positive i-line photoresist is referred to as a *photoactive compound (PAC)*, the most common being a diazonapthoquinone (DNQ). Before exposure to light, the DNQ serves as a strong dissolution inhibitor, reducing the dissolution rate. After exposure to UV light, DNQ promotes dissolution in the developer by a factor of about 100 or more (see Figure 13.20).[9] This i-line resist is often called *DNQ-novolak*, representing the mixture of diazonapthoquinone (DNQ) and novolak resin. It originally evolved from materials used to make blueprints. Note the mask in Figure 13.20 is the direct opposite of the mask used in Figure 13.19.

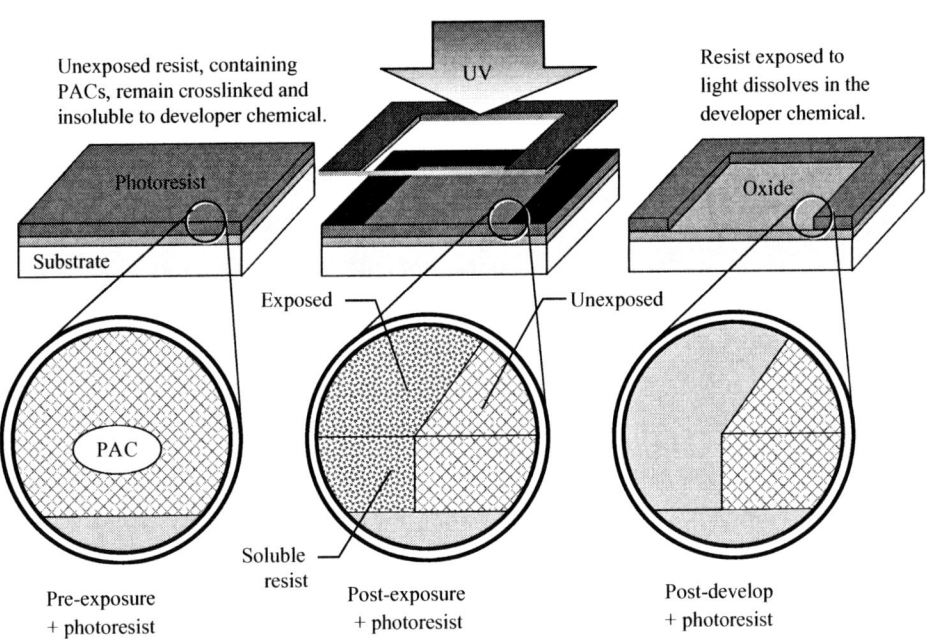

FIGURE 13.20 PAC as Dissolution Inhibitor in Positive I-Line Resist

When exposed to light energy of the proper wavelength, the diazonapthoquinones (DNQs) photochemically decompose in the resist. The exposed DNQ now becomes a dissolution promoter in exposed regions of the resist. This change is accomplished when decomposition of the DNQ creates carboxylic acid, which is highly soluble in the developer solution. In this manner, exposed positive resist changes from an insoluble to a soluble form and dissolves in the developer solution. This is the dissolution of the positive resist (see Table 13.4 on page 354). With this process, areas of the resist that correspond to transparent mask areas will dissolve, thus creating a positive image. In the DNQ resist, regions exposed to light are about 100 times more soluble in the developer than unexposed resist regions.[10]

TABLE 13.4 Steps to Positive I-line Resist Dissolution in Developer

1. Resin is phenol-formaldehyde polymer (novolak) suspended in solvent.
2. Sensitizer photoactive compound DNQ added to novolak as strong dissolution inhibitor (insoluble in developer).
3. During exposure, DNQ photochemically decomposes and creates carboxylic acid.
4. Carboxylic acid promotes dissolution of novolak in exposed areas of resist (resist becomes soluble in developer).

A major advantage of a positive resist is that the unexposed areas of the resist are not affected by the developer, since the resist is initially insoluble and remains as such. Thus the pattern of fine lines that is transferred to the resist will maintain the linewidths and shape during the photolithography process, producing good linewidth resolution. Older positive resists had marginal adhesion strength, but modern positive resists have been designed with more acceptable adhesion.

A reason why positive resist has good resolution is contrast. The positive resist can distinguish better between the light and dark areas of the mask, making a sharp transition from exposure to nonexposure in the resist. Positive photoresist has good contrast characteristics, which produces improved line resolution due to sharp resist sidewalls (see Figure 13.21).

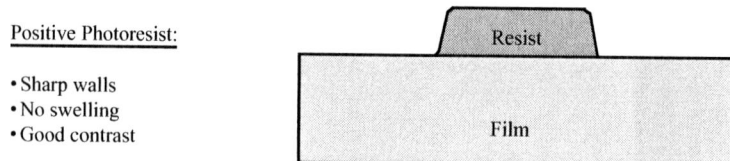

FIGURE 13.21 Good Contrast Characteristics of Positive I-Line Photoresist

Deep UV (DUV) Photoresists

For optimum pattern resolution, a goal in microlithography has been for the wavelength of the exposure light to be proportional to the critical linewidth dimension (this is explained in Chapter 14). For exposure with an i-line wavelength of 365 nm in the mid-1990s, a CD of 0.35 μm was achievable. In order to achieve a CD of 0.25 μm on critical layers in the latter 1990s, it was necessary to reduce the wavelength of the exposure light source to a value of around 250 nm (0.25 μm). This value corresponds to the 248 nm UV wavelength referred to as *deep UV (DUV)*. This wavelength is shown in Figure 13.22.

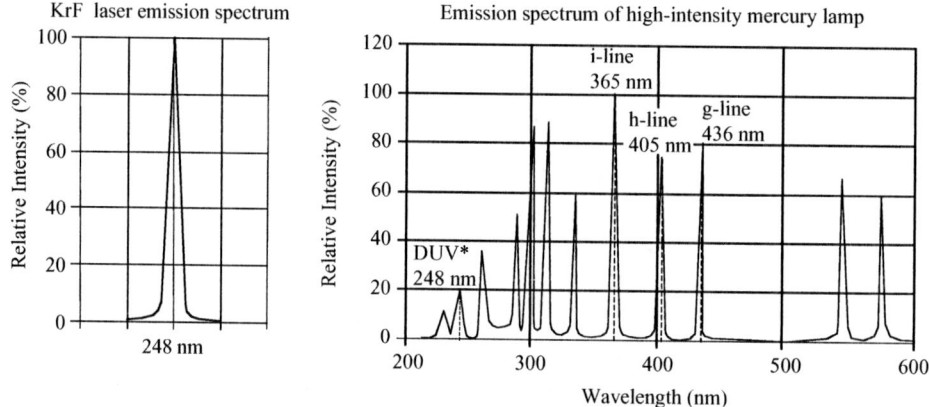

Intensity of mercury lamp is too low at 248 nm to be usable in DUV photolithography applications. Excimer lasers, such as shown on the left, provide more energy for a given DUV wavelength.

FIGURE 13.22 DUV Emission Spectrum Mercury Lamp Spectrum
(Used with permission from USHIO Specialty Lighting Products)

Standard i-line photoresists are generally not acceptable at the DUV wavelengths necessary for CDs of 0.25 μm and below due to lack of sensitivity to smaller wavelengths. Resists have a sensitivity to the minimum amount of light energy required to transfer an acceptable pattern. The DNQ i-line resist system has excessive light absorption and lacks sensitivity at the low light intensity of the shorter DUV wavelength.[11] If the i-line resist has excessive light absorption, then the light never penetrates the full thickness of the resist, leading to incomplete exposure and poor image formation. The lower light sensitivity of the i-line resist would require longer exposure time, causing a decrease in wafer throughput. This low sensitivity is a serious problem when the light source is a mercury lamp with its limited output at the DUV wavelength of 248 nm. For this reason, i-line photoresists are used on noncritical wafer layers of 0.35 μm linewidths and above.

Chemical Amplification for DUV Resists ■ A major change in photoresist technology occurred with the introduction of a new type of photoresist for DUV wavelengths based on chemical amplification. *Chemical amplification (CA)* is a means to increase the sensitivity of the resist substantially over that of the DNQ-novolak i-line resists. All positive and negative tone 248-nm resists are chemically amplified.[12] The researchers who developed chemically amplified resist in the 1980s at IBM (International Business Machines) realized that the light absorption and sensitivity of the i-line resist system was unacceptable for shorter DUV wavelengths, creating the need for a new photoresist that offered the benefits of chemical amplification.[13]

Chemically amplified DUV resists are formulated to react at an accelerated rate to the exposing DUV light source based on an acid-catalyzed reaction. This action is accomplished by boosting the resist sensitivity with the use of a sensitizer known as a *photoacid generator (PAG)*. The PAG produces an acid upon exposure to radiation from deep UV light. This acid is only produced in the resist regions exposed to light. In the areas of resist where there is no light exposure, the acid is not created. Most well-known acid generators used in DUV resists are onium salts, such as iodonium and sulfonium salts.[14]

CA DUV resist technology has undergone significant development over the years. The basic principle for all CA DUV resist resins is that the resin requires a chemical protecting group that makes it insoluble in aqueous (water-based) developer. The acid generated by the PAG in the DUV-exposed areas of the resist then removes the protecting group (referred to as *deprotection*) through a catalytic reaction while heated during the post-exposure bake step. After heating the DUV resist, the exposed resin becomes soluble in the developer.

An early CA DUV resin developed and used at IBM in the 1980s to manufacture DRAMs using DUV lithography was t-BOC (tert-butoxycarbonyl).[15] The t-BOC is the protecting group for the resin known as PHS or PHOST (poly[hydroxystyrene]). Once the photo-generated acid removes the t-BOC-protecting group (referred to as an acid-catalyzed t-BOC deprotection reaction), then the PHS resin becomes soluble in developer.

PHS resin is well-established for use in CA DUV positive resists (note that all major resist manufacturers are actively researching "new generation" CA DUV resists).[16] PHS resin is a phenolic copolymer that has protecting groups that make it dissolve poorly in an aqueous-based solution. During deep UV exposure, the PAG acid is photochemically generated in all exposed resist regions. This acid is the chemical catalyst that removes (cleaves) the protecting group during post-exposure thermal bake. This operation renders the exposed PHS highly soluble in the aqueous developer (see Figure 13.23 on page 356). Essentially, these chemical reactions transform the exposed regions of the DUV resist from an insoluble resist into a highly soluble resist in an aqueous-based developer. Table 13.5 on page 356 outlines the four steps to developing a CA DUV resist.

The first commercial CA DUV resist was a negative system based on crosslinking a phenolic resin with an acid catalyst. All negative CA resists are built on this phenolic resin crosslinking. Negative CA resists have poor process latitude in imaging small contact holes because of the clear-field image during exposure. At the same time, negative DUV resists are resistant to thermal flow for temperatures below 150°C, due to the crosslinking of the phenolic resin polymer.[17] This condition can be beneficial in some etch applications.

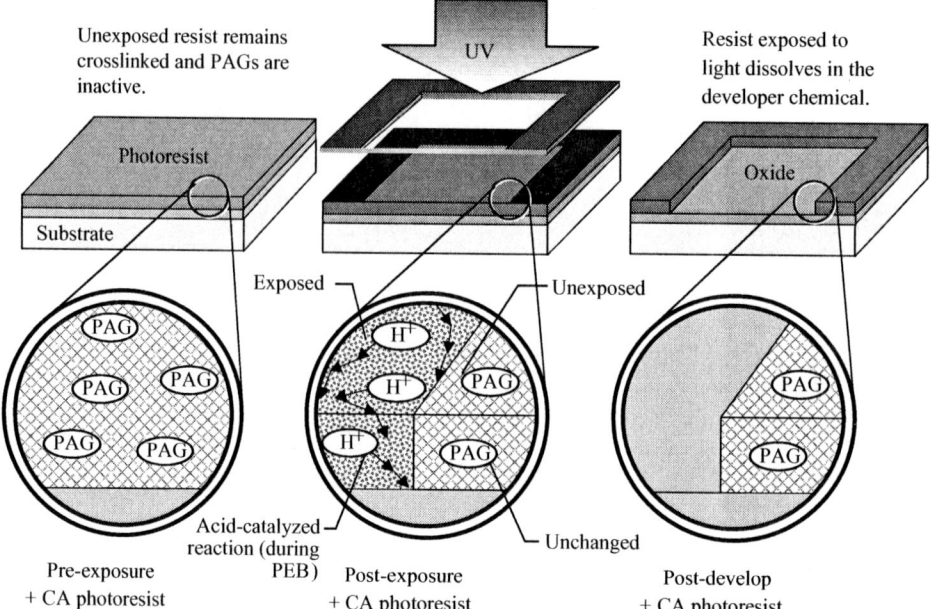

FIGURE 13.23 Chemically Amplified (CA) DUV Resist

TABLE 13.5 Exposure Steps for Chemically Amplified DUV Resist

1. Resin is phenolic copolymer with protecting group that makes it insoluble in developer.
2. Photoacid generator (PAG) generates acid during exposure.
3. Acid generated in exposed resist areas serves as catalyst to remove resin-protecting group during post-exposure thermal bake.
4. Exposed areas of resist without protecting group are soluble in aqueous developer.

Positive CA resists are the dominant type for patterning wafers with 0.18-μm generation devices using DUV wavelengths. One of the main benefits of CA DUV resists is the speed of resist exposure. CA chemistry is capable of providing a 10-fold improvement in the exposure speed over DNQ-novolak-based resist systems without a major compromise in any of the resist lithographic performance areas.[18] Chemical amplification in DUV resist improves the sensitivity of the resist for low-intensity light sources. In addition, DUV offers high-contrast imaging with vertical sidewall profiles and high resolution suitable for resists used on wafers with critical dimensions of 0.25 μm and below.[19]

193 nm DUV Resists. Suitable chemically amplified DUV resists are being introduced into production at this time for 193 nm wavelength exposure light capable of imaging a CD resolution of 0.18 μm. The biggest change is the unacceptability of resists based on phenolic polymers (DNQ i-line resist) or hydroxystyrene (248 nm DUV resist). Some early 193 nm resists are based on polyvinyl phenol and polymethylmethacrylate (PMMA).[20] Another chemical platform under investigation is cyclic olefin polymer materials, which offer superior etch properties.[21] These resists may be used with multilayer resist technology using top surface imaging (see the advanced lithography section in Chapter 15).

DUV Process Requirements ■ DUV resists are very sensitive to contamination, specifically amines (an organic compound such as ammonia) that are found in the ambient atmosphere. Less than several parts per billion (ppb) exposure to amines commonly found in air will lead to undesirable CD variation in the top of the resist profile. This resist variation leads to unacceptable linewidths after etch. Amine sensitivity in CA DUV resists requires that the air in all photolithography equipment be chemically filtered.[22] Monitoring equipment used in photolithography can detect down to 500 parts per trillion amines.

CA resists for DUV are also dependent on the bake temperature at the post-exposure bake (discussed in Chapter 15) for reducing CD variation. Early DUV resists had significant temperature sensitivities, whereas newer resists have less temperature sensitivity. In addition, early DUV resists required the time between resist exposure and post-exposure bake processing to be limited to a few minutes. Recent DUV resists can be delayed slightly longer, up to about 30 minutes.

Photoresist Dispensing Methods

For semiconductor microlithography, the most widely used method to apply the liquid photoresist to achieve a uniform coating on the wafer surface is *spin coating*. There are four basic steps to applying resist by spin coating (see Figure 13.24):[23]

1. **Dispense.** The resist is dispensed onto the wafer while it is stationary or spinning very slowly.
2. **Spin-up.** Quickly accelerate the wafer rotation to a high rpm (revolutions per minute) spin speed to spread the resist over the entire wafer surface.
3. **Spin-off.** Throw off excess resist to obtain a uniform resist film coating on the wafer.
4. **Solvent evaporation.** Continue to spin the coated wafer at a constant rpm until the solvent evaporates and the resist film is nearly dry.

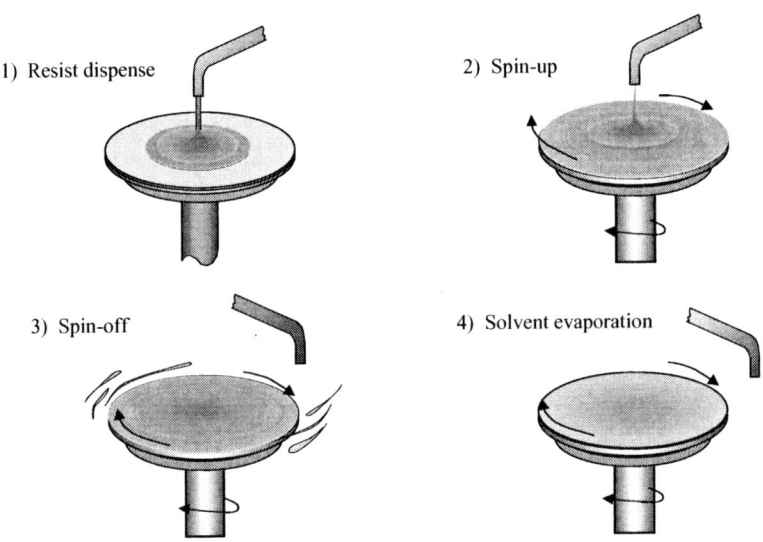

FIGURE 13.24 Four Steps of Photoresist Spin Coating

The two goals in resist spin coating are to have a uniform film coating on a wafer and to achieve a repeatable resist thickness that is maintained from wafer to wafer over extended periods. The target resist thickness is specified for a particular application and is typically on the order of 1 μm. Thickness variation of the resist film on a wafer should be measured less than 20 to 50 Å across the entire wafer surface. Wafer-to-wafer resist thickness control is typically less than 30 Å measured over many wafers.[24]

Spin Coating Equipment ■ Wafer spin coating takes place in an automated wafer track system with wafer handling equipment to move the wafers from operation to operation (see Figure 13.25 on page 358). Robotic handling is preferred over conveyors to minimize particle generation and wafer damage.

The complexity of the automated wafer track is that there are many different operations being performed simultaneously. At any one time, the wafer track could be processing 15 to 20 wafers through different photolithography operations, such as vapor prime resist spin coat, develop, baking, and chilling. Track systems are connected to the wafer stepper and must interface with the stepper so that wafers are delivered and picked up at the correct time with limited waiting.

To spin coat the wafer, robotic handlers position the wafer on a vacuum chuck at the spin coat station in the wafer track. The chuck is a hollow, flat metal disc with holes in its surface that is

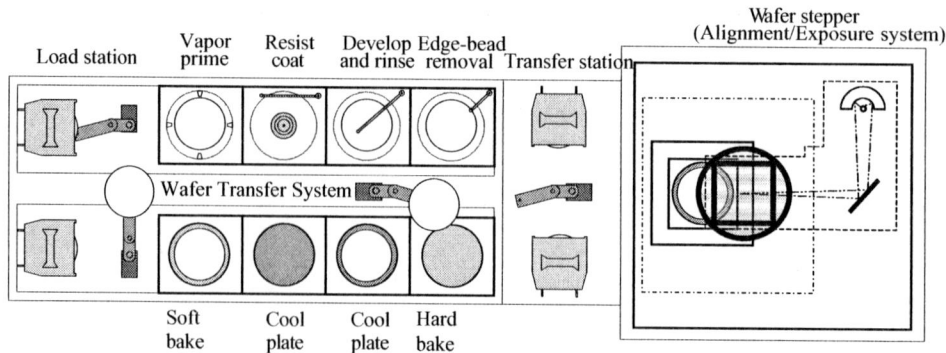

FIGURE 13.25 Automated Wafer Track System for Photolithography

connected to a vacuum. When the wafer is placed on the chuck surface, the vacuum pulls the wafer into intimate contact with the chuck to hold it while spinning.

The resist is dispensed on the wafer through a dispenser nozzle that should deliver a steady stream of resist (see Figure 13.26). The nozzle has a suckback feature to prevent after-dispense drips on the wafer surface. The nozzle is controllable in the X, Y, Z, and θ directions, giving the process engineer flexibility to position the nozzle in a location to optimize thickness, uniformity, and the amount of resist dispensed. A particular dispense method is called radial dispense, where the arm is actually moved across the wafer while the resist is dispensed. This action minimizes the amount of resist dispensed. The movement is computer controlled and can be from edge-to-center or center-to-edge.

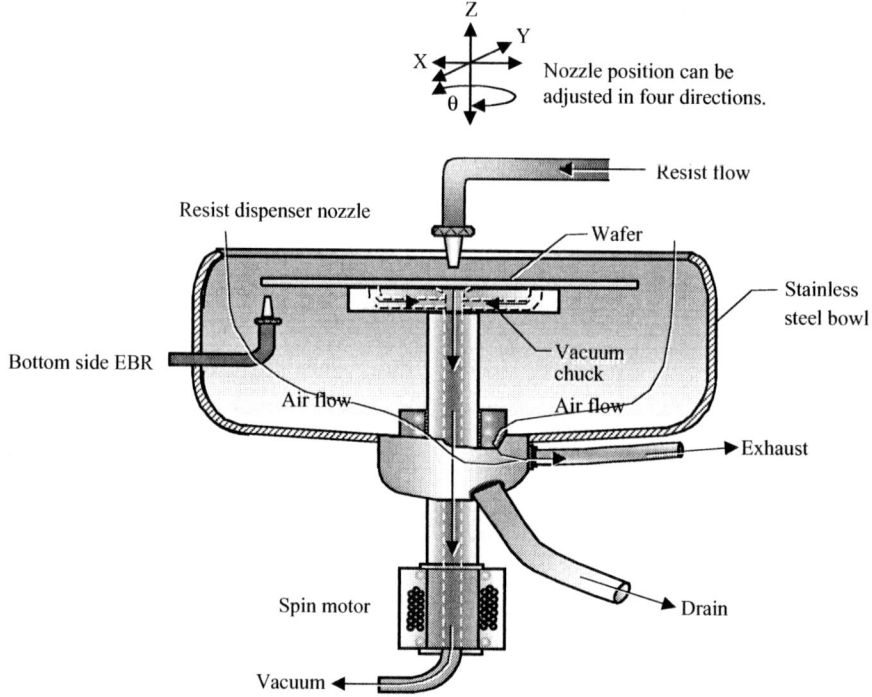

FIGURE 13.26 Photoresist Dispense Nozzle

Spin Coating Parameters ■ The manner of dispensing the liquid resist varies depending on the parameters defined for the application by the process engineer. The resist can be dispensed on a wafer while it is not rotating, known as *static dispense*. After the static dispense, the wafer is first spun at a low rpm to uniformly spread the resist. Once the resist approaches the wafer edge, the rpm is accelerated to the final spin speed (e.g. a typical final speed may be 4,000 rpm). Another approach is to dispense the resist on a wafer that is spinning slowly (e.g., 100 to 200 rpm) in order to uniformly coat the wafer, followed by acceleration to the final spin speed. This is referred to as a *dynamic dispense*.

The amount of resist dispensed depends to a large degree on the viscosity of the resist. For example, a high-viscosity resist with a viscosity of 55 cps may require 2.4 cc of resist for a good coat. On the other hand, a low-viscosity resist with an 8 cps viscosity may need only 1.3 cc of resist. Lower viscosity resists typically have a thinner target thickness.[25]

Resist thickness and uniformity on a wafer are critical quality measures. Thickness is not controlled by the amount of resist deposited since most of the resist flies off the wafer (less than 1% remains on the wafer). The most critical parameters for resist thickness are spin speed and resist viscosity. Many resist manufacturers publish thickness versus spin speed data for their resist formulation (see Figure 13.27). This data assists in determining the optimum final spin speed. Higher viscosities and slower spin speeds will produce thicker layers of photoresist. In general, the resist thickness has been found to vary with spin speed as:[26]

$$\text{Resist thickness} \propto \frac{1}{(\text{RPM})^{1/2}}$$

Where RPM is the spin speed in revolutions per minute.

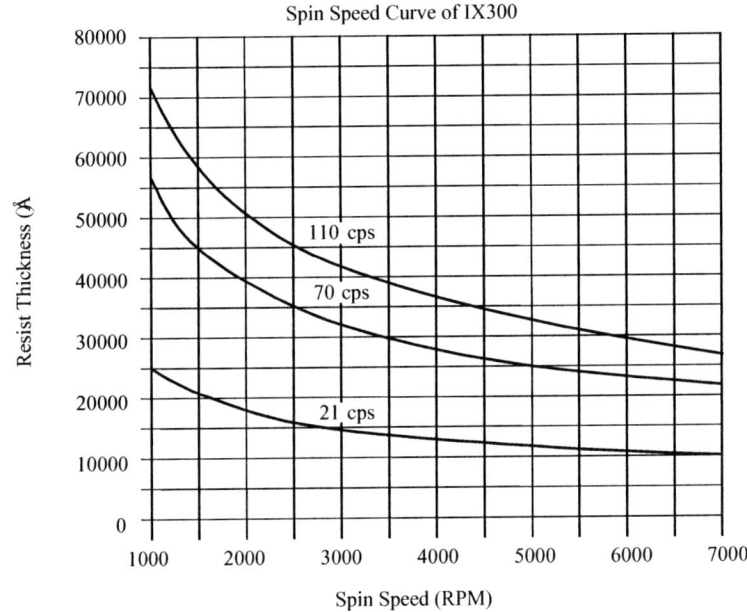

FIGURE 13.27   Resist Spin Speed Curve
(Used with permission from JSR Microelectronics, Inc.)

Different parameters affect resist thickness and uniformity. The acceleration ramp rate to attain the final spin speed is controllable on most spin coaters. High ramp rates usually produce better film uniformity than slow ramp rates. Another parameter is the height of the nozzle and its location during the dispense.

Another important parameter for resist dispense is environmental control. The wafer track system is a closed environment in order to control temperature, humidity, exhaust, and particulate contamination during application of the photoresist coating. Temperature and humidity strongly affect resist uniformity across the wafer. The minienvironment of a wafer track system for photolithography is often controlled for particulate contamination to a class 0.1 level. In addition, the exhaust air flow rate around the spin coater must be properly controlled to avoid negatively impacting resist thickness and uniformity. Too much air flow due to strong exhaust can cause excessive drying of the resist, while poor exhaust flow can cause particles to deposit on the wafer instead of being carried away by the air flow.

Edge-Bead Removal ■ During the wafer spin process, resist flows outward due to centrifugal force toward the wafer edge and onto the backside. This ridge of resist on the wafer edge and backside is called edge bead. When dry, the resist flakes off and causes particles. These particles can land on active circuit areas, on wafer carriers, and inside process equipment, leading to increased defect densities on the wafer. Furthermore, the resist on the back of the wafer can create problems by causing it to stick to wafer chucks.

Resist spin coat stations are equipped with an *edge-bead removal* (*EBR*) device (see Figure 13.26 on page 358). A common approach is to use a spray nozzle assembly to spray a small amount of solvent on the underside of the spinning wafer. The solvent laps up over the beveled edge to the topside, with careful control to ensure the solvent does not reach the topside of the resist (this control is important for defect density reduction). Typical solvents to remove the edge bead are propylene glycol monomethyl ether acetate (PGMEA) or ethylene glycol monomethyl ether acetate (EGMEA).

Some photolithography tools use a laser to expose the edge of wafers right after the normal wafer exposure has taken place. The exposure softens the resist and allows the edge-bead resist to be removed during the normal develop step or by a designated solvent nozzle. This step can be performed in the same track system that was used to spin coat the wafers.

SOFT BAKE

Following the application of resist on the wafer by spin coating, wafers are subjected to an elevated temperature step referred to as *soft bake* (also called *pre-bake*). The reasons for the soft bake of the resist are:[27]

1. Drive off the solvent from the coated resist on the wafer.
2. Improve the adhesion of the resist so that it adheres adequately during the development step.
3. Relieve stresses in the resist film that occurred during the spin process.
4. Prevent resist from getting all over the equipment (keep the tool clean).

The soft bake temperature and time depend on the particular resist and process conditions. The starting point for setting the soft bake parameters is the process recommended by the resist manufacturer. After that, the process is optimized to achieve the adhesion and dimensional control required for the product. The soft bake temperature is usually in the range of 85 to 120°C. The duration of the soft bake varies for different resists, but 30 to 60 seconds is a typical time.

If the resist film were not soft baked and continued directly to alignment and exposure after resist apply, the following problems would occur:[28]

1. The resist film would be tacky and susceptible to particulate contamination.
2. Inherent stresses in the resist film from spin coating would lead to adhesion problems.
3. The high solvent levels would cause inadequate discrimination between exposed and unexposed resist for proper dissolution during development.
4. Outgassing from the resist (from heat during exposure) could contaminate the lens of the optical system.

Before spin coating, photoresist typically contains between 65 to 85% solvents. After spinning, the solvent has been reduced to 10 to 20%, yet the film should still be considered in a liquid state.[29] The ideal amount of solvent after soft bake is about 4 to 7%. Because the solvent is reduced in the soft bake process, the thickness of the resist film also decreases.

Soft Bake Equipment

The preferred method for resist soft bake is heat conduction from a wafer on a vacuum hot plate[30] (see Figure 13.28). In this method, heat is quickly conducted from the hot plate through contact with the backside of the wafer to the resist. The resist is heated from the wafer-resist interface outward, which minimizes the potential for solvent entrapment. Because of the short cycle time (e.g., 30 to 60 seconds), this single-wafer hot plate method is suitable for the flow of multiple wafers through the process steps of an automated wafer track system. In the wafer track process flow, the heating step is usually followed by a cooldown step on a cooling plate. This step rapidly cools the wafer for the next operation. The vacuum hot plate design is the same type as that used for dehydration bake. There is also the option of infrared (IR), microwave, and convection heating for soft bake, but these methods are not commonly used.

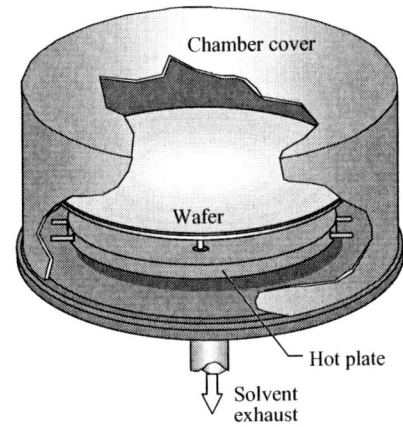

Purpose of Soft Bake:
- Partial evaporation of photoresist solvents
- Improves adhesion
- Improves uniformity
- Improves etch resistance
- Improves linewidth control
- Optimizes light absorbance characteristics of photoresist

FIGURE 13.28 Soft Bake on Vacuum Hot Plate

Process Characterization

The soft bake process requires characterization to determine optimum settings. Reasons to optimize soft bake settings are to have better control over critical dimensions, improve the resist sidewall angle, increase resolution, improve contrast, and achieve wider process latitude.

The solvent content of the resist can be measured as a function of temperature with a *thermogravimetric analysis* (*TGA*). TGA uses the principle of volatization, or polymer weight loss during heating, to represent the amount of solvent loss and thermal decomposition as a function of temperature.[31] To perform this test, a sample is coated with resist and placed immediately into an oven without soft baking. While heating, the sample's weight is measured and plotted (see Figure 13.29).

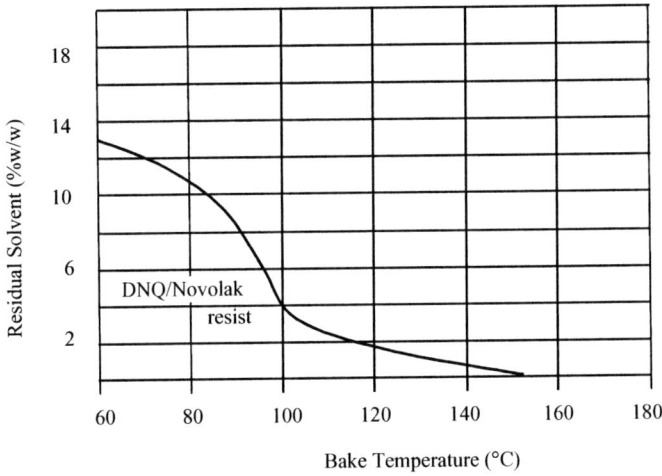

FIGURE 13.29 Solvent Content of Resist versus Temperature During Soft Bake

Another method to characterize the amount of solvent removal is to measure the change in resist thickness as a function of temperature. As solvent evaporates, the resist will become thinner. The temperature at which the resist thickness stabilizes is where most solvents are removed.

PHOTORESIST QUALITY MEASURES

Quality measures for photoresist processing are provided in Table 13.6.

TABLE 13.6 Key Quality Measures for Photoresist

Quality Parameter	Types of Defects	Remarks
1. Resist adhesion.	A. Resist dewetting, which is also called lift-off. The resist does not adhere to the substrate, causing problems during subsequent etch or ion implantation processing.	Possible causes of resist lift-off are: • Contamination on wafer surface. • Inadequate HMDS priming or dehydration bake resulting in moisture on wafer surface (measured by contact angle meter as described in Chapter 7). • Excessive HMDS priming can cause resist "popping," which is failure of poorly adhered resist. • SiO_2 is a difficult surface to achieve good adhesion because it is a hydrophilic surface (water attracting).*
2. General quality of resist coating on wafer.	A. Pinholes (very small holes) in the resist.	• Particulate contamination on mask/reticle (evident after exposure) or on wafer. Check surface preparation cleaning.
	B. Splashback (drops of resist fall on resist coating).	• Improper exhaust level in spin coater. • Adjust vacuum suckback on dispense nozzle to stop drops from forming and dripping on resist.
	C. Resist skinning (thin layer of insoluble resist dried on top of resist coating).	• Spin rate is too high. High spin speeds can cause striations (lines) in the resist. • Exhaust rate too high in spin coater. • For DUV resists, ensure coated resist is not exposed to amines inside track system (check carbon filtration).
3. Thickness of resist coating.	A. Coated resist thickness is out of control. Resist thickness must be uniform: wafer-to-wafer mean thickness requirement is often <30 Å total indicated runout (TIR, which is maximum minus minimum thickness).	Parameters that affect thickness of resist are: • Check spin acceleration and speed. Higher spin speeds produce thinner resist (recall that thickness is inversely proportional to the square root of spin speed). Verify spin speed versus thickness characteristic curve of resist manufacturer. At low spin speeds, irregular solvent loss leads to thickness nonuniformity. • Verify correct spin time. Resist reaches stable thickness in several seconds and requires additional time for thickness uniformity. • Check for correct resist type and viscosity. • Verify no mechanical vibration or air turbulence during high speed spin dry.

* B. Smith, "Resist Processing," *Microlithography, Science and Technology*, ed. J. Sheats and B. Smith, (New York: Marcel Dekker, 1998), p. 529.

PHOTORESIST TROUBLESHOOTING

Common troubleshooting problems for photoresist are provided in Table 13.7.

TABLE 13.7 Common Photoresist Troubleshooting Problems

Problem	Probable Cause	Corrective Actions
1. Excessive resist usage during spin coating (DUV resist costs $2,000 to $5,000 per gallon).	A. Sub-optimum setting of process variables during spin coating process.*	Optimize process variables for spin coating: • First, determine minimum volume of resist needed to coat the wafer (referred to as cut-off volume). • Second, evaluate the following critical parameters using design of experiments: dispense spin speed, exhaust flow rate during dispense and drying steps, dispense rate, resist temperature, cool plate temperature, and ambient air temperature.
	B. Improper spin coater tool setup.	Check the spin coater setup for the following: • Incorrect nozzle size, height, or position. • Calibration of coater equipment. • Wrong software process recipe for wafer.
2. Wafers damaged (broken or chipped) during normal processing in the wafer track.	A. Improper setup of different tools used in track for processing wafers.	Check the following equipment for proper setup: • Improper calibration of robot arms during wafer handoff from one track tool to another. • Loss of chuck vacuum during spin. • Spin coat machine is not leveled properly and has mechanical vibration. • Excessive spin speed. • Verify calibration of cassette indexer to index wafers in cassette for robot pickup. • Warped cassettes cause equipment malfunction.
3. DNQ-novolak resist becomes unstable and contaminated with articles.	A. DNQ-novolak resist has aged due to shelf life.	Excessive shelf life of DNQ-novolak resist (> several months, with maximum of six months to one year) can lead to the following:** • Increase in absorption at longer wavelengths. • Susceptible to thermal degradation of the DNQ, leading to crosslinking and increase of high molecular-weight resist components. • Formation of undesirable acids in the resist. • Precipitation (falling out of solution) of sensitizer to form crystallized particles that contaminate resists. This especially occurs at high-temperature storage. • Point-of-use filtration is common practice for production applications to control resist consistency. Typical filter size is 0.05 μm for 0.25 μm feature size.

* B. Lorefice et al., "How to Minimize Resist Usage During Spin Coating," *Semiconductor International* (June 1998): p. 182.
** B. Smith, "Resist Processing," *Microlithography, Science and Technology,* ed. J. Sheats and B. Smith, (New York: Marcel Dekker, 1998), p. 517.

SUMMARY

Photolithography transfers a pattern from a mask to a UV light-sensitive photoresist on the wafer surface. Resolution is the smallest feature that can be patterned. Photolithography requires process latitude to consistently pattern wafers that meet the requirements. Negative lithography prints a pattern in the wafer that is opposite the mask pattern, whereas positive lithography prints the same mask pattern in the wafer. Photolithography can be divided into eight basic steps: (1) vapor prime, (2) spin coat, (3) soft bake, (4) alignment and exposure, (5) post-exposure bake, (6) develop, (7) hard bake, and (8) develop inspection. The wafer is prepared for resist application by cleaning, dehydration bake, and HMDS vapor prime. Negative photoresist hardens (crosslinks) with UV exposure, while positive resist softens (dissolves). Soft areas are washed away in developer solution. Positive resist is used for submicron lithography. Conventional i-line resist responds to 365 nm UV light for a CD of 0.35 μm and is common on noncritical layers. Positive i-line resist has DNQ-novolak chemistry and is activated by a photo-active compound (PAC). Wafer critical layers (CD of 0.25 μm and below) are patterned with a deep UV (DUV) resist that is chemically amplified. The DUV wavelength is 248 nm and is being reduced to 193 nm. A chemically amplified resist has a chemical protecting group that makes it insoluble in aqueous developer. A photoacid generator (PAG) generated in the DUV-exposed areas deprotects the resist when heated and makes the exposed resist soluble in developer. Liquid photoresist is applied to the wafer by spin coating, followed by edge-bead removal. A soft bake is used to drive off solvents and improve resist adhesion.

KEY TERMS

photolithography (also photo, lithography, masking, or patterning)
microlithography
reticle
photomask (also mask)
optical lithography
resolution
overlay accuracy
overlay budget
runout
process latitude
negative lithography
positive lithography
negative resist
positive resist
dark-field mask
clear-field mask
coater/developer track system (also tracks)
hydrophilic (also hydrated)
hydrophobic (also dehydrated)
dehydration bake
hexamethyldisilazane (HMDS)
vapor prime coating
photoresist
contrast
sensitivity
exposure dose
resist viscosity
resist adhesion
etch resistance
surface tension
conventional i-line photoresists
novolak
photoactive compound (PAC)
DNQ-novolak
deep UV (DUV)
chemical amplification (CA)
photoacid generator (PAG)
spin coating
static dispense
dynamic dispense
edge-bead removal (EBR)
soft bake (also pre-bake)
thermogravimetric analysis (TGA)

REVIEW QUESTIONS

1. What is photolithography?
2. Describe the difference between a reticle and a photomask.
3. List the different UV wavelengths used in photolithography and the name for each wavelength used between 436 and 157 nm.
4. Define resolution.

5. What is overlay accuracy, and how does this contribute to the mask overlay budget?
6. Discuss process latitude.
7. Explain the difference between negative and positive lithography.
8. Describe a clear-field mask.
9. Explain what a dark-field mask is.
10. List the eight steps of photolithography, and give a short explanation of each step.
11. What is one of the major effects of contamination on the wafer surface?
12. Explain the difference between a hydrophilic and a hydrophobic wafer surface.
13. Why is a dehydration bake done?
14. What is HMDS and what purpose does it serve?
15. Describe the most common method of applying HMDS.
16. Define a photoresist.
17. Give two purposes of a photoresist in wafer fabrication.
18. How has photoresist been improved since the early days of wafer fabrication?
19. List and describe the two major types of photoresist.
20. What is the resolution limit of negative resist? Which resist is used for submicron lithography?
21. Define contrast.
22. Explain sensitivity and discuss how this condition is related to the exposure dose.
23. Describe resist viscosity and explain why it is important.
24. Explain resist adhesion.
25. Why is etch resistance an important resist property?
26. What is surface tension, and why is it important for a photoresist?
27. Explain why storage and handling are important for resist.
28. List and describe the four components to an i-line resist.
29. What are two disadvantages of a negative photoresist?
30. What is the most common photoresist used in IC photolithography?
31. State and describe the resin used in i-line positive photoresist.
32. Describe the sensitizer used in i-line positive photoresist.
33. List the four steps to dissolution of i-line positive resist.
34. Give a reason why positive i-line resist has good resolution.
35. Why are i-line resists unacceptable at the DUV wavelengths?
36. What does chemical amplification accomplish in a resist?
37. Describe the purpose of a photoacid generator (PAG).
38. List and describe the four exposure steps for CA DUV resists.
39. What is one of the main benefits of CA DUV resists for wafer fabrication?
40. In what manner are DUV resists sensitive to contamination?
41. How is photoresist applied to a wafer?
42. List and describe the four basic steps to spin coating.
43. Explain the difference between static dispense and dynamic dispense.
44. What does resist thickness vary with? State the formula that describes this relationship.
45. Describe edge-bead removal.
46. State the four reasons for soft bake.
47. What is the ideal amount of solvent remaining in the resist after soft bake?
48. What problems would occur if the resist were not soft baked?
49. Describe the preferred method for performing soft bake.
50. What does TGA stand for and why is this analysis done?

PHOTORESIST MATERIALS AND EQUIPMENT SUPPLIERS' WEB SITES

Allied Signal	http://www.electronicmaterials.com/
Arch Chemicals (aka Olin)	http://www.olinmicro.com/default.asp
Ashland Specialty Chem.	http://www.ashland-act.com/
Clariant Corporation	http://www.azresist.com/
Dainippon Screen Mfg. Co.	http://www.screen.co.jp./eed/index_E.html
DuPont	http://www.dupont.com/semiconductor/
Eastman Chemical	http://www.eastman.com/
EKC Technology	http://www.ekctech.com/ekctech.nsf
FSI International	http://www.fsi-intl.com/
JSR Microelectronics, Inc.	http://www.jsrusa.com/index2.html
J.T. Baker	http://www.jtbaker.com
Karl Suss Inc.	http://www.suss.com/
Olin Microelectronics	http://www.olinmicro.com/
Rite Track	http://www.ritetrack.com/
SEMI	http://www.semi.org/

Shipley Company   http://www.shipley.com/
Silicon Valley Group   http://www.svg.com
TEL, Tokyo Electron Ltd.   http://www.teainet.com
USHIO   http://www.ushio.com/index2.html

REFERENCES

1. S. Campbell, *The Science and Engineering of Microelectronic Fabrication,* (New York: Oxford University Press, 1996), p. 152.
2. P. Castrucci, W. Henley, and W. Liebmann, "Lithography at an Inflection Point," *Solid State Technology* (November 1997): p. 127.
3. S. Campbell, *The Science and Engineering of Microelectronic Fabrication,* p. 178.
4. S. Wolf and R. Tauber, *Silicon Processing for the VLSI Era,* Vol. 1, 2nd ed., *Process Technology* (Sunset Beach: Lattice Press, 2000), p. 510.
5. Ibid.
6. C. Chang and S. Sze, *ULSI Technology,* ed. C. Chang and S. Sze (New York: McGraw-Hill, 1996), p. 290.
7. S. Wolf and R. Tauber, *Silicon Processing for the VLSI Era,* Vol. 1, *Process Technology,* 2nd ed., p. 500.
8. G. Gruetzner et al., "New Negative-Tone Photoresists Avoid Swelling and Distortion," *Solid State Technology* (January 1997): p. 79.
9. R. Dammel, *Diazonaphthoquinone-Based Resists,* (Bellingham, WA: SPIE Optical Engineering Press, 1993), p. 10.
10. S. Wolf and R. Tauber, *Silicon Processing for the VLSI Era,* vol.1, *Process Technology,* 2nd ed., p. 500.
11. H. Ito, "Deep-UV Resists: Evolution and Status," *Solid State Technology* (July 1996): p. 164.
12. D. Seeger, "Chemically Amplified Resists for Advanced Lithography: Road to Success or Detour?" *Solid State Technology* (June 1997): p. 115.
13. C. G. Willson et al., "Approaches to the Design and Radiation-Sensitive Polymeric Imaging Systems with Improved Sensitivity and Resolution," *Journal of Electrochemical Society,* vol. 133, no. 1, (Pennington, NJ: Electrochemical Society, 1986): p. 181.
14. T. Ueno, "Chemistry of Photoresist Materials," *Microlithography: Science and Technology,* ed. J. Sheats and B. Smith, (New York: Marcel Dekker, 1998), p. 451.
15. H. Ito, "Deep-UV Resists," p. 165.
16. G. Amblard, "Lithographic Evaluation of Deep UV Photoresists for 0.25 µm and 0.18 µm Technologies Design Rules," *Advances in Resist Technology and Processing* XV Proceedings of SPIE vol. 3333, February 23–25 1998, Santa Clara, CA, p. 890.
17. H. Ito, "Deep-UV Resists," p. 165–170.
18. M. Toukhy et al., "Chemically Amplified Resist Technology for I-Line Applications," *Advances in Resist Technology and Processing* XV Proceedings of SPIE vol. 3333 (Santa Clara, CA: February 23–25, 1998): p. 1212.
19. H. Ito, "Deep-UV Resists," p. 164.
20. N. Rizvi et al., "A 193 nm Excimer Laser Microstepper System," *Microlithographic Techniques in IC Fabrication,"* Proceedings of the SPIE vol. 3183 (June 25–26, 1997), p. 35.
21. P. Burggraaf, "Optical Lithography to 2000 and Beyond," *Solid State Technology* (February 1999): p. 38.
22. M. Preil and H. Levinson, "Yield-Limiting Issues in Deep-UV Lithography," *Microlithography World* (Spring 1998): p. 24.
23. B. Lorefice et al., "How to Minimize Resist Usage During Spin Coating," *Semiconductor International* (June 1998): p. 179.
24. Ibid.
25. Ibid.
26. S. Campbell, *Science and Engineering of Microelectronic Fabrication,* p. 191.
27. S. Wolf and R. Tauber, *Silicon Processing for the VLSI Era,* vol. 1, *Process Technology,* 2nd ed., p. 515.
28. B. Smith, "Resist Processing," *Microlithography: Science and Technology,* ed. J. Sheats and B. Smith, (New York: Marcel Dekker, 1998), p. 529.
29. Ibid.
30. Ibid., p. 530.
31. D. Elliott, *Microlithography: Process Technology for IC Fabrication,* (New York: McGraw-Hill, 1986), p. 91.

CHAPTER 14
PHOTOLITHOGRAPHY:
ALIGNMENT AND EXPOSURE

Optical lithography is similar to photography with a camera. A camera uses light and lens to transfer an image of an object to a negative film. The develop process then transfers the film's image to paper from which a picture is obtained. In photolithography, the photomask and its pattern are the objects to be photographed. Optics and a light source are used to project the image of the pattern of the mask onto the resist-coated wafer. Once the resist is developed, just as with a negative film for a camera, the mask pattern appears in the resist. Subsequent processing, such as etch, creates a permanent pattern of the mask image on the wafer surface.

Many variables affect the quality of a photograph, such as the type of film, the lighting conditions, whether the object is in focus, and the type of camera lens. A wide range of variables also affect photolithography in wafer fabrication. Examples are the physical conditions of the material on the wafer surface, the type of photoresist, the resolution power of the optics, the nature of the light, and the focusing accuracy of the system. To optimize the photolithography process for submicron critical dimensions, the production team must understand how all factors affect the final image. Team members have an important role because of their daily interaction with the equipment and product. They also should know where the wafers come from and the condition of the wafers prior to processing them in photolithography.

The alignment and exposure process represents a major equipment subsystem for modern photolithography. The wafer is first positioned within the focus range of the optical system. Wafer alignment features are aligned to similar matching features on the photomask, and the UV light is projected through the optics and the mask pattern. The mask pattern appears as light and dark features on the wafer; thus, exposing the photoresist. The numerous variables associated with alignment and exposure are reviewed in this chapter to understand their contribution toward achieving a high-quality, deep submicron patterning process.

OBJECTIVES

After studying the material in this chapter, you will be able to:

1. Explain the purpose of alignment and exposure in photolithography.
2. Describe the properties of light and exposure sources important for optical lithography.
3. State and explain the critical aspects of optics for optical lithography.
4. Explain resolution, describe its critical parameters, and discuss how it is calculated.
5. Discuss each of the five equipment eras for alignment and exposure.
6. Describe reticles, explain how they are manufactured, and discuss their use in microlithography.
7. Discuss the optical enhancement techniques for subwavelength lithography.
8. Explain how alignment is achieved in lithography.

INTRODUCTION

Modern photolithography equipment is based on *optical lithography*, which uses optics to accurately project and expose a mask pattern onto a resist-coated wafer. Basically it consists of an ultraviolet (UV) light source, an optical system, a reticle with the die pattern, an alignment system, and a wafer covered with a light-sensitive photoresist. Optics is at the heart of photolithography's ability to pattern deep-submicron features. Photolithography is at the center of the wafer fabrication process, with wafers spending more time in this process area than any other operation (by some estimates, up to 60% of their fabrication time). Table 14.1 reviews the eight steps of photolithography and highlights the alignment and exposure step covered in this chapter.

TABLE 14.1 Eight Steps of Photolithography

Step	Chapter Covered
1. Vapor prime	13
2. Spin coat	13
3. Soft bake	13
4. Alignment and exposure	**14**
5. Post-exposure bake	15
6. Develop	15
7. Hard bake	15
8. Develop inspect	15

We learned in Chapter 13 about the photoresist material and its importance to successful photolithography. The critical equipment for photolithography is the step-and-repeat aligner (referred to as aligner, but commonly called stepper). Steppers align and expose successive reticle patterns by stepping from one exposure site to another on the resist-coated wafer surface. The industry has recently converted to the step-and-scan system for DUV resists, called step-and-scan technology (discussed later in this chapter).

In photolithography, wafer steppers have three basic purposes—all of which must meet the customers' specifications for accuracy and repeatability:

1. Focus and align the wafer surface to the quartz plate reticle (containing the patterns).
2. Reproduce a high-resolution reticle image on the wafer through exposure of photoresist.
3. Produce an adequate quantity of acceptable wafers per unit time to meet production requirements.

The alignment and exposure operation of the wafer stepper commences once the wafer surface has been coated with the photoresist and soft baked. Resist-coated wafers are automatically loaded onto a wafer stage in the stepper. At this stage the wafer is raised or lowered as needed to bring it into the focus range of the stepper optics. The wafer is aligned to the reticle so that the pattern can be transferred to the proper location on the wafer surface. Once the best focus and alignment are obtained, a shutter opens to allow UV light to pass from the illuminator to the reticle through a projection lens and then onto the resist-coated wafer (see Figure 14.1). The entire focusing, wafer alignment, and exposure operation is done by the stepper. Once a pattern is exposed, the stepper will step to the next location on the wafer and repeat the alignment and exposure (which explains the name step-and-repeat). The stepper is typically attached to an automated wafer track machine, which processes wafers for all other basic photolithography operations.

Importance of Alignment and Exposure

Integrated circuits are fabricated with semiconductor devices in the upper few microns of the silicon wafer, followed by successive deposition and patterning of material layers to form the circuits that interconnect the devices. Circuit designers, using computers with special design software, define the layout of the devices, metal lines, via connections, and other special circuit designs necessary for a chip to function. This is done layer by layer and structure by structure within a layer for the entire wafer. Like creating a stencil, the circuit design pattern for one or more die is transferred to a reticle, with multiple reticles needed to attain the final structure on the wafer surface (see Figure 14.2).

Photolithography: Alignment and Exposure **369**

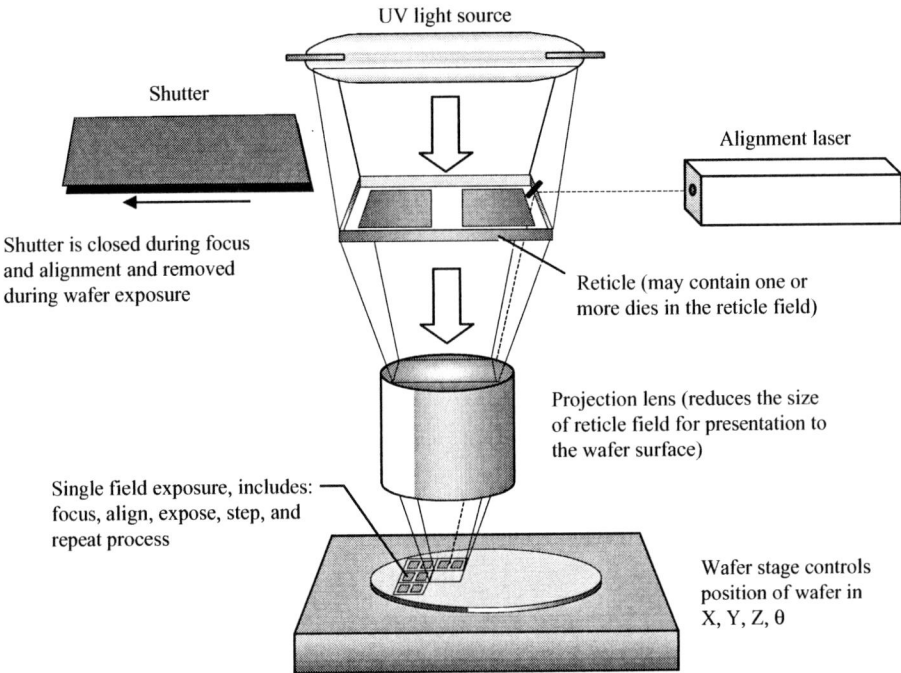

FIGURE 14.1 Reticle Pattern Transfer to Resist

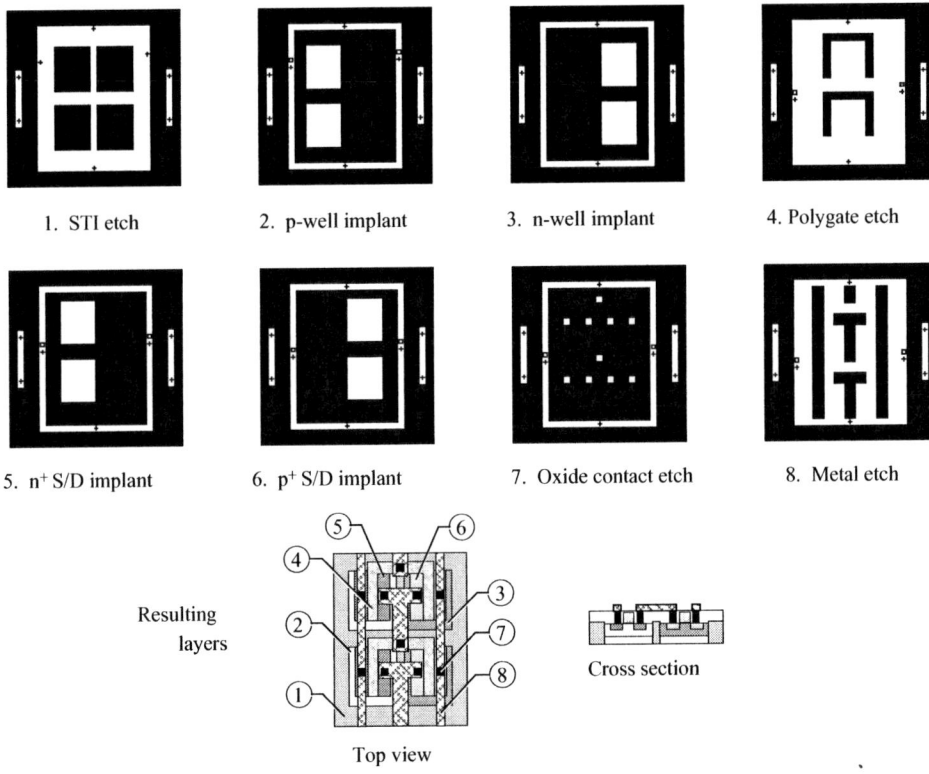

FIGURE 14.2 Layout and Dimensions of Reticle Patterns

The circuit design follows a set of design ground rules that specify the reticle pattern and its required dimensions. The design rules specify parameters such as linewidth, spacing between lines, contact size, via diameter, and the spacing between patterns defined on a reticle. Violation of design rules lead to poor production yield or unreliable product performance.

Each image defined in a reticle pattern has a particular function, such as a contact opening or metal lines. Patterns relate to a distinct material (e.g., oxide, aluminum, and so on) with dimensions and tolerances. Reticles overlay these patterns on top of one another during photolithography to form the devices and circuits on the wafer. The reticle overlay process has alignment specifications, previously referred to as the overlay budget. It is this overlay of reticle patterns onto existing features that poses the special alignment challenge for photolithography—how to accurately overlay patterns with submicron dimensions. Overlay accuracy must be met while maximizing product reliability and producing wafers in high volume.

Optical Exposure ■ During exposure, light from an illumination source passes through the aligned reticle. The reticle has opaque and transparent areas that define a pattern to be transferred to the photoresist on the wafer surface. The goal of exposure is for the reticle pattern to be accurately replicated (within specification) in the resulting image in the resist.

An aspect of exposure is, with all other factors being equal, the shorter the wavelength of the exposing light, the smaller the feature size that can be exposed. This fact has been a driving force for the continual reduction in feature sizes on wafers. Furthermore, the exposing light produces a certain amount of energy that is necessary to produce the photochemical reaction in the resist. This light energy must be uniformly distributed across the exposure field. Photolithography requires an intense exposure at the desired short wavelength to achieve the critical dimensions of today's microlithography.

OPTICAL LITHOGRAPHY

Optical lithography has traditionally been the major limiting factor to the continual reduction in microchip feature sizes since the beginning of wafer fabrication. The demise of optical lithography has been regularly predicted over the years. For instance, in 1985 it was predicted that optical lithography would be incapable of resolving a critical dimension smaller than 0.5 μm.[1] It is now believed that optical lithography is capable of resolutions down to a CD of 0.1 μm and beyond. Lithography is often considered the engine that is driving the performance improvements of Moore's law.[2]

Photolithography for wafer fabrication has been based largely on optical lithography. The longevity of optical lithography is attributed to the basic improvements that have been made to the equipment and process. We will now review the fundamental variables in optical lithography and the basis for these ongoing improvements. At the end of Chapter 15, we will study other next-generation lithography systems that are under investigation for advanced lithography applications and that may become viable for wafer fabrication in the future: extreme UV, e-beam, X-ray, and ion-beam lithography. Nevertheless, it appears that optical lithography will continue to dominate photolithography for the near future.

Light

A light source is needed in optical lithography to project the reticle pattern on the photoresist and cause a photochemical reaction. A practical description of *light* is an electromagnetic wave that is visible to the eye. Light is also radiant energy. These two descriptions reflect the dual nature of light as a wave and a particle. Light travels in waves similar to sound waves. Since light is a wave, it can be described by wavelength (λ) and frequency (f). The relationship of the two is given by the formula shown in Figure 14.3, where v is the velocity of light.

Interference of Light Waves ■ Waves are sinusoidal in nature. Sinusoidal waves of any type (e.g., light, electrical, or sound) that have the same frequency (monochromatic) can have interference between the individual waves. For example, wave interference may be two water ripples interacting with each other and partially canceling each other out. There are two types of interference based on whether the waves are in phase or out of phase (see Figure 14.4):

Constructive interference: Two waves in phase that add to each other.
Destructive interference: Two waves out of phase that subtract from each other.

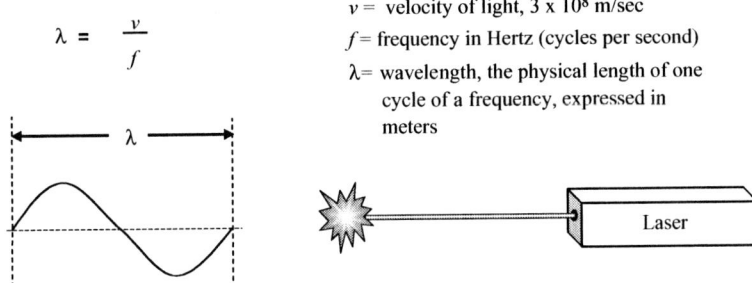

FIGURE 14.3 Light Wavelength and Frequency

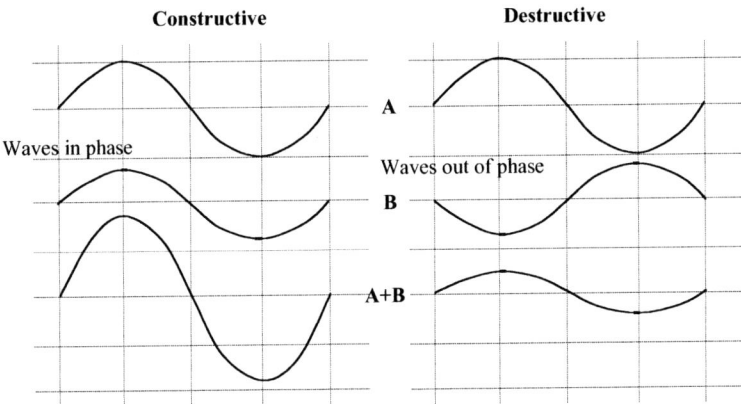

FIGURE 14.4 Wave Interference

Optical Filters. *Optical filters* use light interference to block unwanted incident light through either reflection or interference to obtain a light with a particular wavelength (see Figure 14.5). Optical filters will typically be made of glass and will have one or more layers of thin film coatings. The type of coating and its thickness will determine which wavelengths are blocked through destructive interference or passed through the glass.

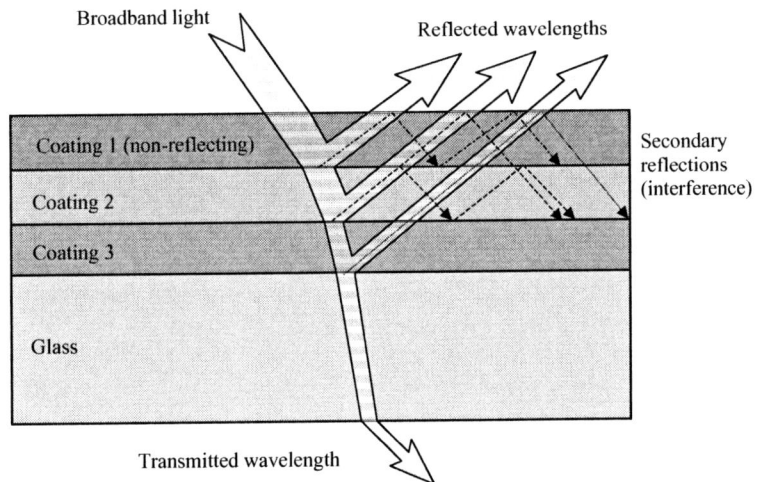

FIGURE 14.5 Optical Filtration

Electromagnetic Spectrum ■ The entire group of visible and nonvisible electromagnetic waves is called the *electromagnetic spectrum,* consisting of radiant energy that varies from very short to very long wavelengths. The visible portion of the spectrum has a wavelength between 390 nm and

780 nm. White light is composed of all the wavelengths of light in the visible spectrum. The UV spectrum ranges from 4 nm to about 450 nm, with a small overlap with the visible spectrum (see Figure 14.6). The deep ultraviolet (DUV) wavelengths are presently used for patterning critical layers in production. The vacuum ultraviolet (VUV) wavelength of 157 nm is the most likely UV successor after the DUV wavelengths of 248 nm and 193 nm. The short wavelength and high photon energy of VUV are strongly absorbed by oxygen absorption bands; therefore, operation is necessary in a vacuum or inert gas ambient (thus the name VUV). Research and development (R & D) is underway in the semiconductor industry into the future technology of using 13 nm UV wavelength, referred to as extreme UV (EUV).

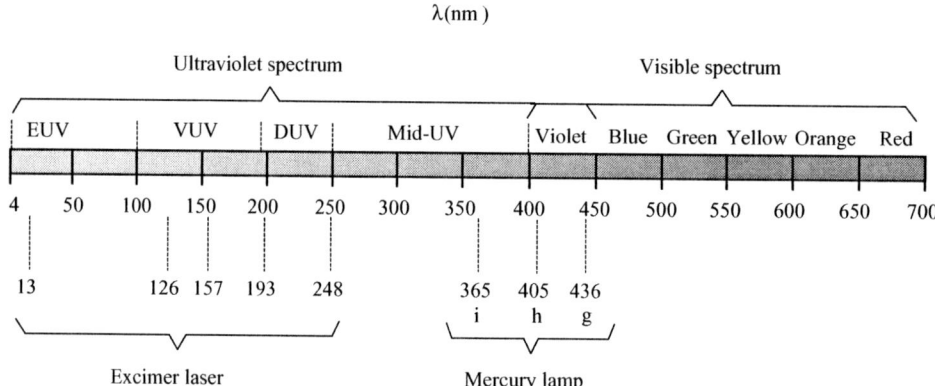

FIGURE 14.6 Ultraviolet Spectrum

Since the UV spectrum overlaps a portion of the visible spectrum, white light also contains a portion of UV (e.g., sunlight is visible and is white light, but it has UV wavelength also). Yellow light is commonly used in the photolithography area of the wafer fab because it is in the portion of the visible spectrum with little UV and, therefore, does not affect photoresists.

Exposure Sources

During exposure of the resist, photochemical transformations occur in the resist material to transfer the pattern from the mask. This is a critical step in photolithography. It must occur in the shortest amount of time and be repeatable for high-volume production of wafers.

Ultraviolet (UV) light is used to expose resist because the resist material reacts to light at this particular wavelength. Wavelength is also important because light with a smaller wavelength allows for the resolution of smaller features on the photoresist (explained in detail later in this chapter). The two types of UV exposure sources most often used today for optical lithography are:

- Mercury arc lamp
- Excimer laser

Besides these more commonly used light sources, other sources used to expose resist in advanced or special applications are X-ray, electron beam, and ion beam. These sources and their unique resist material will be discussed in Chapter 15 in sections on advanced technologies.

Mercury Arc Lamp ■ The high-pressure *mercury arc lamp* is used as the UV illumination source in all conventional i-line steppers. In this type of lamp, electrical current is passed through a tube of mercury-xenon gas to create a discharge arc. This arc emits a characteristic light spectrum, with the useful UV wavelength emissions for a mercury arc lamp between about 240 nm and 500 nm in length (see Figure 14.7).

The mercury arc lamp light spectrum has several intensity peaks. Some have been given letter names that come from the early days of spectrometry. The major intensity peaks are shown in Table 14.2.

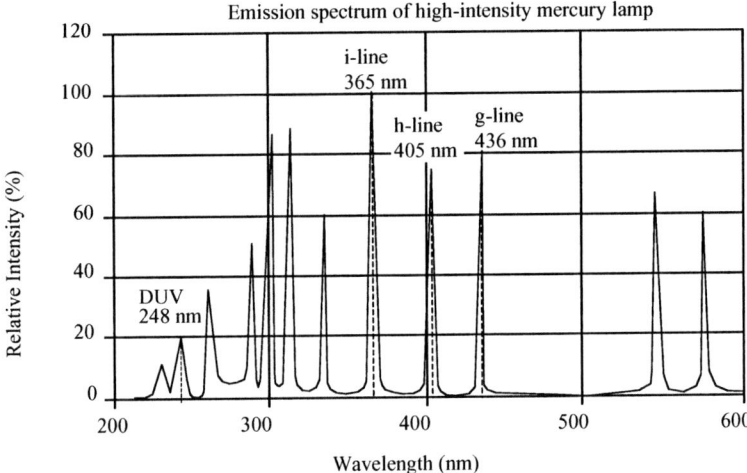

FIGURE 14.7 Emission Spectrum of Typical High Pressure Mercury Arc Lamp (Used with permission from USHIO Specialty Lighting Products)

TABLE 14.2 Mercury Arc Lamp Intensity Peaks

UV Light Wavelength (nm)	Descriptor	CD Resolution (μm)
436	g-line	0.5
405	h-line	0.4
365	i-line	0.35
248	Deep UV (DUV)	0.25

A conventional photoresist will have a certain spectral response that corresponds to a particular UV wavelength during exposure. For instance, a DNQ-novolak i-line resist, which is used for CD feature sizes down to 0.35 μm, responds to the UV i-line wavelength of 365 nm. To expose a resist to the proper UV wavelength, a set of filters is used to block the unwanted wavelengths and any infrared wavelengths. The exposure wavelength is chosen to match the critical feature size on the wafer.

An important aspect of the exposure source is the *light intensity,* which is defined by power per unit area (mW/cm^2), and measured on the surface of the resist. Another way to interpret intensity is the quantity of light per unit area, or brightness. Energy is the product of power and time. If the light intensity (which is power per unit area) is multiplied by the time of exposure, this represents the amount of exposure energy, or exposure dose, in (mJoules/cm^2, or mJ/cm^2) received by the surface of the photoresist.

A typical i-line resist requires an exposure dose of 100 mJ/cm^2 for its exposure.[3] Consider the emission spectrum shown in Figure 14.7 for a mercury arc lamp. The DUV emission at 248 nm is approximately five times less intense than the i-line emission at 365 nm. For an equivalent i-line photoresist at 248 nm, the low mercury arc lamp light intensity at the DUV wavelength would require five times more exposure time. In other words, if light intensity decreases, then the exposure time must increase by a proportional amount. This new exposure time is too long for acceptable wafer production and was a primary reason for the development of chemically amplified DUV photoresists and laser light sources with more power (see Figure 14.8 on page 374).

An undesirable parameter for the resist resin is excessive absorbance of incident radiation. With excessive resist absorption, the light intensity at the bottom of the resist is considerably less than that received at the top. This discrepancy produces image profiles with sloping sidewalls (see Figure 14.9 on page 374). To achieve straight-wall images, the resist must absorb only a small portion of the incident radiation, typically less than 20%.[4] Minimizing resist absorbance requires optimization between the wavelength, light source dose, and the type of resist.

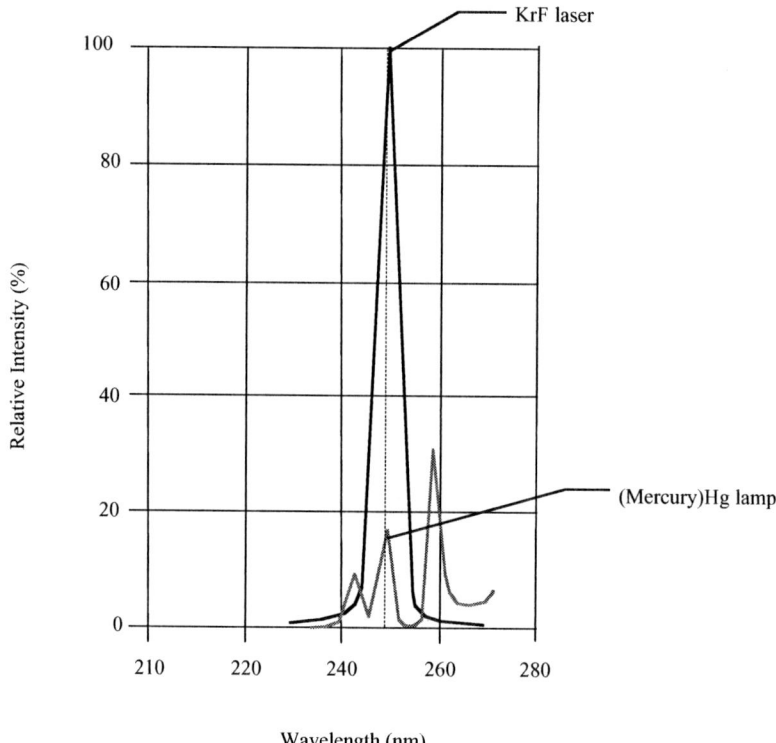

FIGURE 14.8 Spectral Emission Intensity of 248 nm Excimer Laser Versus Mercury Lamp

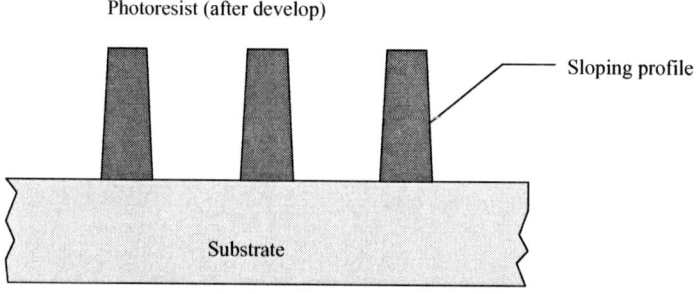

FIGURE 14.9 Excessive Resist Absorption of Incident Light

Excimer Laser ■ Laser light sources for photolithography have been available since the mid-1980s, but reliability and performance concerns delayed their implementation in wafer production until the mid-1990s. Their main benefit has been to provide more light intensity at the DUV wavelengths of 248 nm and below, since the mercury arc lamp is an inefficient emitter at these wavelengths.

The only laser light source used thus far for photolithography exposure is the *excimer laser*. An excimer is an exotic molecule formed from an atom of a noble gas and halogen, such as argon fluoride (ArF), where the molecule exists only in a quasi-stable, excited state.[5] The word excimer actually comes from the expression *exc*ited d*imer*, which is a molecule formed from two identical atoms, such as F_2. It also is used to represent the noble gas and halogen molecules.

Most excimer lasers today contain a high-pressure mixture of two or more elements that are bound in the excited state. Laser light emission occurs when the excited state decays and the exotic molecule falls apart into its two constituent atoms. The laser maintains a condition of more molecules in the excited state than in the ground state by using a high-voltage (10 to 20 kV) pulse discharge across two flat-plate electrodes to excite a high-pressure mixture of the noble gas and halogen.[6]

The most common excimer laser used for DUV resist is krypton-fluoride (KrF) with a wavelength of 248 nm. The KrF laser typically has a power range of 10 to 20 W at a frequency of 1 kHz, which produces high-power pulses of radiant light energy to expose the resist. Table 14.3 highlights excimer laser sources for wafer fabrication photolithography.[7] The argon-fluoride (ArF) laser, at a wavelength of 193 nm, is also envisioned to achieve DUV exposure. Notice that the fluorine (F_2) laser at the 157-nm wavelength has low output energy, which makes it less desirable for future production use because of the longer exposure time required. There has been recent improvement in the F_2 laser output energy to make it equivalent to the KrF and ArF lasers.[8]

TABLE 14.3 Excimer Laser Sources for Semiconductor Photolithography

Material	Wavelength (nm)	Maximum Output (mJ/pulse)	Frequency (pulses/sec)	Pulse Length (ns)	CD Resolution (μm)
KrF	248	300 to 1500	500	25	≤0.25
ArF	193	175 to 300	400	15	≤0.18
F_2	157	6	10	20	≤0.15

As seen in Table 14.3, the excimer laser produces extremely short pulses of light. The peak power of each short pulse is high and can cause damage to optical materials and lens coatings. To minimize this damage, longer laser pulse lengths are desirable and lenses are designed with features to avoid the high concentrations of light from the laser source. Another problem with early excimer lasers was excessive pulse-to-pulse energy variations. To overcome this variation, a sufficiently large number of pulses is required. Improvements to the excimer laser pulse stability continue, which reduces the number of pulses required to achieve acceptable exposure dose uniformity. With fewer pulses needed, the scanning speed of the step-and-scan system can increase, which ultimately increases productivity.[9]

The excimer laser will probably be used as an exposure source for at least another technology generation with the argon-fluoride (ArF) laser at a wavelength of 193 nm. There are no fundamental laser changes required to change from a 248-nm KrF laser to a 193-nm ArF laser. However, optical materials have undesirable absorbance and are more sensitive to laser damage from factors such as optical system heating at the 193-nm wavelength.[10] This condition requires lens systems that are laser damage resistant. An industrial grade F_2 excimer laser at a wavelength of 157 nm has been demonstrated as a potential light source for 0.15 μm critical dimension.

Spatial Coherence. Recall that light is an electromagnetic wave. A light beam has *spatial coherence* when points on different waves remain in phase to one another as the waves propagate. The light waves move in unison (see Figure 14.10). A standard room light bulb has no spatial coherency (completely incoherent). Light beams from excimer lasers have a low amount of spatial coherence, which differs from the beams of conventional lasers. Spatial coherence is

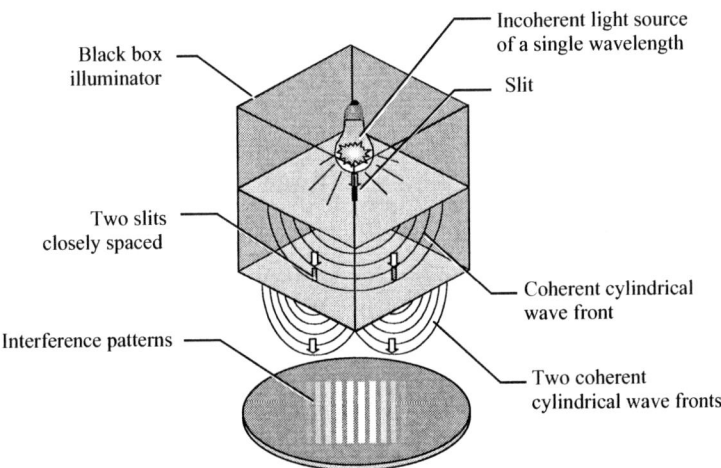

FIGURE 14.10 Spatial Coherence

controlled through optics to minimize interference patterns that can be formed in the image. If not controlled, the interference may appear as a grainy pattern of bright and dark spots on the resist, referred to as speckle.[11]

Exposure Control. ■ A uniform dose of UV light is critical for resist exposure. The exposure dose must be repeatable from exposure to exposure and from wafer to wafer. The exposure latitude for a DUV resist is around 1% in dose variability.[12] The variation in photolithography processes, equipment, and materials requires stringent exposure control.

Exposure control in photolithography is achieved by measuring the intensity of the UV light source at the surface of the wafer with a *dose monitor*. The exposure dose is measured at different locations in the exposure field, permitting the calculation of the percent uniformity of the dose. A photodetector and an electronics circuit are used on automated steppers and step-and-scan systems to monitor and control the exposure dose through use of a shutter or by varying the speed of the scan.

Optics

Optics is the study of the physical properties and composition of light. Optics is important because lithography is based on an optical imaging process used to transfer the reticle pattern to the photoresist. The quality of the resist pattern can be limited by poor-quality optics. All photolithography equipment currently used for high-volume wafer fabrication is based on optical lithography.

Reflection of Light ■ The *law of reflection* describes the relationship between an incident light ray and the corresponding reflected ray, stating that the angle of incidence is equal to the angle of reflection (see Figure 14.11). This law is true whether the reflecting surface is rough or smooth.

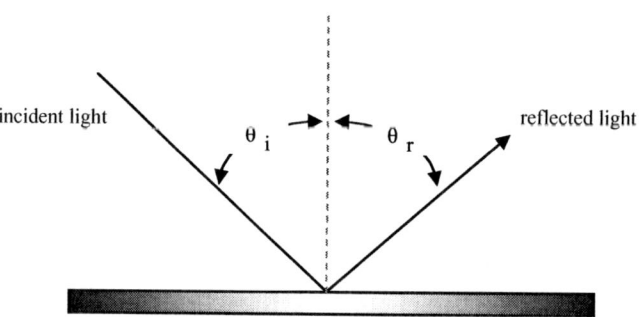

FIGURE 14.11 Law of Reflection

An example of the law of reflection is a plane mirror. A mirror reflects light from its surface and creates a virtual reversed image of an object. The image is said to be virtual because it appears to the viewer as though the object exists behind the mirror, whereas the image on the mirror surface is actually a reversed image reflection of the real object. Mirrors and reflective optics have many applications in steppers and step-and-scan systems, such as for beam orientation or for illuminators that shape and focus light (see Figure 14.12).

Refraction of Light ■ When light passes from one transparent medium into another, such as from air through a glass window, the light changes direction. The change in direction of a light ray when passing from one transparent medium into another is called *refraction*. Refraction of light is caused by the difference in the speed of light in two different media. Light traveling through a uniform medium, such as air, travels at a certain speed. As light enters into and passes through a new medium, such as glass, its speed actually decreases. This decrease occurs because the glass is optically a more dense material. The *relative index of refraction, n,* represents how much the light ray

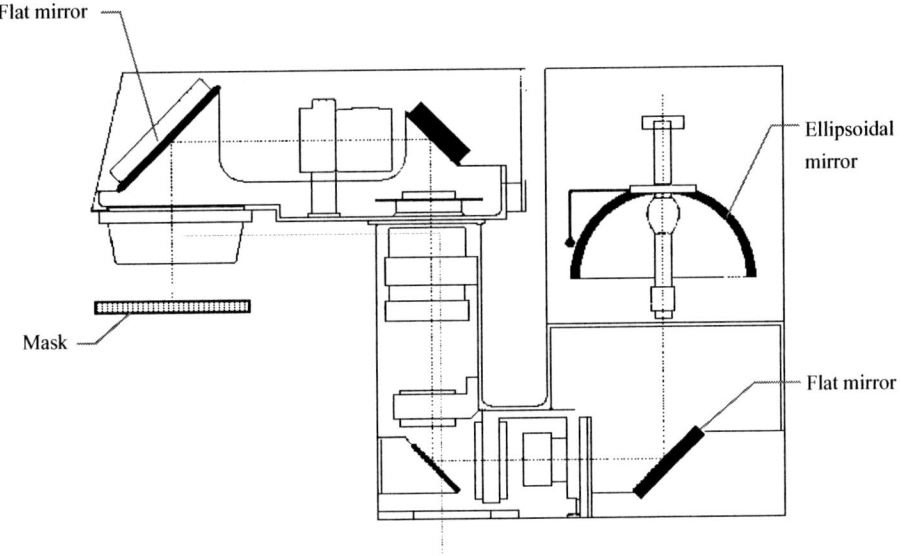

FIGURE 14.12 Application of Mirrors
(Used with permission from Canon USA, illuminator for mask aligner)

bends as it passes through the interface of the two media based on its change in velocity (see Figure 14.13). The *absolute index of refraction* compares the speed of light in a vacuum to the speed of light in the chosen medium. Therefore it is only stated for one medium and the vacuum is assumed. Examples of absolute index of refraction for different media are given in Table 14.4.[13] The refraction of light is also dependent on the wavelength, λ, of the light ray that passes from one medium into another medium.

- Snell's Law: $\sin \theta_i = n \sin \theta_r$
- Index of refraction, $n = \sin \theta_i / \sin \theta_r$

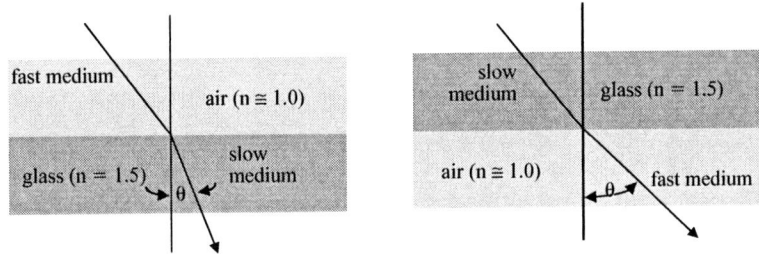

FIGURE 14.13 Refraction of Light Based on Two Mediums

TABLE 14.4 Absolute Index of Refraction for Select Materials

Material	Index of Refraction (n)
Air	1.000293
Water	1.33
Fused silica (amorphous quartz)	1.458
Diamond	2.419

Lens ■ A *lens* is defined as an optical element that refracts light from an object passing through it to form an image of the object. We will consider an individual lens, but optical systems in photolithography consists of many different lenses (refractive) and mirrors (reflective) in the optical system (see Figure 14.14.) Lenses are important in the optical system of photolithography equipment because of the need to project the image of the reticle pattern onto the photoresist. This projection must be done with the proper resolution, dimensional control, and alignment. Lenses and refractive optics in photolithography steppers and step-and-scan systems play a critical role in achieving the resolution needed to achieve the CD linewidths in modern wafer fabrication.

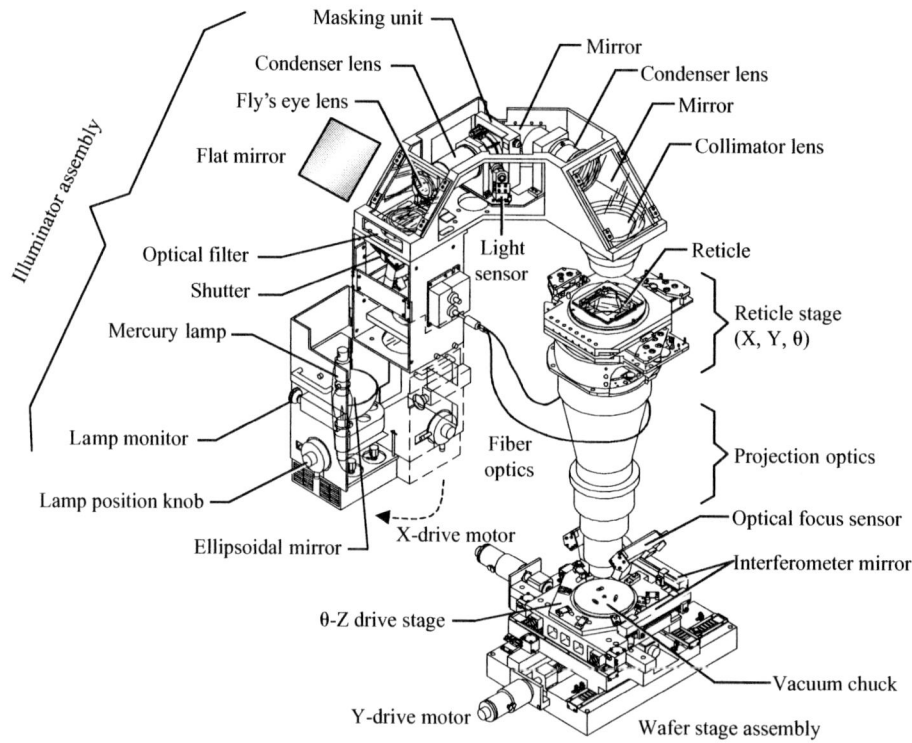

FIGURE 14.14 Optical System of Lenses
(Used with permission from Canon USA, FPA-2000 il exposure system)

A lens has either one or both surfaces as spherical-shaped, which defines whether the lens is converging (convex) or diverging (concave). Lenses refract light in such a way that light rays converge (see Figure 14.15) or diverge (see Figure 14.16) on a principal focus. The principal focus, or *focal point,* is where the light rays are refracted and will converge—the image carried by the light is sharp and in focus. The distance from the center of the lens to the focal point is called the *focal length.*

Lens Material. Lenses have traditionally been made of glass. The lens material is an important variable as the exposure wavelength is continually reduced. For 248-nm wavelength DUV, a suitable lens material is fused silica. It has less light absorption at the DUV wavelengths. At the 193-nm DUV and 157-nm VUV wavelength, calcium fluoride (CaF_2) is being investigated as a possible candidate for lens material.

Traditional optical materials such as glass have greater absorbance and are more sensitive to laser damage at the 248-nm KrF wavelength.[14] Absorption causes a loss in exposure power and induces heat in the optics, which leads to refractive index changes and imaging problems. Thermal effects can also cause focus changes due to optical system heating from DUV laser light sources.

Another concern regarding lens materials is that a laser beam can create compaction damage. *Lens compaction* is a structural rearrangement of the lens material that causes densification of the lens material (see Figure 14.17). Compaction occurs in lens materials, including fused silica. It is not well understood, but it occurs due to the total cumulative laser beam exposure and peak power density.

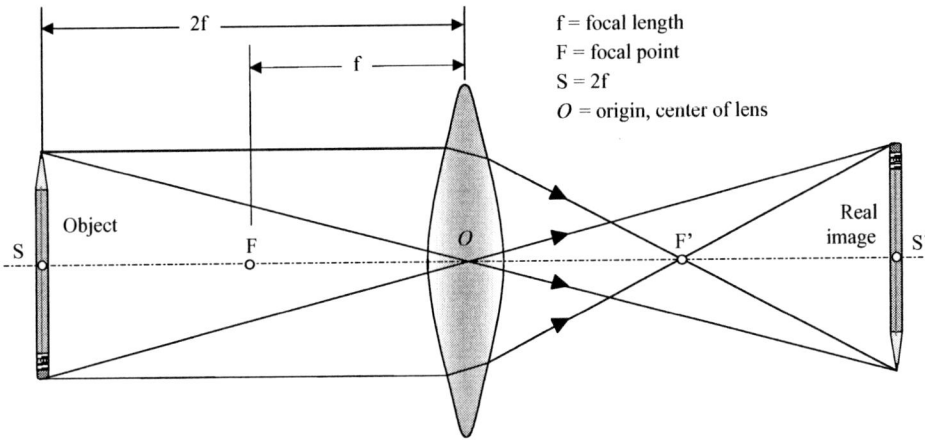

FIGURE 14.15 Converging Lens with Focal Point

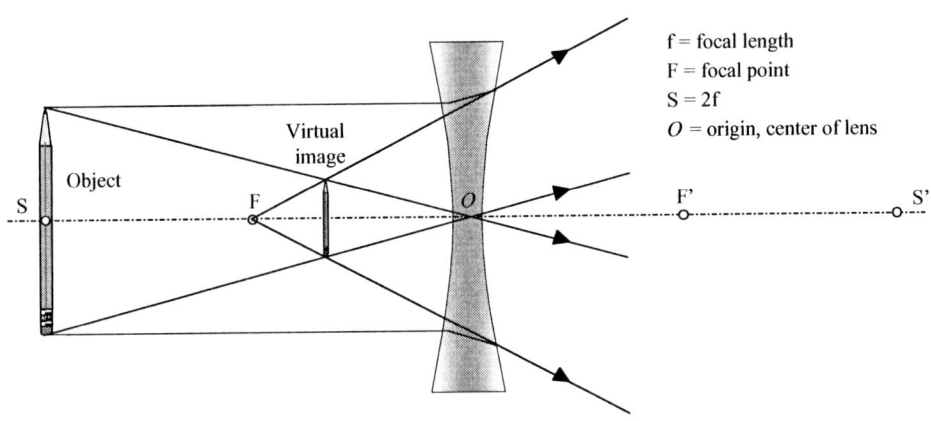

FIGURE 14.16 Diverging Lens with Focal Point

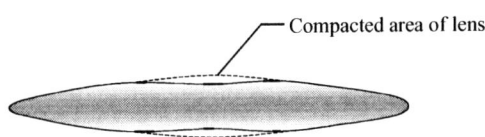

FIGURE 14.17 Laser-Induced Lens Compaction

It can increase the lens material's refractive index in the region traversed by the laser beam. If the index of refraction changes even slightly, then there could be a change in the light wave path, which would lead to loss of image quality. The control of compaction for DUV wavelengths is being researched.

Lenses used in photolithography optical systems are of the finest quality materials and workmanship, but they are not perfect. Deviations from the ideal lens behavior that result from design, fabrication, or usage flaws are referred to as *aberrations*. There are different types of aberrations, such as inaccurate lens surfaces. For the most part, the lens system designer designs the optical system to minimize individual lens aberrations.[15]

Diffraction ■ Light travels in straight lines. When light passes through a narrow opening or past a sharp edge, interference patterns occur along the edge of the opening. The effect is a fuzzy image rather than the expected sharp edge that occurs between light and shadow (see Figure 14.18 on page 380). The light appears to bend around the slit edges. This phenomenon is referred to as *diffraction*.

- Light travels in straight lines.
- Diffraction occurs when light hits edges of objects.
- Diffraction bands, or interference patterns, occur when light waves pass through narrow slits.

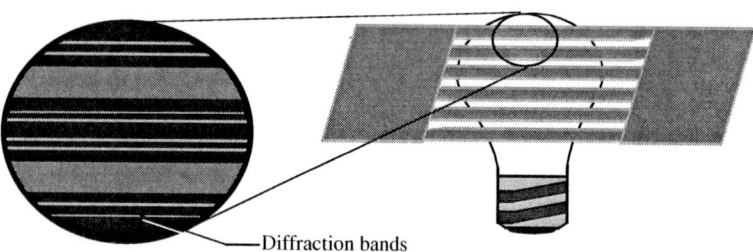

FIGURE 14.18 Interference Pattern from Light Diffraction at Small Opening

You can experience diffraction by putting two fingers close together and looking through them toward a light. The edge of your fingers is not sharp. Diffraction of light passing through a small opening causes an unexpected light intensity profile. In essence, regions outside the boundary are illuminated and the light intensity profile does not have a sharp edge. The central spot is bright and bounded by bands of decreasing intensity which are called *diffraction orders*. The amount of diffraction depends on the width of the opening and the wavelength of the light.

Light diffraction is a concern in photolithography because of the extremely small patterns of sharp edges and narrow spaces on reticles. Light during exposure must pass through these patterns (see Figure 14.19). Diffraction patterns rob exposure energy and scatter it, leading to exposure of unwanted areas of the photoresist. The problem is worse in small holes, such as small contact openings of 200 nm. The interference patterns caused by diffraction can make small contact holes and small lines difficult to print.

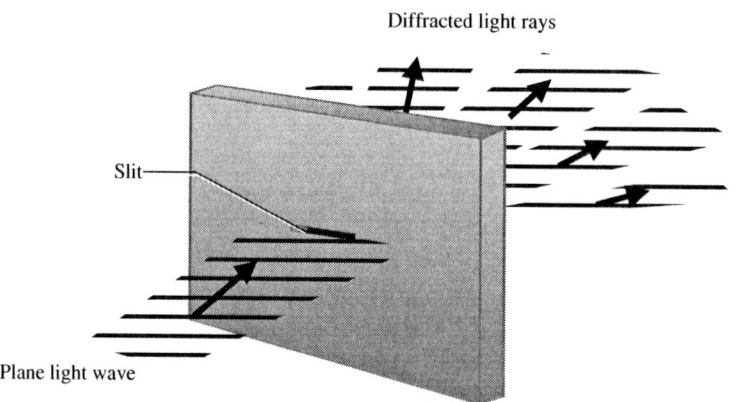

FIGURE 14.19 Diffraction in a Reticle Pattern

Numerical Aperture ■ A lens is able to capture some diffracted light (see Figure 14.20). The ability of a lens to collect diffracted light is referred to as the *numerical aperture* (*NA*) of the lens. For a given lens, the NA is a measure of how much diffracted light the lens can accept and image by converging the diffracted light to a single point.

The numerical aperture is defined by the following formula:

$$NA = (n) \sin \theta_m \approx (n) \frac{\text{radius of lens}}{\text{focal length of lens}}$$

Where, n = index of refraction of the image medium ($n \approx 1$ for air)
 θ_m = angle between the optical principal axis and the marginal ray at the edge of the lens

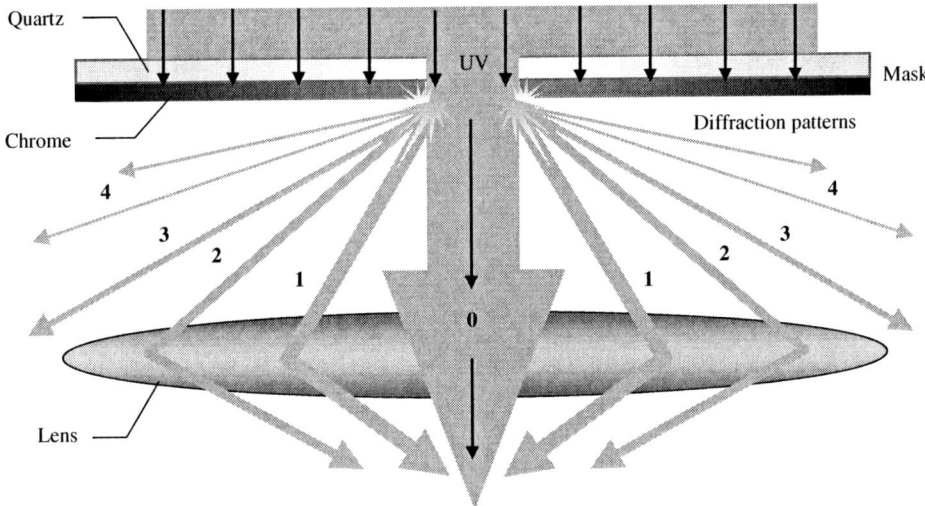

FIGURE 14.20 Lens Capturing Diffracted Light

Note that the value for $\sin \theta_m$ can be approximated by using its trigonometric relationship. This approximation highlights how increasing the radius of the lens will increase the NA and capture more diffracted light. With an increased NA, more of the diffracted light can be converged to a single point for imaging (see Figure 14.21). However, increasing the radius of the lens to increase the NA also means more complicated and costly optical systems.

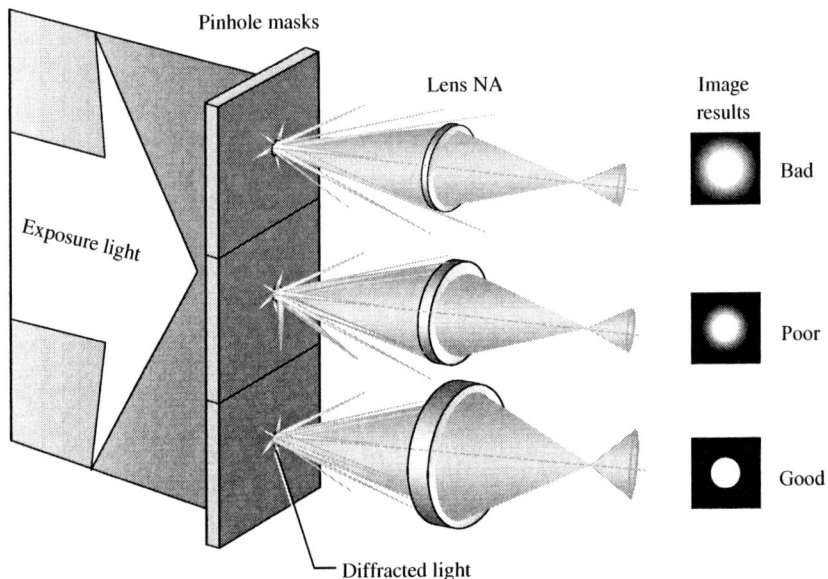

FIGURE 14.21 Effect of Numerical Aperture on Imaging

NA values are typically specified for photolithography steppers and step-and-scan systems. NA has improved considerably over the years. For an air or vacuum medium, the NA value must be less than or equal to 1. Some typical NA values for photolithography equipment are provided in Table 14.5.

TABLE 14.5 Typical NA Values for Photolithography Tools

Type of Equipment	NA Value
Scanning projection aligner with mirrors (1970s technology)	0.25
Step-and-repeat	0.60 to 0.68
Step-and-scan	0.60 to 0.68

Antireflective Coatings ■ Exposure light passes through a reticle to pattern the resist. Below the resist is the underlying layer that will ultimately be etched and patterned. If this underlying film is reflective, as with metal and polysilicon layers, then light rays reflect off this film and potentially damage the adjacent resist. This damage can adversely affect CD control. The two primary light reflectivity problems are reflective notching and standing waves. *Reflective notching* occurs when vertical surfaces on the sides of etched structures reflect light into the resist where exposure is not intended (see Figure 14.22).

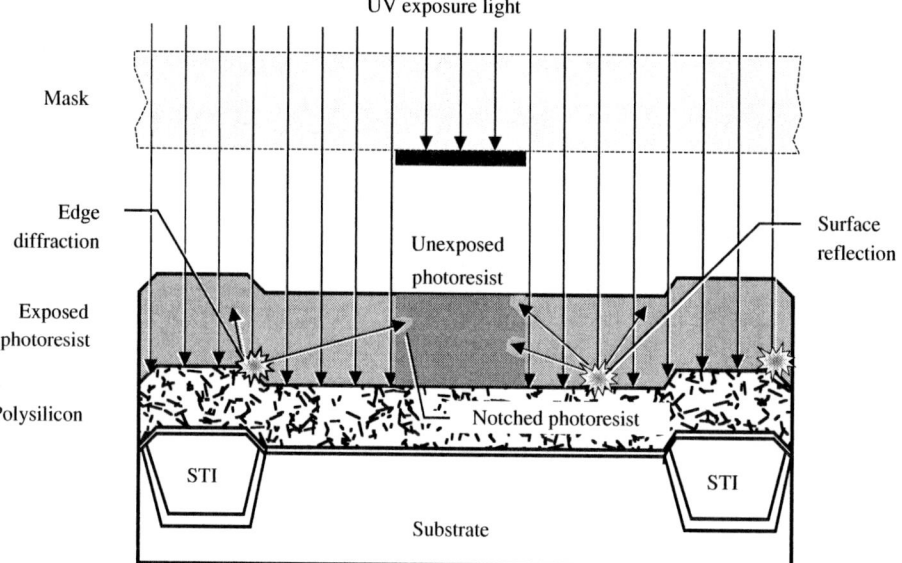

FIGURE 14.22 Photoresist Reflective Notching Due to Light Reflections

Standing Waves. An example of wave reflection and interference in photolithography is the phenomenon of standing waves. If a layer of photoresist is applied to a reflective wafer surface and exposed to monochromatic light, then the incident light wave striking the resist passes through the resist layer and is reflected from the surface of the wafer (see Figure 14.23). Standing waves represent interference between the incident light waves and reflected light waves, which causes nonuniform exposure along the thickness of the photoresist film. The occurrence of standing waves is more pronounced with DUV resists because many wafer surfaces (e.g., oxide, nitride, and polysilicon) are more reflective at the shorter DUV wavelengths.[16] After exposure, the sides of the resist features have striations of overexposed and underexposed areas. Standing waves essentially degrade the resolution of the image in the resist.

Using an *antireflective coating* (*ARC*) applied directly to the reflective substrate surface reduces standing wave effects in photoresist (see Figure 14.24). ARCs reduce unintended light reflection by suppressing the exposure light, with the latest ARCs capable of suppressing 99%

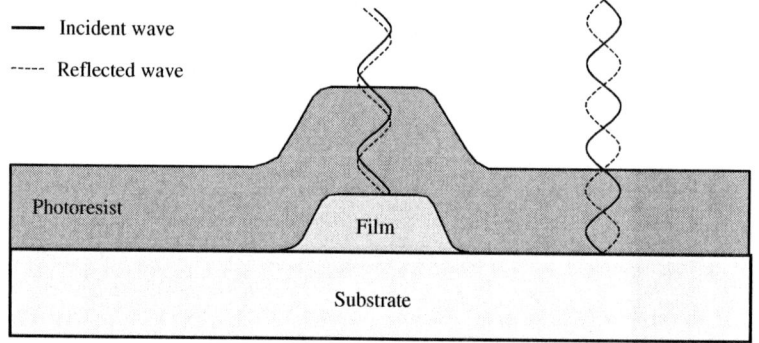

Standing waves cause nonuniform exposure along the thickness of the photoresist film.

FIGURE 14.23 Incident and Reflected Light Wave Interference in Photoresist

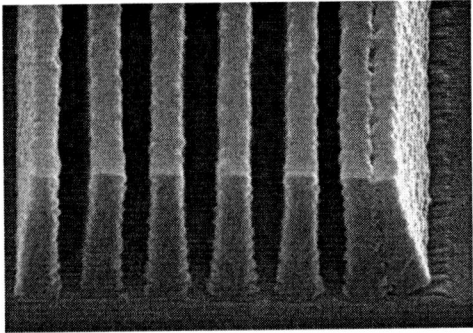

Effect of Standing Waves in Photoresist
(Photo courtesy of Grant Willson's research group at the University of Texas at Austin)

of the substrate-reflected light. They are deposited on the wafer as a thin layer, typically from 200 to 2000 Å, depending on the type of ARC and material used.[17] Dyes can also be added to the resist to help prevent light wave interference. In addition, a post-exposure bake (PEB) between exposure and development can reduce the extent of standing wave striations in conventional i-line resists. The PEB redistributes the photoactive compound (PAC) in the resist and allows for straighter resist sidewall profiles by reducing standing waves.

There are two basic types of ARCs: bottom antireflective coating beneath the photoresist to reduce substrate reflections, and top antireflective coating deposited over the resist to reduce secondary reflections from the resist surface. The bottom ARC has emerged as the most effective method in reducing reflections and problems such as standing waves and, therefore, will be the focus of our discussion.

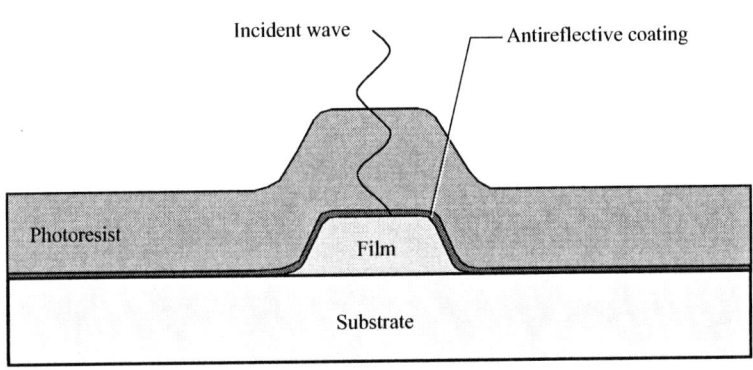

The use of antireflective coatings, dyes and filters can help prevent interference.

FIGURE 14.24 Antireflective Coating to Prevent Standing Waves

Bottom ARCs. A *bottom antireflective coating* (*BARC*) is used to suppress unintended light reflection from a reflective layer that is below the resist, as shown in Figure 14.25 on page 384. The BARC material is an organic or inorganic dielectric material that is applied to the wafer before the photoresist.

Organic ARCs reduce reflection by absorbing light and are typically spin-coated on the wafer in the same manner as the photoresist. Inorganic ARCs are deposited by plasma-enhanced chemical vapor deposition (PECVD). Inorganic ARCs do not absorb light but instead work by phase-shift cancellation of specific wavelengths based on the refractive index, film thickness, and other parameters (see Figure 14.26 on page 384). Successful phase cancellation of light requires very tight control of process parameters, such as a thickness tolerance for the BARC of 15 Å.[18] TiN, used as a diffusion barrier for interconnect metal, is also a good antireflective coating. However, the reflectivity of materials changes for shorter wavelengths, making interference effects more difficult to control.

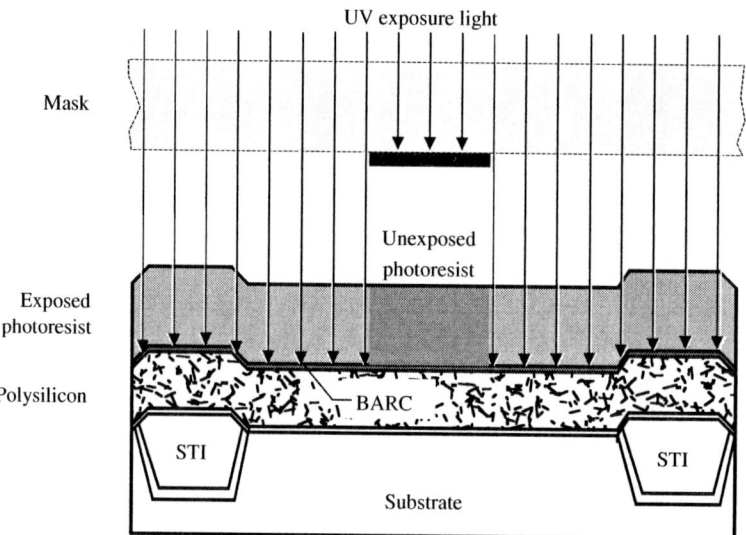

FIGURE 14.25 Light Suppression with Bottom Antireflective Coating

One factor in selecting an ARC is its ability to be removed after the completion of the photolithography process step. In some cases organic ARCs (primarily top ARCs) are aqueous-based and relatively easy to remove by rinsing during the development step. Inorganic ARCs are more difficult to remove, especially if their chemistry is similar to the underlying layer. This ARC layer is sometimes left on the wafer surface and becomes a part of the device.

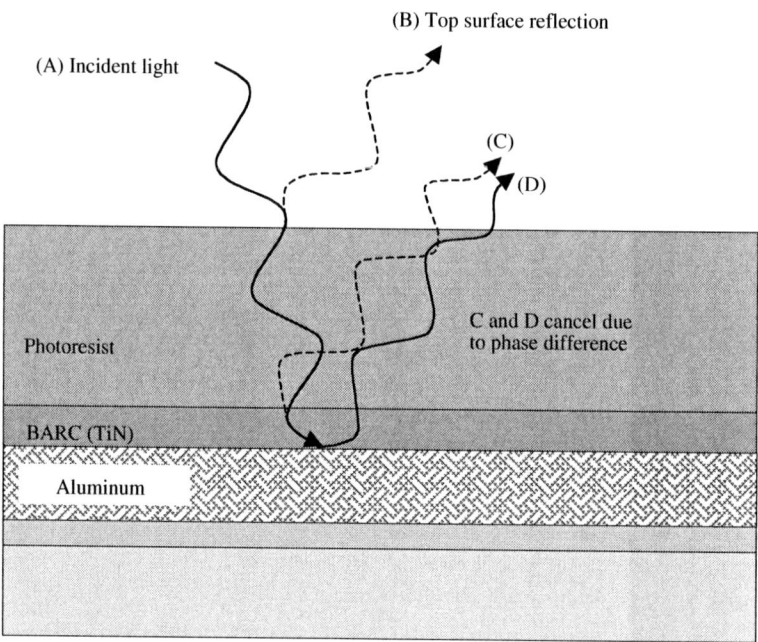

FIGURE 14.26 BARC Phase-Shift Cancellation of Light

Top ARCs. The *top antireflective coating* (*TARC*) reduces reflection at the interface between the resist surface and air (see Figure 14.27). TARC materials do not absorb light, but instead act as a transparent thin-film interference layer that uses destructive interference between light rays to eliminate reflectance.[19]

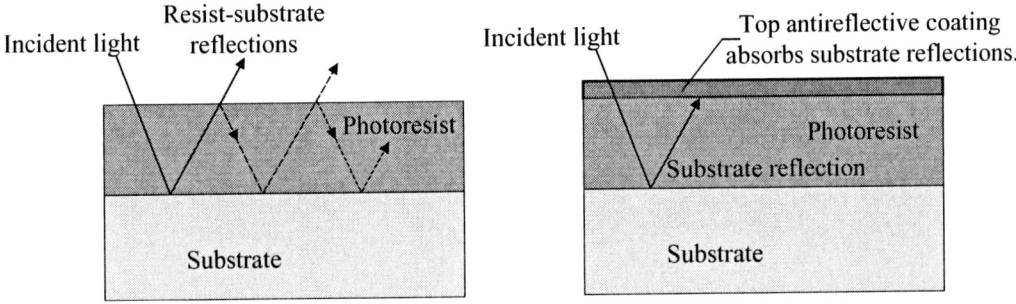

FIGURE 14.27 Top Antireflective Coating

Resolution

In photolithography, *resolution* is defined as the ability to discretely discern pairs of closely spaced features on the wafer (e.g., equal lines and spaces). This quality is shown in Figure 14.28. Resolution is critical given the high packing density for wafer features in advanced IC semiconductor manufacturing. Resolution is an important parameter for any optical system and is critical for photolithography because of the need to print extremely small features sizes for wafer fabrication.

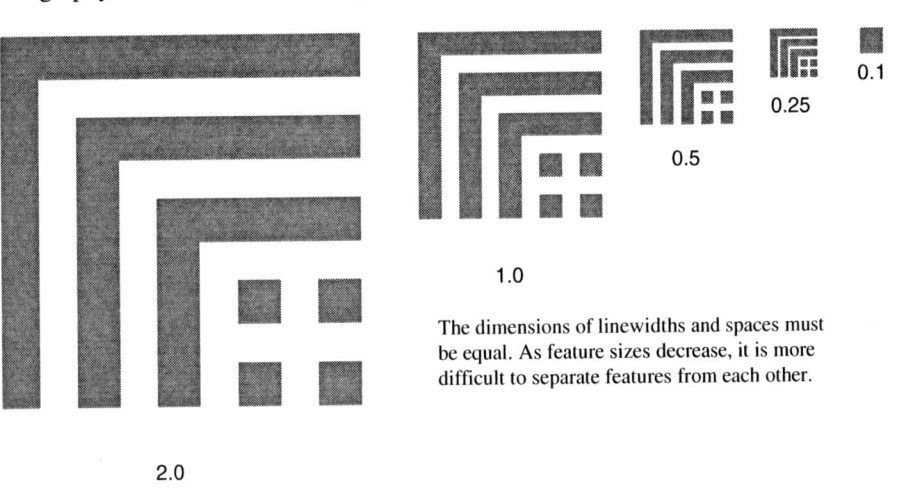

The dimensions of linewidths and spaces must be equal. As feature sizes decrease, it is more difficult to separate features from each other.

FIGURE 14.28 Resolution of Features

The formula for resolution, R, is:

$$R = \frac{k\lambda}{NA}$$

Where, k = factor that represents specific applications, with a range between 0.6 to 0.8
λ = wavelength of the light source
NA = numerical aperture of the exposure system

This formula shows that there are three parameters in photolithography that affect the ability to resolve fine geometry patterns in photoresist:

1. wavelength, λ
2. numerical aperture, NA
3. process factor, k

It should now be obvious why decreasing the wavelength, λ, of the exposing light source is important for improving resolution. A decrease in λ will improve the resolution capability of the lithography system. In other words, smaller wavelength is better. The other important parameter for improving resolution is increasing the NA of the projection lens. Lenses with a high NA also have a high resolution. However, recall that to increase NA also requires a larger diameter lens, which becomes costly.

The third parameter, k, represents process factors of the optical system and can affect resolution. However, there are practical limitations to decreasing k much below 0.6. There are some resolution enhancement techniques available, such as phase-shift masks (PSM) and optical proximity correction (OPC) that are becoming important for reducing k to improve pattern resolution. These techniques are discussed later in this chapter.

Calculating Resolution ■ When given the wavelength, λ, numerical aperture, NA, and process factor, k, it is possible to calculate the predicted resolution, R, of an optical system (see Figure 14.29). Consider the following example:

$$\lambda = 193 \text{ nm}$$
$$NA = 0.60$$
$$k = 0.6$$
$$R = \frac{k\lambda}{NA} = \frac{(0.6)(193 \text{ nm})}{0.6} = 193 \text{ nm}$$

For this optical system, the resolution of the smallest printable pattern is predicted to be 193 nm. If the wavelength of the exposure source is decreased, then the resolution will decrease. As the NA increases, the resolution will decrease. However, as explained in the following section, there is a price to pay for increasing the NA of the system, and that is decreased depth of focus.

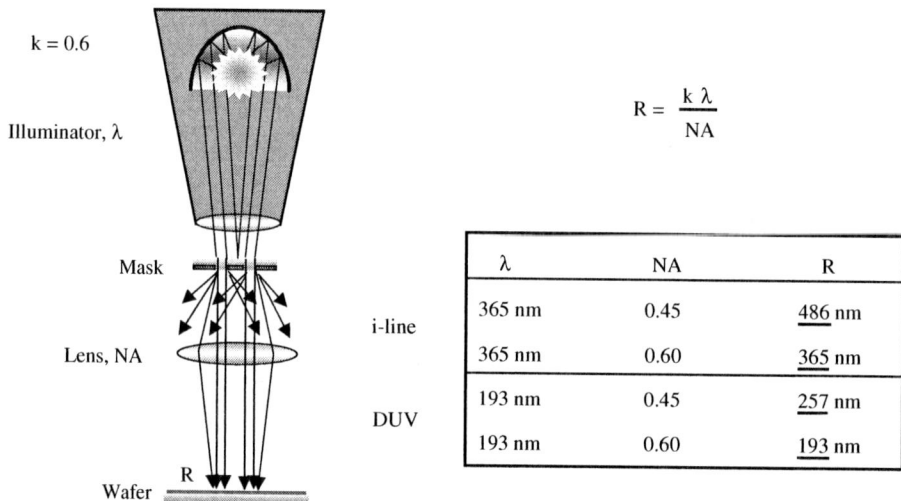

FIGURE 14.29 Calculating Resolution for a Given λ, NA, and k

Depth of Focus ■ The range around the focal point over which the image is continuously in focus is called the *depth of focus*, or *DOF* (see Figure 14.30). The *center of focus* (*COF*) is the point from the center of the lens where the best imaging occurs. The depth of focus is the range above and below the center of focus where the exposure energy is relatively constant. The COF may not be exactly at the middle of the resist layer, while the DOF should extend beyond the upper and lower surface of the resist coating. The actual usable DOF of any exposure system should be confirmed through tests conducted with the relevant processing parameters and environmental conditions. The goal is to find and maintain the best focus across the wafer and from wafer to wafer.

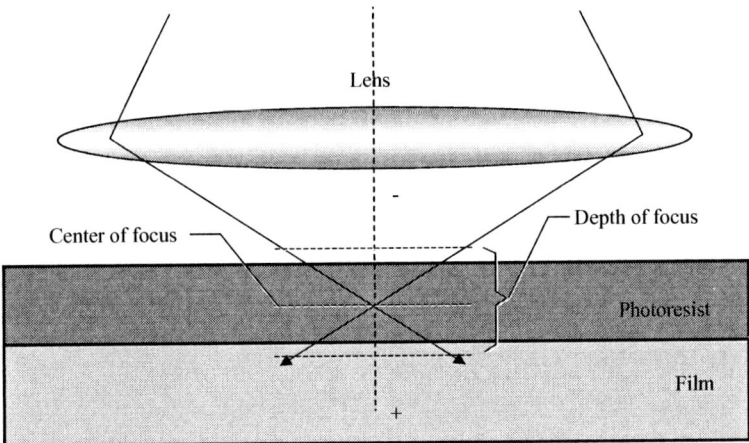

FIGURE 14.30 Depth of Focus (DOF)

The equation to describe the DOF is:

$$\text{DOF} = \frac{\lambda}{2(\text{NA})^2}$$

Where, λ = wavelength of the exposing light
NA = numerical aperture of the optical system

Depth of focus, also known as *depth of field (DOF)*, is commonly used in photography work to indicate the acceptable focus range of a camera's lens. Have you ever seen a picture of someone in which the background is out of focus? The photograph was taken with a camera of reduced DOF. Whereas a depth of focus of 30 cm for handheld cameras is common, in semiconductor photolithography DOF values of 1 μm or less are the norm.

The DOF is calculated using the λ of the illumination source and the NA of the projection lens. As the numerical aperture of a lens increases, more optical detail is captured and the resolution capability of the system increases. The significance of the DOF equation is that if resolution is increased, then the depth of focus will decrease (see Figure 14.31). Improving patterning resolution is imperative for submicron feature sizes. However, the resulting decrease in the DOF severely reduces the process latitude of the optical system.

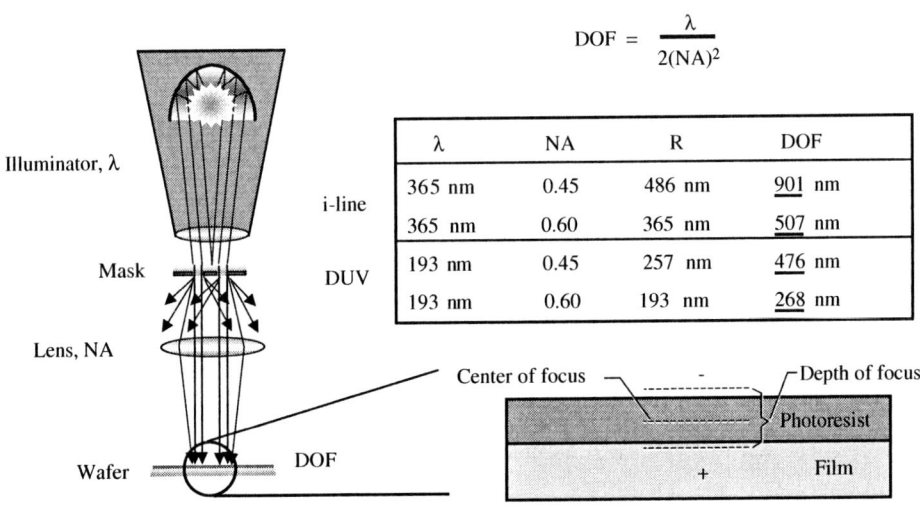

FIGURE 14.31 Resolution Versus Depth of Field for Varying NA

Resolution Versus Depth of Focus ■ In optical lithography, two critical parameters for the image quality are resolution and depth of focus. The semiconductor industry has always been challenged by the competing goals of achieving better resolution to pattern critical dimensions while simultaneously maintaining acceptable depth of focus. Any aspect of the photolithography tool or process that interferes with the ability of the wafer to lay flat, stable, and parallel to the focal plane of the lens will affect the focus quality of the exposure process. If the image plane moves away from the plane of best focus in the resist, image quality suffers. Imagine a stepper having a DOF = 1 μm. Suppose a human hair (diameter ≈ 100 μm) were to fall between the wafer stage and the wafer. The wafer would be lifted roughly 99 μm beyond the focal point of the stepper's lens. The tool would not be able to achieve focus nor be able to successfully align or expose the wafer to the reticle. Photolithography tools have introduced techniques in their equipment that compensate for factors such as poor wafer flatness and equipment vibration but cannot compensate for gross conditions of contamination and damaged parts.

The wafer surface has traditionally been nonplanar due to surface topography introduced during deposition and etching steps. In the 1980s, this wafer surface topography limited resolution because the optical system was not able to adequately focus on all wafer surfaces. Many different techniques were introduced to try to reduce the surface topography during this era.

Surface Planarity. The most important technique to planarize (make flat) wafer surfaces is a planarization process known as chemical mechanical planarization (CMP). CMP was introduced into wafer fabrication in the 1990s and has made a significant contribution toward planarizing the wafer surface to meet reduced depth of focus requirements (which permits increased pattern resolution). CMP has been a major factor in reducing nonplanar wafer surface topography for 0.25-μm critical dimensions and below. The significance of CMP for photolithography is that it makes possible the reduction of the depth of focus to achieve a higher pattern resolution.

PHOTOLITHOGRAPHY EQUIPMENT

The history of photolithography since the early days of wafer fabrication can be separated into five major equipment eras. Each era is roughly based on the type of equipment needed to achieve the CD resolution necessary at the time. The five microlithography equipment eras are:

- Contact aligner
- Proximity aligner
- Scanning projection aligner (scanner)
- Step-and-repeat aligner (stepper)
- Step-and-scan system

Contact Aligner

The *contact aligner* was the primary method of photolithography during the SSI era until the early 1970s. It was used in production mode for linewidth dimensions of about 5 μm and above. Although linewidths of 0.4 μm are possible, the contact aligner is not widely used today.

The mask for a contact aligner has the complete array of all die patterns to be photographed on the wafer surface. The wafer is coated with photoresist and mounted on a stage that has manual knobs to control left, right, and rotational position (a stage is a generic name for a positioning device with x, y, and rotational, or θ, positioning). The mask and wafer are both viewed simultaneously through a microscope having split vision optics (see Figure 14.32). The mask pattern is then aligned to the pattern on the wafer through manual operator control of the stage position.

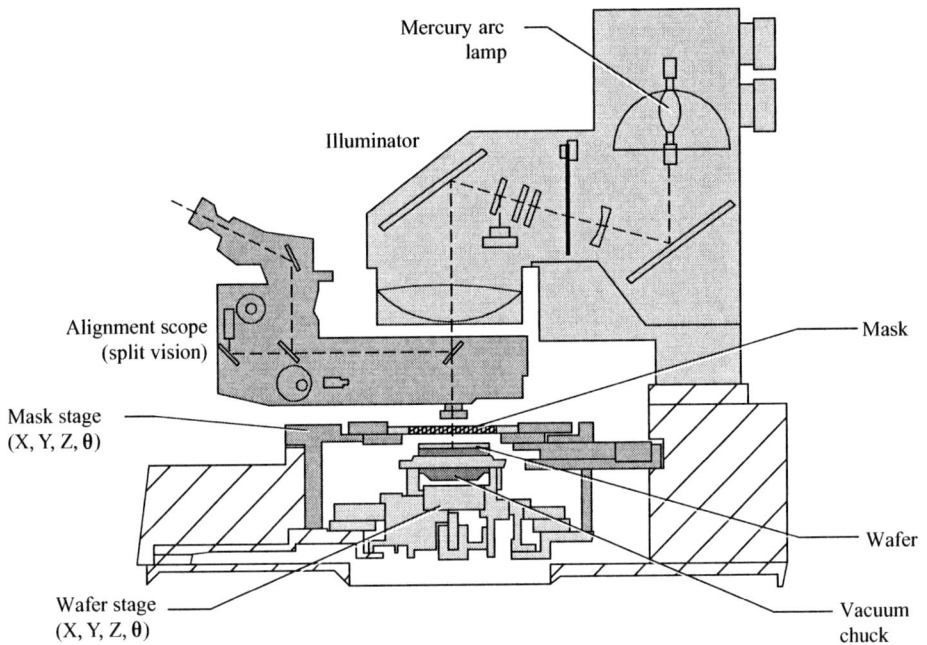

FIGURE 14.32 Contact/Proximity Aligner System
(Used with permission from Canon USA)

Once the mask is aligned to the wafer, then the mask is brought into direct contact with the resist coating on the wafer surface, which is why the equipment is called a contact aligner. At this time, the wafer and mask are exposed to ultraviolet (UV) rays. The UV rays pass through the transparent portions of the mask and the mask pattern is transferred to the resist.

Contact aligner systems were operator dependent and prone to contamination because the mask was placed in direct contact with the resist. Particulate contamination damaged the soft resist layer, the mask, or both, requiring the mask to be replaced every 5 to 25 operations. There were also resolution problems in the immediate area of any particle. As wafer sizes increased, there were overlay accuracy problems from attempting to pattern the entire wafer with one mask because alignment tolerances had to be held over wider areas.

Contact aligner printing actually can produce good image resolution on the wafer surface because the mask pattern and wafer are as close to each other as possible. This proximity reduces image distortion. However, contact alignment is very operator dependent, which introduces problems with repeatability and control.

Proximity Aligner

The *proximity aligner* evolved from the contact aligner and was common in the early 1970s during the SSI and early MSI era. These aligners are still used today in low-volume laboratory applications or older wafer fabs for discrete components where retooling is not economically feasible. They are suitable for linewidth dimensions of 2 to 4 μm, depending on factors such as the reflectivity of the substrate surface.

In proximity alignment the mask, which continues to transfer the entire wafer pattern, does not make direct contact with the resist. It is positioned in close contact to the resist surface, with a gap of approximately 2.5 to 25 μm between the mask and the resist on the wafer surface. The light originating from the source is collimated, which means the light rays are forced to be parallel with each other.

The proximity aligner was an attempt to alleviate the contamination problems associated with the contact aligner by creating a gap between the resist surface and mask that would not necessarily trap particles. Even though the magnitude of the gap was controlled, the performance of the proximity aligner was reduced because of the light scattering as the UV rays passed through the transparent regions of the mask and the air (see Figure 14.33 on page 390). This condition impaired the system's resolution capability, which was a major problem due to the need to reduce the linewidth critical dimension.

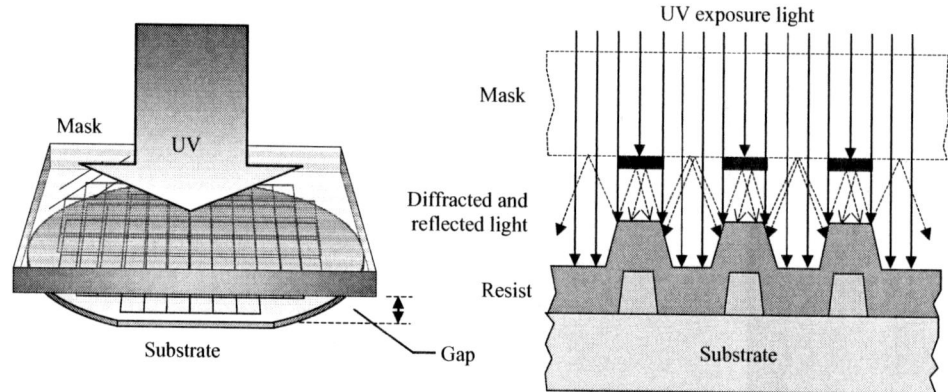

Diffraction of light on edges results in reflections from underside of mask causing undesirable resist exposure.

FIGURE 14.33 Edge Diffraction and Surface Reflectivity on Proximity Aligner

Scanning Projection Aligner

The industry recognized the need to move away from any form of contact or near contact alignment system due to its contamination problems, edge diffraction, resolution limitations, and operator dependency. The development of the *scanning projection aligner* (also referred to as *scanner*) in the early 1970s attempted to address these problems. Scanning projection aligners were the dominant lithography exposure tool in the late 1970s and early 1980s.[20] These aligners are still used today in older wafer fabs. They are suitable for noncritical layers containing linewidths greater than 1 μm.

The *scanning projection aligner* concept is to project a full wafer mask with a 1:1 image onto the wafer surface using a mirror system (i.e., based on reflective optics). Since the mask is 1X, there is no magnification or reduction involved and the pattern on the mask has the same dimensions as the pattern on the wafer.

The UV light rays are focused onto the wafer through a narrow slit, allowing a uniform source of light (see Figure 14.34). The mask and resist-coated wafer are mounted on a scanning carriage and moved in unison across the narrow beam of UV light to expose the resist on the wafer. As the scanning motion takes place, the image of the mask is eventually printed on the wafer surface.

A major challenge of scanning projection aligners was making a good 1X mask that contained all the chips on the wafer. If the chip had submicron feature sizes, then the mask also had submicron dimensions. As submicron feature sizes were introduced, this photolithography approach made it difficult for the mask maker to have no defects in the mask.

Step-and-Repeat Aligner (Stepper)

The mainstay of the 1990s microlithography equipment for wafer fabrication is the *step-and-repeat aligner* (commonly referred to as a *stepper*). Steppers have their distinct name because the tool projects only one exposure field (which may be one or more chips on the wafer), and then steps to the next location on the wafer to repeat the exposure. Steppers were first commercialized in the early 1980s.[21] Though not universally accepted at first, optical lithography steppers have dominated IC fabrication since the later 1980s, primarily for applications with critical dimensions down to 0.35 μm (conventional i-line photoresist) and some 0.25 μm (DUV photoresist).

A stepper uses a reticle, which contains the pattern in an exposure field corresponding to one or more die. A mask is not used in a stepper since a mask contains the entire die matrix. The optical projection exposure system of steppers uses refractive optics to project the reticle image onto the wafer (see Figure 14.35).

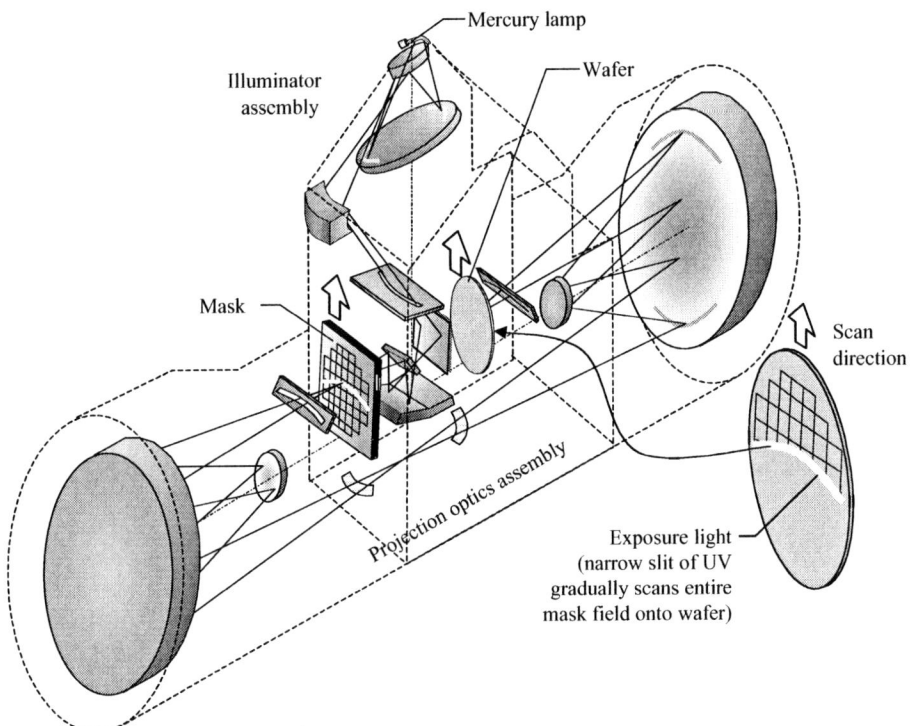

FIGURE 14.34 Scanning Projection Aligner
(Redrawn with permission from Silicon Valley Group Lithography Systems, Perkin-Elmer 500 Micralign)

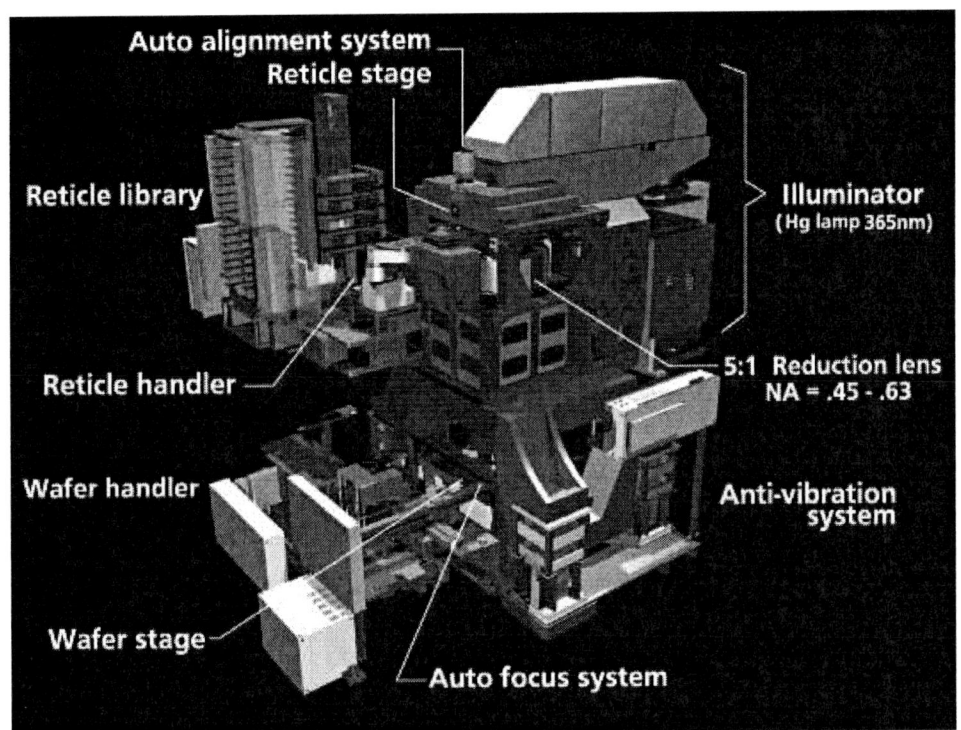

FIGURE 14.35 Step-and-Repeat Aligner (Stepper)
(Used with permission from Canon USA, FPA-3000 i5)

An advantage of optical steppers is their ability to use a reduction lens. Traditionally, i-line stepper reticles are sized 4X, 5X, or 10X larger than the actual image to be patterned (initially steppers used 10X reticles and then later 5X and 4X). To further explain the purpose of a reduction lens, a stepper with a 5X reticle requires a 5:1 reduction lens to transfer the correct image size to the wafer surface. This demagnification factor makes it easier to fabricate the reticle because the features on the reticle are five times larger than the final image on the wafer. At least one stepper manufacturer produces a stepper that provides a 1:1 projected image. The advantage of a nonreduction stepper is a low-cost stepper that can be used on noncritical patterning layers.

At each step in the exposure process, the stepper will focus the wafer and the reticle to the projection lens, align the wafer to the reticle, expose the resist with UV light that passes through the transparent regions of the reticle, and then step to the next location on the wafer to repeat the entire sequence. By following this process, the stepper will ultimately transfer the full die array onto the wafer in a sequence of exposure steps (see Figure 14.36). Because the stepper exposes only a small portion of the wafer at one time (e.g., typically a field size of one die on the wafer for a larger microprocessor die), compensations for variations in wafer flatness and geometry can be easily performed.

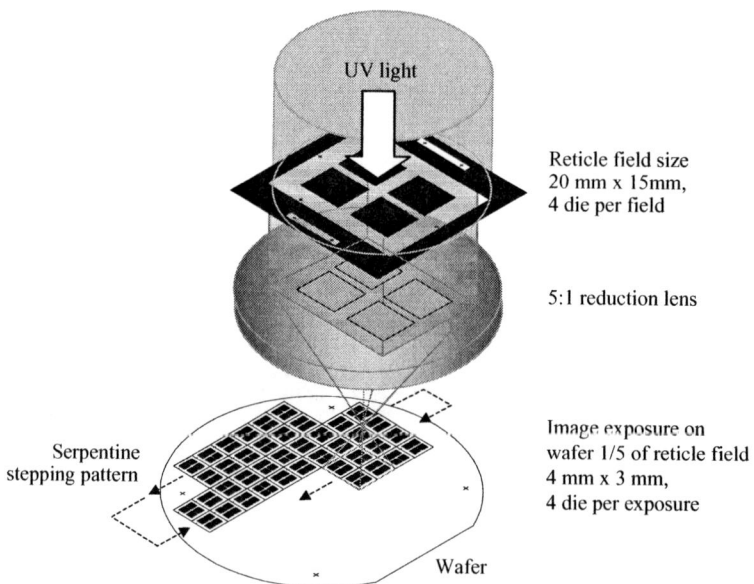

FIGURE 14.36 Stepper Exposure Field

Steppers have used conventional mercury arc lamp illumination sources (for g-line of 436 nm, h-line of 405 nm, and i-line of 365 nm) with a CD down to 0.35 μm. To obtain a 248 nm DUV wavelength source, the mercury arc lamp source is replaced with a KrF (krypton-fluoride) excimer laser. This equipment permits patterning a 0.25-μm CD. Typically, DUV stepper technology is used for patterning critical layers while conventional i-line exposure methods are used for noncritical layers. This mix-and-match approach to lithography is used to reduce production costs.

The growth in the physical size of semiconductor chips along with smaller critical dimensions creates a need for a larger exposure field size and improved lens optics for steppers. This in turn dictates more complex lithographic lens design and fabrication, with the cost of lens systems alone in steppers exceeding $1 million. This cost limits a conventional step-and-repeat exposure field to 22 × 22 mm. To overcome these problems, an evolutionary stepper known as step-and-scan has become predominant for DUV photolithography at technology nodes of 0.25 μm and below.

Step-and-Scan System

A recent development in lithographic exposure equipment employs a technique referred to as a *step-and-scan system*. The step-and-scan optical lithography system is a hybrid tool that combines the technology from scanning projection aligners and step-and-repeat steppers by using a reduction lens to scan the image of a large exposure field onto a portion of the wafer. A focused slit of light is scanned simultaneously across the reticle and wafer (see Figure 14.37). One standard exposure field size for this type of aligner is 26 × 33 mm for a six-inch-square reticle size. Once the scan and pattern transfer is completed, then the wafer is stepped to the next exposure field and the process is repeated.

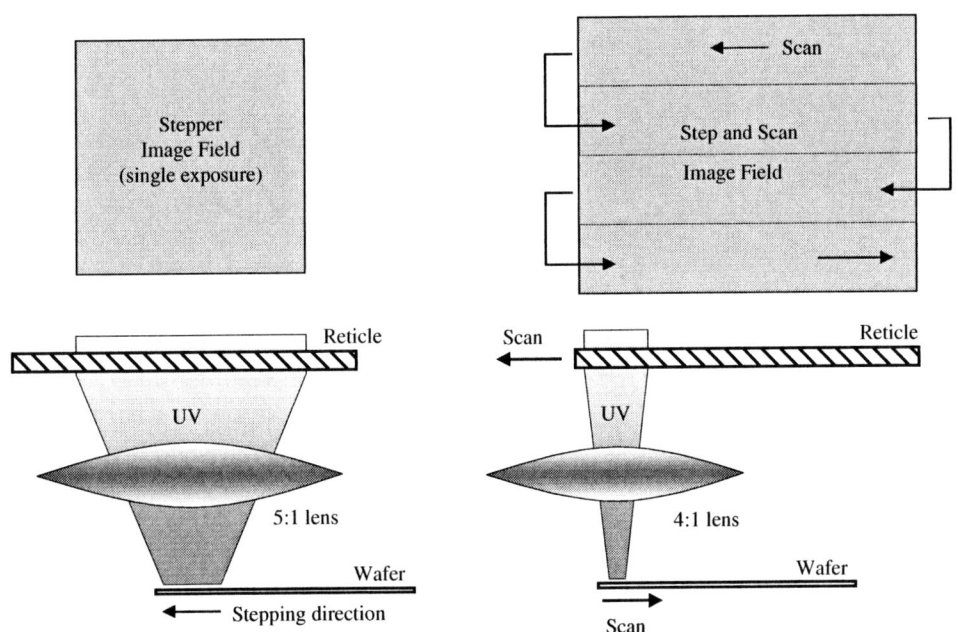

FIGURE 14.37 Wafer Exposure Field for Step-and-Scan (Used with permission from ASM Lithography)

A benefit from using a step-and-scan system to expose wafers is the increased exposure field for larger chip sizes. The lens field only has to be a narrow slit, as in the older full-wafer scanning projection aligners. It scans a reduction reticle (typically a 4X) through a small, well-corrected image field of 26 × 33 mm before stepping to the next location.[22] The optical system can now be designed for this much smaller field, permitting a smaller lens system (see Figure 14.38 on page 394). Another major advantage of a larger field size is the opportunity to put more die on a reticle, thus exposing more die in a single exposure.

Another significant step-and-scan system advantage is the ability to adjust focus throughout a scan (called on-the-fly focus), permitting compensation for lens defects and changes in wafer flatness. This improved control of focus during a scan yields improved CD uniformity control across the exposure field.

The major challenge with step-and-scan systems is the increased demand on mechanical tolerance control due to the motion of the stage controls for the reticle and wafer. Whereas steppers only had to move the wafer rapidly to a new position, a step-and-scan system has to precisely move both the wafer and reticle simultaneously and in opposite directions. These scanning and stepping motions are performed while holding position tolerances to within a few tens of nanometers during the scan.

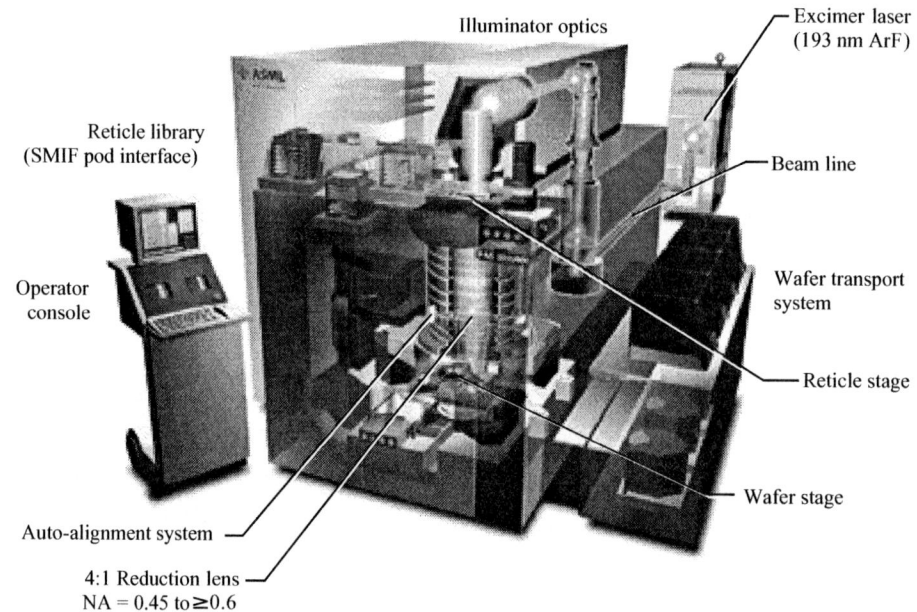

FIGURE 14.38 Step-and-Scan Exposure System
(Used with permission from ASM Lithography, PAS 5500)

Reticles

A *reticle* is a transparent plate that has a pattern image that will be transferred to the photoresist coating on the wafer. Two terms are commonly used in wafer fabrication: reticle and mask. A reticle contains the pattern image for only part of the wafer (e.g., 4 die) that must be stepped and repeated across the entire substrate. Reticles are used for step-and-repeat steppers and step-and-scan systems found in wafer fabrication. A *photomask,* or *mask,* contains the pattern image for a complete wafer die array and the pattern is transferred in a single exposure (1:1 image transfer).[23] Masks are used for older proximity printers and scanning alignment projectors. Table 14.6 compares reticles versus masks to provide insight into why the photolithography industry changed from mask to reticle.

TABLE 14.6 Comparison of Reticle versus Mask

Parameter Number of Exposures	Reticle Multiple Exposures	Mask Single Exposure
Critical Dimension	Easier to pattern submicron dimensions on wafer due to larger pattern size on reticle (e.g., 4:1, 5:1).	Difficult to pattern submicron dimensions on mask and wafer without reduction optics.
Exposure Field	Small exposure field that requires step-and-repeat process.	Exposure field is entire wafer.
Mask Technology	Optical reduction permits larger reticle dimensions—easier to print.	Mask has same critical dimensions as wafer—more difficult to print.
Throughput	Requires sophisticated automation to step-and-repeat across wafer.	Potentially higher (not always true if equipment is not automated).
Die alignment and focus	Adjusts for individual die alignment and focus.	Global wafer alignment, but no individual die alignment and focus.
Defect density	Improved yield but no reticle defect permitted. Reticle defects are repeated for each field exposure.	Defects are not repeated multiple times on a wafer.
Surface flatness	Stepper compensates during initial global prealignment measurements or during die-by-die exposures.	No compensation, except for overall global focus and alignment.

Photolithography Reticle (Photo courtesy of Advanced Micro Devices)

A reticle must be perfectly manufactured. All wafer circuit features ultimately come from patterns on the reticle; therefore, the quality of the reticle plays a key role in achieving high-quality imaging during submicron photolithography. If reticle pattern defects such as distortion and incorrect image placement are not detected, they will be reproduced in the resist on the wafer. Reticles undergo extensive automated testing for defects and particles once they are fabricated.

Reticle Materials ■ The primary reticle substrate material for submicron lithography is fused silica. It is always used for DUV lithography because it has high optical transmission in the DUV portion of the spectrum (248 and 193 nm). Fused silica is the most expensive material for reticles and has very low thermal expansion. A low thermal expansion means the reticle is dimensionally stable with changes in temperature. Because of this, there is only minor thermal run-out, which is an increase in a dimension due to a change in the thermal environment. Other desirable properties for reticle material are high optical transmission and no defects on the surface or within the material.

The most common opaque material deposited on the reticle substrate is a thin layer of chromium (referred to as chrome). This chrome layer will undergo patterning to create the necessary design shapes to define the circuitry for the wafer layer (e.g., holes, lines, pads, and so on). The chrome thickness is usually less than 1,000 Å and is sputter deposited. There sometimes is an antireflective layer of chromium oxide (about 200 Å thick) placed on the chrome.

Reticle Reduction and Size ■ Reticles are used in stepper and step-and-scan system applications that require a reduction lens to reduce overlay accuracy during patterning. A stepper commonly uses a reticle reduction ratio of 5:1 or 4:1, whereas a step-and-scan system uses a reticle reduction ratio of 4:1. Each reticle layer usually contains the necessary pattern for one or more die. The small field exposure size on steppers and step-and-scan systems permits close tolerance control during reticle alignment.

Most reticles are currently 6 × 6 inches (152 mm) square, although 5 × 5 inches is still common. Reticles are usually 0.090" to 0.250" thick. A six-inch reticle with a 4X reduction defines a usable field size at the wafer surface, which on current photolithography step-and-scan equipment is about 25 × 25 mm (see Figure 14.39 on page 396). Multiple exposures are required as the reticle is stepped over the wafer. The required exposure field size is forecasted to increase to accommodate the increasing size of DRAM and microprocessor chips. The industry is investigating a changeover to a nine-inch reticle to replace the current six-inch reticles. This will increase the number of die exposed per field, which permits fewer steps in the stepper or step-and-scan to complete the exposure of a complete wafer.[24] However, there is substantial cost to implementing this change, which may negate the benefit.

Reticle Fabrication ■ A common method for creating a pattern on a reticle is by the use of an electron beam (e-beam). This technique uses direct writing of the pattern from electronically stored original patterns. There is an immense amount of data involved in this process, and the total time to write a pattern on a reticle can be several hours. The basic steps for patterning a reticle are similar to those for a wafer.

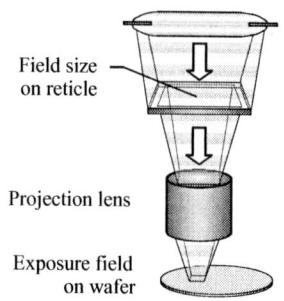

Lens Type	10:0	5:1	4:1	1:1
Reticle Field Size (mm)	100 x 100	100 x 100	100 x 100	30 x 30
Exposure Field on Wafer (mm)	10 x 10	20 x 20	25 x 25	30 x 30
Die per Field of Exposure (assume 5mm x 5mm die size)	4	16	25	36

FIGURE 14.39 Comparison of Reticle Reduction Versus Exposure Field

E-Beam Lithography. The direct-write method of *electron beam* (*e-beam*) lithography is capable of transferring a high-resolution pattern directly to the reticle surface. In e-beam lithography, an electron source produces many electrons that are accelerated and focused in the shape of a beam toward the reticle (see Figure 14.40). The electron beam is focused either magnetically or electrostatically and is scanned in the desired pattern across a special e-beam resist on the reticle surface. The beam can be scanned across the entire reticle (raster scan) or scanned only over the printed areas (vector scan) to eventually transfer the pattern to the reticle.

To apply the e-beam resist on the reticle, the reticle is cleaned and the chrome film is spin coated with a suitable photoresist, then soft baked. The standard e-beam resist is a positive-tone poly(butene 1 sulfone), or PBS. However, this resist is not suitable for submicron linewidths. Alternative resists are being developed, such as chemically amplified resists.[25] Advanced e-beam mask writers are capable of patterning 0.36-µm minimum feature sizes on a reticle.[26]

After exposing and developing the resist, the final patterned surface is etched into the chrome film with a wet- or dry-etch process (advanced reticle fabrication uses dry etch). Recall that the mask features are four or five times larger than the wafer images (due to lens reduction in steppers and step-and-scan tools). However, the dimensional tolerances are very tight on mask features in order to produce acceptable critical dimensions on wafers.

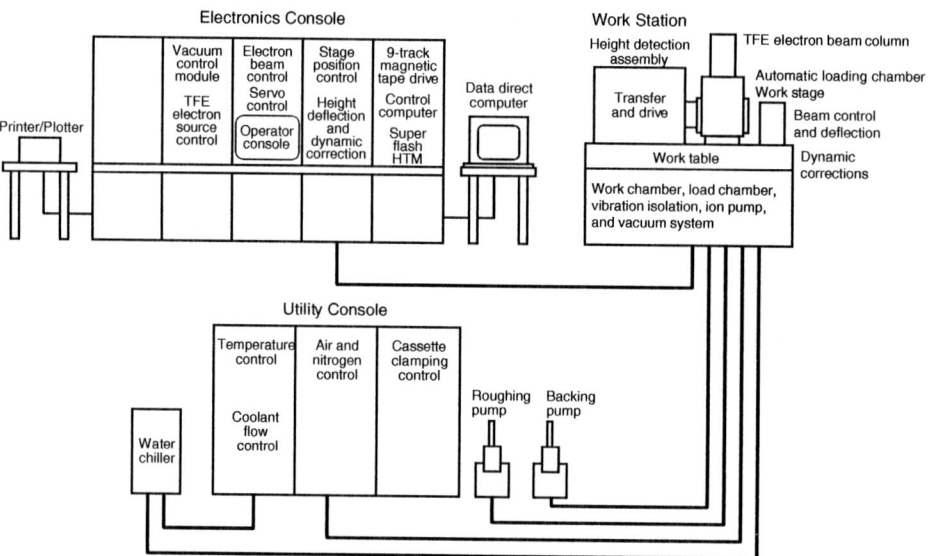

FIGURE 14.40 Principles of Electron Beam Lithography
(Used with permission from Etec Systems, Inc., MEBES 4500 System)

E-beam replaced optical resist patterning on reticles because it has a shorter wavelength and higher exposure speed compared to UV sources. These qualities produce better dimensional control with increased production throughput.[27] E-beam is a reliable method with high resolution and flexibility, but speed and system complexity are disadvantages for its use to pattern wafers.[28] It does have applications in very specialized wafers, such as application-specific integrated circuits (ASICs) that benefit from no reticle to achieve a quick turnaround time from development to production.

Sources of Reticle Damage ■ Reticles on advanced steppers and step-and-scan systems are moved into and out of the tool with automated storage and handling. A reticle must be meticulously clean for it to create perfect images during its many exposures to pattern a wafer layer. Recall that a stepper will step-and-repeat one image over the entire wafer. If there is even one foreign particle on a reticle circuit image, then it is repeated at every site on the wafer.

There are actually many possible sources of damage during use of a reticle, such as dropping the reticle, scratches on the surface, electrostatic discharge (ESD), and particles of dirt. ESD causes a problem if the reticle is handled by an improperly grounded technician. This condition could potentially discharge a small surge of current through the micron-sized chromium lines on the reticle surface, melting a circuit line and destroying the pattern.

If an airborne particle of dirt lands on a critical region of the reticle, this could damage the circuit and create an imaging defect. A solution to particulate contamination on reticles is to protect the surface with a thin, optically transparent membrane known as a *pellicle*. The pellicle is tightly stretched on a sealed frame about 5 to 10 mm above the mask surface. No dirt particles are able to reach the mask surface. If a dust particle lands on the pellicle, it is far removed from the focal plane and is invisible to the projection optics (see Figure 14.41).

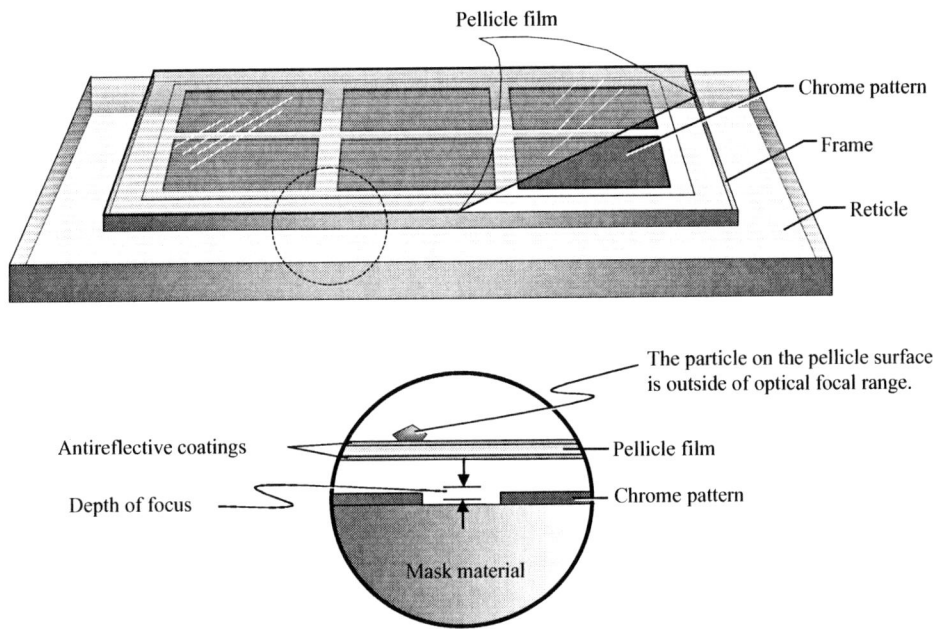

FIGURE 14.41 Pellicle on a Reticle

The pellicle material is transparent to the exposing light energy. There are different materials and thickness used for pellicles, such as nitrocellulose acetate that is typically 0.7 μm thick, or a 12 μm thick Mylar fluorocarbon material. The pellicle is fragile; therefore, any action that would scratch the mask surface will rupture the pellicle membrane (which alerts the user to a potential source of reticle damage). When used for deep UV exposure wavelengths, the optical transparency of the pellicle must be carefully evaluated.

Optical Enhancement Techniques

As wafer critical dimensions decrease to feature sizes of 0.15 µm and below, factors such as diffraction and light scattering prevent reticles from effectively transferring a replica of the pattern to the wafer. Optical enhancement techniques are being used on the reticle to improve the quality and definition of the image. This has become an important area of optical lithography known as *subwavelength lithography*, which permits a wafer to be patterned with a resolution slightly below the light exposure wavelength.

Phase-Shift Mask (PSM) ■ *Phase-shift mask (PSM)* is a method developed in 1982 to overcome problems associated with light diffraction through small openings patterned on the reticle. With PSM, the reticle is modified with an additional transparent layer so that alternating clear regions cause the light to be phase-shifted 180° (see Figure 14.42). As we have learned about interference between out-of-phase waves, there is now destructive interference. The light diffracted into the nominally dark area on the left will encounter destructive interference with the light diffracted from the right clear area. The light diffraction underneath the opaque area is reduced. Phase-shifting techniques on reticles improve image contrast, and has become a critical factor for microlithography at CDs of 0.18 µm and below. There are many approaches to PSM that use the same basic concept, but they are all based on the principle of destructive interference.

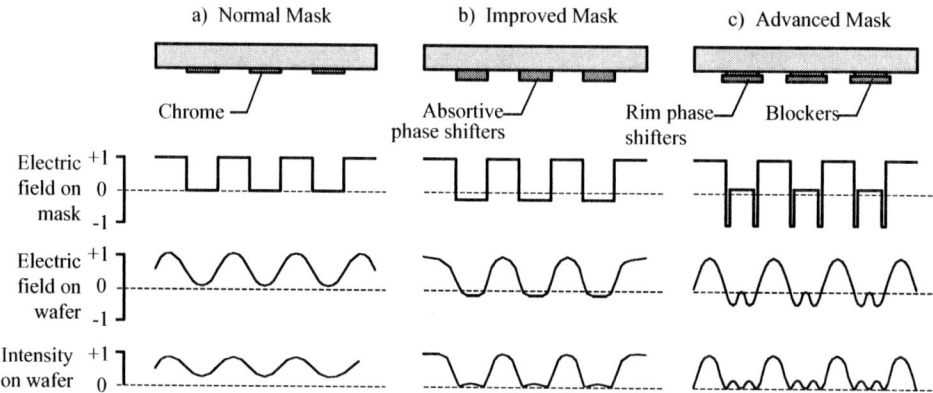

FIGURE 14.42 Phase-Shifting Mask
(Reprinted from the January 1992 edition of *Solid State Technology*, copyright 1992 by PennWell Publishing Company)

Optical Proximity Correction (OPC) ■ Uniformity of feature size dimensions is critical in high-performance ICs, especially in the transistor gate region where varying linewidths will affect the speed of the device. Because of optical proximity effects due to light diffraction and interference between closely spaced features on the reticle, the widths of lines in the lithographic image are affected by other nearby features (see Figure 14.43). Tightly grouped lines will print with a different size than isolated lines, although both have the same linewidth dimension on the reticle. This particular example is referred to as iso-dense bias.

It is possible to introduce selective image size biases (alterations) into the reticle pattern to compensate for optical proximity effects. This is called *optical proximity correction (OPC)*. Computer algorithms are available to the reticle designer for generating optical proximity corrections on the reticle for smaller feature sizes. However, producing reticles with this level of control is a challenge in manufacturing, especially since CD feature sizes are already extremely small. As the critical dimension is reduced to less than 0.18 µm, aggressive scaling of dimensions will require increased use of OPC, which will make reticle fabrication more complicated.[29]

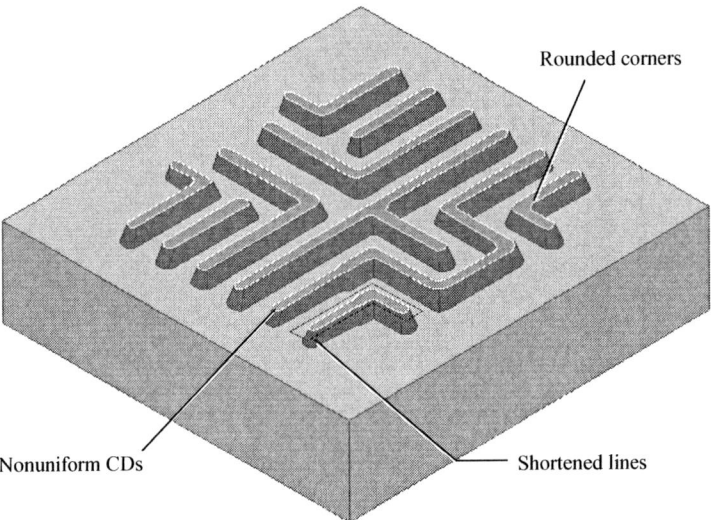

FIGURE 14.43 Optical Proximity Effects

Off-Axis Illumination ■ The conventional form of illumination for lithographic lenses has the exposing light centered on the centerline of the projection optics (see Figure 14.44). However, light diffraction creates a problem as the reticle openings are made smaller and smaller for submicron features sizes, to the point that the imaging lens will not transfer the pattern. A solution is to have the incident light strike the mask at an angle in order to align the diffraction fringes with the lens, which produces a symmetrical intensity profile that corresponds to the reticle pattern. This is referred to as *off-axis illumination* or *(OAI)*. This technique reduces the resolution limit and increases the depth of focus for imaging.

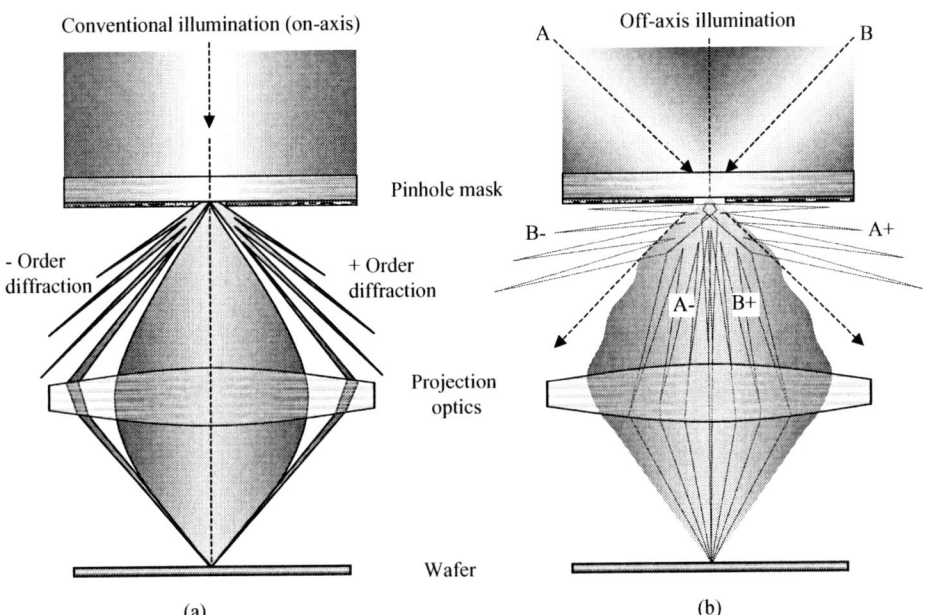

FIGURE 14.44 Off-Axis Illumination

Bias ■ *Print bias* is the difference between a feature dimension on a reticle and the same printed dimension on the wafer. The amount of print bias may be different for varying sized features. An example of print bias is a square contact with dimensions near the optical system resolution that prints with rounded corners. One way to address this problem is by adding serifs to the features on the reticle pattern, which help square features such as contact corners during printing on the wafer (see Figure 14.45 on page 400).[30]

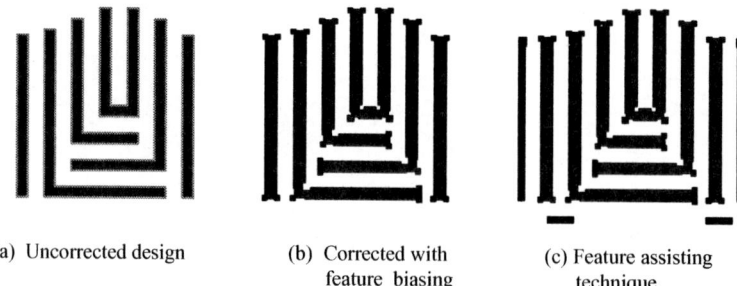

(a) Uncorrected design (b) Corrected with feature biasing (c) Feature assisting technique

FIGURE 14.45 Serifs to Minimize Rounding of Contact Corners
(Redrawn from "Optical/Laser Microlithography," *SPIE Proceeding VIII* 2440 (February 1995)

Alignment

The alignment process begins with the proper alignment of the reticle to a fixed reference mark on the body of the stepper or step-and-scan system. Once the reticle is aligned to the body of the exposure tool, then the wafer stage positioning is measured with respect to the reticle. This positioning provides any baseline correction data that the alignment software will use to compensate for variations in reticle characteristics. This process is referred to as *baseline compensation (BLC)*.

To successfully pattern the wafer, the wafer pattern must be correctly aligned to the existing patterns on the reticle pattern. The IC will function properly only if each successive projected image is correctly matched to the wafer patterns. To this end, *alignment* is the process of determining the position, orientation, and distortion of the patterns already on the wafer and then placing them in correct relation to the projected image from the reticle. Alignment should be fast, repeatable, accurate, and precise. The outcome of the alignment process, or how accurately each successive pattern is matched to the previous layer, is known as *registration*.[31]

Overlay accuracy (also known as registration) is the measure of the alignment system's ability to overlay the reticle pattern onto the wafer pattern. Overlay budget describes the maximum relative displacement between a patterned layer and the previously defined layer (see Figure 14.46). In general, the overlay budget is about one-third of the critical dimension. For 0.15 μm design rules, the overlay budget is expected to be 50 nm.[32]

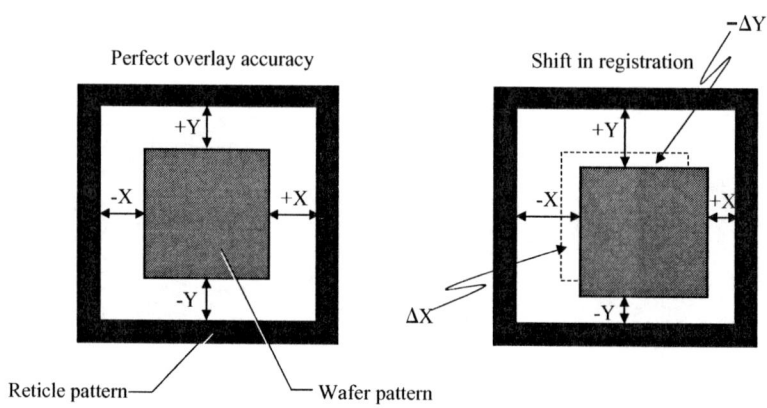

FIGURE 14.46 Overlay Budget

For steppers and step-and scan systems, each reticle pattern is aligned and exposed at multiple locations as the aligner steps across the wafer. Each field corresponds to the reticle pattern of a single large chip or several smaller chips. A *grid* is the particular path that the photo tool follows to step across the wafer and expose the individual fields (see Figure 14.47).

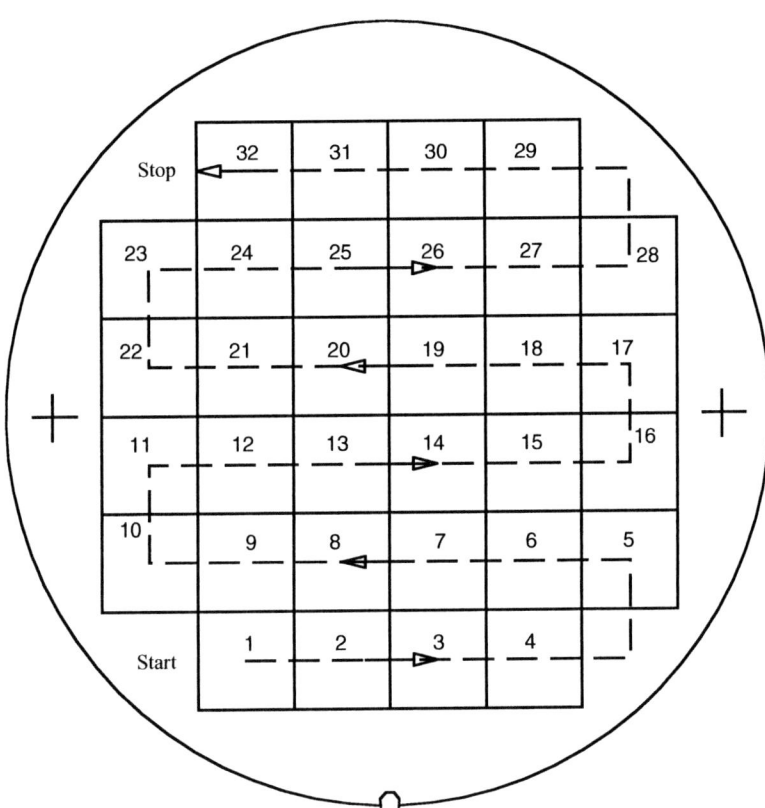

FIGURE 14.47 Grid of Exposure Fields on Wafer

Steppers and step-and-scan systems have a sophisticated automatic alignment system that detects the position and orientation of both the wafer and the reticle and then aligns the wafer to the reticle before exposure (see Figure 14.48 on page 402). The tool's alignment system includes alignment marks on both the reticle and wafer, an alignment detection system, and the electromechanical positioning devices such as motors and drive mechanisms. The goal of alignment detection is to determine as rapidly as possible the position of the different alignment marks in the exposure tool coordinate system. Alignment software algorithms within the tool's process controller calculate offset values and the direction of wafer stage movements necessary to bring the wafer within the overlay budget specified for the tool.

Alignment Marks ■ *Alignment marks* are visible patterns placed on the reticle and the wafer to determine their position and orientation. Marks, also referred to as targets or fiducial marks, may be one or more lines on the reticle, which then become trenches when printed on the wafer. Marks may also be a shape on the reticle which overlays on wafer marks (see Figure 14.49 on page 402). Reticle alignment (RA) marks appear on both the left and right sides of the reticle. The RA marks are aligned to fiducial marks mounted on the stepper's body. Global alignment (GA) marks are printed on the left and right sides of the wafer during first mask exposure. GA marks are used for making coarse alignments of each wafer during subsequent layers. Fine alignment (FA) marks are printed during each field exposure. FA marks are used for making final alignment adjustments between each wafer exposure site and the reticle. The shape and location of marks varies depending on the equipment manufacturer. Once an alignment mark is aligned, it is assumed that the remainder of the reticle patterns is also correctly aligned.

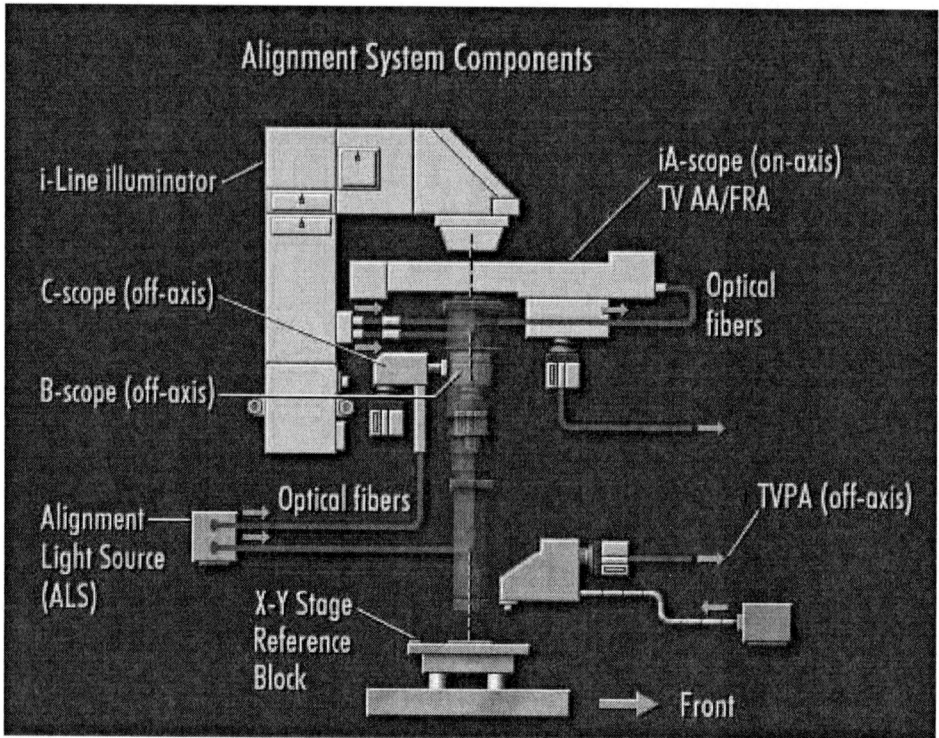

FIGURE 14.48 Step-and-Repeat Alignment System
(Used with permission from Canon USA, FPA-2000 il)

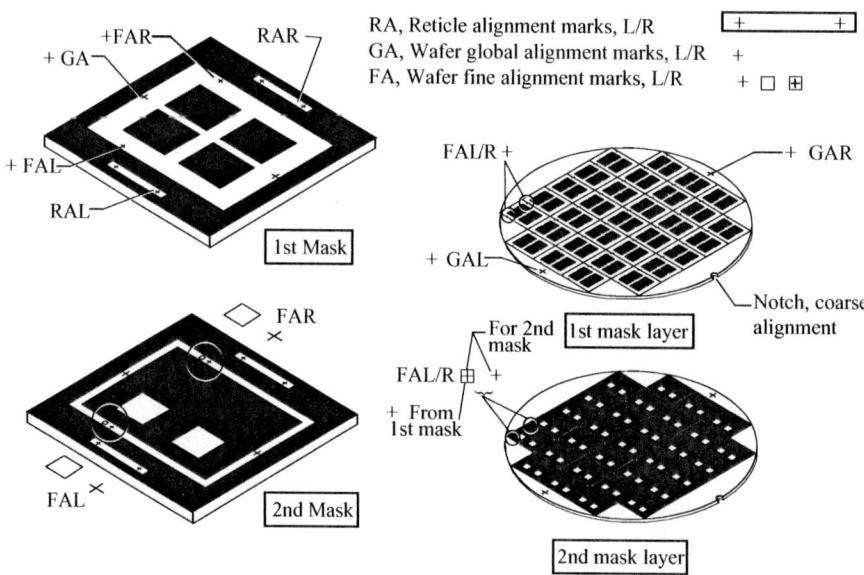

FIGURE 14.49 Alignment Marks

The computer-controlled automatic alignment system identifies the alignment marks. An alignment illumination system projects light through the reticle marks and onto the wafer surface. A common light source is the stepper UV light, with filtering to block unwanted wavelengths that could activate the resist. Another source is a laser such as a helium-neon laser. This laser emits light at 633 nm, which will not activate the resist.

Light illuminates the alignment marks and then photo detectors are used to optically detect the reticle and wafer targets. The alignment illumination system can use the main stepper's projection optics to illuminate the marks (known as on-axis or through-the-lens), or use its own alignment optics (known as off-axis) (see Figure 14.50). There are many different types of alignment methods in use. Laser beams are also used to precisely control the levelness and position of the X-Y stage that holds the wafer chuck. Laser interferometry is used to measure the position of the wafer stage at all times. Once the position data are measured, the data are fed into the system computer with a software interface to the electromechanical system used to make the correct adjustments to align the wafer to the reticle.

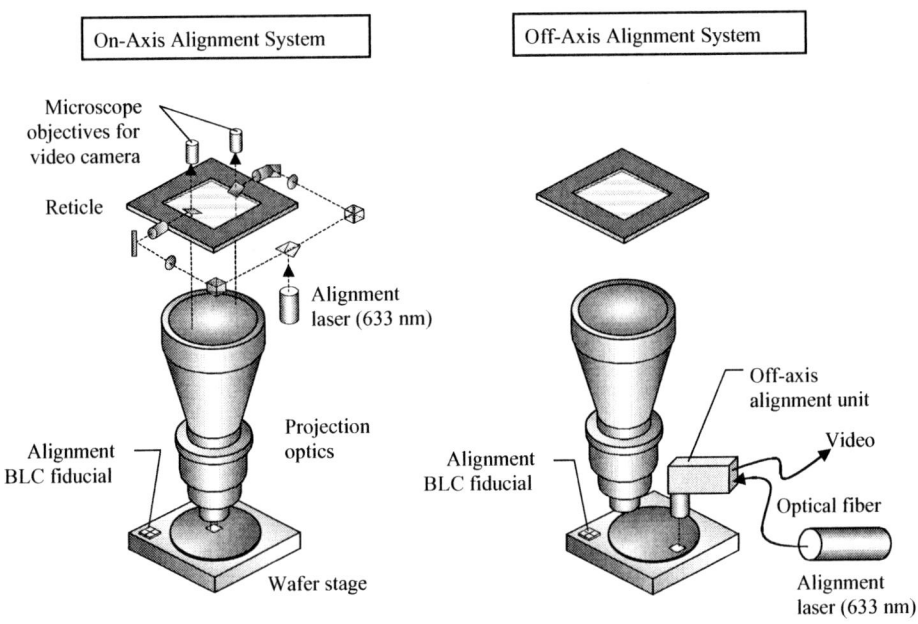

FIGURE 14.50 On-Axis Versus Off-Axis Alignment System
(Used with permission from Canon USA, redrawn after FPA-2000 il schematics)

Step-and-Scan Alignment Marks. An improvement for a step-and-scan system is that multiple alignment marks can be placed in each exposure field, permitting alignment to occur with each field exposure. This means that position parameters such as wafer rotation and X-Y alignment can be adjusted at each scan site, which improves overlay accuracy and reduces the overlay budget.

Alignment Mark Deterioration. The increasing number of film layers on high-performance ICs require greater flatness from one layer to the next. This flatness is achieved by using chemical mechanical planarization (CMP) to planarize the wafer (see Chapter 18). However, CMP also deteriorates alignment marks on the wafer, which makes them more difficult to detect.[33] Stepper and step-and-scan equipment manufacturers have developed alternative approaches to align the wafer. In one case, a laser beam is used to detect a phase change in the beam reflection off the alignment mark relative to a reference signal. Slight phase changes are relatively easy to detect, even with layers that are too flat to measure with the standard light reflection system.

Types of Alignment ■ Each reticle must be aligned to the exposure tool's body when the reticle is mounted on the reticle stage. The reticle alignment marks are illuminated by a laser beam (an example is HeNe) through a fixed referenced mark. Once aligned, the wafer stage is then aligned to the reticle.

The wafer is initially prealigned by holding it in a chuck at the wafer stage and locating the notch or flat. For a *first mask exposure,* the wafer flat or notch is the only alignment feature on the wafer. The blank wafer is "blindly" exposed to the first reticle layer without any further alignment. The only consideration is for ensuring that the lens is properly focused to the wafer prior to each exposure.

For subsequent exposures the wafer and reticle position data are measured with respect to a coordinate system defined for the exposure tool and are then used in a global or field-by-field manner to perform alignment. *Global alignment* (or coarse alignment) uses several marks to align all the exposure fields on the wafer. The goal is to quickly align the entire wafer relative to the reticle.

Field-by-field alignment data (also called fine alignment) is used to align a single field. The stepper steps to each die on the grid, focuses, aligns, and then exposes. This method is the slowest but achieves excellent registration, if necessary, for a critical layer. The specific details of stepper alignment systems are unique to each manufacturer due to the extreme tolerance control required to align the photomask and wafer pattern to within a few nanometers.

Environmental Conditions

The environmental conditions for photolithography steppers and step-and-scan systems are critical for high-yield wafer fabrication. Minor variations in environmental conditions can lead to device defects. Photolithography tools have an enclosed environmental chamber to control conditions such as temperature, humidity, vibration, atmospheric pressure, and particle contamination.

Temperature ■ It is critical to control temperature during photolithography. Temperature can adversely affect all aspects of aligners: mask stage, optical elements, light sources, wafer stage, and alignment systems. Temperatures are typically controlled to within a tenth of a degree Celsius. For this reason, the high-temperature illuminator and heat generating power supplies are generally located away from the main body of the alignment and exposure tool.

Humidity ■ Humidity is carefully controlled inside the fab and within the photolithography environmental chamber. Excessive humidity can affect the resist adhesion to the wafer. Humidity also affects the density of the air, which can adversely affect light passing through it; thus producing adverse effects on interferometer positioning, lens NA, and focusing.

Vibration ■ The occurrence of vibration would create problems for positioning, alignment, focusing, and exposure. Problems could be misregistration, misalignment, defocus, and nonuniform exposure. There are different methods to minimize vibration. In some cases, floors that support the lithography equipment are isolated by large shock absorbers from the other areas of the fab. The photo tools also have their own design features to isolate vibration, such as pneumatic isolation devices and momentum absorption structures.

Atmospheric Pressure ■ Atmospheric pressure changes can affect the refractive index of air in the projection optics and laser interferometers used for stage positioning of steppers. This condition leads to nonuniform CDs and poor registration. For this reason, equipment manufacturers usually include pressure sensors to monitor the atmospheric pressure inside the environmental chamber. The pressure measurement data is fed into a computer for monitoring and control. In some cases, the lens elements are sealed in an airtight housing with a constant air flow and pressure inside the housing.

Particle Contamination ■ Photo tools are designed so that a class 1 or better clean environment is maintained inside the tool chamber. All materials and hardware used in the equipment construction are selected for their low particle generation. Lubricants are avoided. If they are required, then a low vapor pressure lubricant is specified to minimize outgassing. Wafer and reticle handling is done by mechanisms that hold the product from underneath, with automated handling systems used on modern equipment. Air solenoids are vented externally from the tool to avoid scattering particles inside the tool.

Comparison of Photo Tools

Steppers and step-and-scan systems based on optical lithography are the photolithography workhorses in semiconductor manufacturing for VLSI/ULSI deep submicron feature sizes. A comparison of the various features of steppers and step-and-scan systems is listed in Table 14.7.

TABLE 14.7* Comparison of Photolithography Tools

Model	Wavelength (nm)	Stepper/ Step-and-scan	Illumination	Exposure Field Size (mm)	N/A	Resolution (μm)	Overlay Accuracy (nm)
ASML							
PAS5500/22	365	Stepper	Mercury arc	22 × 27.4	0.4	0.70	≤70
PAS5500/60B	365	Stepper	Mercury arc	18 × 24.2	0.54	0.45	≤70
PAS5500/200B	365	Stepper	Mercury arc	22 × 27.4	0.48 to 0.60	0.32	≤50
PAS5500/300B	248	Stepper	KrF laser	22 × 27.4	0.40 to 0.57	0.25	≤45
PAS5500/500	248	Step-and-scan	KrF laser	26 × 34	0.40 to 0.63	0.22	≤45
PAS5500/900	193	Step-and-scan	ArF laser	26 × 33	0.45 to >0.6	0.13 to 0.15	
Canon							
FPA-3000i5	365	Stepper	Mercury arc	22 × 22	0.45 to 0.63	0.35	≤80
FPA-3000EX3	248	Stepper	KrF laser	22 × 22	0.40 to 0.60	0.25	≤80
FPA-4000ES1	248	Step-and-scan	KrF laser	25 × 33	0.40 to 0.63	0.25	≤70
FPA-5000ES2		Step-and-scan		26 × 33	0.68	0.18	
ISI Lithography							
XLS 7500	365	Stepper	Mercury arc	20 × 20	0.55	0.5	≤90
XLS 7800	248	Stepper	KrF laser	22 × 22	0.35 to 0.53	0.35	≤80
Isis 202I	365	Stepper	Mercury arc	22 × 22	0.35 to 0.60	≤0.35	≤60
Isis 302k	248	Stepper	KrF laser	22 × 22	0.35 to 0.55	≤0.25	≤50
ArF MicroStep	193	Stepper	ArF laser	1.5 × 1.5	0.40 to 0.60	≤0.18	
Nikon							
i12D	365	Stepper	Mercury arc	22 × 22	Var. to 0.63	≤0.35	≤55
EX12B	248	Stepper	KrF laser	22 × 22	Var. to 0.55	≤0.28	≤55
S201A	248	Step-and-scan	KrF laser	25 × 33	Var. to 0.60	≤0.25	≤50
SVGL							
MS II+	248	Step-and-scan	KrF laser	22 × 33.5	0.50	0.3	≤70
MS III	248	Step-and-scan	KrF laser	26 × 33.5	0.40 to 0.60	0.25	≤55
MS III+	193	Step-and-scan	ArF laser	26 × 34	0.68	0.18	
Ultratech							
2244I	355 to 375	Stepper		22 × 44	0.32	0.75	≤350
Saturn	355 to 375	Stepper		22 × 44	0.365	0.65	≤250

*Adapted from R. DeJule, "Wafer Stepper Trends," *Semiconductor International* (February 1997), p. 88.

Step-and-Scan System
(Photo courtesy of Silicon Valley Group Lithography Systems)

MIX AND MATCH

Most attention for advancements in wafer fabrication technology is given to leading-edge areas like chemically amplified DUV resists capable of deep submicron CD linewidth control. However, recognize that there can be up to 20 to 30 layers in a high-performance IC. CDs and overlay accuracy are critical quality measures on layers with sub-quarter micron structures. For the subcritical layers, novolak-based DNQ resists with i-line steppers are acceptable and cost-justified.

The total cost of owning and operating equipment for wafer production is referred to as the *cost of ownership,* or *COO.* COO costs include the up-front capital expenditures plus the hidden costs from areas such as maintenance, training, downtime, and reduced process latitude. There is a high COO for operating a high-resolution DUV resist on a subcritical layer where i-line DNQ-novolak resist would suffice. Chip manufacturers have learned it is not cost-effective to use an advanced tool and resist system for patterning subcritical layers.

A *mix-and-match* approach to photolithography matches the photolithography resist and equipment technology to the criticality of the wafer layer. For instance, analysis of a 0.22 μm DRAM device demonstrated that a 0.28 μm i-line stepper with a 70 nm overlay accuracy could image 13 noncritical layers of the total 20 layers on the device. The remaining 7 layers were imaged with a leading-edge DUV step-and-scan system that could achieve the 0.22 μm resolution. In this study, it was estimated that the cost of the i-line product averaged about $5 per layer, vs. $7.50 for the DUV layers.[34] This is the benefit of the mix and match strategy for photolithography technology—to reduce cost of ownership.

ALIGNMENT AND EXPOSURE QUALITY MEASURES

Typical quality measures for alignment and exposure are provided in Table 14.8.

TABLE 14.8 Key Quality Measures for Exposure

Quality Parameter	Types of Defects	Remarks
1. Focus-exposure dose.	A. Incorrect focus-exposure for system. Conduct focus-exposure optimization test while measuring CD linewidth.	• Verify uniform and optimum exposure from illumination source. • Make CD measurements of line versus a series of exposure dose values for a nominal focal position (e.g., 30% from top surface). • With nominal focal position, find optimum dose for producing CD. • Modify focal position and conduct CD measurements. Wide range of acceptable dose exposure at optimum focal position. • Verify the bulk resist meets all quality parameters.
2. Light intensity of illumination source.	A. Non-uniform light intensity in the exposure field.	• Check light intensity for specified energy and uniformity at several locations on a wafer. Most aligners have built-in photodetector (measures in mW/cm^2). • Evaluate resist to ensure it is not outgassing and condensing on optical elements. This will degrade lens transmission and field uniformity (important for DUV resists).*
3. Reticle alignment in stepper or step-and-scan tool.	A. Reticle alignment targets will not align properly with wafer alignment targets.	• Verify appropriate process recipe is loaded for the specific mask layer. • Verify appropriate wafers and reticle are loaded for a specific job. • Check rotation of reticle on reticle stage or wafer on wafer stage due to vacuum leak at chuck or electro-mechanical problem with stage. • Problem with internal optics of aligner. Possible cause is temperature or pressure change that affects the NA of the lens.
4. Pattern resolution.	A. Poor resolution for CDs on wafer: linewidths and holes do not meet specification.	Resolution is often a focus problem: • Perform focus-exposure test. • Check environment (temperature, pressure). • Wafer is not flat on chuck, possibly due to backside contamination or chuck problem. • Look for possible incoming process-related problems or improper process parameters or reticle. • Look for optics problems (e.g., lens aberrations).

Quality Parameter	Types of Defects	Remarks
5. Reticle quality.	The following are reticle defects: A. Dirt or scratches on reticle. B. Pattern defects on reticles: • Break in line. • Bridge between features. • Missing geometry. • Opaque spot from isolated chrome. • Pinhole in chrome line. C. Glass fracture. D. Lifted chrome (poor adhesion). E. Reticle plate flatness.	• Scratches can remove chrome and cause defect in resist. • A break in line extends completely across the chrome feature. • Bridge will join two chrome features across a clear space on the reticle. • Missing geometry is a pattern not on reticle, such as a missing contact. • Opaque spot is an area of chrome that should not be on reticle. • Pinhole is a hole in a chrome pattern. • Reticles should not have fractures and be controlled for flatness and warpage.

*O. Nalamasu et al., "Single-Layer Resist Design for 193 nm Lithography," *Solid State Technology* (May 1999), p. 29.

ALIGNMENT AND EXPOSURE TROUBLESHOOTING

Troubleshooting problems encountered during alignment and exposure are shown in Table 14.9.

TABLE 14.9 Common Alignment and Exposure Troubleshooting Problems

Problem	Probable Cause	Corrective Action
1. Excessive overlay error.*	A. Incorrect alignment system measurement of reticle and wafer alignment marks	Possible measurement error sources are: • Verify correct process recipe and reticle is used for a specific mask layer. • Verify the calibration and stability of the alignment system to determine the position of the alignment marks. • Verify calibration of alignment mark-to-pattern relationship, including thermal and/or mechanical effects. • If error is within one tool, then optical distortion is probably not source of problem. If error is tool-to-tool, check difference in optical distortion of two different projection optics.
	B. Problem with reticle.	• Verify there is no problem with reticle mounting and/or reticle heating that could change alignment mark-to-pattern position. • Check there is no particulate contamination on alignment marks that makes the alignment system incorrectly determine the mark position.

Quality Parameter	Types of Defects	Remarks
	C. Error in wafer or reticle stage that holds and positions wafer and reticle.	• Wafer and/or reticle stage has errors in position and rotation during exposure that contributes to overlay errors. • Excessive vibration of wafer or reticle stage. Tools have built-in vibration and shock isolation. • Unacceptable wafer or reticle stage heating that distorts wafer or reticle. • Chucking errors that vacuum clamp wafers differently and cause wafer distortion.
	D. Problem with projection optics.	• Calibration error in the lens magnification adjustment causes pattern mismatches. • Focus errors and/or unacceptable field flatness that causes image distortion or shifts. • Unacceptable heating causes optics distortion.
2. Drift in KrF laser parameters.**	A. Laser properties that can change are: • Laser spectral bandwidth and energy distribution. • Wavelength stability. • Output energy and repetition rate. • Pulse-to-pulse energy stability.	All laser measurements and calibrations should be performed after specialized training by the supplier, including: • Use special wavemeter to measure wavelength. Drift in wavelength affects focus at the wafer plane. • Assess background optical noise levels and electronic offsets to determine impact on energy distribution. • Measure bandwidth of laser output using procedure specified by supplier.

*G. Gallatin, "Alignment and overlay," *Microlithography, Science and Technology* ed. J. Sheats and B. Smith (New York: Marcel Dekker, 1998), p. 318.

**P. Das and U. Sengupta, "Krypton Fluoride Excimer Laser for Advanced Microlithography," *Microlithography Science and Technology* ed. J. Sheats and B. Smith (New York: Marcel Dekker, 1998), p. 299.

SUMMARY

Modern photolithography is based on optical lithography. Alignment and exposure are critical in lithography to meet critical submicron resolution requirements. There have been five photolithography equipment eras. The step-and-repeat and step-and-scan tools are the preferred optical lithography tools used today for advanced IC fabrication. The continued success of optical lithography is based on improvements to all subsystems. Light is important to project the pattern on the resist. UV light at certain energy peaks creates the photochemical reaction in the resist. UV exposure sources are the mercury arc lamp for conventional lithography and the excimer laser for deep UV. Reflection and refraction are important parameters for manipulating the light through lens systems for image projection. Diffraction describes how the light bends as it passes through the narrow patterns on the reticle. The ability of a lens to capture light is the numerical aperture (NA). Special antireflective coatings are placed on the wafer surface to diffuse reflected light and avoid damage to the light-sensitive resist. Resolution is the smallest feature printable on the wafer and is limited by wavelength, NA, and process factors. Reducing pattern resolution also leads to reduced depth of focus, which requires a planar wafer surface. This is achieved by chemical mechanical planarization (CMP). Quartz reticles contain the pattern to be imaged, using a 4X or 5X reduction optics to simplify reticle fabrication with e-beam lithography. Optical enhancement techniques such as phase-shift mask and optical proximity correction permit subwavelength lithography. Alignment of the reticle to the wafer is critical for meeting overlay budget requirements. Environmental conditions must be tightly controlled for optimum lithography. A mix-and-match approach uses conventional lithography for noncritical layers and DUV lithography for critical layers.

KEY TERMS

optical lithography
light
optical filters
electromagnetic spectrum
mercury arc lamp
light intensity
excimer laser
spatial coherence
dose monitor
optics
law of reflection
refraction
relative index of refraction, n
absolute index of refraction
lens
focal point
focal length
lens compaction
aberrations
diffraction
numerical aperture (NA)
reflective notching
standing waves
antireflective coating (ARC)
bottom antireflective coating (BARC)
top antireflective coating (TARC)
resolution
depth of focus (DOF)
depth of field
center of focus (COF)
contact aligner
proximity aligner
scanning projection aligner (scanner)
step-and-repeat aligner (stepper)
step-and-scan system
reticle
photomask or mask
electron beam (e-beam)
pellicle
subwavelength lithography
phase-shift mask (PSM)
optical proximity correction (OPC)
off-axis illumination (OAI)
print bias
baseline compensation (BLC)
alignment
registration
overlay accuracy
overlay budget
grid
alignment marks
first mask exposure
global alignment
field-by-field alignment
cost of ownership (COO)
mix and match

REVIEW QUESTIONS

1. Describe optical lithography.
2. What are the three basic purposes of a stepper?
3. Explain why design ground rules are important for alignment.
4. What is the alignment challenge for photolithography?
5. Describe the relationship between light exposure wavelength and image resolution.
6. Define light. Why is it needed in optical lithography?
7. List and explain the two types of light wave interference. What is an optical filter?
8. What is the electromagnetic spectrum, and what is the UV range?
9. List and describe the two UV exposure sources used in optical lithography.
10. What is light intensity and exposure dose, and how are they related? Why is exposure dose important?
11. What happens to resist sidewalls if there is excessive resist light absorbance?
12. Describe how an excimer laser functions.
13. What excimer laser is used as a 248 nm light source? As a 193 nm light source?
14. What is spatial coherence? Why is this controlled in photolithography?
15. What is a typical dose exposure latitude for a DUV resist? How is this measured?
16. What is optics, and why is it important in lithography?
17. State the law of reflection and give an example.
18. What is refraction? Explain the relative index of refraction and the absolute index of refraction.
19. What is a lens? Explain the focal point and focal length.
20. What lens material is used at the DUV exposure wavelength? Why?

21. Explain how lens compaction occurs and what problem it creates.
22. Describe lens aberration.
23. What is diffraction? Why is it a concern in optical lithography?
24. What is numerical aperture (NA)? State its formula, including the approximate formula.
25. What happens to NA if the radius of the lens is increased?
26. List and explain the two primary problems of light reflection from the wafer surface.
27. What is an antireflective coating, and how does it reduce standing waves?
28. State the formula for resolution. What three lithography parameters affect resolution?
29. What happens to resolution if the light wavelength decreases? If NA goes up?
30. Calculate the resolution of a scanner if $\lambda = 248$ nm, NA = 0.65, and k = 0.6.
31. Define depth of focus and center of focus. Write the equation to calculate DOF.
32. What happens to depth of focus as resolution inreases?
33. List the two important parameters for image quality in optical lithography.
34. Why is surface planarity important for pattern resolution?
35. Explain the contact alligner. Does it use a mask or reticle?
36. Explain how a proximity aligner works. What problem did it attempt to resolve?
37. Explain how a scanning projection aligner works. What problems did the scanning projection aligner try to resolve?
38. Explain the basic functions of a step-and-repeat stepper.
39. What benefits are gained from step-and-scan technology for lithography?
40. Define a reticle. What is the difference between a reticle and a photomask? What reticle reduction is usually used in a stepper? What reduction is used in a step-and-scan system?
41. What material is used to make a reticle? What opaque material is patterned on a reticle?
42. What is a typical usable field size for a 4:1 reticle?
43. What lithography technology is used to pattern reticles? Why?
44. Explain why a pellicle is used on reticles.
45. Explain phase-shift mask (PSM).
46. Discuss optical proximity correction (OPC).
47. How does off-axis illumination increase the pattern resolution?
48. What is print bias and what is one method to correct it?
49. Explain baseline compensation for alignment.
50. What is alignment? Whjat is registration?
51. Define overlay accuracy. What is the overlay budget? About how much of the critical dimension is the overlay budget?
52. What is a grid?
53. What are alignment marks? Describe RA, GA, and FA alignment marks.
54. How does a step-and-scan system improve alignment?
55. Explain global alignment and field-by-field alignment.
56. List and discuss five environment conditions that must be controlled in steppers and step-and-scan systems.
57. Explain the mix-and-match approach to photolithography. Why is it beneficial?

PHOTORESIST MATERIALS AND EQUIPMENT SUPPLIERS' WEB SITES

ASML	http://www.asml.com/
Canon Semiconductor	http://www.usa.canon.com/indtech/semicondeq/
Charles Evans and Associates	http://www.cea.com
Cymer Inc.	http://www.cymer.com/
Dupont Semiconductor Products	http://www.dupont.com/semiconductor/lith.html
ETEC Systems Inc.	http://www.etec.com/semiprod_frame.html
KLA-Tencor	http://www.kla-tencor.com/product/photo_frame.html
Karl Suss	http://www.suss.com/
Nikon	http://www.nikon.com/
SEMI	http://www.semi.org/
Silicon Valley Group	http://www.svg.com/
SPIE, International Society of Optical Engineering	http://www.spie.org/
Ultratech Stepper	http://www.ultratechstepper.com/
USHIO	http://www.ushio.com/index2.html

REFERENCES

1. H. Levison and W. Arnold, "Optical Lithography," *Microlithography, Micromachining and Microfabrication* vol. I, ed. P. Rai-Choudhury (Bellingham, WA: SPIE, 1997): p. 13.
2. K. Derbyshire, "Issues in Advanced Lithography," *Solid State Technology,* (May 1997): p. 133.
3. M. Hibbs, "System Overview of Optical Steppers and Scanners," *Microlithography Science and Technology,* ed. J. Sheats and B. Smith (New York: Marcel Dekker, 1998), p. 16.
4. B. El-Kareh, *Fundamentals of Semiconductor Processing Technologies* (Boston: Kluwer Academic Publishers, 1995), p. 212.
5. M. Hibbs, "System Overview of Optical Steppers and Scanners," p. 18.
6. B. El-Kareh, *Fundamentals of Semiconductor Processing Technologies,* p. 174.
7. S. Campbell, *The Science and Engineering of Microelectronic Fabrication* (New York: Oxford University Press, 1996), p. 162.
8. R. DeJule, "Extending Optical Lithography to 157 nm?" *Semiconductor International,* (February 1997): p. 84.
9. M. Hibbs, "System Overview of Optical Steppers and Scanners," p. 20.
10. J. Shamaly, "A Full-Field ArF Exposure Tool for 180-nm Lithography," *Microlithography World* (Summer 1996): p. 23.
11. M. Hibbs, "System Overview of Optical Steppers and Scanners," p. 20.
12. C. Cromer et al., "Improved Dose Metrology in Optical Lithography," *Solid State Technology* (April 1996): p. 76.
13. E. Hecht, *Optics,* 2nd ed. (Reading, MA: Addison-Wesley Publishing Co., 1990). p. 143.
14. J. Shamaly, "Full-Field ArF Exposure Tool," p. 23.
15. C. Mack, "Optical Lithography Modeling," *Microlithography: Science and Technology,* ed. J. Sheats and B. Smith (New York: Marcel Dekker, 1998), p. 118.
16. B. El-Kareh, *Fundamentals of Semiconductor Processing Technologies,* p. 206.
17. C. Bencher et al., "Dielectric Antireflective Coatings for DUV Lithography," *Solid State Technology* (March 1997): p. 109.
18. Ibid.
19. B. Smith and M. Hanratty, "Optical Lithography Modeling," *Microlithography: Science and Technology,* ed. J. Sheats and B. Smith (New York: Marcel Dekker, 1998), p. 587.
20. W. Arnold, "Is a Scanner Better than a Stepper?" *Solid State Technology* (March 1999): p. 77.
21. M. Hibbs, "System Overview of Optical Steppers and Scanners," p. 9.
22. W. Arnold, "Is a Scanner Better than a Stepper?" p. 77.
23. S. Wolf and R. Tauber, *Process Technology,* vol. 1, *Silicon Processing for the VLSI Era* (Sunset Beach: Lattice Press, 1986), p. 476.
24. R. Singh, S. Vu, and J. Souze, "Nine-Inch Reticles: An Analysis," *Solid State Technology* (October 1998): p. 83.
25. R. DeJule, "Resists for Next-Generation Masks," Lithography Technology News, *Semiconductor International* (October 1998): p. 46.
26. "Rapid Innovation in Spite of Slow Market," *Solid State Technology* (September 1998): p. 42.
27. B. El-Kareh, *Fundamentals of Semiconductor Processing Technologies,* p. 176.
28. M. Sasago, "Aggressive Optical Lithography," *Microlithography World* (Autumn 1998): p. 5.
29. R. DeJule, "Resists for Next-Generation Masks," p. 46.
30. H. Levison and W. Arnold, "Optical Lithography," *Microlithography, Micromachining and Microfabrication* vol. 1, ed. P. Rai-Choudhury (Bellingham, WA: SPIE, 1997): p. 69.
31. G. Gallatin, "Alignment and Overlay," *Microlithography: Science and Technology,* ed. J. Sheats and B. Smith (New York: Marcel Dekker, 1998), p. 318.
32. R. DeJule, "Lithography," *Semiconductor International* (September 1998): p. 50.
33. R. DeJule, "Wafer Stepper Trends," p. 88.
34. R. DeJule, "More Productivity at Subcritical Layers," *Semiconductor International* (April 1998): p. 50.

CHAPTER 15

PHOTOLITHOGRAPHY: PHOTORESIST DEVELOPMENT AND ADVANCED LITHOGRAPHY

At this stage of the photolithography process, wafers have been coated with resist and exposed to UV light to produce a photochemical reaction in specified areas of the resist. The resist contains a precise coded image of the reticle pattern and is ready for the development step of photolithography. Photoresist development is where the three-dimensional pattern is physically produced in the resist. This step determines whether the resist image is an accurate replication of the reticle pattern. In deep submicron wafer fabrication, success at this step is critical for subsequent processing.

The advanced lithography section summarizes the next generation lithography tools. Fundamental research into alternative lithography tools is necessary so that the semiconductor industry will be ready to replace optical lithography when it has reached its resolution limit. Optical lithography appears to be the optimum patterning process for at least the next several technology nodes below 0.18 micron.

OBJECTIVES

After studying the material in this chapter, you will be able to:

1. Explain why and how a post-exposure bake is done for conventional and chemically amplified DUV resist.
2. Describe the negative and positive resist development process for conventional and chemically amplified DUV resist.
3. List and discuss the two most common resist development methods and the critical development parameters.
4. State why a hard bake is done after resist development.
5. Explain the benefits of a post-develop inspection.
6. List and describe the four different alternatives for advanced lithography, including the challenges for introducing each alternative into production.
7. Describe and give the benefit for the advanced resist process of top surface imaging.

INTRODUCTION

Following the sequential steps of the eight-step photolithography process, wafers have completed alignment and exposure. The reticle pattern has been aligned to each die on individual wafers and exposed to UV light for image transfer to the resist. Wafers with exposed resist leave the stepper or step-and-scan system and return to the automated track system. The next processing step is often a post-exposure bake and then wafers are ready for development. Photoresist development is the chemical dissolution of the soluble resist in the developer chemical. Development removes those areas of the resist rendered soluble by the exposure step. The insoluble resist remaining on the wafer should be an exact copy of the pattern on the reticle (for a positive-tone resist). Photochemistry renders the exposed regions soluble for positive-tone resists and insoluble for negative resists.

The goal of resist development is to accurately replicate the reticle pattern in the resist while maintaining acceptable resist adhesion. The ability to develop resist with deep submicron geometries is essential for advanced ICs. Resist development and inspection is the intermediate pattern transfer step before the wafer moves into another workbay to undergo etch or ion implantation. An improperly developed resist will cause a high reject rate, whereas a properly developed resist is the foundation for a high-yield process. The quality of the resist pattern determines the success of the subsequent process steps.

In early days of wafer fabrication, resist development was a stand-alone process step, with its own tool and workstation. Cassettes of wafers were manually moved from the exposure tool to the develop operation. The manual processing and lack of tool controls produced excessive variability that is unacceptable for submicron lithography. In modern wafer fabrication automated wafer tracks have integrated the development operation into a comprehensive process of photolithography. This represents steps one through seven of the eight-step photolithography process.

The eight steps to photolithography are in Table 15.1. The highlighted steps are covered in this chapter.

TABLE 15.1 Eight Basic Steps of Photolithography

Step	Chapter Covered
1. Vapor prime	13
2. Spin coat	13
3. Soft bake	13
4. Alignment and exposure	14
5. Post-exposure bake	**15**
6. Develop	**15**
7. Hard bake	**15**
8. Develop inspect	**15**

The procedure used to develop resist depends on whether it is positive-tone or negative-tone, and whether it is a conventional i-line resist (DNQ-novolak) or chemically amplified deep UV (CA DUV) resist. Positive resists are the norm for submicron lithography. Noncritical layers typically use conventional i-line resist (making up the major share of photolithography processing), while critical layers with CDs of 0.25 µm and below use DUV resists.

Advanced Lithography

Optical lithography has thrived while contributing to the continual reduction of device feature sizes. It is now expected by some industry experts that optical lithography will continue as the main lithography process until after the arrival of the 0.1 µm technology node sometime around 2006.[1] The reason optical lithography has continued to succeed in wafer fabrication is the ongoing introduction of improvements in equipment, process, and design. We reviewed new exposure techniques in Chapter 14, such as optical proximity correction (OPC) and phase-shift masks (PSM). At the end of this chapter we will review next generation microlithography technologies.

POST-EXPOSURE BAKE

After the wafer with exposed resist exits the exposure system, it enters the wafer track system and undergoes a short *post-exposure bake* (*PEB*) step. A thermal PEB is necessary for chemically amplified DUV resists to catalyze (initiate) critical resist chemical reactions. For conventional i-line resists based on DNQ chemistry, PEB is done to improve adhesion and reduce standing waves. Resist manufacturers include recommended time and temperature specifications for PEB in their product literature.

DUV Post-Exposure Bake (PEB)

It is during the post-exposure bake (PEB) that the exposed regions of a chemically amplified DUV resist become soluble in the developer. Recall that for a chemically amplified DUV resist, a protecting chemical (e.g., t-BOC) makes the resist insoluble in the developer. During UV exposure, a photoacid generator (PAG) generates an acid in the exposed regions. To make the exposed resist soluble to the developer, the post-exposure bake (PEB) heats the resist, which causes the acid-catalyzed deprotection reaction. The acid removes the protecting group from the resin and the exposed resist is now soluble in the developer solution (see Appendix E).

PEB is an important step in resist processing for chemically amplified DUV resists. The requirement for PEB to catalyze the chemical deprotection reaction must be considered nearly as critical as exposure in terms of control, uniformity, and latitude requirements.[2]

Temperature Uniformity ■ The temperature uniformity and duration of PEB for DUV resist is important. Excessive variations will affect the kinetics of the acid catalytic reaction in the resist. Most often, PEB is done on a hot plate in an automated wafer track system. The actual temperature and time that the coated wafer is heated by the hot plate depends on the type of resist used. A representative PEB temperature and time is 90 to 130°C for 1 to 2 minutes. The PEB temperature is usually 10 to 15°C hotter than the soft bake. The hot plate temperature variation and the bake latitude of the particular DUV resist is critical and can affect the amount of CD variation in the resist during the develop step. For commercially available DUV resist systems, a representative post-exposure bake latitude for CD variation is about 5 nm /°C.[3] To reduce CD inconsistencies, hot plates are set at 130°C and controlled within ± 0.1°C.[4]

PEB Delay ■ Early DUV resists were very sensitive to the delay time from UV exposure to PEB. A delay greater than a few minutes permitted the acid to neutralize due to amine contamination from the ambient air. The neutralization occurred on the top surface of the resist, creating a thin, less-soluble inhibition layer that led to the formation of a "T-top" after development (see Figure 15.1). More recent DUV resists permit a delay of up to about 30 minutes. However, it is still desirable to develop the resist as soon as possible after exposure to minimize any potential chemical reactions that can inhibit the development process.

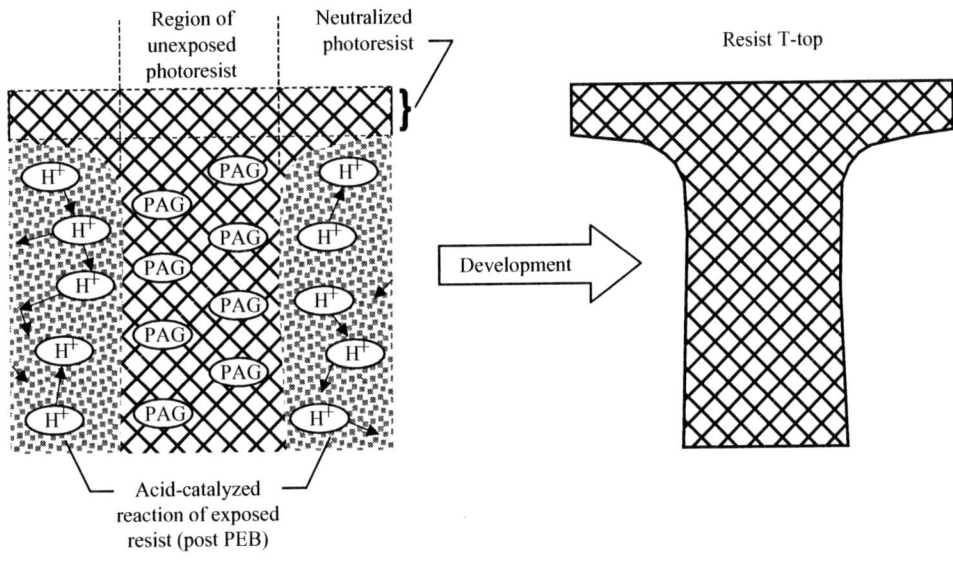

FIGURE 15.1 Amine Contamination of DUV Resist Leading to T-top Formation

Conventional I-Line PEB

It is common practice to perform a PEB for conventional DNQ-novolak i-line resist. The PEB extends some of the same benefits gained by the soft-bake step that occurred before exposure. PEB helps drive out remaining solvents that are left in the resist, reducing solvent content from 7 to 4% (before exposure) to about 5 to 2%.[5] The most significant benefit from PEB for i-line resist is a reduction in the standing wave defect from reflective surfaces during exposure.

Recall from Chapter 14 that standing waves result from light exposure of a resist that is coated on a reflective substrate. Coherent light reflects off the substrate and interference produces a nonuniform intensity in the resist film that appears as standing waves in the sidewall. The PEB minimizes the standing wave effect because the increased temperature causes the PAC sensitizer in the resist to diffuse through the novolak polymer matrix, essentially producing an averaging effect across the standing wave boundary (see Figure 15.2).

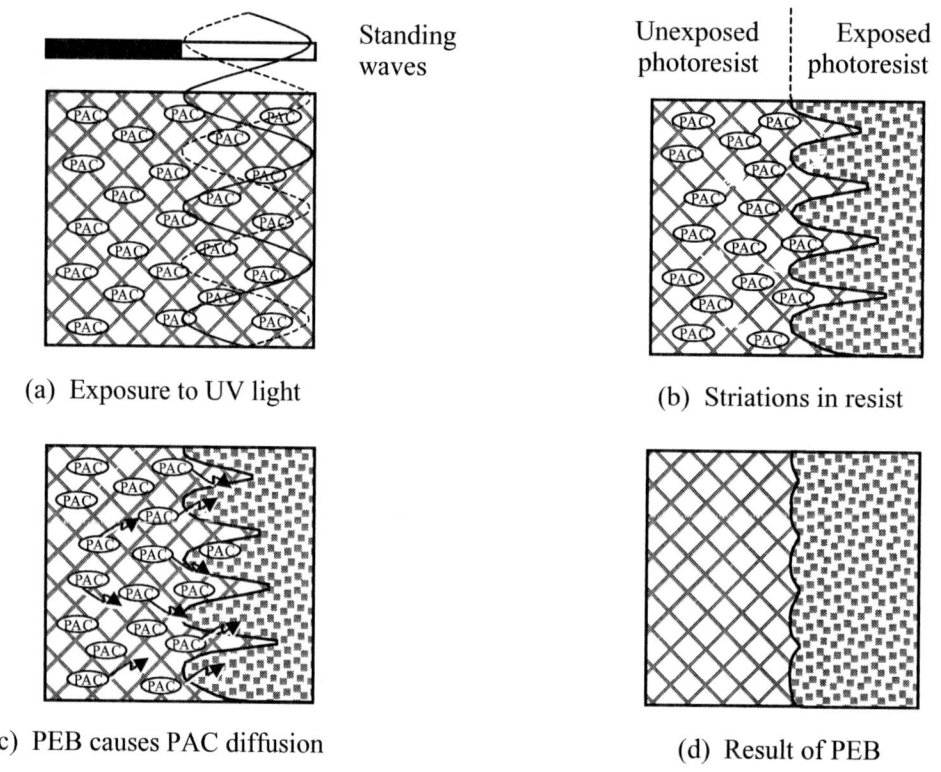

FIGURE 15.2 Reduction of Standing Wave Effect Due to PEB

DEVELOP

Photoresist development (develop) uses a liquid chemical developer to dissolve the soluble regions of the resist that were formed during the mask exposure. The primary goal of resist development is to accurately replicate the reticle pattern in the resist material. The emphasis is on producing CD features that meet the required specifications. If the CDs meet the specifications, then all other features are assumed acceptable since the CD is the most difficult structure to develop.

Resist patterning problems occur if the development process is not properly controlled (see Figure 15.3 on page 417). These resist problems can negatively affect production yield, showing up as defects in the subsequent etch process. The three primary types of development problems are underdevelopment, incomplete development, and overdevelopment. The photoresist lines in Figure 15.3 illustrate these types of problems as compared to a normal developed line. An underdeveloped line appears wider than a normal line and will have sloping sidewalls. The incomplete developed line has residual resist on the substrate that should have been removed during develop. Overdevelopment removes too much resist causing features to appear narrower and poorly-defined.

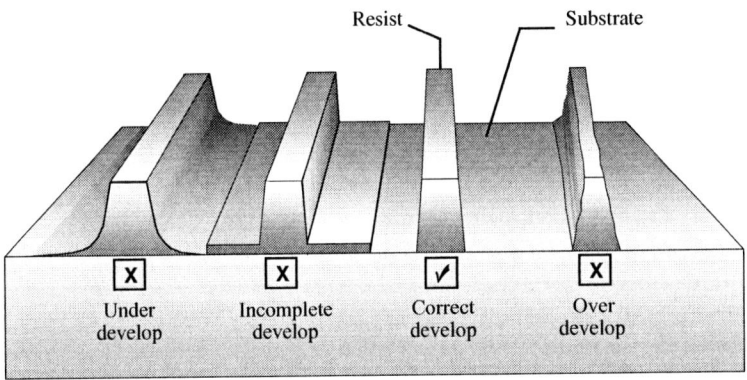

FIGURE 15.3 Photoresist Development Problems

Negative Resist

Negative resist is crosslinked (hardened) by exposure to UV light. This makes the exposed resist nonsoluble in the developer solution (see Figure 15.4).

Very little chemical reaction is necessary for *negative resist development* in the developer solution. This process consists mainly of a solvent wash of the unexposed resist, which is not crosslinked and therefore soft and soluble. The developer is typically an organic solvent such as xylene that is sprayed on the resist while the wafer is spinning on a vacuum chuck. Developer spray may be followed by another organic solvent sprayed on the wafer to stop the develop process.

An organic solvent rinse cycle is then introduced to clean the wafer and remove partially crosslinked pieces of resist. In addition, rinse removes any remaining developer to ensure the development process stops. Rinse typically consists of an organic solvent such as n-butylacetate or an alcohol or trichloroethylene. There is finally a spin-dry step to dry the wafer.

A significant problem with negative resist is swelling and distortion of the crosslinked exposed resist due to absorption of the developer solution during the wash. Because of this, the sidewalls of remaining negative resist on the wafer become jagged and swollen. This absorption is a primary reason why traditional negative resists are not suitable for geometries below about 2 μm. There is ongoing research into improving negative resists to minimize swelling.

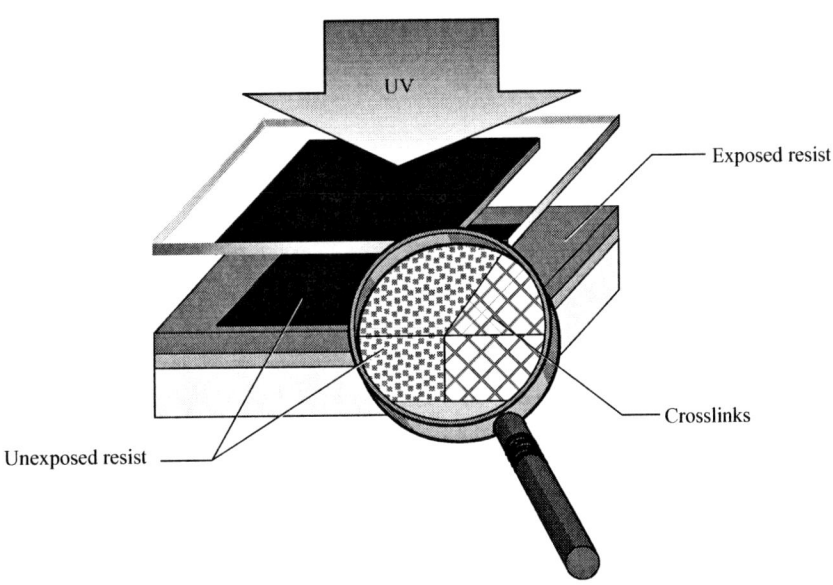

FIGURE 15.4 Negative Resist Crosslinking

Positive Resist

Positive-tone resists are the most common resists used in submicron wafer fabrication due to improved linewidth resolution. The two general types of positive resists used in wafer microlithography are conventional DNQ i-line resists (noncritical layers down to 0.35-μm linewidth) and chemically amplified (CA) DUV resists (critical layers with 0.25-μm linewidth and below). Both of these types of resists use a phenolic-based resin. Conventional i-line resists consist of a novolak resin, while CA DUV resists commonly use a PHS resin. The type of resin is important for resist development because this is the material that the developer will remove in order to pattern the resist. Although these two resists are very different chemically (see Chapter 13), it is generally true that phenolic resin is soluble in a base solution.

Developer ■ *Positive resist development* involves a chemical reaction between the developing solution and the resist to dissolve the exposed resist (see Figure 15.5). The rate at which a developer dissolves the resist is termed the *dissolution rate* (also referred to as the speed of the developer). A fast dissolution rate is desirable for productivity, but too fast a rate can also be bad for resist performance. Developers also have selectivity. High *developer selectivity* means the developer reacts quickly with the exposed resist (fast removal rate) relative to the slow reaction with the unexposed resist (slow removal rate). A developer with high selectivity produces sharper and cleaner resist sidewalls, which is desirable for high-density patterning.

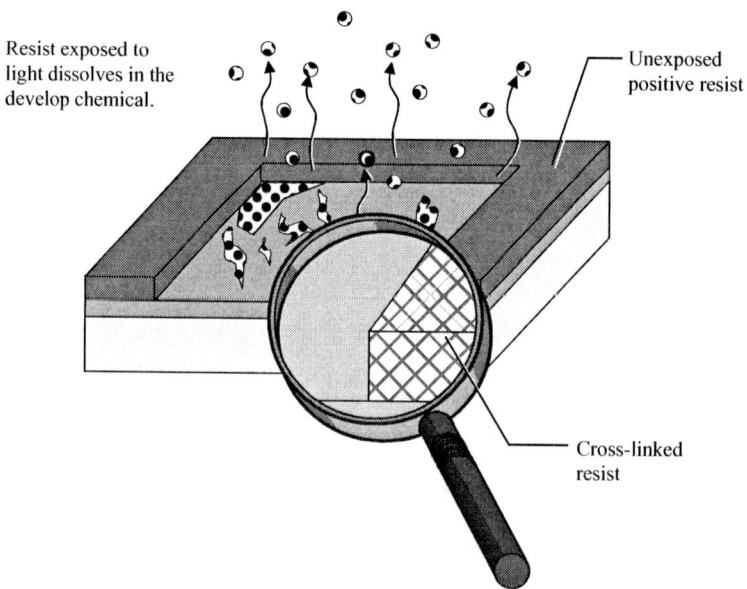

FIGURE 15.5 Development of Positive Resist

The chemical reaction of developer with positive resists is significantly different than the solvent wash needed to dissolve negative-tone resists. Positive resist developer solution is based on a strong alkaline diluted with water (aqueous) developer. Early developer solutions were an alkaline-water mixture of sodium hydroxide or potassium hydroxide. However, both of these solutions contain mobile ionic contaminants (MICs), which are unacceptable for processing high-performance IC circuits sensitive to corrosion. The most common developer today for positive-tone resists is tetramethyl-ammonium hydroxide (*TMAH*). This developer has a very low concentration of metal ions, particularly sodium ions, to avoid introducing MICs on the wafer surface. An advantage of aqueous developers is that the rinse only requires DI water.

Standardized TMAH developer formulations are common in the industry. These developers are more economical and they improve developer repeatability for better quality control. TMAH developer concentrations in the range of 0.2 to 0.3 normality (N) allow sufficient resist selectivity with high contrast between exposed and unexposed resist regions. *Normality* refers to the exact chemical makeup of a particular developer (or any chemical solution); it has a direct effect on developer dissolution rates. A 0.26 N TMAH solution is becoming a standard for resist processing.[6] The objective is to have a high selectivity toward removing the exposed portions of the resist while leaving the unexposed resist intact on the wafer.

Exposed positive resist dissolves into the TMAH developer. The unexposed resist does not dissolve into nor absorb any of the developer. For this reason positive unexposed resist does not swell as with the negative resist. The lack of swelling makes positive resist desirable for producing submicron geometries.

TMAH requires careful control of the developer strength as measured with pH. It is also important to control the developer solution temperature and DI water rinse temperature. The dissolution rate varies as a function of developer temperature, with a result of faster resist dissolution occurring at lower developer temperatures.[7] Small ppm levels of surfactants are often added to the TMAH developer. A surfactant reduces surfaces tension and increases wetting, which makes a liquid flow across the surface better. This condition improves the dissolution of small geometry features such as contacts. Surfactants also minimize any residual film left on the resist surface.

TMAH developer chemistry for positive resist is highly caustic, meaning it destroys or erodes by chemical action. Alkaline chemicals erode metal as well as human tissue. Proper safety precautions (e.g., chemical protective clothing and eyewear) must be used by technicians when working around develop stations, such as on the automated wafer tracks.

Conventional I-Line Resists. Recall from Chapter 13 that positive i-line resists create carboxylic acid during UV exposure. When TMAH developer is applied to exposed resist, a chemical reaction occurs that neutralizes the carboxylic acid by the alkaline TMAH developer. The neutralized carboxylic acid of the exposed resist rapidly goes into solution. The unexposed (hard) resist does not react with the developer and remains intact on the substrate surface.

Chemically Amplified DUV Resists. To review, chemically amplified (CA) DUV resists often consist of phenolic resin in the form of PHS (poly[hydroxystyrene]). PHS has a protection group (such as t-BOC) that makes the PHS insoluble in the base developer. The PAG (photoacid generator) in the CA DUV generates an acid during exposure. During post-exposure bake, this acid removes (or "cleaves") the protection group away from PHS in exposed regions. The PHS is now highly soluble in the base developer.

The developer does not actually react with the PHS during the development operation. This is because there are hydroxyl OH groups that exist in a "corkscrew" configuration along the PHS polymer chain. These OH groups create effective diffusion paths and bring about extremely high dissolution rates for the PHS in the basic developer solution.[8]

Development Methods

Static immersion of a cassette of wafers into developer solution was an early method for resist development. However, it is no longer used today in high-volume production. Batch production is not suitable for automated wafer tracks that efficiently transport single wafers simultaneously between photolithography operations. Static immersion consumes a large quantity of developer chemicals. Furthermore, batch immersion is not conducive to producing uniform development for high-density ICs across large diameter wafers.

The two most common development techniques used to remove exposed resist on spin-coated wafers today are:

- Continuous spray development
- Puddle development

Continuous Spray Development ■ The dissolution of exposed resist with a *continuous spray develop* tool and solution is a similar process to resist dispense systems. In the past, batch development was done using spray systems similar to a spin-rinse-dryer. Development can also be done in a wafer track system after the wafer has completed post-exposure bake. A single wafer is positioned on a vacuum chuck and spun at a slow speed (e.g., 100 to 500 RPM) while one or more nozzles dispense developer on the resist-coated wafer surface (see Figure 15.6). The developer is dispensed in a fine mist, with some processes using ultrasonic atomization to allow for low-velocity dispersion. A low-velocity exit minimizes adiabatic (constant heat transfer) cooling effects during dispense, where the

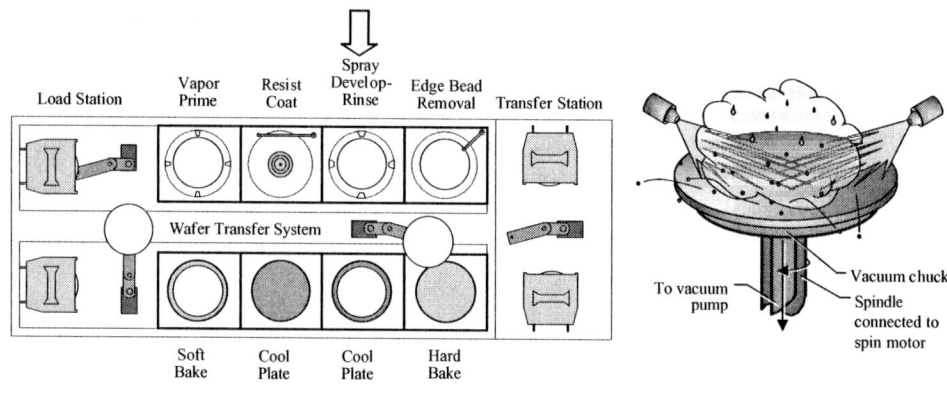

FIGURE 15.6 Resist Development with Continuous Spray

temperature of the developer drops due to its expansion from a high pressure region to a low pressure region. Early nozzle designs required a heating system for the developer to minimize the cooling effect.

The nozzle spray pattern and speed of the wafer rotation are critical variables to achieve repeatability in the resist dissolution rate and uniformity across the wafer. In recent years the spray process method has been largely replaced by puddle development because the latter method allows more process latitude for these variables.

Puddle Development ■ The *puddle develop* approach uses the same basic equipment as the spray method to develop the resist. A small amount of developer is dispensed onto the wafer and forms a puddle (see Figure 15.7). Adequate developer volume is needed to form a puddle meniscus over the entire wafer. Excessive developer is avoided to minimize backside wafer wetting. The wafer can be stationary or slowly rotating on a heated chuck.

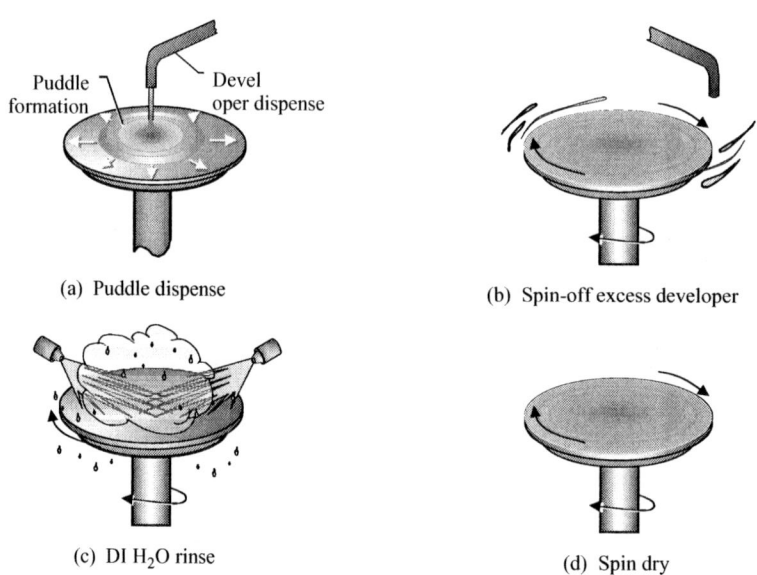

FIGURE 15.7 Puddle Resist Development

There are variations as to whether the wafer is static or rotating after the initial developer is puddled on the wafer. In all cases, the developer must be left on the resist for sufficient time to allow the soluble resist areas to become completely dissolved. As an example, a multiple-puddle method is used where the first puddle is left on the wafer for a predetermined time (such as 10 to 30 seconds, depending on the type of developer). It is then spun off and a new puddle is dispensed and left on the wafer for a defined time. This second puddle replenishes the developer chemicals and rejuvenates the chemical reaction between the developer and the resist. Some manufacturers also spray the developer onto the wafer during the second puddle application.

After the resist has dissolved in developer, the wafer is rinsed with DI water and spun dry. It is important to rinse all remaining chemicals after development from both the topside and backside of the wafer.

For optimal performance, the flow of developer is kept low to reduce variations in development rate at the edge of the wafer. This reduction is an advantage of puddle development over spray because a minimal amount of developer chemical is used. At the same time, adequate developer volume is needed to achieve full and uniform coverage over the entire wafer. Puddle development introduces fresh chemicals for every wafer, which improves wafer-to-wafer uniformity. The puddle development method minimizes temperature gradients and permits control of the variables affecting uniformity of single wafer resist development.

Resist Development Parameters

Critical parameters must be controlled during the development process. These parameters are:

- Developer temperature
- Developer time
- Developer volume
- Wafer chuck
- Normality
- Rinse
- Exhaust flow

Developer Temperature ■ The optimum temperature for developer mixtures is between 15 to 25°C. The temperature must be maintained to within ± 1°C or better from the optimal setpoint once it has been determined. The temperature of the developer has a direct effect on resist dissolution rate. As previously noted, faster resist dissolution for positive resists occurs at lower developer temperatures. For solvent developers (negative resist), the dissolution rate increases with an increase in temperature.

Developer Time ■ Developer continues to react with resist until it is removed by a rinse process. This condition can lead to overdevelopment of the resist and cause unacceptable CD variations. Off-line analysis of CD resist linewidth can be done on a sample of wafers from a batch to determine optimum develop time. However, this sampling gives only limited information and is costly.

An in-situ *dissolution rate monitor* (*DRM*) is used in automated wafer track systems for real-time monitoring of the development process.[9] A DRM uses interferometric signal data collected by an optical sensor that measures the phase difference of light reflected off the baked-on resist film surface and the wafer surface underneath. As the resist dissolves, the phase difference between the two surfaces decreases to zero at the breakthrough point for endpoint (EP) detection. Monitoring each wafer with a DRM provides more useful information about the develop process and permits more accurate control of CDs.

Developer Volume ■ The amount of developer volume dispensed on the wafer is critical for successful resist development. The use of inadequate developer volume results in *scumming*, which is a residue film left on the wafer surface if it has been underdeveloped (such as from inadequate developer volume) or improperly rinsed. Excessive developer is undesirable because the goal is to reduce chemical usage in the fab. Excessive developer is a cost issue, but this condition will not necessarily ruin the resist on the wafer.

Normality ■ *Normality* (*N*) is a concentration unit that refers to the quantity of substances (stoichiometry) in a solution. For a particular developer, normality represents the exact chemical makeup, which in turn indicates how strong or weak the developer is. Normality has a direct effect on resist dissolution rates. It is primarily altered by the addition of moisture, so the chemical distribution system must be carefully controlled to minimize leaks.

Rinse ■ Rinsing is typically done with DI water, which is sprayed onto the wafer, and then the wafer is spun dry. Rinse serves to stop the development process and also remove developer from the wafer surface.

Exhaust Flow ■ The develop module is enclosed on the wafer track machine. The exhaust flow is critical so as not to disturb the developer that is applied in spray form. Too low an exhaust can leave residual developer mist in the developer module, which builds up and reacts with subsequent wafers.

Wafer Chuck ■ The wafer must be held perfectly flat on the wafer chuck during the static or spin development cycle. This placement ensures coverage uniformity during puddle development.

HARD BAKE

A post-development thermal bake, referred to as *hard bake,* is done to evaporate any residual solvent and harden the resist. This action improves the resist's adhesion to the substrate and prepares it for subsequent processing, such as becoming more resistant to etch. Hard bake also removes any residual developer and any water.

Since all light exposure is complete, the baking temperature for hard bake can be elevated toward the solvent boiling point, effectively evaporating the solvent and achieving maximum resist densification. For DNQ-novolak resist, there is also the need to evaporate any remaining DNQ to avoid nitrogen diffusion during subsequent high-energy processes. The nitrogen causes localized resist popping in the densified resist, which disperses resist particles on the wafer surface.[10]

The starting point for the hard bake temperature is the recommended setting determined by the resist manufacturer. After that, the process can be fine-tuned to meet the adhesion and dimensional control necessary for the product. Nominal hard bake temperatures are 130°C for positive resist and 150°C for negative resist. Hard bake is typically done on an in-line hot plate in a wafer track system, or in an in-line oven. Resist is a material that softens and flows when sufficiently heated (see Figure 15.8). Higher hard bake temperatures could cause the resist to flow slightly, which can deform the pattern.

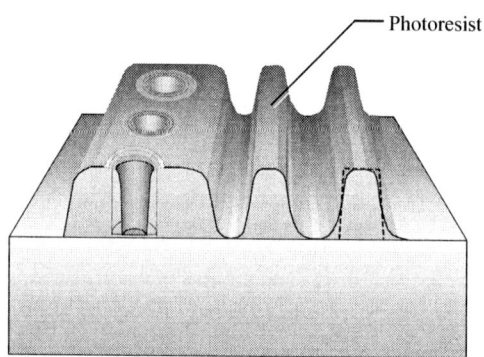

FIGURE 15.8 Softened Resist Flow at High Temperature

Resist Hardening With Deep UV. Resist hardening of DNQ-novolak resist is also achieved by exposing it to deep UV. This exposure causes the positive resist resin to crosslink and form a thin surface crust, which increases the resist thermal stability. The crosslinked resist can now withstand thermal processes up to 210°C without significant resist flow.[11] This condition is beneficial for subsequent processing such as plasma etching and ion implant (see Chapters 16 and 17, respectively), where temperatures can rise up to 125 to 200°C. If the resist is not sufficiently hardened, it risks flowing, which reduces the resolution of the original pattern.

DEVELOP INSPECT

The imaged pattern in the resist undergoes a *post-develop inspection* to find defects. This step is necessary to identify and remove defective wafers before they continue to the following processes of etch or implant. A wafer with a defective resist pattern that is etched or implanted will become scrap. The post-develop inspection is used to characterize the photo process, providing information to the photolithography production team for corrective action.

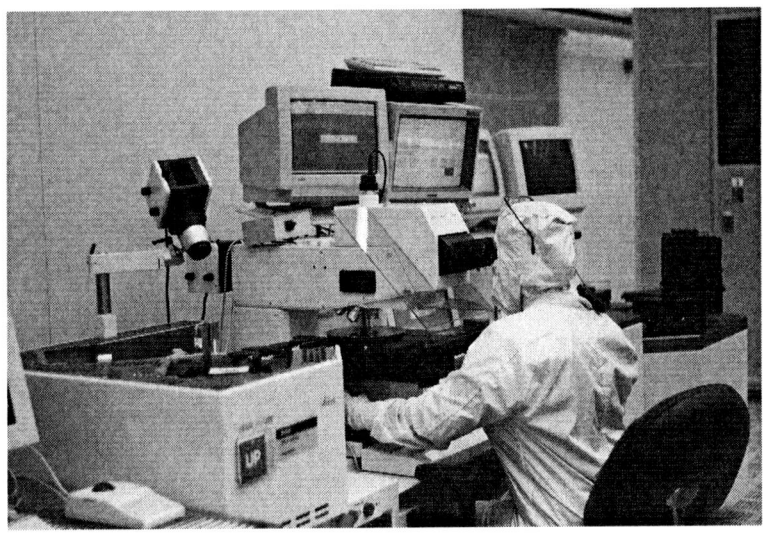

Automated Inspect Tool for Develop Inspect
(Photo courtesy of Advanced Micro Devices)

Most post-develop defects are relatively large and belong to a wide range of defect types (see the Quality Measure section at the end of this chapter). The defects may occur during all previous photolithography steps plus during the processing before lithography. Post-develop inspection requires a flexible system. Traditionally, this inspection has been a manual, operator-intensive inspection process with an optical microscope. Depending on the criticality of the resist layer, the inspection may be done on a sample basis. Automated inspection equipment for post-develop inspection is becoming more common in advanced fabs, especially for deep submicron lithography where the defects are not detectable with optical microscopes.

There are two possible outcomes for a wafer that fails post-develop inspection. If the wafer has a problem from a previous operation that makes it unacceptable, then the wafer is scrapped. If the problem is associated specifically with the quality of the pattern in the resist film, then the wafer is reworked. Wafer rework is done by stripping the photoresist off the wafer surface and inserting the wafer back into the photolithography process (see Figure 15.9).

This is one of the few operations in the fabrication process where rework of rejected parts is possible. The percent of wafers reworked is monitored and action plans are put in place for rejects. The goal is zero defects, while many fabs run typically less than 2% rejects. If the rework rate is greater than about 4%, then there is a quality problem that requires corrective action.

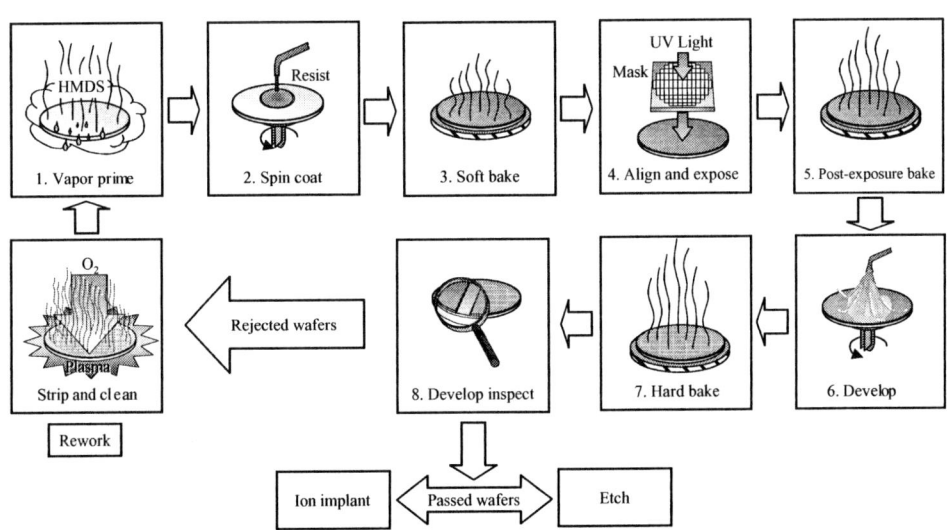

FIGURE 15.9 Develop Inspect Rework Flow

ADVANCED LITHOGRAPHY

The resolution capability of optical lithography has been extended by many equipment and process improvements, especially in the areas highlighted in Table 15.2. These improvements are the basis for a recent development in optical lithography referred to as subwavelength printing. *Subwavelength lithography* images CDs that are significantly less than the light source wavelength. An example is a 248 nm DUV light source used to print CD linewidths at 0.2 µm (200 nm) or below. Recent research activities have demonstrated a 50 nm CD pattern using a 248 nm DUV light source, chemically amplified DUV resist, and phase-shift mask. Subwavelength printing further extends the versatility of optical lithography while lowering its cost of ownership.

TABLE 15.2 Photolithography Improvements

1. Reduction in wavelength of the UV light source.
2. Increase in numerical aperture of optical lithography tools.
3. Chemically amplified DUV resists
4. Resolution enhancement techniques (e.g., phase-shift masks and optical proximity correction).
5. Wafer planarization (chemical mechanical planarization, or CMP) to reduce surface topography.
6. Advances in photolithography equipment (e.g., stepper and step-and-scan).

Next-Generation Lithography

Predicting the actual resolution limit of optical lithography has been a futile exercise for many years. There is some future limit where the extension of optical lithography will not be feasible. Wafer patterning must then transition to an alternative lithography process referred to as *next-generation lithography*. Major research activity is underway to investigate the type of next generation lithography tool that will replace optical lithography. Advanced study of this nature requires many years of research and development effort and sustained funding. There are four major lithography technologies considered as possible successors to optical lithography:[12]

- Extreme UV (EUV)
- SCALPEL
- Ion projection lithography (IPL)
- X-ray

Each of these technologies has demonstrated patterning capability down to 70 nm or below. At the time of this writing, International SEMATECH has selected EUV and SCALPEL for research funding as the next generation lithography tool.

Extreme UV (EUV) ■ *Extreme UV* lithography builds on the production knowledge of optical lithography. It uses a laser-produced plasma source to produce UV wavelengths of ~13 nm and is expected to print images down to about 30 nm.[13] The source operates in a vacuum environment to produce the EUV radiation, which is collected by condenser optics and formed into a beam (see Figure 15.10 on page 425). The beam is reflected off a reflective reticle that scans the pattern through the beam. An all-reflective (meaning mirrors instead of lens) 4X projection optics will image the UV beam onto a resist-coated wafer. The wafer is scanned at one-fourth the reticle speed in the opposite direction.

There are a number of challenges for bringing EUV into wafer production.[14] The precision optics will be difficult to achieve given the stringent requirements for manufacturing the high-quality surfaces. Mirror reflectivities will be optimized by using precision multilayer coatings on the surfaces. The reflective reticle is four times larger than the feature size; therefore, a 100 nm CD has a minimum reticle feature size of 400 nm (at 4X demagnification). The reticle consists of a thin metal absorber patterned on a multilayer coating that is patterned with conventional electron beam mask tools. The short wavelengths and penetration depths will require the use of top surface imaging resists (see the following section) or bi-layer resists with a very thin top layer. For alignment, the total overlay budget is approximately 35 nm for devices at the 0.1 µm design rules. All equipment will be in a vacuum environment.

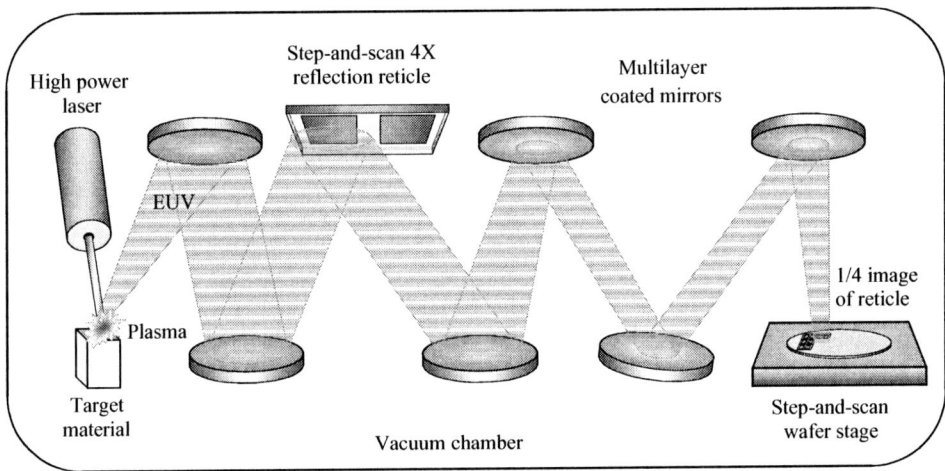

FIGURE 15.10 Concept for Extreme Ultraviolet Lithography
Redrawn from International SEMATECH's Next Generation Lithography Workshop brochure

SCALPEL ■ Feasibility studies are in progress for an electron beam imaging approach referred to as *SCALPEL* (SCattering with Angular Limitation Projection Electron Beam Lithography).[15] SCALPEL, in development since the late 1980s, uses an established electron beam source instead of a light source to image a wafer pattern (see Figure 15.11). A multilayer membrane mask is used that absorbs few electrons. The pattern is formed as the electron beam passes through a high atomic number layer in the mask that scatters the electrons for a high-contrast image at the wafer plane.

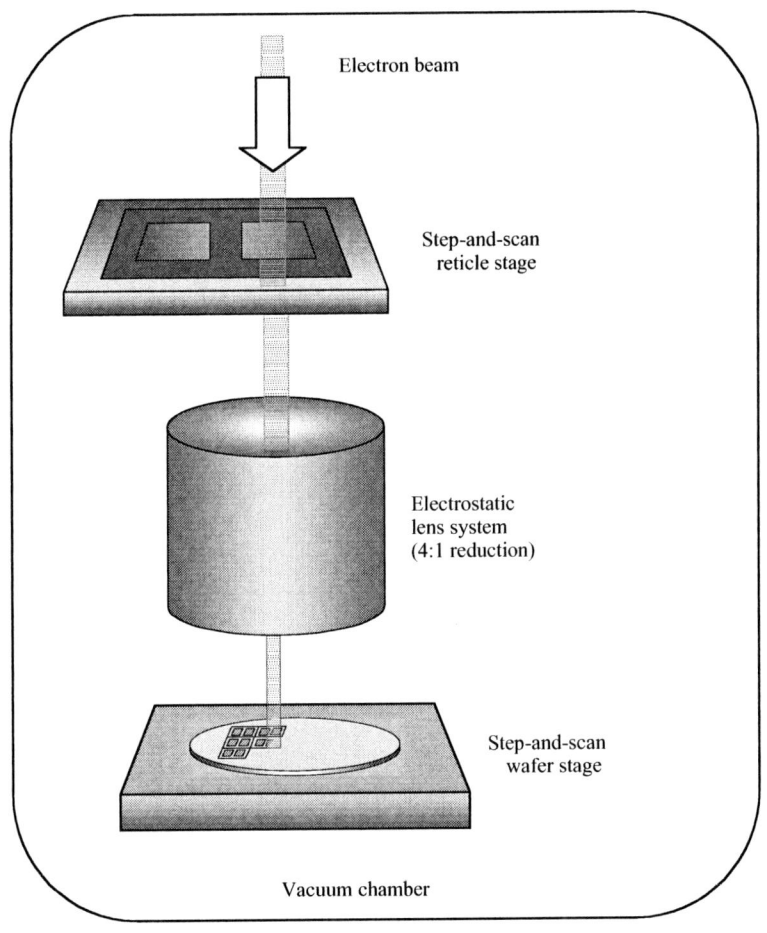

FIGURE 15.11 Concept of SCALPEL
Redrawn from International SEMATECH's Next Generation Litihography Workshop brochure

Increased beam current permits faster exposure but there is also a problem with beam blur caused by space charge effects at higher beam currents. Beam blur occurs when like-charged particles repel one another and cause the image to blur. The mask for SCALPEL is a 4X mask that does not require complicated resolution enhancement techniques, nor does the system need expensive optics.

SCALPEL prints linearly, using a step-and-scan writing strategy that produces stripes of exposed resist. To expose an entire chip, several stripes are stitched together per chip. This approach will require good overlay accuracy so that the stitching does not exceed CD error budgets. The first commercial SCALPEL tools are targeted for 2002.

Ion Projection Lithography (IPL) ■ *Ion projection lithography (IPL)* is a technique in early development that uses ion beams to expose resists, either through a mask or by serially writing on the resist with a finely focused beam (see Figure 15.12). If a mask is used, then a stitched approach is needed, using a broad beam to create small subfields of exposed resist on the wafer surface. IPL uses multielectrode electrostatic optics to direct hydrogen or helium ions to the wafer. Ions transfer their energy more efficiently to the resist than electron beams because ions have a larger mass. There are also fewer secondary electrons produced by ion bombardment and those produced have very low energy, which makes for less backscatter. Backscatter creates proximity effects that determine the limit to the minimum feature size for wafer patterning.[16] Very high resolution is achievable from ion projection lithography, with 50 nm feature sizes already demonstrated in a research environment.

X-Ray ■ *X-ray lithography* is an established technology for imaging patterns on wafers with CDs below 100 nm (0.1 μm). An X-ray source projects X-rays onto a special mask that forms a pattern

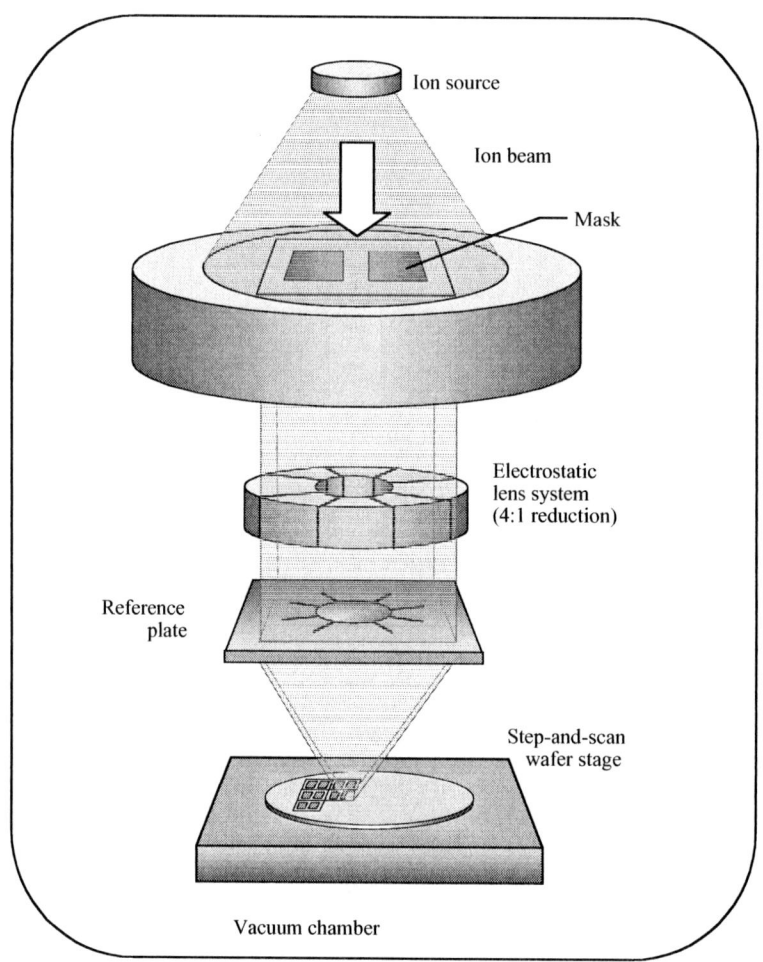

FIGURE 15.12 Ion Projection Lithography
Redrawn from International SEMATECH's Next Generation Lithography Workshop brochure

on a resist-coated wafer. This technique is not widely used for wafer fabrication, primarily due to its high cost of ownership in comparison to optical lithography.[17] One major chipmaker has used X-ray lithography for patterning commercial chips throughout the 1990s, including advanced microprocessors on 200-mm diameter silicon wafers. The system components for X-ray lithography are: (1) a mask that has a pattern of X-ray absorbing materials on a material that transmits X-rays, (2) an X-ray source, and (3) an X-ray resist.

The X-rays used in lithography are known as soft X-rays. They operate in a region of the electromagnetic spectrum from about 0.1 nm to about 10 nm (see Figure 15.13).[18] These differ from the familiar X-rays used in medicine, which have a shorter wavelength and are referred to as hard X-rays.

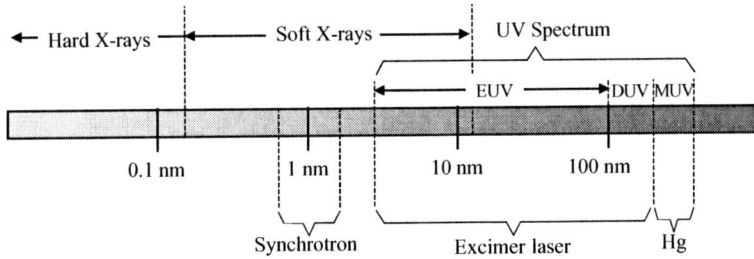

FIGURE 15.13 X-ray Spectrum

The most common X-ray source for lithography is known as a synchrotron (an electron storage ring). High-energy electrons are forced into closed curved paths by magnetic fields and made to accelerate which causes them to emit radiation. This action produces X-ray radiation that is very intense and reasonably collimated (parallel). A projection system based on special mirrors will project the X-rays onto the mask and resist coated wafer. Each synchrotron ring provides radiation for multiple X-ray steppers (e.g., 15 or more).

A special X-ray photomask is used to define the pattern. Because X-ray lithography has such a short wavelength, there are no diffraction interference effects from the mask, producing wide process latitude. An X-ray mask has a membrane substrate (e.g., polyimide) that transmits X-rays and a patterned material that absorbs X-rays, such as gold, tungsten, or tantalum (see Figure 15.14). However, the ability to fabricate masks is one of the many challenges for X-ray lithography. X-ray patterning uses a 1X mask that has the same dimensions as the wafer CDs. This feature eliminates the benefit from having a 4X mask with four times larger feature sizes than the wafer. Mask makers

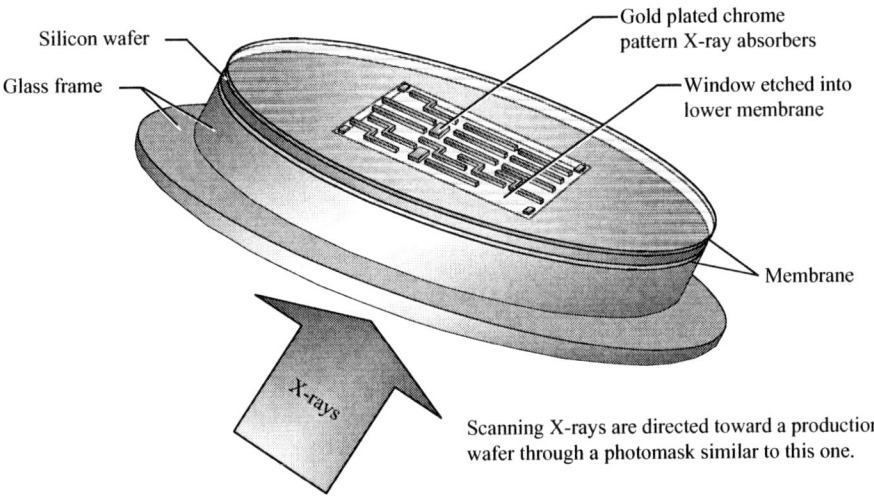

FIGURE 15.14 Concept of X-ray Photomask
Redrawn from K. Nalcamura, Lithography, *ULSI Technology*, ed. by C. Chang and S. Sze (New York: McGraw-Hill, 1996), p. 314.

are challenged to control pattern placement and CDs and minimize mask defects on X-ray masks with existing e-beam mask writers.[19]

Advanced Resist Processing

Photoresist has undergone two basic turning points in semiconductor manufacturing: (1) changing to positive resist, and (2) using chemically amplified deep UV resist (see Figure 15.15). Because resist is the medium that transfers the reticle pattern to the wafer surface material, it is a critical contributor to next-generation lithography.

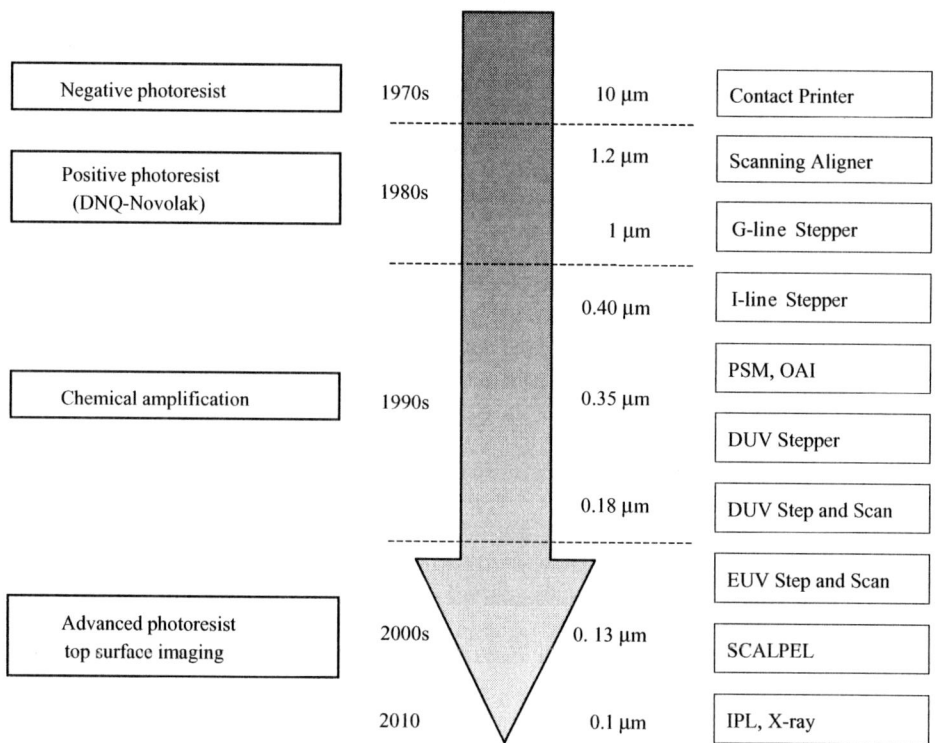

FIGURE 15.15 Development Trends of Photoresist and Lithography

An alternative resist technology that is being evaluated for sub-0.1 μm feature sizes is top-surface imaging resist. In *top-surface imaging,* the photoresist is imaged only at its top surface. This feature action provides a wider process latitude for focus issues than that found on typical resist systems since conventional resist must have its entire resist thickness in focus. At the same time, top-surface imaging permits resist materials to be opaque to the exposure light.[20] This feature could be beneficial since 248 nm DUV resist materials become opaque when the UV wavelength is reduced to 193 nm.

DESIRE Process ■ There are different ways to achieve surface imaging on resists. One method is through a process referred to as DESIRE (diffusion-enhanced silylated resist), which uses *silylation* to selectively place a thin layer of silicon on a resist. This process effectively transfers the reticle pattern to the resist-coated wafer in the form of a silicon pattern on the top surface of the resist.

The silicon is applied by first exposing the spin-coated resist to UV light per the standard process (see Figure 15.16 on page 429). The unexposed resist has the PAC (photoactive compound) sensitizer that is a diffusion inhibitor, while the exposed resist is more susceptible to silicon diffusion. There is often a presilylation bake to crosslink unexposed resist and increase the contrast between the exposed and unexposed region. The silicon is then incorporated into the resist by treating it with a gas-phase silylating agent (such as hexamethyldisilazane, or HMDS) at elevated

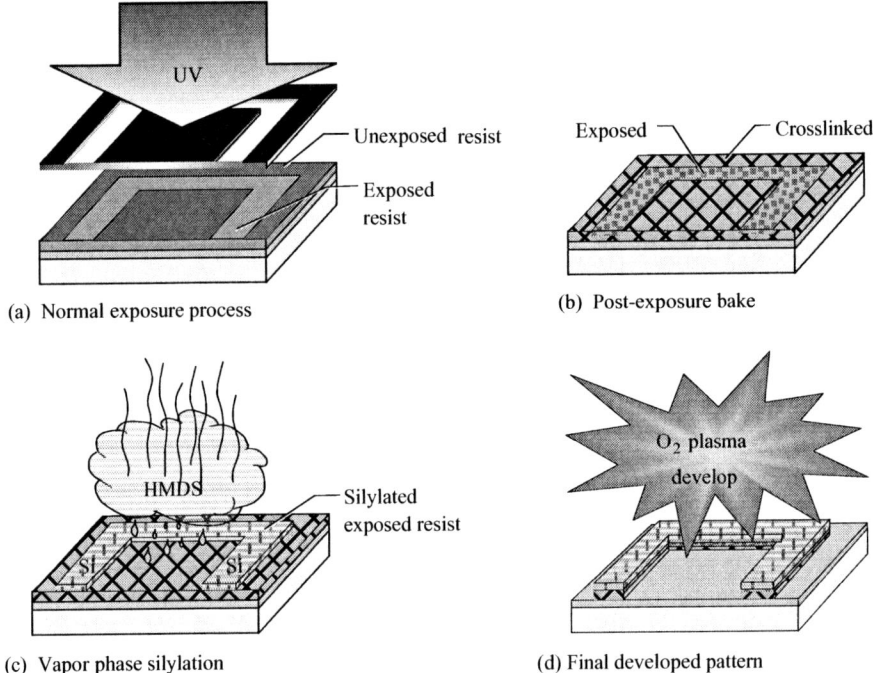

FIGURE 15.16 Top-Surface Imaging

temperatures. The exposed areas of the resist selectively bond chemically to the silicon to a depth of 100 to 200 nm.[21] The silicon-bonded areas of the resist represent the pattern transfer from the reticle to the substrate.

The final step is to dry-develop the wafers in oxygen plasma to remove the unprotected resist. During this plasma etch the silicon is converted into silicon dioxide which forms a thin protective mask that stops the etch process in the resist directly underneath the SiO_2. This is just one example of many silylation-based processes. They all essentially use a single-layer patterning technique that depends on selectively incorporating a silicon-containing compound into the photoresist after the exposure step.

DEVELOP QUALITY MEASURES

There are different types of defects found at post-develop inspect. The quality measures are shown in Table 15.3.

TABLE 15.3 Key Measures at Post-Develop Inspection

Quality Parameter	Types of Defects	Remarks
1. Critical dimensions.	A. Wider CDs than normal.	• Improper stepper focus. • Not enough time or energy during exposure. • Not enough develop time or weak developer solution. • Incorrect process recipes during exposure or develop steps.

Quality Parameter	Types of Defects	Remarks
	B. Narrower CDs than normal.	• Excessive time or energy during exposure. • Excessive develop time or strength of developer solution. • Incorrect process recipes during exposure or develop steps.
2. Contamination.	A. Particles and foreign contamination on the resist surface.	• Equipment needs to be kept clean, with special focus on the track equipment. • Inadequate cleaning and rinsing of wafers. • Developer chemicals and rinse water require point-of-use filters to remove contaminants.
3. Surface defects.	A. Scratches in the resist surface.	• Wafer handling errors or tool misadjustments related to cassette indexers and robotic handling systems.
	B. Particles, spots and stains.	• Chamber exhaust flow, dispenser alignment, dispense pressure, wafer leveling, splashbacks, drips, spin speed, are all possible contributors.
	C. Missing resist, extra resist, or scumming.	• Incorrect puddle time. • Incorrect dispense volume or position. • Improper rinsing after the develop process. • Improper or uneven baking.
	D. Striations in photoresist along the sidewalls of the resist features.	• Standing waves or reflective notches (unacceptable CD variation). • Improper or no antireflective coating (ARC) was used.
4. Overlay registration.	A. Improper alignment or overlaying of one layer over a previous layer.	• This is not a problem caused by the develop process. • It is more likely a stepper-induced problem. • Wrong process recipe or reticle was used. • Poor temperature and humidity control.

DEVELOP TROUBLESHOOTING

Various troubleshooting problems encountered during develop are shown in Table 15.4.

TABLE 15.4 Common Develop Troubleshooting Problems

Problem	Probable Cause	Corrective Actions
1. Linewidths and holes do not meet CD requirements.	A. Underdeveloped or underexposed positive resist across entire wafer.	• Ensure correct stepper process recipe was used. • Check for insufficient exposure time and energy settings. • Check stepper illuminator system. • Verify dose meter (light integrator) is functioning properly. • Ensure correct recipe on developer was used. • Check for insufficient puddle time and quantity settings. • Check developer equipment. • Check bake temperatures.
	B. Overdeveloped or overexposed positive resist across entire wafer.	• Ensure correct stepper process recipe was used. • Check for excessive exposure time and energy settings. • Check stepper illuminator system. • Verify dose meter (light integrator) is functioning properly. • Ensure correct recipe on developer was used. • Check for excessive puddle time and quantity settings. • Check developer equipment. • Check bake temperatures.
	C. No measurable CDs.	• Check exposure or develop operation. • Check reticle or process recipe. • Wafer may have skipped coat, expose, post-exposure bake, or develop step. • Rework wafer.
2. Resist scumming.	A. Resist residue remains on the wafer after the develop operation is completed.	• Check develop equipment process recipe. • Verify puddle time and puddle quantity are correct. • Check the rinser operation. • Check bake oven times and temperatures. • Rework wafer.

Problem	Probable Cause	Corrective Actions
3. Contamination and defects.	A. Possible causes may include the chemicals, rinse water, and process chamber.	• Verify puddle time and puddle quantity are correct. • Clean the develop process chamber, then recheck a test wafer for further contamination. • Check, and if necessary, replace line filters for the developer and rinse water.
	B. Misting or backsplashing from the dispenser can cause contamination.	• Check chamber exhaust level. • Check alignment of dispenser relative to the wafer. • Check for drips from dispenser assembly. • Rework wafer.
4. Collapsing of resist pattern after develop.	A. High aspect ratio (>5:1) will lead to collapsing of resist lines.*	• Verify that the resist is not excessively thick, since this will increase the aspect ratio and make the resist more likely to collapse. • Check that the resist has proper adhesion to the wafer. • The problem resolution may require a material or process change (e.g., more rigid resist).
5. Unacceptable CD variation at top of CA DUV resist profile.	A. Amine contamination of resist after exposure.	• Check integrity of environmental chamber filtering system. • Check to see if coated-wafers may have been exposed to external chemical contamination. • Asses whether resist is best selection for length of delay (newer resists withstand a longer time before PEB).

*J. Yu, et al., "Analysis of Resist Pattern Collapse and Optimization of DUV Process for Patterning Sub-0.20 mm Gate Line," *Advances in Resist Technology and Processing XV, Proceedings of SPIE,* Vol. 333, (Bellingham, WA: SPIE, 1998): p. 880.

SUMMARY

Resist development removes those areas of the resist rendered soluble by the exposure step. After exposure, a post-exposure bake (PEB) is performed for chemically amplified DUV resists to catalyze critical resist chemical reactions. The temperature uniformity, duration, and time delay for the PEB of DUV resist is important. For conventional DNQ-novolak resist, a PEB drives out additional solvent for better adhesion and reduces the standing wave effect. Negative resist development is mainly a solvent wash of unexposed resist. The exposed negative resist tends to swell during development, which limits its use in submicron lithography. Positive resist development involves a chemical reaction between the developer and the exposed resist, with parameters of develop speed and selectivity. The most common positive-tone developer is TMAH. The two most common development methods are continuous spray and puddle, which require the control of critical parameters for an optimum process. Hard bake is a post-development thermal bake used to drive out any residual solvent and harden the resist. A post-develop inspection is performed to characterize defects for corrective action and remove defective wafers from the process before etch or implant.

Lithography improvements have produced sub-wavelength lithography, where patterning achieves a CD below the exposure light wavelength. Next-generation lithography is evaluating the lithography technology for eventually replacing optical lithography. There are four primary alternatives: Extreme UV (EUV), SCALPEL, ion projection lithography (IPL), and X-ray. EUV uses a UV wavelength of about 13 nm to achieve a pattern resolution of 30 nm. SCALPEL uses an established electron beam for lithography. IPL uses ion beams to expose resist. X-ray lithography projects X-rays with a wavelength from about 0.1 to 10 nm onto a special mask to pattern a resist. Top-surface imaging is a technique usedto image only the top surface of the resist for reduced depth-of-focus and increased resolution.

KEY TERMS

resist development
post-exposure bake (PEB)
negative resist development
positive resist development
dissolution rate
developer selectivity
continuous spray develop
puddle develop
dissolution rate monitor (DRM)
scumming
normality

hard bake
post-develop inspection
subwavelength lithography
next-generation lithography
extreme UV (EUV)
SCALPEL
ion projection lithography (IPL)
X-ray lithography
top surface imaging
silylation

REVIEW QUESTIONS

1. Explain resist development. What is its goal?
2. Why is a post-exposure bake done for chemically amplified DUV resist? Be specific about deprotection.
3. Why is temperature uniformity important for PEB?
4. Describe the benefits of a post-exposure bake for an i-line conventional resist.
5. Explain negative resist development. What is the primary problem for submicron patterning?
6. Why is positive-tone resist the most commonly used resist?
7. What is the developer dissolution rate? Is it desirable for the rate to be high or low?
8. Explain resist selectivity and whether it should be high or low.
9. What is the name for the most common positive-tone developer?
10. Why is a surfactant added to the developer?
11. Explain how the TMAH developer functions for a conventional i-line resist.
12. Is there a chemical reaction between PHS and the developer for chemically amplified DUV resists?
13. List the two methods of resist development.
14. Explain continuous spray development.
15. Describe puddle development.
16. List the seven resist development parameters.
17. Explain why hard bake is done.
18. Describe UV resist hardening.
19. Why is a post-develop inspection performed?
20. What is subwavelength lithography?
21. List four alternative lithography methods under evaluation for next-generation lithography.
22. Explain extreme UV (EUV).
23. Describe SCALPEL lithography.
24. Discuss ion projection lithography (IPL).
25. Explain X-ray lithography. What is one of the major challenges for X-ray lithography?
26. Describe top-surface imaging. Why is silylation used in this process?

PHOTOLITHOGRAPHY MATERIALS AND EQUIPMENT SUPPLIERS' WEB SITES

Allied Signal — http://www.electronicmaterials.com/
Applied Materials — http://www.appliedmaterials.com/products/
Arch Chemicals (aka Olin) — http://www.olinmicro.com/default.asp
Ashland Specialty Chem. — http://www.ashland-act.com/
ASML — http://www.asml.com/
Canon Semiconductor — http://www.usa.canon.com/indtech/semicondeq/
Charles Evans and Associates — http://www.cea.com
Clariant Corporation — http://www.azresist.com/
Cymer Inc. — http://www.cymer.com/
DuPont — http://www.dupont.com/semiconductor/
Eastman Chemical — http://www.eastman.com/

EKC Technology	http://www.ekctech.com/ekctech.nsf
ETEC Systems Inc.	http://www.etec.com/semiprod_frame.html
FSI International	http://www.fsi-intl.com/
International SEMATECH	http://www.sematech.org/
JSR Microelectronics. Inc.	http://www.jsrusa.com/index2.html
J.T. Baker	http://www.jtbaker.com/
Karl Suss Inc.	http://www.suss.com/
Lucent Technologies	http://www.bell-labs.com/project/SCALPEL/
MICRO Magazine	http://www.miciromagazine.com/
Nikon	http://www.nikon.com/
Olin Microelectronics	http://www.olinmicro.com/
Photronics Inc.	http://www.photronics.com/
SEMI	http://www.semi.org/
Semiconductor International	http://www.semiconductor.net/
Shipley Company	http://www.shipley.com/
Silicon Valley Group	http://www.svg.com/
Solid State Technology	http://sst.pennet.com/home/home.cfm
SPIE	http://www.spie.org/
TEL, Tokyo Electron Ltd.	http://www.teainet.com
Ultratech Stepper	http://www.ultratechstepper.com/

REFERENCES

1. P. Burggraaf, "Optical Lithography to 2000 and Beyond," *Solid State Technology* (February 1999): p. 31.
2. B. Smith, "Resist Processing," *Microlithography: Science and Technology,* ed. J. Sheats and B. Smith (New York: Marcel Dekker, 1998), p. 542.
3. D. Seegar, "Chemically Amplified Resists for Advanced Lithography: Road to Success or Detour?" *Solid State Technology* (June 1997): p. 115.
4. A. Braun, "Track Systems Meet Throughput and Productivity Challenges," *Semiconductor International* (February 1998): p. 63.
5. B. Smith, "Resist Processing," p. 541.
6. B. Smith, "Resist Processing," p. 551.
7. C. Mack et al., "New Model for the Effect of Developer Temperature on Photoresist Dissolution," *Advances in Resist Technology and Processing* XV Proceedings of SPIE vol. 3333, (Bellingham, WA: SPIE, 1998): p. 1218.
8. B. Smith, "Resist Processing," p. 548.
9. D. Velikov et al., "Endpoint Detector Monitors Photoresist Develop Process," *Semiconductor International* (August 1996): p. 144.
10. B. Smith, "Resist Processing," p. 561.
11. B. Smith, "Resist Processing," p. 562.
12. R. DeJule, "Next-Generation Lithography Tools: The Choices Narrow," *Semiconductor International* (March 1999): p. 48.
13. Ibid.
14. Challenges summarized from A. Hawryluk, N. Ceglio, and D. Markle, "EUV Lithography," *Microlithography Word* (Summer 1997): pp. 18–20.
15. R. DeJule, "Next-Generation Lithography Tools," p. 48.
16. S. Ghandhi, *VLS Fabrication Principles: Silicon and Gallium Arsenide,* 2nd ed., (New York: Wiley, 1994), p. 693.
17. T. Ueno and J. Sheats, "X-ray Lithography," ed. J. Sheats and B. Smith (New York: Marcel Dekker, 1998), p. 403.
18. T. Ueno and J. Sheats, "X-ray Lithography," p. 405.
19. R. DeJule, "Next-Generation Lithography Tools," p. 48.
20. S. Postnikov et al., "Top Surface Imaging Through Silylation," *Advances in Resist Technology and Processing* XV Proceedings of SPIE vol. 3333, (Bellingham, WA: SPIE, 1998): p. 997.
21. K. Nakamura, "Lithography," *ULSI Technology,* ed. C. Chang and S. Sze (New York: McGraw-Hill, 1996), p. 294.

CHAPTER 16
ETCH

A wafer is fabricated into many functional microchips by creating electronic devices in the silicon and then sequentially depositing layers of dielectric and conductive materials to interconnect the devices. This is the concept of planar technology for wafer fabrication, used since the early days of semiconductor manufacturing.

In general, interconnect materials are deposited on the wafer surface and then selectively removed to form the circuit patterns that were defined by photolithography. This selective material removal process, known as etch, follows post-develop inspection. It is critical that etch be performed correctly or else the chip will not function. More importantly, once material is removed by etch, it is difficult to correct a mistake. Improperly etched wafers are most likely scrapped and discarded at a loss to the company.

Etch requirements depend on the type of feature being fabricated, such as the aluminum alloy metal stack, a polysilicon gate, a silicon trench for isolation, or a dielectric via. IC structures are complex, with a range of materials that require different etch parameters. Shrinking feature sizes drive tighter etch dimensional control and make inspections more difficult.

An example of an etch application is aluminum. The traditional metallization process deposits an aluminum alloy layer on the wafer surface. This action is followed by photolithography and etch to form the interconnects. The different metal layers are electrically connected by previously-formed tungsten plugs in interlayer dielectric (ILD) vias.

With the introduction of damascene processing for copper metallurgy, the metallization process becomes a dielectric etch to form a trench in the ILD. Copper is blanket-deposited into the dielectric patterns and polished back to the ILD height using chemical mechanical planarization. With damascene processing, there will be less emphasis on metal etch and more importance placed on dielectric etch.

OBJECTIVES

After studying the material in this chapter, you will be able to:

1. List and discuss nine important etch parameters.
2. Explain dry etch, including its advantages, and discuss how etching action takes place.
3. List and describe the equipment systems for seven dry plasma etch reactors.
4. Explain the benefits of high-density plasma (HDP) etch, and discuss the four types of HDP reactors.
5. Give an application example for dielectric, silicon, and metal dry etch.
6. Discuss wet etch and its applications.
7. Explain how photoresist is removed.
8. Discuss etch inspection and its related important quality measures.

INTRODUCTION

Etch is the process of selectively removing unneeded material from the wafer surface by using either chemical or physical means. The fundamental goal of etching is to accurately reproduce the mask features on the resist-coated wafer. The patterned resist layer is not attacked significantly by the etchant. This mask layer is used to protect specific regions of the wafer surface while permitting selective etching through openings in the photoresist layer (see Figure 16.1 on page 436). The etch process follows photolithography in the general CMOS process flow (see Figure 16.2 on page 436). In this manner, etch can be viewed as the last major pattern transfer process step necessary to replicate the desired pattern on the wafer surface.

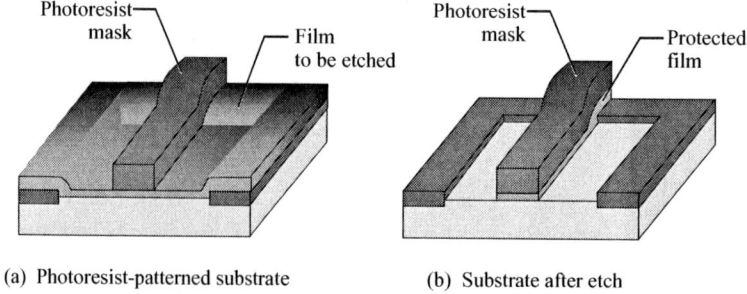

(a) Photoresist-patterned substrate   (b) Substrate after etch

FIGURE 16.1 Applications for Wafer Etch in CMOS Technology

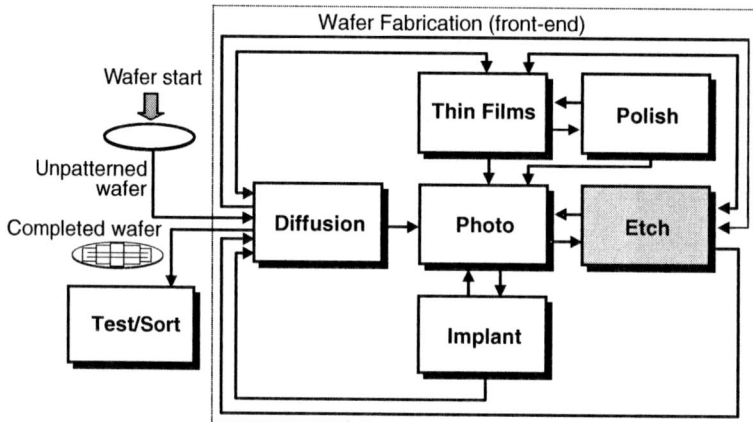

FIGURE 16.2 General CMOS Process Flow
(Used with premission from Advanced Micro Devices)

Etch Processes

Two basic types of etch processes are used in semiconductor manufacturing: dry etch and wet etch. *Dry etch* exposes the wafer surface to a plasma created in the gaseous state. The plasma passes through the openings in the patterned resist and interacts physically or chemically (or both) with the wafer to remove the surface material. Dry etching is the primary method used for etching devices with submicron geometries. Thus, dry etching will receive the most coverage in this text. In *wet etch*, liquid chemicals such as acids, bases, and solvents are used to chemically remove wafer surface material. Wet etch is generally applicable only for larger geometries (>3 μm). It is still used to remove some layers from the wafer and to remove dry etch residues.

Dry etch can also be categorized by the type of material removed. The three major material categories for etching are metal etch, dielectric etch, and silicon etch. *Dielectric etch* is used for applications with dielectric material, such as silicon dioxide. The formation of contact and via structures require a dielectric etch to form openings in the ILD. This is a challenge for etching high-aspect ratio openings (ratio of height to width for an opening). *Silicon etch* (including polysilicon) is used for applications requiring silicon removal, such as etching polysilicon transistor gates and silicon trench capacitors. *Metal etch* is primarily used for removal of aluminum alloy stacks to form interconnect wiring on metal layers. At the time of this writing, there is no acceptable method for etching copper metal with submicron feature sizes (which is an important reason for the introduction of dual-damascene processing into the wafer fab).

Etch can also be categorized as either patterned or unpatterned etching. *Patterned etching* uses a masking layer (patterned photoresist) to define areas of the surface material that are to be etched. Only selected portions of this wafer surface layer are removed during the etch process. Patterned etching is used to form the many different features on the wafer surface, including gates, metal interconnects, vias, contact holes, and trenches. *Unpatterned etching, etchback,* or *stripping,* occurs when the entire wafer surface is etched with no mask present. This etching process is used to strip masking layers (e.g., STI nitride strip and the titanium strip after the salicide process used to form spacers for transistor implants). Etchback is used when it is desirable to reduce the overall thickness of a specific layer of film (e.g., when planarizing a surface to reduce topographical features). Photoresist is another example of a material stripped off the wafer. In summary, patterned and unpatterned etch processes can be performed using either dry-etch or wet-etch techniques.

ETCH PARAMETERS

Etch has specific requirements it must meet in order to replicate the mask pattern on the wafer surface material. Important parameters for etch are:

- Etch rate
- Etch profile
- Etch bias
- Selectivity
- Uniformity
- Residues
- Polymer formation
- Plasma-induced damage
- Particle contamination and defects

Etch Rate

Etch rate is the speed at which material is removed from the wafer surface during etching (see Figure 16.3). It is usually measured in Å/minute. The depth of the etched opening is known as the *step height*. It is desirable to have a high etch rate in order to keep wafer throughput high. This feature is even more important when using single-wafer processing in cluster tools. The etch rate is determined by process and equipment variables such as the type of material being etched, reactor configuration, gases used for the etch, and the process parameter settings. The etch rate is calculated by the following formula:

$$\text{Etch Rate} = \frac{\Delta T}{t} \quad (\text{Å/minute})$$

Where, ΔT = amount of material removed (Å or μm)
t = time elapsed during etch (typically minutes)

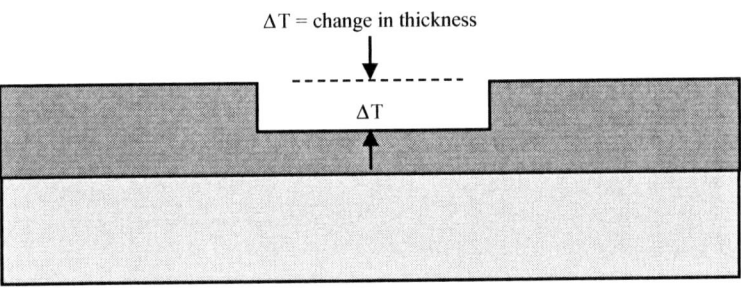

FIGURE 16.3 Etch Rate

Etch rate is generally proportional to the concentration of the etchant. Factors such as wafer surface geometry can affect the wafer-to-wafer etch rates. Wafers with significant surface area for etching will deplete the etchant concentration and etch slower, whereas wafers with small surface areas for etching will etch faster. This condition is referred to as *loading effects*. The decrease in etch rate is caused by the plasma etching reactions consuming most of the available etchant species in the gas phase. The change in etch rate due to loading effects is a primary reason why effective endpoint detection is critical (see the following section).

Etch Profile

Etch profile refers to the shape of the sidewall of the etched feature. There are two basic etch sidewall profiles: isotropic and anisotropic. An *isotropic etch profile* etches at the same rate in all directions (laterally and vertically), leading to undercutting of the etched material under the mask (see Figure 16.4). This action results in an undesirable loss of linewidth. Wet chemical etching is usually isotropic in nature, which is the primary reason why wet etching is not used for selective patterned etching of submicron devices. Some dry plasma systems are also capable of providing an isotropic etch profile. There are instances where isotropic etching may be desirable depending on the specific needs of the material being etched and the requirements of subsequent processing steps.

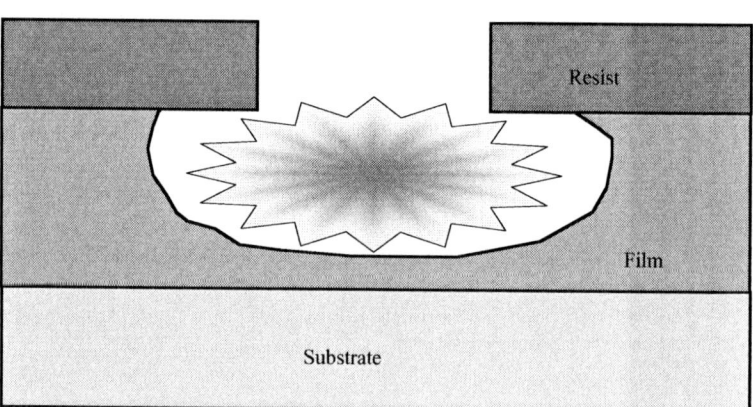

FIGURE 16.4 Wet Chemical Isotropic Etch

The desirable etch sidewall profile for submicron feature sizes is an *anisotropic profile,* where the rate of etching is in only one direction perpendicular to the wafer surface (see Figure 16.5). There is very little lateral etching activity. This leaves vertical sidewalls, permitting a higher packing density of etched features on the chip. Anisotropic etch is critical for the patterning of submicron devices with small linewidths and features. Advanced IC applications usually require 88 to 89° vertical sidewall profiles. Anisotropic etch is achieved with most dry plasma etching. Table 16.1 shows the profile for wet etch and the various profiles achievable with dry etch.

The amount of anisotropic etch can be moderate (slight sidewall angle) or highly anisotropic (vertical sidewalls). Etch profile refers to the shape of the etched film wall. A vertical profile is the result of a highly anisotropic etch.

With smaller geometries, the etch profiles have higher aspect ratios. It is difficult to get etchant chemicals in and reaction by-products out of the high-aspect ratio openings. To overcome this, it is desirable to have *directionality* to drive the plasma into the high-aspect ratio openings. If plasma ions are directional (perpendicular to the wafer surface), then only the surface is bombarded, not the feature's sidewalls. This action forces etchant chemicals into high-aspect ratio openings with little undercutting. For advanced ICs with sub-0.25 μm CDs, directionality is achieved with a high-density plasma source capable of generating enough etchant species to achieve acceptable etch rates.

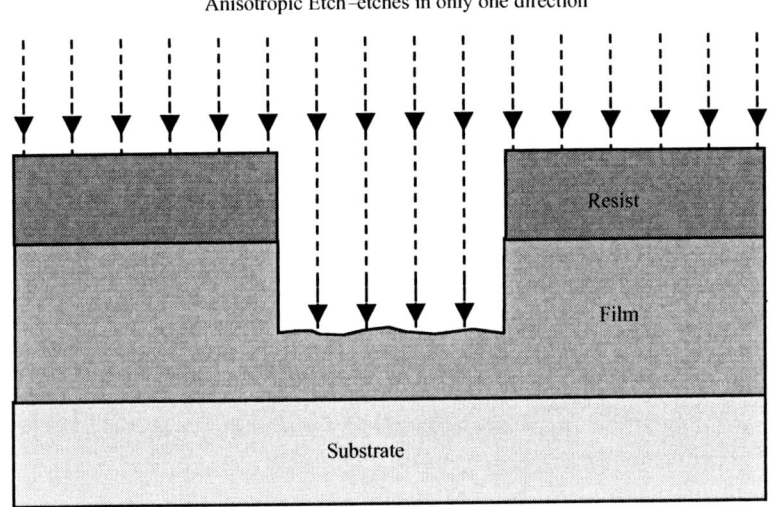

FIGURE 16.5 Anisotropic Etch with Vertical Etch Profile

TABLE 16.1 Sidewall Profiles for Wet Etch Versus Dry Etch

Type of Etch	Sidewall Profile	Diagram
Wet Etch	Isotropic	
Dry Etch	Isotropic (depending on equipment and parameters)	
	Anisotropic (depending on equipment and parameters)	
	Anisotropic—Taper	
	Silicon Trench	

Etch Bias

Etch bias is a measure of the change in linewidth or space of a critical dimension (CD) after performing an etch process (see Figure 16.6 on page 440). It is usually caused by undercutting (see Figure 16.7 on page 440), but can also be the result of an etch profile. Undercutting occurs when the etch process removes excessive material below the mask, causing the top surface of the etched film to be recessed from the resist edge. The formula to calculate etch bias is:

$$\text{Etch bias} = W_b - W_a$$

Where, W_b = the original linewidth in photoresist before etch.
W_a = the final linewidth of the etched material after resist removal.

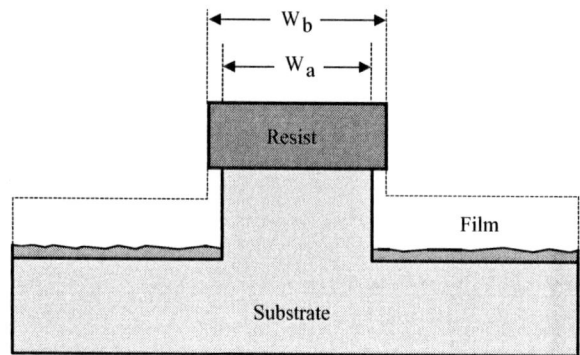

FIGURE 16.6 Etch Bias

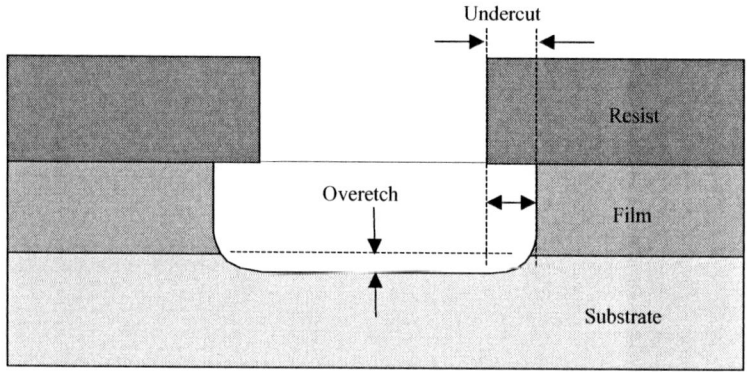

FIGURE 16.7 Etching Undercut and Slope

Selectivity

Selectivity represents how much faster one film etches than another film under the same etch conditions. It is defined as the etch rate of the material being etched relative to the etch rate of another material (see Figure 16.8). High selectivity means that etching only occurs on the desired layer. A high selectivity etch process does not etch the underlying film (etching stops at the right depth) and the protective photoresist is not etched. Shrinking geometries require thinner layers of resist. High selectivity is necessary in most advanced processes to ensure critical dimension and profile control. Specifically, the smaller the critical dimension then the higher the selectivity must be.

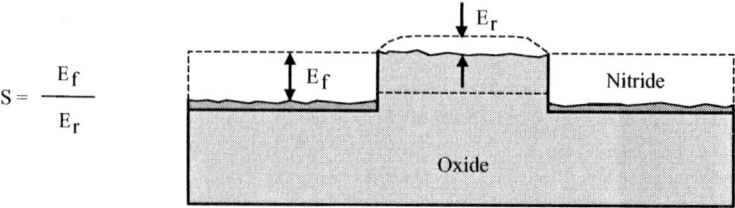

FIGURE 16.8 Etch Selectivity

The selectivity, S_R, for the film undergoing etch and the mask layer (e.g., photoresist) can be calculated from the following formula,

$$S_R = \frac{E_f}{E_r}$$

Where, E_f = the etch rate of the film undergoing etch
E_r = the etch rate of the masking layer (e.g., photoresist)

Based on this formula, selectivity is often expressed as a ratio. A process with poor selectivity may be 1:1 (meaning the film being etched is removed as quickly as the photoresist mask), whereas a good selectivity could be 100:1. Interpret this as the etched film is removed 100 times faster than the film not being etched (e.g., photoresist).

Dry etching frequently does not provide adequate etch selectivity to the layer underneath. In this case, a plasma reactor should be equipped with an endpoint detection system that signals when it is time to stop the etch process with minimal overetching. The endpoint detector notifies the etch equipment controller when the underlying film is beginning to be exposed in order to stop the etch process.

Uniformity

Etch uniformity is a measure of the capability of the process to etch evenly across the entire surface area of the wafer, across the entire wafer lot, and from lot to lot. Uniformity is closely related to selectivity, since nonuniformity results in additional overetch. Maintaining uniformity across the wafer surface is key to ensuring consistent manufacturing performance. The challenge is that etching must be uniform in different types of wafer surface pattern density, such as densely populated wafer areas, large open spaces, and within high-aspect ratio features. Some problems in uniformity occur because etch rates and profiles depend on feature size and pattern density.[1] The etch rate is slower in small openings, to the point that etching can actually stop in small geometries with high aspect ratios. For instance, silicon trenches with a high-aspect ratio opening etch slower than trenches with a small-aspect ratio opening (see Figure 16.9). This phenomenon is known as *aspect ratio dependent etching (ARDE)*, also referred to as *microloading*. The objective is to minimize ARDE across the wafer surface in order to improve uniformity.

Randomly select 3 to 5 wafers in a lot

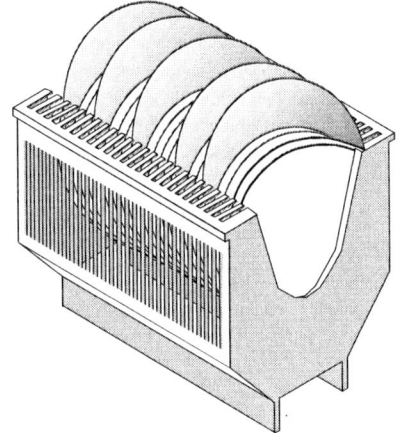

Measure etch rate at 5 to 9 locations on each wafer, then calculate etch uniformity for each wafer and compare wafer-to-wafer.

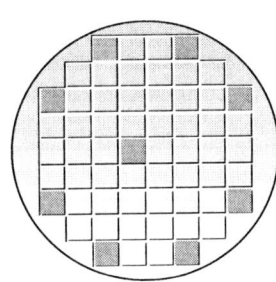

FIGURE 16.9 Etch Uniformity

Residues

Etch residue is the unwanted material remaining on the wafer surface after etch. It often coats the walls of the chamber and the bottom of the feature being etched. It can occur for reasons such as contaminants in the etched film, improperly chosen etch chemistry (e.g., etching too fast), contaminants in the chamber, and nonuniform dopant distribution in the film. There are different names for the residues left after etch, including stringers, veils, crowns, and fences. Stringers are small residues of the etched material that are not totally removed, are electrically active, and can form an undesirable connecting short between features. Residues are a source of wafer contamination for IC fabrication and can cause problems during resist stripping. An overetch is sometimes done at the end of the etch process to remove residues. In some cases the etch residues can be removed by a resist strip process or by wet chemical etching.

Polymer Formation

A *polymer formation* is sometimes intentionally deposited on the sidewalls of the etch feature to form an etch-resistant film that prevents lateral etching (see Figure 16.10). This produces highly anisotropic features because it blocks the etching of the sidewall, thus increasing the etch directionality. The result is better control of critical dimensions of patterned structures. The polymers come from photoresist carbon converted into polymers during etching and combines with etching gases (i.e., C_2F_4) and etch by-products to form this sidewall polymer.[2] The need for sidewall polymers depends on the type of etching gas used.

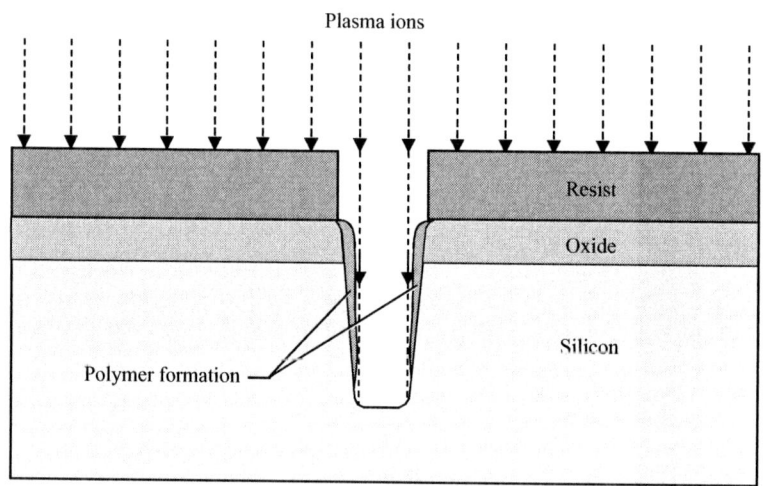

FIGURE 16.10 Polymer Sidewall Passivation for Increased Anisotropy

These sidewall polymers are complex, containing etchant elements and reaction by-products such as aluminum, titanium from barrier layers, oxides, and other inorganic materials. The polymer chains have strong carbon-fluorine bonds that are difficult to oxidize and remove.[3] However, the polymers must be removed after the etch process or device yield and reliability is affected. The cleaning of the sidewall often requires special gas chemistry for a plasma stripping process or possibly a wet clean using strong solvents followed by DI-water rinsing (see the following section).

Unfortunately, an undesirable side effect of polymer deposition is that the internal parts of the process chamber also become coated with polymer. Etch process chambers require periodic cleaning to remove the polymer and to replace parts that can not be cleaned.

Plasma-Induced Damage

Plasma consists of energized ions, electrons, and excited molecules, which can cause *plasma-induced damage* to sensitive devices on the wafer surface. A major type of damage is nonuniform plasma that creates trapped charges on the gate electrode of a transistor, causing a breakdown of the thin gate oxide.[4] Plasma becomes nonuniform due to poor equipment design or from operating the plasma reactor outside of the optimum process window. Another type of device damage is from energetic ion bombardment to gate oxide that is directly exposed to the ions. This damage could occur at the edge of a gate electrode during etching. Plasma damage is sometimes removed from the wafer through anneals or chemical wet etching.

Particle Contamination

Wafer damage from plasma can also come from particle contamination generated by the plasma near the wafer surface. Studies have shown that particles are trapped near the plasma/sheath interface because of the electrical potential difference.[5] When the plasma is turned off, these particles fall onto the wafer surface. Plasma based on fluorine gas chemistries produces fewer particles than chlorine or bromine gas chemistries because fluorine generates etch by-products with a higher vapor pressure. Control of particle contamination is done by an optimized tool design, proper tool operation and shutdown, and use of the appropriate gas chemistry for the film being etched.

DRY ETCH

Dry etch is the primary etching method used to remove surface material in semiconductor manufacturing. The goal of dry etch is to reproduce the image of a mask on the wafer surface with a high degree of integrity. The advantages to using dry etch over wet etch are listed in Table 16.2.

TABLE 16.2 Advantages of Dry Etch Over Wet Etch

Description of Advantages
1. Etch profile is anisotropic with excellent control of sidewall profiles.
2. Good CD control.
3. Minimal resist lifting or adhesion problems.
4. Good etch uniformity within wafer, from wafer-to-wafer, and from lot-to-lot.
5. Lower chemical costs for usage and disposal.

There are disadvantages to using dry etch. The primary disadvantages are poor selectivity to the underlying layer, risk for device damage from plasma, and expensive equipment.

In the dry etching process, a low-pressure plasma discharge is used to remove material in small feature sizes of integrated circuits (the creation of a plasma was discussed in Chapter 8). The plasma interacts with the wafer surface to cause etching action and the subsequent removal of the surface material. The major actions that occur for plasma etching of a substrate material are shown in Figure 16.11.[6]

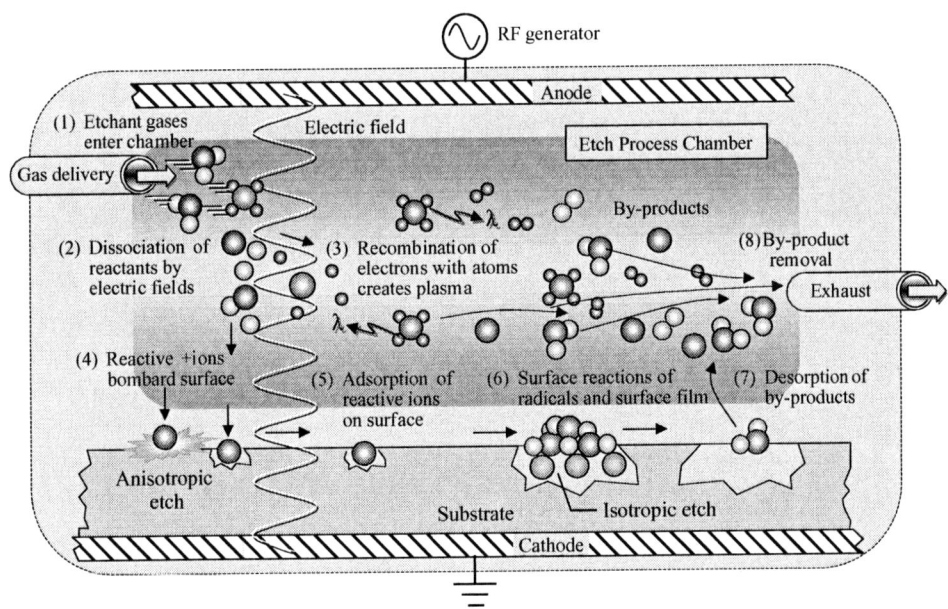

FIGURE 16.11 Plasma Etch Process of a Silicon Wafer

Etching Action

Etching action in dry etch systems is achieved by either a chemical or a physical technique or a combination chemical/physical technique (see Figure 16.12). In a purely *chemical mechanism,* the plasma creates reactive species (free radicals and reactive atoms) that chemically react with the materials on the wafer surface. The gases introduced into the chamber (typically containing chlorine and/or fluorine) are carefully selected to attain high selectivity (that is, to minimize the chemical reaction with the photoresist mask and underlying wafer layer). The chemical etching component of the plasma produces poor CD control because of its isotropic profile. Volatile by-products of the reaction are removed by the low-pressure pumping system.

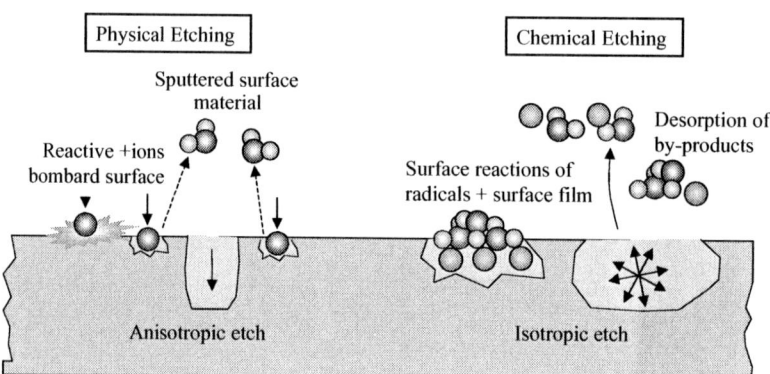

FIGURE 16.12 Chemical and Physical Dry Etch Mechanisms

To achieve etching action with a *physical mechanism,* the plasma provides energetic species (bombarding positive ions) that are accelerated toward the wafer surface by strong electric fields. The ions physically remove the unprotected wafer surface material by a sputter-etch action. Typically a nonreacting gas such as argon (Ar) is used. A benefit to this mechanical etching approach is the strong directionality of the etch, making it possible to achieve highly anisotropic profiles (near vertical sidewalls) for good CD control. Sputter etching produces a high etch rate; however, it does have poor selectivity. Another problem is that the species removed by sputtering are not volatile and may redeposit back on the wafer, causing particulate and chemical contamination.

There is also a *combined physical and chemical mechanism* where ion bombardment improves the chemical etching action. The etch profile is varied from isotropic to anisotropic by adjusting the plasma conditions and gas composition. Combined physical and chemical etch produces good CD control with fair selectivity and is often preferred for most dry etch processes. Table 16.3 summarizes the different etch parameters for chemical, physical, and combined chemical/physical etching. The advantages from each type depend on the objectives of the etch process.

Note that dry etch systems may be designed to operate either as isotropic or anisotropic etchers depending on the direction of the RF electric field relative to the wafer surface. This means that positive ion sputtering can occur respectively on the wafer surface or on the edge of the wafer. If the field is perpendicular to the wafer surface, then the etching is done by a combination of heavy positive ion sputtering and some radical chemical reactions. If the field is parallel to the surface of the wafer, very little physical sputtering occurs, so the etching is done primarily by chemical reactions between radicals and the surface material.

Potential Distribution

The *plasma potential distribution* in the glow discharge region of a plasma has a strong effect on the etching capability of the system. This is because the amount of particle energy that bombards the etched surface depends on the electrical potential distribution. Figure 16.13 shows a reactor glow discharge between two electrodes. The power electrode has RF power applied, while the ground electrode is at ground potential. The potential of the plasma is positive relative to the grounded electrode (which is also connected to the reactor walls, making them grounded as well). The plasma region has the most positive potential in the system.

TABLE 16.3 Chemical Versus Physical Dry Plasma Etching

Etch Parameter	Physical etch (RF field perpendicular to wafer surface)	Physical etch (RF field parallel to wafer surface)	Chemical etch	Combined Physical and Chemical
Etch Mechanism	Physical ion sputtering	Radicals in plasma reacting with wafer surface*	Radicals in liquid reacting with wafer surface	In dry etch, etching includes ion sputtering and radicals reacting with wafer surface
Sidewall Profile	Anisotropic	Isotropic	Isotropic	Isotropic to anisotropic
Selectivity	Poor/difficult to increase (1:1)	Fair/good (5:1 to 100:1)	Good/excellent (up to 500:1)	Fair/good (5:1 to 100:1)
Etch Rate	High	Moderate	Low	Moderate
CD Control	Fair/good	Poor	Poor to nonexistent	Good/excellent

*Used primarily for stripping and etchback operations.

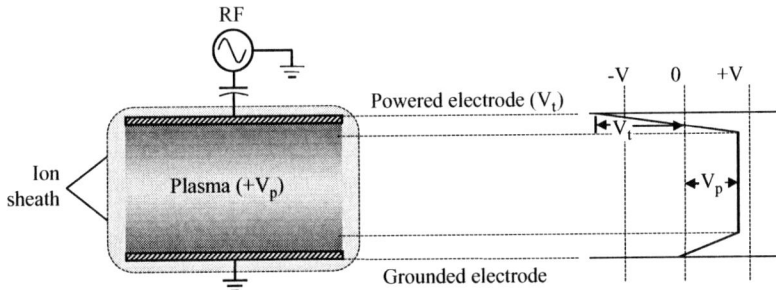

FIGURE 16.13 Schematic View of Reactor Glow Discharge with Potential Distribution

The powered electrode in an etch reactor develops a negative self-bias (DC self-bias or DC bias) voltage relative to ground because fast-moving electrons leave the plasma to strike the electrode. After a certain amount of negative charge, electrons are repelled from the electrode, thus creating the dark space (ion sheath) region with positive ionic charge. The magnitude of the powered electrode self-bias voltage depends on the amplitude and frequency of the RF voltage applied to the electrodes.[7] If the electrodes are of similar area, the potential difference across the dark space of each electrode will be the same. Since the powered electrode develops a negative self-bias, the plasma must assume a positive potential to produce a potential of equivalent magnitude at the grounded electrode. In essence, the plasma forms a compensating positive potential.

There are etch trends that are affected by parameters of the etch process. If the RF frequency is reduced, then ions efficiently cross the plasma dark space in a small fraction of an RF cycle. This action increases the ion energy and etch rate. The amount of ion bombardment depends on the size of the electrodes. For asymmetrical electrode size, if the area of the powered electrode is small, then the positive plasma potential is small. This condition results in a larger dark space potential at the powered electrode and a high-energy ion bombardment of the powered electrode surface.[8] Some basic trends for etching process parameters are provided in Table 16.4 on page 446.

TABLE 16.4 Effects of Changing Plasma Etch Parameters

Increase (↑) or Decrease (↓) in Etch Control Parameters		Ion Energy	DC Bias	Etch Rate	Selectivity	Physical Etch
RF Frequency	↑	↓	↓	↓	↑	↓
	↓	↑	↑	↑	↓	↑
RF Power	↑	↑	↑	↑	↓	↑
	↓	↓	↓	↓	↑	↓
DC Bias	↑	↑	↑	↑	↓	↑
	↓	↓	↓	↓	↑	↓
Electrode Size	↑	↓	↓	↓	↑	↓
	↓	↑	↑	↑	↓	↑

PLASMA ETCH REACTORS

The basic components of a plasma dry etch system include: a reaction chamber where the etching takes place, an RF power supply to ignite the plasma, a gas flow control system, and a vacuum system to remove etch by-products and gases. The etch system includes sensors, gas flow control units, and an endpoint detector. A wide range of gas chemistries is used in etch (see the following section), but in general, fluorine etches SiO_2; chlorine and fluorine etch aluminum; chlorine, fluorine, and bromine etch silicon; and oxygen removes photoresist. Different parameters controlled in dry plasma etching are: vacuum operation, gas mixture, gas flow rate, temperature, RF power, and the wafer position relative to the plasma. The interaction of these different parameters is a function of the dry etch process controller.

Prior to the 1980s, most plasma etch equipment used a barrel reactor designed for batch processing (many wafers processed simultaneously). This type of reactor is no longer common in advanced IC production and is only used in noncritical applications. The current trend in semiconductor manufacturing is for single-wafer processing in integrated cluster tools. This technique achieves manufacturing efficiency by reducing the batch size for single-piece flow. Integrated cluster tools also reduce exposure to contamination between process steps and achieve better wafer uniformity because the reactor parameters are optimized for one wafer. The most important development in dry etch systems for sub-0.25 μm geometries is the high-density plasma reactor.

The different types of dry plasma reactors are:

- Barrel plasma etcher
- Parallel plate (planar) reactor
- Downstream etch systems
- Triode planar reactor
- Ion beam milling
- Reactive ion etch (RIE)
- High-density plasma etchers

Barrel Plasma Etcher

The *barrel reactor* is a cylindrical design with almost pure chemical isotropic etching at a pressure of about 0.1 to 1 torr (see Figure 16.14). Wafers are mounted vertically in a quartz boat with a small separation between wafers. The RF power is applied by placing electrodes on both sides of the cylinder. There is typically a perforated metal cylindrical etch tunnel that confines the plasma to the outer region between the etch tunnel and chamber wall. The wafers are placed parallel to the electric field to minimize physical etching. The etchant species in the plasma diffuse to the etch tunnel, while the energetic ions and electrons of the plasma do not enter this region. The etching is almost purely chemical with isotropic etching and high selectivity.[9] There is minimal plasma-induced damage because there is no physical sputtering on wafers. The barrel plasma reactor has been used primarily for stripping photoresist from wafers. Oxygen is the primary reactant for stripping photoresist.

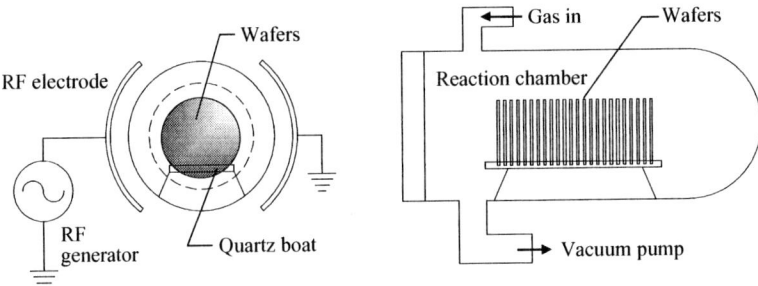

FIGURE 16.14 Typical Barrel Reactor Configuration

Parallel Plate (Planar) Reactor

The *parallel plate (planar) reactor* has two parallel plates that are symmetrical in size and position in the reactor (see Figure 16.15). A wafer can be placed backside down on the grounded electrode cathode with the RF signal applied to the upper electrode. This is the plasma etch mode, with energetic ion bombardment since the plasma potential is always above ground potential. If a wafer is placed directly on the RF-powered electrode, the wafer is in direct contact with the plasma and energetic ions. This contact can result in high-energy ion bombardment and is said to be in the reactive ion etch mode. Physical and chemical etch mechanisms occur in both the plasma etch mode and the reactive ion etch mode. However, energies of the bombarding ions are about ten times higher in the reactive ion etch mode.[10] A reactive gas is needed to form the plasma, such as fluorine (F_2) or oxygen (O_2). The reactor operates at a pressure of about 0.1 to 1 torr, with high RF power to control the etch rate. The planar etcher was one of the early reactors. Single-wafer parallel plate etchers are commonly used today.

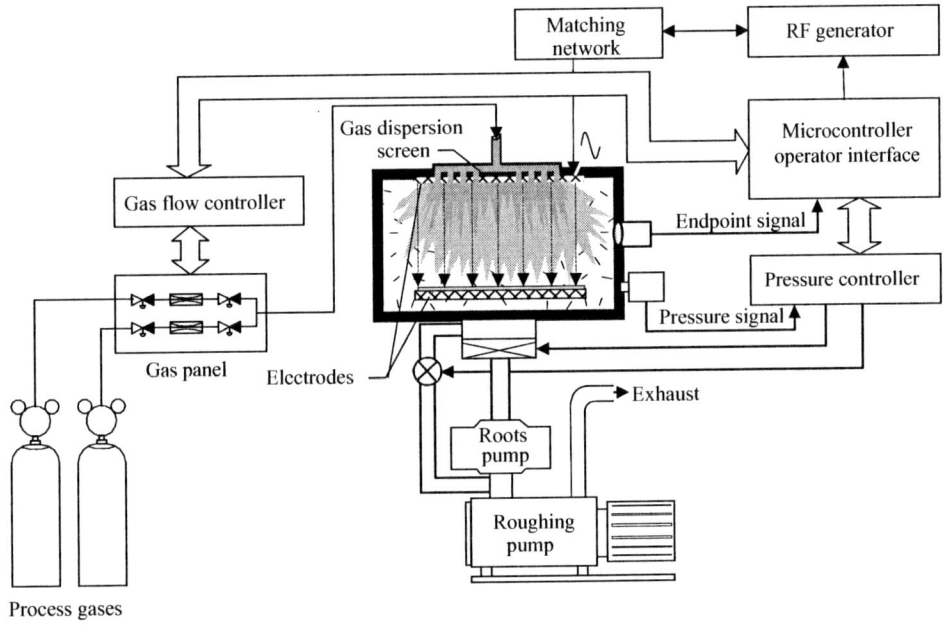

FIGURE 16.15 Parallel Plate Plasma Etching

Downstream Etch Systems

Repeated exposure to ion bombardment at the wafer surface increases the probability of device damage. A means of reducing damage to the wafer surface as well as heat buildup due to ion bombardment is to locate the wafer etch region away from the plasma in a *downstream reactor*. The plasma is formed in a separate source at a pressure of about 0.1 to 1 torr, transferred to the process chamber, and uniformly distributed over the heated wafer surface (see Figure 16.16). Since there are no ions to create directional etching, downstream reactors employ chemical etching, are isotropic, and are often used to remove resists or other noncritical layers (see the following section). A microwave source (2.45 GHz) for exciting the plasma for downstream etching is common because it produces the maximum concentration of atomic oxygen and the lowest concentration of ionic oxygen, which minimizes device damage and yields a high strip rate.

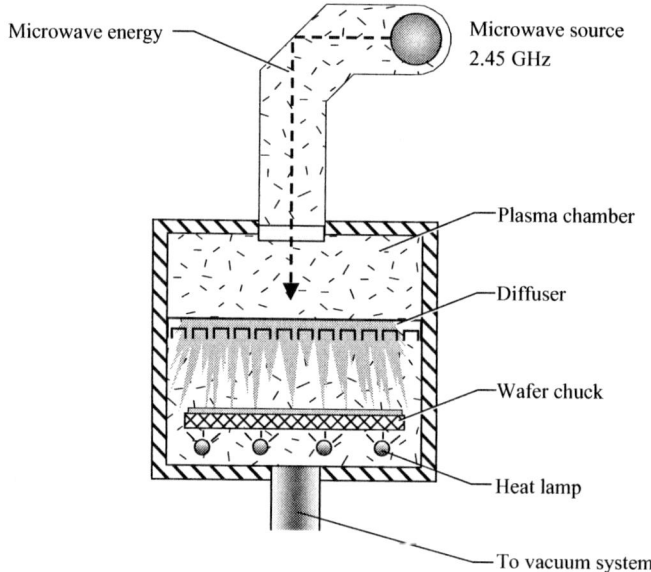

FIGURE 16.16 Schematic of a Downstream Reactor

There are variations of the downstream etching concept used with different etch systems discussed in the following section. The objective is to minimize wafer exposure to plasma ions in order to reduce or eliminate plasma-induced damage.

Triode Planar Reactor

The *triode planar reactor* adds a third electrode to attain control over the amount of ion bombardment. Figure 16.17 shows a setup with two power supplies. The inductively-coupled RF generates the plasma to create the ions and reactive species at a pressure of about 10^{-3} torr. The low-frequency generator controls the ion bombardment. A typical use is in single-crystal silicon trench etching.

Ion Beam Milling

Ion beam milling, also called *ion beam etching* (*IBE*), has a physical etch mechanism with a strongly directional plasma. It is capable of anisotropic etching of small features. The plasma is commonly generated using an RF inductively-coupled source or a microwave source. Fast-moving electrons are emitted by a hot filament. Argon atoms enter the plasma chamber through a diffuser screen. An electromagnet surrounds the plasma chamber. The magnetic field causes electrons to travel in circular paths. This cyclical motion produces a high number of collisions with argon atoms, which results in a high number of positive argon ions. Positive argon ions are drawn out of the plasma source with grid electrodes and formed into a high-density beam using a set of collimated electrodes (see Figure 16.18). A high-voltage accelerator grid boosts ion energy as high as 2.5 keV.

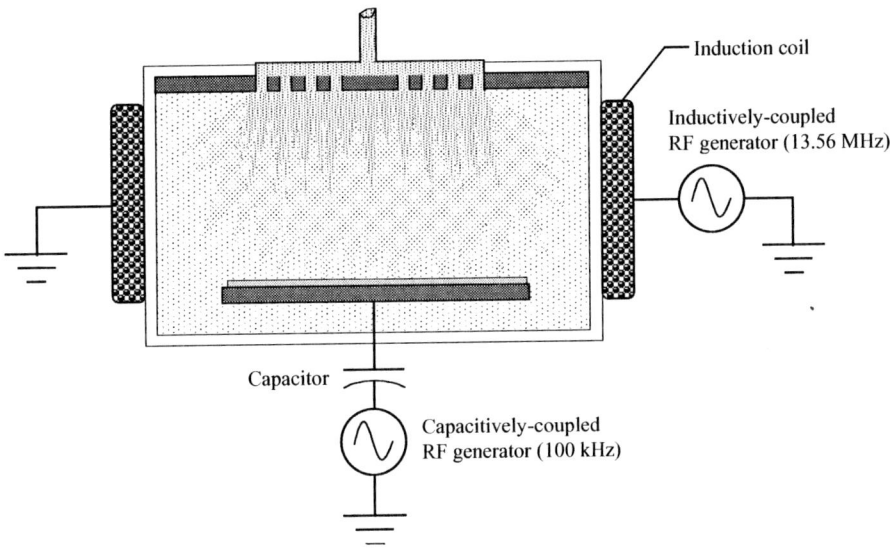

FIGURE 16.17 Triode Planar Reactor

A neutralizing filament emits electrons to recombine with argon atoms to prevent charging of the wafer with the positive ions. Ion beam etchers operate with argon (Ar) gas in the low-pressure range of 10^{-4} torr, which is lower than that commonly used for high-density plasma etching. Ion beam etching is used to etch difficult materials, such as gold, platinum, and copper.[11] The wafer can be tilted to produce variable sidewall geometries. The major problems inhibiting widespread use of ion beam etchers in semiconductor processing are low selectivity (usually below 3:1) and low etch rates, which lead to low wafer throughput.

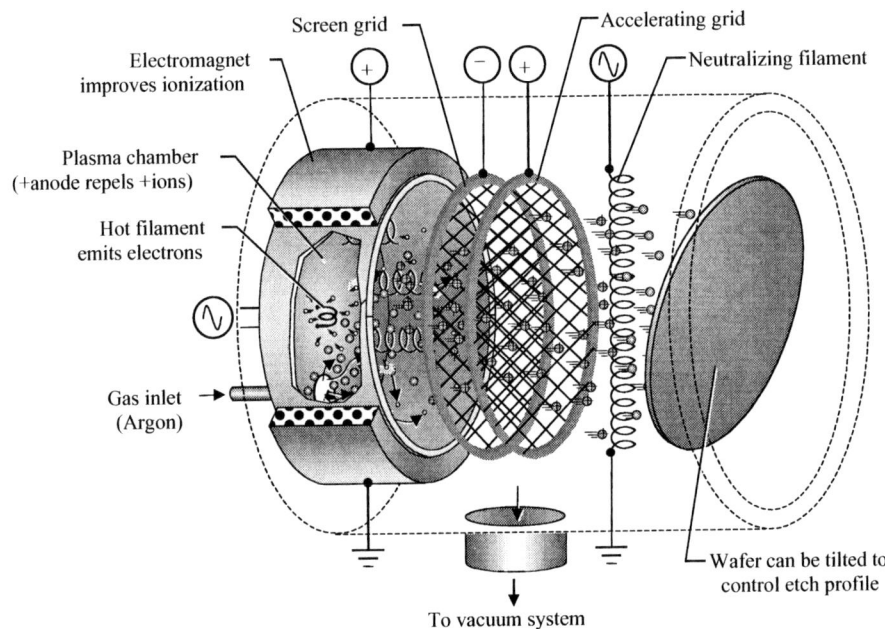

FIGURE 16.18 General Schematic of Ion Beam Etcher
Redrawn from *Advanced Semiconductor Fabrication Handbook,* Integrated Circuit Engineering Corp., pp. 8–12.

Reactive Ion Etch (RIE)

Reactive ion etch (RIE) is a technique for removing material from the wafer surface with both a reactive chemical process and a physical process using ion bombardment. RIE is similar to the standard parallel-plate plasma etcher, except the wafers are placed on the RF-powered electrode (cathode) and the powered electrode size is greatly reduced relative to the grounded electrode size (see Figure 16.19). In this manner a DC self-bias develops on the cathode and the wafers acquire a large voltage difference with respect to the plasma. This condition creates directionality for the ionized species moving toward the wafer, creating improved anisotropic sidewalls for features. There is no sputtering on the anode. The pressure is relatively low at <0.1 torr.

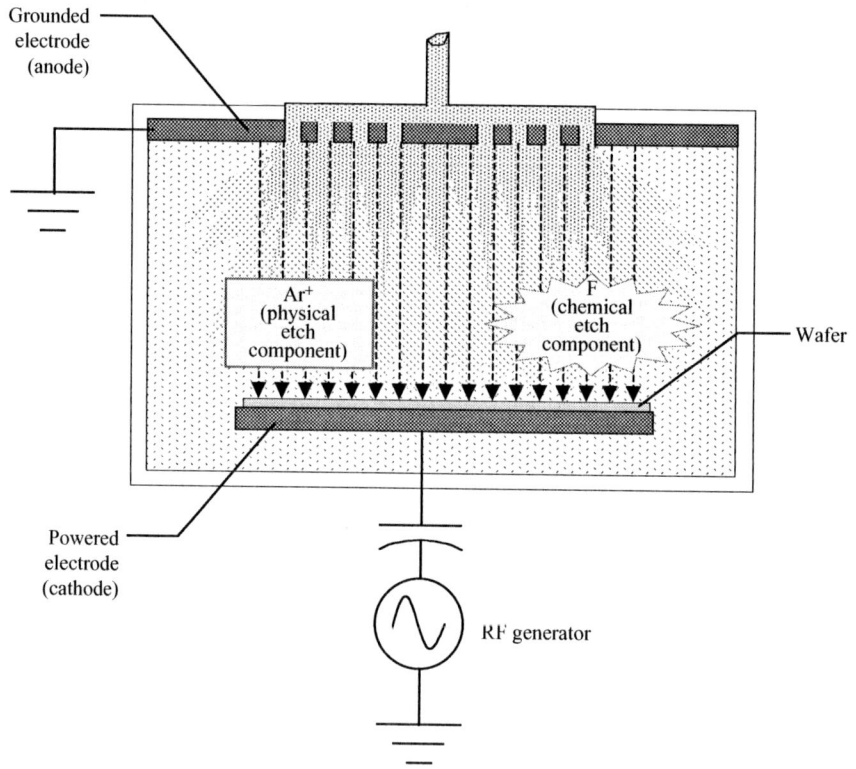

FIGURE 16.19 Parallel-Plate RIE Reactor

High-Density Plasma Etchers

The most predominant dry etching methodology in use for critical layers of advanced ICs is *high-density plasma etch* with single-wafer processing in a cluster tool.[12] Standard plasma systems previously used in wafer fabrication operated in a vacuum regime of a few hundred mtorr, where it is relatively straightforward to create a plasma (such as by applying an RF signal across two parallel plates). However, for 0.25-μm geometries and below, it has become too difficult to get etchant ions into and etch by-products out of high-aspect ratio features. The etching action slows down and will actually stop altogether at the bottom and lower sides of the feature profile.

The solution is to lower the system pressure to about 1 to 10 mtorr to increase the mean free path lengths of gas molecules and ions. This condition effectively reduces collisions that cause loss of profile control. However, a drawback to reduced pressure is the lower etch rate due to the rapid reduction in ion density as pressure decreases. To overcome this concern, a high-density plasma is needed to generate enough ions for an acceptable etch rate with a lower pressure. High-density refers to the number of active species in the plasma relative to conventional plasma at the same process pressure. In conventional plasma, the degree of ionization is typically on the order of 0.01% to 0.1%. High-density plasma more efficiently couples input power with the plasma, resulting in greater dissociation of etch species to achieve a degree of ionization as high as 10%. This technique yields highly directional, low-energy ions that produce anisotropic etching in high aspect ratio openings.

High-Density Plasma Etcher
(Photo courtesy of Applied Materials, Inc.)

High-density plasma etchers generally immerse the plasma in a magnetic field. The reasons for using a magnetic field during plasma etch are: (1) plasma generation is more efficient for creating highly-directional, low-energy ions entering high-aspect ratio openings, yet causes less wafer damage; (2) plasma is denser, with more reactive species and charged particles that increases the etch rate; and (3) the DC bias on the wafer can be reduced, which leads to less wafer bombardment (or damage).

Electron Cyclotron Resonance (ECR) ■ The *electron cyclotron resonance* (*ECR*) reactor was one of the earliest high-density plasma reactors to be developed commercially, first introduced in the early 1980s. It is still used in today's modern wafer fabs for etching 0.25-μm feature sizes and below. The ECR reactor produces very dense plasmas at operating pressures of about 1 to 10 mtorr. ECR etching uses microwave excitation (2.45 GHz) in the presence of a magnetic field to generate high-density plasma (see Figure 16.20 on page 452). A key feature of the ECR reactor is a magnetic field parallel to the direction of reactant flow that causes free electrons to move in a spiral path due to the magnetic force. An efficient transfer of energy occurs from the electric field to the plasma electrons when the electron orbit frequency (known as electron cyclotron resonance) equals the frequency of the applied electric microwave field. This resonance condition increases the probability of electron collisions, which generates a very dense plasma to create a large flux of ions. The reactive ions move toward the wafer surface and react with the surface layer to cause etching.

A low-power RF bias (13.56 MHz) or DC bias can be applied to the electrode holding the wafer to control the energy of the ions striking the surface. This action permits the ECR to operate as a combined chemical and physical etch process and produces an anisotropic etch profile. The main drawback to the ECR reactor for semiconductor manufacturing is its equipment complexity.[13] A variation of the ECR is a system with pairs of magnets and microwave antennas distributed around a central reactor, known as distributed ECR (DECR).

Inductively-Coupled Plasma (ICP) ■ Another high-density, low-pressure etch reactor that has plasma decoupled from the wafer is the *inductively-coupled plasma* (*ICP*) reactor. This reactor is less complicated and less expensive than the ECR and widely used in the United States. The ICP generates plasma by means of a spiral coil separated from the plasma by a dielectric plate or quartz tube (see Figure 16.21 on page 452).[14] The wafer is located away from the coil so it is not affected by its electromagnetic field. The wafer can be biased to have both chemical and physical etching. This reactor can achieve anisotropic sidewall profiles in high-aspect ratio openings.

Another inductively-coupled reactor that generates a high-density plasma is the *helicon wave*. This system receives power from an RF signal (13.56 MHz) that is inductively coupled into the plasma from a double-loop antenna located outside a quartz source tube.

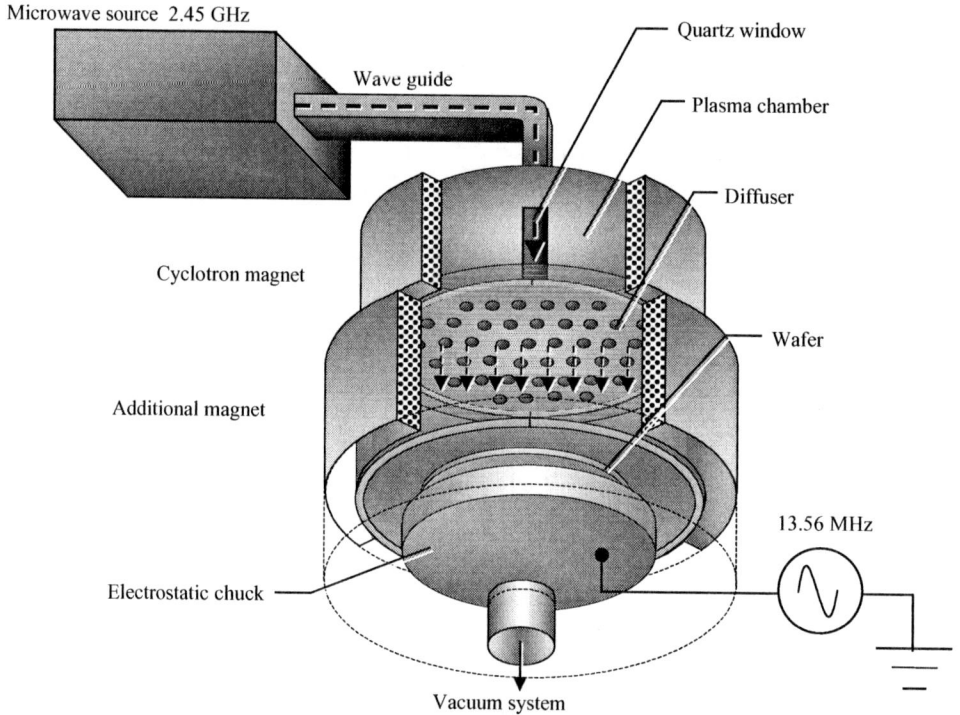

FIGURE 16.20 Schematic of Electron Cyclotron Reactor
Redrawn from Y. Lii, "Etching," *ULSI Technology,* ed. by C. Chang and S. Sze (New York: McGraw-Hill, 1996), p. 349.

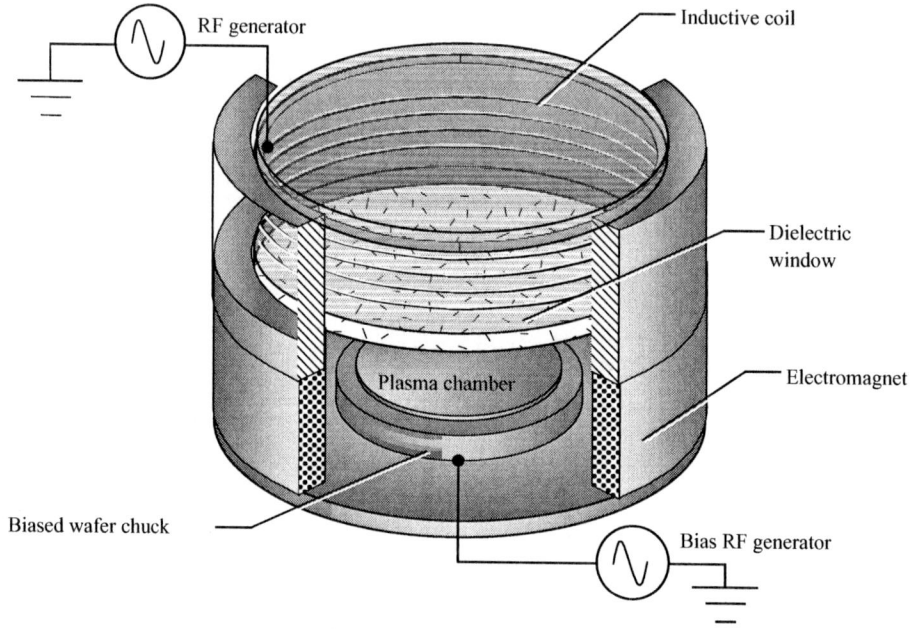

FIGURE 16.21 Inductively-Coupled Plasma Etch
Redrawn from Y. Lii, "Etching," *ULSI Technology,* ed. by C. Chang and S. Sze (New York: McGraw-Hill, 1996), p. 351.

Dual Plasma Source (DPS) ■ Figure 16.22 is a schematic of a *dual plasma source (DPS)* etch chamber (also referred to as decoupled plasma source). It has four major components: a source power unit, an upper chamber, a lower chamber with the wafer, and a movable electrode.[15] As with the previous high-density plasma systems, there are two sources of RF power used. The source power unit has an inductive coil for transferring RF power to the plasma in order to generate the reactive ions and neutral species. This is referred to as the source power. The source power unit has a temperature control system. RF power is also supplied to the wafer electrode to bias the wafer substrate and is referred to as the bias power. The wafer cathode is movable in the vertical direction. Only the upper chamber is exposed to the plasma and process gases, which keeps the lower chamber clean and makes it much easier to maintain.

A key aspect of the DPS plasma is the decoupling of the source plasma power from the bias power. This arrangement permits greater control over the ion density and ion energy, which results in a larger process window for physical and chemical etching. The result is improved critical dimension control and less etch residue.

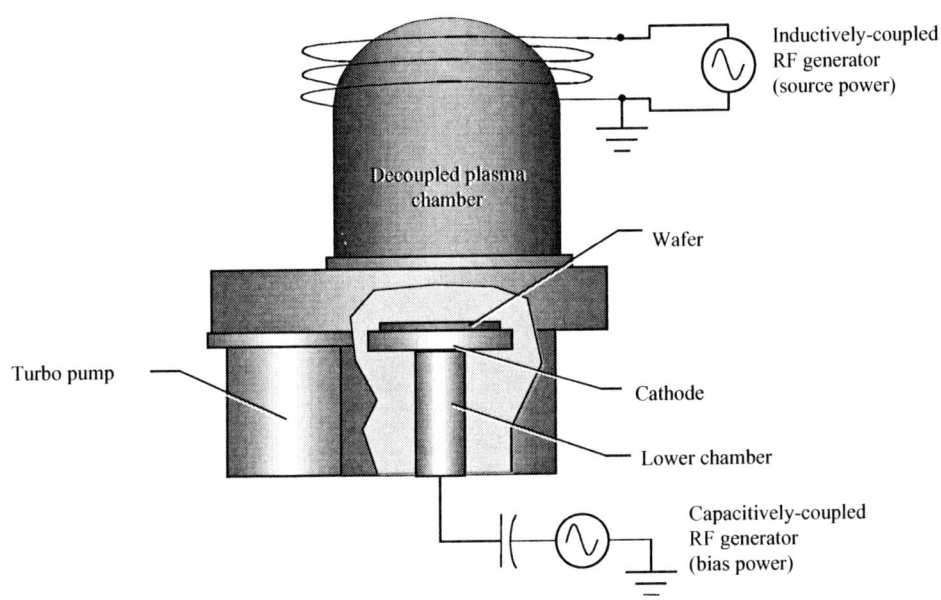

FIGURE 16.22 Dual Plasma Source (DPS)
Redrawn from Y. Ye et al., *Proceedings of Plasma Processing XI*, Vol. 96-12, ed. G. Mathod and M. Meyyoppan (Pennington, NJ: The Electrochemical Society, 1996): p. 222.

Magnetically Enhanced RIE (MERIE) ■ The *magnetically enhanced RIE (MERIE)* reactor (also referred to as a *magnetron*) is a combined physical and chemical etch system. It is similar to the RIE reactor except that now there is a magnetic field that holds the plasma away from the chamber walls and increases the electron and ion concentration near the wafer to create a high-density plasma (see Figure 16.23 on page 454). The magnetic field can be rotated electrically by its three-phase AC power supply or physically rotated in a dipole ring magnet system. Confining the plasma with the magnetic field creates a high-density plasma and permits lower pressure, effectively maintaining etch directionality and uniformity, especially when etching high-aspect ratio features.

Etch System Review

Shrinking feature sizes and the introduction of new wafer materials place stringent demands on etch performance. Dry etch systems use physical or chemical or a combination of these two mechanisms to etch material. Some equipment is anisotropic while other equipment is purely isotropic. One can appreciate the wide variety of etch equipment that is available for wafer manufacturing. The capability and control of a particular etch system is critical for successful wafer fabrication. Table 16.5 on page 454 summarizes the important characteristics for each equipment system.

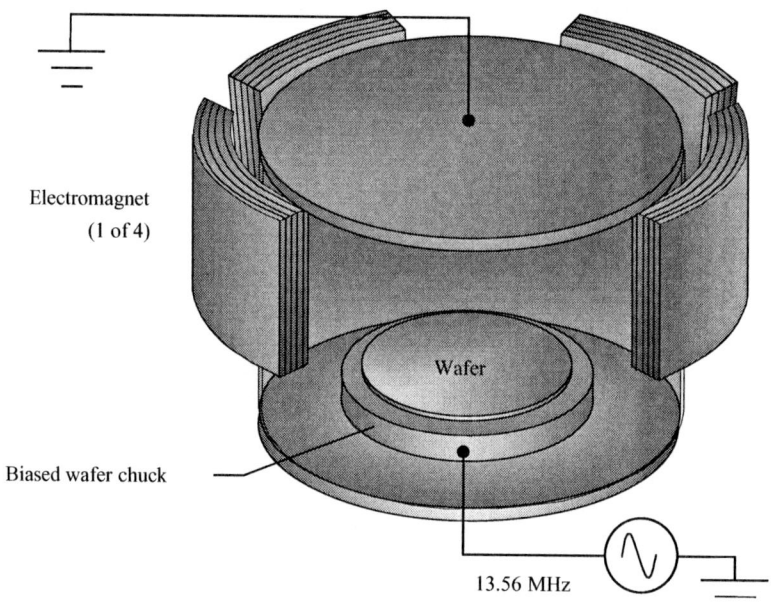

FIGURE 16.23 Magnetically Enhanced Reactive Ion Etch (MERIE)
Redrawn from *Wet/Dry Etch* (College Station, TX: Texas Engineering Extension Service, 1996), p. 165.

TABLE 16.5 Dry Etcher Configurations

Configuration	Etch Mechanism	Pressure (Torr)	Arrangement	High-Density Plasma	Biasing	Bias Source	Profile
Barrel	Chemical	10^{-1} to 1	Coil or electrodes outside vessel	No	In cassette (bulk)	RF	Isotropic
Parallel Plate (Planar)	Physical and Chemical	10^{-1} to 1	Planar diode (two electrodes)	No	On powered electrode (anode or grounded electrode)	RF	Anisotropic and isotropic
Downstream Plasma	Chemical	10^{-1} to 1	Coil or electrodes outside vessel	No	In cassette (bulk) downstream of plasma	RF or Microwave	Isotropic
Triode Planar	Physical	10^{-3}	Triode (three electrodes)	No	On powered electrode		Anisotropic
Ion Beam Milling	Physical and Chemical	10^{-4}	Planar triode	No	On powered electrode (anode)		Anisotropic
Reactive Ion Etch (RIE)	Physical	<0.1	Planar or cylindrical diode	No	On cathode		Anisotropic
Electron Cyclotron Resonance (ECR)	Physical	10^{-4} to 10^{-3} (low)	Magnetic field in parallel with plasma flow	Yes	On cathode	RF or DC	Anisotropic
Distributed ECR	Physical	(low)	Magnets distributed around central plasma	Yes	On cathode	RF or DC	Anisotropic

Configuration	Etch Mechanism	Pressure (Torr)	Arrangement	High-Density Plasma	Biasing	Bias Source	Profile
Inductively-Coupled Plasma (ICP)	Physical	(low)	Spiral coil separated from plasma by dielectric plate	Yes	On cathode	RF or DC	Anisotropic
Dual Plasma Source	Physical	(low)	Independent plasma and wafer biasing	Yes	On cathode	RF or DC	Anisotropic
Magnetically Enhanced RIE (MERIE)	Physical	(low)	Planar diode with magnetic field confining plasma	Yes	On cathode	RF or DC	Anisotropic

Endpoint Detection

Dry etch differs from wet etch in that it usually does not have good etch selectivity to the underlying layer. For this reason, a form of *endpoint detection* is required to monitor the etch process and stop etching to minimize overetching of the underlying layer. Endpoint detection systems measure different parameters, such as a change in the etch rate, the types of etch products removed from the etch process, or a change in the active reactants in the gas discharge (see Figure 16.24). A method used for endpoint detection is optical emission spectroscopy. This measurement tool is integrated into the etch chamber for real-time monitoring of the etch process.

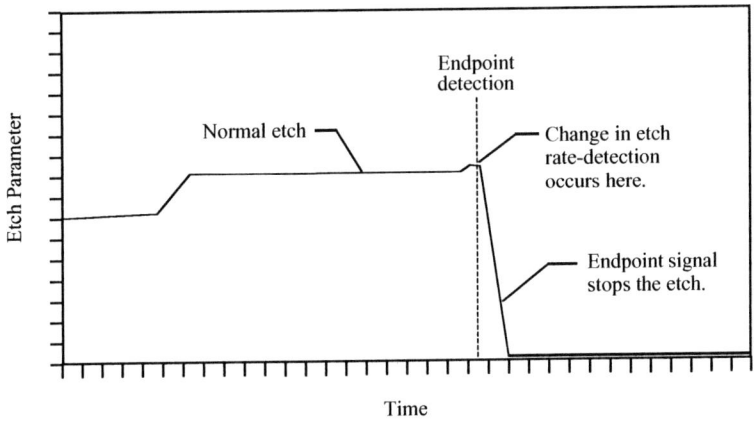

FIGURE 16.24 Endpoint Detection for Plasma Etching

Optical Emission Spectroscopy ■ An excited species emits light at a wavelength that corresponds to a specific material (see Table 16.6 on page 456).[16] The light emitted by excited atoms and molecules in a gas discharge can be analyzed with optical emission spectroscopy to identify the element. The emitted light is passed through a detector with a filter to let light of a specific wavelength pass through that is used to identify the etched materials. The emission intensity is directly related to the relative concentration of a species in a plasma. In this manner the endpoint detector can determine when the etch process has removed the desired material and proceeded into the underlying layer. Optical emission spectroscopy is the most common method for etch endpoint detection because it is easy to implement with high sensitivity.[17]

Optical emission can also be used to perform etch reactor diagnostics. If a system has leaks to the atmosphere, the presence of nitrogen is identified. Water moisture can be detected, which could be caused by inadequate vacuum pumpdown after chamber cleans. Absolute concentrations of species can not be obtained by optical emission spectroscopy.

TABLE 16.6 Characteristic Wavelengths of Excited Species in Plasma Etch

Material	Emitting Species of Etchant Gas	Some Products	Wavelength (nm)
Silicon	CF_4/O_2	SiF	440; 777
	Cl_2	SiCl	287
SiO_2	CHF_3	CO	484
Aluminum	Cl_2	Al	391; 394; 396
	BCl_3	AlCl	261
Photoresist	O_2	CO	484
		OH	309
		H	656
Nitrogen	N_2 (used as a purge gas prior to and after etch)	N_2	337
		NO	248

Vacuum for Etch Chambers

The vacuum system in an etch process is critical because it affects plasma parameters associated with gas flow and pressure. A typical high-density plasma vacuum system for etch will achieve a chamber pressure in the 1 mtorr range or less with gas flows as high as 800 sccm at the wafer surface. The vacuum components must be capable of handling the corrosive by-products produced during etching. The pumps used for an etch process are typically a turbopump, a roots-type blower and a dry backing pump. Typically the turbopump is located close to the chamber to maximize the pump speed at the wafer surface, while the other pumps are located in a service bay or sub fab area.[18] Additional dry pumps are used to evacuate the loadlock and transfer chamber of an integrated cluster tool.

The fluorine, chlorine, and bromine gas chemistries used in etch (see the following section) are effective at removing the wafer surface films because of their reactivity. Unused reactants as well as reaction by-products are removed at a high rate. The high reactivity of etch by-products creates corrosion problems for vacuum system components, requiring special designs such as nitrogen purge of bearings and magnetically levitated (maglev) pumps that only rely on bearings during start-up and shutdown.[19]

DRY ETCH APPLICATIONS

There are many different types of plasma dry etching applications needed to fabricate integrated circuits. These applications are found in all materials used in the fab. For our purposes, we will review dry etch by the type of material to be etched: dielectric, silicon, and metal. The move to smaller critical dimensions, higher aspect ratio openings, and new materials in wafer fabrication creates challenges in etch processing for all three material types. Optimizing etch conditions leads to competing objectives for the production team. In general, the requirements for successful dry etch are:

1. High selectivity to avoid etching materials that are not to be etched (primarily photoresist and underlying materials).
2. Fast etch rate to achieve an acceptable throughput of wafers.
3. Good sidewall profile control.
4. Good etch uniformity across the wafer.
5. Low device damage.
6. Wide process latitude for manufacturing.

Critical etch parameters are determined through optimization activities for each specific dry etch application. Some of these parameters are shown in Figure 16.25. Note that in many cases optimization is being done through process equipment modeling using computer software. This is due to the high cost of conducting prototype hardware testing with actual wafers.

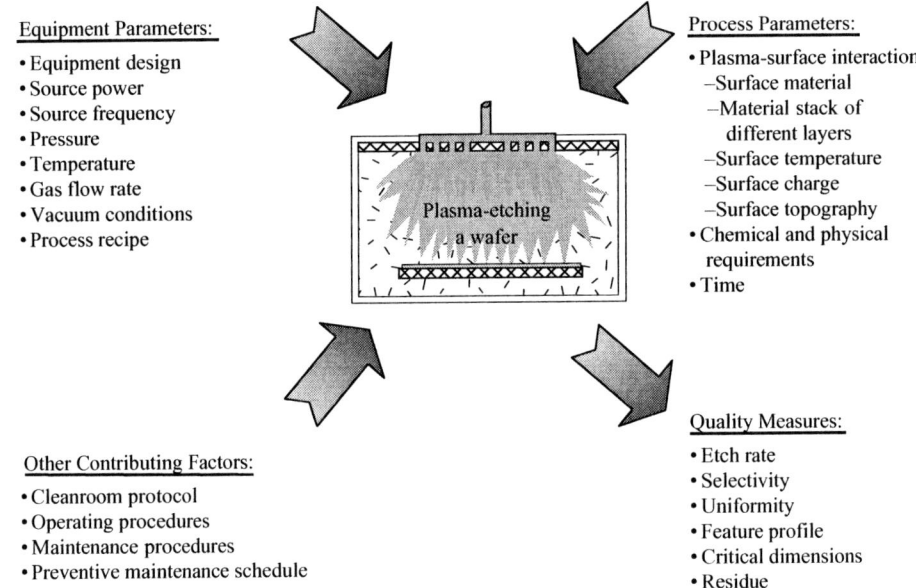

FIGURE 16.25 Dry Etch Critical Parameters

Table F.1 in Appendix F lists the typical gas chemistries used for plasma etching. Dielectric and silicon materials are usually etched with fluorine-based gas etchants such as CF_4. Aluminum etching is typically done with chlorine-based gas chemistries such as BCl_3.

Dielectric Dry Etch

Dielectric dry etch is the most complicated etch process for 200 mm wafers and will remain the biggest etch challenge for 300 mm wafers.[20] The challenges for oxide etching are tighter process specifications because of shrinking feature sizes, new trench etch processes for interlayer dielectrics (ILD) used in the dual-damascene approach for copper metallization, and the need to etch new ILD low-k materials.

Oxide ■ Common reasons for etching oxide are for forming contact holes and vias. These features are critical applications, with requirements such as etching high-aspect ratio openings in the oxide. Aspect ratios for DRAM applications are expected to be at 6:1 for 0.18 μm feature sizes. The required selectivity to underlying silicon and polycide is about 50:1.[21] There are new oxide etch applications such as the dual-damascene structure, with new trench requirements and high-aspect ratio etch. There are also noncritical oxide etch applications such as a low-aspect ratio via etch.

The oxide plasma etching process is commonly based on fluorocarbon chemistry. Fluorocarbon is a fluorinated hydrocarbon. It has one or more of the hydrogen atoms in a hydrocarbon molecule replaced by fluorine (F). In this manner, many gases are sources of fluorine, such as CF_4, C_3F_8, C_4F_8, CHF_3, NF_3, and SiF_4. Common gases are tetrafluoromethane (CF_4), which has a high etch rate but poor selectivity with polysilicon, and trifluoromethane (CHF_3), which has a high rate of polymer formation. Fluorocarbon gases are chemically stable in their ground state (nonplasma state) and will not attack silicon or its oxide because their bonds are stronger than SiF bonds. There are also buffer gases such as Ar and He added to the etch chemistry. Argon has a relatively high mass that is used for physical etching (sputtering). Helium has a low mass and is used to diminish the etchant gas concentration (referred to as a diluent) to enhance plasma uniformity.

The desired fluorocarbon feed gas, which may have a primary gas component such as CF_4, is fed into the plasma and dissociates (see Figure 16.26 on page 458). This reaction forms many different types of reactive ions (radicals) and neutrals.[22] Examples of reactive species are CF_4, CF_2, CHF, HF, and F. It is the radical F species that attacks the oxide and causes etching. The F reactive ions easily form volatile by-products, which are then pumped out of the etch reactor by the vacuum system. The CF_x radicals also serve to passivate the sidewall surfaces with a polymer. A higher carbon-to-fluorine ratio generally means more polymer formation, lower etch rate, and higher oxide-to-silicon selectivity. In some cases, a silicon source is used above the plasma to getter excess fluorine atoms, thereby increasing the ratio of carbon to fluorine.

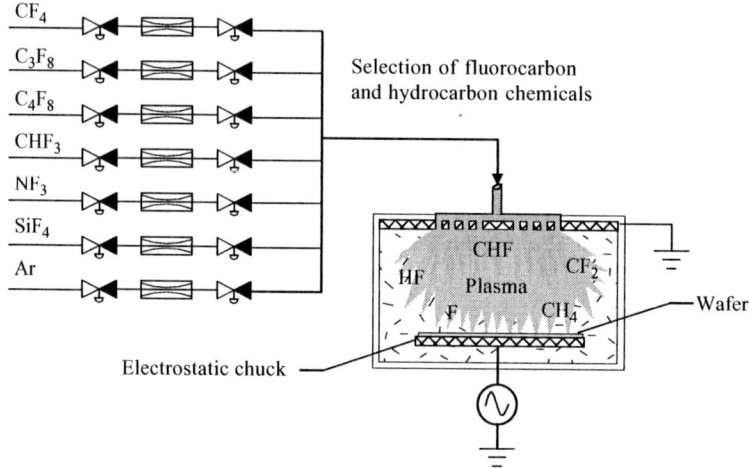

FIGURE 16.26 Oxide Etch Reactor

Underlying Material Selectivity. One of the major challenges for oxide etching is obtaining high selectivity to the underlying material, usually silicon, nitride, or an antireflective coating. For example, high selectivity to silicon is critical during contact etch through the LI oxide dielectric to avoid etching into the source/drain region. High selectivity to silicon is required when etching oxide sidewall spacers on the gate structure (see Chapter 9). For via etch, high selectivity to TiN, W, or Al is a requirement.

One method of achieving selectivity to silicon is by adding oxygen to the gas chemistry to control the selectivity between oxide and silicon. Small concentrations of O_2 improve the etch rate of both oxide and silicon. At greater concentrations of up to about 20% O_2, there is more rapid etching of oxide over silicon, improving the selectivity to minimize etching of the underlying silicon. Another way to improve selectivity is to add hydrogen to the gas mixture as the etch rate of silicon decreases until it reaches almost zero for a H_2 volume of about 40%. At the same time, the oxide etch rate is not affected by H_2 at concentrations below 40%.[23]

Selectivity during oxide etch can be maintained to the underlying material by use of a hard "etch stop" masking layer of silicon nitride, as with etching the contact hole (see Figure 16.27). This action requires good oxide etch selectivity to the nitride layer. A hard mask approach to selectivity adds processing steps and is done only when necessary.

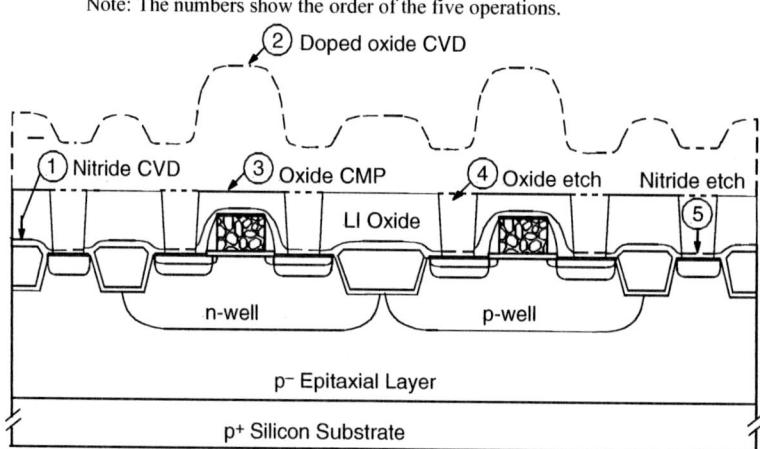

FIGURE 16.27 Etch Stop Hard Mask Layer

Oxide/silicon selectivity is also achieved by forming a passivating polymer layer on top of the silicon as part of the etch process. The polymer residue is introduced into the gas chemistry to inhibit lateral silicon etching. The polymer tends to form more easily on silicon rather than SiO_2, which is thought to occur because the carbon combines with the oxygen from the SiO_2 to form volatile CO or CO_2, which is then pumped away.[24] This approach to selectivity permits continued etching of the oxide while inhibiting the silicon etch. Unfortunately the polymer also deposits on the chamber surfaces during etch and becomes a source of particulate contamination, requiring frequent cleaning of the etch system.

Photoresist Selectivity. Achieving high selectivity to photoresist during oxide etch is important for avoiding tapered sidewalls. The photoresist defines the pattern that is being etched. In the case of contacts and vias, millions of holes are being etched simultaneously, each requiring the removal of a precise amount of surface material, often at different depths (see Figure 16.28). One factor that reduces photoresist selectivity (an undesirable situation) is the effective formation of aggressive fluorine atoms in a high-density plasma. Free fluorine will etch the organic photoresist. This lowers selectivity from about 10:1 from conventional plasma etchers to a range of 4:1 to 7:1 in the high-density plasma tools.[25] In addition, the need to etch through antireflective coating layers, which lengthens the etch time, further reduces the resist thickness. In general, deep UV resists are less resistant than i-line resists to plasma processing, which also lowers resist selectivity.

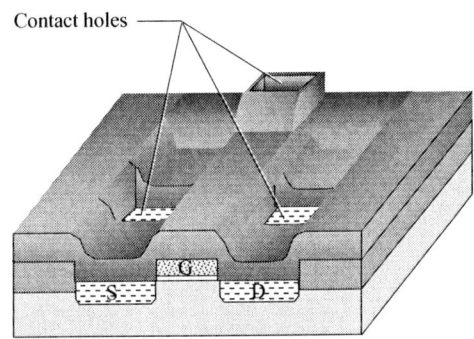

FIGURE 16.28 Contact Etching to Varying Depths

Sidewall Profile. Contact windows in the local interconnect (LI) oxide dielectric layer generally have dimensions equivalent to the smallest feature size with a high-aspect ratio. For this type of application, a high degree of anisotropy with a vertical sidewall profile is needed. The most significant factor is the highly directional ion bombardment of high-density plasma. A factor in obtaining optimum sidewall profile is the amount of resist selectivity. Low resist selectivity can lead to tapered sidewalls in an anisotropic etch process because of gradual photoresist erosion during the etch process.

Silicon Nitride ■ There are two basic types of silicon nitride used in wafer fabrication. One type is deposited with LPCVD at 700° to 800°C, which produces a stoichiometric compound of Si_3N_4 film. The other nitride film is deposited by plasma-enhanced CVD at <350°C that produces a film having lower density.[26] The etch rate is faster for the PECVD film because of its lower density.

Silicon nitride is etched by different gas chemistries. A common primary gas is carbon tetrafluoride (CF_4) that is mixed with O_2 and N_2. Increasing the amount of O_2/N_2 dilutes the concentration of fluorine species and lowers the etch rate of any underlying oxide. A high etch rate of 1200 Å/min and a high selectivity to oxide would be about 20:1 for a nitride film deposited with LPCVD.[27] Selectivity to oxide is important in an application with a thin pad oxide used as an etch-stop, requiring the etch process to quickly slow down when oxide is reached. Other possible primary gases used for nitride etch are SiF_4, NF_3, CHF_3, and C_2F_6.

Silicon Dry Etch

Plasma dry etch of silicon is a critical process in wafer fabrication. The two major layers for plasma etching are (1) polysilicon gate length formation for the MOS gate structure, and (2) single-crystal silicon trench creation used for either device isolation or capacitor structures in dynamic random-access memories (DRAM).

Polysilicon etch gas chemistries have traditionally been fluorine-based gases, including CF_4, CF_4/O_2, SF_6, C_2F_6/O_2, and NF_3. The fluorine atoms produce a very fast etch reaction but with an isotropic profile and average resist selectivity. The sidewall profile is improved by reducing the fluorine atoms and increasing the ion energy, but this also reduces the poly/oxide selectivity. The energy of the bombarding ions must be low enough to avoid sputtering away an underlying oxide layer.

To overcome these difficulties, polysilicon plasma etch gases are often based on chlorine, bromine, or chlorine/bromine chemistries. Chlorine (Cl_2) produces an anisotropic silicon sidewall profile with good oxide selectivity (>10 selectivity for poly/oxide and poly/nitride). Bromine gas chemistry, as either Br_2 or HBr, etches silicon anisotropically with very high selectivity (>100) to oxide and nitride. This quality is important for applications such as poly gate etching over the gate oxide.[28] Bromine gas chemistry also has better selectivity to photoresist than Cl_2. Another silicon etch chemistry is a blend of chlorine and bromine, such as HBr and Cl_2 with the addition of O_2. Adding O_2 increases the etch rate and the selectivity to oxide. Polymers form during etching with chlorine, and fluorine gas chemistries to deposit on sidewalls and contribute to sidewall profile control. SiF_4, $SiCl_4$, and $SiBr_4$ are some volatile etch by-products for fluorine, chlorine, and bromine gas chemistries.

Poly Gate Etch ■ In MOS devices, doped LPCVD polysilicon is used as the gate conductor material. The linewidth of the doped polysilicon conductor defines the gate length in the active device and therefore affects transistor performance (see Figure 16.29). For this reason, CD control is critical. The poly gate etch process must have high selectivity to the underlying gate oxide with excellent uniformity and repeatability. A high degree of anisotropy is also required since the polysilicon line also serves as an implant barrier during the source/drain doping step. Sloped sidewalls would yield partial doping of the implant species under the polysilicon gate structure.

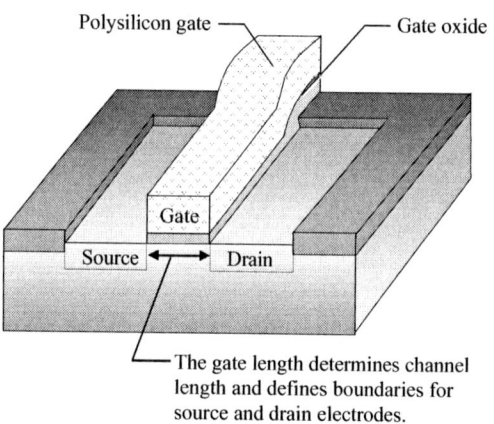

FIGURE 16.29 Polysilicon Conductor Length

Etching polysilicon (and silicon in general) is often a three-step process. This permits optimization for either anisotropy or selectivity during the different etch steps. These three steps are:

1. The first step is a *breakthrough step* that removes the native oxide, hard mask (e.g., SiON) and surface contaminants to achieve a uniform etch (this minimizes etching surface defects from contaminants that act as a micro-mask).

2. This is followed by a *main-etch step* to endpoint. This step removes most of the polysilicon film without damaging the gate oxide and achieves the desired anisotropic sidewall profile.

3. The final step is an *overetch step* that removes the remaining residues and stringers of polysilicon while maintaining high selectivity to the gate oxide. This etch process should be done while avoiding any microtrenching (formation of small trenches) of the gate oxide around the periphery of the polysilicon (see Figure 16.30).

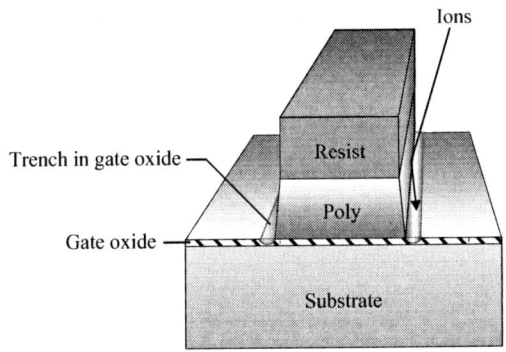

FIGURE 16.30 Undesirable Microtrenching During Polysilicon Gate Etching

The polysilicon gates are difficult structures to etch—requiring care and precision in the etch process. Devices with a 0.15 μm-feature size will have a projected gate oxide thickness of 20 to 30 Å (equivalent thickness of 6 to 10 monolayers of oxide atoms). The process specifications for selectivity of polysilicon etch to oxide typically require no more than a 5 Å loss of oxide thickness (about 1.5 monolayers) across a 70,686 mm² surface area on a 300 mm wafer. This will require development of a selectivity >150:1 for the main polysilicon etch process to prevent punchthrough of the gate oxide and an overetch selectivity of >250:1 to remove etch residue and stringers.[29]

Polysilicon etch gas chemistries have traditionally been fluorine-based gases, including CF_4, CF_4/O_2, SF_6, C_2F_6/O_2, and NF_3. The fluorine atoms are instrumental in the etching of silicon, however, this chemical-type of etch results in etch with an isotropic profile. This profile is improved by reducing the fluorine atoms and increasing the ion energy, but this also reduces the polysilicon-to-SiO_2 selectivity. The energy of the bombarding ions must be low enough to avoid sputtering away the gate oxide. To overcome these difficulties, polysilicon plasma etch gases often contain both chlorine and fluorine. Chlorine produces an anisotropic polysilicon sidewall profile with good selectivity to SiO_2. Another common gas chemistry for etching polysilicon is bromine because it gives higher etch selectivity of polysilicon to gate oxide than chlorine. Volatile etch by-products for fluorine, chlorine, and bromine gas chemistries are SiF_4, $SiCl_4$, and $SiBr_4$, all of which are evacuated from the chamber after etching.

Single-Crystal Silicon Etch ■ Single-crystal silicon is etched to form trenches for applications such as device isolation or for the fabrication of vertical capacitors in high-density DRAM ICs. Silicon trench fabrication permits a decrease in the surface area needed for these technologies, which is important as feature sizes continue to decrease on advanced ICs. For STI device isolation in both bipolar and MOS technology, the shallow silicon trenches (depths less than 1 μm to several μms in depth) are filled with oxide dielectric. Capacitors are fabricated by etching a deep trench (a depth greater than 5 μm), oxidizing the sidewalls, and then filling the trench with polysilicon. The vertical capacitor trench design is advantageous because it takes up less surface area and provides higher capacitance values.

Silicon Trench Etching. Silicon trench etching requires precise dimensional control for each trench on the IC. This is a challenge, considering there are millions of capacitor trenches on a high-density IC, with each trench required to have identical smooth, nearly vertical sidewalls, the correct depth, and rounded corners at the bottom and top of the trench. A multistep etch process (as outlined above) is needed, with a modification for the final step where the bottom of the trench is rounded to remove any silicon damage (see Figure 16.31 on page 462).

Sidewall passivation formed by adding carbon to the gas is often used to protect the trench sidewalls from lateral etch attack. The sidewall shape is a function of the wafer temperature—increasing the temperature results in less deposition of sidewall passivation and more lateral etching. Wafer temperature can be controlled by backside cooling of the vacuum chuck with helium.[30]

For plasma dry etching of shallow trenches, fluorine gas chemistry is sometimes used because it has a high silicon etch rate with adequate selectivity to the photoresist that is used as the mask material. For deep trenches (e.g., several microns depth), a chlorine-based or bromine-based gas chemistry is often used because this gas chemistry has a high silicon etch rate and high selectivity

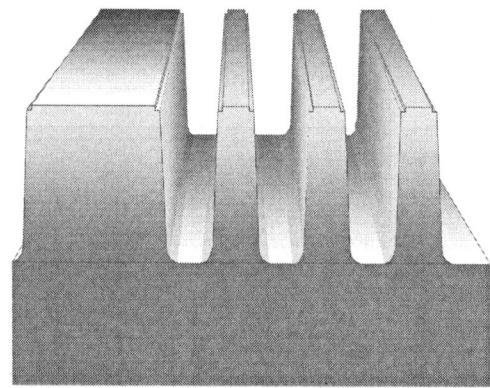

FIGURE 16.31 Silicon Trench Etching

to the oxide mask. Bromine gas chemistry is becoming common in high-density plasma reactors because it is an aggressive etchant that does not require extensive use of carbons for sidewall passivation (which reduces contamination problems). The primary problem with bromine gas chemistry is that it is extremely corrosive to the gas delivery system and reactor. The gas suppliers and equipment manufacturers are addressing this issue.

Metal Dry Etch

A major application in metal etch is aluminum alloy for interconnect wiring. Recall from Chapter 12 that copper alloyed with aluminum reduces electromigration and hillock formation in aluminum, while silicon is added to aluminum to minimize spiking at contact interfaces. Silicon poses no additional etch resistance. Tungsten, barrier metals, and the contact metal are also common metals etched during wafer fabrication. However, copper metal is not dry etched because it does not easily form volatile compounds, which makes plasma etching difficult. Due to the density of interconnect lines on the critical layers of ULSI wafers, an anisotropic dry etch is necessary to attain the desired edge profile and narrow lines. The major requirements for metal etching are:[31]

1. High etch rates (>1000 nm/min).
2. High selectivity to the masking layer (>4:1), interlayer dielectric (>20:1), and the underlying layers.
3. High uniformity with excellent CD control and no microloading (<8% at any location on the wafer).
4. No device damage from plasma-induced electrical charging.
5. Low residue contamination (e.g., copper silicon residue, developer attack, and surface defects).
6. Fast resist strip, often in a dedicated cluster tool chamber, with no residual contamination.
7. No corrosion.

Aluminum and Metal Stacks ■ A chlorine-based chemistry process is commonly used for aluminum etch. Pure chlorine etches aluminum isotropically. To achieve a directional etch process with anisotropic etching, sidewall passivation is needed by adding polymers to the etch gas, such as with CHF_3 or carbon obtained from the photoresist. BCl_3 is also used for aluminum etch, which has heavy molecules for a physical etch and good sidewall profile control. A small amount of N_2 is often added to the gas chemistry to minimize microloading and assist sidewall passivation. Aluminum etching with fluorine-based gas chemistry is unacceptable because of the low vapor pressure of the etch by-product, nonvolatile AlF_3, which means that the etch rate is extremely limited. Bromine-based chemistry has been investigated for metal etch.

A challenge in metal etch for interconnect wiring is the complexity of the multilayer stack that is common in VLSI/ULSI technology (see Figure 16.32). There is often an antireflective coating (ARC) of TiN or other materials and an underlying adhesion barrier (e.g., Ti), which complicates the etch process. To remove ARC films, possible etch gas chemistries are CCl_4/O_2 or SF_6/Cl_2. TiN films are removed in CCl_4/N_2 or Cl_2/Ar or $BCl_3/Cl_2/CHF_3$.[32] Another factor is that aluminum oxidizes almost instantly when exposed to air and the resulting aluminum oxide (Al_2O_3) inhibits the chlorine etch reactions.

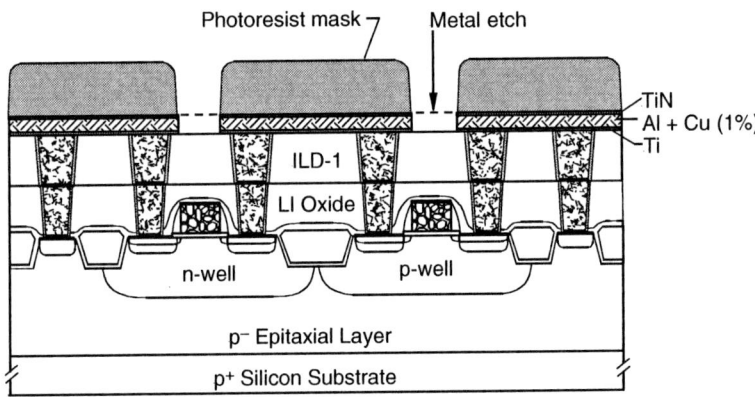

FIGURE 16.32 Metal Stack for VLSI/ULSI Integration

To etch a metal stack structure, multiple steps are used. This includes the first step to remove the native oxide layer (e.g., Al_2O_3 for aluminum) on the surface. The typical steps for etching a metal stack are:

1. Breakthrough step to remove native oxide.
2. ARC layer etch (may be combined with above step).
3. Main etch step of aluminum.
4. Overetch step to remove residue. It may be a continuation of the main etch step.
5. Barrier layer etch.
6. Optional residue removal process to prevent corrosion.
7. Resist removal (see section below).

Thorough corrosion control is essential to device performance after metal etching. Any remaining corrosive by-products of the etch process must be quickly neutralized or removed from the wafer surface. For aluminum etching, the major corrosive by-products are $AlCl_3$ or $AlBr_3$. These by-products react with water to form highly corrosive HCl or HBr, which are damaging acids to aluminum. Control of water vapor and oxygen content is critical in the etch process, which is best achieved in single-wafer etch reactors with a loadlock chamber to isolate the wafers from atmospheric contaminants and moisture. Another approach is to use the resist strip process to remove corrosive compounds while removing the resist, thus reducing their potential for corrosion.[33] The need to minimize the time delay and exposure to moisture are reasons for an etcher to have an integrated photoresist removal chamber.

Tungsten ■ Tungsten (W) is an important metal commonly used for via fill in a multilevel metal structure (see Chapter 12). Tungsten etch can be done with fluorine- and chlorine-based chemistry. However, fluorine-based chemistry, such as SF_6 and CF_4, has poor oxide selectivity. Chlorine-based gas chemistry has good oxide selectivity. N_2 is often added to the gas mixture to give better selectivity to the photoresist mask. Sometimes an O_2 forming gas is added to reduce carbon deposits. A chlorine gas such as Cl_2 or CCl_4 can be used to etch tungsten with improved anisotropy and selectivity.

Tungsten Etchback. *Tungsten etchback* is used as one step in the formation of tungsten plugs. Via openings are first etched in the SiO_2 interlayer dielectric (ILD). The W is blanket-deposited (CVD) into the via openings on top of a TiN barrier layer. This operation is followed by a dry plasma etchback step to remove the excess blanket layer and finish with a tungsten-filled via (see Figure 16.33 on page 464). The etch involves a two-step process. In the first step, 90% of the tungsten is etched at a high rate with excellent uniformity. In the second step, the etch rate is reduced and a gas chemistry is used with a high selectivity to the TiN barrier layer.[34] The etch reduction serves to decrease gas pressure and wafer temperature to reduce loading effects that may cause a recess in the W plug. Anisotropic etching is not needed, but minimal residue and plug loss are important. This etchback process has been largely replaced for advanced wafer fabrication by chemical mechanical planarization (CMP) to remove excess W and planarize the top surface of the contacts and vias (see Chapter 18).

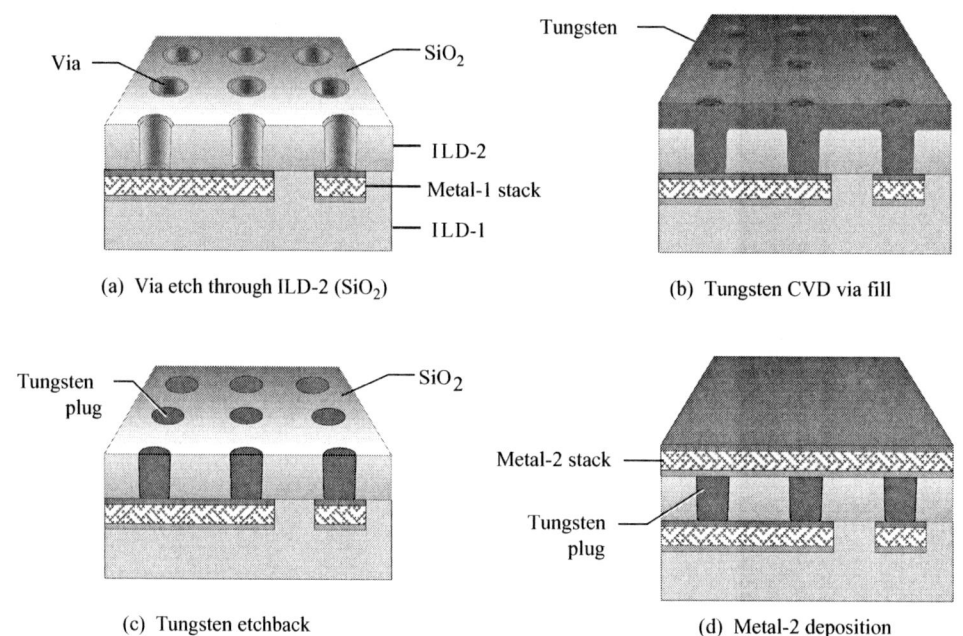

FIGURE 16.33 Tungsten Etchback

Contact Metal Etch ■ Refractory metals alloyed with silicon form the silicides common in wafer fabrication, including $CoSi_2$, WSi_2, $TaSi_2$, and $TiSi_2$ (see Chapter 12). A polycide is a composite of doped polysilicon and a refractory metal. Silicon dioxide does not form an alloy with the refractory metals, and it is the unreacted metal that must be removed during the contact metal etch process. Contact metal etch is critical in MOS device fabrication because the dimensional control will affect the channel length of the device. Contact metal etch with dry plasma can be done with fluorinated and chlorinated gas chemistries. Fluorine-based gas chemistry, such as NF_3 and SF_6, is used because of its good dimensional control with increased etch rate. Since the contact formation is a self-aligning process, no photoresist mask is required for this etch process.

WET ETCH

Wet etching has been associated with wafer fabrication since the beginning of the semiconductor industry. It has been replaced to a large degree by dry etching but still plays an important role in oxide cleaning, residue removal, the stripping of surface layers, and etching for larger geometry applications. Wet cleaning (see Chapter 6) can actually be considered a wet etch process. An example is the residue layer and damaged silicon layer formed by reactive ion etching of a contact hole. The contact residue is removed by downstream microwave ashing followed by wet etching as a form of cleaning to remove residues and silicon damage.[35] The benefits from wet etching over dry etching are good selectivity to the underlying film, no risk of plasma damage to devices, and use of simple equipment. Basic wet-etch parameters (see Table 16.7) must be controlled during the etching process. These parameters are applicable to all wet etching applications.

Wet etching is done by exposing a batch of wafers, usually 25 but sometimes up to 50 wafers, to an appropriate acid bath, either by immersion or spray. Immersion is the simplest method, while spray etching requires less volume of chemicals and is faster than immersion. Wet etching produces an isotropic sidewall with substantial undercutting of the film underneath the edge of the mask material, making it unsuitable for feature sizes <3 μm. Isotropic etching with undercutting is the main

TABLE 16.7 Wet-Etch Parameters

Parameter	Explanation	Difficulty to Control
Concentration	Solution concentration (e.g., ratio of $NH_4F:HF$ to etch oxide).	Most difficult parameter to control because the bath concentration is continually changing.
Time	Time of wafer immersion in the wet chemical bath.	Relatively easy to control.
Temperature	Temperature of wet chemical bath.	Relatively easy to control.
Agitation	Agitation of the solution bath.	Moderately difficult to properly control.
Number of Runs	Solution must be replaced after specified number of runs to reduce particles and to ensure proper solution strength.	Relatively easy to control

reason why the majority of semiconductor etching is done with dry plasma etch. Other disadvantages to wet etch are the chemical safety concerns for wet baths, its tendency to cause resist lifting and bubbles, the difficulty in controlling the bath parameters for uniformity, and the high cost of chemical disposal.

Types of Wet Etch

Wet etching can provide an alternative for high etch selectivity that might not be available with dry etching. In addition, wet etching eliminates plasma damage. As previously stated, wet etch has been widely replaced by plasma dry etching in advanced IC wafer fabrication. Materials such as silicon and aluminum are nearly always dry etched for submicron fabrication. General safety information for wet etch chemicals is provided in Appendix A.

Wet Oxide Etch ■ Oxide can be wet etched by hydrofluoric acid (HF). Selective removal of oxide is done by spraying or immersing the wafer in a dilute solution of HF that is frequently buffered with ammonium fluoride (NH_4F), known as a *buffered oxide etch* (*BOE*) or *buffered HF* (*BHF*). A chemical buffer is a solution that resist changes in pH when small amounts of strong acid or base are added. Buffering the HF with NH_4F provides a well-controlled etch solution that slows and stabilizes the etch and does not appreciably attack the photoresist. The BOE is rinsed from the wafer using DI water and quick dump or cascade overflow rinse equipment. The quick dump rinse is faster and uses less DI water, but cascade rinsing can significantly reduce particulate contamination levels.

SiO_2 is an amorphous material and etches equally well in all directions when exposed to BOE. If a 1 μm thick oxide layer is etched, there is also a lateral etch of 1 μm under the mask material. This lateral etching and the subsequent undercutting is undesirable from a dimensional standpoint, limiting the density of lines and spaces that can be achieved.

The oxide etch rate also depends on whether it is thermal or deposited oxide (see Table 16.8 on page 466). Since dry grown oxide is denser than wet grown oxide, dry grown oxide etches at a slower rate. In addition, doped oxides etch differently and usually faster than undoped oxides.

Wet Chemical Strips ■ Due to the high selectivity of wet etch, *wet chemical strips* are sometimes used to remove surface layers, including photoresist (see the following section) and masking layers. Silicon nitride (Si_3N_4) is widely used in wafer fabrication as a mask material in the isolation technologies of shallow trench isolation (STI), local oxidation of silicon (LOCOS), and self-aligned contact structure. The removal of the nitride masking layer is often done by a wet chemical strip with hot phosphoric acid (H_3PO_4).[36] The wet acid bath solution is typically maintained at around 160°C and has the desirable property of high selectivity to exposed oxide layers. Control of the hot phosphoric nitride removal bath is difficult and it is done as a timed operation (no endpoint detection) with the use of monitor wafers. There is usually some oxynitride formation on the exposed nitride layer that requires a short hydrofluoric acid (HF) deglaze before the nitride removal. If the oxynitride layer is not removed, then there is a risk of nonuniform nitride removal. Another material often

TABLE 16.8* Approximate Oxide Etch Rates in BHF^a Solution at 25°C

Type of Oxide	Density (g/cm^3)	Etch Rate (nm/s)
Dry grown	2.24 to 2.27	1
Wet grown	2.18 to 2.21	1.5
CVD deposited	<2.00	1.5^b to 5^c
Sputtered	<2.00	10 to 20

a) 10 parts of 454 g NH$_4$F in 680 ml H$_2$O and one part 48% HF
b) Annealed at approximately 1000°C for 10 minutes
c) Not annealed
*B. El-Kareh, *Fundamentals of Semiconductor Processing Technology* (Boston: Kluwer Academic, 1995), p. 277.

removed by wet strip is the titanium layer deposited to form the titanium silicide at the contacts. Ammonium hydroxide (NH$_4$OH) and hydrogen peroxide (H$_2$O$_2$) are diluted with DI water for this process.

HISTORICAL PERSPECTIVE

Etch technology has undergone many changes throughout the years of wafer fabrication. The initial barrel plasma etcher was a simple tool with limited operator controls. The modern plasma etcher generates a high-density plasma, has independent RF power supplies for plasma generation and wafer bias, endpoint detection, gas pressure and flow control, and is integrated with software control of the etch parameters. Table 16.9 details the advances for polysilicon etching from wet etch technology through reactive ion etch (RIE) and up to modern high-density plasma etch tools for subquarter micron geometries.

PHOTORESIST REMOVAL

One of the last steps after etching is the removal of the photoresist mask. The photoresist serves as the pattern transfer medium from the reticle to the wafer surface and blocks films from being etched or ions from being implanted. Once etch or implantation is complete, the resist serves no further purpose on the wafer surface and must be completely removed. In addition, any trace residues from the etch process are also removed.

Photoresist stripping is the general term that describes the wet removal of the resist. Stripping is sometimes used for resist residues that are difficult to remove. In most applications, wet resist stripping is not cost effective because of the required handling and disposal of wet chemicals. Furthermore, if the resist has previously undergoine dry etch processing, its top surface could be hardened from exposure to fluorinated and chlorinated gas chemistry. This renders the resist insoluble in most wet chemistry stripping solutions. In this case, plasma stripping is required to remove at least the top resist layer.

A major challenge for photoresist removal is that resists are designed and processed to achieve maximum adhesion to the wafer surface. As discussed in Chapter 13, this is required for the resist to withstand the rigors of the etch process and ion implantation. Maximum resist adhesion creates an obvious challenge for photoresist removal. Another challenge is the need for a high resist strip rate to attain a higher wafer throughput. However, a high strip rate generally tends to leave more resist residues, which places more burden on the removal process to effectively remove all resist and residues.

Plasma Ashing

Plasma ashing is the dry removal of resist with oxygen, and is the dominant technique for bulk resist removal.[37] The first plasma ashers in the 1970s used barrel reactor technology, with oxygen

TABLE 16.9* Polysilicon Etch Technology Evolution

Geometry Requirements	Time Frame and Reactor Design	Chemistries	Strengths	Limitations and Problems	Controls
4 to 5 µm, isotropic etch	Pre-1977: wet etch	HF/HNO_3 buffered with acetic acid or H_2O	Batch process	Resist lift; bath aging; temperature sensitive	Operator judgement for endpoint
3 µm	1977: barrel etcher	CF_4/O_2	Batch process	Nonuniformity; isotropic etch; large undercut	Manometer and timer
2 µm	1981: single wafer etch	CF_4O_2	Single wafer; individual etch endpoint, improvement in repeatability	Low oxide selectivity; isotropic process	Endpoint detection
1.5 µm	1982: single wafer RIE	SF_6/Freon 11, SF_6/He	MFCs; independen pressure and gas flow control, improvement in repeatability	Low oxide selectivity; profile control	MFCs; separate gas flow and pressure control
To 0.5 µm	1983: variable gap; load-locked	CCl_4/He, Cl_2/He, Cl_2/HBr	Load-locked chamber; variable gap; improvement in repeatability	Microloading in high-aspect ratios; profile control	Control of electrode gap; computer controls
To 0.25 µm and below	1991: inductively-coupled plasma (ICP)	Cl_2, HBr	High-density plasma; low pressure; simple gas mixtures; improvement in repeatability	Complex tool; many variables	Independent RF control for plasma generation and wafer bias

*Adapted from C. Almgren, "The Role of RF Measurements in Plasma Etching," *Semiconductor International,* (August 1997): p. 100.

plasmas to strip the resist. Barrel reactors have been widely replaced with downstream plasma reactors that strip the resist with an oxygen plasma in single-wafer chambers.

Asher Overview ■ A *photoresist asher* removes the resist layer by reacting atomic oxygen atoms with the resist material in a plasma environment (see Figure 16.34 on page 468). Atomic oxygen (O) is created by using microwave or RF energy to dissociate molecular oxygen (O_2). There are usually forming gases such as N_2 or H_2 added to improve the ash performance and enhance polymer residue removal.[38] Thus, a typical asher gas chemistry would be O_2/N_2. Recall from Chapter 13 that resist is basically a hydrocarbon polymer. The atomic oxygen atoms rapidly react with the resist material to create volatile carbon monoxide (CO), carbon dioxide (CO_2), and water (H_2O) as the main by-products. The asher by-products are pumped away by the vacuum system.

Plasma Damage ■ A concern during the ashing process is plasma damage to devices on the wafer surface from ion bombardment or wafer charging. As gate oxides are scaled thinner, this concern is more critical due to the sensitivity of thin oxides to plasma degradation. Wafer damage from plasma has largely been remedied by using single-wafer downstream ashers that position the wafer away from the damaging plasma ions, allowing only reactive chemical species to reach the wafer. Microwave (2.45 GHz) frequencies are common for source plasma generation in downstream systems. This is because a microwave glow discharge produces more atomic oxygen reactant and a lower proportion of ionic species. This yields lower risk of ion-induced damage.

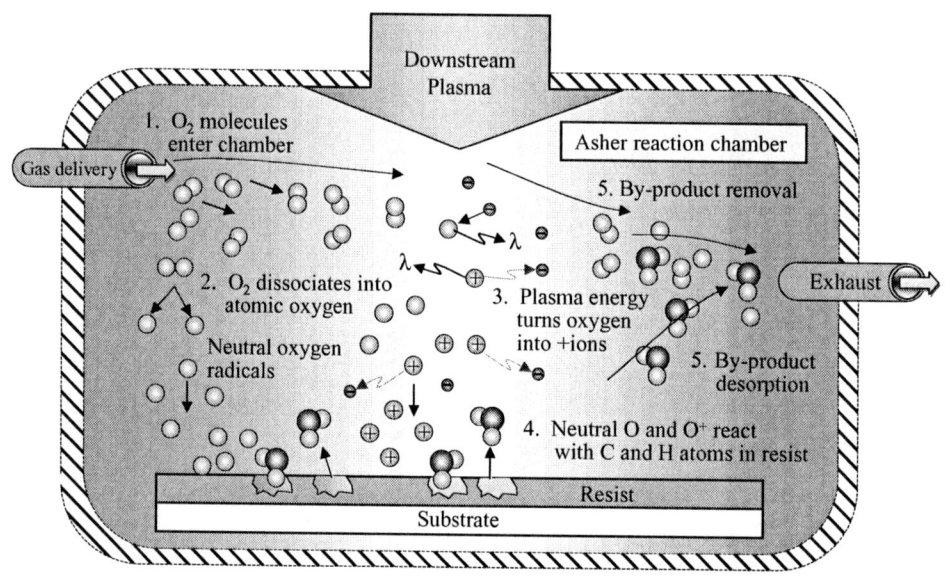

FIGURE 16.34 Atomic Oxygen Reaction with Resist in Asher

Residue Removal ■ Standard ashing has evolved to include the removal of post-etch residues such as sidewall polymers and via veils (see Figure 16.35). As previously discussed, these residues are complex and can contain plasma etching and ashing by-products such as aluminum, titanium, oxides, and silicon. A complicating factor is that ashing may occur at an elevated temperature (e.g., 200°C), which can harden residues and make them harder to remove. If not removed, residues are a source of particles and contaminants that increase the wafer defect density. To fully remove etch residues, especially the inorganic residue materials (e.g., silicon oxides, metal oxides, aluminum, and so on), some ashers are based on alternative plasma chemistries to oxygen, such as NO or N_2O. Other asher systems add a small amount of fluorine to the gas chemistry in the form of CF_4 or NF_3 to more effectively remove residues containing oxide and silicon by making them more water soluble.[39] The ash process is typically followed by a DI water rinse to remove residual particles on the wafer surface.[40] In some cases a strong sulfuric acid (H_2SO_4) mixture with hydrogen peroxide (such as the piranha mixture discussed in Chapter 6) is used to thoroughly clean the remaining stripped photoresist.

Wet cleaning for residue removal still is found in backend processes, particularly for the removal of sidewall passivating residues that have high-density inorganic contaminants. Dry ashing alone is not sufficient to strip and clean residues because of the multitude of inorganic materials present that are not volatized by the plasma. Until recently, wet chemistry residue strippers were commonly based on a hydroxylamine compound. However, there is a corrosion risk to metal and barrier layers from this stripper and new hydroxylamine-free formulations are in use based on a chelation chemistry to strip residues (a chelate is a complex ion that involves metal cations).[41] This new formulation dissolves inorganic residues at low temperature with no corrosion.

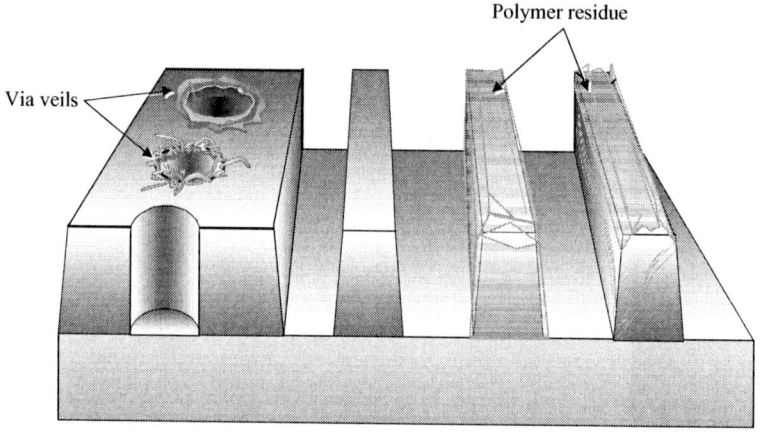

FIGURE 16.35 Post-Etch Via Veil Residue

ETCH INSPECTION

The final step in the etch process is an *etch inspection* to verify quality measures. This inspection occurs on the patterned wafer after all etching and photoresist stripping is completed. It has traditionally been done with a manual microscope inspection that used incident white or ultraviolet light to inspect for defects such as stains and large particulate contamination. Manual microscopes have been largely replaced by automatic inspection systems, especially on critical layers with deep submicron patterning. Advanced metrology instruments are able to automatically inspect patterned wafers for patterned defects and distortions (see Chapter 7). One of the most important inspections that occurs at final inspection is the automated measurement to verify critical dimensions (CD) on a particular mask level. The quality of the etch process is also verified by inspecting for etch problems such as overetching, underetching and undercutting.

Etch inspection is similar to the inspection at develop inspect, except that there is no photoresist and defects usually cannot be reworked. Almost any defect, if numerous enough, will cause the wafer to be scrapped. An exception to this rule is surface particle contamination, which can be cleaned. Representative quality measurements for etch final inspection are described in Table 16.10.

ETCH INSPECTION QUALITY MEASURES

Typical quality measures for etch inspection are provided in Table 16.10.

TABLE 16.10 Quality Measures for Etch Final Inspection

Quality Parameter	Types of Defects	Remarks
1. Critical dimension bias.	A. Linewidth change: excessive difference between photoresist linewidth and final feature linewidth after etch.	• CD bias for linewidth change is measured by comparing pre-etch linewidth in photoresist with post-etch linewidth of the same feature on the wafer. Measurements can be done with an SEM. • Excessive CD bias requires optimization of the etch process for undercutting or slope. • The photoresist profile has an effect on CD bias. Vertical resist profile produces best CD bias.
2. Metal corrosion.	A. Corrosion or attack of metal film after etch.	• Corrosion often results from residual HCl on the wafer that is exposed to water vapor in the air. • Corrosion may be visible as small bubbles along the side of metal lines. This defect is looked for by optical microscope or SEM.
3. Sidewall contaminants after etching.	A. Residual sidewall passivants that remain after etch, including remaining photoresist.	• Minimize formation by optimizing the pre-metal etch photoresist hard bake. • Residues are usually removed by post-etch solvents or dilute buffered HF.*
	B. Contaminants backsputtered onto the sidewall of the metal line or a via hole.	• Backsputtering can occur due to polymer used to passivate sidewall, leaving a "veil" residue (thin overhang along metal line or via sidewall).
4. Loading effects.	A. Microscopic nonuniformity of etch process.	• Etch rate between narrow openings is slower than open-field areas due to reduced density of reactive radicals in small space. • Balance pressure and power of etch process.**

Quality Parameter	Types of Defects	Remarks
5. Shorts after metal etch.	A. Bridging of metal lines after etch leading to electrical short.	• Reduce pattern density effects from microloading.
6. Excessive post-etch residue.	A. The following types of residue may exist after etch: • Stringers (small strings of residue). • Veils (thin residue overhang). • Crowns • Rails • Corrosion after metal etch.	The causes of residue after etch are: • Nonuniform etch process. • Films deposited over topography. • Nonuniform dopant distribution in film. • Contaminants in the film (in addition to intentional alloys such as Cu in Si). • Contaminants in gas or chamber. Check supply filters or clean system. • Incorrect process parameter, such as high etch rate.

*K. Mautz, *Optimization of Single Wafer and Batch Metal Etch Manufacturing Processees* vol. 96-12 (Pennington, NJ: The Electrochemical Society, 1996), p. 283.

**S. Gonzales, J. Quijada, and G. Grivna, *Submicron Metal Etch Integration Study* vol. 2875 (Bellingham, WA: SPIE, 1996), p. 302.

DRY ETCH TROUBLESHOOTING

Problems associated with dry etch equipment and process are provided in Table 16.11.

TABLE 16.11 Common Dry Etch Troubleshooting Problems

Problem	Probable Cause	Corrective Actions
1. Incorrect etch rate.	A. Change in RF power. B. Incorrect temperature. C. Problem with pressure. D. Endpoint detection not functioning properly. E. Improper wafer spacing. F. Improper gas-flow dynamics. G. Improper maintenance. H. Incorrect process recipe.	• Check and troubleshoot RF generator and matching unit. • Check backside wafer cooling system. • Calibrate vacuum gauges (e.g., capacitive manometer) and pressure control system. • Check endpoint detection system. • Check wafer-to-electrode gap. • Verify gas distribution system. • Perform chamber wet clean. • Verify process recipe and parameters.
2. Inadequate selectivity.	A. Etch rate too high. B. Improper gas flow or pressure. C. Endpoint detection problem. D. Wrong wafer temperature. E. Incorrect process recipe.	• Verify etch rate. • Calibrate MFCs and vacuum gauges. • Check/calibrate endpoint detection. • Check wafer cooling system. • Confirm process recipe and parameters.
3. Improper sidewall profile angle.	A. Contamination on sidewall. B. Temperature of wafer. C. System pressure. D. Incorrect process recipe (misprocess).	• Check for polymer buildup in chamber. • Backside contamination of wafer causing nonuniform heating. • Check/calibrate MFCs and perform leak test to check for contamination.

Problem	Probable Cause	Corrective Actions
4. Etch nonuniformity across wafer.	A. Depletion of etchant gas concentration due to ARDE. B. Improper gas flow. C. Temperature of wafer. D. Improperly positioned wafer in chamber. E. Chamber configuration. F. Improper film thickness. G. Improper maintenance.	• Verify acceptable design for dense and nondense areas of wafer. • Check/calibrate gas distribution system. • Check thermocouples and wafer cooling. • Check robotics, wafer handling system, and vacuum chuck • Check reactor plate spacing. • Measure and verify film thickness. • Perform chamber wet clean.
5. Plasma damage.	A. Nonuniform plasma. B. Excessive ion bombardment of gate oxide. C. Excessive RF power. D. Improper maintenance.	• Poorly designed or maintained plasma equipment. • Suboptimum process conditions. • Check recipe and RF generator.
6. Particle Contamination.	A. Leak/contamination from gas lines. B. Tool operation. C. Improper gas chemistry.	• Leaks or faulty MFCs. • Improper tool shutdown, operation, or maintenance. • Wrong process recipe. • Perform wet clean.
7. Metal Corrosion.	A. Moisture. B. Gas flow. C. Contaminants from etch process. D. Wrong maintenance procedure.	• Excessive time delay for post-etch residue cleanup. • Check MFCs for correct process gases. • Control time to resist strip. • Check maintenance procedure.

SUMMARY

Etch selectively removes material from the wafer surface by chemical or physical means. Dry etch uses a plasma, whereas wet etch uses liquid chemicals. Etch is generally divided into categories including dielectric, silicon, and metal etch. There are nine important parameters: etch rate, etch profile, etch bias, selectivity, uniformity, residues, polymer formation, plasma-induced damage, and particle contamination. Plasma etch has many advantages over wet etch and is the most common etch process. Sidewall profiles are generally divided into isotropic and anisotropic. Anisotropic, or vertical sidewalls, are produced by dry plasma etching. Oxide etch requires high selectivity to the underlying material. Silicon etch is used for critical applications such as polysilicon gate formation and trench formation. A main application of metal etch is aluminum interconnect wiring, with the need to etch through complex metal stacks. Wet etch has been largely replaced by dry etch for advanced IC fabrication but is still used for oxide removal, wet cleans, and stripping. The removal of photoresist at the completion of etch or ion implant is known as photoresist ashing. An asher uses a downstream process to minimize plasma-induced damage and is usually followed by a wet etch to remove residues. Etch inspection is the last operation in the etch process and is necessary to ensure defects are identified and corrected.

KEY TERMS

etch
dry etch
wet etch
dielectric etch
silicon etch
metal etch
patterned etching
unpatterned etching (etchback or stripping)
etch rate
step height
loading effects
etch profile
isotropic etch profile
anisotropic etch profile
directionality
etch bias
selectivity
etch uniformity
aspect-ratio-dependent etching (ARDE)
microloading
etch residue
polymer formation
plasma-induced damage
chemical mechanism
physical mechanism

combined physical and chemical mechanism
plasma potential distribution
barrel reactor
parallel plate (planar) reactor
downstream reactor
triode planar reactor
ion beam milling or ion beam etching (IBE)
reactive ion etch (RIE)
high-density plasma etch
electron cyclotron resonance (ECR)
inductively coupled plasma (ICP)
dual plasma source (DPS)
magnetically enhanced RIE (MERIE) or magnetron
endpoint detection
breakthrough step
main-etch step
overetch step
tungsten etchback
buffered oxide etch (BOE) or buffered HF (BHF)
wet chemical strips
photoresist stripping
plasma asher
photoresist asher
etch inspection

REVIEW QUESTIONS

1. Define etch. What is the goal of etching?
2. What are the two types of etch processes? Give a short description of each type.
3. List the three major material categories for dry etch.
4. Explain the difference between patterned and unpatterned etching.
5. List nine important parameters for etch.
6. Define etch rate and state its formula. Why is it desirable to have a high etch rate?
7. Explain loading effects and how this is related to the etch rate.
8. Describe isotropic and anisotropic etch profiles and what are the desirable and undesirable aspects of each profile.
9. Is a dry etch profile isotropic, anisotropic or both? What about a wet etch profile?
10. What is directionality and why is it desirable in etching?
11. What is etch bias and what is it caused by? State and explain the etch bias formula.
12. Define selectivity. Does dry etching have good or poor selectivity? What does high selectivity mean? State and explain the selectivity formula.
13. What is etch uniformity? What is the challenge for etch uniformity? Explain ARDE and discuss how it is related to etch uniformity. What is another name for ARDE?
14. Discuss etch residues, why they occur, and how they are removed.
15. Why is a polymer formation deposited sometimes on feature sidewalls during etch? What is an undesirable side effect of polymer formation?
16. What is plasma-induced damage from etching and what problems occur from this damage?
17. Explain how plasma can cause particle contamination on the wafer surface.
18. What is the goal of dry etch? List the advantages of dry etch over wet etch. What are the disadvantages of dry etch?
19. List the three ways to achieve etching action during dry etch.

20. Explain the chemical mechanism and the physical mechanism for achieving etching action.
21. Describe the combined physical and chemical mechanism for dry etch.
22. Describe the plasma potential distribution and why it is important for etch.
23. What are the etch results if the RF frequency is decreased? RF power is decreased? DC bias is increased? Electrode size is decreased?
24. Describe the basic components of a plasma dry etch system. What gas chemistries are used for silicon dioxide, aluminum, silicon and photoresist?
25. Describe the barrel plasma etcher.
26. Describe the parallel plate (planar) reactor.
27. Why is downstream etching beneficial? Describe a downstream reactor.
28. How does a triode planar reactor function?
29. Explain ion beam milling. For what materials is it used?
30. Describe reactive ion etch (RIE).
31. Explain the principle for how high-density plasma improves etching of high-aspect ratio gaps.
32. Give three reasons for using a magnetic field with a high-density plasma etcher.
33. Describe electron cyclotron resonance (ECR).
34. Explain inductively-coupled plasma (ICP).
35. Describe dual plasma source (DPS).
36. Discuss the MERIE technology for high-density plasma.
37. What is endpoint detection and why is it necessary for dry etch? What is the most common type of endpoint detection?
38. Describe optical emission spectroscopy for endpoint detection.
39. List six requirements for successful dry etch.
40. List three challenges for oxide etching.
41. Oxide plasma etching is usually based on what gas chemistry? Give a specific gas chemistry common for oxide etch.
42. Give an example why oxide etch selectivity to the underlying material is important. Provide two methods for achieving oxide etch selectivity to silicon.
43. What is one factor that reduces photoresist selectivity? Is this desirable or undesirable?
44. Explain the effect of low resist selectivity on contact via etch in the LI oxide.
45. Is the etch rate faster for silicon nitride deposited with PECVD or LPCVD? Why?
46. What gas chemistries are usually used for etching polysilicon and why has these chemistries replaced fluorine chemistries?
47. Give three requirements for polysilicon gate etch.
48. List and describe the three steps for etching polysilicon.
49. Discuss the polysilicon etch gas chemistries.
50. How are silicon trench sidewalls protected during plasma etch from lateral etch attack?
51. What gas chemistry is sometimes used for shallow trench etching? Deep trench etching? Why are these gas chemistries used?
52. List the seven major requirements for metal etching.
53. List the seven steps for etching a metal stack.
54. What control is necessary after metal etching?
55. What gas chemistry has good oxide selectivity during tungsten dry etch?
56. Describe the tungsten etchback process.
57. What is the main reason why dry etch has largely replaced wet etch?
58. List four wet etch parameters, explain each parameter and state how difficult it is to control.
59. How is oxide wet etched? What are the two names of the FF solution buffered with ammonium fluoride?
60. What is the wet etch rate difference for dry grown versus wet grown oxide?
61. Describe the silicon nitride wet chemical strip.
62. What is photoresist stripping?
63. What is plasma ashing? What is the purpose of a photoresist asher?
64. What is a concern from the ashing process? How is this concern resolved?
65. How are post-etch residues removed?
66. What is an important aspect of performing an etch inspection?

ETCH EQUIPMENT SUPPLIERS' WEB SITES

Applied Materials — http://www.appliedmaterials.com/products/
Eaton Corporation — http://www.semiconductor.eaton.com/
Gasonics International — http://www.gasonics.com/
Hitachi — http://www.hitachi.com/semiequipment/products.html
International SEMATECH — http://www.sematech.org
Lam Research Corp. — http://www.lamrc.com/
Leybold-Inficon — http://www.leyboldinficon.com/
MRC, Materials Research Corp. — http://www.materialsresearch.com/
Plasmos — http://www.plasmos.com/
SEMI — http://www.semi.org/
Tegal Corporation — http://www.tegal.com
TEL, Tokyo Electron Ltd. — http://www.teainet.com

REFERENCES

1. Y. Lii, "Etching," *ULSI Technology,* ed. C. Chang and S. Sze (New York: McGraw-Hill, 1996), p. 342.
2. P. Singer, "New Frontiers in Plasma Etching," *Semiconductor International* (July 1996), p. 153.
3. R. DeJule, "Managing Etch and Implant Residue," *Semiconductor International* (August 1997): p. 60.
4. C. Gabriel, "Measuring and Controlling Gate Oxide Damage from Plasma Processing," *Semiconductor International* (July 1997): p. 151.
5. Y. Lii, "Etching," p. 364.
6. Adapted from G. Oehrlein and J. Rembetski, "Plasma-Based Dry Etching Techniques in the Silicon Integrated Circuit Technology," *IBM Journal of Research & Development* vol. 36, no. 2 (Armonk, NY: March 1992): p. 140.
7. S. Wolf and R. Tauber, *Silicon Processing for the VLSI Era*—Vol. 1, *Process Technology,* (Sunset Beach: Lattice Press, 1986), p. 545.
8. Ibid.
9. Ibid., p. 569.
10. Ibid., p. 700.
11. P. Singer, "New Frontiers in Plasma Etching," *Semiconductor International* (July 1996): p. 153.
12. Ibid., p. 154.
13. Y. Lii, "Etching," p. 349.
14. Ibid.
15. Y. Ye et al., "0.35-Micron and Sub-0.35 Micron Metal Stack Etch in a DPS Chamber—DPS Chamber and Process Characterization," *Proceedings of the Eleventh International Symposium on Plasma Processing* vol. 96 12 ed. G. Mathad and M. Meyyappan (Pennington, NJ: The Electrochemical Society, 1996), p. 222.
16. G. Oehrlein, "Reactive Ion Etching," *Handbook of Plasma Processing Technology* ed. S. Possnagel, J. Cuomo, and W. Westwood (Park Ridge, NJ: Noyes Publishing, 1990).
17. S. Wolf and R. Tauber, *Silicon Processing for the VLSI Era,* Vol. 1—*Process Technology* p. 697.
18. J. Baliga, "Vacuum Pump Designs Adjust to Harsher Conditions," *Semiconductor International* (October 1997): p. 87.
19. P. Singer, "Vacuum Pumping in Etch and CVD," *Semiconductor International* (Septem-ber 1995): p. 78.
20. S. Tandon, "Challenges for 300 mm Plasma Etch System Development," *Semiconductor International* (March 1998): p. 78.
21. Ibid.
22. P. Singer, "The Many Challenges of Oxide Etching," *Semiconductor International* (June 1997): p. 110.
23. B. El-Kareh, *Fundamentals,* p. 326.
24. Ibid., p. 327.
25. P. Singer, "Many Challenges of Oxide Etching," p. 110.
26. S. Wolf and R. Tauber, *Silocon Processing for the VLSI Era*—Vol. 1, *Process Technology,* p. 556.
27. Y. Wang et al., "High-Selectivity Silicon Nitride Etch Process *Semiconductor International* (July 1998): p. 238.
28. Y. Lii, "Etching," p. 357.
29. S. Tandon, "Challenges for 300 mm,"p. 80.
30. Y. Lii, "Etching," p. 355.
31. Adapted from S. Tandon, "Challenges for 300 mm," p. 80.
32. B. El-Kareh, *Fundamentals,* p. 330.
33. R. DeJule, "Managing Etch and Implant Residue," p. 58.
34. Y. Lii, "Etching," p. 360.
35. D. Taylor, "Wet-Etch Process Improvements Through SPC," *Solid State Technology* (July 1998): p. 119.
36. J. Rembetski, W. Rust, and R. Shepherd, "The Removal of Hard Masks in Semiconductor Processing," *Solid State Technology* (March 1995): p. 68.
37. P. Singer, "Plasma Ashing Moves into the Mainstream," *Semiconductor International* (August 1996): p. 84.
38. Ibid.
39. C. Cheng and J. Oncay, "A Downstream Plasma Process for Post-Etch Residue Cleaning," *Semiconductor International* (July 1995): p. 185.
40. G. Herdt, P. Gillespie, and Y. Wasserman, "Characterization of Damage from Dry Ashing and Residual Removal," *Semiconductor International* (November 1997): p. 80.
41. R. DeJule, "Managing Etch," p. 58.

CHAPTER 17
ION IMPLANT

Intrinsic silicon crystal structure is formed by silicon covalent bonds. As explained in Chapter 2, intrinsic silicon is a poor conductor. Silicon only becomes useful as a semiconductor when its structure and conductivity are altered by adding small amounts of impurities called dopants. This process is referred to as doping. The doping of silicon is fundamental to pn junction fabrication in semiconductor devices. Ion implantation is the primary method for doping wafers.

Doping is widely used throughout wafer fabrication. Dopants are precisely and uniformly placed into the device to alter its electrical performance. Silicon chips require doping with Group IIIA and Group VA dopants for many different reasons, which will be discussed in this chapter.

The continual shrinking of chip critical dimensions and the subsequent increase in chip packing density force the scaling of all device elements. An example of this is the reduction of the polysilicon gate length to obtain a narrower channel region. A shorter channel length requires shallower junction depths for the source and drain doped regions. Junction depths are ultrashallow for chips designed at the 0.18 μm and below technology nodes. The junction depth requirements decrease to 30 ± 10 nm for devices with a CD of 0.1 μm.[1]

OBJECTIVES

After studying the material in this chapter, you will be able to:

1. Explain the purpose and applications for doping in wafer fabrication.
2. Discuss the principles and process of dopant diffusion.
3. Provide an overview of ion implantation, including its advantages and disadvantages.
4. Discuss the importance of dose and range in ion implant.
5. List and describe the five major subsystems for an ion implanter.
6. Explain annealing and channeling in ion implantation.
7. Describe different applications of ion implantation.

INTRODUCTION

Doping is the introduction of a dopant into the crystal structure of a semiconductor material to modify its electronic properties (e.g., electrical resistivity). As discussed in Chapter 2, certain Group IIIA and VA elements are used as dopants in wafer fabrication. For review, Table 17.1 on page 476 highlights the four principal dopants used in semiconductor manufacturing.

Dopant species are also referred to as impurities but should not be confused with contaminating impurities. Dopants are introduced into the silicon and other fabrication materials for a variety of reasons. For instance, dopants such as boron and phosphorus are used to form the majority carriers in silicon devices, to form conductive layers in the wafer, or to alter material properties (e.g., doping SiO_2 to form borophosphosilicate glass, or BPSG). The polysilicon gate electrode is doped for conductivity.

There are two techniques in wafer fabrication for introducing dopant elements to the wafer materials: thermal diffusion and ion implantation. *Thermal diffusion* uses high temperature to move the dopant through the silicon lattice structure. This method is dependent on time and temperature. *Ion implantation* introduces dopants into the substrate through a high-voltage ion bombardment. The dopants are implanted into the wafer through high-energy collisions at the atomic level. In the beginning of semiconductor manufacturing, thermal diffusion was the primary method for doping the wafer. However, due to shrinking critical dimensions and the corresponding scaling requirements, nearly all dopant processes in modern wafer fabrication are now done

TABLE 17.1 Common Dopants Used in Semiconductor Manufacturing

Acceptor Dopant Group IIIA (p-Type)		Semiconductor Group IVA		Donor Dopant Group VA (n-Type)	
Element	Atomic Number	Element	Atomic Number	Element	Atomic Number
Boron (B)	5	Carbon	6	Nitrogen	7
Aluminum	13	Silicon (Si)	14	Phosphorus (P)	15
Gallium	31	Germanium	32	Arsenic (As)	33
Indium	49	Tin	50	Antimony	51

Note: Common dopants are shaded.

with ion implantation (see Figure 17.1). Table 17.2 outlines some critical processes commonly doped with ion implantation versus the limited doping done by diffusion. Figure 17.1 shows these doped regions on a cross section drawing of a CMOS inverter. Note the dopant type (p or n) and concentration (− or +) are also shown. The different doped regions labeled A to O are defined in Table 17.2.

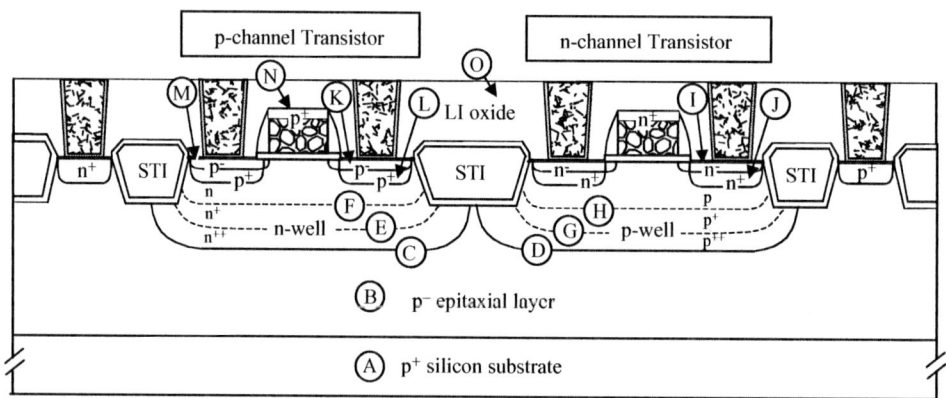

FIGURE 17.1 CMOS Structure with Doped Regions

TABLE 17.2* Common Dopant Processes CMOS Fabrication

Process Step	Common Dopant Species	Ion Implant or Diffusion	Remarks
A. p+ Silicon Substrate	B	Diffusion	The monosilicon crystal is doped during the crystal growing process.
B. p− Epitaxial Layer	B	Diffusion	The epitaxial layer is doped by diffusion during epitaxial silicon growth.
C. Retrograde n-well	P	Ion Implant	Retrograde well has dopant profile with peak concentration buried at a certain depth. The dopant concentration decreases as it approaches the surface.
D. Retrograde p-well	B	Ion Implant	Retrograde well has dopant profile with peak concentration buried at a certain depth.
E. p-Channel Punchthrough	P	Ion Implant	Phosphorus implanted more laterally than source/drain implants to keep electric field of drain from punching through p-channel and reaching source region.

Process Step	Common Dopant Species	Ion Implant or Diffusion	Remarks
F. p-Channel Threshold Voltage (V_T) Adjust	P	Ion Implant	Implant phosphorus to adjust MOS V_T threshold voltage.
G. n-Channel Punchthrough	B	Ion Implant	Boron is implanted more laterally than source/drain implants to keep electric field of drain from punching through n-channel and reaching source region.
H. n-Channel V_T Adjust	B	Ion Implant	Implant boron through oxide layer to adjust MOS threshold voltage.
I. n-Channel Lightly Doped Drain (LDD)	As	Ion Implant	Low-dose implant of arsenic in region adjacent to the n-channel to improve electrical performance by reducing the peak electrical field and therefore hot carriers, which minimizes interface charges trapped in the gate oxide.
J. n-Channel Source/Drain (S/D)	As	Ion Implant	Higher implant dose of arsenic to form n-channel source and drain.
K. p-Channel LDD	BF_2	Ion Implant	Low-dose implant of boron in region adjacent to the p-channel to improve electrical performance between drain and channel region.
L. p-Channel S/D	BF_2	Ion Implant	Higher implant dose of boron to form p-channel source and drain.
M. Silicon	Si	Ion Implant	Implant with nondopant atom to amorphize silicon to reduce transient enhanced diffusion (TED) and channeling effects.
N. Doped Polysilicon	P or B	Ion Implant or Diffusion	Doping of polysilicon gate electrode to make conductive.
O. Doped SiO_2	P or B	Ion Implant or Diffusion	Doped oxide to obtain material benefits (e.g., create better flow and serve as getter).

*Adapted from E. Rimini, *Ion Implantation: Basics to Device Fabrication,* (Boston: Kluwer Academic Publishers, 1995).

Ion implantation typically follows photolithography in the general process flow (see Figure 17.2 on page 478); although, there may be some processes that call for the use of an etched-oxide layer to serve as a mask during either diffusion doping or ion implant. In this case, wafers could flow from etch to diffusion or ion implant. Ion implantation plays a crucial role in chip performance improvement because of the need for smaller dimensions. Transistor performance is increasingly dependent on precise doping profiles in the silicon that are achievable only with ion implant.

Doped Regions

Recall from Chapter 4 that a wafer is uniformly doped with dopant atoms during the crystal growth process to create a p-type or n-type wafer. The type of dopant depends on the manufacturer's specification. During wafer fabrication, dopants are selectively introduced to create the devices on the individual wafers. The selective introduction of dopants into masked openings on the wafer forms a *doped region* where dopants reside within the silicon crystal structure (see Figure 17.3 on page 478). The doped region is characterized by its *dopant profile,* which determines the amount of dopant actually added in the doped region as a function of depth.

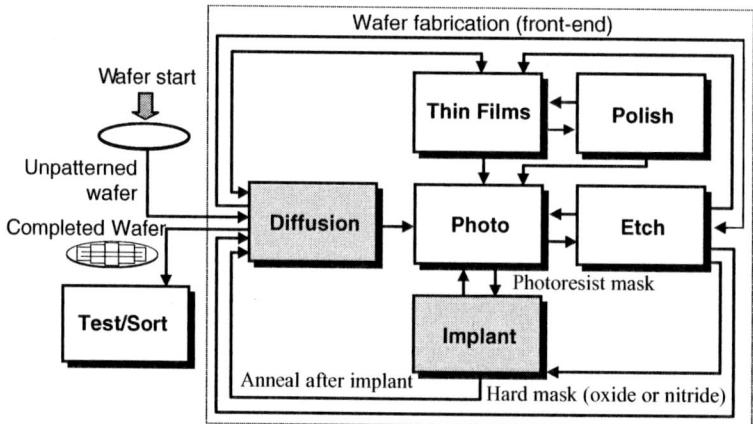

FIGURE 17.2 Ion Implant in Wafer Process Flow
(Used with permission from Advanced Micro Devices)

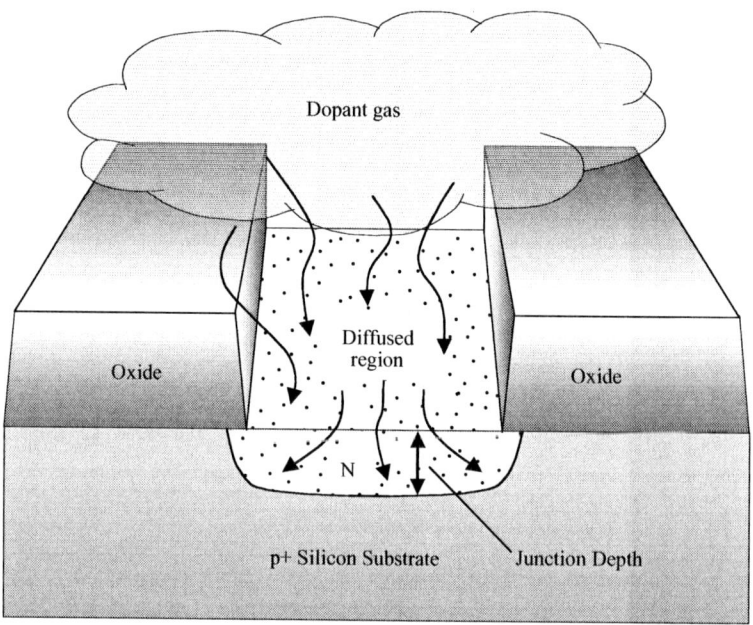

FIGURE 17.3 Doped Region in a Silicon Wafer

The dopant atoms enter the solid silicon through an oxide mask opening. In ion implantation, this mask is often photoresist, but a variety of masks can be used since it is a low-temperature process. Doping by diffusion requires an oxide or nitride mask because this technique is done in a high-temperature furnace.

The doped region can be of the opposite type as the silicon substrate, such as an n-type doped region in a p-type wafer, or it can be the same type with a different dopant concentration (e.g., a heavily doped p$^+$ region in a p-type wafer). The location where the conductivity of the doped region makes a change from p-type to n-type, or vice versa, is the pn junction. The exact depth in the wafer where the p-type dopant meets the n-type dopant is the *junction depth*, x_j. At the precise location of the junction depth, the concentration of electrons equals the concentration of holes. Another way of looking at this feature is the net dopant concentration at x_j is zero.

The ultimate interest in wafer doping is related to the individual dopant profiles for the devices after all processing is complete. A wafer undergoes many thermal excursions during processing, such as oxide growth, CVD, and so on, all contributing to undesirable dopant diffusion through the silicon. Dopant diffusion can alter critical parameters (e.g., junction depth and concentration) of the initial doped region and affect device performance. This is an important reason for minimizing the wafer thermal budget.

The main focus of this chapter will be doping by ion implantation. We will first review diffusion to analyze basic concepts and explain how intentional and unintentional diffusion occurs in wafer fabrication.

DIFFUSION

Diffusion is a basic property of matter that describes the movement of one material through another. Diffusion occurs from a region of relatively higher concentration into a region of lower concentration, resulting from the motion of atoms, molecules, or ions. In semiconductor fabrication, high-temperature diffusion is the mechanism used to move a dopant through the silicon crystal lattice. Diffusion can occur as a gas-state, liquid-state or solid-state. An example of liquid-state diffusion is a drop of food colorant in a glass of water. The highly concentrated drop of colorant immediately starts to diffuse into the glass of water and will continue until the entire volume of water is the same color as the drop of colorant. The same type of diffusion occurs in the solid-state when a high concentration of dopant material is introduced to a wafer. The dopant material diffuses through the silicon atoms of the solid crystal lattice.

Diffusion Principles

Thermal diffusion of a solid-state dopant in silicon requires three steps: predeposition, drive-in, and activation. During *predeposition,* or *predep,* wafers are loaded into a high-temperature diffusion furnace and a quantity of dopant atoms is transferred from the source cabinet to the diffusion furnace. The furnace temperature setting is typically 800 to 1100°C for 10 to 30 minutes. The dopants are introduced into a very thin layer of the wafer while maintaining a constant concentration of dopants at the surface. A thin oxide layer, referred to as a *cap oxide,* is grown on the surface to prevent dopant atoms from diffusing out of the silicon. The total number of dopant atoms deposited is referred to as Q. Note that for ion implantation (discussed later in this chapter), Q is referred to as the dose and also represents the number of dopant atoms implanted.

The predeposition step establishes the concentration gradient for the entire diffusion process. The concentration of dopants is highest at the surface and reduces when moving away from the surface, forming a gradient (or slope). This slope establishes the dopant profile, which can be mapped using four point probes (see Chapter 7). *Fick's laws,* first postulated in 1855, mathematically describe the diffusion process. These laws state in part that the number of particles moving through a cross-sectional area is proportional to the concentration gradient.[2] Fick's laws are used to predict the dopant concentration at some distance, x, from the surface.

The second part of thermal diffusion is the *drive-in* step. This is a high-temperature process (1000 to 1250°C) used to move the deposited dopants through the silicon crystal to the desired junction depth in the wafer. No additional dopants are added to the wafer. However, the wafer surface oxidizes in the high-temperature environment, which affects the diffusion of the dopant atoms during drive-in. Some dopants, such as boron, tend to move into the growing oxide layer while others, such as phosphorus, are pushed away from the SiO_2. This modification of dopant concentration due to the silicon surface oxidation is termed *redistribution.*

The third part of thermal diffusion is the *activation* step. Here the temperature of the furnace is raised slightly higher to enable the dopant atoms to bond with the silicon atoms in the lattice structure. This action activates the dopant atoms, which changes the conductivity of silicon.

Dopant Movement ■ Each dopant has a particular *diffusivity* in silicon, which represents the rate (or speed) that the dopant moves in the wafer. The higher the diffusivity, the faster the dopant moves. Diffusivity increases with temperature, which is reflected in a term known as the *diffusion coefficient,* D, of the dopant being diffused. Elements have different diffusion coefficients within the temperature range of the drive-in step that reflect whether they are fast or slow diffusers in silicon. The diffusion coefficient is an equation variable in Fick's laws used to predict the final dopant concentration in silicon.

Within the wafer, dopant atoms move by two different mechanisms: interstitial and substitutional (see Figure 17.4 on page 480). Dopants with high diffusivity, such as gold (Au), copper (Cu), and nickel (Ni), use primarily *interstitial movement* to move between the interstitial space between regular crystal sites of the silicon lattice. Slower moving dopants, such as arsenic and phosphorus that are common in semiconductor doping, typically use *substitutional movement* where atoms fill empty crystal positions in the lattice.

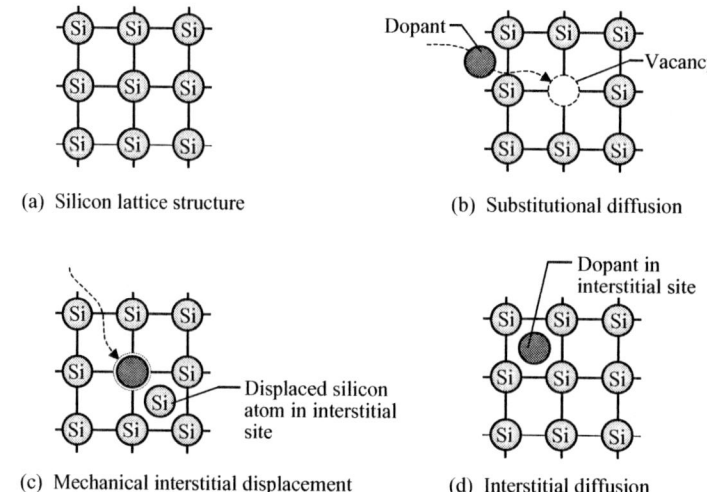

FIGURE 17.4 Dopant Diffusion in Silicon

Dopants are useful to form semiconductor silicon only if they are *activated dopants*, meaning they are part of the silicon lattice structure. An activated dopant can act as a donor or acceptor of electrons and is an n-type or p-type dopant with respect to silicon. If the dopant occupies interstitial space, then it is not activated and is ineffective as a dopant. Thermal energy moves dopants into the regular crystal sites, a process known as *crystal activation*. Activation is a regular part of diffusion because it occurs at a high temperature. For ion implant, crystal activation is done with an anneal step.

Solid Solubility ■ At a given temperature, there is a limit to how much dopant can be absorbed by the silicon. This is referred to as the *solid solubility limit*. This is true for most materials, such as pouring salt into a container of water and watching it dissolve. If salt is continually added, at some point the solubility of the salt in the water reaches a limit where it no longer dissolves. Each particular dopant has a solid solubility limit (see Table 17.3). Note that only a fraction of the dopants in silicon are actually activated and contribute electrons or holes for conduction (about 3 to 5%).[3] Most dopants remain in interstitial sites and are electrically inactive.

TABLE 17.3* Solid Solubility Limits in Silicon at 1100°C

Dopant	Solubility Limit (atoms/cm^3)
Arsenic (As)	1.7×10^{21}
Phosphorus (P)	1.1×10^{21}
Boron (B)	2.2×10^{20}
Antimony (Sb)	5.0×10^{19}
Aluminum (Al)	1.8×10^{19}

*SEMATECH "Diffusion Processes," *Furnace Processes and Related Topics* (Austin, TX: SEMATECH, 1994), p. 15.

Lateral Diffusion ■ The diffusion mask is either silicon dioxide (SiO_2) or silicon nitride (Si_3N_4) since a photoresist mask could not withstand the high-temperature process. When atoms diffuse into the wafer, they move in all directions: down into the silicon, laterally, and back out of the wafer. *Lateral diffusion* occurs when dopant atoms travel in a direction along the surface of the wafer. Typically, lateral diffusion during a thermal diffusion process is 75 to 85% of the vertical junction depth.[4] Lateral diffusion is undesirable in advanced MOS circuits because it can cause a reduction in the channel length, affecting device density and performance.

Diffusion Process

The objective of the diffusion process is to bring the diffusing impurity in contact with a wafer and maintain the specified time and temperature for diffusion to occur. Diffusion takes place in high-temperature diffusion furnaces, discussed in Chapter 10. Typically, one furnace is used for the pre-deposition, drive-in, and activation steps. Quartz and other furnace parts used in diffusion processes should be segregated from other furnace processes to avoid cross-contamination.

Diffusion should produce reproducible results from run-to-run and wafer-to-wafer. The eight steps required to properly perform diffusion in wafer fabrication are:

1. Run qualification test to ensure the tool meets production quality criteria.
2. Verify the wafer properties using a lot control system.
3. Download the process recipe with the desired diffusion parameters.
4. Set up the furnace, including a temperature profile.
5. Clean the wafers and dip them in hydrofluoric (HF) to remove native oxide.
6. Perform predeposition: load wafers into the deposition furnace and diffuse the dopant.
7. Perform drive-in: increase the furnace temperature to drive-in and activate the dopant bonds, then unload wafers.
8. Measure, evaluate, and record the junction depth and sheet resistivity.

The first six steps are done to prepare wafers for diffusion in the drive-in operation of step 7. A qualification test (or "qual") is performed on the furnace. This test determines the tool's production worthiness in terms of particles and other specified criteria. Once the tool is proven production worthy, the appropriate wafer lots are identified and the proper process recipe is downloaded to the furnace. A thermal profile should always be done to verify furnace temperatures are correct for the process to be performed. In a production environment, a given furnace is dedicated to a specific process step and is maintained at the precise temperature profile.

Wafer Cleaning ■ Wafer cleaning is critical because contaminants can block the diffusion of the dopant atoms into the wafer. Wafers should be cleaned immediately prior to loading the wafers into the furnace to minimize wafer contamination (e.g., native oxide). Cleaning typically includes an immersion in an acid and oxidizer, followed by an etch in an HF solution to remove any remaining oxides (see Chapter 6 for wet cleaning). The use of a dilute HF solution is referred to as a *deglaze* step to remove the poor quality native oxide on the wafer surface.

Dopant Sources ■ Although they may have been used in the early days of the semiconductor industry, it is impractical to use pure elements as a dopant source for submicron IC manufacturing. For instance, boron and phosphorus are solids at room temperature with low vapor pressures, making them difficult to melt or vaporize. Dopant concentration is also difficult to control when using a solid dopant source. Dopants are typically supplied as a gaseous or liquid source in the form of a compound. Some of the most common dopants are listed in Table 17.4 on page 482.

A liquid source (bubbler) has a carrier gas (e.g., nitrogen) bubbled through it and is delivered to the furnace tube as a vapor. Oxygen is also supplied for the dopant source to react and form oxides. Using boron as an example, the boric oxide undergoes a second reaction with silicon to form a boron-rich silicon dioxide layer on top of the wafer. This layer serves as the local source of boron for the predeposition step. The following two reaction equations illustrate how the diborane source is converted into a boron dopant:[5]

1st Reaction: B_2H_6 (gas) + $3O_2$ (gas) → B_2O_3 (solid) + $3H_2O$ (liquid)
 diborane *oxygen* *boric oxide* *water*

2nd Reaction: $2B_2O_3$ (solid) + $3Si$ (solid) → $4B$ (solid) + $3SiO_2$ (solid)
 boric oxide *silicon* *boron (dopant)* *silicon dioxide*

TABLE 17.4* Typical Dopant Sources for Diffusion

Dopant	Formula of Source	Chemical Name
Arsenic (As)	AsH_3	Arsine (gas)
Phosphorus (P)	PH_3	Phosphine (gas)
Phosphorus (P)	$POCl_3$	Phosphorus oxychloride (liquid)
Boron (B)	B_2H_6	Diborane (gas)
Boron (B)	BF_3	Boron tri-fluoride (gas)
Boron (B)	BBr_3	Boron tri-bromide (liquid)
Antimony (Sb)	$SbCl_5$	Antimony pentachloride (solid)

*SEMATECH "Diffusion Processes," *Furnace Processes and Related Topics,* (Austin, TX: SEMATECH, 1994), p. 7.

Gaseous sources are delivered to the furnace directly from the gas cylinder through a mass flow controller. The dopant is diluted in the cylinder with an inert gas to help prevent system corrosion and control the flow of the gas. A point-of-use filter should always be used to prevent fine particulates from getting into the furnace. Some diffusion sources are highly toxic, especially the gaseous sources. The most toxic of the commonly used dopant sources are arsine (AsH_3) and diborane (B_2H_6). Proper storage and usage is required at all times, especially to avoid a gas leak.

ION IMPLANTATION

Ion implantation is a method for introducing controlled amounts of dopants into the silicon substrate to change its electronic properties. It is a physical process—there are no chemical reactions. A number of applications use ion implantation in modern wafer fabrication (see Table 17.2 on page 476). The main application of ion implantation is the doping of semiconductor materials. Each doping application has specific requirements for the concentration of the dopant material and its depth. Ion implantation is preferred over diffusion in nearly all applications because of its ability to repeatedly control dopant concentration and depth. It has become a critical process to meet the challenges of wafer fabrication with sub-quarter micron feature sizes and larger diameter wafers (see Figure 17.5).

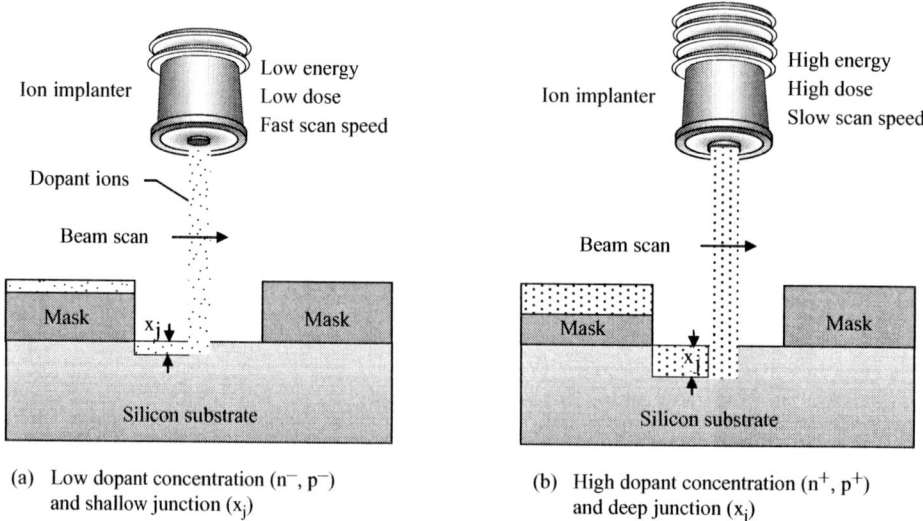

(a) Low dopant concentration (n^-, p^-) and shallow junction (x_j)

(b) High dopant concentration (n^+, p^+) and deep junction (x_j)

FIGURE 17.5 Controlling Dopant Concentration and Depth

Overview

Ion implant processing is done in one of the most complex semiconductor processing tools, called the *ion implanter* (see Figure 17.6). The implanter has an ion source component that creates positive-charged dopant ions from a source material. The ions are extracted and then separated in a mass analyzer to form a beam of the desired dopant ions. The number of ions in the beam is related to the desired concentration of dopants to be introduced into the wafer. The beam of ions is accelerated in a voltage field to attain a high velocity (on the order of 10^7 cm/sec).[6] Because of the high velocity, the ions have kinetic energy that is used to implant the dopants into the silicon crystal lattice structure of the target wafer. The beam scans the wafer to provide uniform doping across the wafer surface. Implantation is followed by a thermal anneal step to activate the dopant ions in the crystal structure. All implanter processing is done in a high vacuum.

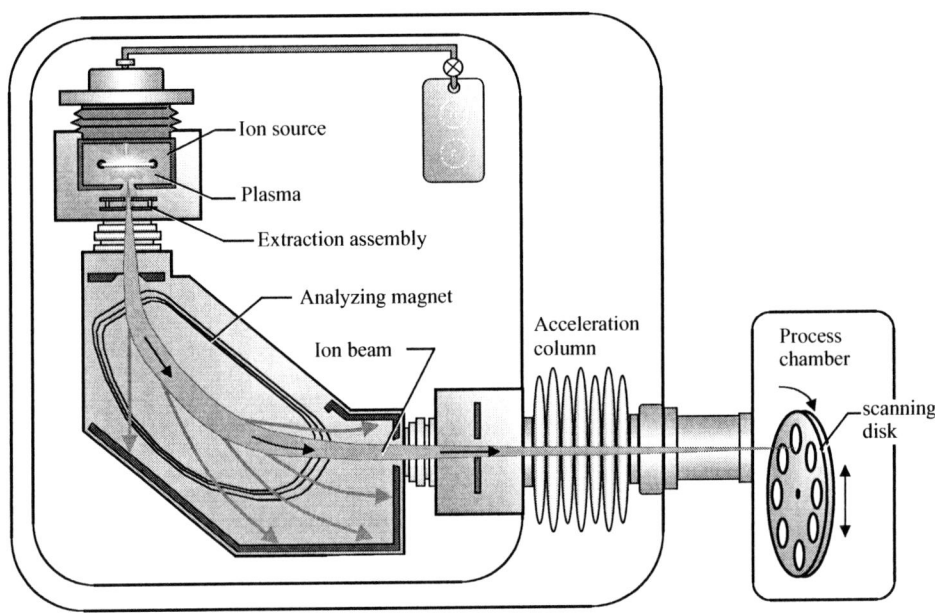

FIGURE 17.6 General Schematic of an Ion Implanter

Ion Implanter
(Photo courtesy of
Varian Semiconductor
Equipment, VIISion 80)

There are two major goals for sub-0.25 μm doping requirements:

1. To introduce a uniform, controlled amount of a specific dopant into the wafer
2. To place the dopants at a desired depth.

Advantages of Ion Implant ■ The primary advantages of ion implantation are listed in Table 17.5. These benefits effectively increase the process flexibility for fabricating wafers with sub-quarter micron technology.

TABLE 17.5 Advantages of Ion Implantation

Advantages	Description
1. Precise control of dopant concentration.	Ability to precisely control the number of implanted dopants over a wide range of concentrations, from 10^{10} to 10^{17} ions/cm^2, and controlled to within ±2% over this range. In contrast, diffusion can control dopant concentration to within 5 to 10% at high concentrations and becomes less effective at low concentrations.
2. Good dopant uniformity.	Controls dopant uniformity by scanning the ion beam over a large wafer area, which is critical for smaller feature sizes on larger wafers.
3. Good control of dopant penetration depth.	Controls the ion energy during the implant to achieve good control of dopant penetration depth. This permits design flexibility, such as buried layers where the maximum dopant concentration is at the buried layer and the minimum concentration is at the wafer surface.
4. Produces a pure beam of ions.	Mass separation techniques produce a pure beam of ions free of contamination. Different impurities can be selected for implantation. It is done in a high vacuum that minimizes contaminants.
5. Low temperature processing.	Implantation is performed at moderate temperatures (<125°C) which permits the use of a variety of masks, including photoresist.
6. Ability to implant dopants through films.	Dopants can be implanted through films, such as an oxide or nitride. This condition allows adjustment of the threshold voltage of MOS transistors after the gate oxide is grown. It increases the flexibility to implant at various stages of the process.
7. No solid solubility limit.	The dopant concentration during implantation is not limited by the solid solubility limit of the silicon wafer.

Disadvantages of Ion Implant ■ The major disadvantage of ion implantation is the damage done to the crystal structure when the energetic dopant ions bombard the silicon host atoms. As the high-energy ions enter the crystal and collide with substrate atoms, energy is transferred and some silicon atoms are displaced from the lattice. This action is referred to as *radiation damage*. Most or all of this crystal damage is corrected with a high-temperature anneal step.

Another drawback of implantation is the complexity of the implanter equipment. However, this drawback is offset by the implanter's capability for dose and depth control and overall process flexibility.

Ion Implant Parameters

Ion implant is a flexible process that must meet stringent chip design and productivity requirements for wafer fabrication. The important ion implant parameters are:

♦ Dose
♦ Range

Dose ■ The *dose* (Q) is the number of implanted ions per unit area of wafer surface, with units of atoms/cm^2 (also stated as ions/cm^2). To count the implanted ions, Q is calculated by the following dose equation:

$$Q = \frac{It}{enA}$$

Where, Q = dose in atom/cm^2
I = beam current in coulombs/sec (amperes)
t = implant time in seconds
e = electronic charge equal to 1.6×10^{-19} coulombs
n = charge per ion (for instance, B$^+$ is equal to 1)
A = area implanted in cm^2

One of the main reasons why ion implantation is so important in wafer fabrication is its ability to repeatedly put the same dose into each wafer. The implanter does this by virtue of the positive charge on each ion. When formed into a beam, the flux of positive dopant ions represents a *beam current* measured in milliamps (mA). A low-to-medium beam current is from less than 0.1 mA up to about 10 mA, while a high beam current is from about 10 mA up to approximately 25 mA. As seen in the equation above, the magnitude of the beam current is a key variable for defining the dose. If the beam current increases, then more dopant atoms are implanted per unit time. A high beam current is often desirable to increase wafer throughput (more ions are implanted per unit time of production), but can also create uniformity problems.

Range ■ The ion *range* is the total distance an ion travels in the silicon during ion implantation. To characterize range it is necessary to understand energy. When the ions are accelerated due to electric potential difference to a high velocity, they gain energy. The ion energy is kinetic energy (KE) due to its motion, and it typically is expressed in units of joules. However, for ion implantation, the energy is usually stated in terms of the number of the electronic charge times the difference in potential, or electron volts (eV). The equation that describes this energy is:

$$KE = nV$$

Where, KE = energy in electron volts (eV)
n = charge state of the ion (e.g., "+" = 1, "++" = 2)
V = voltage difference in volts

For example, if an ion with a single positive-charge state moves in an electric field with a voltage difference of 100,000 volts (100kV), then its energy is:

$$\begin{aligned} KE &= nV \\ &= (1)(100\text{kV}) \\ &= 100 \text{ keV} \end{aligned}$$

Higher implanter energy means the dopant atom will have an increased range and travel deeper into the wafer. Energy is important in implanters because of the need to control the range and, therefore, junction depth of the dopants. High-energy implanters have an energy greater than 200 keV and may be as high as 2 to 3 MeV. High-energy implant is used in applications such as deep retrograde wells and retrograde triple wells (see Table 17.6 on page 486). A retrograde well has higher doping concentrations deeper in the well than at the surface. Ultralow-energy implanters have energies presently down to about 200 eV to enable doping at very shallow depths for source/drain applications.[7]

There is a *projected range,* R_p, which is how far the implanted ions travel into the wafer, depending on the ion mass and energy, the target mass and the beam direction with respect to the wafer crystal structure (see Figure 17.7 on page 486).[8] At the same time, not all ions come to stop simply at the projected range; some stop at a shorter distance and others at a greater distance. Ions will also move in a lateral direction. Combining all these ion movements produces a distribution of distances traveled by the dopant atoms implanted in the wafer, referred to as *straggle,* or ΔR_p. The R_p (projected range) indicates how shallow or deep the junction depth can be formed, whereas the ΔR_p (straggle) represents the spread of the implanted species around R_p. As the implant energy of

TABLE 17.6 Classes of Implanters

Class of Implanter System	Description and Applications
Low/Medium Current	• Highly pure beam currents <10 mA. • Beam energy is usually <180 keV. • Most often the wafer is stationary and the ion beam is scanned. • Specialized applications of punchthrough stops.
High Current	• Generate beam currents >10 mA and up to 25 mA for high dose implants. • Beam energy is usually <120 keV. • Most often the ion beam is stationary and the wafer does the scanning. • Ultralow energy beams (<4keV down to 200 eV) for implanting ultrashallow source/drain junctions.
High Energy	• Beam energy exceeds 200 keV up to several MeV. • Place dopants beneath a trench or thick oxide layer. • Able to form retrograde wells and buried layers.
Oxygen Ion Implanters	• Class of high current systems used to implant oxygen in silicon-on-insulator (SOI) applications.

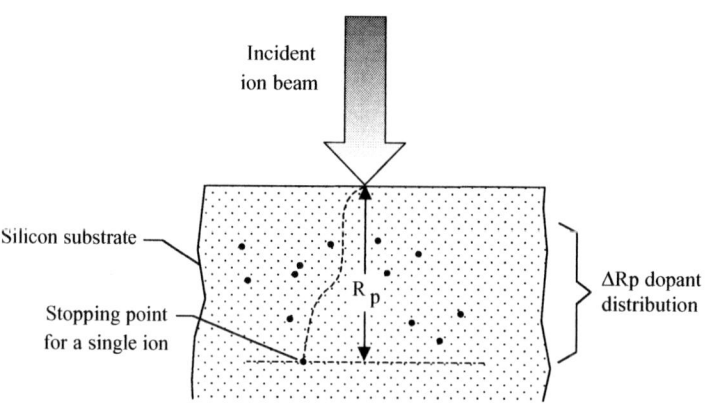

FIGURE 17.7 Range and Projected Range of Dopant Ion

dopant atoms increases, the projected range increases but the peak concentration of dopants decreases because the range straggling increases. Projected range charts are available that predict the projected range for a given implant energy (see Figure 17.8).

The implanted ions will travel some depth into the target wafer before they are brought to rest by energy loss due to collisions with silicon atoms (see Figure 17.9). The two primary energy loss mechanisms are electronic stopping and nuclear stopping.[9] *Electronic stopping* of dopant atoms is caused by interactions with the target electrons, similar to stopping a projectile in a thick medium, such as a child jumping into a pile of plastic balls. *Nuclear stopping* of implanted ions is caused by collisions between atoms that cause a displacement of silicon atoms. It can be visualized as the collision between two hard spheres, such as billiard balls. Depending on the ion mass and energy, an implanted atom can displace as many as 10^4 silicon atoms by nuclear collisions before coming to rest.[10]

The dopant moving through the silicon atoms creates a damage path in the crystal lattice, with the nature of the damage depending on whether the dopant ion is light or heavy (see Figure 17.10). A light dopant atom glances off silicon atoms with little energy transfer and is deflected through a large scattering angle. A heavy ion has a large energy transfer each time it strikes a silicon atom and is deflected through a relatively small scattering angle. Each displaced silicon atom also produces a large number of displacements.

Ion Implant **487**

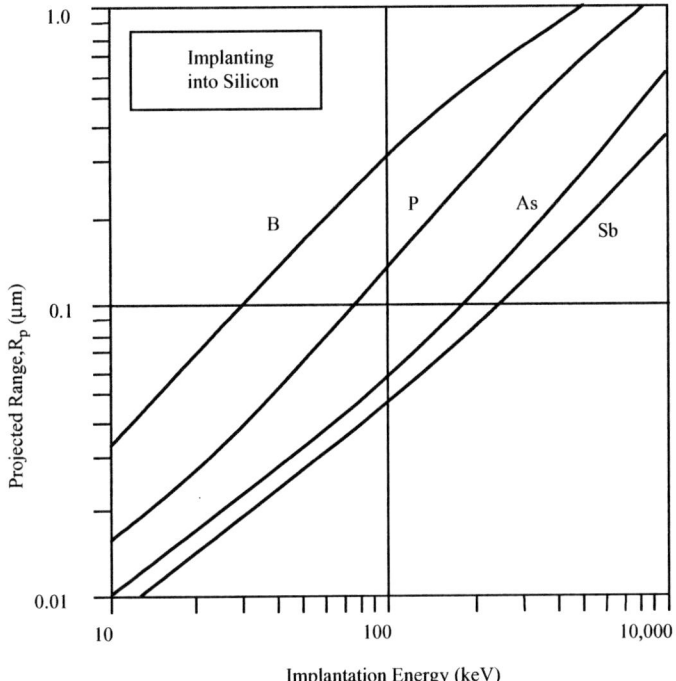

FIGURE 17.8 Projected Range Chart for Implant Energy
Redrawn from B. El-Kareh, *Fundamentals of Semiconductor Processing Technologies,* (Boston: Kluwer, 1995), p. 388.

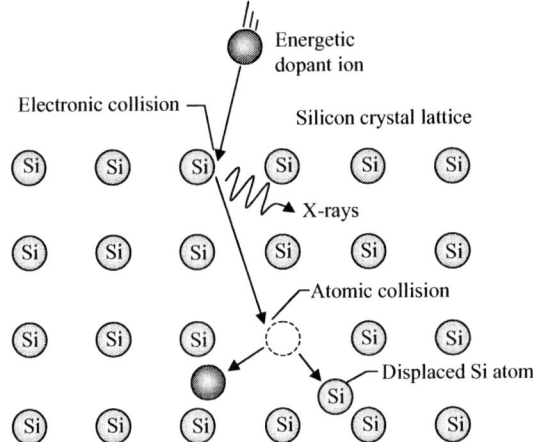

FIGURE 17.9 Energy Loss of an Implanted Dopant Atom

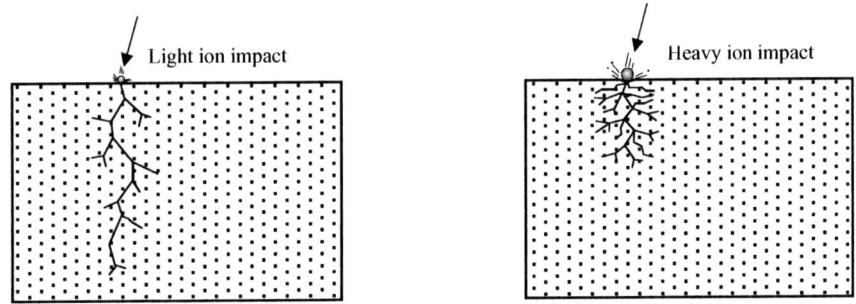

FIGURE 17.10 Crystal Damage Due to Light and Heavy Ions

ION IMPLANTERS

An ion implant tool consists of the following five major subsystems:

- Ion source
- Extraction electrode and ion analyzer
- Acceleration column
- Scanning system
- Process chamber

The ion source and extraction assemblies are generally mounted in the same vacuum chamber (see Figure 17.11). The ion source generates positive ions from a specific dopant gas or solid. The circular suppression electrode attached to the extraction assembly uses the electric field of a negative high voltage to draw the positive ions out of the ion source. Electromechanical controls on the extraction assembly aid in positioning the extraction electrode directly in front of the ion source plasma chamber. On older ion implanters, these controls could be manually adjusted from the operator's control panel. Today these controls are set by a process recipe and maintained by the tool's internal computer software.

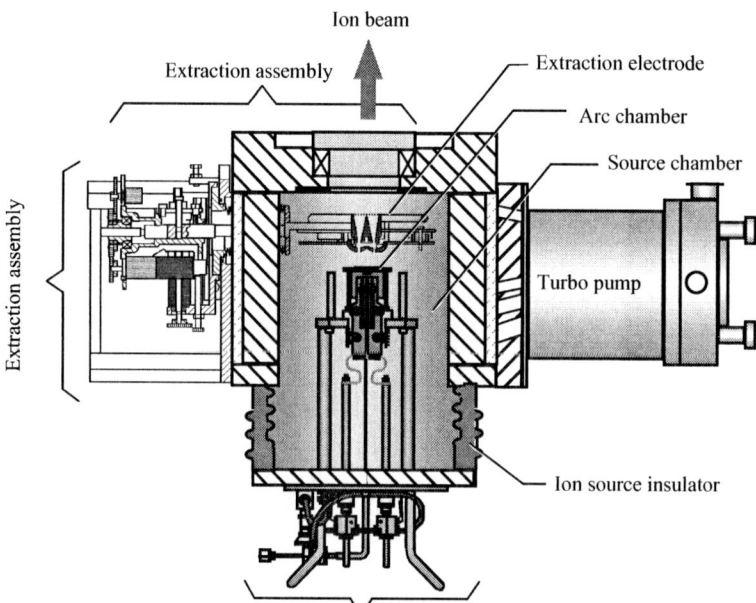

FIGURE 17.11 Ion Source and Extraction Assembly Housing
(Used with permission from Applied Materials, Inc., Precision Implanter 9500)

Ion Source

The species to be implanted must be present in the beam as a charged particle, or ion. Because of their electrical charge, ions can be controlled and accelerated with electric and magnetic fields. Ions for implantation are generated in an *ion source*. A positive ion is formed from a dopant gas source or by vaporizing a solid. Typically B^+, P^+, As^+, and Sb^+ are produced by ionization of atoms or molecules. The most common feed material for a source of dopant atoms is a gas, such as B_2H_6, BF_3, PH_3, and AsH_3. The gas is packaged in a relatively small cylinder (0.4 to 2.2 liters in size) and diluted with hydrogen to reduce the risks associated with inadvertent gas releases.

Another method of supplying the feed material is by heating and vaporizing a solid material, which is sometimes used to obtain arsenic (As) or phosphorus atoms from solid pellets.[11] The solid material is vaporized at about 900°C and the volatile dopant atoms are transported to the ion source chamber. No dilute hydrogen is added, which permits higher maximum beam currents. A disadvantage of a solid source is the long setup time, or tune time, which can be 40 to 180 minutes. Most of this time is required to heat and stabilize the vaporizer. There is also the need for more frequent maintenance due to the buildup of sputtered materials on the beam-line components. However, many IC manufacturers prefer to use solid dopant sources because it is easier to deal with them from an environmental and safety standpoint.

Ion Creation ■ Ions are generated in the ion source by bombarding the feed gas atoms with electrons. Electrons are often produced by a simple and robust hot tungsten filament source. The *Freeman ion source* is one of the most common electron sources.[12] In this source the rod-shaped cathode filament is held inside an arc discharge chamber with a feed gas inlet. The walls of the arc chamber are the anode. When feed gas is introduced, a plasma is created around the filament in the arc chamber by passing a high current through the filament and applying a voltage of about 100V between the cathode and anode. Positive ions are produced by the collisions between energetic electrons and the feed gas molecules. An external source magnet applies a magnetic field parallel to the filament to increase ionization and to stabilize the plasma. During operation the current between the cathode and anode is adjusted, along with the feed gas, to maintain a stable plasma. A variation of the Freeman ion source is the Bernas ion source that uses a pigtail type of filament and a negative-charged reflector plate to improve electron confinement and efficiency (see Figure 17.12).

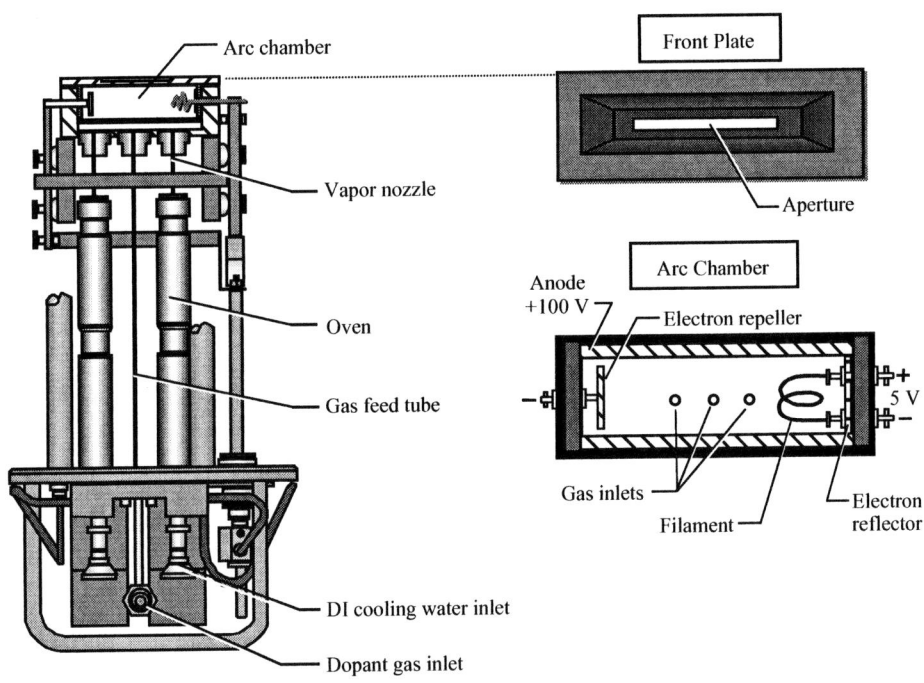

FIGURE 17.12 Schematic of Bernas Ion Source Assembly
(Used with permission from Applied Materials, Inc., Precision Implanter 9500)

The bombardment of electrons breaks up the feed gas molecule into different ion species. Boron trifluoride (BF_3) is often used as a feed gas because boron in its elemental form has a very low vapor pressure, requiring a temperature of 2000°C to vaporize it. BF_3 also has greater mass and lower diffusivity than boron, making it a preferred p-type dopant for forming shallow junctions. When BF_3 is used as the source of boron, many different ion species are generated in the ion source. These include B^+, B_{10}^+, B_{11}^+, BF^+, BF_2^+, F^+, and F_2^+. The B^+ ion is the atom desired to be implanted into the wafer and will be selected when the beam passes through the analyzer magnet.

There are other designs for ion sources, such as an RF (radio frequency) ion source, cold cathode source, and microwave ion source. An RF source excites the feed gas molecules in a magnetic environment, generating a cooler plasma with the possibility of higher beam currents with a longer ion source lifetime.[13]

Extraction and Ion Analyzer

The traditional implanter extraction system collects all the positive ions created inside the ion source and forms them into a beam. The ions are extracted through a slit in the ion source. They are repelled by the positive bias arc chamber (anode) and are attracted to the negative bias (cathode) of the *extraction assembly* (see Figure 17.13). Since positive ions are like little magnets, each with a positive charge, they are attracted to a negative electric field. The higher the electric field, then the faster the ion will move. At the same time, the faster the ion moves, then the higher is its kinetic energy and the farther it will penetrate into the wafer when it strikes the surface. The negative bias of the extraction electrode also repels any stray electrons in the source plasma and forms the positive ions into a beam. A negatively biased *suppression electrode* is used to focus the ion beam into a parallel beam for efficient beam transport through the implanter. This condition is important for delivering high beam currents to wafers.

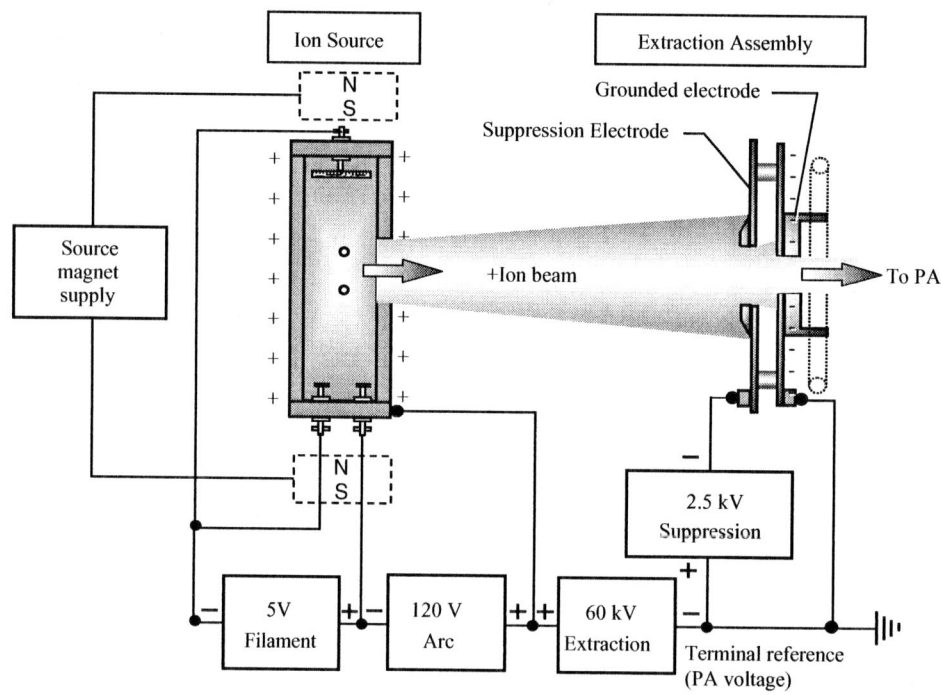

FIGURE 17.13 Interaction of Ion Source and Extraction Assemblies
(Used with permission from Applied Materials, Inc., Precision Implanter 9500)

Mass Analyzer Magnet ■ The ions extracted from the source contain many different ion species and travel at a relatively high speed due to the acceleration provided by the extraction voltage. The species in the ion beam have different atomic mass units (amu). In ion implanters, a magnetic *ion analyzer* separates the desired dopant ion from the main body of ion species. As shown in Figure 17.14, the analyzer magnet is shaped in a 90° angle. The magnetic field of the analyzer magnet causes the ion species to be deflected into an arc. A spectrum of ions is formed based on their amu. For a given field strength, ions with heavy masses are not able to bend at the appropriate angle, while lighter ions bend too easily. There is one ion species, however, that bends just enough at the appropriate field strength to allow it to pass through the center of the analyzer magnet. This is the dopant species that eventually gets implanted into the wafer.

The radius of the arc formed by the ion depends on the mass of the individual species, its speed, the strength of the magnetic field, and the ion's electronic charge. The magnetic field strength is adjusted to match the desired path of the dopant ion. This desirable dopant ion passes through a slit at the end of the analyzer while all other ions strike the walls of the analyzing magnet.

Ion Implant **491**

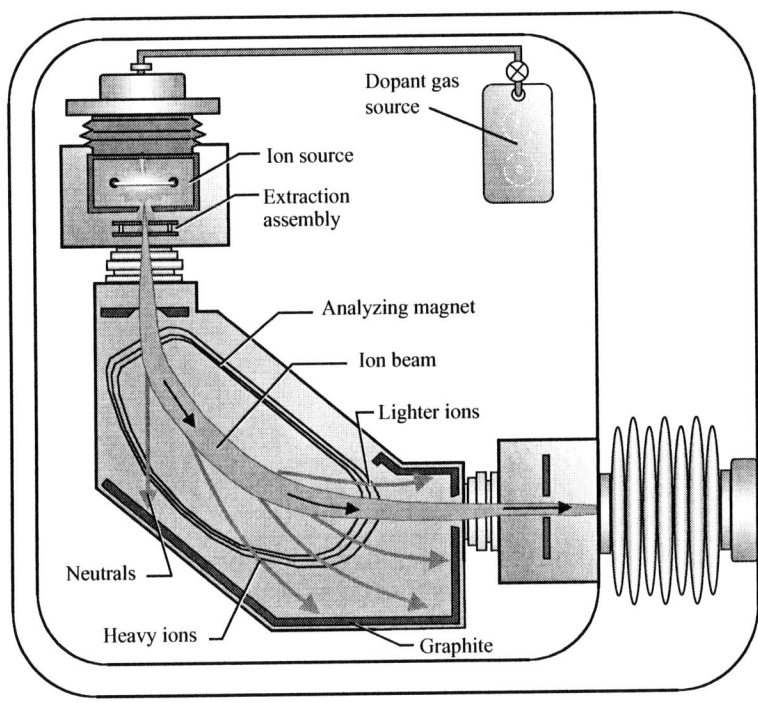

FIGURE 17.14 Analyzing Magnet

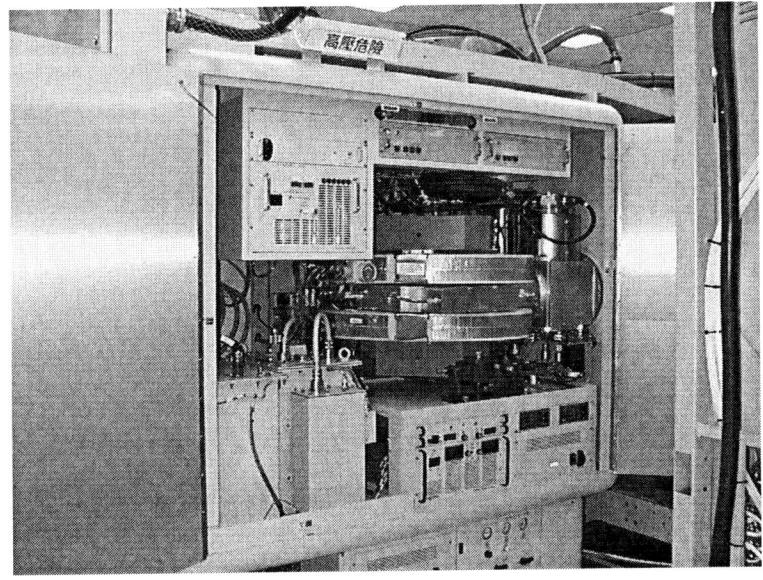

Ion Implanter
Analyzing Magnet
(Photo courtesy
of Varian
Semiconductor
Equipment,
VIISion 80)

Acceleration Column

To achieve additional ion acceleration (and therefore energy) beyond the analyzer magnet, positive ions are accelerated in an electric field inside an *acceleration column* (see Figure 17.15 on page 492). This acceleration column is a linear design made of a series of electrodes separated by insulators, each with an increasing negative charge. As the positively charged ions enter the column, they start accelerating. In this manner the total voltage difference in the electrode assembly and the acceleration column accelerates the dopant ions. The higher the total voltage, then the higher the velocity and, therefore, energy. Higher beam energy means the dopant ions are implanted deeper into the wafer, while lower beam energy is used for ultrashallow implants. Different beam energies relative to the dose are shown in Figure 17.16 on page 492.

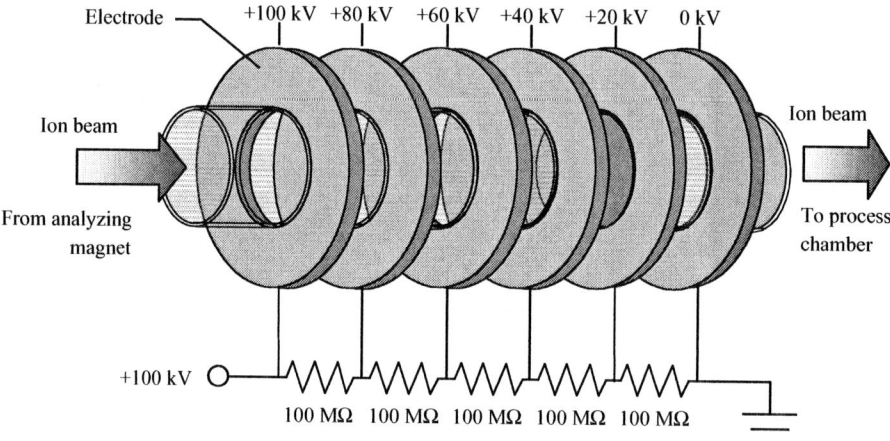

FIGURE 17.15 Acceleration Column

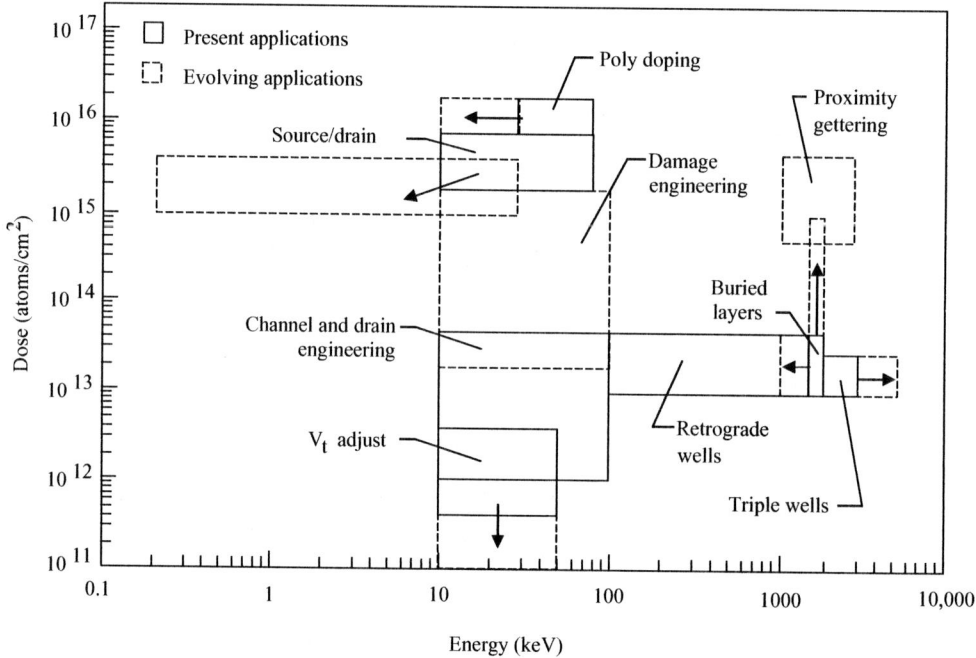

FIGURE 17.16 Dose Versus Energy Map
(Used with permission from Varian Semiconductor Equipment)

The dose versus energy map highlights the importance of energy (range) and dose (concentration) for ion implantation.[14] Beam energy defines the projected range that the ions are implanted in the wafer surface and varies from high energy (>200 keV and into multiple MeV) to low energy (<120 keV and down to 200 eV). Dose is directly related to the concentration of dopants in the wafer and is represented by the beam current, or number of ions in the beam, ranging from low, medium, to high current (>15 mA). Implanters are often classified according to their maximum operating beam current and acceleration energy.

Post-Accelerator ■ There is additional beam focusing that takes place in the *post-accelerator* component of the acceleration column. It typically uses quadrupole lens focusing based on four cylindrical poles in a two-lens system, employing electrostatic or magnetic repulsion to form the ion beam into a focused, circular beam.

High-Current and High-Energy Beam ■ High-energy implanters are used to implant buried dopant layers, such as for retrograde and triple wells. Most commercial high-energy ion implanters employ *linear accelerator* technology to accelerate ions to high velocity and obtain beam energies in excess of 200 keV and into the MeV range.[15] High beam current (up to 25 mA) is desirable to reduce the implant time and increase wafer throughput through the implanter. A linear accelerator is constructed as an alternating series of high-voltage (several 10 kV) electrodes and grounded quadrupole focusing lenses (see Figure 17.17). The timing of the voltage between each electrode and grounded quadrupole is set to match the arrival of the dopant ions at each gap. The ions are accelerated through the gaps, with the final energies up to 20 times higher than the maximum acceleration in a traditional beamline acceleration column. This condition avoids the use of extremely high voltages. A traditional bending magnet is located at the end of the linear accelerator that serves as an energy analyzer to ensure that a pure, monoenergetic beam is formed for wafer implant. This added analysis magnet also aids in the removal of beam contaminants by removing masses other than the desired species.

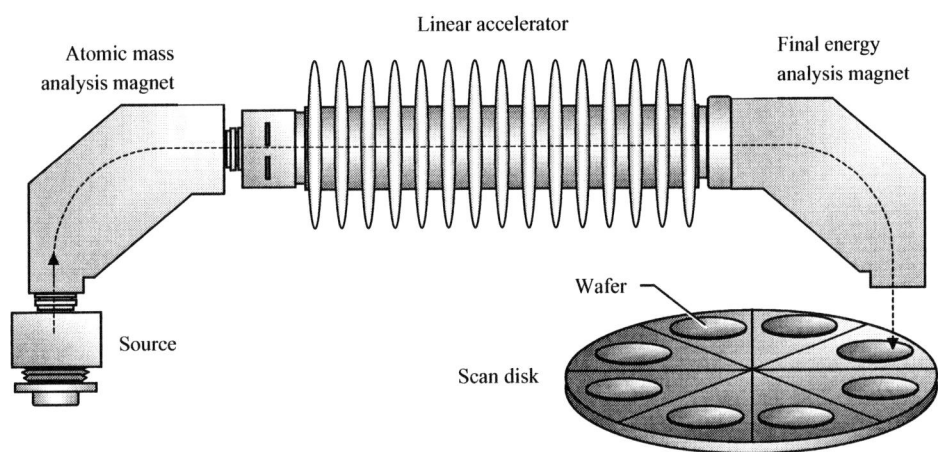

FIGURE 17.17 Linear Accelerator for High-Energy Implanters

High-Current and Low-Energy Beam ■ The need to form ultrashallow implants has promoted the development of high-current, low-energy beams. A low-energy beam accelerates ions to a low velocity so that their projected range is shallow in the target wafer. Maintaining a well-focused beam with low-energy acceleration is difficult because beam size is hard to control and usually increases with high current and low energy. Visualize this action by imagining a bowling ball rolled slowly down a bowling alley lane. It tends to veer to the side and into the gutter much more easily than a bowling ball rolled straight and fast. A method used to achieve a high beam current in a low-energy beam is *beam deceleration,* or *decel.*[16] With this approach, the beam is extracted from the ion source at a high energy and then decelerated down to the desired energy without loss of beam current. An electrode positioned farther down the beamline is used to decelerate the ions from the initial extracted energy. This approach may cause energy contamination of the beam if high-energy ions manage to pass through to the wafer. A variation to decel is the *differential lens,* which combines a short beam path and optimized optics to minimize beam current loss.

Space Charge Neutralization ■ A beam of only positively charged ions is inherently unstable, since like charges repel one another with mutual repulsion. This condition causes *beam blow-up,* meaning the diameter of the beam increases as the ions pass down the beamline, resulting in nonuniform implantation. Beam blow-up is minimized by ensuring that the positive ions in the beam are neutralized with secondary electrons, a condition referred to as *space charge neutralization.*

Secondary electrons are generated when high-energy dopant ions strike a surface (e.g., aperture plates or beam guide assembly) as they pass along the beamline. The secondary electrons become trapped within the beamline due to the negative-biased electrodes along the beam path, including the extraction electrode and the acceleration. The secondary electrons bounce back and forth between the negatively biased plates, unable to leave their specific region in the beamline. The trapped electrons serve to neutralize the beam and prevent or reduce beam blow-up (see Figure 17.18).

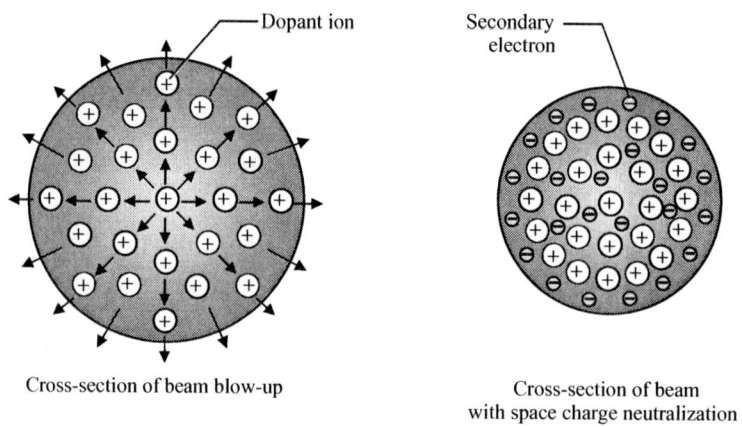

FIGURE 17.18 Space Charge Neutralization

Neutral Beam Trap ■ The positively charged ion beam is formed in a vacuum of less than 10^{-6} torr, but there are always residual gas molecules. Neutral ions are formed in the beam when a dopant ion gains an electron by colliding with a residual gas molecule. They cannot be deflected because of their lack of charge and, if not removed, they will be implanted. A *neutral beam trap* is created by placing a bend in the beamline a short distance before the target chamber using beam-bending electrodes. Since the neutral ions will not be deflected by the electrodes at the bend, they travel straight and collide with a grounded collector plate (see Figure 17.19).

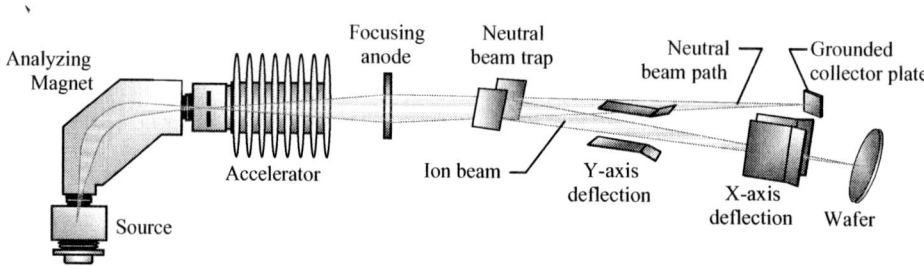

FIGURE 17.19 Neutral Beam Trap
(Used with permission from Varian Semiconductor Equipment)

Scanning System

The goal of the implanter is to form a highly pure ion beam from both a contamination and energy standpoint. This pure beam is implanted into the target wafer. That is, the beam has only the desired dopant ions at a predetermined energy level. Because the focused ion beam is typically small, on the order of 1 cm^2 for medium implanters and 3 cm^2 for high-current implanters, it must be scanned to cover the entire wafer. Wafer scanning is done either by moving the beam over a stationary wafer or moving the wafers through a stationary beam. Scanning plays a critical part in dose uniformity and repeatability.

Traditionally, low- to medium-current implanters keep the wafer stationary, while high-current implanters will keep the ion beam stationary. The different types of scanning systems used in implanters are:

- Electrostatic scanning
- Mechanical scanning
- Hybrid scanning
- Parallel scanning

Electrostatic Scanning ■ *Electrostatic scanning* deflects the ion beam across a stationary wafer by applying a specific controlling voltage to a set of X-Y electrodes (see Figure 17.20). When the polarity of one side of the electrodes is made negative, the positive ion beam will deflect toward it. By properly positioning the two sets of electrodes and constantly varying the voltage, the beam is made to deflect across a wafer. This type of scanning could be compared to spray painting a surface, where the spray needs to be worked back and forth many times to uniformly coat the surface. An electrostatic scan moves the beam across the wafer (x-axis) 15,000 times in one second while it scans up and down (y-axis) 1,200 times per second. Special care is made to ensure the scan is uniform at the outer edges of the wafer where the scan actually has to stop and reverse directions. In electrostatic scanning systems the wafer can be twisted and tilted relative to the ion beam to achieve desired junction characteristics, and to reduce channeling effects.

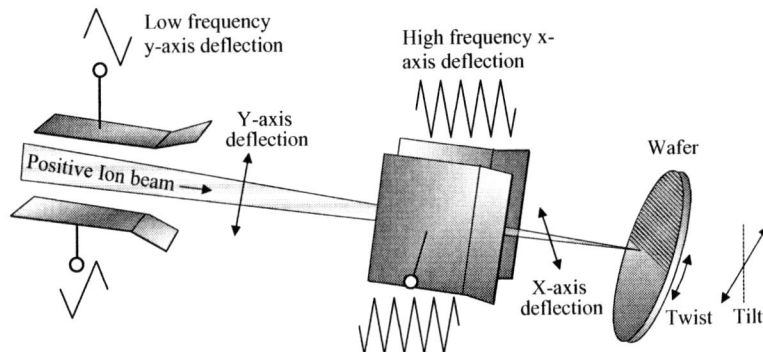

FIGURE 17.20 Electrostatic Ion Beam Scanning of Wafer

Implanting a single wafer at a time is referred to as a serial process and is generally used for low- to medium-current implant applications. In a typical low-current implant operation, a wafer might be scanned for 7 to 10 seconds to obtain uniform dose. Note that this same method of beam deflection can also be accomplished using an electromagnetic field approach instead of an electrostatic approach.

Because the wafer is held fixed during electrostatic scanning with no elaborate mechanical assembly, there is less possibility of particulate contamination. Another advantage to this type of scanning is that electrons and neutrals are not deflected and are eliminated from the beam. The major disadvantage is that the ion beam does not strike perpendicular to the wafer, which leads to undesirable shadowing from the mask material that partially blocks implantation of the ion beam (see Figure 17.21).

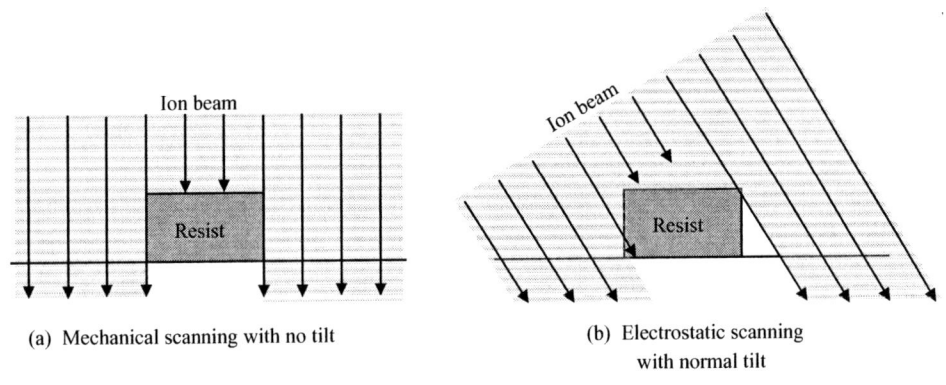

FIGURE 17.21 Implant Shadowing

Mechanical Scanning ■ With *mechanical scanning,* the ion beam is fixed and the wafers are mechanically moved through the beam. This method is generally used for high-current implanters because electrostatic beam deflection is difficult at high currents and energies. The beam size may be 1 cm wide by 3 cm high. Mechanical scanning is done by rotating multiple wafers (e.g., up to 25 200-mm wafers) fixed on the outer circumference of a large wheel assembly disk that is simultaneously moving up and down while rotating at 1000 to 1500 rpm (see Figure 17.22). The disk diameter may be 5 feet or larger. While the disk is spinning, it is also moved up and down so that the beam scans from the outer edge to the inner edge of the wafer. The disk can also be tilted relative to the ion beam to prevent channeling through the interstitial spaces of the silicon lattice (see the following section). Mechanical scanning is a batch process that effectively averages the beam power over a very large area. An advantage is the reduced wafer heating from the ion beam energy. However, there are potentially more particles generated from the mechanical assembly.

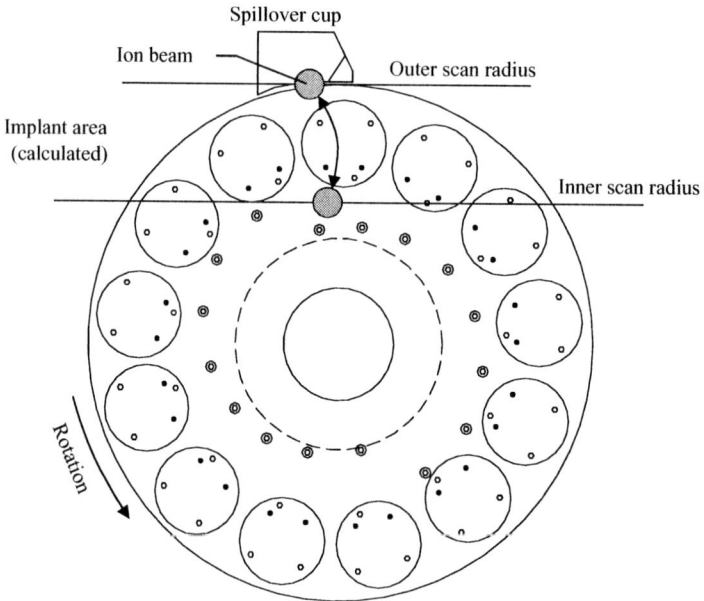

FIGURE 17.22 Mechanical Scanning of Implanted Wafers
(Used with permission from Varian Semiconductor Equipment, VIISion 80 Ion Implanter)

Hybrid Scanning ■ In the *hybrid scanning* system, the wafers are placed on a disk and rotated to scan in the y axis. The beam is then electrostatically (or electromagnetically) scanned in the x-direction. This method is generally used in a serial process for low- to medium-current with a single wafer.

Parallel Scanning ■ In electrostatic scanning, the beam is not scanned perpendicular to the wafer surface, which causes shadowing. *Parallel scanning* is a method used to reduce shadowing and channeling (see the following section) by scanning the ion beam to within <0.5° of the wafer surface.[17] In the parallel scanning method, the beam is first scanned electrostatically. Then the beam passes through another set of magnets which corrects the beam angle and forces it to strike perpendicular to the wafer surface.

Wafer Cooling ■ The energy of the ion beam striking the wafer is converted into heat and will cause the wafer temperature to increase. Wafer cooling is used to control the temperature and prevent problems caused by heating, usually controlling the wafer temperature to <50°C. Photoresist will blister and flake off if exposed to temperatures above 100°C and will become difficult to remove during photoresist stripping.[18] This feature is especially critical during high-dose implantation. If the wafer temperature exceeds 300°C, the electrical properties of the device are affected and partial annealing could occur that could alter the sheet resistivity of the wafer. Factors that affect how much the wafer temperature rises during implantation include ion beam energy, implant time, scan speed, and wafer size.

The two techniques in wide use for wafer cooling in ion implantation are gas cooling and elastomeric cooling.[19] For *gas cooling* the wafer is sealed against a platen (a cooled plate, often with internal circulating water) and a gas such as helium is introduced behind the wafer. The gas provides a thermal conduction path to carry heat from the wafer to the platen. In *elastomeric cooling*, the metal platen is covered with a thin layer of elastomer material. This elastomer is in contact with the backside of the wafer and conforms to the wafer surface to maximize heat transfer between the wafer and platen.

Wafer Charging ■ During implantation, as the ion beam strikes the wafer, some positive ions accumulate in the masking layer, causing *wafer charging*. This condition can lead to a significant electric charge buildup on the wafer, especially for high-current implanters. Such wafer charge buildup can change the charge balance in the ion beam and lead to beam blow-up, causing substantial dose variations across the wafer. Wafer charging can also damage a surface oxide, including damage to the gate oxide leading to device reliability problems.

Control of wafer charging has traditionally been done by flooding the surface of the wafer with low energy electrons using an electron shower known as *secondary electron flood* (see Figure 17.23). In this method, primary electrons with the energy of a few hundred eV are directed to a target located close to the ion beam path. When these primary electrons strike the target, a cloud of low-energy (<20 eV) secondary electrons is generated.[20] These secondary electrons are trapped by the ion beam and are used to neutralize positive charge buildup on the wafer surface. It is important that no high-energy primary electrons reach the wafer or damage may occur to gate oxides.

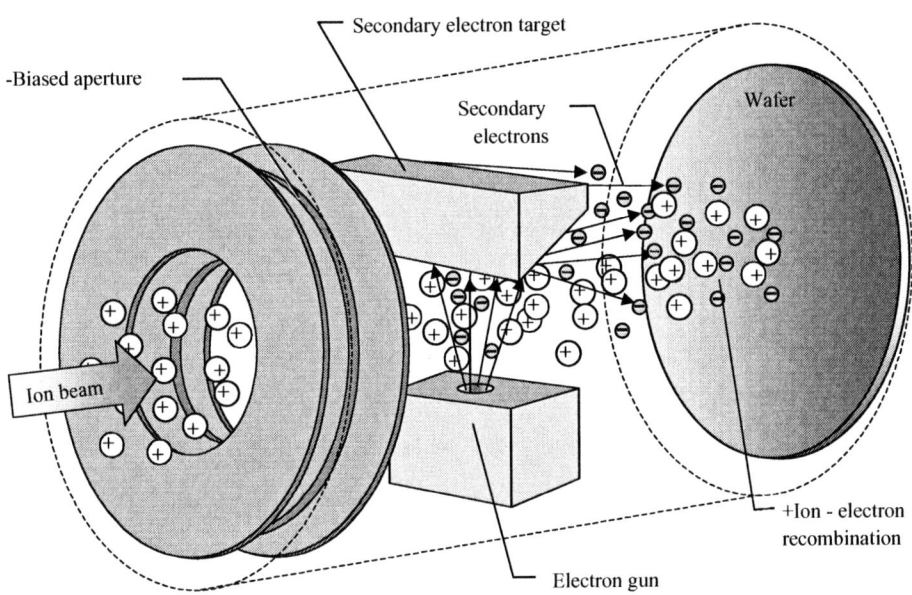

FIGURE 17.23 Electron Shower for Wafer Charging Control
(Used with permission and adapted from Eaton NV10 Ion Implanter, circa 1983)

Control of wafer surface charging is often done now by immersing the ion beam and wafer in a stable, high-density plasma environment, referred to as a *plasma electron flood* system.[21] It is based on the extraction of electrons from a plasma (usually Ar or Xe) maintained in an arc chamber located close to the beam path and wafer (see Figure 17.24 on page 498). The plasma is filtered so that only secondary electrons reach the wafer surface and neutralize the positive charge buildup. The main advantage of plasma flood over the electron shower is that no high-energy electrons are generated in the plasma, which means only low-energy electron energy is used. Plasma flood effectively reduces wafer charge buildup and damage.

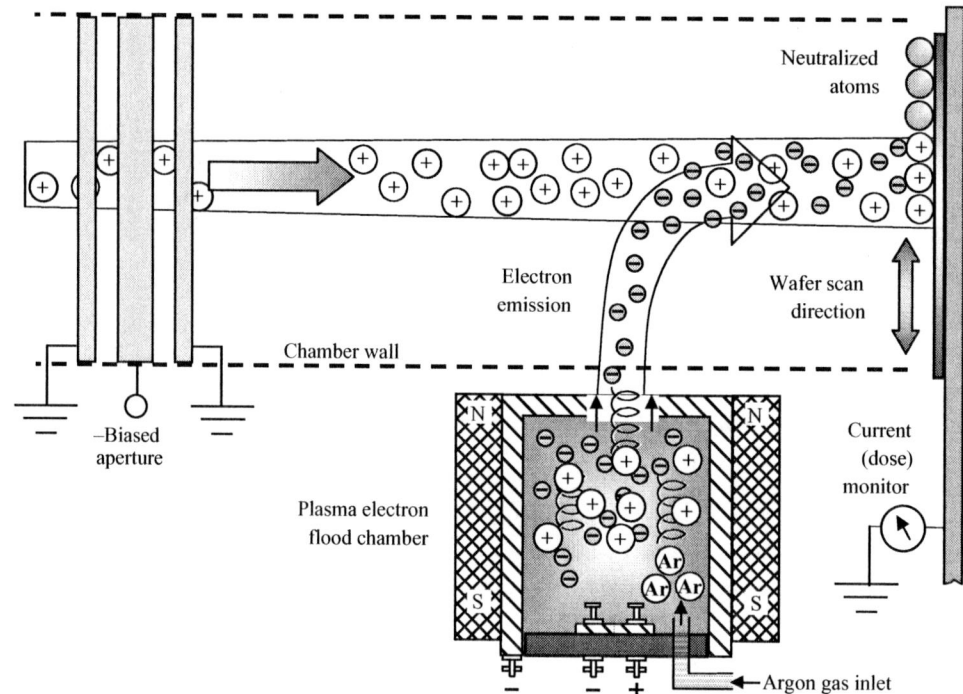

FIGURE 17.24 Plasma Flood to Control Wafer Charging

Process Chamber

The implantation of the ion beam into the wafer takes place in the *process chamber*. The process chamber is a major implanter subassembly that includes the scanning system (discussed above), *end station* with vacuum load locks for loading and unloading wafers, the wafer handler system, and the computer control system. There are also methods for dose monitoring and channeling control (see the following section). The end station can be large if used for mechanical scanning. It is pumped down to the vacuum required for implantation with multiple roughing, turbopumps, and cryopumps to reach base pressure (e.g., typically 10^{-6} torr).

End Station for Ion Implanter (Photo courtesy of International SEMATECH)

Load/unload systems in the end station use robotic handling to move wafers between an input station and the scanning disk in the target chamber (see Figure 17.25). Cassettes are loaded into the input rack and load locks are used to initially seal the input chamber. A roughing pump lowers the pressure around the cassettes. When pressure is low enough, a turbopump continues the pumpdown until reaching high vacuum. At this point, an isolation valve opens and a robotic mechanism moves the cassette into the main target chamber. When the cassette is in the process chamber, a robotic wafer handler system moves the wafers from the cassette and places them on the scanning disk. The wafer usually undergoes initial alignments using the wafer flat or notch to ensure proper wafer orientation in the disk.

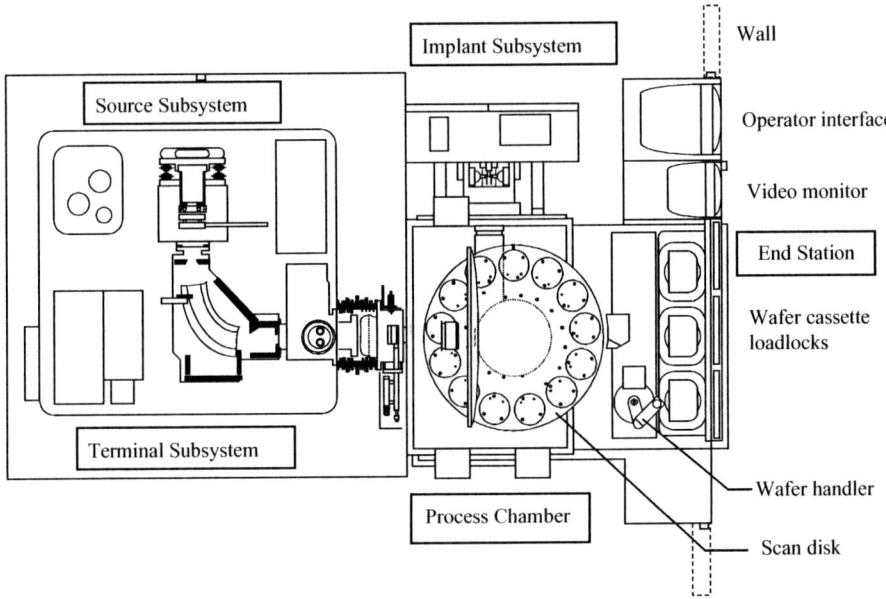

FIGURE 17.25 Wafer Handler for an Implant Process Chamber
(Used with permission from Varian Semiconductor Equipment, VIISion 200 Ion Implanter)

All subsystems are accessed by the technician through a central computer system. System status and information are obtained through the computer, including diagnostics and maintenance information.

Dose Control ■ In ion implanters, real-time dose control is done by measuring the ion beam impinging upon the wafers. The beam current is measured using a sensor called the *Faraday cup*. In a simple Faraday system, a current sensing target is placed in the ion beam path to measure the beam current. However, this placement poses a problem because the ion beam interacts with the target and creates secondary electrons, which give a false beam current reading. The Faraday system uses a design that suppresses the secondary electrons using either an electrical bias (see Figure 17.26 on page 500) or magnetic field to obtain a true ion beam current reading. The measured beam current from the Faraday cup is input into the electronic *dose controller*, which functions as a current integrator (it continuously sums the measured ion beam current). The dose controller correlates the summed beam current to the elapsed implant time and calculates the total time needed to reach a set dose value.

Annealing

Ion implantation damages the silicon lattice by knocking atoms out of the lattice structure. With high doses, the implanted layer actually becomes amorphous. Furthermore, the implanted ions rarely enter the lattice structure sites of the silicon, instead stopping in interstitial sites outside the lattice. These interstitial dopants are electrically inactive until activated by a high-temperature *annealing* step. Annealing heats the implanted silicon substrate to repair crystalline damage and electrically activate dopants by moving the atoms into crystal lattice sites (see Figure 17.27 on page 500). Crystal damage repair is done at about 500°C and dopant activation occurs at about 950°C. Electrical activation of dopants occurs as a function of time and temperature, with longer

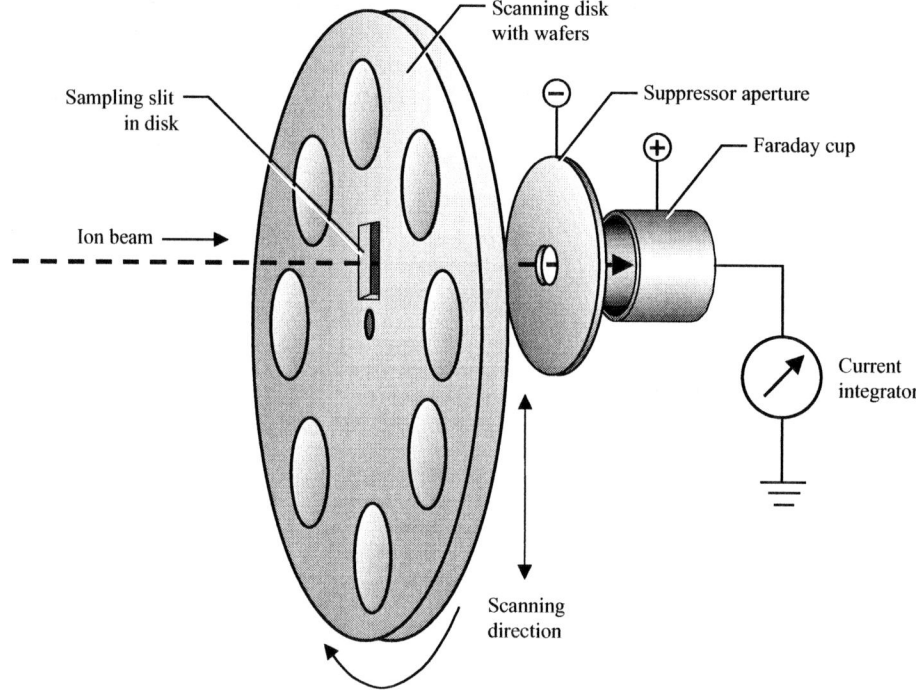

FIGURE 17.26 Faraday Cup Beam Current Measurement
Redrawn from S. Ghandhi, *VLSI Fabrication Principles: Silicon and Gallium Arsenide*, 2nd ed., (New York: Wiley, 1994), p. 417.

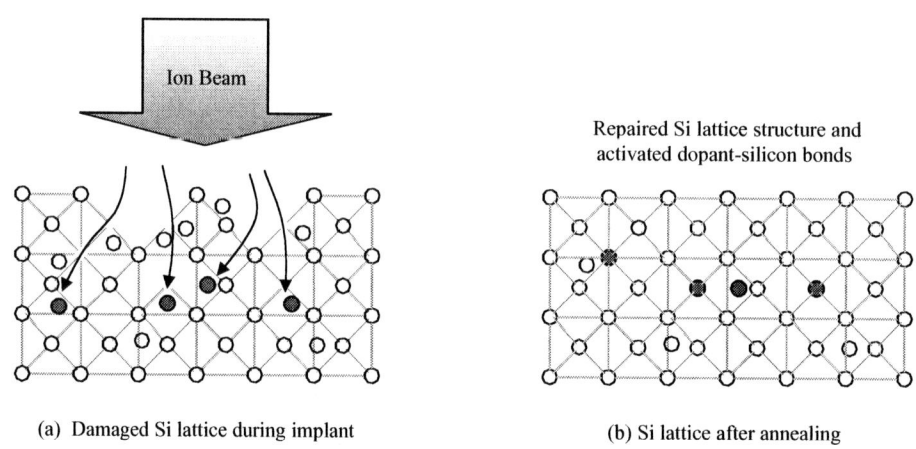

(a) Damaged Si lattice during implant   (b) Si lattice after annealing

FIGURE 17.27 Annealing of Silicon Crystal

times and higher temperatures increasing the dopant activation. There are two basic methods for heating the implanted wafer for anneal:

- Furnace anneal
- Rapid thermal anneal (RTA)

Furnace Anneal ■ *Furnace anneal* is the traditional method of annealing by heating the wafer in a hot-wall furnace in the range of 800 to 1000°C for approximately 30 minutes. At this temperature, the silicon atoms move back into the lattice sites and the dopant atoms enter substitutional sites in the lattice. However, annealing cycles at this time and temperature can cause extensive dopant diffusion and is undesirable for advanced IC wafer fabrication.

Rapid Thermal Anneal (RTA) ■ *Rapid thermal anneal (RTA)* anneals the wafer by using an extremely fast ramp and short dwell time at the target temperature (typically 1000°C). Annealing of implanted wafers is usually done in a single-wafer rapid thermal processor (RTP) with argon (Ar) or nitrogen (N_2) flowing into the chamber. A fast ramp rate and short dwell time is optimized to anneal the wafer to restore lattice damage and electrically activate the dopant while minimizing dopant diffusion in the silicon. RTA also minimizes a phenomenon known as transient enhanced diffusion (see the following section). RTA is the optimal anneal method to achieve acceptable junction depth control in shallow implants.

Investigations have been done into spike anneals for RTA. The ramp-up rate is very fast at up to 150°C/sec. This speed permits a soak time reduction to about 1 second. Uniform ramp rates during a spike anneal are critical because much of the thermal exposure occurs during the ramp.[22]

The RTA equipment design is similar to the rapid thermal processor previously discussed in Chapter 10. There are different types of heating, with a common method being tungsten halogen lamps positioned on both sides of the wafer. Multiple optical pyrometers are often used to measure temperature across the wafer to ensure uniformity.

Transient Enhanced Diffusion. There is a recent discovery of a dopant diffusion referred to as *transient enhanced diffusion* (*TED*) that occurs in the anneal step after implantation. It is caused by extra interstitial atoms from the dopant ion implanted in the silicon and is not directly related to silicon damage. It has significance when forming ultrashallow junctions because of the need to minimize all dopant diffusion.

Channeling

Single-crystal silicon atoms exhibit long range order, lying in rows and planes throughout the crystal. *Channeling* occurs when the implanted ions are not slowed by collisions with silicon atoms and instead pass through the interstitial areas of the crystal (see Figure 17.28). The undesirable aspect of channeling is that the expected projected range has a very large spread, especially for shallow implants at low ion beam energies. Dopant ions passing through lattice channels do not encounter electronic or nuclear stopping (i.e., have little energy loss) and therefore penetrate deeper than those ions that are stopped by the lattice atoms. There are four ways that channeling is controlled during implant: (1) wafer tilt, (2) screen oxide layer, (3) preamorphization of the silicon, and (4) using dopants with greater amu's (atomic mass units).

FIGURE 17.28 Silicon Lattice Viewed Along <110> Axis
(Used with permission from Edgard Torres Designs)

Wafer Tilt. The most widely used method to reduce channeling during ion implantation is *wafer tilt*. This method orients the wafer at an angle relative to the incident ion beam. A common angle for (100) wafers is 7° from the perpendicular and ensures that a collision occurs within a short distance of the dopant ion entering the silicon. The tilt angle is set during the scanning process. In this manner, the crystal lattice presents a dense orientation to the incident beam to obtain better control of the projected range on the implanted ion (see Figure 17.29 on page 502). Note that channeling behavior is different at low energies (<1 keV) for ultrashallow junctions, where tilt has little effect on reducing channeling.[23] Furthermore, tilt angles increase shadowing effects and may lead to asymmetric device behavior. It is also necessary sometimes to twist (or rotate in the vertical plane) the wafer surface. For the <110> crystal orientation, twisting the wafer 15° to 35° can help reduce channeling.

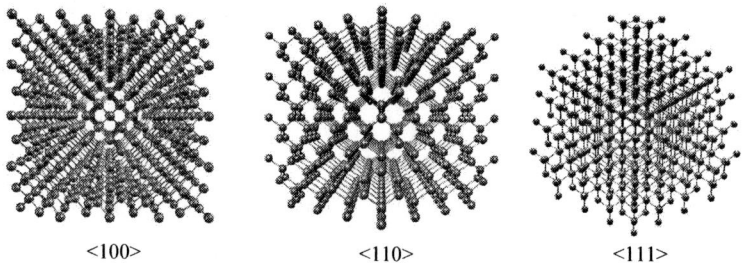

<100>   <110>   <111>

FIGURE 17.29 Ion Entrance Angle and Channeling
(Used with permission from Edgard Torres Designs)

There are situations where high-current implantation of shallow source/drain extensions (e.g., lightly doped drain) may require high-angle implants under the photoresist mask to obtain a shallower depth. Modern serial process (single-wafer) implanters are designed to permit wafer tilt of ±60° or greater in two axes.[24] This permits implants underneath wafer features and onto trench sidewalls. Trench sidewall doping is required for applications such as a DRAM capacitor cell in order to form a conductive capacitor plate along the sidewall.

Screen Oxide Layer. In some cases a thin oxide layer (10 to 40 nm) is grown or deposited as a *screen oxide layer* on the wafer surface prior to implantation. It is also sometimes referred to as a *sacrificial oxide* since it is deposited for the implantation process and then removed. Implanting through this amorphous oxide layer randomizes the directions of the ions as they enter the silicon lattice to reduce channeling effects. Randomization occurs due to the dopant ion impinging on the SiO_2 molecules. Investigations indicate that a screen oxide is not always effective at reducing channeling and can lead to dose uniformity problems. Variables that affect how successful a screen oxide is to control channeling are the implant energy, dopant species, the screen oxide thickness, and the orientation of the incident ion beam.[25]

Preamorphization. A technique for reducing channeling is *preamorphization* of the single-crystal silicon lattice with an electrically inactive species, usually Si^+. This step is done prior to implantation and serves to destroy the single crystal structure in a thin layer of the silicon surface. The subsequent implant with the dopant ion into the amorphous crystal structure leads to minimal channeling. The implantation is followed by a thermal anneal to correct the lattice damage. Investigations show that preamorphization is especially important for shallow profiles at energy levels below 1 keV.[26] The biggest disadvantage of preamorphization is that it involves an additional implant step that increases production cycle times.

Particles

Ion implantation is sensitive to particulate contamination. A particle located on the wafer surface can block the incident ion beam and cause improper implantation (see Figure 17.30). High-current implanters create more particles due to erosion of beamline components. Implanter suppliers design the target chamber to reduce the high temperature from the high-current beam and minimize particle formation. Sources of particles are reviewed in a later section of this chapter. Particles are usually monitored on unpatterned wafers to calculate the particles per wafer per pass and take corrective action when necessary (see Chapter 7).

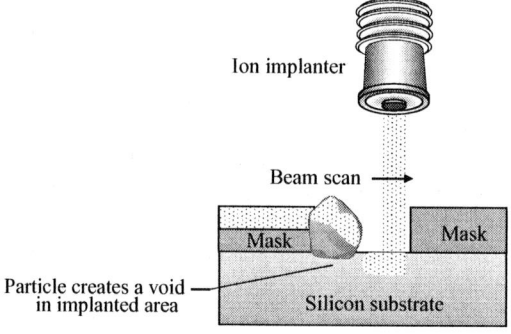

FIGURE 17.30 Implantation Damage from Particulate Contamination

ION IMPLANT TRENDS IN PROCESS INTEGRATION

Ion implantation requirements vary widely because of the different dopant requirements necessary for leading edge processes. Examples of different implant process requirements for advanced MOS wafer fabrication are:

- Deep buried layers
- Retrograde wells
- Punchthrough stoppers
- Threshold voltage adjustment
- Lightly doped drain (LDD)
- Source/drain implants
- Polysilicon gate
- Trench capacitor
- Ultrashallow junctions
- Silicon-on-insulator (SOI)

Deep Buried Layers

Deep buried layers implanted in silicon with high-energy ion implanters (>200 keV) are used in CMOS circuits for various reasons, such as the creation of retrograde wells (Figure 17.31). A *triple well* has a buried implanted well layer beneath the standard n- and p-retrograde wells for improved device performance and packing density. An important reason for buried layers is to control latch-up, a reliability problem in CMOS circuits. As discussed in Chapter 3, latch-up occurs when parasitic transistors cause the device to turn on unintentionally, leading to complete failure of the chip. Latch-up is typically controlled in wafer fabrication by the use of an epitaxial silicon layer (epilayer) on the surface of the wafer. The low-resistance epilayer effectively shunts current to the ground plane. The epilayer also serves as an impurity getter. Because of the high cost of epilayer substrates, there is increased interest in high-dose buried layers formed by ion implantation to replace the current shunting and impurity gettering of the epilayer.[27]

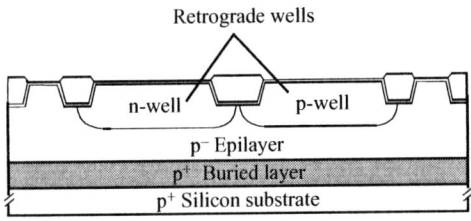

FIGURE 17.31 Buried Implanted Layer

Retrograde Wells

Recall from Chapter 3 that a well is a diffused area where active devices (e.g., MOSFETs) are constructed. The n-type or p-type MOS transistor is placed in a well (or tub) of opposite conductivity type to form the semiconductor junction. Also recall from earlier in this chapter that, with diffusion, the dopant profile always has the maximum concentration at the silicon wafer surface. An important design alternative for MOS devices is a *retrograde well,* which has the peak implanted dopant profile buried at a certain depth (e.g., several microns below the wafer surface). This feature is shown in Figure 17.32 on page 504. Another term for a retrograde well is a *vertically modulated well.* A retrograde well is formed by a high-energy implant that produces high dopant concentration deep in the retrograde well to improve transistor immunity to problems such as latch-up and punchthrough.[28] There are also some applications for having a triple well structure to achieve additional device isolation, such as in a DRAM or EEPROM circuit.

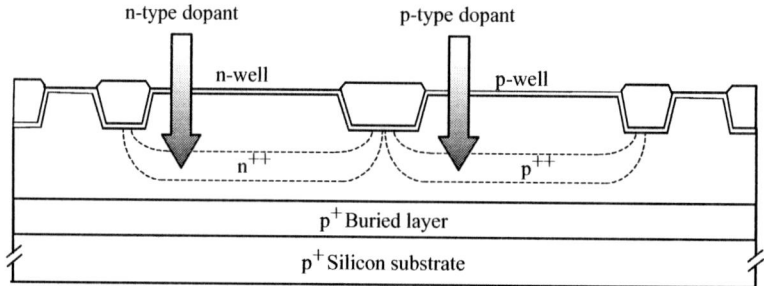

FIGURE 17.32 Retrograde Well

Punchthrough Stoppers

In submicron devices with short channel lengths, punchthrough stopper implants are necessary. *Punchthrough* is a condition where the channel is shorted (punched through) with undesirable leakage current and leads to device failure. It occurs when the channel length is reduced for scaling (known as a short channel) and high electric fields occur at the drain end of the channel. In the case of an n-channel, the electrons moving from source to drain are accelerated by this high electric field and by energetic collision and create a free electron and hole pair. This action can cause the drain field to extend too far into the lightly doped channel region and contact the source region field, causing punchthrough.[29] The punchthrough implant places dopants just below the active channel adjacent to the source and drain in order to carefully modify the well doping and prevent expansion of the drain depletion region into the lightly doped channel when the device is biased for operation (see Figure 17.33). Precise control of the placement and dose of a punchthrough implant is necessary, with boron implant used for n-channel devices and phosphorus used for p-channel devices.

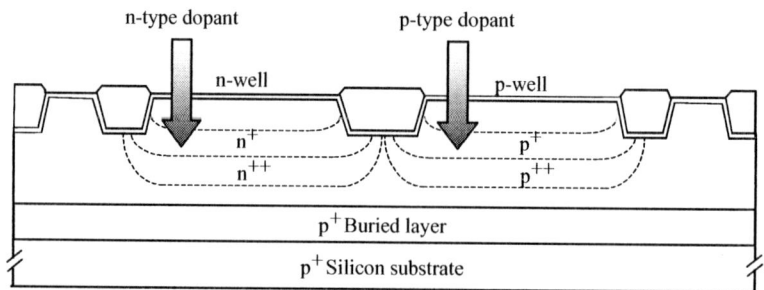

FIGURE 17.33 Punchthrough Stop

Threshold Voltage Adjustment

An MOS transistor consists of three distinct parts: source, drain, and gate. No current flows between the source and drain until the channel region underneath the gate becomes conductive when triggered by a voltage applied at the gate. The threshold voltage, V_{TH}, is the amount of voltage required to form the conductive channel between the source and drain that causes enhancement mode transistors to turn on. V_{TH} is very sensitive to the dopant concentration in the channel region. For optimum device performance, a dopant is implanted beneath the silicon layer to adjust the channel region to the required dopant concentration. This is called the MOS gate *threshold voltage implant adjustment* and is of fundamental importance to device performance. To illustrate this adjustment, take the case of a p-type dopant (e.g., boron) implanted in an n-channel region to adjust V_{TH} (see Figure 17.34). The increase in the p-type dopant concentration will causes a change in the necessary V_{TH} (e.g., V_{TH} increases from 0.9 V to 1.0 V). Threshold voltage adjustment with ion implantation was the first widespread use of implanters in wafer fabrication due to the implantation's ability to produce uniform, repeatable dopant concentrations.

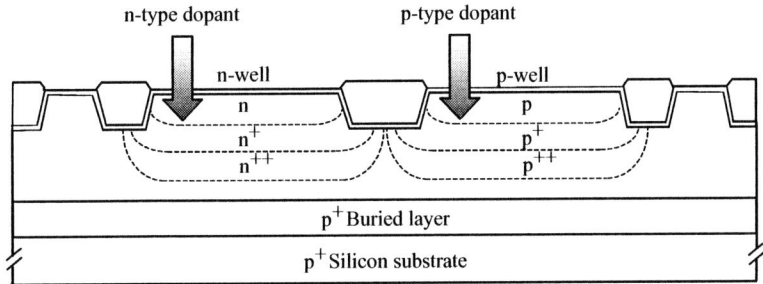

FIGURE 17.34 Implant for Threshold Voltage Adjustment

Lightly Doped Drain

A *lightly doped drain (LDD)* implant is used when defining the source and drain regions of the MOS transistors (see Figure 17.35). This type of region is often referred to as a source/drain extension. It is needed to achieve dimensional reductions for the scaling of submicron devices. The implant places the LDD dopant just to the edge of the channel region under the gate to provide a gradual dopant concentration to the source/drain regions. The LDD creates complex lateral and vertical doping profiles in the interface region at the channel edge. The LDD implants for nMOS and pMOS transistors occur in two separate masking and implant operations. The gate of each MOSFET is implanted as the shallow junctions of the source and drain are also formed.

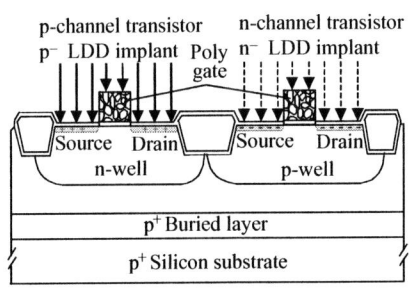
a) p⁻ and n⁻ lightly-doped drain implants
(performed in two separate operations)

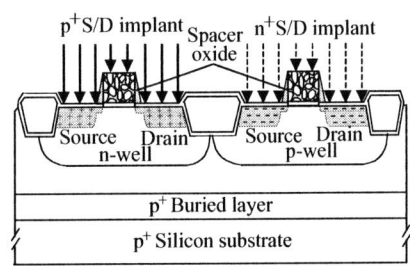
b) p⁺ and n⁺ source/drain implants
(performed in two separate operations)

FIGURE 17.35 Source/Drain Formations

The LDD structure is formed by a medium- to low-dose implant aligned with the gate structure (this is the n^- or p^- implant), followed by a high-dose (called the n^+ or p^+) source/drain implant. The source/drain (S/D) implant is aligned to the gate by an oxide sidewall spacer formed between the two implants (see Figure 17.35). If an LDD is not formed, then high electric fields are present between the junction and channel regions during normal transistor operation. Electrons exposed to a high electric field during their movement from source to drain (for an n-channel) are accelerated by this high field. Energetic electron collisions create a free electron and hole pair (called *hot carriers* or *hot electrons*).[30] Hot electrons acquire energy from the electric field and cause electrical performance problems, such as becoming trapped in the gate oxide layer and interfering with the gate threshold voltage control of the device.

The LDD creates a gradual lateral dopant concentration gradient between the high concentrations in the S/D regions (10^{20} to 10^{21} atoms/cm^3) and the low concentrations in the channel region (10^{16} to 10^{17} atoms/cm^3) under the gate.[31] The reduced doping of the LDD decreases the electric field between the junction and channel regions. This technique separates the maximum current path in the channel from the maximum electric field location in the junction to avoid creating hot carriers.

Source/Drain Implants

Source/drain (S/D) implants form highly doped regions (10^{20} to 10^{21} ions/cm^3) that interface with the lightly doped active channel and well regions (10^{16} to 10^{17} ions/cm^3) of an MOS transistor. S/D regions are doped to have the opposite conductivity type as the well that surrounds them. Arsenic implants are often used to form n-type source/drains for n-channel (nMOS) transistors, and boron or BF_2 implants are usually used to form p-type source/drains for p-channel (pMOS) transistors. By using implantation with its excellent control of doping placement, the lateral diffusion of the source/drain dopant ions into the channel region is minimized.

Polysilicon Gate

The *polysilicon gate,* or simply *poly gate,* must be doped to render it conductive and this is done with either diffusion or ion implantation. For ease of fabrication, the deposited polysilicon film has traditionally been doped with an n^+ dopant for both n-channel and p-channel devices prior to the gate patterning process. This doping can pose some electrical performance problems in sub-0.25 μm devices with short channel lengths. An approach that is becoming more widespread to resolve this problem is to create a separate p^+ doped gate for the p-channel devices to complement the n^+ dopant for the n-channel, known as *dual-polysilicon gate structure.*[32] To produce such a device, undoped poly gates are first patterned and then doped when the source and drain of each device type are implanted. For this design to function properly, the doped poly gates must be sufficiently activated or else a depletion layer may form at the poly/oxide interface and adversely affect device performance. Lateral diffusion of dopants between the two different dopant regions is also a concern for a dual-polysilicon gate structure, making thermal budget more critical.

Trench Capacitor

The *trench capacitor* has replaced the planar storage capacitor because of the shrinking DRAM memory cell size. The trench capacitor is a three-dimensional device formed by dry etching the trench into silicon. In order to obtain sufficient capacitance, a dopant level of about 10^{19} atoms/cm^3 is placed in a very shallow layer into the vertical sidewalls of the capacitor, with a dielectric deposited in the trench (see Figure 17.36).[33] Ion implantation done with an implant angle is used to dope the capacitor sidewalls. The actual angle of implant depends on the trench aspect ratio. There are other three-dimensional structures that may use implantation, such as certain applications for trench isolation.

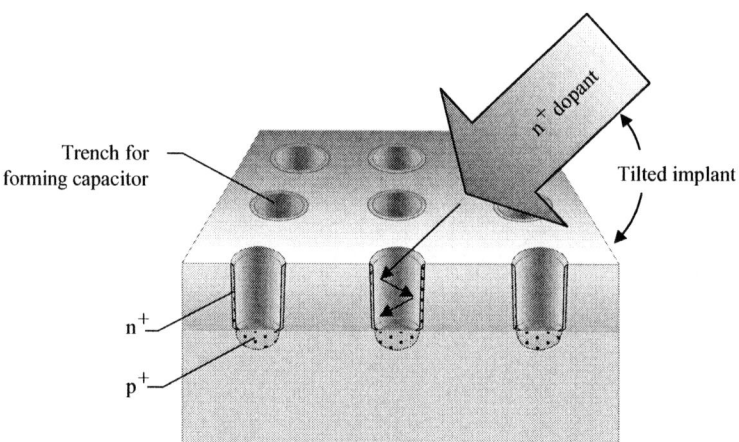

FIGURE 17.36 Dopant Implant on Vertical Sidewalls of Trench Capacitor

Ultrashallow Junctions

The high level of scaling required for high speed and increased transistor packing densities on chips requires a reduction in MOS device channel lengths. To maintain the device electrical performance, key device elements must be scaled (reduced in size) appropriately with the channel length. *Ultrashallow junctions* are formed so that the depth of the source and drain (S/D) junctions are in scale with the short channel length (see Figure 17.37). Ultrashallow junction depths range from 54 ± 18 nm for the 0.18 μm technology node to 30 ± 10 nm for the 0.10 μm technology node.[34] It is also

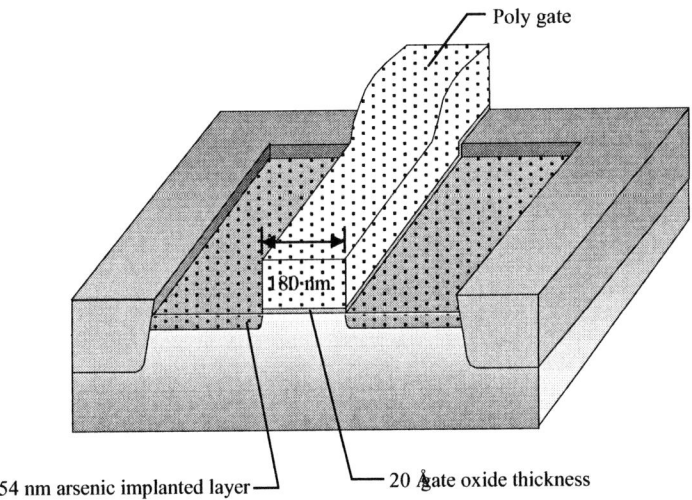

FIGURE 17.37 Ultrashallow Junctions

necessary to carefully control the lateral dopant profile. Ultrashallow junctions are implanted using a high beam current, low-energy implanter.

Shallow n-type source and drain junctions are formed relatively easily due to the large mass of the arsenic dopant ion (see Table 17.1 on page 476). The large mass forms an amorphous layer to assist in depth control by minimizing channeling. Shallow p-type junctions are more difficult to form because the low mass of boron does not easily form an amorphous layer, leading to more channeling effects and difficulty in controlling junction depth. To overcome this problem, the silicon may undergo preamorphization (e.g., with a silicon implant) or a heavier molecular ion such as BF21 is implanted.35

Silicon-On-Insulator (SOI)

A form of vertical device isolation that is becoming more important in advanced wafer fabrication is *silicon-on-insulator* (*SOI*). SOI technology is based on an insulator layer (oxide) buried within the silicon that effectively isolates the devices on the silicon surface. It was developed in the 1970s, but has not been widely used due to process complexity and cost. SOI offers several advantages for deep submicron CMOS applications, including the complete elimination of latchup, reduced electric fields to minimize hot carriers, and reduced parasitic capacitance.[36] The concept of forming single-crystal silicon on an oxide layer (or other insulator material) is difficult because the dielectric material's crystalline properties are so different from pure silicon. If not properly controlled, this difference in crystal structure can lead to crystal defects in the silicon that affect device performance. The most widely used SOI technology is SIMOX.

SIMOX ■ At the time of this writing, SIMOX is the leading commercial process for SOI. In the *SIMOX* (Separation by IMplanted OXygen) process, a well-defined horizontal oxide layer is buried in the silicon wafer. This is done by implanting a high concentration of oxygen atoms into the wafer, traditionally using a high-energy implanter (e.g., a 200 keV oxygen implanter.) The implant step is followed by a high-temperature thermal anneal (e.g., 1300°C) to react the buried oxygen with silicon and form a continuous SiO_2 layer under the thin silicon surface (see Figure 17.38 on page 508). This buried oxide (referred to as BOX) layer is typically about 50 to 500 nm thick and serves as an excellent device isolation layer. The anneal process also regenerates the crystalline quality of the silicon layer remaining over the oxide.[37] It should also be noted that there are new SIMOX techniques in development using low-energy, low-dose oxygen implanters that produce buried layers with improved dielectric properties.

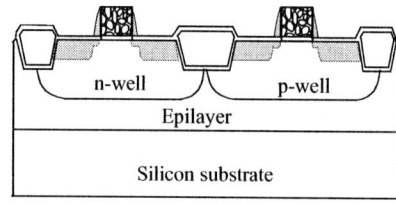
(a) Common CMOS wafer construction

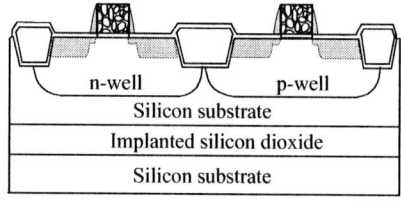
(b) CMOS wafer with SIMOX buried layer

FIGURE 17.38 CMOS Transistor With and Without SIMOX Buried Oxide Layer

ION IMPLANT QUALITY MEASURES

Quality measures used for ion implantation are shown in Table 17.7.

TABLE 17.7 Key Quality Measures for Ion Implantation

Quality Parameter	Types of Defects	Remarks
1. Unacceptable particle contamination on the wafer surface.	A. Particle measurement on monitor wafer surface (particles per wafer per pass).	Possible particle sources are: • Microdischarge from dirty electrodes. • Excessive wear in moving mechanisms. • Improper cleaning of implanter (e.g., ion source, extraction assembly, and any part of the beam line). • Breakdown of photoresist from excessive temperature or sputtering. • Heatsink elastomer on wafer backside. • Condensation in loadlock. • Wafer handling. • Machine movements: opening valves, load/unload indexers, etc.
2. Dose control (incorrect dose).	A. Dose defects are detected by incorrect sheet resistance measurement. Wafer with underdose exhibits high sheet resistance (addition of dopants increase conductivity) and overdose exhibits low sheet resistance.	Unacceptable uniformity exhibits itself as regions of low and high dopant concentration on the wafer surface. Possible causes of this problem are: • Wrong process recipe. Check the process recipe against the wafer lot. • Improper beam current measurement. Check Faraday system for leakage current and proper operation. • An overdose could be caused by electrons entering the beam. The electrons displace beam ions and cause an overdose because the beam counter is counting fewer ions. Check the electron suppression system. • Improper annealing. For RTA, check: (1) improper temperature or time during anneal, (2) incorrect ramp rates, (3) nonuniform heating, or (4) incorrect N_2 flow rates.

Quality Parameter	Types of Defects	Remarks
3. Junction depth for ultrashallow junctions using low energy implant.	A. Incorrect junction profile as indicated by spreading sheet resistance measurement.	Check the following possible causes for junction depth: • Verify that the anneal ramp rate and soak time in the RTA is correct. Excessive thermal exposure will increase the junction depth due to dopant diffusion. • Investigate ion channeling. The wafer tilt may be incorrect. At low energies, the wafer may require an amorphous implant with a heavy ion to destroy the local crystal structure, followed by an anneal. • If implanting boron (low mass element), verify there is no TED problem during anneal. Ultralow energy implant may be required.

ION IMPLANTATION TROUBLESHOOTING

Common ion implant troubleshooting problems are reviewed in Table 17.8.

TABLE 17.8 Common Ion Implant Problems

Problem	Probable Cause	Corrective Actions
1. Dose uniformity problems causing regions of low and high dopant concentrations on the wafer surface (detected by sheet resistance or thermal wave contour maps).	A. Charge neutralization system.	• Wafer charging shows on contour maps as high resistivity values in central wafer region. • Check secondary electron flood or plasma flood for proper operation.
	B. Scan system problems (e.g., channeling).	• For high current, check for mechanical problems in the drive system of the target disk. • For low and medium current with fixed targets, inspect the scan system for improper x and y direction scans.
	C. Leakage currents in implanter.	Beam current leakage in the implanter system create a risk of dose error and nonuniformity. • Verify all insulators are clean and have no sputtered material build-up. Sputtered material comes from the ion beam striking materials in the beam path. • All cabling insulation must be of high quality.
2. Contaminants in the ion beam detected by SIMS analysis.*	A. Ion source contaminants.	Possible sources of the ion source components are: • Check for dirty ion source components. • Leak-test the vacuum system to verify no leaks. • Verify source material meets purity requirements.
	B. Contaminants in beam line before mass analysis.	The region after the extractor and before the mass analyzer is not a significant source of contamination, but check for vacuum leaks.

Problem	Probable Causes	Corrective Actions
2. Contaminants in the ion beam (continued).	C. Contamination created during mass analysis.	Mass resolving slit is set too wide to gain a higher beam current, thus not excluding all ions. An example is hydrogen contamination during a P implant from a phosphine source.
	D. Contamination in beam line after mass separation.	Residual gas species interact with the ion beam to create species with incorrect energy (energy contamination). Corrective action includes: • Leak-check vacuum system. • Check energy filtration systems. • Investigate for metallic contaminants from components in system.
	E. End station contamination.	Possible contamination sources are: • Sputtered metals and dopants from the end station materials (e.g., ion beam striking electrode). • Alkali-element contamination from photoresist. • Aluminum from the Faraday cup assembly. • Cross-contamination from other dopant elements used in the same implanter.

*F. Stevie et al., "Using SIMS to Detect Contamination Sources from Ion Implanters," *Solid State Technology* (May 1995): pp. 51–58.

SUMMARY

Doping introduces a dopant into silicon to modify its electronic properties. This is done through ion implantation or diffusion. Diffusion has been largely replaced by ion implantation due to its many advantages. Diffusion is the movement of one material through another due to a concentration gradient. Ion implantation is a physical process that uses an implanter tool. Two important ion implant parameters are dose and range. The beam current is used to define the dose. The range is the total distance a dopant travels in the wafer and is related to energy and the mass of the dopant ions. The implanted ions are eventually stopped in the wafer due to collisions or interactions with silicon atoms. There are six major subsystems of an implanter tool: ion source, extraction assembly and ion analyzer, acceleration column, scanning system, and target chamber. The ion source generates ions for implantation. The extraction assembly extractsions from the source and the analyzer magnet separates the desired dopant from the ion species. The accelerator accelerates the dopants, which are scanned on the wafer with the scanning system in the target chamber. Wafers are annealed in either a hot-wall furnace or RTP to repair crystal damage and electrically activate dopants. Channeling is the passage of dopants through interstitial crystal areas, which leads to nonuniform junction depth. It is controlled by wafer tilt, preamorphization, or screen oxide layer. Implanters are sensitive to particles. Implantation is widely used in wafer fabrication, including MOS gate threshold adjustment, retrograde wells, source/drain implants, ultrashallow junctions, lightly doped drains, polysilicon gates, deep buried layers, punch-through stoppers, trench capacitors, and SIMOX (a silicon-on-insulator technology).

KEY TERMS

doping
thermal diffusion
ion implantation
doped region
dopant profile
junction depth, x_j
diffusion
predeposition or predep
cap oxide
Fick's laws
drive-in
activation
redistribution
diffusivity
diffusion coefficient, D
interstitional movement
substitutional movement
activated dopants
crystal activation
solid solubility limit
lateral diffusion
deglaze
ion implanter
radiation damage
dose, Q
beam current
range
projected range, R_p
straggle, ΔR_p
electronic stopping
nuclear stopping
ion source
Freeman ion source
extraction assembly
suppression electrode
ion analyzer
acceleration column
post-accelerator
linear accelerator
beam deceleration or decel

differential lens
beam blow-up
space charge neutralization
neutral beam trap
electrostatic scanning
mechanical scanning
hybrid scanning
parallel scanning
gas cooling
elastomeric cooling
wafer charging
secondary electron flood
plasma electron flood
process chamber
end station
Faraday cup
dose controller
furnace anneal
rapid thermal anneal (RTA)
rapid thermal processor (RTP)
transient enhanced diffusion (TED)
channeling
wafer tilt
screen oxide layer
sacrificial oxide
preamorphization
deep buried layers
triple well
retrograde well or vertically modulated well
punchthrough
threshold voltage implant adjustment
lightly doped drain (LDD)
hot carrier or hot electrons
source/drain (S/D) implants
polysilicon gate or poly gate
dual-polysilicon gate structure
trench capacitor
ultrashallow junctions
silicon-on-insulator (SOI)
SIMOX

REVIEW QUESTIONS

1. What is doping?
2. List the four common dopants and state whether each is a p-type or n-type dopant.
3. Give a short description of thermal diffusion.
4. Give a short description of ion implantation.
5. List five common dopant processes in wafer fabrication.
6. Ion implantation typically follows what process step?
7. Describe a doped region and a dopant profile.
8. What is the junction depth?
9. Describe diffusion.
10. List and explain the three steps of diffusion.
11. What is redistribution in diffusion?
12. Explain diffusivity and the diffusion coefficient.
13. Explain interstitional movement and substitutional movement in diffusion.
14. Why does a dopant need to be activated?
15. What is the solid solubility limit for dopants?
16. Explain lateral diffusion and why it is undesirable.
17. List the eight steps to properly perform diffusion in wafer fabrication.
18. What is a deglaze step?
19. How are dopant sources typically supplied to the diffusion furnace?
20. State the two reaction equations that illustrate how a diborane source is converted into a boron dopant.
21. Give a general overview of an ion implanter.
22. State the two major goals for sub-0.25 micron doping requirements.
23. List seven advantages of an ion implanter over diffusion.
24. What is the major disadvantage of ion implantation and how is it corrected?
25. Define dose. State and explain the dose equation.
26. What is the beam current? What is the relationship between beam current and dose?
27. What is range? Describe the relationship between energy and range.
28. If an ion with a single positive charge moves in an electric field with a voltage difference of 200kV, what is its energy?
29. List four classes of ion implanters, and give a short description of each one.
30. Describe projected range.
31. Define straggle.
32. Describe the two primary energy loss mechanisms during implantation.
33. Describe the silicon crystal damage depending on whether the dopant is a light or heavy ion.
34. List the five major subsystems of an ion implant tool.
35. What is the purpose of the ion source? What is the most common source feed material?
36. Describe the Freeman ion source.
37. Discuss the purpose and operation of the extraction assembly.
38. What is the purpose of the suppression electrode?
39. What is the purpose of the mass analyzer magnet? Describe how the mass analyzer functions.
40. How does the acceleration column increase ion beam energy?
41. Using Figure 17.16, give a process step for a high-current implanter and a high-energy implanter.
42. What does the beam energy define? What is dose related to in ion implant?
43. How high is the energy for a high-energy implanter? Why is high beam current desirable?
44. Why is a high-current and low-energy beam desirable? How is a high beam current achieved in a low-energy beam?
45. Explain beam blow-up and space charge neutralization.
46. What is the reason for a neutral beam trap?
47. List and give a short explanation of the four types of scanning systems.
48. Which type of scanning is used for low-current applications? High-current applications?
49. Why is it necessary to cool the wafer?
50. Discuss wafer charging, secondary electron flood, and plasma electron flood.
51. Describe the process chamber.
52. How is real-time dose control performed?
53. What is the purpose of annealing? Is it preferred to perform a furnace anneal or an RTA?
54. Describe transient enhanced diffusion (TED).
55. Describe channeling. List and give a short explanation of the three mechanisms for controlling channeling.
56. List the ten different doping processes that us ion implant. Give a short explanation of each implant process.
57. What is a dual-polysilicon gate structure, and why is it needed?
58. Why is SIMOX a potentially beneficial device fabrication technology? What is SOI?

ION IMPLANTER EQUIPMENT SUPPLIERS' WEB SITES

Advanced Energy Industries	http://www.advanced-energy.com
Applied Materials	http://www.appliedmaterials.com/products/
Axcelis Technologies (formerly Eaton)	http://www.axcelis.com/
Charles Evans and Associates	http://www.cea.com
CVC Technologies	http://www.cvc.com/
Eaton Corporation	http://www.semiconductor.eaton.com/
Epion Corporation	http://www.epion.com/
High Voltage Engineering Europa BV	http://www.highvolteng.com/
Ion Implant Services	http://www.ionimplant.com/
Implant Sciences Corp.	http://www.implantsciences.com/
International SEMATECH	http://www.sematech.org/public/index.htm
SEMI	http://www.semi.org/
Therma-Wave	http://www.thermawave.com/index.htm
Varian Semiconductor	http://www.vsea.com/
Veeco Instruments, Inc.	http://www.veeco.com/

REFERENCES

1. M. Foad and D. Jennings, "Formation of Ultra-Shallow Junctions by Ion Implantation and RTA," *Solid State Technology* (December 1998): p. 43.
2. S. Wolf and R. Tauber, *Silicon Processing for the VLSI Era,* Volume 1—*Process Technology,* 2nd ed., (Sunset Beach, CA: Lattice, 2000), p. 325.
3. SEMATECH, "Diffusion Processes," *Furnace Processes and Related Topics,* Austin: (SEMATECH, Inc., 1994), p. 15.
4. S. Ghandhi, *VLSI Fabrication Principles, Silicon and Gallium Arsenide,* 2nd ed., (New York: Wiley, 1994), p. 182.
5. SEMATECH, "Diffusion Processes," p. 8.
6. J. Mayer and S. Lau, *Electronic Material Science: For Integrated Circuits in Si and GaAs,* (New York: Macmillan, 1990), p. 222.
7. M. Foad and D. Jennings, "Formation of Ultra-Shallow Junctions by Ion Implantation and RTA," *Solid State Technology* (December 1998): p. 44.
8. E. Rimini, *Ion Implantation: Basics to Device Fabrication,* (Boston: Kluwer, 1995), p. 79.
9. S. Wolf and R. Tauber, *Silicon Processing for the VLSI Era,* Volume 1—*Process Technology,* p. 377.
10. B. El-Kareh, *Fundamentals of Semiconductor Processing Technologies,* (Boston: Kluwer, 1995), p. 381.
11. T. Romig, J. McManus, K. Olander and R. Kirk, "Advances in Ion Implanter Productivity and Safety," *Solid State Technology* (December 1996): p. 69.
12. E. Rimini, *Ion Implantation: Basics to Device Fabrication,* p. 36.
13. R. DeJule, "New Designs in High-Current Ion Implanters," *Semiconductor International* (April 1998): p. 61.
14. Ibid., p. 61.
15. D. Duff and L. Rubin, "Ion Implant Equipment Challenges for 0.18 µm and Beyond," *Solid State Technology* (June 1998): p. 90.
16. R. DeJule, "New Designs in High-Current Ion Implanters," p. 61.
17. B. El-Kareh, *Fundamentals of Semiconductor Processing Technologies,* p. 368.
18. E. Rimini, *Ion Implantation: Basics to Device Fabrication,* p. 59.
19. M. Mack, "Wafer Cooling and Wafer Charging in Ion Implantation," *Handbook of Ion Implantation Technology,* (Amsterdam: Elsevier, 1992), p. 613.
20. Y. Erokhin, R. Reece and R. Simonton, "Charge Control for High-Current Ion Implant," *Solid State Technology* (June 1997): p. 104.
21. Ibid.
22. M. Foad and D. Jennings, "Formation of Ultra-Shallow Junctions by Ion Implantation and RTA," p. 51.
23. Ibid., p. 44.
24. R. DeJule, "New Designs in High-Current Ion Implanters," p. 62.
25. R. Simonton and A. Tasch, "Channeling Effects in Ion Implantation," ed. J. Ziegler (Amsterdam: Elsevier, 1992), p. 206.
26. M. Foad and D. Jennings, "Formation of Ultra-Shallow Junctions by Ion Implantation and RTA," p. 44.

27. L. Rubin and W. Morris, "High-Energy Ion Implanters and Applications Take Off," *Semiconductor International* (April 1997): p. 77.
28. L. Rubin and W. Morris, ibid, p. 86.
29. R. Simonton and F. Sinclair, "Ion Implantation Applications," *Handbook of Ion Implantation Technology,* ed. J. Ziegler (Amsterdam: Elsevier, 1992), p. 282.
30. J. Yue, "Reliability," *VLSI Technology,* ed. C. Chang and S. Sze (New York: McGraw-Hill, 1996), p. 658.
31. R. Simonton and F. Sinclair, "Ion Implantation Applications," p. 282.
32. C. Lu and W. Lee, "Process Integration," *ULSI Technology,* ed. C. Chang and S. Sze (New York: McGraw-Hill, 1996), p. 494.
33. R. Simonton and F. Sinclair, "Ion Implantation Applications," p. 318.
34. M. Foad and D. Jennings, "Formation of Ultra-Shallow Junctions by Ion Implantation and RTA," p. 43.
35. E. Rimini, *Ion Implantation: Basics to Device Fabrication,* p. 274.
36. R. Simonton and F. Sinclair, "Ion Implantation Applications," p. 339.
37. R. Simonton and F. Sinclair, "Applications in CMOS Process Technology," *Handbook of Ion Implantation Technology,* ed. J. Ziegler (Amsterdam: Elsevier, 1992), p. 339.

CHAPTER 18

CHEMICAL MECHANICAL PLANARIZATION

The ULSI era with sub-0.25 μm design rules represents an advanced level of chip integration, with tens of millions of transistors and an estimated 50 million connections on a few square centimeters of chip surface.[1] Multiple metal layers is the enabling technology that permits interconnecting the millions of transistors and supporting components on individual ICs. Multilayer metal promotes higher device density because of its efficient use of vertical space on the chip surface.

Multilayer metal technology existed since the 1970s, but excessive surface topography became a limiting factor in submicron patterning. As additional layers were added, the wafer surface became nonplanar. A nonplanar wafer surface has undesirable topography. This leads to other problems, the foremost being the inability to pattern the surface because of the limited depth of focus in stepper lenses in optical lithography. Various planarization techniques were developed to reduce or minimize process-induced topography, most were ineffective for overcoming severe wafer surface topography.

In the late 1980s, a global planarization method known as chemical mechanical planarization (CMP) was developed at IBM. It became the standard for planarization in high-density semiconductor manufacturing during the 1990s. Without CMP, chips fabricated to ULSI integration would not be feasible. For dual-damascene copper metallization, CMP is the enabling process that integrates the different layers.

OBJECTIVES

After studying the material in this chapter, you will be able to:

1. Describe the terminology for planarization.
2. List and discuss three traditional types of planarization.
3. Discuss chemical mechanical planarization (CMP), the issues of wafer planarity, and the advantage of CMP.
4. Describe the slurry and pad for both oxide and metal CMP.
5. Discuss CMP equipment, including endpoint detection and wafer carriers.
6. Explain the post-CMP clean procedure.
7. List and describe seven different CMP applications.

INTRODUCTION

Wafer fabrication involves processes that deposit or grow films, followed by repeated patterning to form device and interconnect structures. Advanced ICs require multiple levels of metallization, with at least six or more metal layers—each separated by an interlayer dielectric (ILD). A natural result of building device structures and the multilevel interconnect wiring is the creation of steps in the different layers. *Topography* describes the nonplanar surface of wafer layers during fabrication. Wafer topography becomes more pronounced with additional layers. Acceptable step coverage and gap-fill is critical for chip yield and long-term reliability. To illustrate wafer topography, a single metal layer IC from the 1970s is shown in Figure 18.1 on page 516.

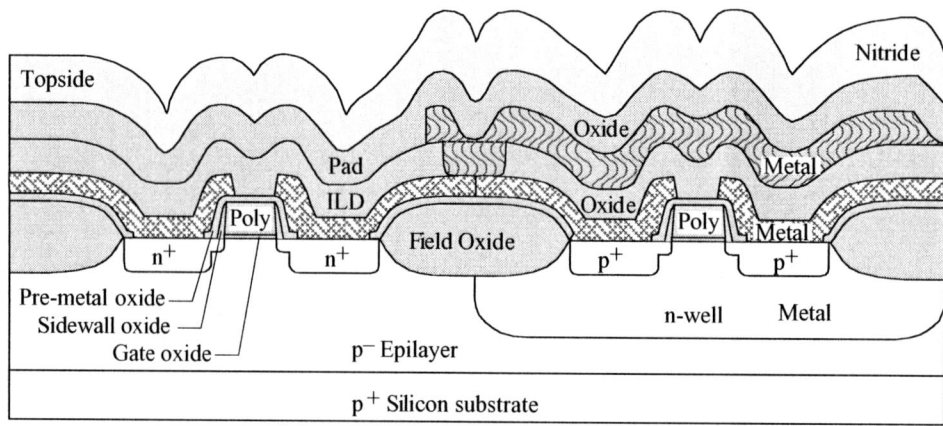

FIGURE 18.1 Single Metal Layer IC with Topography

Increased topography is a consequence of higher chip packing density. The trend toward multilevel metal technology in IC design is compounded by the ever-present need to decrease the device size and interconnection dimensions. Surface topography on advanced ICs have deeper steps and higher aspect ratios for gaps, making step coverage and gap-fill more difficult. A major negative consequence of topography is loss of linewidth control during photolithography. The photoresist thickness variations due to topography are a major factor inhibiting subquarter micron lithography. This variation is due to the depth-of-focus limitations of optical steppers. It is also difficult to pattern nonuniform photoresist thickness over etched steps.

A *planarized wafer* has a flat surface with minimal layer thickness variations (meaning minimal topography) on each layer. Filling in low features or removing high features are two ways to planarize a wafer surface. Planarizing the wafer surface is critical for follow-on process steps (e.g., lithography) and can actually serve to increase device yield by removing undesirable foreign material on the wafer surface during the actual planarization process. Qualitative planarization terminology is outlined in Table 18.1.[2] A wafer can have *local planarization,* associated with certain closely spaced features on a die, or *global planarization* of widely spaced steps that encompass all die across the entire wafer surface (see Figure 18.2). Most traditional planarization processes developed during the 1970s and 1980s resulted in the smoothing of step corners and local planarization (see the following section for an explanation of traditional planarization methods).

The primary planarization process used since the mid-1990s for multilevel metal technology has been *chemical mechanical planarization* (*CMP*). An IC cross section that has been polished using CMP is shown in Figure 18.3. It achieves global planarization across the wafer surface. CMP, commonly referred to as *chemical mechanical polish,* or *polish,* has been used for many years for optical glass polishing and wafer polishing during silicon wafer production. It was developed for

TABLE 18.1 Terminology for Wafer Planarization

Type of Planarization	Description
Smoothing	Step height corners rounded and sidewalls sloped, but the height is not significantly reduced.
Partial Planarization	Smoothing plus a local reduction in step height.
Local Planarization	Complete filling of smaller gaps (1 to 10 μm) or local areas within a die. The total step height to flat areas across the wafer is not significantly reduced.
Global Planarization	Achieves local planarization plus a significant reduction in the total step height across the entire wafer surface. This is also referred to as uniformity.

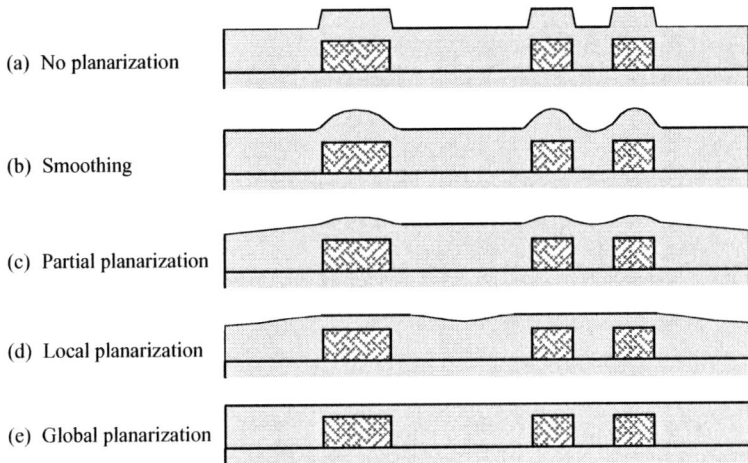

FIGURE 18.2 Qualitative Definitions of Planarization

semiconductor wafer planarization in the fabrication process at IBM in the late 1980s. The first application in the wafer fab was dielectric material planarization. This was followed with planarization for tungsten plug metallization. CMP is now used for most films and multiple materials on the same wafer surface. Its flexibility has made it an essential part of chip fabrication (see Figure 18.4). For copper metallurgy, CMP becomes an enabling technology for implementing multiple metal levels and interconnects in wafer fabrication.[3] This chapter reviews traditional planarization techniques to understand their limitations, followed by an in-depth review of planarization using CMP.

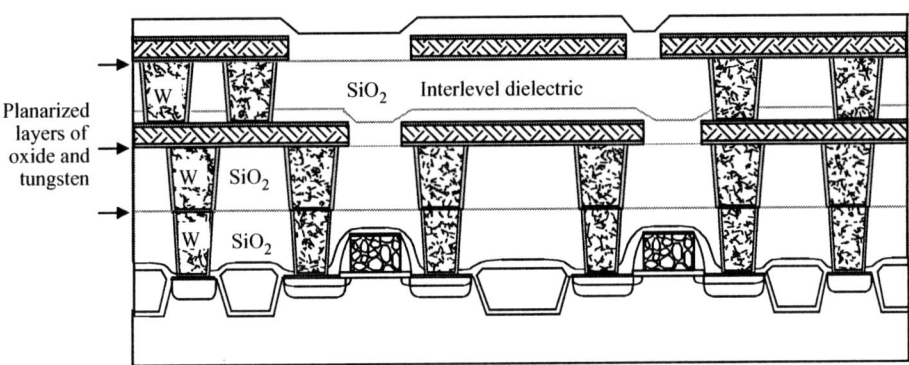

FIGURE 18.3 Multilayer Metallization with Chemical Mechanical Planarization (CMP)

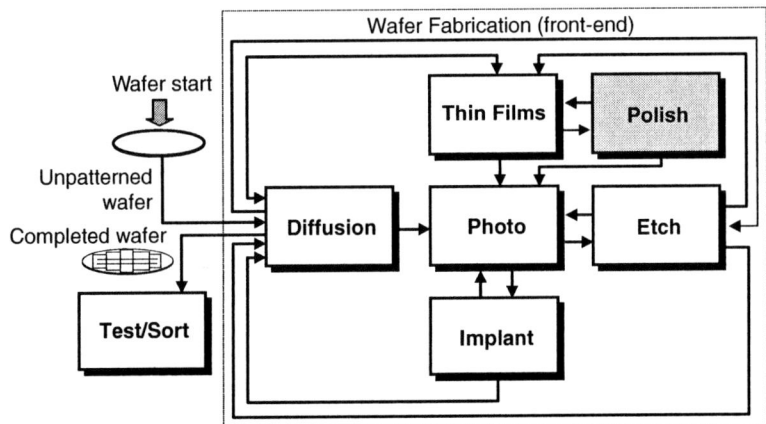

FIGURE 18.4 Wafer Process Flow with CMP
(Used with permission from Advanced Micro Devices)

518 CHAPTER 18

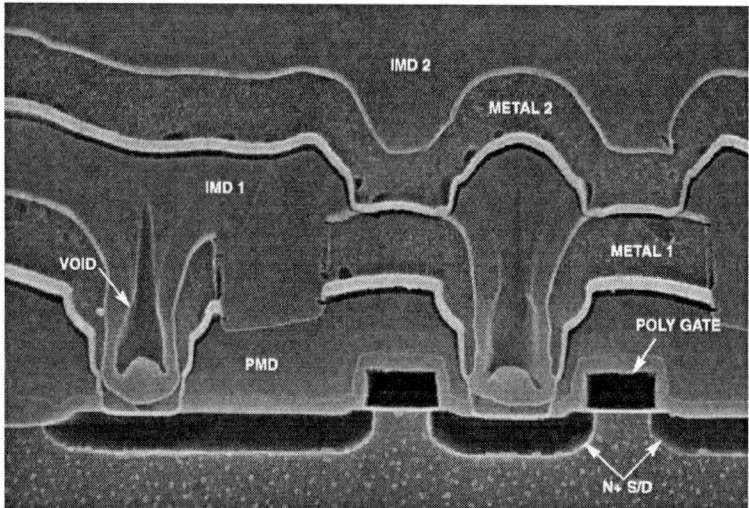

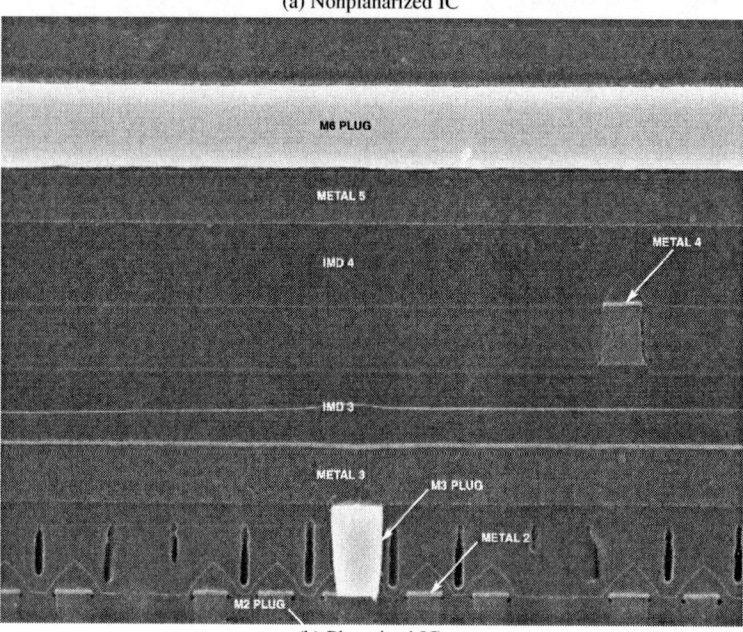

Multilayer Metallization with Nonplanarized and Planarized Surfaces (Micrographs courtesy of Integrated Circuit Engineering)

(a) Nonplanarized IC

(b) Planarized IC

TRADITIONAL PLANARIZATION

A short discussion of some of the traditional planarization methods is in order. This serves to highlight their contributions and weaknesses for wafer planarization and illustrate why chemical mechanical planarization (CMP) has become so important to advanced IC fabrication. The traditional planarization methods covered are:

- Etchback
- Glass reflow
- Spin-on films

Etchback

Topography created by surface features can be smoothed by applying a thick layer of a dielectric or other material followed by the application of a sacrificial layer of planarizing material (e.g., photoresist or spin-on glass). The sacrificial material fills voids and low spots on the surface. Etching of the sacrificial layer is then done using a dry etch to smooth the surface features by removing high features at a faster rate than low features. This process is known as *etchback planarization* (see

Figure 18.5). The etching is continued until the dielectric reaches a final thickness, with the planarizing material still filling the low-lying areas. There are different variations of etchback depending on the features, metal level, and so on. It is a local planarization of topography that smooths the surface over closely spaced steps. Etchback does not provide global planarization.

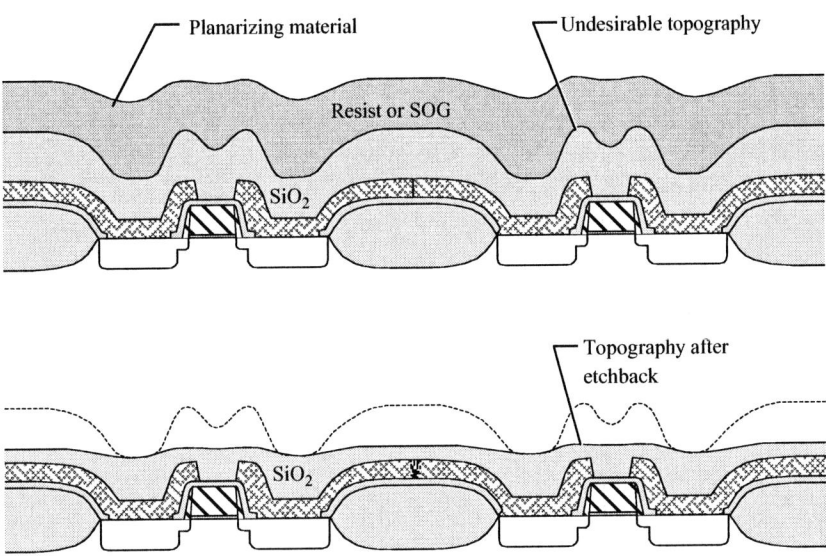

FIGURE 18.5 Etchback Planarization

Glass Reflow

Borophosphosilicate glass (BPSG) and other doped oxides have been used for interlayer dielectric (ILD) applications, and are often deposited with atmospheric pressure chemical vapor deposition (APCVD). *Glass reflow* is the heating of a doped oxide to cause the material to flow at elevated temperatures. For instance, BPSG annealed in a high-temperature furnace at 850°C for 30 minutes in nitrogen causes the BPSG to flow, resulting in BPSG flow angles of about 20° over step coverage (see Figure 18.6).[4] This flow property can be used to achieve BPSG conformality in step coverage or gap-fill, thus achieving some measure of partial planarization around the feature. BPSG reflow for conformality over features attains partial planarization, but is not adequate for deep submicron ICs with multilayer metallization.

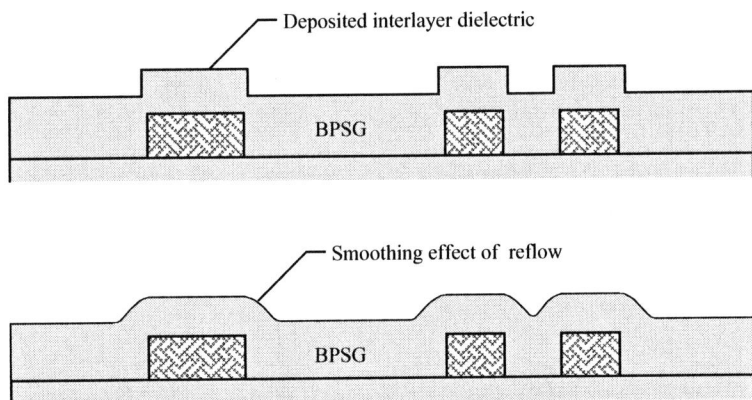

FIGURE 18.6 BPSG Reflow Planarization

Spin-On Films

Spin-on films is the application of different types of liquid materials on wafer surfaces and then spinning to achieve planarization, primarily for interlayer dielectric (ILD). This technology was most common for planarization and gap-fill at 0.35 μm and larger device generations. Spinning uses centrifugal force to attain a smoothing effect over topography by filling in low features. The planarizing capability of a spin-on solution depends on many factors, such as the chemical composition of the solute, its molecular weight, and its viscosity (tendency to flow). The spin-on film materials were either organic or inorganic, including photoresists, spin-on glass (SOG), and various types of resins.[5] SOG has different formulations, such as 80% solvent and 20% SiO_2, or an organic SOG formulation (e.g., polysiloxane). A bake step after spinning evaporates solvents and leaves oxide to fill the low-lying gaps. An oxide layer is deposited by CVO to further fill in the gaps in the topography (see Figure 18.7).

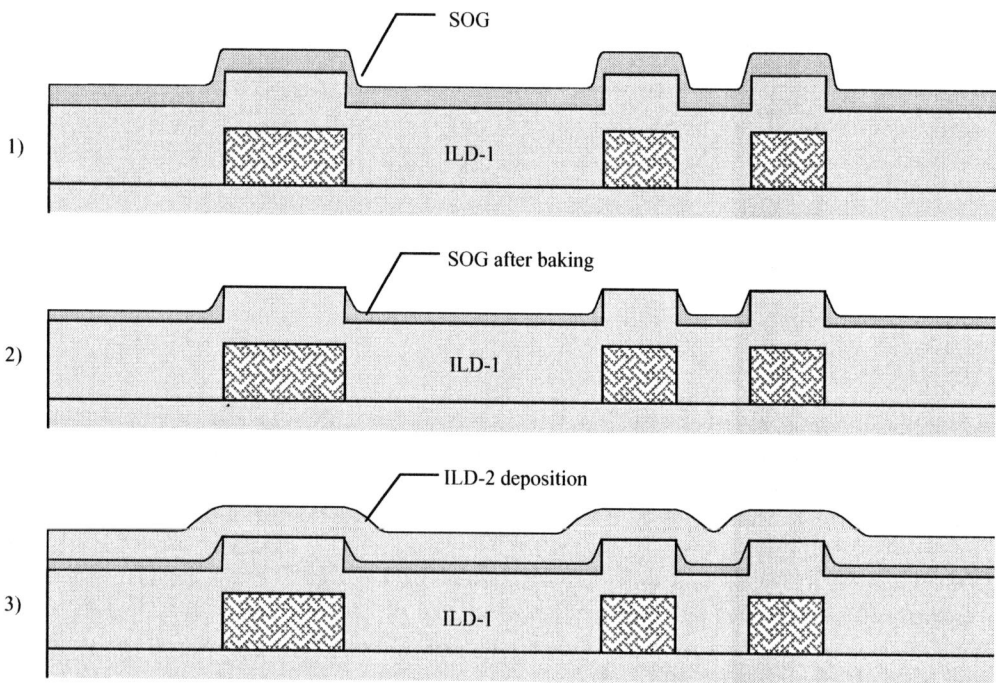

FIGURE 18.7 Spin-On Film Followed by ILD-2 Oxide Layer

As with the other traditional planarization methods, spin technology is limited in its ability to achieve global planarization for multilayer technology in advanced ICs. In a controlled study conducted to compare the planarization capability of a spin-on film with etchback to CMP, the maximum step height after etchback was about 7,000 Å, whereas CMP planarized the same material to a maximum step height of 50 Å.[6] This 140X improvement in surface planarity demonstrates why CMP has become the dominant planarization technology in ULSI fabrication. It should be pointed out that spin-on technology for applying films is potentially an important process for future wafer fabrication. It is envisioned that some of the future low-*k* dielectric films for ILD may be applied by spin-on techniques.

CHEMICAL MECHANICAL PLANARIZATION

Chemical mechanical planarization (CMP) is a global surface planarization technique. It planarizes the wafer surface by relative motion between a wafer and a polishing pad in the presence of a slurry while applying pressure (see Figure 18.8). The CMP tool is often referred to as a *polisher*. In a polisher, the wafer is positioned in a wafer holder, or carrier, and held against a polishing pad on a flat surface known as the platen. The motion between the wafer and polishing pad is controlled differently depending on the tool manufacturer. Most polishers use either rotary or orbital motion. In some instances, the platen is powered and moves (e.g., rotates), while on other polishers the carrier is powered while the platen rotation is driven only by the carrier motion.

CMP achieves wafer planarity by removing high features on the surface more quickly relative to the low features. It has become the most widely used method for accurately and uniformly polishing a wafer to a required thickness and planarity. One of the unique aspects of CMP is its ability to polish both dielectric and metal layers in multilevel metallization interconnect structures using an appropriately designed slurry and pad system.

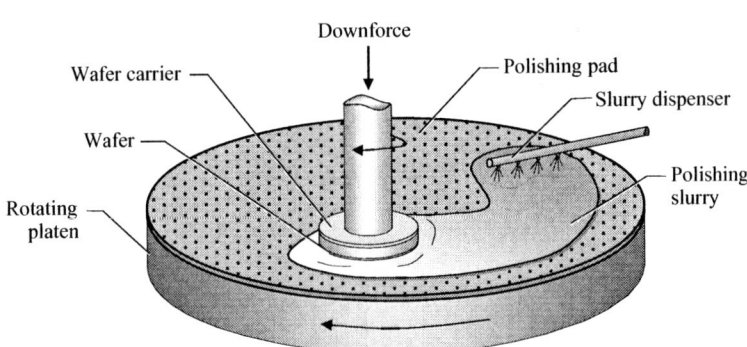

FIGURE 18.8 Schematic of Chemical Mechanical Planarization (CMP)

CMP Planarity

CMP is used to reduce the effects of wafer thickness variations and surface topography for fabrication. The concepts of wafer planarity and uniformity are important for describing the effectiveness of CMP. *Planarity* describes surface topography variations for wafer distances ranging from microns to millimeters. *Uniformity* is measured over millimeters or centimeters and represents a film thickness variation across the entire wafer.[7] In this manner, a wafer can be planar but nonuniform, or vice versa. One way to look at this condition is to consider how uniform are the different areas on a wafer that have been planarized. Two specific regions on a wafer could have been planarized very flat, but with each wafer surface region planarized to a different thickness. Each wafer region has good planarity when compared to itself, but poor uniformity when compared to one another.

The *degree of planarization* (*DP*) indicates how flat the wafer surface is at a particular step height location after CMP, relative to the initial step height prior to CMP (see Figure 18.9 on page 522).[8] In this manner, DP is a function of a specific feature. DP is calculated as:

$$DP\,(\%) = \left(1 - \frac{SH_{post}}{SH_{pre}}\right) \times 100$$

Where, DP = Degree of planarization
SH_{post} = Distance between the maximum and minimum step heights (thickness variation) at a specific location on the wafer surface after CMP
SH_{pre} = Distance between the maximum and minimum heights at a specific location on a wafer surface before CMP

If the topography measured on a wafer is perfectly flat after CMP, then $SH_{post} = 0$ and the DP is 100%. This means the planarization from CMP is perfect. For another example, if a wafer has $SH_{pre} = 20\ \mu m$ and $SH_{post} = 1\ \mu m$, then DP is,

$$DP\,(\%) = \left(1 - \frac{SH_{post}}{SH_{pre}}\right) \times 100$$

$$= \left(1 - \frac{1}{20}\right) \times 100$$

$$= 95\%$$

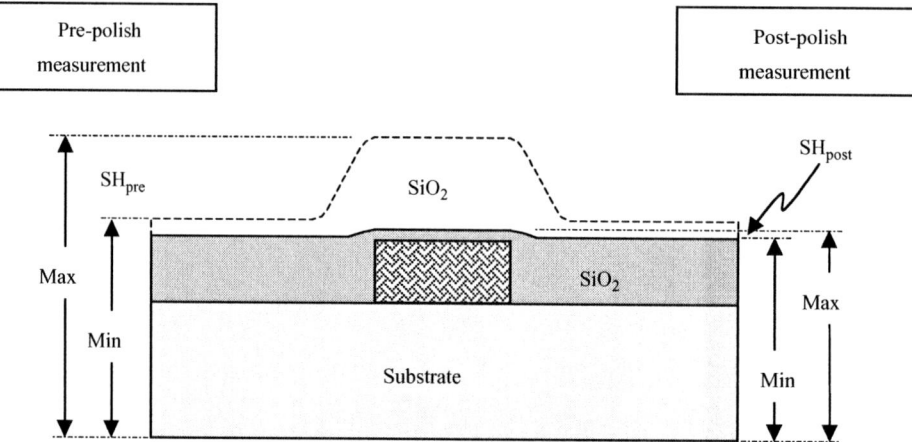

FIGURE 18.9 Wafer Measurements for Degree of Planarization

Two expressions are used to describe wafer nonuniformity: *within-wafer nonuniformity* (*WIWNU*) and *wafer-to-wafer nonuniformity* (*WTWNU*). WIWNU is a measure of the amount of film thickness variation across an individual wafer and is obtained by measuring numerous sites on a wafer (e.g., >9 sites). WTWNU represents film thickness variation across multiple wafers. These terms are often used to describe global planarization.

Advantages of CMP

There are many advantages to CMP, which has made it the most widely used planarization method for ULSI-era integration. The major CMP advantages for wafer fabrication are shown in Table 18.2.[9]

TABLE 18.2 Advantages of CMP

Benefits	Remarks
1. Planarization	Achieves global planarization.
2. Planarize different materials	Wide range of wafer surfaces can be planarized.
3. Planarize multimaterial surfaces	Useful for planarizing multiple materials during the same polish step.
4. Reduce severe topography	Reduces topography to allow for fabrication with tighter design rules and additional interconnection levels.
5. Alternative method of metal patterning	Provides an alternate means of patterning metal (e.g., damascene process), eliminating the need to plasma etch difficult-to-etch metals and alloys.
6. Improved metal step coverage	Improves metal step coverage due to reduction in topography.
7. Increased IC reliability	Contributes to increasing IC reliability, speed, and yield (lower defect density) of sub-0.5 μm and circuits.
8. Reduce defects	CMP is a subtractive process and can remove surface defects.
9. No hazardous gases	Does not use hazardous gases common in dry etch process.

These advantages distinguish CMP from the traditional planarization techniques that merely smooth over local topography with no effect on global planarization. There are also challenges to using CMP. Given the importance of CMP to integrating the metal and dielectric layers and interconnects in dual-damascene processing, the control of the CMP process is critical. The disadvantages to CMP are listed in Table 18.3.[10]

TABLE 18.3 Disadvantages of CMP

Disadvantages	Remarks
1. New technology	CMP is a new technology for wafer planarization. There is relatively poor control over the process variables with a narrow process latitude.
2. New defects	New types of defects from CMP can affect die yield. These defects become more critical for sub-0.25 μm feature sizes.
3. Need for additional process development	CMP requires additional process development for process control and metrology. An example is the endpoint of CMP is difficult to control for a desired thickness.
4. Cost of ownership is high	CMP is expensive to operate because of costly equipment and consumables. CMP processes materials require high maintenance and frequent replacement of chemicals and parts.

CMP Mechanisms

There are two CMP mechanisms that explain how planarization takes place on a wafer surface: (1) a chemical reaction by the slurry chemistry forms a wafer surface layer that is relatively easy to remove, and (2) this reacted wafer surface layer is mechanically removed by the slurry's abrasive component and the applied pressure and relative velocity of a polishing pad.[11] The microscopic CMP action required to planarize a wafer is both chemical and mechanical. A purely mechanical process, such as using sandpaper to smooth a board, cannot be used because such an abrasive process would damage the surface with gouges and scratches.

Oxide Polish ■ *Oxide polish* was the first and most widely used CMP planarization process in semiconductor wafer fabrication. Oxide polish is used to planarize the deposited ILD between metal layers to achieve global planarity. The rate of oxide polishing, or speed of removal, is given by *Preston's equation*, which shows that a number of parameters affect the removal rate. If pressure or velocity increases, then the rate of removal increases. Preston's equation is:[12]

$$R = kPv$$

Where, R = Rate of removal (thickness of glass removed per unit time)
P = Applied pressure
v = Relative velocity between the wafer and polishing pad
k = Constant for given tool and process conditions, including parameters such as the oxide hardness, polishing slurry, and polishing pad

The basic mechanism for oxide CMP is known as *Cook's theory*, and is the same polishing mechanism as for polishing optical glass. In a basic slurry, the water in the slurry reacts with the oxide to form a hydroxyl bond (Si bonded with OH, as shown in Figure 18.10 on page 524). This reaction is referred to as surface hydration. Surface hydration of the oxide decreases its hardness, mechanical strength, and chemical durability. There is also heating of the wafer surface due to friction during the polishing process, which additionally decreases the hardness of the oxide. The removal of the softer hydrated surface layer takes place by the mechanical action of the abrasive slurry particles.[13] In the higher regions of the wafer, the local pressure is greater than it is in the lower regions. By Preston's equation, the removal rate of oxide is greater for the higher regions, resulting in planarity.

Metal Polish ■ The mechanism for *metal CMP* is not as well understood as oxide polish. One simplistic model explains metal polishing using both a chemical oxidation and mechanical abrasion mechanism.[14] The slurry comes in contact with the metal surface and oxidizes it. For example, in the case of copper CMP, copper oxides (CuO or Cu_2O) and copper hydroxides ($Cu(OH)_2$) are formed. The metal oxide layer is then removed by mechanical abrasion from the particles in the slurry. Once the oxide layer is removed, the chemicals in the slurry oxidize the newly exposed metal surface and the process repeats itself (see Figure 18.11 on page 524). Recent investigations into the metal CMP mechanism have found that the chemical oxidation and dissolution of the oxidized

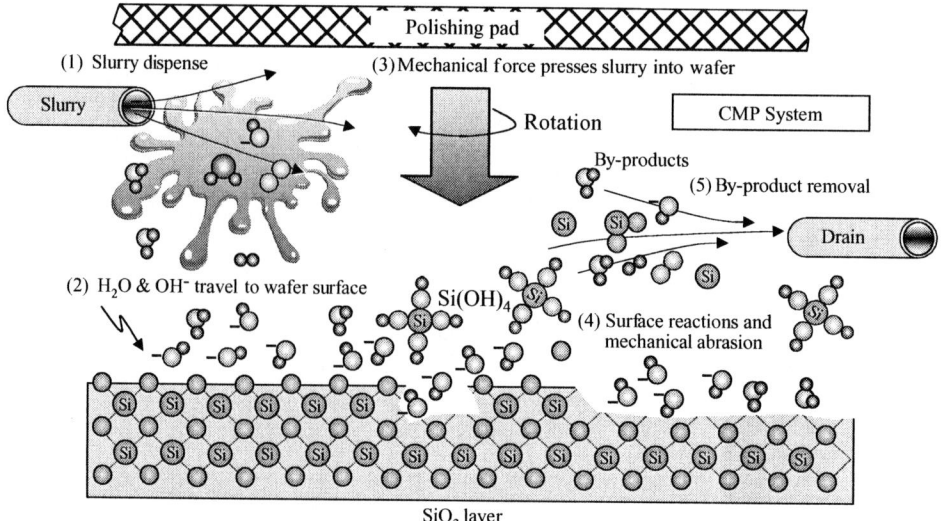

FIGURE 18.10 CMP Oxide Mechanism

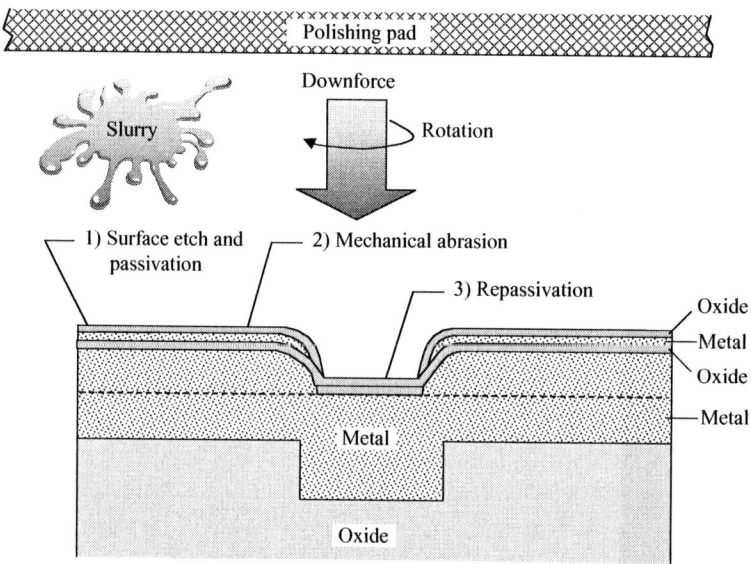

FIGURE 18.11 Mechanism for Metal CMP

metal are more important than mechanical abrasion. This means that careful control of the slurry chemistry is important for metal CMP.[15]

Most CMP processes have become a two-step polishing operation. The first polish is the primary removal step. The second polish is a buff clean that uses only DI water (or possibly a unique slurry). The main reasons for this second buff step are to remove microscratches and particles on the wafer surface and prepare the wafer to undergo the post-CMP clean process.

Pattern Density Effects ■ The degree of planarity that can be achieved with CMP is governed by the *pattern sensitivity* of this planarization technique.[16] Regions of narrowly spaced features, referred to as high pattern density, often polish at a greater rate than widely spaced features. Small, isolated raised features encounter greater pressure during planarization and polish at a higher rate. On the other hand, low features experience less pressure and polish at a slower rate. Variation in polishing rate can significantly affect CMP results for high-performance ICs. Local interconnect and dual-damascene metal levels require pattern densities that vary across a chip surface from very dense arrays of metal interconnect to areas with virtually no patterns (or no metal lines).

In situations where metal lines are densely packed (see Figure 18.12), undesirable erosion of the metal structures may occur during CMP).[17] *Erosion* is the thinning of oxide and metal in a patterned area. It is defined as the difference in the SiO_2 thickness before and after the polish step. One reason why it occurs is the slight overpolish into the underlying SiO_2 when planarizing a blanket metal layer. SiO_2 erosion is greater in areas of high pattern density.[18] To minimize this erosion, the overpolish step could be shortened (e.g., ideally by decreasing surface topography). In another approach, an oxide buff step may be added to the polish step to planarize the elevated regions of the oxide. The target oxide removal for the buff step is about 300 Å.[19] If not corrected, erosion in the surface can become so severe that it causes problems such as incomplete via etch due to oxide thickness variation (see Figure 18.13).

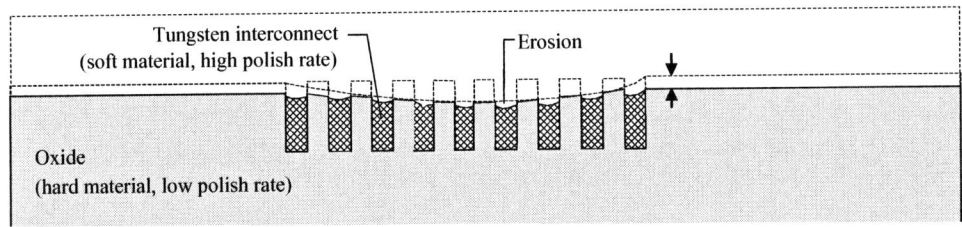

FIGURE 18.12 CMP Erosion in High Wiring Density

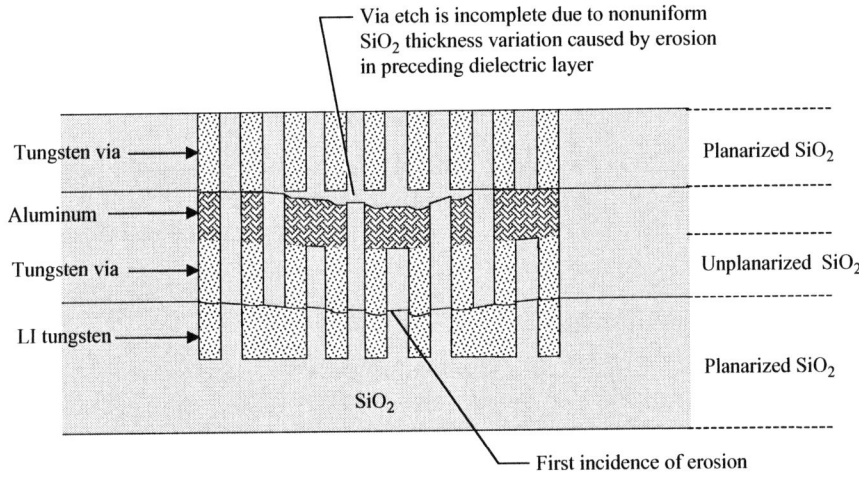

FIGURE 18.13 Incomplete Via Etch due to Erosion

Another undesirable side effect of CMP is dishing. *Dishing* is a reduction in the thickness of a material toward the center of the feature (see Figure 18.14 on page 526). It is defined as the difference in height between the center of the metal line (which is the lowest point of the dish) and the point where the SiO_2 levels off (the highest point of the SiO_2). The amount of dishing is dependent on the width of the line being polished, with wider lines having more tendency to dish. The hardness of the pad also affects dishing. Softer polishing pads bend into the soft metal line and exert pressure to cause dishing.

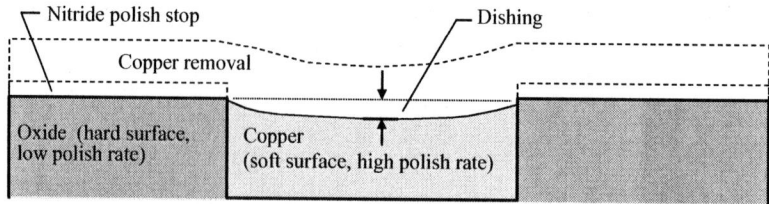

FIGURE 18.14 CMP Dishing in a Large Feature

CMP Slurry and Pad

The slurry and polishing pad are referred to as *consumables* because of their continual usage and replacement in the CMP process. They are critical components for CMP and must be carefully controlled.

Slurry ■ *Slurry* is a mixture of fine abrasive particles and chemicals that are used to remove specific material from the wafer surface during CMP. It is an important component of CMP because it contains the chemicals and polishing particles needed for planarization. Slurry is sometimes dispensed through a slurry dispense nozzle above the platen. Precise slurry mixing and consistent batch blends are critical to achieving the necessary repeatability from wafer to wafer, lot to lot, and day to day. It is also important for the slurry to be uniformly distributed across the wafer surface during polishing. The quality of the slurry is a factor in avoiding surface scratches during polish.

Oxide Slurry. One common type of slurry for oxide dielectrics is ultrafine silica colloidal (uniformly suspended) particles in an alkaline medium of aqueous potassium hydroxide (KOH) or ammonium hydroxide (NH_4OH). The KOH-based slurry is the most widely used slurry for oxide CMP because of its stable colloidal suspension. The K^+ ion is a mobile ionic contaminant (MIC), but it is easily gettered by the local interconnect (LI) oxide such as borophosphosilicate glass (BPSG). The NH_4OH-based is free of MICs, but it has unstable colloidal suspension and a higher cost.[20] The slurry pH is usually around 10.0 to 11.0. As previously explained, the presence of water in the slurry is critical for surface hydration and subsequent oxide planarization.

Tungsten Metal Slurry. The slurry for tungsten metal CMP is based on fine alumina (Al_2O_3) powder or silica as the abrasive component. The silica powder is softer than alumina with less possibility of surface scratching and is more commonly used. A slurry chemistry used today is a mixture of hydrogen peroxide (H_2O_2) and silica or alumina abrasive. During polishing, the H_2O_2 decomposes into H_2O and dissolved O_2. The O_2 reacts with and converts tungsten into tungsten oxide (WO_3). The tungsten oxide is softer than tungsten, so the WO_3 is polished and removed.[21]

Copper Metal Slurry. There is active research on the optimum copper metal slurry, without industry consensus at the time of this writing. Alumina powder slurry is sometimes used for copper CMP. However, in aqueous solutions copper behaves differently than tungsten. One difference is that copper oxidizes in a basic pH range of 7 to 12.5 while tungsten does not. For this reason, the slurry solution that works best for tungsten is not desirable for copper.[22] One approach for copper slurry is to use a basic solution of NH_4OH in the slurry that oxidizes the surface to form native copper oxide films, especially in the low-lying feature areas. At the same time, an ammonia complexing agent is added to rapidly dissolve the copper oxides in high feature areas for abrasive removal by the slurry particles. This action produces a high polish rate and a good surface finish.[23]

Application-Specific Slurry. Different slurries tailored for specific wafer surface materials are sometimes used on the same CMP equipment. For instance, with a copper CMP process, there is the copper layer and the barrier metal of tantalum (Ta) or tantalum nitride (TaN). If the same slurry is to be used in one approach for both materials, then both should polish at the same rate (the selectivity should be minimal). Depending on the slurry chemistry, there is a significant difference in chemical reactivity (or selectivity) between the copper and tantalum. This difference causes a slow polishing rate on the barrier metal, which leads to quality problems. An alternative approach is the use of a two-slurry system. The first slurry planarizes the copper and stops on the barrier, while the second slurry removes the barrier metal with minimum effect on the copper.[24] This approach may require a different polishing pad for each slurry, which can be done on the same tool with different platens.

New slurries are in development that will have chemistries and abrasive particles tailored for specific CMP applications. For example, a new slurry under development is a cerium oxide slurry where the particles are very uniform in size, with an average particle diameter of 0.2 μm. There is also development work on a slurry-free pad that has the abrasive integrated into the pad and uses DI water.[25]

Polishing Pad The *polishing pad* attaches to the top surface of the platen. It is an important part for determining the polish rate and planarization ability during CMP.[26] The pad is typically made of polyurethane because of its mechanical characteristics and is porous, like a sponge, in order to hold slurry. The pad has small perforations to help transport the slurry and promote uniform polishing (see Figure 18.15). The pad also removes the reacted products away from the wafer surface. After polishing a number of wafers, the pad surface becomes flattened and smooth, causing a condition referred to as *glazing*. A glazed pads has less ability to hold polishing slurry, which significantly decreases the polishing rate.

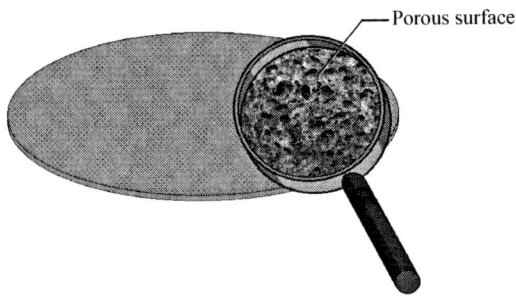

FIGURE 18.15 CMP Polishing Pad

CMP Polishing Pad
(Photo courtesy of
Speedfam-IPEC)

Polishing pads require regular *conditioning* to slow the effects of glazing. The purpose of conditioning is to provide consistent polishing performance throughout the pad life. Pads are conditioned by techniques such as mechanical abrasion or DI water jet spray to regenerate a rough surface. Another approach is to use a diamond wheel that is applied to the pad while it rotates. The conditioning process removes pad surface material and is a significant factor in the pad life. Some CMP tools have in-situ (real-time) conditioning, where a conditioning wheel is applied to one portion of the pad while the polishing of the wafer is occurring on another portion of the pad. In ex situ (external) pad conditioning, the conditioning is not done during polishing but only after a sequence of specified number of wafers are polished.

Polishing pads are also changed regularly with use. On older CMP tools, pads lasted around 100 wafers, depending on the application. Today, because of pad, slurry, and tool advancements, pads are lasting 800 wafers.[27] It is also common to stack multiple polishing pads on a platen to obtain the right combination of hardness and softness. One stacked pad approach for oxide polishing is to have a very soft sub pad on the platen with a hard pad in contact with the wafer surface.

Controlling the properties of polishing pads to maintain a repeatable removal rate over time is one of the biggest challenges in CMP. The type of motion is a factor in polishing pad control, with most approaches based on rotary or orbital motion. There are variations, with one manufacturer making a polishing tool that moves with linear motion (the pad motion is similar to a belt sander). Linear polishing permits a simpler CMP tool design, but there is still a need for stacked pads and backpressure in order to control for nonuniformity. Linear polishing also produces high removal rates at low pressures.[28] However, modern CMP polishers based on a rotary or orbital approach can achieve comparable downforce and removal rates.

Polish Rate and Uniformity ■ The *polish rate* is the speed at which the material is removed during planarization, with typical units of nm/minute or µm/minute. A hard polishing pad generally promotes local wafer planarization by evenly polishing over topography with minimal erosion in areas with dense patterns. It bridges over the low spots of the wafer and removes material from the high spots, thereby making the wafer surface more planar at specific locations.[29] However, hard pads generally have greater within wafer nonuniformity (WIWNU).[30] A soft polishing pad will reduce the formation of surface scratches.

Higher pressure (downforce) and rotational speed will increase the removal rates, but possibly at the expense of uniformity. Higher pressure also causes more severe surface damage (e.g., scratches) and contamination when used with hard polishing pads.[31] Low pressure may improve planarity but wafer-to-wafer nonuniformity (WTWNU) increases. In many cases, the best planarity results occur with a hard polishing pad and low pressure.

The movement of the slurry during polish affects the polish rate. In tools with rotary motion, slurry is moved primarily through rotational forces along the surface of the pad. The edge of the wafer may get more slurry than the center of the wafer. This gives the edge of the wafer a higher polishing rate than the center, referred to as a *center slow* tool (see Figure 18.16). To compensate, some manufacturers will flow N_2 into the carrier to produce backpressure on the backside of the wafer. This action gives the wafer a convex shape and helps reduce center slowness by slightly increasing the pressure in the center of the wafer. The carrier may also have a slight contour to improve polishing in the center.[32]

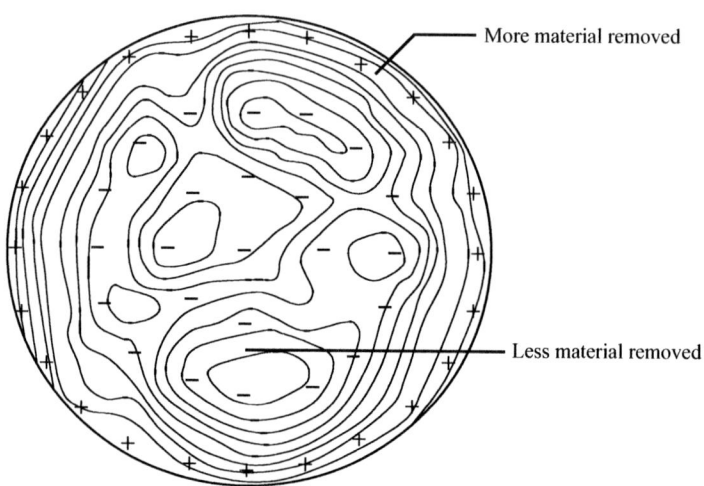

FIGURE 18.16 CMP Wafer Contour Plot for Center Slowness

Selectivity. The polishing rate of the different materials undergoing CMP is an important factor that affects wafer planarity and uniformity. *Selectivity* is a ratio that compares the polishing rate for blanket-coated materials. Selectivity is defined as:

$$\text{Selectivity} = \frac{\text{polish rate of material \#1}}{\text{polish rate of material \#2}}$$

If a soft material polishes at 3,000 nm/min and a hard material polishes at 100 nm/min, then the selectivity is 30. In an application that requires simultaneous polishing of multiple materials, it may be ideal for the selectivity to be one. This condition will minimize problems such as dishing and erosion. On the other hand, for metal polishing, a high selectivity is often desirable between the metal and the dielectric to minimize erosion. If a barrier film is present between the metal and interlayer dielectric (ILD), a low selectivity to the barrier film may be desirable so that it can be removed during the main CMP step without the need for a subsequent barrier removal step.

Polish Time and Layer Thickness. The amount of time needed to polish a layer is directly related to the amount of material to be removed. *Overburden* is the amount of excess material deposited on the wafer surface prior to polishing. The goal is to reduce the amount of polish time (and film deposition time) by minimizing the overburden thickness. At the same time, there must be adequate film thickness to achieve planarity. For example, consider the case of an ILD polish. The post-polish film thickness must be greater than the step height of the underlying metal feature by an amount determined by electrical performance considerations. This state is achieved by requiring a film overburden thickness to be equal to the step height plus some additional amount.

CMP Variables ■ Control of the CMP process is difficult because of the influence and interaction of many different parameters that affect planarization and uniformity. Parameter optimization is required for different applications. A summary of these parameters is given in Table 18.4.[33]

TABLE 18.4 CMP Parameters

Parameter	Planarization Results on Wafer
Polish time	• Amount of material removed • Planarity
Pressure on wafer carrier (downforce)	• Removal rate • Planarization and nonuniformity
Platen speed	• Removal rate • Nonuniformity
Carrier speed	• Nonuniformity
Slurry chemistry	• Material selectivity • Removal rate
Slurry flow rate	• Affects amount of slurry on the pad and the lubrication properties of the system
Pad conditioning	• Removal rate • Nonuniformity • Stability of CMP process
Wafer/slurry temperature	• Removal rate
Wafer back pressure	• Center slowness/nonuniformity • Wafer breakage

CMP Equipment

A CMP polish tool planarizes using a polishing pad that is bonded to the surface of the platen (see Figure 18.17 on page 530). A wafer carrier holds the wafer in place during polishing. Most production polishers have multiple platens and polishing pads to provide flexibility for polishing different layers. Multiple platens operate simultaneously with multistep processing using different slurries. As discussed earlier, the final polishing step is often a buff to remove a few hundred angstroms of material.

Variations of present-day CMP equipment have been used in the semiconductor industry for decades as wafer lapping machines in silicon wafer production. However, lapping machines did not have to address the wide range of materials currently used in IC wafer fabrication. Due to the increased number of metal layers and the sensitivity of depth-of-focus in lithography, the planarization capability of the CMP equipment is critical. An important equipment issue is the ability to detect when the proper amount of material has been removed. The device is known as endpoint detection.

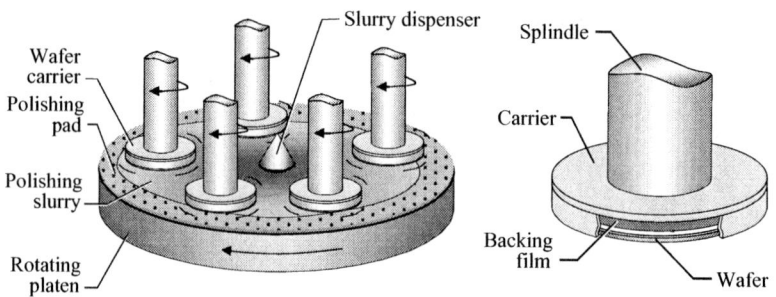

FIGURE 18.17 CMP Tool with Multiple Wafer Carriers

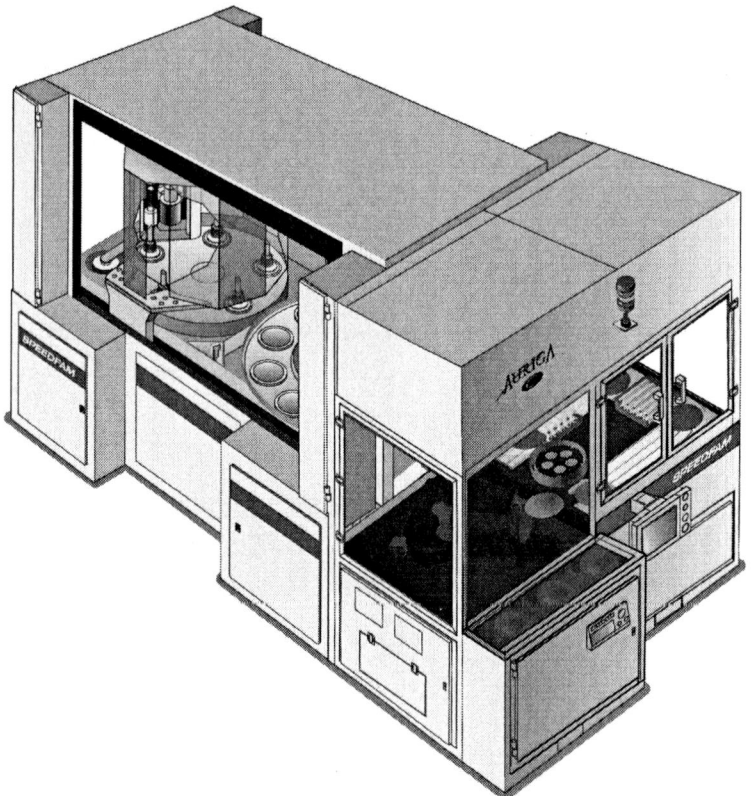

CMP Tool
(Photo courtesy of Speedfam-IPEC)

Endpoint Detection. *Endpoint detection* is the ability of the CMP tool to detect when the planarization process has polished the material to the correct thickness. Some CMP applications are straightforward for endpoint detection, such as polishing blanket tungsten and stopping on the underlying dielectric oxide film with high selectivity. However, other applications pose challenges, such as polishing ILD oxide with no polish stop material or damascene copper with a requirement to have 1000 Å of copper remaining after a specific CMP step. The wide range of CMP process variables also makes an accurate prediction of the correct polish time difficult.

The first CMP tools in the early 1990s used monitor wafers for endpoint detection to estimate the duration of the wafer polish. This process was crude and often required a repolishing step to get within the target thickness. The latest models of all CMP tools have some form of in-situ endpoint detection device that permits the user to stop at the remaining thickness without repolishing. Endpoint detection has traditionally been limited to sampling and averaging the remaining film across the entire wafer. Recent equipment improvements permit endpoint detection at a specific location on the wafer. In an ideal situation, in-situ endpoint detection would compensate for changes in the polishing rate and provide early detection in polishing nonuniformities. This is an area of active research in the CMP equipment industry. The two most common in-situ methods are based on either motor current or optical measurements.

Motor Current Endpoint Detection. Motor current endpoint technology monitors the magnitude of the electrical current supplied to the motor used to rotate the wafer carrier or platen. Since the wafer carrier rotates at a constant speed to maintain the proper polishing rate, the drive current is varied to compensate for any load changes on the motor. In this manner, the motor current is sensitive to changes at the wafer surface due to friction or surface roughness. Frictional changes occur when the polisher finishes with one material, exposing a material with different polishing and frictional characteristics. An example is passing from polishing a blanket tungsten film to the underlying Ti/TiN barrier metal and then oxide layer. The change in the drive motor current easily detects each of these material layer changes (see Figure 18.18).

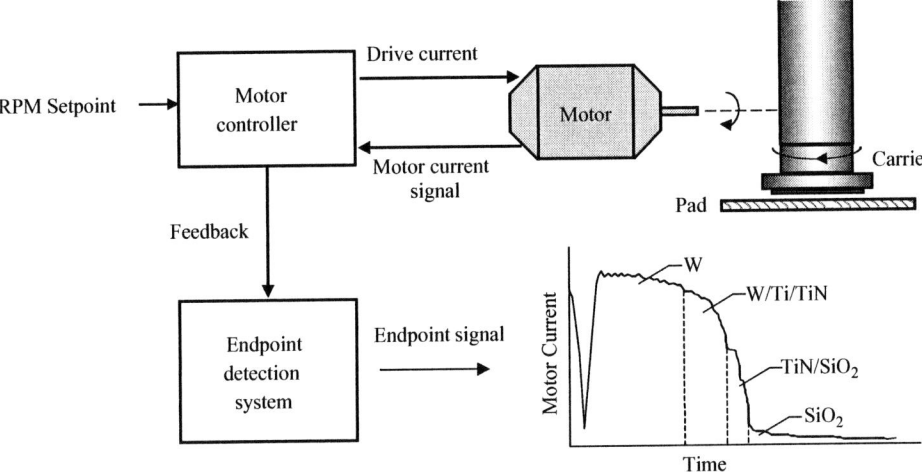

FIGURE 18.18 Motor Current Endpoint Detection

An application where motor current monitoring is not appropriate for endpoint detection is the CMP of interlayer dielectric (ILD). The goal is to leave a predetermined oxide thickness on the surface. Because of this, there is no layer breakthrough (or stopping layer) that produces a change in surface friction and motor drive current. Endpoint detection for this application requires a different technique, with most equipment using optical detection.

Optical Endpoint Detection. A method widely used on CMP equipment is *optical interferometry endpoint* detection. This technique is based on light reflectance. Recall from Chapter 8 that in reflection spectroscopy, light reflects at different angles off film layers depending on the type of film and its thickness. Optical endpoint detection measures the constructive and destructive interference between UV or visible light rays reflected off the polished layer as the film changes from one material interface to another. By continually sensing the film thickness during polishing, this technique can determine the removal rate during the polish (see Figure 18.19 on page 532). Signal processing algorithms are used to analyze the reflected light and reduce electrical noise from the pattern on the wafer surface. A representative accuracy for endpoint detection of STI (shallow trench isolation) oxide over nitride is within ± 100Å of the nitride layer.[34]

Wafer Carrier ■ The *wafer carrier* (also known as the *polishing head*) holds the wafer above the polishing pad on the platen surface. The downward force and rotational motion of the wafer carrier relative to the platen can affect uniformity. Vacuum often is used to hold the wafer in the carrier during transportation and polishing, although sometimes the vacuum is turned off during polishing or even reversed to apply a backpressure. Some carriers include a *backing film* sandwiched between the wafer and carrier to provide a compliant support that compensates for flatness or particles on the wafer backside. The backing film is spongy and contains holes for the vacuum.

The design of the wafer carrier affects the amount of nonuniformity wafers exhibit at their outer edge. The distance from the wafer edge where polishing is less controlled is referred to as *edge exclusion*. For high-density ICs, edge exclusion is currently about 3 mm.[35] This size was reduced from 6 to 7 mm a few years ago and will be reduced again in the future. The source of

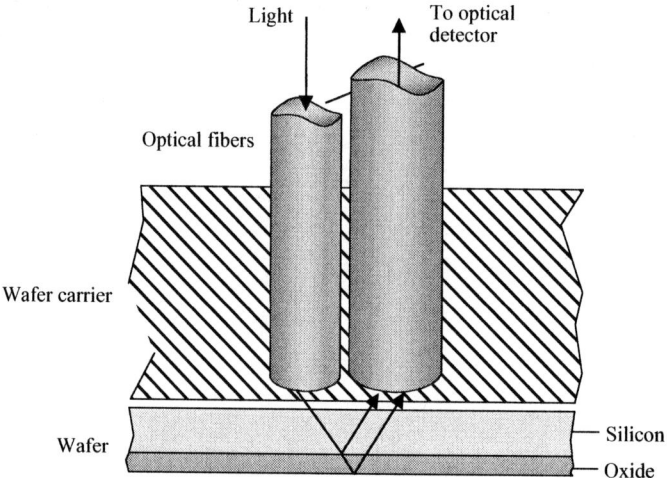

FIGURE 18.19 Optical Interferometry for Endpoint Detection
Redrawn from H. Litvak and H. M. Tzeng, "Implementing Real-Time Endpoint Control in CMP," *Semiconductor International* (July 1996): p. 262.

the nonuniformity is a slight buckling of the polishing pad at the wafer edge. The conventional head-on older tools were not able to apply pressure on the retaining ring that held the wafer in place (see Figure 18.20). On more recent tools, independent pressure control is applied to the retaining ring of the wafer carrier to optimize the removal rate at the wafer edge with the removal rate across the wafer, thus achieving better uniformity.

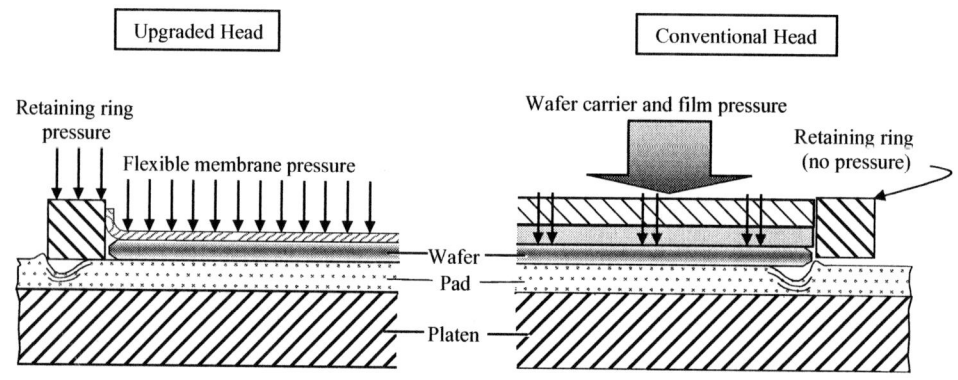

FIGURE 18.20 CMP Head Carrier Design and Wafer Edge Nonuniformity
Redrawn from K. Wijekoon, et al., "Tungsten CMP Process Developed," *Solid State Technology* (April 1998): p. 55.

A new carrier design, sometimes referred to as a *bladder carrier*, is based on applying constant pressure to the back of the wafer. The pressure is applied through the use of pneumatic or hydraulic pressure. A flexible membrane, or bladder, uniformly couples the pressure across the wafer during polishing to achieve improved global planarization.[36]

CMP Clean

The front-end-of-line (FEOL) cleaning process in a wafer fab is driven mainly by gate oxide performance. Cleaning for the back-end-of-line (BEOL) has traditionally addressed contaminants related to thin film layering, contact formation, and more recently, CMP. With the incorporation of CMP for a wide range of applications such as interlayer dielectric (ILD), tungsten plug formation, and dual-damascene processing, BEOL cleaning processes have become more stringent.

The focus for CMP cleaning is to remove all contaminants from the polishing process. These contaminants include slurry particles, any particles from the polished material, and chemical contamination from the slurry. Special cleaning effort is made to remove particulate contamination from slurry and polished material. Particles become either mechanically embedded in the wafer surface due to the pressure applied during CMP or physically adhered to the polished wafer surface due to electrostatic or atomic (van der Waals) forces. The electrostatic forces are surface charges that can be attractive or repulsive depending on the zeta potential (see Chapter 6). In an acidic slurry, such as that used for tungsten metal, the surface charges on the slurry alumina and the silicon wafer have opposite charges and therefore lead to particulate contamination.[37]

The cleaning processes used in post-CMP employ various cleaning tools: a brush scrub, acid spray clean, megasonic rinse, and spin rinse dryers. These cleaning systems were discussed in Chapter 6. The major CMP process steps for post-CMP clean are post-oxide, post-STI (shallow trench isolation), post-poly, post-tungsten, and post-copper.

Post-CMP Clean ■ Post-CMP cleaning has evolved since CMP's wafer fab beginning in the early 1990s (see Figure 18.21). Post-CMP cleaning initially used DI water with megasonics. It evolved to physical scrubbing of the wafer using a double-sided scrubber (DSS) brush and DI water. The brushes rotate and press against the wafer surface to mechanically remove particles. However, for cleaning with a DSS brush and only DI water, the brush quickly becomes loaded (contaminated) with particles. A *loaded brush* condition with particle contamination easily transfers particles to other wafers. To resolve loaded brushes, post-CMP clean is typically done using a DSS brush scrub with dilute ammonium hydroxide (NH_4OH) that flows through the brush core to flush the brush (see Figure 18.22).[38] The outward liquid flow through the brush stems serves to continually displace particles.

	Wet bench with megasonics	Double-sided scrubber DSS+ DI water	DSS + NH_4OH	DSS + NH_4OH and HF	DSS + Additional Chemistries
Oxide CMP	✓	✓	✓	✓	
Tungsten CMP			✓	✓	
Copper CMP					✓

FIGURE 18.21 Evolution of Post-CMP Cleaning

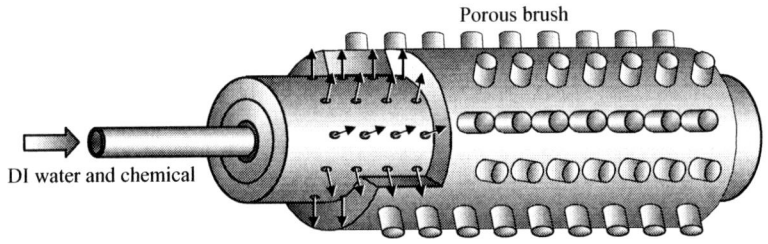

FIGURE 18.22 Through-the-Brush Chemical Delivery for Post-CMP Cleaning
Redrawn from D. Hymes, et al., "Brush Scrubbing Emerges as Future Wafer-Cleaning Technology," *Solid State Technology* (July 1997): p. 210.

In some cases, hydrogen peroxide (H_2O_2) is added to the post-CMP cleaning solution. This alkaline solution is used to control the pH, and therefore zeta potential, so that the particles and wafer surface are electrostatically repelled.[39] For oxide and tungsten metal cleaning, this step is done by cleaning in a basic solution with a pH above 8. Only when the pH is 8 or above do the slurry particles have the same surface charge sign as that of the tungsten metal and SiO_2. This condition causes particles to be repelled by the wafer surface, making them more easily removed during cleaning.

There are variations in post-CMP cleaning. A short cleaning step in diluted HF is often done to remove trace metal contaminants by etching away a few angstroms of the surface. This process is most effective on an oxide surface but is also done on tungsten metal if the time is very short (e.g., 15 seconds). Evaluations have also shown citric acid is effective at removing alumina particles from SiO_2 surfaces.[40]

A spray acid tool or megasonic cleaner is used in some applications for applying the cleaning solutions instead of the scrub brush. However, a DSS brush supplies much higher energy to the wafer surface than megasonic energy, which produces better cleaning results and shorter processing times. A spin rinse dryer is used after wet cleaning to rinse and dry the wafers.

Post-Copper CMP Clean. Copper diffuses very quickly through silicon and silicon dioxide. To prevent degradation in device electrical performance, all copper must be removed from wafer surfaces (front, back, and bevel edge). The NH_4OH cleaning chemistry that avoids brush loading for traditional CMP clean is not acceptable for copper cleaning, as it causes nonuniform etching of the copper layer, leading to local surface roughening. It is anticipated that new cleaning chemistries will be required for copper that not only control electrostatic forces to repel particles but also prevent copper corrosion.[41] The cleaning solution should be compatible with the copper surface. It is critical that the residual copper contamination be removed from the wafer, especially in dielectric regions and areas of high pattern density. At the time of this writing, the chemistry of the cleaning solution for copper is an area of ongoing research.

Dry-In/Dry-Out ■ The latest trend in CMP equipment is integrated CMP processing and cleaning, commonly referred to as a *dry-in/dry-out* process. This approach improves productivity by reducing the time needed for polishing and clean. Wafers are planarized, cleaned, dried, measured, and returned to the cassette within the single tool. However, for Cu CMP, studies also indicate there is a 50% reduction in particles with an integrated dry-in/dry-out clean process. Another benefit is that the quick removal of the slurry from the Cu surface creates less potential for corrosion.[42]

CMP Equipment Manufacturers

Various manufacturers offer commercial CMP equipment (see Table 18.5). The production issue of wafer throughput is important. The equipment must account for the mismatches in process time between the different materials to be polished on a layer, which is why multiple platen and wafer carriers are available on CMP equipment. Efficient production requires the use of simultaneous wafer processing. This result is achieved with flexible equipment hardware and software (e.g., the ability to use different process recipes for each material, and so on).

TABLE 18.5 Examples of Some Commercial CMP Equipment Systems

Supplier/Model	Type of Motion	No. of Platens/ Diameter (in.)	# of Wafer Carriers	Dry-In/ Dry-Out	Endpoint Detection
Applied Materials Mirra 3400	Rotary	3 / 20"	4	Yes	Yes
Ebara EPO-222	Rotary	2 / 23.6"	1	Yes	Yes
Speedfam-IPEC Avanti 472 Avanti 672 IPEC 676/776 Auriga-C	Orbital Orbital Orbital Rotary	2 / 22.5" 3 or 6 / 32" 4 / 16"	1 3 or 6 4	Yes Yes Yes	Yes Yes Yes
Lam Teres	Linear	2 belts	4	Yes	Yes
SpeedFam Auriga	Orbital	2 / 32"	5	Yes	Yes
Strasbaugh Symphony	Rotary	3 / 32"	4	Yes	Yes

CMP APPLICATIONS

CMP has become the dominant planarization technology for multilevel metallization integration for ULSI subquarter micron technology. CMP applications involve a wide range of materials. A short review will be done of some key applications to demonstrate how CMP is an enabling technology for overall process integration. Refer to Chapter 9 for specific details to understand the entire process flow. The CMP applications to be reviewed are:

- STI oxide polish
- LI oxide polish
- LI tungsten polish
- ILD oxide polish
- Tungsten plug polish
- Dual-damascene copper polish

STI Oxide Polish

Shallow trench isolation (STI) is a front-end process that forms isolation regions between devices on the wafer surface. STI has replaced the LOCOS (local oxidation of silicon) technique primarily because of STI's more efficient use of space for device structures. Refer to Chapter 11 for a description of the STI process flow.

The oxide fill layer in STI is planarized with CMP to remove all oxide above the nitride layer (otherwise the hot phosphoric acid strip will not remove the nitride in the operation that follows STI polish). This CMP process is shown in Figure 18.23. The nitride functions as a polish-stop layer for CMP, with endpoint detection to stop the polish process at the transition from oxide to nitride. The thickness of the nitride layer also defines the permissible amount of CMP overpolish before active device areas are exposed and damaged by polishing.

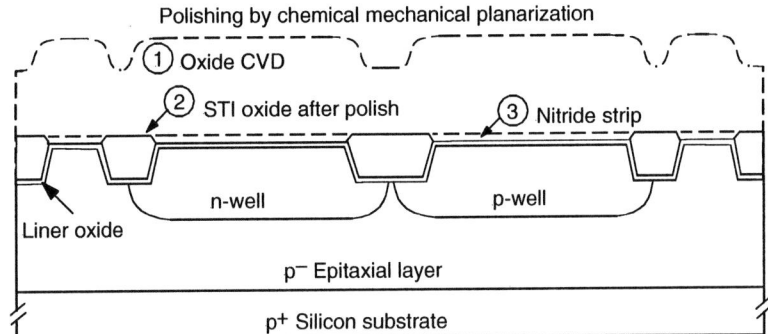

FIGURE 18.23 CMP for Oxide Fill of STI

A challenge for STI polish is to avoid excessive oxide thinning, or dishing, in the trench. This condition is caused by excessive polishing pad flexibility that deflects under pressure into the wide trench region. When the silicon nitride mask layer is exposed during CMP, the pad flexes into the trench opening and forms a concave-up surface that leads to dishing. The amount of dishing is influenced by factors such as the hardness of the pad, width of the trench, and overpolish time.

LI Oxide Polish

LI oxide is polished and later patterned to form the local interconnect (see Figure 18.24 on page 536). The local interconnect (LI) provides metal connecting lines between active devices that pass through the ILD-1. The LI metal is typically tungsten to connect components at the transistor and substrate levels. A doped oxide layer, such as phosphosilicate glass (PSG), is deposited for the LI oxide. The ILD-1 oxide electrically isolates active devices from the LI metal and physically isolates devices from contamination sources such as mobile ionic contaminants.

The LI oxide film includes a thin layer of silicon nitride or oxynitride that serves as device protection and an LI stop layer, followed by a thick layer of doped oxide deposited by plasma-enhanced TEOS or high-density plasma CVD (HDPCVD). Phosphorus-doped HDPCVD oxides (PSG) are most commonly deposited and reflowed at 600 to 650°C for the sub-0.25 μm generation devices, replacing BPSG for thermal budget reasons due to the high-temperature (850°C) reflow required for BPSG.[43] After deposition, the LI oxide film has excessive thickness and topography that conform to the surface features. The LI oxide layer requires planarization with CMP, with a final thickness on the order of 8,000 Å (see Figure 18.24).

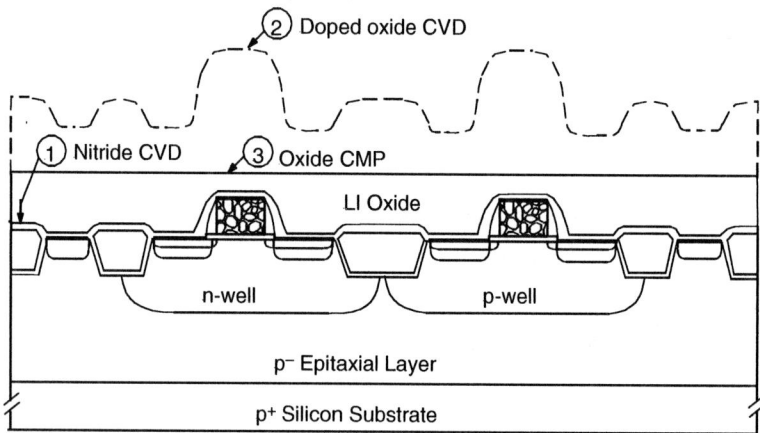

FIGURE 18.24 LI Oxide Before and After CMP Planarization

LI Tungsten Polish

The LI (local interconnect) is formed by patterning openings in the ILD-1 oxide for vias and local wiring to the source/drain contacts. Tungsten metal is deposited into the vias to form plugs and trenches that comprise the local metal connections. Prior to tungsten, a thin Ti/TiN stack layer is deposited. The Ti improves adhesion of tungsten to SiO_2. The TiN serves as a diffusion barrier for tungsten metal and helps improve source/drain contact resistance. Tungsten is blanket deposited over the entire wafer to fill the vias and LI trenches. It is preferred over aluminum for the LI because of its superior gap-fill properties without voids. It is anticipated that even for copper metallurgy, tungsten will remain in use for LI metal. CMP is used to polish the tungsten using the oxide as a polish stop. Tungsten has good polish characteristics. A light buff is used to remove residual debris from the tungsten CMP step and prepare the surface for cleaning.

The tungsten metallization for LI layer has the highest tungsten pattern density with many tungsten features. Dishing and erosion are significant problems for CMP of LI-tungsten. Overpolish must be minimized during this process.

ILD Oxide Polish

The interlayer dielectric (ILD) is the oxide deposited between metal layers to electrically isolate the metal conductors. It usually consists of consecutive deposits of high-density plasma CVD (HDPCVD) followed by PECVD oxide. The HDPCVD is deposited because of excellent gap-fill characteristics in narrow geometries, whereas the PECVD is used for increased wafer throughput and lower cost. The oxide is polished with CMP to a specific target thickness, which is essentially a blind process because there is no stop-polish layer. ILD oxide polish requires effective endpoint detection (see Figure 18.25). This process is different from tungsten CMP, which uses the ILD oxide as an effective stop layer.

Tungsten Plug Polish

The ILD oxide is patterned with photolithography and dry-etched to open vias. A thin blanket layer of Ti is deposited for adhesion, followed by a thin TiN barrier metal. Blanket deposition of tungsten fills all via holes and the top of the ILD oxide. The tungsten is then polished with CMP back to the ILD oxide surface, conveniently using the oxide as a stop layer. This step forms the metal plugs that make electrical connections between adjacent metal layers (e.g., Metal 1 and Metal 2).

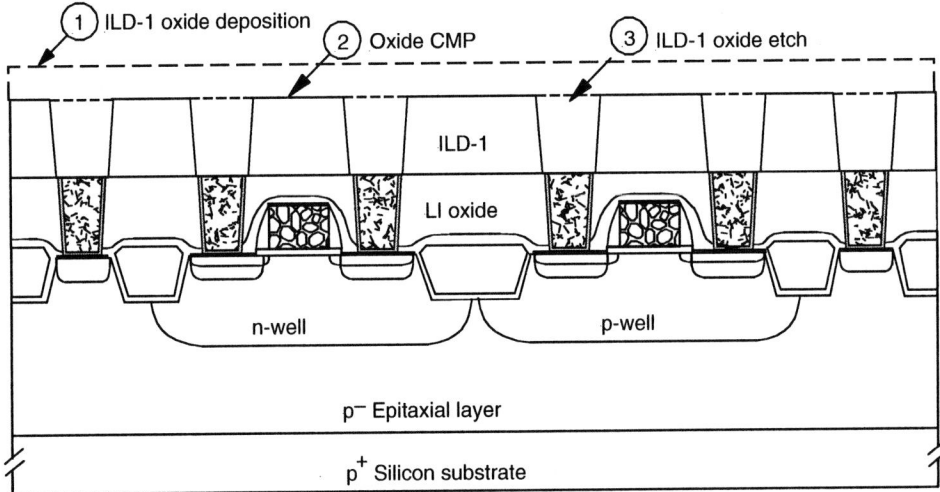

FIGURE 18.25 ILD Oxide Polish

Dual-Damascene Copper Polish

CMP is used to polish the vias and fine lines of copper used in the dual-damascene approach (see Figure 18.26). Via and trench patterns are first formed in the ILD material by photolithography and dry etching. A barrier metal (~75 Å), followed by a thin copper seed layer (~500 Å), is deposited to prepare the surface. Then copper is deposited into the high-aspect ratio patterns using electrochemical deposition (ECD). This copper is planarized using CMP, with the dielectric serving as a stop layer.

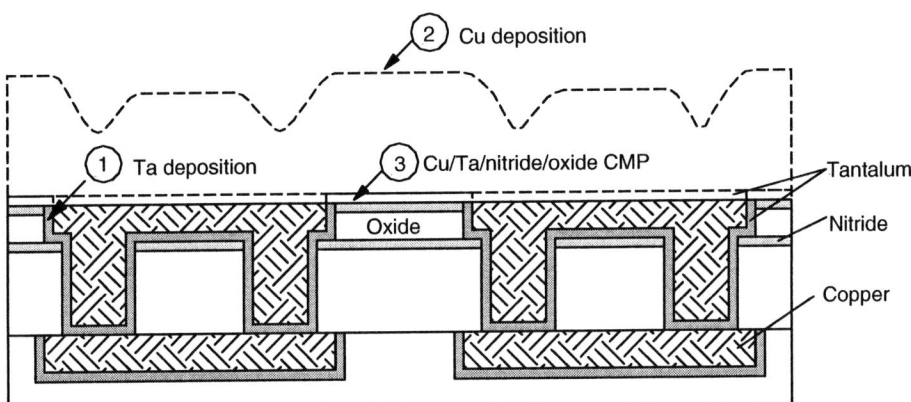

FIGURE 18.26 CMP for Dual-Damascene Copper Metallurgy

Depending on the dual-damascene approach, there may be a challenge for CMP planarization of damascene copper while also removing the barrier metal. This removal would be required at the dielectric/barrier/Cu interface, requiring careful selectivity in the CMP process. An ideal case has a selectivity of nearly 1 for the Ta and Cu materials. At the time of this writing, it appears that Cu CMP will have a two-step process. One slurry removes all or most of the Cu and another slurry is used to remove the Ta layer and remaining Cu with nearly equivalent removal rates (a selectivity close to 1).

The selectivity of Cu and Ta also affects how well planarity quality measures can be met. Higher polishing rates of one material can cause severe dishing of the copper and erosion of the dielectric surface. The most susceptible regions for dishing are wide buses and bond pads on the upper noncritical metal layers. Dual-damascene trench depths are currently about 5,000 Å and less than 20% variation is needed for layer integration. This means there are less than 1,000 Å permissible for surface dishing and erosion from CMP.[44] Uneven surface planarization from dishing and erosion can transfer to subsequent layers, making it more difficult to achieve good planarization. This condition causes electrical shorts from residue because of the poor planarization resulting from underlying topography.[45]

CMP QUALITY MEASURES

A significant quality problem due to CMP is surface microscratches. A small and difficult to find microscratch serves as a hiding place for deposited metal, potentially forming metal-to-metal shorts within a level (see Figure 18.27). The major source of microscratches is unacceptable particle contamination in the slurry. Representative quality measures for CMP are highlighted in Table 18.6.

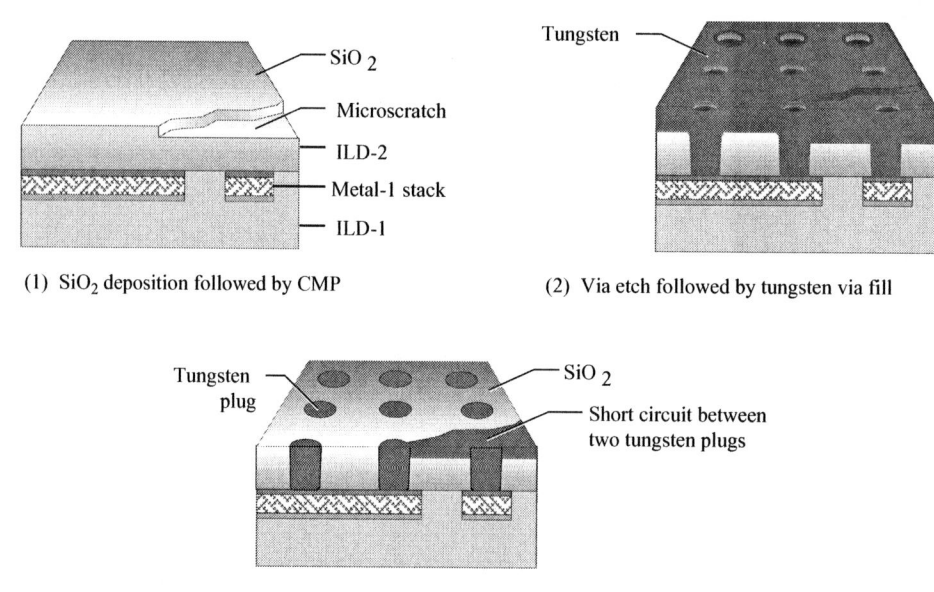

FIGURE 18.27 Results of CMP Microscratch

TABLE 18.6 Key Quality Measures for CMP

Quality Parameter	Types of Defects	Remarks
1. Scratches or gouges on wafer surface	A. Intrametal shorts (metal layer to metal layer)	Microscratches can be extremely difficult to detect with inspection. Possible causes are: • Poorly controlled particle size distribution in slurry. • Long storage or improper usage causes slurry particles to dry up and cause larger particles. • Particulate contamination in slurry. • Nonoptimized or lack of secondary buff operation to reduce microscratches.

Quality Parameter	Types of Defects	Remarks
2. Dishing	A. Depression in high polish rate material	• Dishing typically occurs in high polish rate materials (e.g., STI oxide) in low polish rate surroundings (e.g., nitride used as a polish stop). • More dishing occurs when the high polish rate material has larger surface area. • Increasing the hardness of the pad will decrease dishing (pad bends less into the recesses). • As a temporary solution, some designers are using dummy features as support pads to prevent dishing in large features (e.g., bonding pads). This can only be used where surface area is available.* • For copper, a combination of a hard pad and a slurry with a high degree of chemical removal (in high areas only) should minimize dishing.**
3. Erosion	A. Depression or excessive removal in low polish rate material	• Erosion of low polish rate materials (e.g., nitride). • It is most pronounced when the metal pattern density increases (small spacing between lines), such as for tungsten interconnect with a nitride barrier. • Can cause incomplete via etch that does not make contact with the bottom metal layer. • Investigate tunable polishing pads for local compressibility and optimum slurry movement.
4. Residues	A. Residue in the form of stringers along edges of features	CMP is a relatively clean process (when compared to the RIE etchback process). Properly performed CMP should reduce stringers along feature edges. Stringers could lead to shorts or degrade chip reliability. Possible causes/actions for stringers are: • Stringers occur for different materials, such as tungsten or polysilicon stringers. • Minimize stringers by achieving more uniform polish between different materials by optimizing polishing pad, slurry chemistry, etc.

*J. Damiano et al., "Characterization of the Oxide CMP Process for Shallow Trench Isolation Based Advanced BiCMOS Technologies," *Proceedings* 98-7 (Pennington, NJ: The Electrochemical Society, 1998), p. 53.

**J. Steigerwald, S. Murarka, and R. Gutmann, *Chemical Mechanical Planarization of Microelectronic Materials*, (New York: John Wiley and Sons, 1997), p. 266.

CMP TROUBLESHOOTING

Common problems found during CMP troubleshooting are provided in Table 18.7.

TABLE 18.7 Common CMP Troubleshooting Problems

Problem	Probable Cause	Corrective Actions
1. Excessively large particles in slurry	A. Slurry drying on inside of dispenser walls and falling into slurry mixture. B. Lumping of abrasive particles in incoming slurry. C. Poor slurry preparation by supplier.	• Use point-of-use (POU) filtration for slurry. • Optimize slurry manufacturing and delivery. • Improve stability of slurry with surfactants and stabilizers.
2. Nonuniform polishing of the wafer surface	A. Center fast (center polishes faster than other areas of wafer). B. Center slow (center polishes slower than other areas of wafer).	• Improper pad contour on platen can cause center fast, center slow, or edge problems. Check for excessive pad wear. • Pressure (downforce) on wafer carrier set incorrectly. • Check pad conditioning arm for proper adjustment and wear (no glazing present). • Insufficient slurry flow rate or incorrect viscosity (ability to flow). • Deterioration of backing film on wafer carrier causing wafer to not be held flat. • Table rotational speed set incorrectly. • Backpressure on wafer not set properly.
3. Particles on cleanliness of wafer after copper CMP	A. Excessive particles found on wafers (requirement for 0.18 μm CD is <20 defects/wafer at >0.08 μm size).* B. Build-up of residual slurry in deep features such as photoalignment marks and bond pad areas. C. Removal of residual copper from dielectric region or from between the lines on patterned wafers.	A major concern with CMP is the level of particles on the surface after cleaning. Possible corrective actions are: • Investigate if cleaning of residual copper is done on patterned wafers. This is more difficult for high pattern density. • Check for brush loading on double-sided brush scrubbers (contamination). • Verify the optimum cleaning chemistry is used, especially for copper residues. • Ensure there is no chemical cross contamination from the exposure of the different materials on the wafer surface to the brushes and cleaning chemistry.
4. Glazing of polishing pad (leading to decreased polish rate)	A. Improperly adjusted conditioning arm. B. Pad excessively worn.	Pad break-in and conditioning are necessary to achieve consistent polish performance. Possible actions for glazing are: • Verify proper pad break-in was performed (e.g., running dummy wafers prior to running product wafers). • Adjust conditioning downforce or change conditioning surface. • Replace polishing pad.

*R. DeJule, "CMP Challenges Below a Quarter Micron," *Semiconductor International* (November 1997): p. 58.

SUMMARY

Topography describes the nonplanar wafer surface that results from the fabrication of different features. A planarized wafer has a flat surface with local or global planarization. Three traditional planarization methods are etchback, glass reflow, and spin-on films. These methods are not adequate for global planarization of advanced ICs. Chemical mechanical planarization (CMP) achieves local and global planarity of metal and dielectric films. There are many advantages to CMP for ULSI integration, primarily the ability to achieve global planarity. There is a different CMP mechanism for polishing oxide and metal films. The oxide mechanism is based on surface hydration, which decreases the surface hardness of the oxide to prepare it for abrasive removal. Metal films are oxidized by the slurry, which makes it easier to remove a surface layer by polishing. CMP is sensitive to pattern density, with defects of erosion and dishing occurring with certain patterns. Slurry consists of abrasive particles and chemicals that remove unwanted material from the wafer surface. There is a different slurry chemistry depending on the material being removed. The polishing pad attaches to the top of the platen and is an important factor in planarization. After usage, the polishing pad becomes glazed and requires mechanical conditioning. CMP parameters such as hardness of polishing pad and down pressure will affect the polishing rate and uniformity. Most CMP tools have multiple polishing surfaces (platens) for flexibility. Endpoint detection is used to determine when the CMP has reached the correct thickness. Motor current and optical interferometry are the two major endpoint detection methods. The wafer carrier is optimized to apply uniform pressure across the entire wafer. CMP clean is able to thoroughly clean particles, primarily using double-sided brush scrubbing with ammonium hydroxide chemistry to avoid brush contamination. The latest CMP equipment integrates polishing and cleaning with a dry-in/dry-out approach. There are many CMP applications used throughout multilayer metallization.

KEY TERMS

topography
planarized wafer
local planarization
global planarization
chemical mechanical planarization (CMP)
etchback planarization
glass reflow
spin-on films
polisher
planarity
uniformity
degree of planarity (DP)
within wafer nonuniformity (WIWNU)
wafer-to-wafer nonuniformity (WTWNU)
planarization distance
oxide polish
Preston equation
Cook's theory
metal CMP
pattern sensitivity

erosion
dishing
consumables
slurry
polishing pad
glazing
conditioning
polish rate
center slow
selectivity
overburden
endpoint detection
motor current endpoint detection
optical interferometry endpoint detection
wafer carrier or polishing head
backing film
edge exclusion
bladder carrier
loaded brush
dry-in/dry-out process

REVIEW QUESTIONS

1. Describe topography. What happens to topography with higher chip packing density?
2. Describe a planarized wafer.
3. List and discuss the four terms used for the terminology for wafer planarization.
4. What is the primary planarization process used since the mid-1990s?
5. List and describe three traditional planarization methods.
6. Describe chemical mechanical planarization (CMP). How does CMP achieve planarity?
7. What is another name for a CMP tool?
8. What is a unique aspect of CMP?
9. Explain the difference between planarity and uniformity. How can a wafer be planar but not uniform?
10. What is the degree of planarization (DP)? What is the DP if $SH_{pre} = 10$ µm and the $SH_{post} = 1$ µm?
11. Explain the difference between WIWNU and WTWNU.
12. List and explain nine advantages to CMP.
13. List and explain four disadvantages to CMP.
14. State the two mechanisms that explain how a surface is planarized by CMP.
15. What is oxide polish used for? State and explain the Preston's equation.
16. Describe the mechanism for oxide polish known as Cook's theory.
17. Explain the mechanism for how metal is polished.
18. What is pattern sensitivity and how does it affect CMP?
19. Discuss erosion and dishing and how they affect wafer planarization quality.
20. Define slurry. Why is slurry important for CMP?
21. Discuss the slurry characteristics for oxide, tungsten, and copper.
22. Describe the polishing pad.
23. Explain glazing. What is the purpose of conditioning?
24. Define polish rate.
25. How does the polishing pad affect uniformity?
26. Explain selectivity. State the formula for selectivity. What selectivity do you want for the simultaneous polishing of different materials?
27. What is overburden? How is overburden related to the polish time?
28. For each of the following parameters, describe what results are achieved on the wafer: polish time, pressure on wafer carrier, platen speed, carrier speed, slurry chemistry, slurry flow rate, pad conditioning, wafer/slurry temperature, and wafer back pressure.
29. Why is endpoint detection needed in CMP?
30. Give an example of a straightforward CMP application for endpoint detection, and give a challenging application.
31. List and describe the two types of endpoint detection used in CMP.
32. Describe the wafer carrier and how this affects wafer uniformity.
33. What is the backing film on the wafer carrier?
34. Describe edge exclusion and how this affects nonuniformity.
35. Explain the bladder carrier.
36. What is the focus for CMP cleaning?
37. Why is a brush used for CMP cleaning? What is a loaded brush and why is this a concern for CMP cleaning?
38. Discuss post-copper CMP clean.
39. Explain dry-in/dry-out clean concept for CMP.
40. List and give a short description of six applications for CMP in wafer fabrication.

CMP EQUIPMENT SUPPLIERS' WEB SITES

Applied Materials	http://www.appliedmaterials.com/products/
Arch Chemicals (aka Olin)	http://www.olinmicro.com/default.asp
BOC Edwards	http://www.boc.com/edwards/
Dainippon Screen Mfg. Co.	http://www.screen.co.jp/eed/index_E.html
Dow Corning	http://www.dowcorning.com/
DuPont	http://www.dupont.com/semiconductor/
Ebara Technologies Inc.	http://www.ebaratech.com/
EKC Technology Inc.	http://www.ekctech.com/
FSI International	http://www.fsi-intl.com/
International SEMATECH	http://www.sematech.org/public/index.htm
Intersurface Dynamics Inc.	http://www.isurface.com/
KLA-Tencor	http://www.kla-tencor.com/product/photo_frame.html
Lam Research Corp.	http://www.lamrc.com/
Millipore Corporation	http://www.millipore.com/
Nova Measuring Instruments Ltd.	http://www.nova.co.il/

Peter Wolters CMP Systems	http://www.peter-wolters.com/
Polypros Inc.	http://www.polypros.com/
Prodeo Technologies	http://www.prodeotech.com/
Rippey Corporation	http://www.rippey.com/
SEMI	http://www.semi.org/
Semitool	http://www.semiatool.com/
Sitek Inc. (see Prodeo Technologies)	http://www.sitekgroup.com/
Speedfam-IPEC	http://www.sfamipec.com
Steag Electronic Systems	http://www.steag.com/
Strasbaugh	http://www.strasbaugh.com/
TBW Industries Inc.	http://www.tbw-inc.com/
Verteq	http://www.verteq.com/

REFERENCES

1. R. DeJule, "CMP Challenges below a Quarter Micron," *Semiconductor International* (November 1997): p. 55.
2. R. Liu, "Metallization," *ULSI Technology,* ed. C. Chang and S. Sze (New York: McGraw-Hill, 1996), p. 425.
3. D. Pramanik, M. Weling, and X.-W Lin, *CMP Applications for Sub-0.25 μm Process Technologies,* Proceedings Vol. 98-7, (Pennington, NY: The Electrochemical Society, 1998): p. 1.
4. W. Schaffer and H. Fry, "BPSG Improves CMP Performance for Deep Submicron ICs," *Semiconductor International* (July 1996): p. 205.
5. G. Schwartz and K. Srikrishnan, "Chip Integration," *Handbook of Semiconductor Interconnection Technology,* ed. G. Schwartz, K. Srikrishnan, and A. Bross, (New York: Marcel Dekker, 1998), p. 368.
6. S. Chooi et al., *A Comparison of Spin-On Materials in IMD Planarization,* Proceedings Vol. 2875 (Bellingham, WA: 1996), p. 276.
7. R. DeJule, "Advances in CMP," *Semiconductor International,* SPIE (November 1996): p. 88.
8. M. Pecht, R. Radojcic, and G. Rao, *Guidebook for Managing Silicon Chip Reliability,* (Boca Raton: CRC Press, 1999), p. 82.
9. Adapted from J. Steigerwald, S. Murarka, and R. Gutmann, *Chemical Mechanical Planarization of Microelectronic Materials,* (New York: Wiley, 1997), p. 4.
10. Adapted from J. Steigerwald, S. Murarka, and R. Gutmann, *Chemical Mechanical Planarization of Microelectronic Materials,* p. 32.
11. G. Schwartz and K. Srikrishnan, "Chip Integration," p. 428.
12. M. Tomozawa, "Oxide CMP Mechanisms," *Solid State Technology* (July 1997): p. 169.
13. M. Tomozawa, "Oxide CMP Mechanisms," p. 172.
14. R. Liu, "Metallization," p. 439.
15. T. Bibby and K. Holland, "Equipment," *Chemical Mechanical Polishing in Silicon Processing,* ed. S. Li and R. Miller (San Diego: Academic Press, 2000), p. 8.
16. G. Schwartz and K. Srikrishnan, "Chip Integration," p. 434.
17. M. Rutten et al., "Pattern Density Effects in Tungsten CMP," *Semiconductor International* (September 1995): p. 123.
18. J. Steigerwald, S. Murarka, and R. Gutmann, *Chemical Mechanical Planarization of Microelectronic Materials,* p. 255.
19. K. Wijekoon et al., "Tungsten CMP Process Developed," *Solid State Technology* (April 1998): p. 56.
20. S. Li, B. Tredinnick and M. Hoffman, "Consumables I: Slurry," *Chemical Mechanical Polishing in Silicon Processing,* ed. S. Li and R. Miller (San Diego: Academic Press, 2000), p. 148.
21. Ibid., p. 149.
22. C. Sainio and D. Duquette, *Electrochemical Interactions During the Chemical-Mechanical Planarization of Copper in Ammonia-Based Slurries,* Proceedings Vol. 96-22 (Pennington, NJ: The Electrochemical Society, 1997): p. 110.
23. C. Sainio and D. Duquette, *Electrochemical Characterization of Copper in Ammonia-Containing Slurries for Chemical Mechanical Planarization of Interconnects,* Proceedings Vol. 98-7 (Pennington, NJ: The Electrochemical Society, 1998): p. 127.
24. A. Braun, "Slurries and Pads Face 2001 Challenges," *Semiconductor International* (November 1998): p. 72.
25. Ibid., p. 73.
26. J. Steigerwald, S. Murarka, and R. Gutmann, *Chemical Mechanical Planarization of Microelectronic Materials,* p. 66.
27. A. Brain, "Slurries and Pads Face 2001 Challenges," p. 68.
28. R. DeJule, "Advances in CMP," p. 90.
29. M. Leach, "Local Planarization Process Developed for Higher Yields, *Semiconductor International* (October 1996): p. 137.
30. R. DeJule, "Advances in CMP," p. 56.
31. J. de Larios, "Post-CMP Cleaning for Oxide and Tungsten Applications," *Semiconductor International* (May 1996): p. 122.
32. T. Bibby and K. Holland, "Equipment," p. 20.
33. Adapted from J. Steigerwald, S. Murarka, and R. Gutmann, *Chemical Mechanical Planarization of Microelectronic Materials,* p. 40–46.
34. R. DeJule, "Advances in CMP," p. 92.

35. T. Marbeiter, T. Cleary and K. Sutter, "An Update: Transition to 300 CMP," *Semiconductor International* (November 1998): p. 78

36. T. Bibby and K. Holland, "Equipment," p. 22.

37. L. Zhang et al., *Inhibition of Alumina Particle Deposition Onto SiO_2 Surfaces During Tungsten CMP Through the Use of Citric Acid,* Proceedings Vol. 98-7 (Pennington, NJ: The Electrochemical Society, 1998): p. 161.

38. L. Peters, "Clean Processing," *Semiconductor International* (March 1998): p. 52.

39. B. Fraser et al., *Evaluation of Non-Contact Post-CMP Cleaning Process Utilizing Split-Lot Polishing and Cleaning Comparisons,* Proceedings Vol. 97-35 (Pennington, NJ: The Electrochemical Society, 1998): p. 634.

40. L. Zhang et al., *Inhibition of Alumina Particle Deposition Onto SiO_2 Surfaces During Tungsten CMP Through the Use of Citric Acid,* p. 161.

41. C. Hymes et al., "The Challenges of the Copper CMP Clean," *Semiconductor International* (June 1998): p. 118.

42. R. DeJule, "CMP Grows in Sophistication," *Semiconductor International* (November 1998): p. 62.

43. L. Peters, "Is the 0.18 µm Node Just a Roadside Attraction?" *Semiconductor International* (January 1999): p. 52.

44. S. Selinidis et al., *Development of a Copper CMP Process for Multilevel, Dual Inlaid Metallization in Semiconductor Devices,* Proceedings Vol. 98-7 (Pennington, NJ: The Electrochemical Society, 1998): p. 9.

45. S. Selinidis et al., *Development of a Copper CMP Process for Multilevel, Dual Inlaid Metallization in Semiconductor Devices,* Proceedings, Vol. 98-7, p. 11.

CHAPTER 19
WAFER TEST

Chip fabrication in the ULSI era continues to grow in complexity. IC scaling to increase microchip performance is driving critical dimensions into the deep submicron technology nodes of 0.15 µm and below. Contamination is controlled to the molecular level during chip fabrication to reduce the number of killer defects. The wafer diameter is increasing to put more die on a wafer, which requires new fabrication tools and handling methods.

However, microchips require more than just fabrication. They must first be designed. The appropriate production materials are selected. Early prototype hardware is built and tested to verify the chip meets specifications. When successful, the chip is released to production. Once the design, materials, and fabrication come together in the manufacturing process, chips are tested to ensure functionality. The art of producing a high-density microchip culminates in electrical tests to verify that the chip meets customer requirements.

Electrical tests must be done quickly and accurately. A high reject rate at wafer test can cripple a chipmaker for producing chips in a narrow market timeframe. A low reject rate due to failure of the test procedure to find defects will permit more components to fail during use by the customer, causing customers to buy chips elsewhere. To avoid these situations, the correct test procedures are needed in wafer fabrication to ensure IC functionality and reliability.

OBJECTIVES

After studying the material in this chapter, you will be able to:

1. Discuss the electrical tests done for IC fabrication.
2. Explain the purpose of the in-line parametric test, and describe how it is conducted.
3. Describe the equipment used for in-line parametric tests.
4. State the objectives of the wafer sort test, and explain how it is performed.
5. Outline and discuss the different types of wafer sort tests.
6. Discuss the test issues associated with wafer sort.
7. State and explain the factors that affect yield at wafer sort.
8. Describe three wafer yield models and discuss yield management.

INTRODUCTION

Wafer test is the measurement of electrical parameters on ICs at the wafer level to verify conformance to specifications. The goal of wafer test is to verify acceptable electrical performance. The electrical specifications used in the test procedure vary depending on the purpose of the test. If a defect is found, the production team uses the test data to make sure defective chips are not shipped to the customer and fabrication problems are corrected.

Considered by itself, test does not add value to the chip. If all steps in wafer fabrication were done 100% correctly and all materials used on a chip were perfect, then there would be no need for wafer test during production (assuming all prototype ICs were developed, tested, and released to production properly). Every chip would conform precisely to specification. Such a condition is highly unlikely in today's advanced wafer fabs because of the increasingly complex IC circuits, materials, and processes being rapidly introduced.

Test is a critical operation for verifying a chip's functionality in a continually changing production environment. It is essential that wafer test be able to determine a good die as good and a bad die as defective. Wafers with acceptable die will continue in the process, and wafers with excessive bad die are held for corrective action. Incorrect test data will lead to serious problems for customers and ultimately the chipmaker.

IC Electrical Tests

Electrical tests are done at various stages of the chip process. These tests start in the early design stage, continue at critical steps in wafer fabrication, and finish with testing of the final packaged IC product (see Chapter 20 for a description of assembly and packaging). A summary of all major tests done to a chip, whether at wafer level or packaged IC, is provided in Table 19.1.

TABLE 19.1 Different Electrical Tests for IC Production (from Design Stage to Packaged IC)

Test	Stage of IC Manufacture	Wafer- or Chip-Level	Test Description
1. IC Design Verification	Pre-Production	Wafer level	Characterize, debug and verify new chip design to ensure it meets specifications.
2. In-Line Parametric Test	During wafer fabrication	Wafer level	Production process verification test performed early in the fabrication cycle (at front-end of line) to monitor process.
3. Wafer Sort (Probe)	After wafer fabrication	Wafer level	Product functional test to verify each die meets product specifications.
4. Burn-In Reliability	Packaged IC	Packaged chip level	ICs powered up and tested at elevated temperature to stress product to detect early failures (in some cases, reliability testing is also done at the wafer level during in-line parametric testing).
5. Final Test	Packaged IC	Packaged chip level	Product functionality test using product specifications.

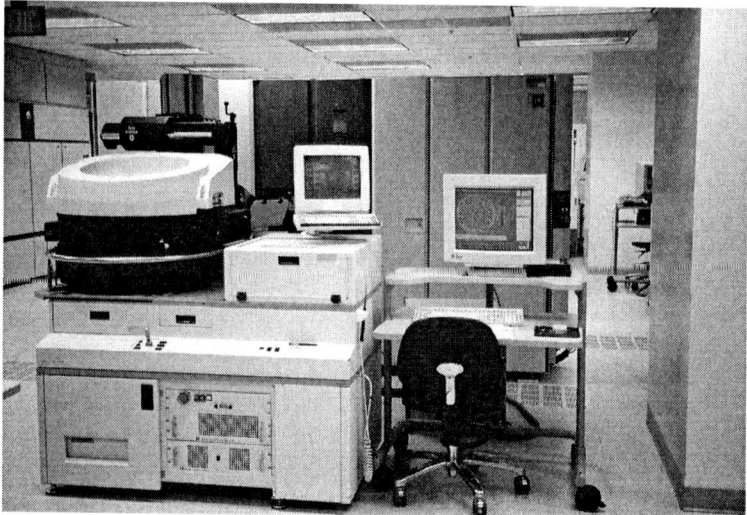

Automated Electrical Tester
(Photo courtesy of Advanced Micro Devices)

Because the subject matter of this book is wafer fabrication, we will study the in-line parametric test and wafer sort (tests listed as items two and three in Table 19.1). These are the wafer-level electrical tests used in wafer fabrication. Wafer sort is the primary wafer level test done to verify that each die functions properly and is done at the completion of wafer fabrication (see Figure 19.1). The other tests listed in Table 19.1 will not be discussed in detail. They are provided for completeness to demonstrate the entire range of tests done during IC manufacturing from early product development until packaged microchip.

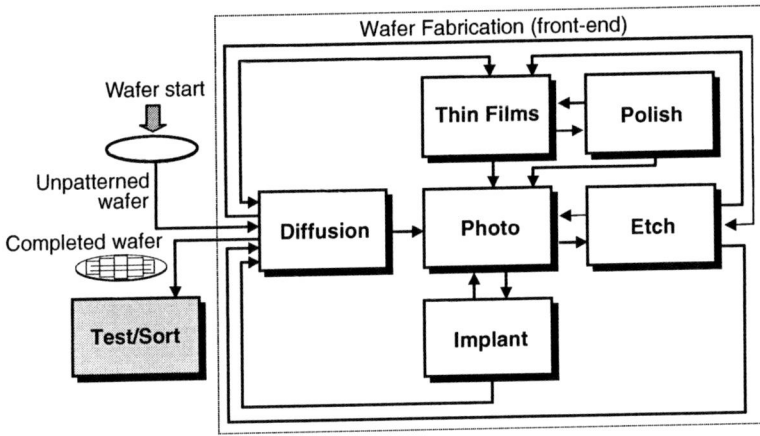

FIGURE 19.1 Wafer Fab Process Flow with Test
(Used with permission from Advanced Micro Devices)

Test defect data is used for yield management activities to reduce wafer defects. Electrical test data categorizes wafers as pass or fail depending on the number of chips per wafer that fail. The pass/fail data is used to calculate the yield, which expresses the percentage of good die on the wafer. The production team performs defect reduction activities by analyzing wafer test data to determine the source of problems and to implement corrective actions (see later in this chapter for an explanation of defect reduction).

Semiconductor fabs can succeed or fail based on the yield of the wafers. It has been estimated that fabs lose from $1 million to $8 million in earnings for each percentage point of yield loss.[1] With the small market window for new product introduction, and intense competition for market share, poor yielding wafers means a chipmaker might not be able to produce adequate chips to supply the market—making their product vulnerable to other competitors with higher yield and adequate chip supply. With this scenario, one can appreciate the importance of having good test data that reflect the IC performance with respect to product specification.

WAFER TEST

There are two types of electrical tests done during wafer fabrication. These tests are wafer tests because they are performed on wafers (instead of packaged die). The tests are:

- In-line parametric test
- Wafer sort

These two electrical tests have different requirements and occur at different stages of the fabrication process (see Figure 19.2 on page 548). The in-line parametric test is performed right after completion of the first metal layer etch (end of front-end-of-line processes) to gain early information about process and device performance. The wafer sort test is a major test stage of IC fabrication. It is done at the completion of wafer fabrication to determine which chips on the wafer meet product specification and are acceptable for shipment to assembly and packaging.

In-Line Parametric Test

The *in-line parametric test* (also known as *wafer electrical test* or *WET*) is an electrical test performed on test pattern structures located on wafers. It is sometimes referred to as a *DC test* because it tests the devices with DC voltage applied to the physical structures. The in-line parametric test is done as early as possible after the completion of the front-end processes (i.e., diffusion, photo, and implant). The test would typically be done after first metal layer has been deposited and etched. This allows contact probes to make electrical contact with the special test structure pads.

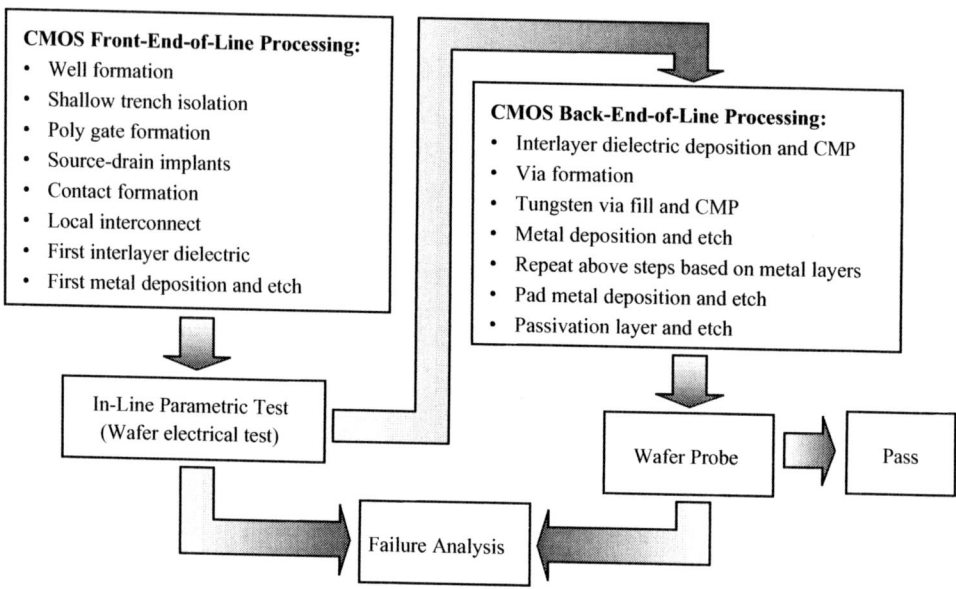

FIGURE 19.2 Location of Wafer Fabrication Electrical Tests

The reasons for the in-line parametric test are:

1. **Identify process problems:** Early identification of process problems during wafer fabrication (instead of waiting until an entire wafer is fabricated before finding out if there is a problem).

2. **Pass/fail criteria:** Determine whether wafers should continue the fabrication process based on pass/fail criteria.

3. **Data collection:** Collect wafer data to assess process trends in order to make process improvement (e.g., channel length variation).

4. **Special tests:** Assess specific performance parameters when required (e.g., special customer requests).

5. **Wafer level reliability:** Perform optional wafer level reliability tests on lots when needed to assure reliability concerns of process conditions.

In-line parametric tests are done as early as possible in the fabrication process. No power supply and signal voltages are applied to the devices on the wafer. Instead current, voltage, and capacitance measurements are made using special parametric test structures to determine the processes' capability. The test is important because it is the first time the wafer undergoes a complete set of tests to verify it was fabricated correctly. The pass/fail data establishes a close link between process conditions and device performance.

With increasing IC complexity, testing early in the process is critical. If the process conditions for fabricating the wafer are unacceptable, then parametric tests will detect this by demonstrating the product fails the electrical limits of the test. The defect reduction team is alerted to the problem and can immediately take corrective action. The data provides ongoing information to monitor and optimize the front-end semiconductor processes.[2]

Wafer Test Structures ■ Parametric testing is not performed on the individual wafer devices. Rather, it is done on special *test structures* (also referred to as *process control monitors*, or *PCMs*) arranged at specific locations on the wafer. Test structures are used because they prevent damage to the actual production die during testing. For early design verification tests, the test structures can be a test pattern on an entire special chip because of the need for many structures and test data. For production wafers, space is a premium; therefore, test structures are commonly located in the scribe line region between the individual die. These are sometimes referred to as *scribe line monitors* (*SLMs*) (see Figure 19.3). The width of the scribe line that the SLM must fit in might be 100 to 150 μm; thus the SLMs are limited in size.

PCM test structures are used to test a wide range of parameters. There is a growing trend in the industry toward standardized test structures for various measurements, mainly due to the

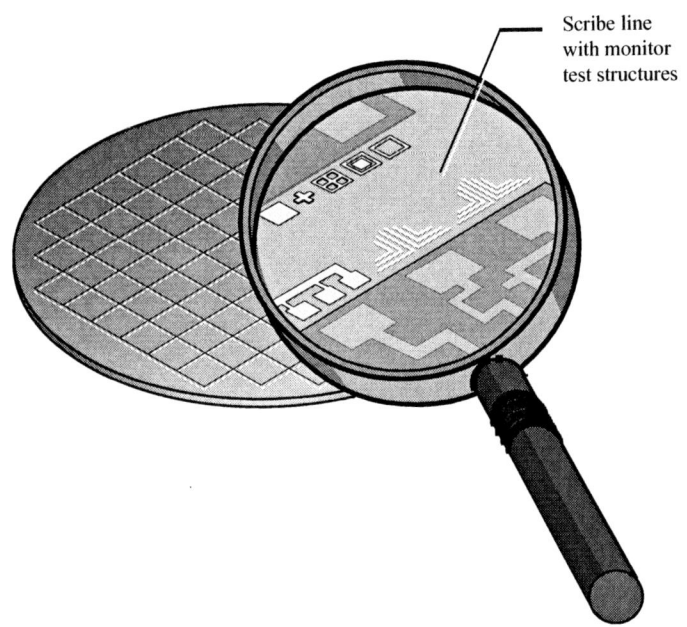

FIGURE 19.3 Scribe Line Monitor Test Structure

structure design complexity and limited development time for new products.[3] Examples of some typical test structures are provided in Table 19.2.[4]

TABLE 19.2 Examples of Test Structures

Test Structure	Fault Measurement
Discrete transistors	Leakage current, breakdown voltage, threshold voltage, and effective channel length
Various linewidths	Critical dimensions
Box in a box	Critical dimensions and overlay registration
Serpentine structure over oxide steps	Continuity and bridging
Resistivity structure	Film thickness
Capacitor array structure	Insulator materials and oxide integrity
Contact or via string	Contact resistance and connections

Figure 19.4 shows a representative PCM test structure used to assess the film thickness of the first metal contacts as a function of sheet resistance. The in-line parametric test measures the series resistance of the structure. A problem with the contacts, such as an improper barrier film thickness, would cause this test structure to electrically fail the parametric test. Since the test structure was fabricated at the same time and under the same process conditions as the wafer, it highlights a problem existing with the actual contacts on individual wafer die. This information is valuable to the production team for defect reduction.

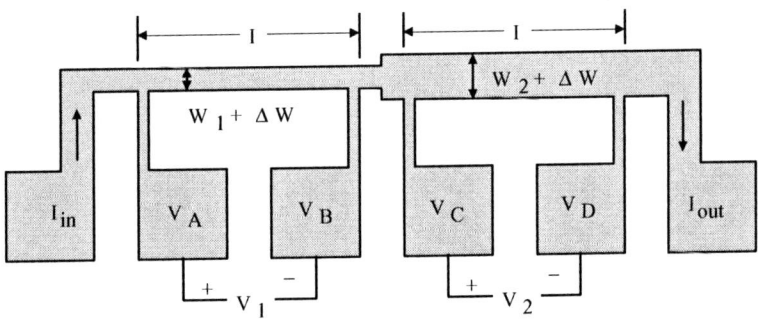

FIGURE 19.4 Test Structure for First Metal Contacts
Redrawn from *Microelectronics Manufacturing Diagnostics Handbook,* ed. A. Lanzberg (New York: Van Nostrand Reinhold, 1993).

Types of Parametric Tests ■ Once test structures are measured during the in-line parametric test, the resulting data is recorded and compared to the expected electrical results. Descriptions of tests and typical measurement levels for conducting the tests are provided in Table 19.3.

TABLE 19.3* Sample Suite of In-Line Parametric Tests

Test Paramater	Description	Typical # of Tests in Program	Typical Measurement Level
Opens/Shorts	Test for open or shorted circuits to check integrity of signal paths. Open and short tests are usually done first because they are fast tests that quickly screen out bad wafers.	2	Go/No-go
Gshorts	Test for shorted gate structures.	1	Go/No-go
Gateleak	Measures gate oxide leakage. Leakage current is a reverse current carrier by minority carriers that goes in the opposite direction of forward current. Small device geometries make leakage current a serious concern.	1	1 pA
BVox	Gate oxide breakdown voltage. This is a quick check for the strength and quality of the gate oxide.	2	10 V
Idsat	Saturation current (negligible channel resistance) flows from drain to source. Apply known gate, drain and substrate voltages. This is a measure of drain current with maximum gate voltage applied.	16	20 mA
Vt	Measures the gate threshold voltage where the transistor starts to draw current from drain to source.	22	0.2 to 1 V
Vtsat	The amount of threshold voltage needed at the gate to drive the drain current into saturation.	16	0.4 to 1 V
Idoff	Device leakage in off mode, drain to source. Gate voltage is not large enough to form a channel.	20	5 to 100 pA
Rds	Vds/Ids at specified drain current (Id) and drain voltage (Vds).	20	25 to 1000Ω
Peakisub	Peak substrate current.	6	5 µA
BVdss	Breakdown voltage from drain to source (punchthrough voltage), measured on minimum channel length transistor with gate grounded to source. Value must be greater than the minimum operating voltage the device will see while functioning.	10	10 V
Pfieldvt	Threshold voltage of a pMOSFET with field oxide as dielectric.	2	12 V
Nfieldvt	Threshold voltage of a nMOSFET with field oxide as dielectric.	2	12 V
Res2t	Determine resistance value using two terminal connections.	21	2 to 1 kΩ
Isolation	Tests leakage characteristics of the isolation structures.	11	100 nA
Diode fvmi	Diode characterization by applying a voltage and measuring current.	2	10 nA
Diodebv	Diode breakdown voltage.	2	3 to 10 V
Res4t	Determine resistance value using 4-point probe connection.	11	2 to 1 kΩ

*Adapted from W. Merkel, "Parametric Testing to Improve Semiconductor Yields," *Semiconductor Online*, http://www.semiconductoronline.com/, (March 12, 1998), p. 1.

Interpreting Parametric Data ■ Failure data from the in-line parametric test is analyzed using the relationship between the different parameters (such as those listed in Table 19.3). Engineers establish the relationship (referred to as correlation) between these parameters and the finished product specification. This information is used to troubleshoot parametric failures. The defect reduction team uses measurement data from failed wafers to assess the state of the fabrication process.

To understand how data interpretation works, it is beneficial to consider an example. A possible in-line parametric test problem is a large variation in the threshold voltage (V_{TH}) that produces only a minor change in the drain current (I_d).[5] This voltage fails the parametric test because I_d does not have the expected current flow that correlates to the magnitude of V_{TH}. One expects that if V_{TH} changes, I_d should have a proportional change (see Figure 19.5). Based on this type of failure, the most likely suspects are the channel implant or the channel length. If channel implant is the source of the problem, then ion implantation must be investigated. If the problem is channel length, that problem is primarily addressed in photolithography. One might also consider other factors such as the implant of the lightly doped drain region or gate oxide thickness.

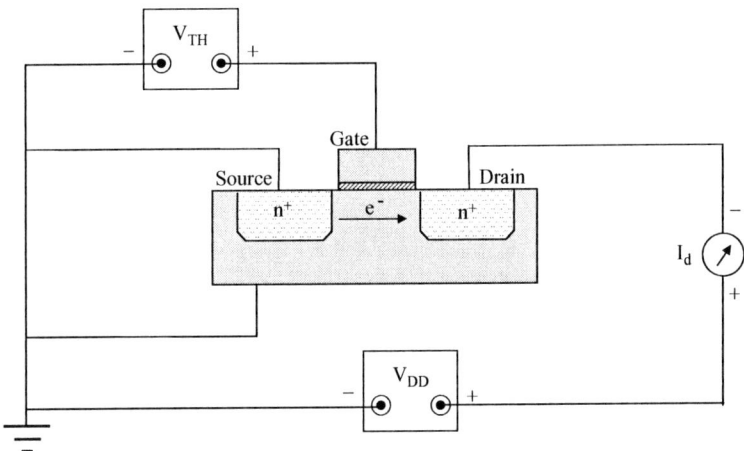

FIGURE 19.5 Threshold Voltage and Drive Current

In-line parametric tests are usually done on a sample basis to statistically assess performance. A possible sample size during normal fabrication could be measuring 100 to 200 parameters at 3 to 5 sites per wafer.[6] If there are excessive failures, then additional wafers in a lot might be tested to verify the defects. In addition, the technician conducts a set of checks to ensure the tester is functioning properly. A correlation wafer is sometimes used to check the set-up of the parametric test equipment and ensure an equipment problem is not causing the failures. If the correlation wafer fails the test or if the wafers will not pass after a retest, then the test engineer is notified.

The in-line parametric test is an early warning about potential problems. Parametric testing supplies information about how the wafers were processed and where problems are in the process flow. The objective is improvement in the fabrication process yield.

Data Trends. An important aspect of parametric testing is detecting data trends. Unacceptable trends are:

1. The same die location keeps failing a parameter on a wafer.
2. The same parameter is consistently failing on different wafers.
3. There is excessive variation (e.g., >10%) in measurement data from wafer to wafer.
4. Lot-to-lot failure for the same parameter, indicating a major process problem.

The reason for repeated die location, parameter failures, or excessive variation must be determined. Lot-to-lot failure for the same parameter indicates there is a major process problem and requires immediate corrective action. Communication within the production team is important for interpreting data trends.

Wafer Level Reliability ■ *Wafer level reliability* (*WLR*) is a specialized form of parametric testing that uses test structures to assess the reliability of the devices on the wafer.[7] WLR, developed in the 1980s and not used by all semiconductor manufacturers, predicts how long the devices on a wafer will last during customer usage. Statistical sampling techniques are used to establish an expected performance for the material and design of the test structure. If WLR test results show a change in test output, then there is possibly a change in the material or its microstructure.

WLR is not a pass/fail test. It uses variables such as voltage, current, temperature, and time to stress the wafer test structure and cause any weak areas to fail during the test. It can also be used to qualify a new process by monitoring process changes to predict their effect on device quality. Examples of different WLR tests are:

1. Stressing metal lines for electromigration failure by applying a high current density.
2. Assessing how much charge an oxide layer can hold and for how long before it is destroyed.
3. Determining how much charge can be trapped in an oxide.
4. Evaluating the effect of a new wet cleaning process on oxide growth.

IC Reliability. What is reliability? *IC reliability* is the probability that devices will perform properly in their usage environment over their intended life. In other words, how long will the IC last during normal usage. Although the general reliability of ICs is extremely high, they are not 100% fail-safe. Reliability tests help ensure that chips shipped to customers will survive their specified usage conditions without failing. Traditionally, reliability testing has been done using burn-in testing with packaged chips. A *burn-in test* powers up and tests chips in a harsh environment (e.g., elevated temperature at 85°C and an elevated bias voltage) with the intent of forcing weak devices to fail so that they will not be shipped to customers.[8] This testing produces a more reliable IC. However, this often requires a very long test, possibly tens or hundreds of hours of continuous testing, which is costly and time consuming.

In-Line Parametric Test Equipment ■ The in-line parametric test equipment is an automated tester designed to interface the test structures on the wafer to sophisticated hardware and software that performs the electrical test. The primary tester subsystems are listed and shown in a system block outline in Figure 19.6.

- Probe card interface
- Wafer positioning
- Tester instrumentation
- Computer as host or server/network

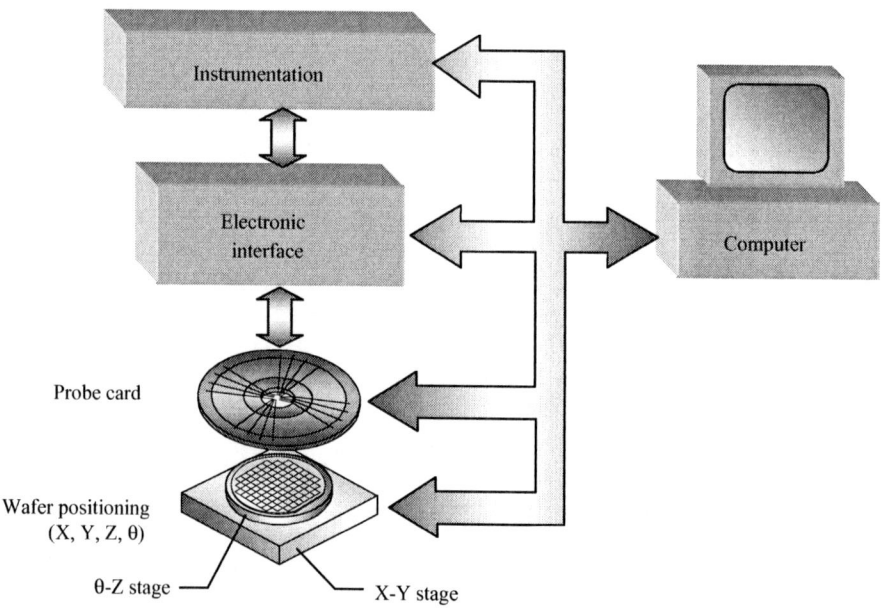

FIGURE 19.6 System Blocks for Automated Parametric Tester System

Probe Card Interface. A *probe card* is the interface between the automatic tester and the device under test (DUT). The probe card is typically a printed circuit board with fine probe needles that make physical and electrical contact to the test structures of the DUT. Probe needles, often made of tungsten, conduct current into and out of the wafer test structure pads for performing the electrical test. An individual probe needle is customized to fit a specific test structure pad, which means each wafer product usually has a specific probe card. There may be hundreds of probe needles on a card that all must remain aligned and planar to make good electrical contact. Individual probe cards are expensive, ranging from several thousand dollars to >$10,000 per probe card.

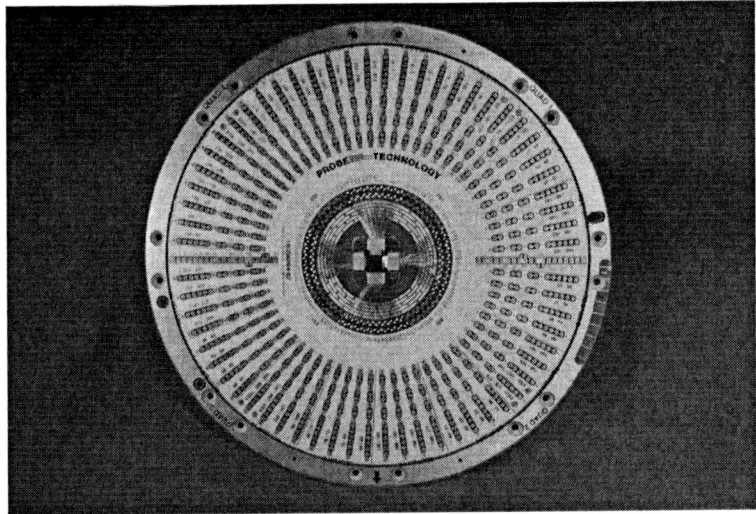

Probe Card of Automatic Tester (Photo courtesy of Probe Technology)

Advanced probe cards are often capable of multisite probing to increase the wafer test throughput. Multisite probing will use a single tester and probe card to test two or more test structures at the same time. Because of more circuitry and needles in a reduced space, multisite probe cards may generate a poorer electrical signal (e.g., reduced signal-to-noise and increased leakage). High-performance ICs may not be testable with multisite probing.

Wafer Positioning. For a wafer to be tested, the *prober* positions the wafer for contact with the probe card interface needles. The wafer is initially moved from the cassette by a wafer handler and mounted on a vacuum chuck. Motors move the wafer in the X, Y, Z, and θ positions using an optical alignment system. The X-Y motion positions the probe card over the test structures. Once the probe card is aligned in the X-Y position, the theta angle of rotation is adjusted so that the needle tips are centered on all the pads (see Figure 19.7). Good alignment of all needles on individual pads is essential.

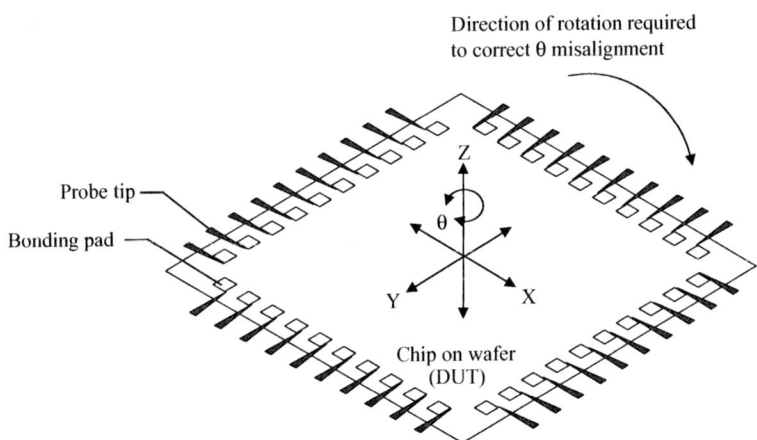

FIGURE 19.7 Probe Card Theta Angle Adjustment

The Z stage position is critical and requires regular system checks for planarization to ensure all components are planar. The prober Z stage positions the wafer vertically to make contact with the probe needles. If the Z stage positions the wafer too high, there is a risk of damaging the wafer and the probe needles due to excessive interference. If the Z stage height is too low, there is insufficient or no contact. There must be a slight amount of interference (referred to as *overdrive*) that causes the needles to pierce (or to scrub) the surface of the probed aluminum pad (see Figure 19.8).[9] Typical probe needle overdrive is 50 to 100 μm (2 to 4 mils). Pad probe marks are an indication that the prober is properly aligned in all axes and has made electrical contact. There is a clear relationship between the condition of the probe mark and whether the prober was set up properly (see Figure 19.9).

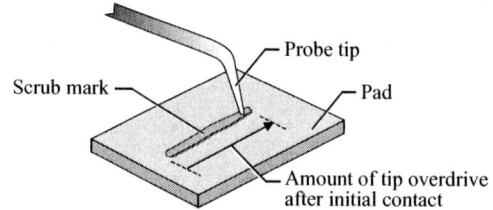

FIGURE 19.8 Probe Needle Overdrive and Scrub

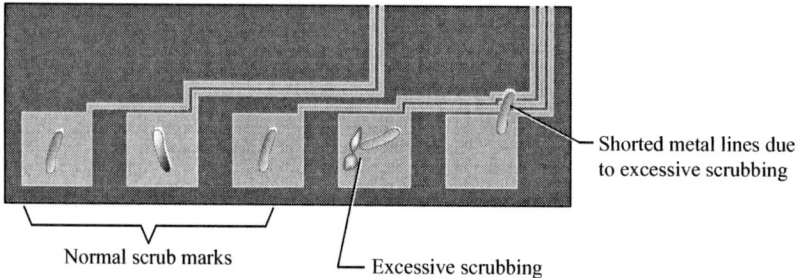

FIGURE 19.9 Types of Probe Marks on Pads

Tester Instrumentation. Advanced ICs require *automated test equipment* (*ATE*) that can measure sub-picoamp currents and picofarad-level capacitances on test structures quickly, accurately, and repeatedly. The ATE controls the test process. A parametric tester has ATE instrumentation that serves as a voltage or current source and is capable of making output voltage and current measurements (see Figure 19.10). This is sometimes referred to as the source-measure unit (SMU).[10] The SMU interfaces to the different probe card needles through a matrix of solid state switches, which have recently replaced reed relays for improved signal control. The force measurement unit (FMU) instrument measures resistance by forcing a specified voltage across a circuit and measuring the resulting current flow. There are also instrument units capable of measuring capacitance (capacitor measurement unit, or CMU) and small current flow in the picoamp range (picoammeter unit, or PAU). The measured analog signals from these instrument units are converted to digital information and transferred to the host computer.

Computer as Host or Server/Network. A computer directs the operations of the test system, including test software algorithms, automated test equipment (instrumentation), prober control software for wafer positioning, test data storage and control, system calibration, and failure diagnostics. A host computer connects to the different subsystems of the tester. Modern test systems are generally connected to a network for ease of access to the computer system for data transfer and control.

A *test algorithm* is a computer program written by engineers that controls the test instrumentation to perform measurements. The program is written for specific test structures and directs the tester hardware to perform the measurement test. The software stores the test data and assigns a unique category of failure, known as a *bin*, to each failed wafer. The computer usually has custom prober software for controlling the wafer positioning. The software must also be capable of acquiring and storing the test data for product disposition and data analysis to determine the source of the problem.

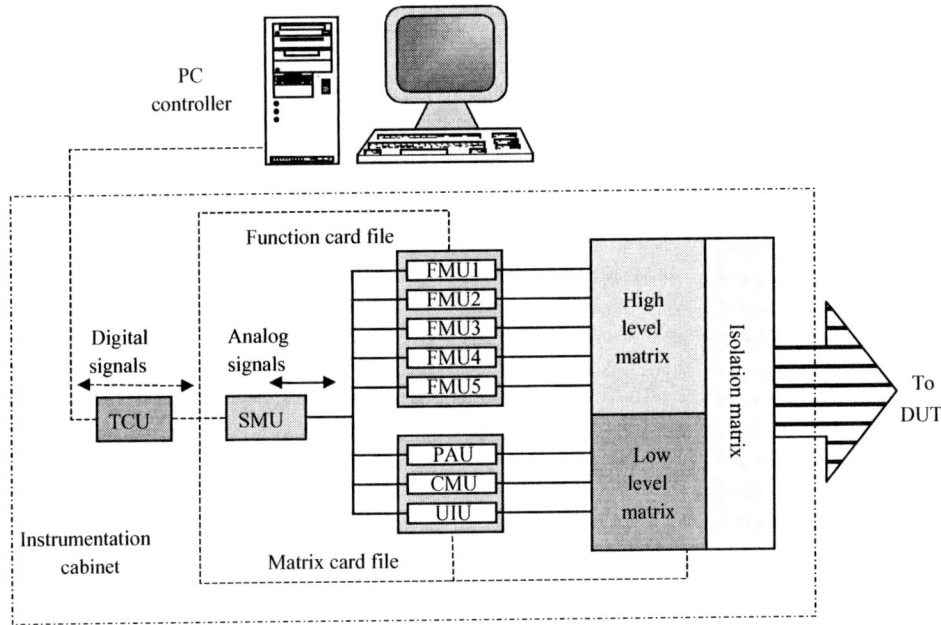

FIGURE 19.10 Tester Instrumentation Block Diagram
(Used with permission from Keithley Instruments)

Test equipment *calibration* is necessary to compensate for variations in tester components.[11] Calibration ensures integrity of the test data so that good wafers continue to be processed and defective wafers are rejected. Most tester software has autocalibration routines that use internal correction factors to quickly perform the calibration steps. The main calibration is for DC levels and timing of the various signals sent to the DUT. Manual calibration is done by a technician at less frequent intervals (e.g., monthly). Diagnostic failure routines are used to analyze the cause for failure and assign bin numbers that keep track of the test results.

Challenges for In-Line Parametric Testing ■ Larger wafers increase the number of tests because there are typically more test sites per wafer. Because in-line parametric testing is done during the fabrication process, speed of testing is critical to achieving adequate wafer throughput. A challenge for parametric testing is the physical distance of the instruments from the DUT. Long cables and connectors in the signal path increase measurement noise and have higher parasitic capacitance. This condition reduces the sensitivity and accuracy of the measurements.[12]

Wafer Sort

At the end of wafer fabrication, wafers undergo 100% *wafer sort* testing of each die, also referred to as *electrical sort, wafer probe,* or simply *probe*. The objective of a wafer sort is to verify which devices on a wafer function properly. This test is a major stage in IC fabrication. Each die on a wafer is tested to all functional product specifications for both DC and AC parameters. The objectives of wafer sort are:

1. **Chip functionality:** verify the operation of all chip functions to ensure only good chips are sent to the next IC manufacturing stage of assembly and packaging.
2. **Chip sorting:** sort good chips based on their operating speed performance (this is done by testing at several voltages and varying timing conditions).
3. **Fab yield response:** Provide important fab yield information to assess and improve the performance of the overall fabrication process.
4. **Test coverage:** Achieve high test coverage of the internal device nodes at the lowest cost.

Wafer sort is a functional test that verifies a device by ensuring it can perform all specified tasks within the limits specified by the IC data book. An example of a functional test that an individual might do is checking a new personal computer to make sure all the different software programs and hardware devices function when it is first installed. Ideally, functional tests cover all faults that could occur during fabrication.

Wafer sort makes an important contribution to wafer fabrication. We have seen how the fab process is continually undergoing process improvements to maintain Moore's law. In this respect, the wafer fabrication process is never completely under control. Wafer sort provides a measure of stability to ensure that process and design changes do not negatively affect the chip's functionality for the customer.

Performing Wafer Sort ■ The procedure for performing wafer sort is similar to in-line parametric test, except now every die on the wafer is tested. In some cases, the same automated test equipment (ATE) is used. However, wafer sort is usually located in an adjacent facility to the fab with a less stringent cleanroom class since wafer fabrication is complete, including the final surface passivation layer. On the other hand, the in-line parametric test is often located in a workbay of the fab to minimize contamination.

Automated wafer handling to perform wafer sort is similar to in-line parametric test. To verify tester setup, a correlation wafer is used. The correlation wafer is a known good wafer that ensures the test system is working properly. The wafers are indexed from the cassette to the prober, followed by mounting on a vacuum chuck with Z-positioning (vertical movement). Probe needle alignment is done automatically with software control. Mechanical probe needles contact the bond pads to create electrical continuity. The probes are interfaced to the ATE to perform the range of AC functional tests based on test algorithms. The type of tests, the number of tests, and their order are defined by the test program in the computer.

Once the test is complete, the defective die are flagged in the computer data base so that they will be rejected before packaging. The traditional method for highlighting defective chips has been *ink mark*, which puts a drop of ink on each unacceptable die. Ink can be messy and introduces possible contaminants to the chip. An approach that is gaining wide acceptance is an *electronic wafer map* that creates a computer image of chip location and test results to categorize good and bad die (see Figure 19.11). Chip assembly downloads the electronic wafer map into its equipment database to reject all defective chips after the wafer has been diced into the individual die.

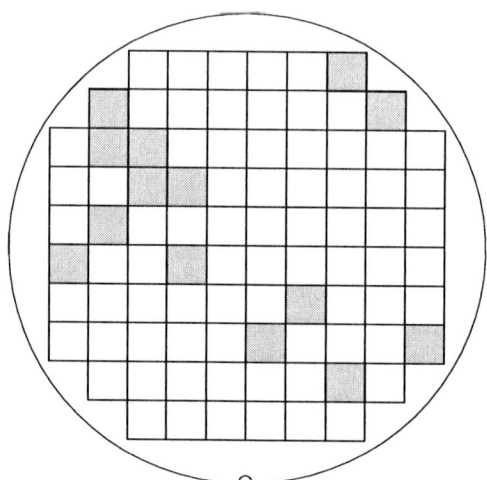

FIGURE 19.11 Wafer Map of Rejected Die

Bincode number. Wafers are often assigned a *bincode number* after wafer sort test to categorize the test results (see Figure 19.12). The bincode number summarizes the test results and groups similar wafers together. Wafers that pass are assigned a bincode number, while wafers with defective die have a bincode number assigned by the type of die defect (e.g., output leakage, open, short, and so on). The bincode numbers are summarized by wafer to show the probe results, such as how many die passed and how many failed a specific test. A *bin map* is often made for individual wafers to provide a visual image that highlights the bincode number of each die on the wafer (see Figure 19.13). Specialized software analysis, referred to as *spatial signature analysis (SSA)*, recognizes unique bin map failure distributions on the wafer surface to pinpoint the source of the

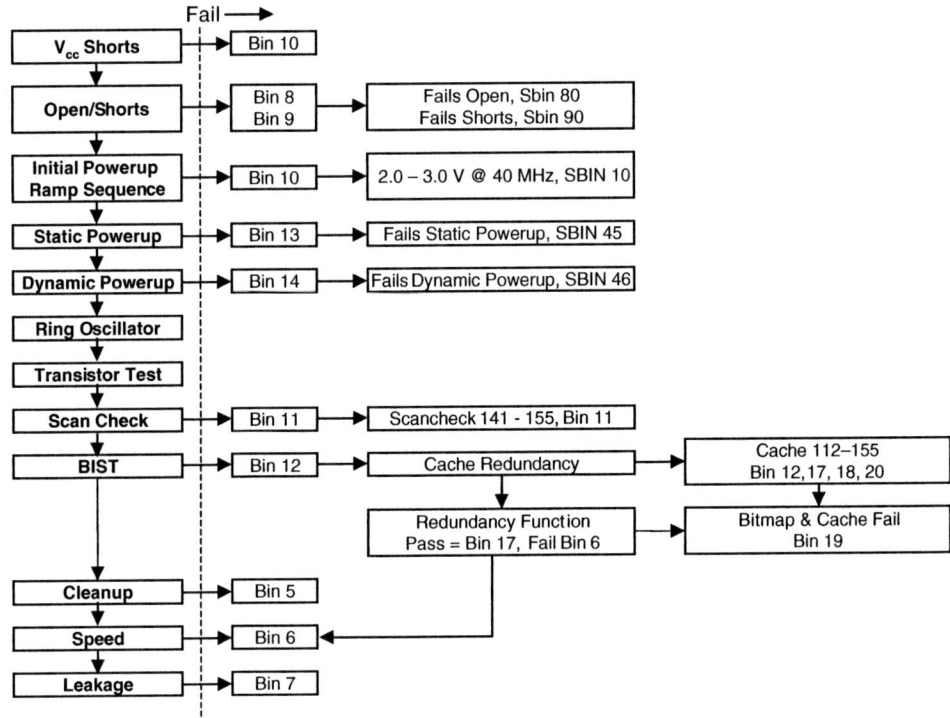

FIGURE 19.12 Bincode Numbers at Wafer Sort

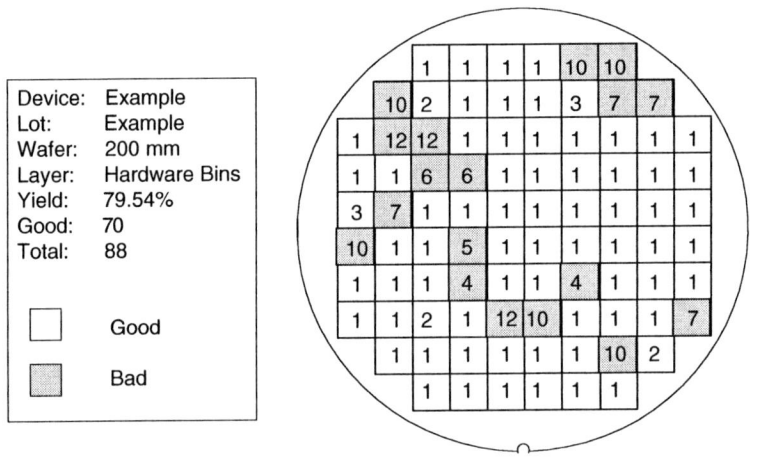

FIGURE 19.13 Wafer Bin Map with Bin Failures

wafer defects.[13] Automatic recognition of failure patterns assists the production team in quickly identifying the source of problems.

Wafer Sort Tests ■ Wafer sort tests are sometimes referred to as AC tests because a system clock and high-frequency input signal are used to verify the chip performance. There are also DC tests at wafer sort to check for continuity, opens/shorts, leakage current, and I_{DD} tests. These DC tests are usually the first tests performed during functional testing to determine if the die should continue for the longer AC tests. Three tests typically done during wafer sort are:[14]

- DC tests (continuity, shorts/opens, and leakage)
- Output checks
- Functional tests

DC Tests. The first electrical check is a continuity test to verify that the probe needles are making good electrical contact with the bonding pads. This check confirms an acceptable tester setup by the technician. The probe needle scrub mark, as previously discussed for in-line parametric testing, is a reliable visual check for good mechanical and, therefore, good electrical contact. Tests for shorts and opens are conducted by applying a forward bias voltage and measuring the voltage drop to check for low (short) or high (open) input resistance. Leakage current has become an important DC test due to shrinking device geometries. When a device is off, one expects no current flow. However, there can be reverse leakage current, which is parasitic to device performance, slowing the transistor's switching time and degrading logic levels. Steps in chip design are taken to minimize leakage current.

Output Checks. To verify acceptable chip performance, wafer sort tests the output signals. This verifies that the correct pattern of bits (logic "1" or high voltage and logic "0" or low voltage) are present and matching an expected output. The test also confirms that the bits are at the correct voltage levels for a logic "1" output high or a logic "0" output low. There are also output verifications for concerns such as signal pattern synchronization and the stability of logic levels.

Functional Tests. *Functional tests* verify that the chip performs as intended by the product data specification. The functional test software program is used to test all aspects of the chip by inputting binary test patterns (a string of 0's and 1's), referred to as *test vectors,* to the devices and verifying the correct output. The test vectors are applied at circuit nodes (device circuit junctions). If the expected output is not obtained during the test, then there is a fault. An example of a functional test for a very small section of an IC is shown in Figure 19.14. There are two steps to this particular example, first forcing node 9 to a logic "1" and then to a logic "0". In each condition there is an expected output at node 11. If the correct output is attained, then the devices that control node 9 are acceptable. If the correct output is not obtained, then there is a functional problem at the input to node 9.

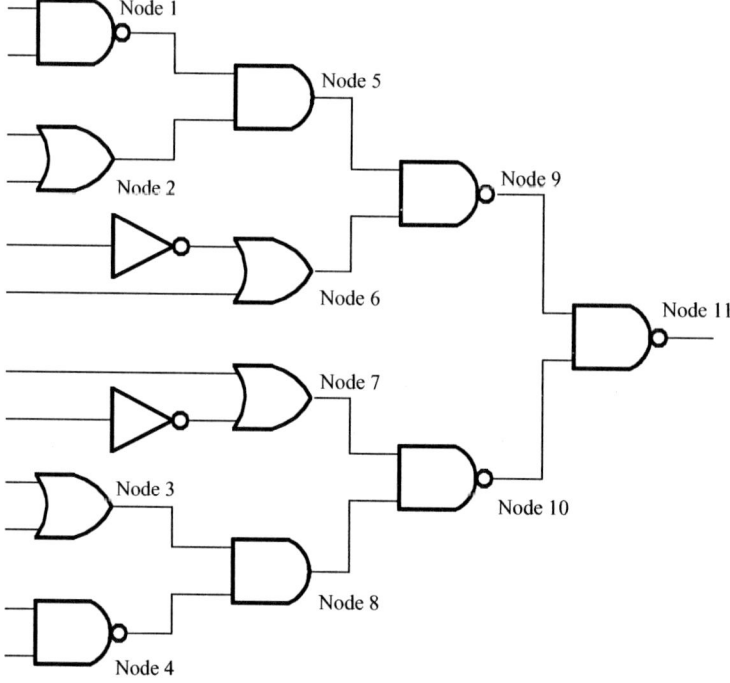

FIGURE 19.14 Example of a Functional Test

The goal in functional test is to test the operation of all devices on the chip. The supply voltage may be varied to make it harder for the part to pass the functional test. For microprocessors, the clock speed is checked to make sure the chip works at a particular speed (e.g., 800 MHz, 1.13 GHz, and so on). However, it is physically impossible in the ULSI era to test for every possible random defect in a chip because the test algorithm cannot duplicate all the different ways customers might use the chip. It also would take a long time to test all failure scenarios.

Wafer sort uses the concept of test coverage. *Test coverage* defines the percent of nodes that are actually tested in wafer sort, such as 99% test coverage. The goal for most manufacturers is at least 95% test coverage, meaning that the remaining 5% of nodes are untested and could possibly have problems that lead to field failures.

The type of functional test varies for different ICs. To functionally test a memory chip, a digital number might be stored in RAM by writing it into certain memory cells and then reading it to verify it is the same number. If a memory cell (an address location in the chip for storing a binary bit) is defective, the digital output will not be the same as the digital input. A traditional memory test is the zero-one algorithm that verifies logic "0" at all cells and then changes the cells to logic "1" and makes sure this is accepted (see Figure 19.15).[15] Memory devices are also tested for parameters such as access time to the cell output, measured in nanoseconds. Memory might also have a thermal bake (e.g., 24 hours at 250°C) to stress the part and verify acceptable memory retention.

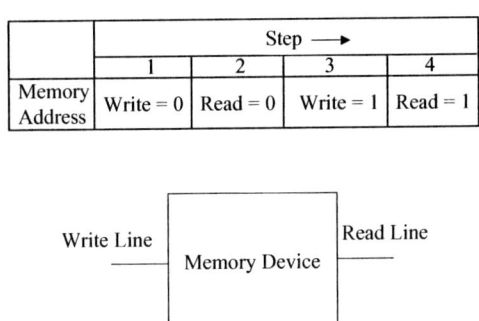

FIGURE 19.15 Zero-One Algorithm for Memory Test

Test Issues at Wafer Sort ■ Wafer sort testing is a complicated task that is under pressure to reduce test time without permitting defects to escape the process. Larger diameter wafers with increased functions on a chip require longer probe times, increased power supplies, more sophisticated test algorithms, improved wafer handling, and larger computer systems to perform the tests and track results. To overcome these challenges, test engineers work with design engineers early in the product development cycle to incorporate test features into the chip and develop built-in-self-tests (BIST) and parallel testing concepts. Several relevant topics that affect chip testing at wafer sort are:

- Total test time
- Fault models
- I_{DDQ} testing
- Guardbanding

Total Test Time. ULSI technology integrates many different functions onto one chip. This integration is highly beneficial for chip performance, but it requires more valuable production time to complete chip testing. To increase tester throughput and quality, *design for test* (*DFT*) strategies are used early in chip design by taking into account the testability of the chip. An example of DFT is scan testing, where special circuits are designed into the IC that bypass normal data paths to apply special test patterns, thus decreasing test time. During production testing, *parallel testing* of devices is sometimes performed, such as parallel testing of mixed-signal analog and digital sections on a chip.[16] Parallel testing can reduce the total test time.

Fault Models. Many different types of IC failures are detected by *fault models* used in the test software algorithm. An example of a widely used model is the single stuck-at-fault (SSAF) model.[17] This fault exhibits a problem by being permanently held, or stuck, at logic "1" or logic "0" level. To test a SSAF, if the circuit line is stuck at logic "1," then the test is to apply a logic "0" to the line. If the logic level does not change, then there is a defect that causes the SSAF failure. Another fault model that represents a common IC failure is bridging faults, which is a failure between two unconnected lines. Delay fault is another type of fault model. In this case, faults occur when increased propagation delays cause gates in the circuits to fail to meet their data specification. A limitation of fault models is their inability to detect many types of physical defects, such as short circuits that change the functionality of the circuit but still permit it to pass the simplistic fault model (see Figure 19.16).

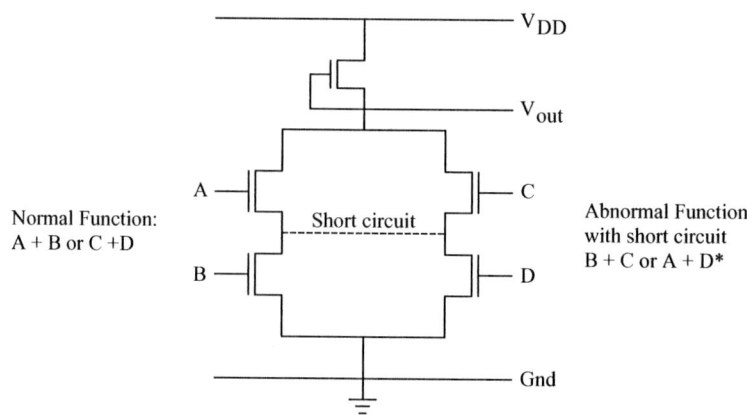

*Short circuit causes change in functionality, but does not cause any electrical node to be "stuck at."

FIGURE 19.16 Limitation of Stuck-At-Fault Model

I_{DDQ} Testing. With the increased use of ICs, the customer demand for higher reliability has also increased. The simplicity of fault models makes them limited in the defects they can detect. One test widely used during the 1990s to increase fault coverage for CMOS ICs is I_{DDQ} testing.[18] I_{DDQ} is the quiescent current, or steady state current, from the source to drain of the transistor when it is off. The principle is that in a fault-free situation, when the transistor is off, the steady state current should be negligible. There is some junction leakage current but this is on the order of nanoamps and can be neglected. However, in the presence of various physical defects, the steady state current in a CMOS circuit no longer remains negligible. The magnitude of I_{DDQ} can increase several orders of magnitude because of the fault. This current increase is detected during the test and highlights the fault. The most common faults detected by I_{DDQ} are physical shorts (bridging), power supply to ground shorts, gate oxide shorts, and punchthrough. The drawback of I_{DDQ} testing is the difficulty in determining the root cause of the fault. Nevertheless, I_{DDQ} testing has significantly reduced CMOS IC defect test escapes, increased IC reliability (lowered early failures at the customer), and reduced the need to perform burn-in testing.[19] It should be pointed out that the benefits of I_{DDQ} testing are being eroded in deep submicron CMOS ICs because of the increased subthreshold current in MOS transistors, making it more difficult to detect the defects.[20]

Guardbanding. The product data sheet published for customers in a chipmaker's IC data book implies that all chips will meet specifications. *Guardbanding* is the practice of testing a device to a more stringent requirement than that specified in the product data book specification.[21] Guardbanding tightens the test limits in the tester in order to ensure that chips passing functional test meet the product specification and customer requirements. Tightened test limits could be in the form of reduced electrical requirements, such as an 8 pA current leakage requirement at the customer, with 7 pA specified at final test (after IC packaging) and 6 pA specified at wafer sort (see Figure 19.17). Guardbanding may also include testing at an elevated temperature during test (e.g., 75°C) that is higher than the customer usage temperature. The benefit from guardbanding is that test limits are set to account for equipment and process variations, including instrument error, measurement error, and product variability. The goal is to ensure that the product data sheet specifications are met and the customer receives good ICs.

Yield

Wafer sort yield is the percentage of acceptable die that pass the wafer sort test. A major goal for wafer fabrication is to maintain a high yield at wafer sort. A low yield means that a large number of chips will be rejected at assembly and packaging. This result is costly and reduces the throughput of the factory. Low yield makes it difficult for a chipmaker to deliver high-quality chips on time

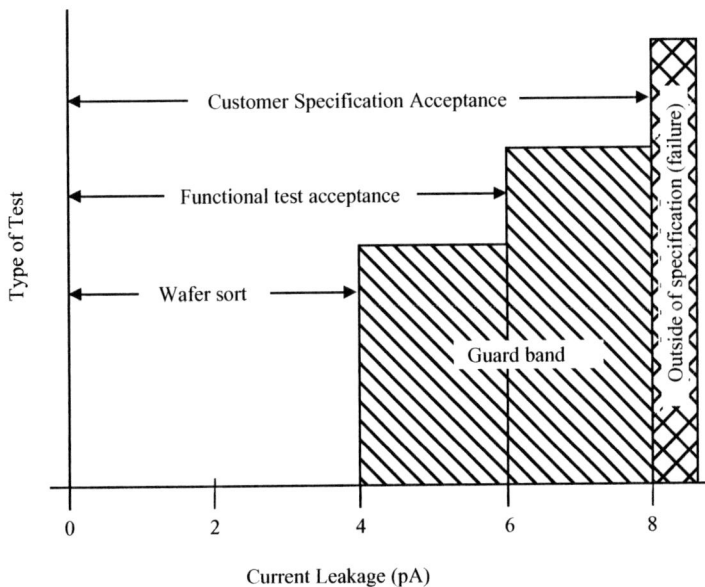

FIGURE 19.17 Guardbanding at Functional Test to Verify Customer Specifications

to the market. Typical wafer sort yields are approximately 60% for the first year of production and 80 to 90% yield in several years, depending on the type of product. For DRAM yields, 98% yields after one to two years of production are common.[22]

Because the wafers have completed the entire fab process, wafer sort yield indirectly measures the overall stability and cleanliness of the wafer fab processes. Wafer sort encompasses all process variability in one test. Wafer sort yield is defined by:

$$\text{Wafer sort yield} = \frac{(\text{\# of good die})}{(\text{Total \# of die started on wafers})}$$

For example, a lot has 25 wafers with 50 chips per wafer. Out of the total 1,250 chips, 1,140 pass wafer sort. The wafer sort yield is:

$$\text{Wafer sort yield (\%)} = \frac{1{,}140}{1{,}250} \times 100 = 0.912 \times 100 = 91.2\%$$

From this lot, 91.2% of the chips are acceptable for assembly into an IC package. The remaining 8.8% of the chips are scrapped (thrown away). Fabrication and design factors that affect wafer sort yield are:

- Larger wafer diameter
- Increased die size
- Increase in number of process steps
- Shrinking feature sizes
- Process maturity
- Crystal defects

Larger Wafer Diameter ■ Wafer diameter has steadily increased since the beginning of semiconductor manufacturing to improve production efficiency. This is because more chips on a larger wafer has a smaller effect on chip costs. Although there is a large up-front capital equipment cost for changing over the fab equipment to accommodate a larger wafer diameter, this cost is justified if there is a need for more chips to be produced in the fab. In essence, up-front costs are amortized over a high volume of chips. A benefit from larger diameter wafers is that each wafer has a smaller proportion of partial-die chips (see Figure 19.18 on page 562). Partial-die chips are nonfunctional; therefore, reducing partial-die chips effectively increases wafer sort yield.

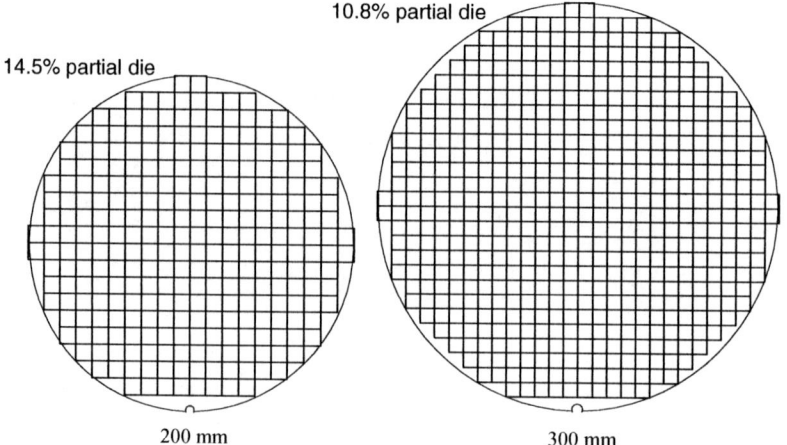

FIGURE 19.18 Reduced Partial Die on Large Wafer

Larger diameter wafers have more die away from the wafer edge and are less affected by edge problems. Process variations occur at a higher rate around the edge of the wafer. For instance, more rapid heating and cooling at the wafer edge causes increased thermal nonuniformity. The wafer edge is also more susceptible to handling and contamination problems.

Increased Die Size ■ Increasing the die size without increasing the wafer diameter results in a smaller percentage of whole die on the wafer. This percentage affects the wafer sort yield, since there are fewer die undergoing probe test. As an example, consider the extreme case of one die on the wafer. If this wafer with one die has a defect, sort yield is 0%. If there are 100 die on the wafer surface and one die has a defect, sort yield is 99%. Defect density (the number of defects per unit area) will also decrease with more die on a wafer, if all other variables are constant. Die size can dramatically affect probe yield. Die size has generally increased over the years to support more functions on a chip. This increase is balanced with larger wafer diameters to maintain an adequate number of die on a wafer.

Increase in Number of Processing Steps ■ The number of IC processing steps is steadily increasing, with currently around 450 process steps needed to fabricate a high-performance microprocessor IC (see Figure 19.19). The increase in process steps is driven by increased chip complexity. More process steps means increased opportunity to contaminate or damage a wafer due to handling or a misprocess. Increased contamination leads to an increase in the defect density and lower wafer sort yield. A higher number of process steps also increases the process cycle time, which is the time it takes to process a part. Long cycle times create product flow bottlenecks throughout the process, increasing the risk of chip contamination.

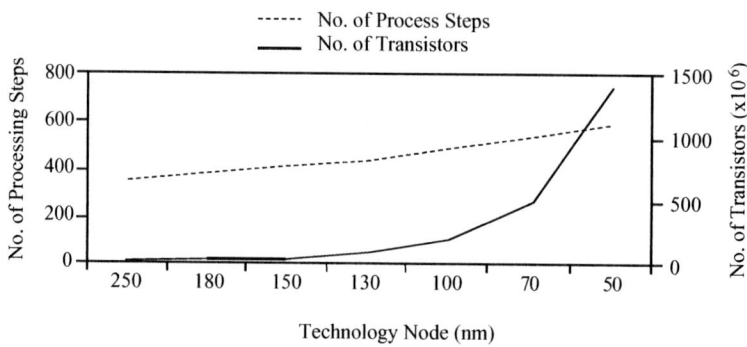

FIGURE 19.19 Growth in Number of Process Steps Due to Chip Complexity
Redrawn from C. Gross et al., "Assessing Future Technology Requirements for Rapid Isolation and Sourcing of Faults," *Micromagazine* (online version) http://www.Micromagazine.com/archive/98/07/jensen.html> (July 1998), p. 6.

Shrinking Feature Sizes ■ Shrinking feature sizes to increase the chip density are a major factor in wafer fab productivity improvements, accounting for 12% to 14% of these improvements since the 1980s.[23] At the same time, reduced critical dimensions make it more difficult to pattern, introducing photolithography defects into the process that affects probe yield. Deep submicron wafers are more susceptible to contamination defects, as well as the overall defect density. Critical wafer layers have a higher probability of causing failure at wafer sort than noncritical layers.

Process Maturity ■ Chipmakers must rapidly develop new products to compete. The continual introduction of new products into a wafer fab leads to process instability that increases defects at wafer sort. The standard product life cycle predicts lower yields early in the product's life, with increased yield after the process matures. A mature process has a stable period of high yield where it repeatedly produces good chips. Competitive pressure often requires the product life cycle of new products to be shortened with a faster ramp-up to a mature process. Figure 19.20 shows the dramatic improvement in reaching product maturity for DRAM devices, ranging from about 5 years for 64kb DRAM to about 1 year for 256 Mb DRAM.[24] An accelerated time to full production increases the pressure on the production team to make rapid yield improvements. This avoids production loss from wafer sort yield.

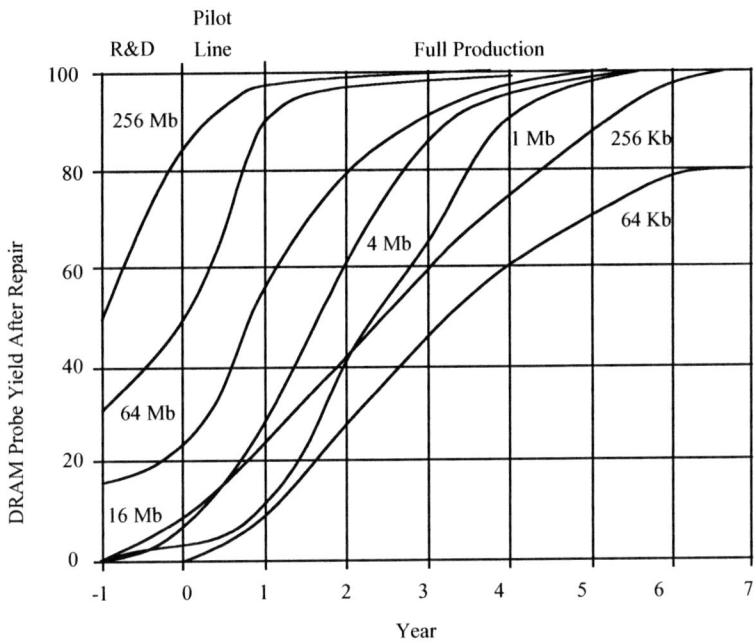

FIGURE 19.20 Reduced Time to Product Maturity for DRAM Production
Redrawn from C. Gross et al., "Assessing Future Technology Requirements for Rapid Isolation and Sourcing of Faults," *Micromagazine* (online version) http://www.Micromagazine.com/archive/98/07/jensen.html> (July 1998), p. 6.

Crystal Defects ■ Crystal defects such as dislocations can affect wafer sort yield. Dislocations can arise from chips and cracks on the wafer edge, resulting from poor handling or automatic handling equipment. The dislocations move toward the wafer center, especially during heat treatments (e.g., oxidation).

Wafer Sort Yield Models

Yield models have developed over the years to predict the wafer sort yield based on the wafer design, prior yield results, and statistical analysis. These models are beneficial for estimating the cost of changing design rules to reduce the chip area or increase the size of the wafer. Yield models are also used when estimating the cost of producing new chips. Three traditional yield models are:[25]

- Poisson's model
- Murphy's model
- Seed's model

Poisson's Model ■ *Poisson's model,* also referred to as the *exponential model,* is the simplest yield model and one of the first developed. The poisson yield model is:

$$Y = \frac{1}{e^{AD}}$$

Where,   Y = yield of functioning die
         A = die surface area
         D = defect density

Note: These variable descriptions are not repeated for the following two models.

This model shows the relationship between die area, defect density, and wafer sort yield. It assumes that the defect density across the wafer is uniform and constant from wafer to wafer, which is a poor assumption for larger diameter wafers fabricated with complex process integration. We have also noted that defects usually occur more often around the edge of the wafer than in the center, which this model does not include. For these reasons, Poisson's model only applies to smaller diameter wafers.

Murphy's Model ■ *Murphy's model* is a widely used yield prediction model that assumes the defect density varies across the wafer and from wafer to wafer. The variation across the wafer has central tendency (normal distribution) with the lowest density at the center and highest at the wafer edge. This yield model has been a good predictor for VLSI and ULSI yield. The yield model equation is:

$$Y = \left[\frac{1 - e^{-AD}}{AD}\right]^2$$

Seed's Model ■ *Seed's model* also assumes varying defect density across the wafer and from wafer to wafer. This model is suitable for VLSI/ULSI technology wafers. The model equation is:

$$Y = \frac{1}{e^{\sqrt{AD}}}$$

Usefulness of Yield Models ■ Yield models are useful when modeling fabrication processes that are stable, which means predictable die failures from random defects. If nonrandom defects occur that reduce wafer sort yield (e.g., a chip design modification) then the models may not be applicable. The predicted yield loss in these models addresses the killer defects that cause the wafer die to fail wafer sort. Defects that cause reliability failures in the field, such as electromigration, are not modeled. Yield models are most beneficial when comparing the yield of an existing process to the projected yield of a new product or when planning major wafer fab improvements.

These three models are examples of models for predicting yield. There are other yield models in use, with many of them tailored to a particular company's product and fabrication process. The development of accurate yield models is an ongoing task, with the goal of factoring into new models the issues of complex process integration and parametric yield losses.[26]

Yield Management Systems ■ The reduction of defects to improve fabrication yield is a major objective of semiconductor manufacturers. This reduction leads to faster yield ramps and higher yields on increasingly complex ICs. The immense amount of measurement and test data collected during wafer fabrication has created the need for *yield management systems.* This is also referred to as *defect reduction.* If used properly, yield management improves yield by linking defect and parametric data back to the individual workstations in the fab process. The defect data is analyzed to determine the problem's root cause, followed by corrective action that may include shutting down, servicing, and correcting tools. The measurement data at the workstation, usually in the form of statistical process control (SPC), can be correlated to the defects found at wafer test. SPC is a statistical method of analyzing data to determine when a process is stable or unstable and in need of corrective action. With the high degree of automation and hundreds of interdependent process steps, an active yield management team can assist the production teams to improve yield. Figure 19.21 gives a schematic representation of yield management in the wafer fab.

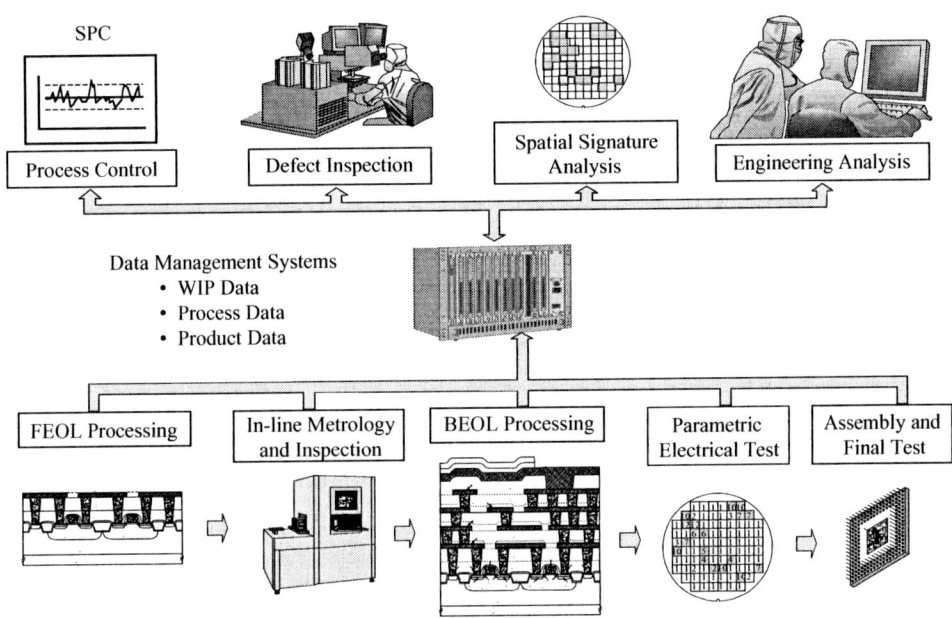

FIGURE 19.21 Yield Management in a Wafer Fab

TEST QUALITY MEASURES

Common quality measures for wafer test are presented in Table 19.4.

TABLE 19.4 Quality Measures for Wafer Test*

Quality Parameter	Types of Defects	Remarks
1. Electrical overstress.	A. Circuit failure (e.g., electrical open) due to overvoltage or overcurrent stress for a long time duration (e.g., >1 ms).	Nearly 50% of all IC failures are from electrical overstress (EOS). Possible causes are: • Latchup (see below). • Power supply switching. • Relay operation. • Power supply line variations. • Improper test procedure (e.g., wrong tester). Most prevention is done through circuit design (e.g., addition of external voltage clamps).

Quality Parameter	Types of Defects	Remarks
2. Electrostatic discharge (ESD).	A. Circuit failure (e.g., electrical open) due to overvoltage: • 100 V electrostatic pulse can damage the gate oxide if there is no protection. • Drain-source shorts are most severe form of ESD damage. • Damage is also done to contacts with a change to input/output leakage current.	Electrical failure is caused by exposure of the circuit to ESD voltage potentials ranging from 100 V to 20 kV. ESD protections are: • Current shunts and voltage clamps at each pin. • ESD protection for sub-0.5 μm features is influenced by lightly doped drain design and hot electron carrier performance. • ESD protection is becoming an integrated part of IC design.
3. Latchup.	IC Circuit failure due to open interconnect lines: • Latchup is typically a sudden and massive increase in the current draw of the IC. • Triggering of latchup is not always destructive but the chip must be reset before normal operation can resume.	The most common form of latchup is from parasitic elements in an IC (see Chapter 3). Methods to control latchup are:** • Fabricating devices in lightly doped epilayers grown on heavily doped substrates. • Trench isolation is effective in reducing susceptibility of latchup, particularly if the trench depth is greater than the well depth.
4. Gate oxide breakdown.	Device failure from two main types of gate oxide damage: • Catastrophic damage from overvoltage such as EOS. • Time-dependent dielectric breakdown during operation with the rated voltage, temperature, and power dissipation.	Maintaining gate oxide integrity (GOI) is critical for device performance. Thinner gate oxides are more sensitive to damage. • Breakdown of gate oxide can cause leakage at the chip input and output pins. • Oxide defects are introduced during the oxide growth process, especially the wafer clean, rinse, and dry process. Ultraclean oxide growth techniques improve yield.
5. Mobile ionic contamination (MIC).	MICs cause failure due to changes in device parameters, including: • Threshold voltages. • Off-state leakage currents. • Transistor drive currents.	Principal ions causing failure are Na^+, Cl^-, and K^+. Na^+ is most mobile due to its small radius. MICs have been steadily reduced in the wafer fabs, including sources such as processing equipment, packaging materials, fabrication environment, and the human body.

*Adapted from E. Amerasekera and F. Najm, *Failure Mechanisms in Semiconductor Devices,* 2nd ed., (New York: John Wiley and Sons, 1997), pp. 71–111.

**S. Campbell, *The Science and Engineering of Microelectronic Fabrication,* (New York: Oxford University Press, 1996), p. 446.

TEST TROUBLESHOOTING

Common test troubleshooting problems are described in Table 19.5.

TABLE 19.5 Common Test Troubleshooting Problems

Problem	Probable Cause	Corrective Action
1. Probe card problem at in-line parametric test.	A. Source of problem is based on interpretation of probe card problem: • Open readings on all measurements. • Open readings on some measurements. • Same electrical parameter (e.g., leakage) fails on every wafer.	• Check probe card needles alignment and planarization. • Verify probe marks on test pads. • Check probe card needles for dirt on needle tip, broken needle, and misaligned needles. • Verify wafer is flat on the chuck and properly aligned. • Leakage failure can result from probe card components. Verify space between needles and needles themselves are clean with no current leakage path.
2. Damaged test pads or bonding pads.	A. Damage from probe needles.	Damaged probe needles can be a significant problem for obtaining reliable test data. Probe needles are damaged by: • Excessive overdrive that puts too much needle force on pads. • Pad scratches from probe needles. • Reprobes that damage test pads. • Careless setup and handling.
3. Tester not functioning properly during test.	A. Hardware error: • Damaged probe needle card. • Faulty disk drive. • Using wrong test station. • Input power problems. • Improper DUT loading in prober.	Perform visual and electrical check of test setup to verify: • Correct test station for product. • Run correlation wafer to ensure test setup is correct. • Position of wafer. • Scrub marks on pads from needles.
	B. Software error: • System out of calibration. • Waveform signal degradation. • Wrong test software loaded, applying wrong forcing functions.	• Run autocalibration for tester. • Run diagnostic or self-test software. In worst case, verify each line of software code. • Verify signal integrity with oscilloscope. • Run correlation wafer to verify test setup. • Install power line conditioner.
	C. Test algorithm error: • Error in design of test algorithm. • Wrong algorithm for wafer, causing test vector or timing error.	• Verify algorithm with design check tool. • Compare two different software models to note difference.

SUMMARY

Wafer test measures electrical parameters on chips to verify acceptability. Electrical tests are done at different stages of the IC fabrication process, from early development chips to final packaged ICs. There are two wafer-level tests: in-line parametric test and wafer sort test. The in-line parametric test is done on a sample of wafer test structures located in the scribe line area of a wafer. This test is often done right after first metal etch. There are a wide range of parameters tested to provide early response on process performance. Reliability test structures are also used to assess reliability at the wafer level. Automated test equipment for in-line parametric test consists of a probe card to contact the wafer, wafer positioning mechanisms, tester instrumentation, and a computer to control the tester. A software test algorithm controls the measurements. Wafer sort performs functional tests on each individual die of a wafer. Defective die are electronically marked (or ink marked) in order to be scrapped once the chips are separated from the wafer. Wafers are assigned a bincode number to categorize the wafer based on the test results. The electrical tests for wafer sort are DC tests, output tests, and functional tests. Functional tests verify the chip performs as intended by the specification. Fault models are used to detect die problems during test. To increase fault coverage, I_{DDQ} testing measures steady state current leakage. Wafer sort yield measures the percentage of acceptable die that pass wafer sort. Different factors affect wafer yield, such as the wafer diameter and process maturity. There are yield models to predict the fab yield depending on wafer parameters. Yield management integrates all fab measurement and process data to improve yield.

KEY TERMS

wafer test

in-line parametric test (also wafer electrical test, WET, or DC test)

test structures (also process control monitors or PCMs)

scribe line monitors (SLMs)

wafer level reliability (WLR)

IC reliability

burn-in test

probe card

prober

overdrive

automated test equipment (ATE)

test algorithm

bin

calibration

wafer sort (also electrical sort, wafer probe, or probe)

ink mark

electronic wafer map

bincode number

bin map

spatial signature analysis (SSA)

functional tests

test vectors

test coverage

design for test (DFT)

parallel testing

fault models

I_{DDQ} testing

guardbanding

wafer sort yield

Poisson's model (also exponential model)

Murphy's model

Seed's model

yield management systems

REVIEW QUESTIONS

1. Define wafer test. What is the goal of wafer test?
2. List and describe the five different electrical tests performed during IC production.
3. List the two wafer-level tests done during wafer fabrication.
4. Explain how wafer fabs can succeed or fail based on the yield of wafers.
5. At what stage of the fab process is the in-line parametric test typically performed?
6. What is another name for in-line parametric test? Is the in-line parametric test a DC test or an AC test?
7. List and explain the five reasons for performing an in-line parametric test.
8. Why is it important to do an in-line parametric test?
9. Explain what is a test structure and how it is used for in-line parametric testing. Give an example of three different test structures.
10. What is a scribe line monitor (SLM)?
11. List and explain five different tests that may be done during in-line parametric testing.
12. Why is it important to interpret in-line parametric test data?
13. List four unacceptable test data trends.
14. Explain wafer-level reliability. Give an example of a wafer-level reliability test.
15. What is IC reliability? Explain burn-in testing.
16. List the four major subsystems for in-line parametric testers.
17. What is a probe card?
18. How is a wafer positioned during in-line parametric testing?
19. Explain overdrive and indicate why it is important.
20. Why is automated test equipment (ATE) required for advanced ICs?

21. What is a test algorithm?
22. Why is test equipment calibration necessary?
23. State a challenge for in-line parametric testing.
24. List and explain the objectives of wafer sort.
25. Describe how wafer sort is performed.
26. Why is an electronic wafer map compiled of defective die?
27. What is the purpose of a bincode number?
28. Explain spatial surface analysis.
29. Why are wafer sort tests sometimes described as AC tests?
30. List and describe three electrical tests typically done during wafer sort.
31. Explain how functional testing is conducted using test vectors.
32. Explain the concept of test coverage.
33. Describe how the zero-one algorithm is used for a functional test.
34. List four test issues that affect wafer sort.
35. Why are design for test (DFT) and parallel testing important for total test time?
36. Explain the single-stuck-at-fault (SSAF) model.
37. Explain the benefit of I_{DDQ} testing. What is its drawback?
38. Discuss guardbanding and how it ensures the product specification is met.
39. What is wafer sort yield?
40. List and explain six factors that affect wafer sort yield.
41. List three yield models and give a short description of each one.
42. Which models are acceptable for ULSI chips?
43. Explain the benefits of a yield management system. What is another name for yield management?

TEST AND PROBER EQUIPMENT SUPPLIERS' WEB SITES

Advantest America Inc. — http://www.advantest.com/
Agilent Technologies — http://www.agilent.com/
Cerprobe Corporation — http://www.cerprobe.com/home.asp
Electroglas Incorporated — http://www.electroglas.com/
Exatron Automatic Test Equipment — http://www.exatron.com/
Integrated Technology Corp. — http://www.inttechcorp.com/
International SEMATECH — http://www.sematech.org/
Keithley Instruments — http://www.keithley.com/
Micro Control Company — http://www.microcontrol.com/
The Micromanipulator Co. Inc. — http://www.micromanipulator.com/
Micro Photonics Inc. — http://www.microphotonics.com/
National Institute of Standards — http://www.nist.gov/
Pacific Western Systems — http://www.pacificwesternsystems.com/
Probe Technology — http://www.probecard.com/
QC Solutions — http://www.qcsolutions.com/
Schlumberger — http://www.1.slb.com/ate/diagsys
SEMI — http://www.semi.org/
Signatone — http://www.signatone.com/
SISA, Semiconductor Industry Suppliers Association — http://www.sisa.org/
TEL, Tokyo Electron Ltd. — http://www.teainet.com
Teradyne — http://www.teradyne.com/

REFERENCES

1. W. Merkel, "Parametric Testing to Improve Semiconductor Yields," *Semiconductor Online* (March 12, 1998): p. 1.
2. G. Pinkerton, "New Parametric-Test Technologies Meet Future Production Challenges," *Solid State Technology* (December 1996): p. 53.
3. T. Turner, "Test Structure Design: An Opportunity for Fab Outsourcing," *Solid State Technology* (April 1998): p. 44.
4. Adapted from E. Hnatek, *Digital Integrated Circuit Testing From a Quality Perspective,* (New York: Van Nostrand Reinhold, 1993), p. 155.
5. W. Merkel, "Parametric Testing to Improve Semiconductor Yields," p. 5.
6. Ibid., p. 3.
7. R. DeJule, "Expanding Applications and Demands on Parametric Test," *Semiconductor International* (June 1996): p. 110.

8. M. Pecht, R. Radojcic and G. Rao, *Guidebook for Managing Silicon Chip Reliability,* (Boca Raton: CRC Press, 1999), p. 184.
9. R. Iscoff, "What's in the Cards for Wafer Probing," *Semiconductor International* (June 1994): p. 77.
10. Keithley Vendor Literature, *S600 DC Parametric Test System,* Keithley Instruments, Inc.
11. S. Max, *Extending Calibration Intervals,* Proceedings of International Test Conference (Piscataway, NJ: IEEE, 1996): p. 118.
12. G. Pinkerton, "New Parametric-Test Technologies meet Future Production Challenges," p. 53.
13. C. Gross, et al., *Assessing Future Technology Requirements for Rapid Isolation and Sourcing of Faults,* Micromagazine, on-line version, (July 1998), p. 7.
14. T. Shao and F. Wang, "Wafer Fab Manufacturing Technology," *ULSI Technology,* ed C. Chang and S. Sze (New York: McGraw-Hill, 1996), p. 631.
15. A. J. van de Goor, *Testing Semiconductor Memories: Theory and Practice,* (New York: John Wiley and Sons, 1991), p. 96.
16. S. Sasho and M. Shibata, *Multi-Output One-Digitizer Measurement,* Proceedings of International Test Conference (Piscataway, NJ: IEEE, 1998): p. 258.
17. E. Hnatek, *Digital Integrated Circuit Testing from a Quality Perspective,* p. 133.
18. R. Rajsuman, I_{ddq} *Testing for CMOS VLSI,* (Boston: Artech House, 1995): p. 21.
19. A. Righter, et al., *CMOS IC Reliability Indicators and Burn-In Economics,* Proceedings of International Test Conference (Piscataway, NJ: IEEE, 1998): p. 194.
20. M. Sachdev, P. Janssen, and V. Zieren, *Defect Detection with Transient Current Testing and its Potential for Deep Sub-Micron CMOS ICs,* Proceedings of International Test Conference (Piscataway, NJ: IEEE, 1998): p. 204.
21. V. Agrawal and S. Seth, *Test Generation for VLSI Chips,* (Washington, D.C.: Computer Society Press, 1988): p. 328.
22. D. Jensen, C. Gross, and D. Mehta, *Mapping the Roadmap, New Industry Document Explores Defect Reduction Technology Challenges,* Micromagazine, on-line version (January 1998), p. 2.
23. D. Scott and R. Pisa, "Can Overall Factory Effectiveness Prolong Moore's Law?" *Solid State Technology* (March 1998): p. 75.
24. C. Gross, et al., *Assessing Future Technology Requirements for Rapid Isolation and Sourcing of Faults,* p. 6.
25. T. Price, *Introduction to VLSI Technology,* (New York: Prentice Hall, 1994): p. 105.
26. D. Jensen, C. Gross, and D. Mehta, *Mapping the Roadmap, New Industry Document Explores Defect Reduction Technology Challenges,* p. 8.

CHAPTER 20
ASSEMBLY AND PACKAGING

The use of wafers is a cost-effective way to simultaneously fabricate many chips. Once all fabrication and tests are complete, chips are separated from the wafer and assembled into the final IC package. The assembly and packaging process takes good electrical devices, places them in a package, and interconnects the device bonding pads to the package leads. Packaging provides a means of protecting the chip and attaching it to a higher level of assembly.

In the early days of semiconductor manufacturing, the cumbersome interconnection of the chip to the metal enclosure surrounding the chip was a manual process. U.S. chipmakers viewed it as highly labor intensive and transferred this process technology to offshore facilities or contractors. In the modern era, IC assembly and packaging has become highly automated and is an increasingly important part of the product price and performance for the end user. High-volume IC assembly is still dominated by Asia-Pacific region companies.

There are many new chip assembly and packaging designs that are in development for ICs. New designs are continually benchmarked for cost and performance against the existing approaches that have been proven over the past 20 to 30 years. As customer demand improves chip performance, new requirements are placed on IC packaging. Critical packaging parameters are the number of input and output (I/O) pins, electrical performance, thermal dissipation, and size.

The general trend is the merging of wafer fabrication technology with assembly and packaging. The size of the package is continually reducing to reflect the chip size. Some chipmakers are packaging ICs at the wafer level. Multiple chips are placed in one package. Ultimately, it is inevitable that the wafer fabrication process and the IC assembly and packaging eventually become part of the same process.

OBJECTIVES

After studying the material in this chapter, you will be able to:

1. Describe the general trends and design constraints of assembly and packaging.
2. State and discuss the traditional assembly methods.
3. Describe the different traditional packaging options.
4. Discuss the benefits and limitations of seven advanced assembly and packaging techniques.

INTRODUCTION

At the completion of the fab process, wafers that pass electrical test are ready for assembly and packaging of individual chips. This is done during *final assembly and packaging,* referred to as the *back-end* of the IC manufacturing process. IC final assembly and packaging is a big business. In 1998, 62 billion IC package units were sold worldwide.[1]

Final assembly and packaging are two distinct processes in the back-end of IC manufacturing. Each has their specific processes and tools. In the traditional process, *IC final assembly* separates each good die from the wafer and attaches the die to a metal leadframe or substrate. For leadframe assembly, the metal bond pads on the chip surface are interconnected with fine diameter wire to an internal leadframe to provide electrical access to the chip. After final assembly, *IC packaging* encloses the die in a protective package. The most common package in use today is the encapsulation of the chip with a plastic molding compound. This plastic package provides environmental protection and formed leads for connection to the next higher level of assembly (e.g., soldered to a circuit board). An outline of the traditional final assembly and packaging process is shown in Figure 20.1.

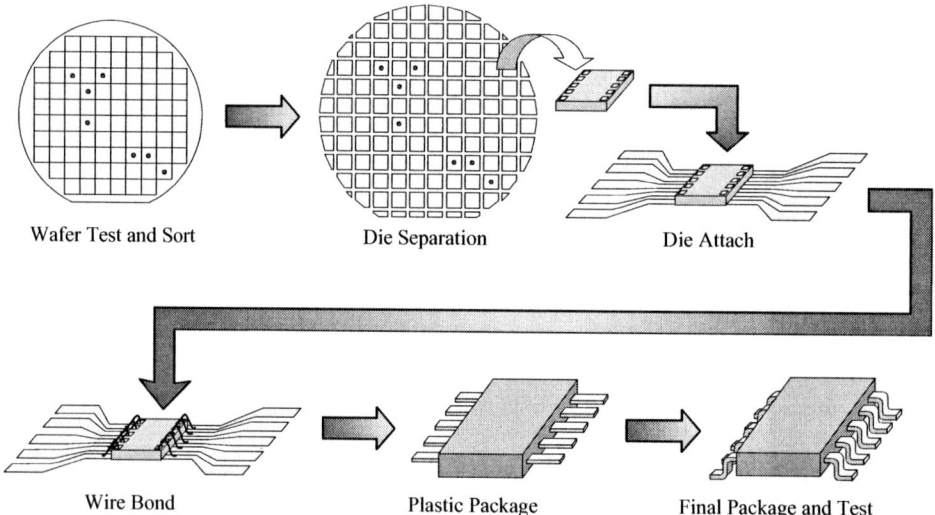

FIGURE 20.1 Traditional Assembly and Packaging

IC packaging has four important functions for all chips:

1. Protection from the environment and handling damage.
2. Interconnections for signals into and out of the chip.
3. Physical support of the chip.
4. Heat dissipation.

Many packaging variations exist in the industry, with some of the most common packages shown in Figure 20.2 (additional information on packages is provided later in this chapter). The package is selected so that the four functions are optimized to meet certain design constraints: performance, size, weight, reliability, and cost objectives (see Table 20.1). Packaging the IC permits the microchip to function in a wide range of customer environments, such as in a notebook computer, in the engine compartment of an automobile, and sandwiched between the plastic layers of a credit card. The various environmental conditions of contamination, moisture, temperature, mechanical vibration, and physical abuse all must be taken into account when the design engineers select the IC package.

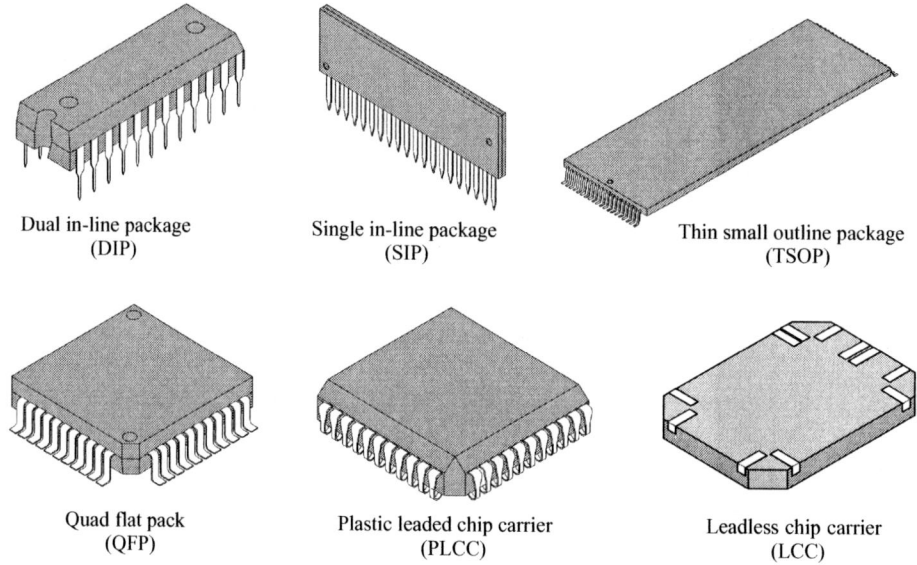

FIGURE 20.2 Typical IC Packages

TABLE 20.1 Design Constraints for IC Packaging

Design Parameters	Design Constraints
Performance	• RC time delay • Number of signal inputs/outputs (IOs) • Wirebond versus bump attachment • Signal rise time • Switching transients • Thermal • Power dissipation • Input and output impedance • Frequency response
Size/weight/form	• Chip size • Package size • Bond pads size and pitch • Package leads size and pitch • Substrate carrier pads size and pitch • Design of heat sink
Materials	• Chip substrate (plastic, ceramic, or metal) • Carrier (organic, ceramic) • Thermal expansion mismatch • Lead metallurgy
Cost	• Integration into existing process • Package materials • Yield
Assembly	• Method of die attach • Package attach (through hole, surface mount, or bumped) • Heat sink assembly • Encapsulation

For high-performance chip applications, such as high-end computers, performance and reliability are critical. For most consumer applications, cost is an important criteria, along with size and weight.

Packaging Levels

Two different packaging levels exist for electronic components (see Figure 20.3 on page 574). The chip assembly and packaging, which is the topic of this chapter, is referred to as *1st level packaging*. Once the chip is packaged into an IC component, the package I/O terminals connect the chip to the next level of assembly. The *2nd level packaging* assembles the IC component into a system with many components and connectors.[2] In most instances of 2nd level packaging, the IC component is assembled onto a printed circuit board using eutectic Sn/Pb solder (melting temperature of 183°C). A *printed circuit board* (*PCB*), also referred to as a *substrate* or *carrier*, has the circuit interconnections to attach the chip-carrying IC components onto the board with solder, while using connectors to interface into the remaining electronic subsystems of the product. Assembled circuit boards are then put into the final product (sometimes referred to as box-line assembly, such as for the assembly of computers).

This chapter covers the traditional method of IC assembly and packaging and then uses this information as a foundation to introduce the advanced assembly and packaging approaches. Many of the traditional methods are still in use in a wide range of products, primarily because of their low cost and proven reliability. Advanced packaging techniques are generally used for high-performance microchip applications.

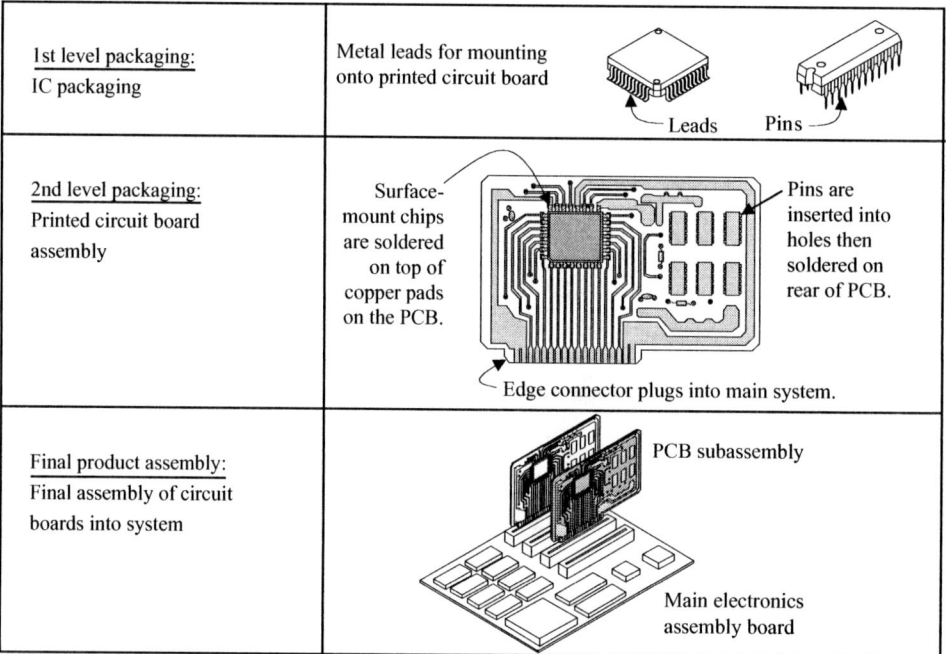

FIGURE 20.3 Levels of IC Packaging

TRADITIONAL ASSEMBLY

Final assembly consists of the operations required to attach the chip to the IC packaging. Manufacturing yield is critical during final assembly, since the majority of the fabrication cost is already in the chip. Traditional final assembly, which has been used for an estimated 95% of all ICs assembled during the latter 1990s,[3] consists of the following four steps:

- Backgrind
- Die separation
- Die attach
- Wire bonding

Backgrind

The first operation in final assembly is *backgrind* (sometimes this is done right after wafer sort before shipment to final assembly). Large diameter wafers are relatively thick to minimize breakage during front-end fabrication (a 300-mm wafer is 775 μm thick, which is roughly $1/32$ of an inch). However, wafers must be thinned before assembly begins. Wafers are usually thinned to a thickness of 200 to 500 μm.[4] Thinner wafers are easier to dice into chips and have improved thermal dissipation, which is beneficial for reducing thermal stresses in thin ULSI packages. Thinner chips also reduce the size profile and weight of the final IC package.

The backgrind operation is done with fully automated machines (see Figure 20.4). Backgrind is carefully controlled to minimize the introduction of stress into the wafer.[5] Stress can cause the wafer to warp, making it more likely to break during dicing and more difficult to mount into the package. In some cases, backside metallization is deposited after backgrinding. The metal is typically a gold film applied to improve electrical conductivity to the substrate and for eutectic die bonding (see the following section).

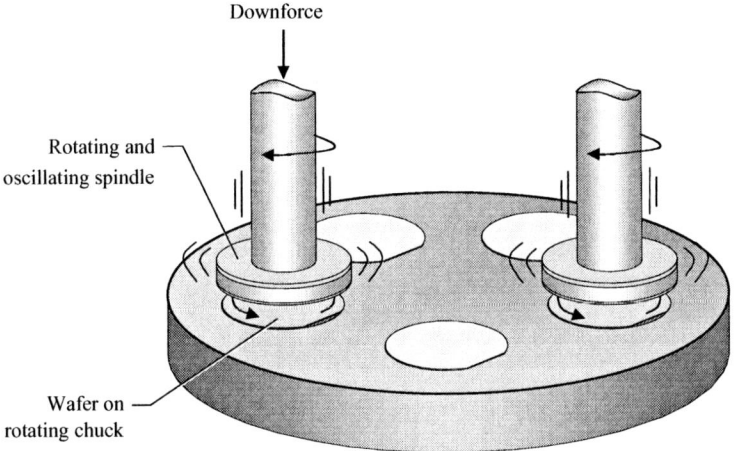

FIGURE 20.4 Schematic of the Backgrind Process

Die Separation

Die separation (also known as *die singulation*) cuts each die from the wafer using a diamond-blade dicing saw. Before dicing, wafers are removed from a cassette and properly oriented and placed onto an adhesive film secured to a rigid frame. The adhesive film holds the wafer intact once all chips are diced. The wafer is transferred into the saw, sprayed with DI water, and then sliced in the X and Y direction by a 25-μm thick diamond saw blade rotating up to 20,000 RPM (see Figure 20.5). The wafer is rinsed with DI water to remove the silicon slurry residue from the cutting action while each individual die is held in place by the adhesive backing. The saw usually cuts the wafer with 90 to 100% saw-through of the scribe lines. Wafer dicing is done with fully automated equipment that uses alignment systems and integrated wafer cleaning. Older dicing methods with 50% saw-through followed by a breaking process are usually not acceptable for ULSI devices.[6]

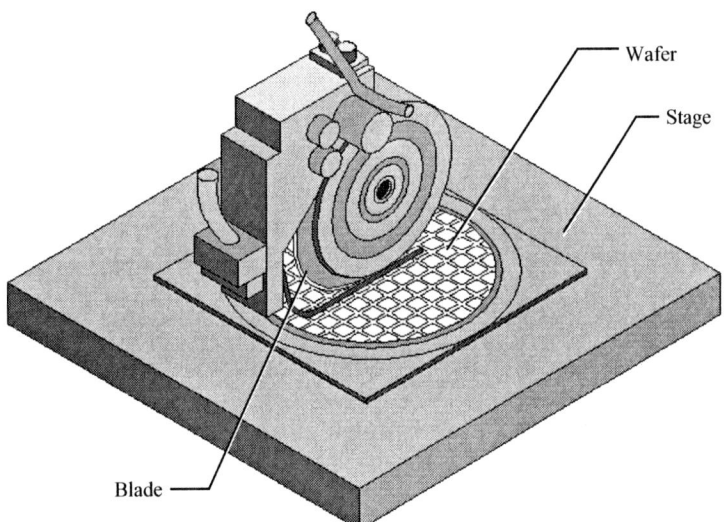

FIGURE 20.5 Wafer Saw and Sliced Wafer

Die Attach

After die separation, the wafers are moved to the *die attach* operation. At die attach, each good die is individually picked from the adhesive backing and physically attached to the substrate or lead-frame. An example of a leadframe is shown in Figure 20.6 on page 576, with the leads fanning out from the internal chip-bonding region to the larger lead pitch needed for higher-level assembly (pitch is the center-line to center-line spacing of leads). Leadframe magazines are small handling racks used to efficiently handle the leadframes and move them from tool to tool. The automated die bonder is a high-speed tool that uses a special gripper, referred to as a collet, to pick up die by their edge (to avoid chip damage) and place them on the substrate or leadframe for assembly. Die bonder tools require flexibility to attach die to a variety of applications, including leadframes, ceramic substrates, and circuit boards. The good die are selected based on a detector that recognizes the absence of an ink mark or by using the computerized wafer mapping data supplied from wafer sort.

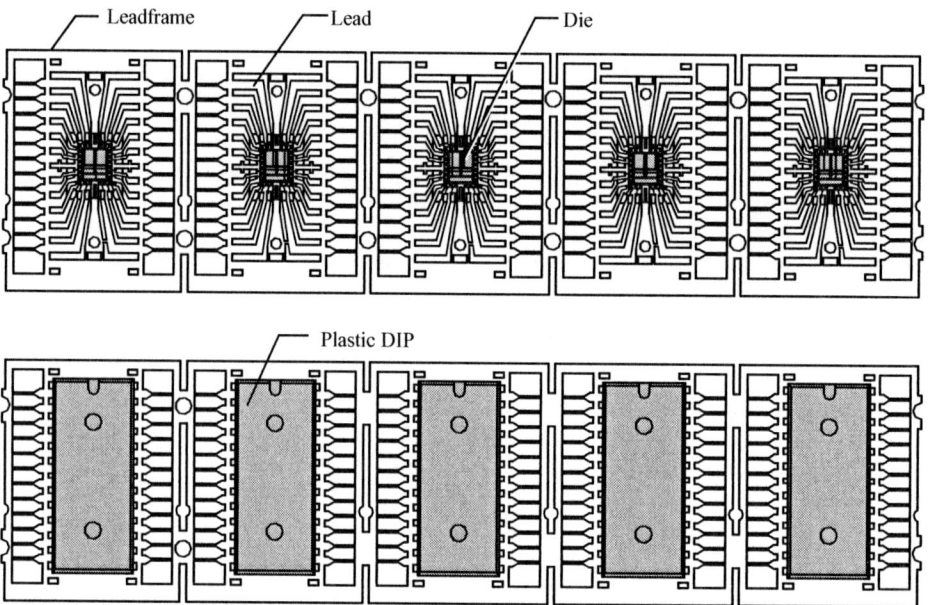

FIGURE 20.6 Typical Leadframe for Die Attach

Chip Bonding ■ The die is physically attached to the leadframe or substrate using one of the following techniques:

- Epoxy attach
- Eutectic attach
- Glass frit attach

Epoxy Attach. *Epoxy attach* is the most common method for bonding the chip to the leadframe or substrate. An epoxy is dispensed in the center region of the leadframe or substrate. The die bonder tool places the back of the chip onto the epoxy (see Figure 20.7), followed by a thermal heat cycle to cure the epoxy (e.g., 125°C for one hour). Most MOS applications use straight epoxy. However, the epoxy can be formulated with silver flakes to make a thermally conductive epoxy if there is a need for heat dissipation between the chip and the rest of the package.

Eutectic Attach. To attach a chip using *eutectic attach,* a thin film of gold (Au) is deposited on the backside of the wafer after backgrind (see Figure 20.8). Recall that eutectic defines a solution composition in order to minimize its melting point. This gold is then joined by alloying to the substrate, which is usually either a metal leadframe (e.g., alloy 42, a Ni-Fe alloy) or a ceramic substrate (e.g., 90 to 99.5% Al_2O_3). Typically, the substrate has a metallized surface of Au or silver (Ag). When heated to 420°C for approximately six seconds, which is slightly above the Au-Si eutectic temperature, this method forms a eutectic alloy interconnection between the chip and leadframe. Eutectic attach provides a good thermal path and mechanical strength. The eutectic attach technique is more common for bipolar ICs.

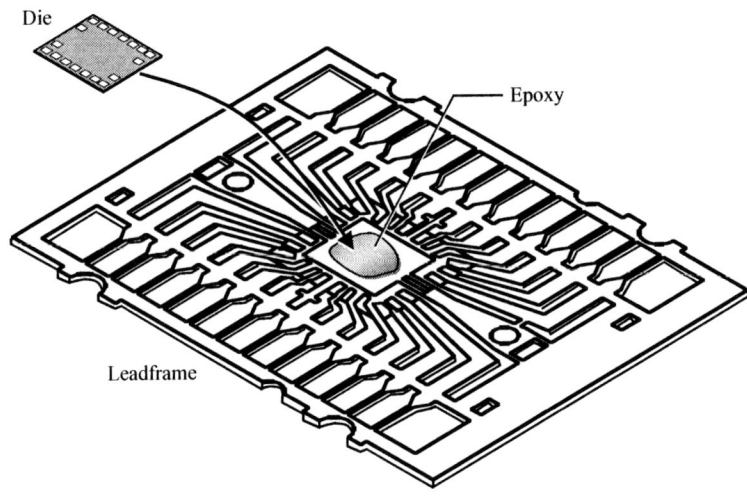

FIGURE 20.7 Epoxy Die Attach

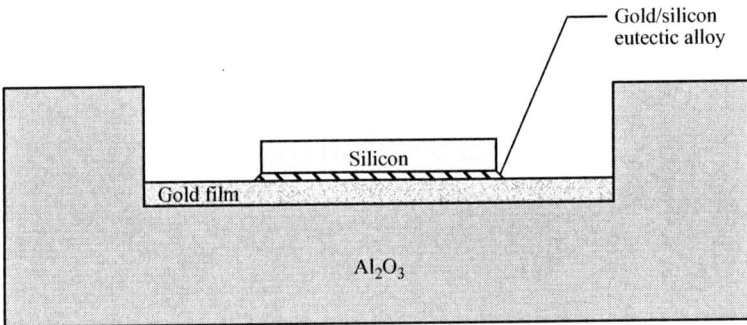

FIGURE 20.8 Au-Si Eutectic Attach

Glass Frit Attach. The *glass frit attach* consists of a mixture of silver and glass particles suspended in an organic medium. It is used to attach chips directly to Al_2O_3 ceramic packages without metallization to attain a hermetic seal. *Hermetic sealing* refers to protecting the silicon device from the external environment, specifically moisture and contamination. The silver and glass used in glass frit soften during cure and form a bond to the ceramic with good thermal conduction. It requires a relatively high temperature to cure the silver-filled glass.

Wirebonding

Wirebonding is the most common method for electrically connecting the aluminum bonding pads on the chip surface to the package inner lead terminals (sometimes referred to as *posts*) on the leadframe or substrate (see Figure 20.9 on page 578). This high-speed operation spools and bonds a fine diameter wire from the chip bond pads to leadframe inner lead pads, with the ability to form multiple bonds per second (e.g., a common speed is ten wire bonds per second). The tool bonds the wire at each die bonding pad or leadframe pad and steps to the next location. Wirebond placement accuracy is often +5 μm.[7] The bond wire is either Au or Al wire because it bonds well to both chip pads and leadframe inner lead pads, with a common wire diameter between 25 to 75 μm. A standard wire diameter of 25 μm is used on a die bonding pad pitch of 70 μm.[8]

The three basic types of wirebonding derive their names from the type of energy used during the wire termination process. The three methods of wirebonding are:

- Thermocompression bonding
- Ultrasonic bonding
- Thermosonic ball bonding

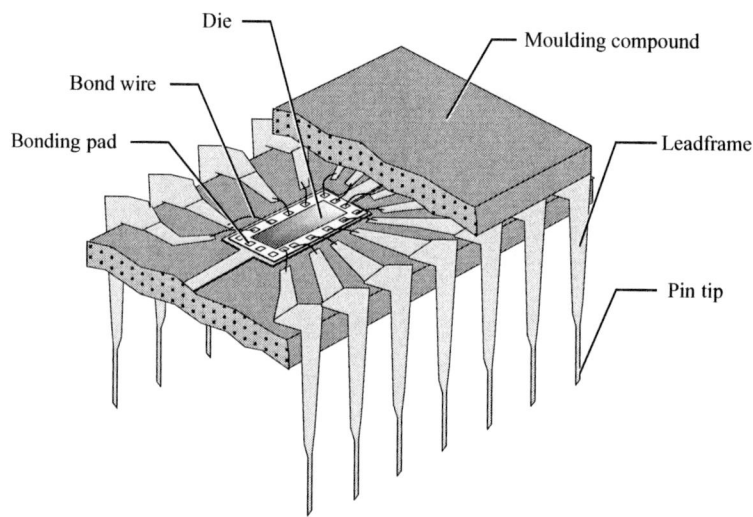

FIGURE 20.9 Wires Bonded from Chip Bonding Pads to Leadframe

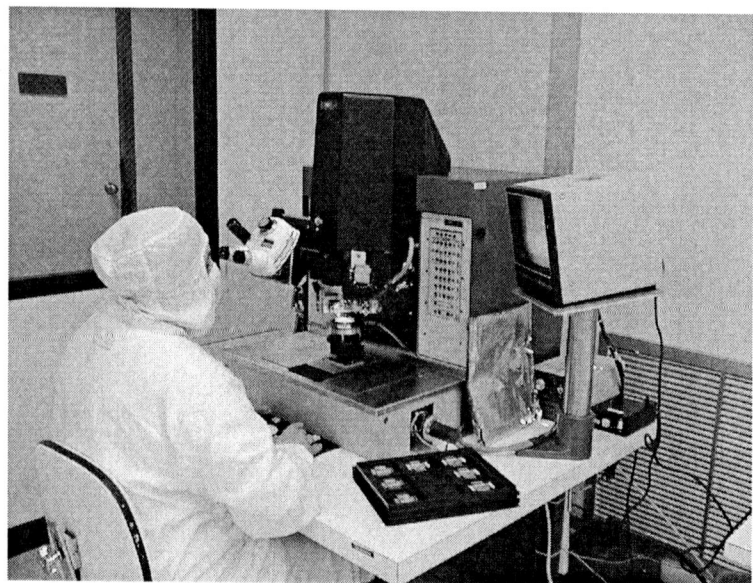

Wirebonding Chip to Leadframe (Photo courtesy of Austin Semiconductor, Inc.)

Thermocompression Bonding ■ In *thermocompression bonding,* thermal energy and pressure are used to form the wirebond of a gold wire to the chip pad and the leadframe inner lead. A bonding mechanism, referred to as a capillary tip, positions the wire on a heated chip bonding pad and applies pressure. The combination of force and heat causes the gold wire and aluminum pad to form a bond, referred to as a *wedge bond.* The capillary tip then feeds additional wire while moving to the leadframe inner lead, where another wedge bond is formed in the same manner (see Figure 20.10). This wirebonding process repeats until all chip bond pads are bonded to their corresponding leadframe inner lead posts.

Ultrasonic Bonding ■ *Ultrasonic bonding* is based on ultrasonic energy and pressure as a means to form a wedge bond between the wire and pad. It can form bonds between similar and dissimilar metals, such as Al wire/Al pad or Au wire/Al pad. The wire is fed through a groove in the bottom of the capillary tip (similar to thermocompression bond) and positioned over the die bonding pad. The capillary tip applies pressure and rapid mechanical vibration, usually at an ultrasonic frequency of 60 kHz (as high as 100 kHz), to form a metallurgical bond. The substrate is not heated in this technique. Once the bond is formed, the tool moves to the inner lead pad of the leadframe, forms

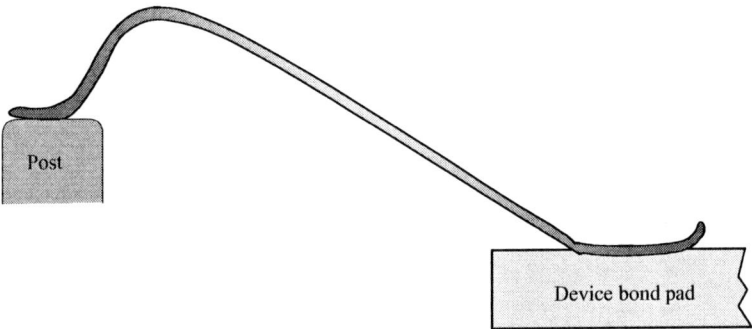

FIGURE 20.10 Thermocompression Bonds

the bond, and breaks the wire (see Figure 20.11). This process repeats until all chip bonding pads are wirebonded to the appropriate leadframe inner lead.

Thermosonic Ball Bonding ■ *Thermosonic ball bonding* is a technique combining ultrasonic vibration, heat, and pressure to form a bond referred to as a *ball bond*. The substrate is maintained at a temperature of approximately 150°C. The thermosonic ball bond has a capillary tip, made of tungsten carbide or a ceramic material, that feeds a fine diameter Au wire vertically through a hole in its center. The protruding wire is heated by a small flame or capacitor discharge spark, causing the wire to melt and form a ball at the tip. During bonding, ultrasonic energy and pressure causes a metallurgical bond to form between the Au wire ball and the Al pad (see Figure 20.12 on page 580). After completion of the ball bond, the bonding mechanism moves to the substrate inner lead pad and forms a thermocompression wedge bond. The wire is broken and the tool continues to the next die bonding pad. This ball bond/wedge bonding sequence has excellent control of the wire loop size between the bonding pad and inner lead pad, which is important for thinner IC packages.

Wirebond Quality Measures ■ It is essential that wire bonding have a consistently high yield. Two primary methods for assessing quality are visual inspection and pull tests. Visual inspection is done by looking at the wedge or ball and verifying that a good bond has formed. For example, a wedge bond should have a flat region where the ultrasonic vibration of the capillary tip came in contact. For a ball bond, there is deformation of the ball from the applied pressure, but it is undesirable to have excessive deformation.

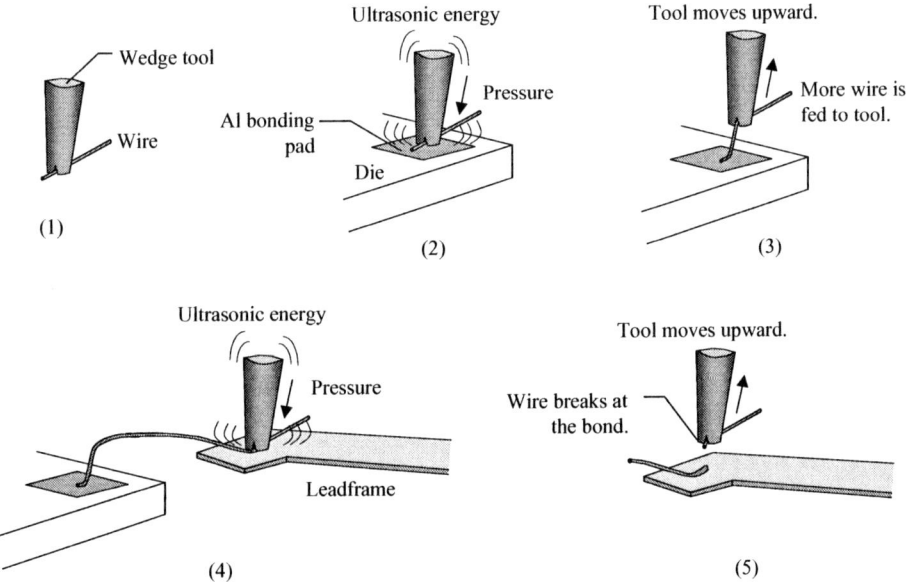

FIGURE 20.11 Ultrasonic Wirebonding Sequence

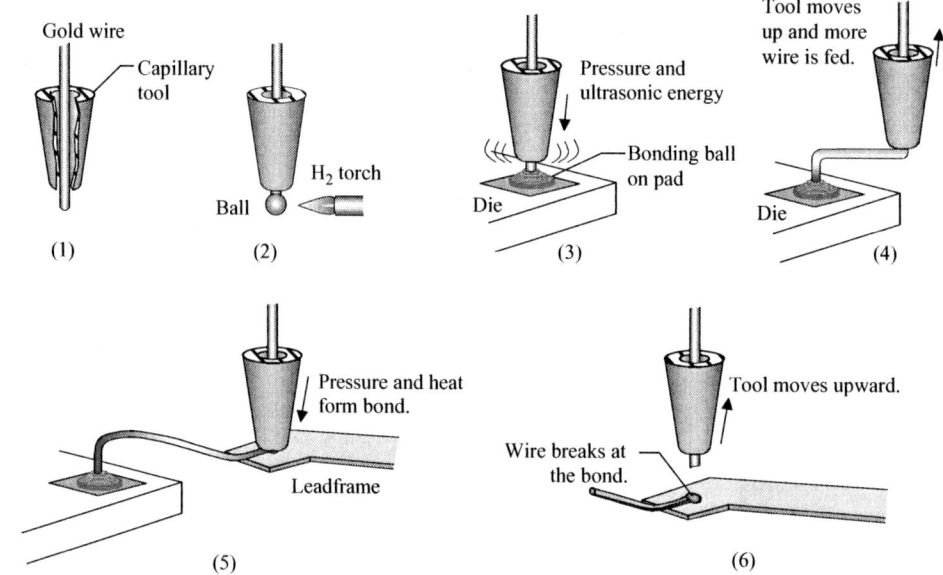

FIGURE 20.12 Thermosonic Ball Bond

Wire pull tests provide a quantitative assessment of the wirebond quality (see Figure 20.13). The pull test measures the strength of individual bonds and highlights where the bond fails, such as at the heel (the interface between the wire and flat region). These numerical measurements can be monitored with statistical process control (SPC) to assess process stability and trends.

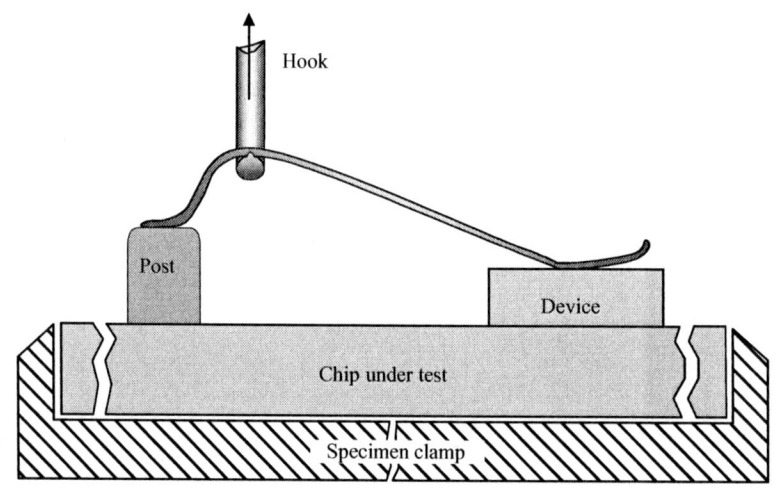

FIGURE 20.13 Wirebond Pull Test

TRADITIONAL PACKAGING

Integrated circuits have a wide array of traditional packaging styles. As previously noted, packaging is necessary to protect the chip from moisture and contaminants in the environment and handling damage. IC packaging forms the leads on the leadframe to interconnect the chip bonding pads with the 2nd level assembly circuit board. Chip bond pads are on a pitch ranging from 60 to 115 μm.

The leadframe leads fan out from this bonding pad pitch to the larger pad pitch used on circuit boards. A relatively large pitch is used on circuit boards, ranging from about 12 mils (300 μm) up to 25 to 50 mils (625 to 1250 μm) for surface mount components. Pin-in-hole components have a 100 mil (2,500 μm) pitch on circuit boards.

Metal packaging was common in the early history of the semiconductor industry. It is still used today for discrete device applications and small-scale integrated circuits. The die is attached to the center of a gold-plated header and wirebonded to the leads. A glass seal is formed around the leads and a metal cover is welded to the base to create a hermetic seal. An example is the metal TO-style (transistor outline) package (see Figure 20.14).

The two most widely-used types of traditional IC packaging materials are:

- Plastic packaging
- Ceramic packaging

FIGURE 20.14 TO-Style Metal Package

Plastic Packaging

Plastic packaging uses an epoxy polymer to completely encapsulate the wirebonded die and leadframe in a molding process. Plastic packaging has been an industry mainstay since its introduction in the 1960s. A key feature of plastic packaging is that the design lends itself to high-volume production techniques. The leadframes (with the die attached and wirebonded) are in strips and move through the process in racks to simplify handling. The racks interface with the different tools used to encapsulate the die and inner lead frame. A major reason why the plastic package has remained popular is flexibility for the shape of the leads, either as pin-in-hole (PIH) leads or surface mount technology (SMT) leads. PIH leads pass through the circuit board, whereas SMT leads attach to the board surface. Components with SMT leads are preferred because of higher density packaging (permitting a higher number of input/output leads, or I/Os) for both the IC component and circuit board. Other beneficial reasons for using plastic packaging are low material cost and low weight.

The cross-linked polymer of plastic packaging is dimensionally stable, ionically clean, and resistant to processing temperatures up to about 250°C. Other important epoxy parameters are low moisture absorption and the addition of fillers to reduce the thermal coefficient of expansion (TCE) to match the leadframe and chip's TCE. Although epoxy plastic molding is not considered a hermetic seal to protect the die from the environment and contamination, substantial improvement has been made in this area.

Once encapsulated, only the leads necessary for 2nd level assembly to the circuit board protrude from the IC package. The molded packages go through a deflashing step, which removes excess molding material from the package enclosure. Deflashing is done typically with a physical abrasion process similar to a sand blaster. Later the plastic package is marked with manufacturing and product information using ink or laser.

Component lead forming is done after molding. The strips of molded ICs are put into a lead trim-and-form tool where the leads are formed into the necessary shape: gull wing and J-lead for SMT and straight form for PIH. Each IC is separated from the leadframe strip by trimming off the tie-bar that holds all the leads straight (see Figure 20.15). Following lead forming, a thin lead finish (usually solder or tin) is applied to inhibit corrosion. The lead finish is typically applied by electroplating deposition. In some cases, the lead finish is applied before forming the lead.

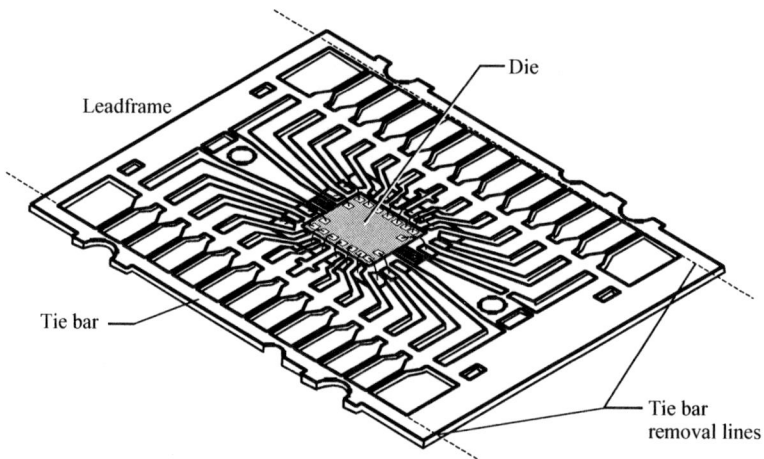

FIGURE 20.15 Tie Bar Removal from Leadframe

Types of Plastic Packages ■ There are many different types of plastic packages. Most are attached to corresponding pads on a printed circuit board by high-volume reflow soldering. A representative sample of the key plastic packages are presented here:

Dual in-line package (DIP), shown in Figure 20.16A, with two rows of PIH leads bent downward to pass through holes in the circuit board. This package was popular in the 1970s and 1980s, but its usage is declining.

Single in-line package (SIP), shown in Figure 20.16B, is an alternative to the DIP and is used to reduce circuit board space occupied by the IC component body, such as for memory application.

Thin small outline package (TSOP), widely used in memory and smart cards, is shown in Figure 20.16C, with gull-wing surface mount technology (SMT) leads along two sides that attach to corresponding pads on the circuit board. TSOPs are often attached to dual in-line memory modules (DIMMs) used as memory cards plugged into the main circuit board of a computer (see Figure 20.16D on page 584). An earlier version of a TSOP was a small-outline IC, or *SOIC*. Small outline packages have been and remain the most widely used type of IC package in the 1990s and early 2000s.[9]

Quad flatpack (QFP), shown in Figure 20.16E on page 584 is a surface mount component with leads on all four sides of the package to attain a high lead density (up to 256 leads or more). The QFP has been manufactured with the finest lead pitch for SMT plastic packages, down to 12 mil (300 µm). A pitch this fine is often a limiting factor for high yield in circuit board assembly. QFP and small outline packages accounted for 80% (approximately 50 billion units) of all ICs shipped during 1998.[10]

Plastic leaded chip carrier (*PLCC*) with J-leads is shown in Figure 20.16F on page 584. This package is used instead of the QFP if high I/O count is not necessary.

Leadless chip carrier (*LCC*) is a plastic package with leads that wrap around the package edge to maintain a low profile (see Figure 20.16G on page 584). The LCC either plugs into a socket or solders directly to the circuit board. A socket is used for ease of removal for field upgrades or repair.

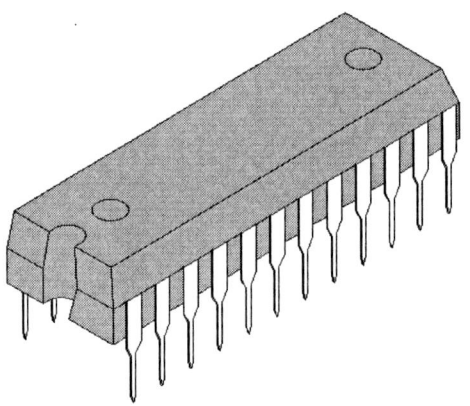

FIGURE 20.16A   Plastic Dual In-Line Package (DIP) for Pin-In-Hole (PIH)

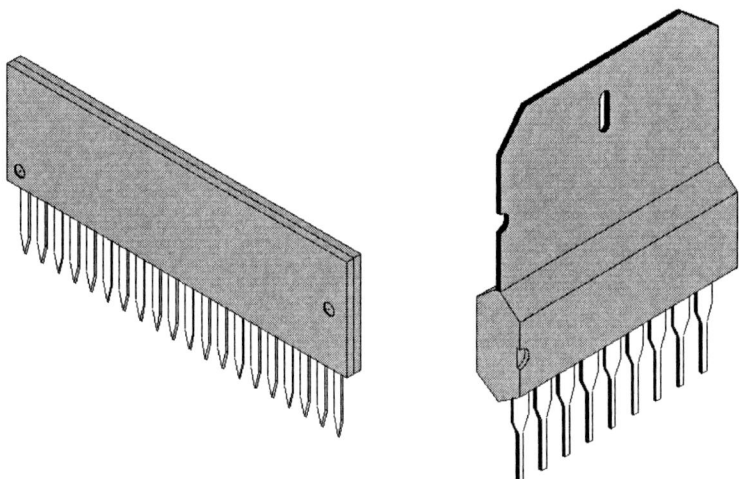

FIGURE 20.16B   Single In-Line Package (SIP)

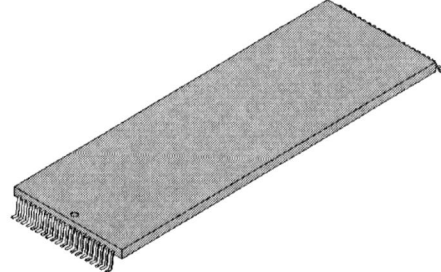

FIGURE 20.16C   Thin Small Outline Package (TSOP) with Gull-Wing Surface Mount Leads

FIGURE 20.16D Dual In-Line Memory Module (DIMM)

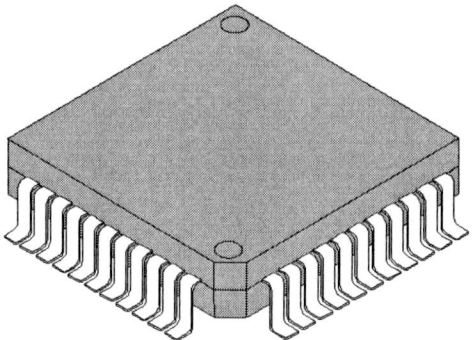

FIGURE 20.16E Quad Flatpack (QFP) with Gull-Wing Surface Mount Leads

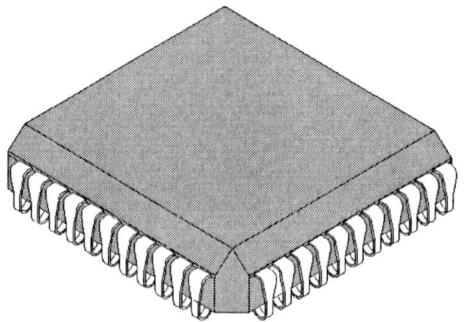

FIGURE 20.16F Plastic Leaded Chip Carrier (PLCC) with J-Leads for Surface Mount

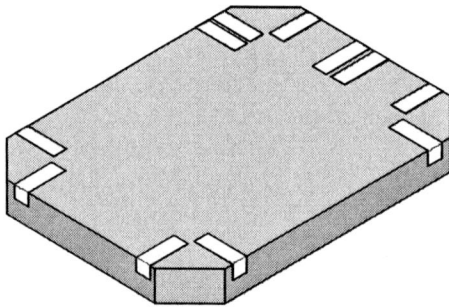

FIGURE 20.16G Leadless Chip Carrier (LCC)

Ceramic Packaging

Ceramic packaging is used for IC packaging, typically in state-of-the-art applications that require either maximum reliability or high-power, with a hermetic seal.[11] Ceramic packaging has two main methods, either a refractory (high melting temperature) ceramic that is processed separately from the chip assembly and packaging, or a ceramic DIP (CERDIP) technology, with lower packaging costs but still maintaining a hermetic seal.

Refractory Ceramic ■ A *refractory ceramic* substrate is common for IC packaging. It is made from alumina (Al_2O_3) powder mixed with an appropriate glass powder and an organic vehicle to form a slurry. The slurry is cast into thin sheets approximately 1 mil thick, dried, and then patterned to build a multilayer ceramic substrate (see Figure 20.17). Custom wiring circuits are deposited on the individual layers, with metallized vias for interconnecting the different layers. Several of the ceramic sheets are precisely laminated together and then fired at 1600°C to form a monolithic (meaning formed into one) sintered body. This is referred to as a *high-temperature cofired ceramic* (*HTCC*). There is also a *low-temperature cofired ceramic* (*LTCC*) that is fired at 850 to 1050°C, which can use circuit fabrication materials that do not withstand the HTCC temperature.[12] Ceramic technology is excellent for constructing complex packages with many signal, ground, power, bonding, and sealing layers. The major challenges for refractory ceramic as a substrate in IC packaging are: (1) high shrinkage, which makes tolerance difficult to control, (2) high dielectric constant, which increases parasitic capacitance and can affect high-frequency signals, and (3) the conductivity of alumina, which becomes an issue for signal delay.[13]

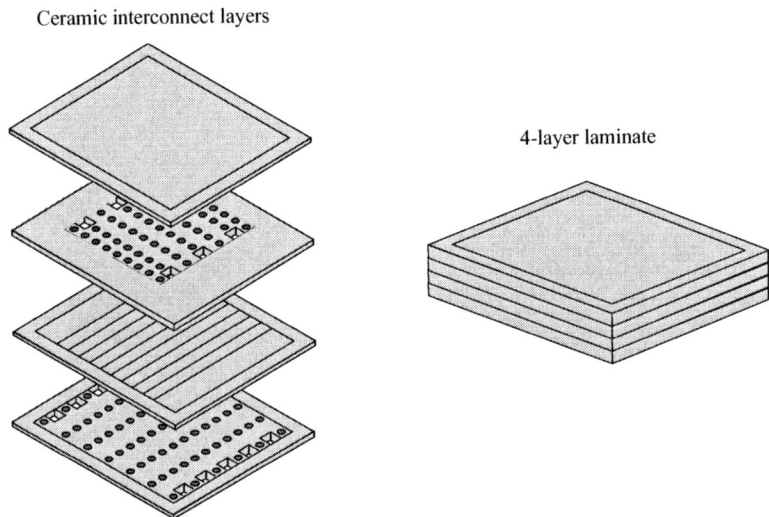

FIGURE 20.17 Laminated Refractory Ceramic Process Sequence

The most common lead format for ceramic packages is brazed pins on a 100-mil pitch that create the *pin grid array* (*PGA*) package. This is a pin-in-hole package for circuit board assembly. The die can be attached and wirebonded to either the bottom or top of the ceramic, followed by a lid to make a hermetic seal. PGAs are used for high-performance ICs such as high-frequency and fast microprocessors with pin counts up to 600 pins. PGA packages often need some form of heatsink or a small fan to remove the heat generated inside the package.

Ceramic Pin Grid Array (PGA)
(Photo courtesy of
Advanced Micro Devices)

Laminated Ceramic ■ A low-cost approach for ceramic packaging technology is to press two ceramic units together (after chip wirebonding) with a leadframe positioned between them (see Figure 20.18). This package is referred to as a *CERDIP*. The ceramic layers are hermetically sealed using a low-temperature glass seal.

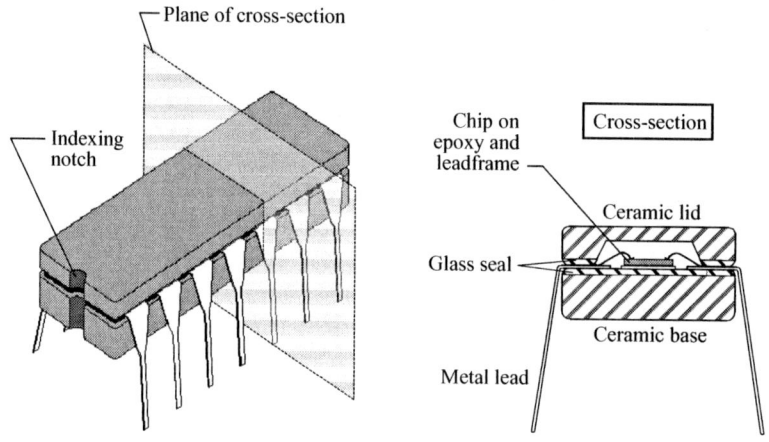

FIGURE 20.18 CERDIP Package

Final Test

All assembled and packaged chips undergo a final electrical test to ensure IC quality. The test is the same functional test performed at wafer sort, except now each die is tested as a final IC package. IC chip handlers are required for production testing of the individual chips in the automated test equipment (ATE). The IC handler quickly inserts each IC into the electrical contacts of the tester. Small, spring-loaded pins, referred to as pogo pins, make electrical contact with the package leads to perform the electrical test. After completion of the test, the IC handler moves the IC back to its final shipping package (i.e., tray, reel, or tube).

Advanced IC packages with a high number of inputs/outputs and a small package footprint pose a challenge for final test. Special test fixtures, often referred to as *contactors* or *sockets,* are used make electrical connection between the leads of the IC package and the contact pins on the automated tester (see Figure 20.19). These contactors must work reliably for millions of IC insertions without significant wear or electrical signal degradation. For advanced ICs, test fixtures are designed to minimize inductance and optimize the impedance of the signal path.

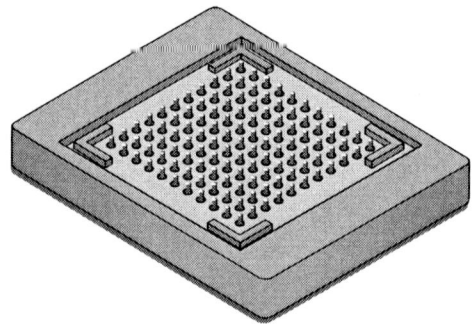

FIGURE 20.19 Test Socket for IC Package

ADVANCED ASSEMBLY AND PACKAGING

Reliable, faster, and higher-density circuits at lower cost are the goals for IC packaging. Wirebond technology has proven to be low cost and reliable. For the future, packaging goals will be met by increasing the density of chips and reducing the number of internal interconnections. Packages with fewer interconnection links decrease potential failure points, reduce the circuit resistance, potentially shorten the circuit length, and reduce inter-electrode capacitance, which affects electrical performance. The need to reduce the IC package to fit end-user applications (e.g., smart cards, palmtop computers, camcorders, and so on) is driving the new packaging designs that reduce size and overall profile. This reduction is offset by the need for handling larger amounts of parallel data lines, thus driving the need to increase package input/output requirements with more leads. Most of the increased I/O lead requirements are occurring in CMOS microprocessors, whereas memory I/O lead requirements will remain relatively low.

New packaging designs are regularly introduced to resolve 2nd level packaging challenges. Standardization groups such as the Joint Electronic Device Council (JEDEC) of the U.S. and the Electronic Industries Association of Japan (EIAJ) produce standards so that all companies use uniform IC package designs. Advanced IC packaging designs include:

- Flip chip
- Ball grid array (BGA)
- Chip on board (COB)
- Tape automated bonding (TAB)
- Multichip modules (MCM)
- Chip scale packaging (CSP)
- Wafer-level packaging

Flip Chip

Flip chip is a packaging technique of mounting the active side of a chip (with the surface bonding pads) toward the substrate (i.e., upside down placement of the bumped die relative to the wire-bonding approach—thus the reason for the term "flip" chip). This is currently the package design with the shortest path from the chip devices to the substrate, producing a good electrical connection for high-speed signals. There is also a reduction in weight and profile since leadframes or plastic packages are often not used. Flip chip technology uses bumps—usually formed from tin/lead solder in a 5% Sn and 95% Pb ratio—to interconnect the chip bonding pads to the substrate (see Figure 20.20). High Pb solder is used because of its increased reliability in this bump application. The most common solder bump process is referred to as C4 (controlled collapse chip carrier), developed by IBM in the 1960s for attaching silicon chips to ceramic substrates. The substrate in use today is ceramic or plastic-based, either as a rigid printed circuit board or flexible polyimide circuit. There has also been active investigation into using conductive adhesives (e.g., silver-filled adhesive polymer) in place of solder for the flip chip bump.[14]

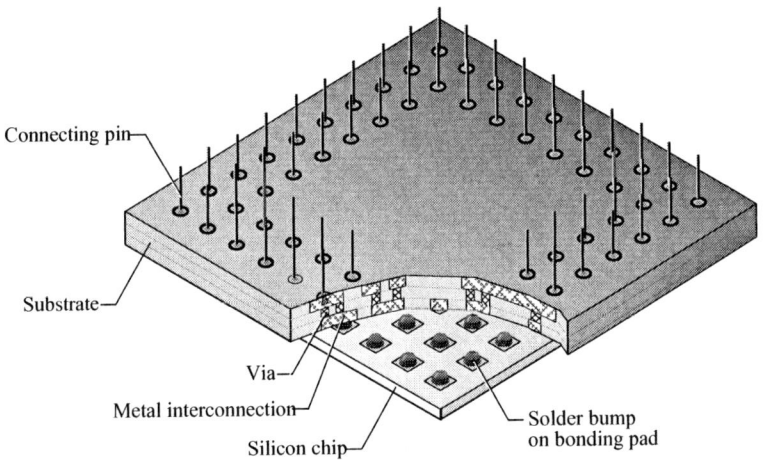

FIGURE 20.20 Flip Chip Packaging

C4 solder bumps are typically deposited on chip bonding pads at the wafer level using evaporation or physical vapor deposition (sputtering). There is a special barrier layer metallurgy (BLM) required at the chip C4 solder pads on the wafer (see Figure 20.21). The BLM provides good C4 solder adhesion to the pad and inhibits diffusion. The diameter of the C4 bump has traditionally been 4 mils on a 10 mil pitch.

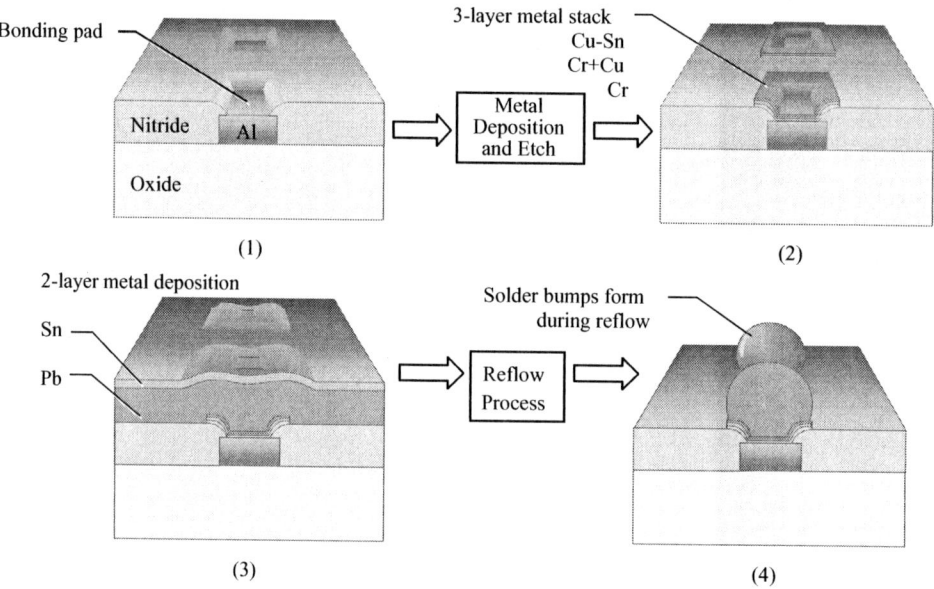

FIGURE 20.21 C4 Solder Bump on Wafer Bonding Pad

The flip chip is attached to the substrate by using an aligner-bonder tool. It uses a vision system to automatically align and place the chip on the substrate. The chip C4 solder bumps are positioned on the corresponding substrate contact pads. Heat, often applied with hot air, and slight pressure then cause the C4 solder to reflow and form an electrical and physical connection between the substrate and the die.

Epoxy Underfill ■ An important issue for flip chip reliability is the coefficient of thermal expansion (CTE) mismatch between the silicon chip and the substrate. Excessive CTE mismatch introduces stress into the C4 solder joints and causes premature failure due to solder cracks. The problem is resolved by using an *epoxy underfill* that flows between the chip and the substrate (see Figure 20.22). The epoxy CTE is matched to the C4 solder joint and effectively reduces the stress applied to the C4 joints. The stress reduction on the C4 solder joints can be more than a 10X improvement with the use of underfill.[15]

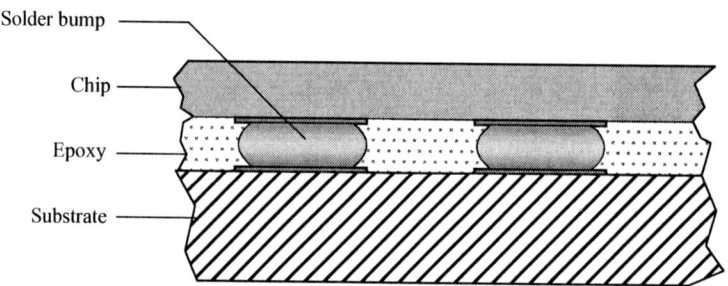

FIGURE 20.22 Epoxy Underfill for Flip Chip

A major challenge with the use of epoxy underfill is that the flip chip can not be removed once the epoxy is applied. This creates problems for rework if the chip is found defective during test. Epoxy is often applied after electrical test. There are also techniques developed that avoid the use of epoxy underfill, such as the addition of an interposer (a compliant polymer material with an interconnect structure) between the die and substrate to decouple CTE stress between the two bodies.

Another aspect of flip chip processing is the ability to clean underneath the chip before applying the epoxy underfill. The reflowed C4 solder bumps leave only a 2 to 3 mil clearance between the chip and substrate. Soldering requires a flux chemical to remove oxidation and create an acceptable solder joint. Flux is sometimes an ionic contaminant that must be thoroughly removed by DI water or solvent cleaning. Advanced no-clean fluxes leave no ionic contaminants and do not require cleaning.

Number of I/O Leads ■ Flip chip technology facilitates the need for more I/O leads in a package because it is an *area array* technology. This means that the C4 solder bumps are placed in an x-y grid across the entire chip surface, efficiently using chip surface area for a greater number of leads (see Figure 20.23). The traditional wirebonding is a *perimeter array* technique, with bonding pads only around the die perimeter (such as for a QFP). Perimeter array limits the number of leads in the package and does not efficiently use available surface area in the center of the die. One of the biggest challenges with area array technology is inspection for C4 bump integrity. Automated inspection systems based on X-rays have been used.

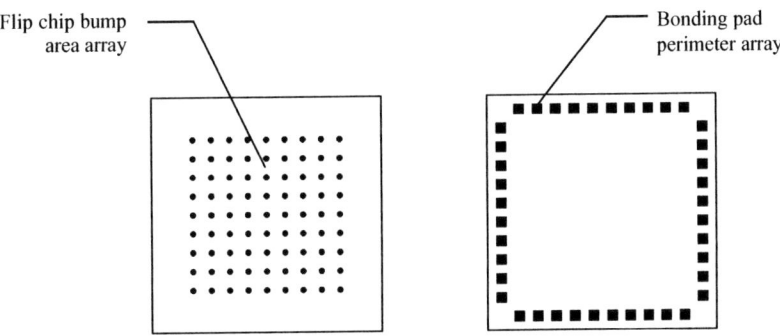

FIGURE 20.23 Flip Chip Area Array Solder Bumps Versus Wirebond

Ball Grid Array (BGA)

The *ball grid array* (*BGA*) package was introduced in the early 1990s and has a similar package design to pin grid array (PGA). BGA consists of a ceramic or plastic substrate that has an area array of eutectic Sn/Pb solder balls used to connect the substrate to the circuit board (see Figure 20.24 on page 590). The silicon chip is attached to the top of the substrate using flip chip C4 or wirebond technology. BGA is an extension of flip chip with a considerably larger pitch between interconnects to simplify 2nd level assembly. As with flip chip, BGA effectively achieves a high lead count in a surface mount package with a small profile. High-density BGA packages have up to 2400 leads. The BGA solder ball pitch is commonly 40, 50, or 60 mils versus the 100 mil pin pitch for PGA. This feature is a main contributor to the high lead count. A recently introduced BGA package has a solder ball pitch of 20 mils, which was the minimum BGA pitch in use in the latter 1990s.

Chip with Ball Grid Array (BGA)

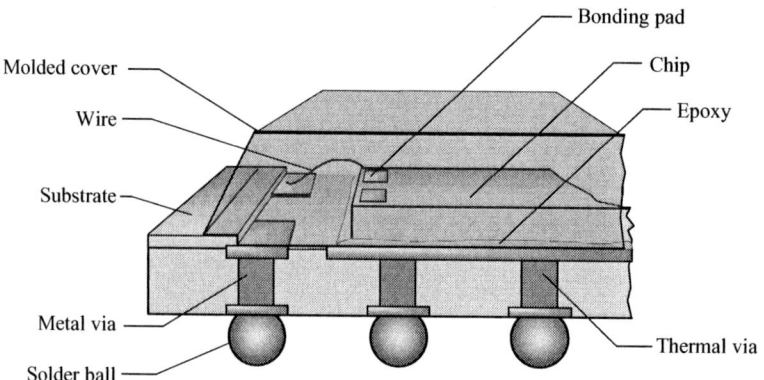

FIGURE 20.24 Ball Grid Array

There has been considerable development work on plastic substrates for BGA packages, also referred to as an organic or laminate carrier. In this case, the silicon die is attached with wirebond or C4 solder bumps to the plastic substrate, which has solder balls for attachment to the circuit board. Plastic substrates have a lower dielectric constant than ceramic, which translates into better high-frequency performance and high-speed switching because of reduced signal transmission delay.

A positive aspect of BGA is that during 2nd level assembly, the BGA component is placed on the circuit board along with other surface mount components (e.g., QFPs, TSOPs, and so on) and solder reflowed. The BGA solder balls reflow and form an interconnection to the board. Integrating the BGA process with existing surface mount components lowers the cost of assembly.

Chip on Board (COB)

During the late 1980s, the *chip on board* (*COB*) process was developed to mount IC chips directly onto a substrate with other SMT and PIH components. It is also referred to as *direct chip attach* (*DCA*). The chip is epoxied and wirebonded to the substrate (often a printed circuit board) using the standard attachment process (see Figure 20.25). There is no package around the silicon chip, but it is covered with an epoxy resin, known as a glob-top. The COB approach reduces the package size over traditional SMT and PIH packaging with minimal process and equipment changes. It became popular in applications where size and cost are important, such as video game cartridges and smart cards.

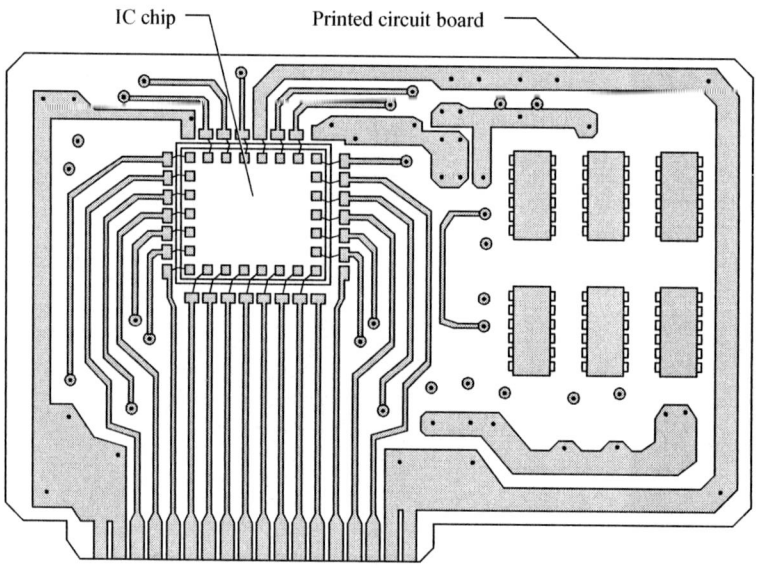

FIGURE 20.25 Chip on Board (COB)

Tape Automated Bonding (TAB)

Tape automated bonding (TAB) is a high-I/O packaging approach that uses a plastic tape as a chip carrier (see Figure 20.26). The tape has a thin copper foil sandwiched between two layers of polyimide dielectric film. The copper is etched to form leads that match the chip bond pads, with an inner lead bond region (ILB) for bump attachment to the chip and an outer lead bond region (OLB) for solder attachment to the circuit board. Once the chip is attached to the ILB, the chip is covered with an epoxy resin known as glob-top for protection and the tape is rolled into a reel. The tape and reel format is used for 2nd level assembly of the TAB chip to the circuit board. During assembly, the chip and leads are excised from the tape and the leads are formed into a gull wing and then bonded by solder reflow to the circuit board. TAB was once thought to be the definitive high-I/O package during the 1980s, but expensive tooling and integration costs for assembly kept it from becoming widely used. It is now applicable only to special niche applications.

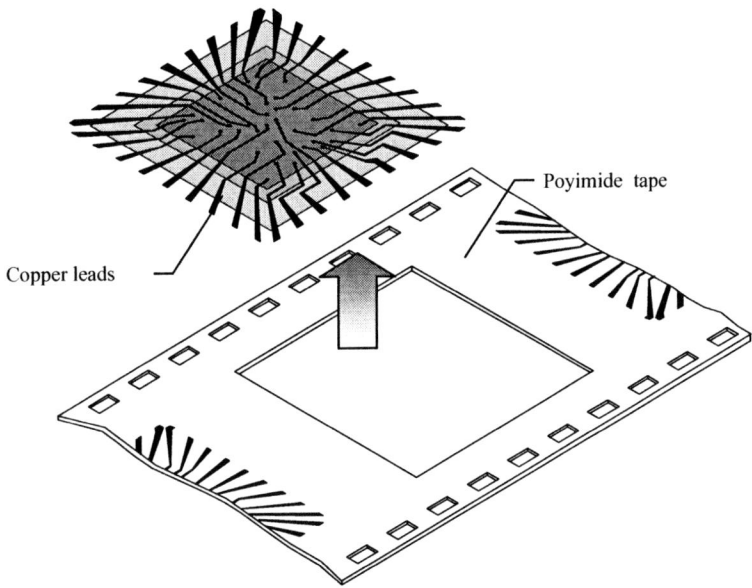

FIGURE 20.26 Tape Automated Bonding (TAB)

Multichip Module (MCM)

A *multichip module (MCM)* is a package that mounts several die onto one substrate (see Figure 20.27). This mounting permits a higher density of silicon chips on the MCM substrate material. An MCM is defined as having a silicon chip module surface area that covers 30% or more of the substrate surface area.[16] The most common MCM substrates are ceramic or advanced printed circuit boards with high chip density. MCM packaging design enhances electrical performance by reducing circuit resistance and parasitic capacitance while reducing the size and weight of the total package. The MCM is an evolution of an older technology known as a hybrid circuit, which had active and passive components mounted on ceramic and connected by conductive thick-film paste that was screened on (similar to silk screening). Hybrid circuits had low die density on the substrate.

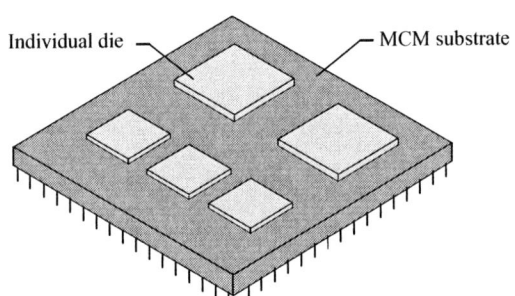

FIGURE 20.27 Multichip Module (MCM)

Chip Scale Packaging (CSP)

IC packaging designers seek lower cost, lower weight, and lower thickness with enhanced electrical performance. The development in the 1990s of a diverse range of IC packages that were approximately the same size as the silicon chip led to the concept of chip scale packages. A general definition of a *chip scale package* (*CSP*) is an IC package that has a footprint (surface area) that is <1.2 times the footprint of the die.[17] Since CSP packages have roughly the same size as the die, it creates efficient use of 2nd level circuit board surface area, especially when chips use area array bump technology. The main CSP packaging technologies today are the flip chip and BGA methods discussed above, since both use bump interconnects. Furthermore, CSP includes multiple chips on a substrate, which encompasses the MCM concept. Flip chip is the fastest growing advanced packaging method (see Figure 20.28). Examples of different chip scale packages are provided in Table 20.2.

It is desirable for CSP is to be compatible with existing surface mount packages for 2nd level assembly to the circuit board. The use of CSP equipment, processes, and materials that are compatible with the existing surface mount infrastructure simplifies a CSP product's introduction into manufacturing and lowers production costs.

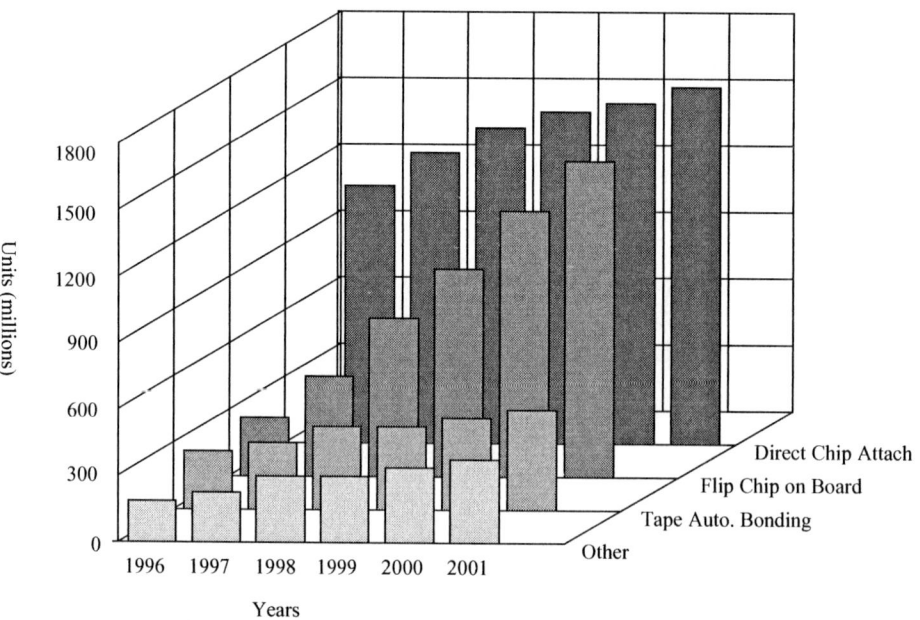

FIGURE 20.28 Trends for Advanced Packaging
Redrawn from S. Winkler, "Advanced IC Packaging Markets and Trends," *Solid State Technology* (June 1998): p. 63.

Wafer-Level Packaging

Until now, all IC assembly and packaging to connect chip bond pads to standard pads on a substrate had been done on die separated from the wafer. This process created a natural separation between the front-end wafer fab processes and the back-end assembly and packaging processes used to produce finished ICs. To attain lower costs with increased production efficiency, wafer-level packaging was developed in the latter 1990s. *Wafer-level packaging* is the formation of the 1st level interconnections and the package I/O terminals on the wafer before it is diced (see Figure 20.29). Many of the package design proposals for wafer-level packaging use flip chip material and processing technologies.

TABLE 20.2* Diversity of Chip Scale Packages

General CSP Approach	CSP Package Name	Company
Custom Leadframe	Area array, bumped CSP	Amkor/Anam
	Small outline no-lead/C-lead (SON/SOC)	Fujitsu
	Bump chip carrier (BCC)	Fujitsu
	Micro-stud-array (MSA)	Hitachi
	Bottom leaded plastic (BLP)	LG Semicon
	Quad flat no-lead (QFN)	Matsushita
	Memory CSP	TI Japan
	Quad outline non-leaded	Toshiba
Interposer (flexible material with interconnets) between die and substrate	Enhanced flex CSP	3M
	FleXBGA	Amkor/Anam
	FBGA	Fujitsu
	Chip-on-flex CSP	GE
	Multichip scale package (MCSP)	Hightec MC AG
	CSP for memory devices	Hitachi
	IZM flexPAC	Fraunhofer Institute
	Molded ball grid array	Mitsubishi Electric
	Chip-on-flex chip size package	Motorola Singapore
	Fine-pitch BGA (FPBGA)	NEC
	MicroBGA	Tessera
Rigid Substrate	Chip array package (CABGA)	Amkor/Anam
	CSP	Cypress Semiconductor
	Ceramic mini-BGA	IBM
	Molded array process CSP	Motorola
	Plastic chip carrier	National
	CSP	Oki Electric
	Transformed grid array package	Sony
	Ceramic/plastic fine-pitch BGA	Toshiba

Adapted from J. Baliga, "Making Room for More Performance with Chip Scale Packages," *Semiconductor International* (October 1998): p. 86.

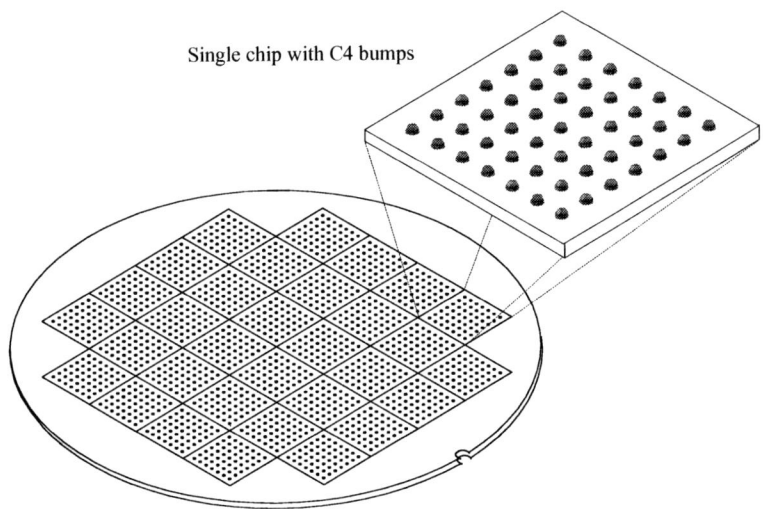

FIGURE 20.29 Wafer-Level Packaging

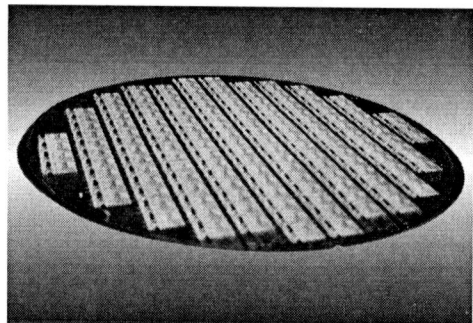

C4 Bumped Wafer
(Photo courtesy of Advanced Micro Devices)

There are competing wafer-level packaging designs under development and it remains to be seen which designs will gain acceptance. The key is developing a reliable interconnect system to interface between the fine pitch geometry of the die bonding pads and the coarser-pitch geometry needed for the 2nd level circuit board assembly. One approach uses thin film layering processes to build an interface between the die bond pads and the larger geometry pads needed to attach the chip to a circuit board. Wirebonding makes an interconnect between the chip and the interface (see Figure 20.30). A BGA solder ball array is used to attach the chip package directly to the 2nd level assembly circuit board. A leadframe is not typically a part of wafer-level packaging. The total height of the package on the circuit board is often 1mm or less.

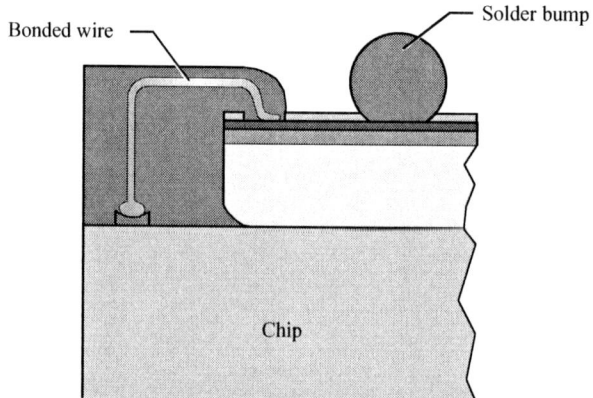

FIGURE 20.30 Design Concept for Wafer-Level Packaging
Redrawn from V. DiCaprio, M. Liebhard, and L. Smith, "The Evolution of a New Wafer-Level Chip-Size Package," *Chip Scale Review* (May/June 1999).

If the entire assembly and packaging process is done at wafer level, then test and burn-in will also be done with wafers. Chipmakers are developing the wafer-level test procedures at this time. In most respects, test after wafer-level packaging is similar to wafer sort. A major change is the use of probe card needles that contact die solder bumps on a wafer instead of aluminum pads. This setup could simplify the probe card because the BGA solder ball pitch is much larger than the bond pad pitch on a wafer. Wafer-level burn-in and test eliminates the need for test sockets and special IC handling procedures that are currently used to test individual IC packages. This would be a significant efficiency improvement and cost savings. Another possible improvement is the elimination of the wafer probe test, since the final functional test is now done at the wafer level (see Figure 20.31).

Cost savings is a big factor in IC assembly and packaging. As chip geometries shrink and assembly and packaging technology remains constant, the cost of packaging becomes a greater percentage of the total IC component cost. In some cases, the cost of the IC package exceeds the cost of the IC itself.[18] The ultimate goal of wafer-level packaging is to provide high-density IC packaging while still at the wafer level, thus unifying the front-end and back-end processes to reduce process steps for a substantial cost savings. This integration is a natural progression that results in a true chip scale package. Benefits of wafer-level packaging are listed in Table 20.3.

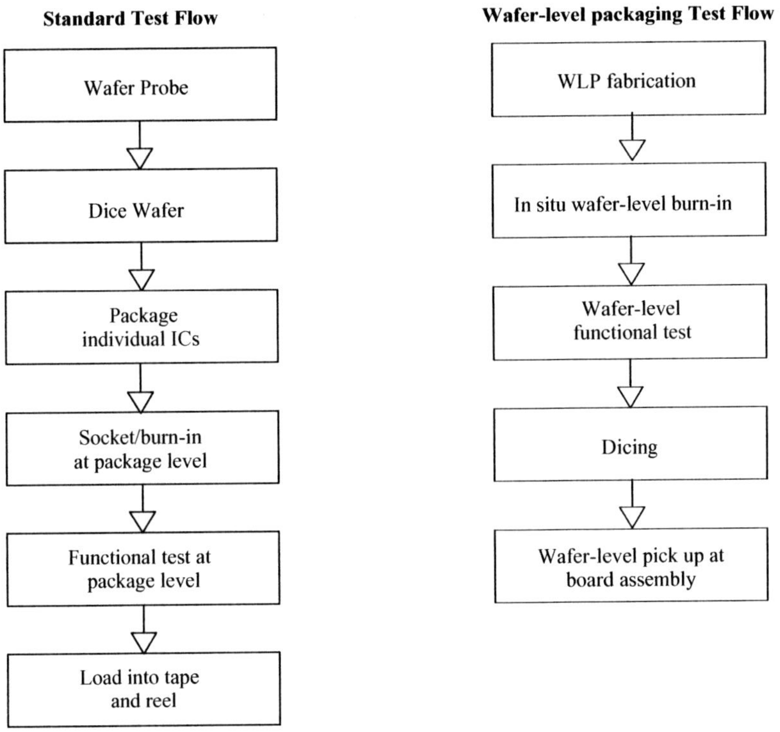

Standard Test Flow

```
Wafer Probe
    ↓
Dice Wafer
    ↓
Package
individual ICs
    ↓
Socket/burn-in
at package level
    ↓
Functional test at
package level
    ↓
Load into tape
and reel
```

Wafer-level packaging Test Flow

```
WLP fabrication
    ↓
In situ wafer-level burn-in
    ↓
Wafer-level
functional test
    ↓
Dicing
    ↓
Wafer-level pick up at
board assembly
```

FIGURE 20.31 Comparison of Standard Test Flow with Wafer-Level Package Test Flow.
P. Elenius, "Wafer-Level Packaging Gains Momentum," *Solid State Technology* (April 1999): p. 46.

TABLE 20.3* Wafer-Level Packaging Features and Benefits

Parameter	Benefits
Package size	The package is equal to the chip size in x and y dimensions. It is the smallest possible IC package and minimizes the package weight.
Mounted package height	It is extremely thin with a total height < 1.0 mm as measured from the circuit board surface after 2nd level assembly.
Component reliability	Test results indicate that wafer-level packaging components pass existing reliability tests for passivated components.
Solder joint reliability	Test results indicated solder joint reliability meets standard thermal cycle (-65 to $125°C$) reliability tests.
Electrical performance	Electrical simulation tests indicates that the die face-down (flip chip) configuration of wafer-level packaging with its short circuit traces results in very good electrical performance for minimizing inductance and parasitic capacitance losses.
Integration with existing SMT infrastructure	The wafer-level package is compatible with existing surface mount technology and uses standard solder balls and ball pitches.
Alpha-particle protection	Radioactive elements occurring naturally in packaging materials emit alpha-particles that can cause voltage loss in memory cells. The use of polyimide tape and film adhesive provides alpha-particle protection for memory chips.
Low system cost	The use of existing materials with wafer integration to reduce handling and a wafer test strategy to minimize duplicate testing provides for a low overall system cost.

*Adapted from V. DiCaprio, M. Lebhard, and L. Smith, "The Evolution of a New Wafer-Level Chip-Size Package," *Chip Scale Review* (May/June 1999): p. 34.

ASSEMBLY AND PACKAGING QUALITY MEASURES

A sampling of quality measures for IC assembly and packaging is provided in Table 20.4.

TABLE 20.4 Key Quality Measures for IC Assembly and Packaging

Quality Parameter	Types of Defects	Remarks
1. Crack or void in flip chip C4 bump.	A. A solder crack or void in a C4 bump is a stress concentration point that will fail prematurely. B. A severe crack with complete delamination of the C4 bump should be detected in final test.	• A solder crack could result from tooling damage during processing. • Voids may result from improper temperature profile during solder preheat which leaves excessive flux residues that vaporize during soldering.
2. Irregularity of C4 bump size in flip chip.	A. Irregular C4 bump size can lead to increased solder stress and electrical failure of the bump.	• Ensure pad surface metallurgy (BLM) is properly cleaned prior to deposition of C4 solder. • Review the C4 solder deposition process to verify correct parameters.
3. Flux contamination residues on C4 solder bump of flip chip.	A. Flux contaminants can produce dendritic growth (a conductive film of contaminants that causes a short) between C4 solder bumps and cause device electrical failure. B. Flux residue prevents epoxy underfill from adhering to the C4 solder column, creating a delamination (halo defect). This leads to a stress buildup and eventual failure in C4 bump.	• Verify post-reflow cleaning process to verify cleaning chemistry is properly formulated to remove flux residues. • Verify the epoxy underfill is completely encapsulating the C4 solder bumps between the chip and substrate (no voids).
4. Flip chip epoxy underfill has voids and delamination between die and substrate.	A. Epoxy voids and delamination can initiate cracks in the C4 solder bump due to CTE mismatch that leads to electrical failure.	• It is critical that the epoxy underfill process be properly controlled to minimize defects. • Rework of the flip chip is difficult if the epoxy is already applied.
5. Wirebond strength of wedge or ball bond.	A. Low pull strength data during pull test (values read below minimum or out-of-control from control chart analysis). B. Bond failure leading to electrical test failure due to open circuit.	Possible causes for bond failures are: • Formation of intermetallics between gold and aluminum at the bond interface, causing voids that degrade the quality of the bond. • Poor wedge or ball bond formation due to contamination or improper cleaning, causing lifted bond. • Improper pressure during bond process. Low bonding pressure forms bond prone to lifting, while high bond pressure is prone to failure from cracking.

IC PACKAGING TROUBLESHOOTING

Common IC packaging troubleshooting problems are listed in Table 20.5.

TABLE 20.5 Common IC Packaging Troubleshooting Problems

Problem	Probable Cause	Corrective Actions
1. Plastic package delamination or cracking.	A. Excessive moisture absorbed into plastic package (molded plastic can absorb up to 0.4% moisture from the air by weight). Exposure to temperature cycling from processing or chip usage can cause expansion and failure. B. Excessive CTE mismatch among package, chip, and interconnect structures.	• If necessary, prevent moisture from reaching the package by using dry storage. • Modules packed for shipping should be dry baked and dry packed with a desiccant to absorb moisture. • Excessive CTE mismatch requires package redesign to minimize mismatch.
2. Corrosion causing die electrical failure due to increase in metallization resistance and eventual open circuit or increase in leakage current.	A. Moisture reaching die in the presence of ionic contaminants: • The silicon nitride passivation layer can crack due to packaging stress and expose underlying metallization to moisture.	Corrective action is needed to reduce level of ionic contaminants and inhibit access of moisture to die, including: • Reduce plastic shrinking during molding by adding fillers, thus minimizing stress on the die and forming good adhesion to all surfaces. • Improve die cleaning process to reduce presence of ionic contaminants.
3. Soft memory error in DRAMs or SRAMs (soft errors are usually correctable, but can become a serious problem if they occur too frequently).	A. Radioactive elements present in packaging materials emit alpha (α) particles with energies up to 8 MeV that can upset memory storage cell (cause a cell storing a "1" to lose memory and change to a "0").	Sources of α-particles are alumina and epoxy. Protection methods against α radiation includes: • Shielding around the device because radioactive impurities are never completely eliminated. • Coating wafer with polyimide. • Building devices in heavily doped silicon epilayer to inhibit movement of charge.
4. Wire bonds failing by open circuit at the chip bond pad.	A. Formation of a purple-colored intermetallic compound (known as purple plague) at interface between gold and aluminum on the bond pad. At elevated temperatures, aluminum from the bond pad diffuses into the purple plague intermetallic and causes voids. Excessive voids lead to an open circuit.	The solutions to purple plague are: • Minimizing exposure to elevated temperature. • Ensure the gold has acceptable purity. • Follow standard design rules for bond pad metallization. Purple plague problems may become more common with devices that operate at higher temperatures.

SUMMARY

Final assembly and packaging assembles each chip into a package for protection and attachment to the higher level of assembly. Traditional assembly has four steps: wafer preparation (backgrind), die separation, die attach, and wire bonding. Die attach is used to physically attach the die to the leadframe or substrate. There are three methods of die attach: epoxy attach, eutectic attach, and glass frit attach. Epoxy attach is the most common. Wirebonding electrically connects the die to the leadframe or substrate and consists of thermocompression, ultrasonic, and thermosonic ball bonding. Traditional packaging is oriented around different types of plastic packages. Ceramic packaging has been used in traditional packaging for high reliability or high power. The ceramic packages are either refractory ceramic or laminated ceramic. All packaged ICs undergo a final electrical test. Advanced packaging methods are flip chip, ball grid array (BGA), chip on board (COB), tape automated bonding (TAB), multichip modules (MCM), chip scale packaging (CSP), and wafer-level packaging. Flip chip, BGA, and COB are the most commonly used techniques. These techniques are all considered chip scale packages. The industry is moving toward wafer-level packaging for process efficiency and cost benefits.

KEY TERMS

final assembly and packaging or back-end
IC final assembly
IC packaging
1st level packaging
2nd level packaging
printed circuit board (PCB) (also substrate or carrier)
backgrind
die separation or die singulation
die attach
epoxy attach
eutectic attach
glass frit attach
hermetic sealing
wirebonding
posts
thermocompression bonding
capillary tip
wedge bond
ultrasonic bonding
thermosonic ball bonding
ball bond
plastic packaging

dual in-line package (DIP)
single in-line package (SIP)
thin small outline package (TSOP)
quad flatpack (QFP)
plastic leaded chip carrier (PLCC)
leadless chip carrier (LCC)
ceramic packaging
refractory ceramic
pin grid array (PGA)
CERDIP
contactor or socket
flip chip
epoxy underfill
area array
perimeter array
ball grid array (BGA)
chip on board (COB)
direct chip attach (DCA)
tape automated bonding (TAB)
multichip module (MCM)
chip scale package (CSP)
wafer-level packaging

REVIEW QUESTIONS

1. What is the name given for the back-end of the IC manufacturing process?
2. Give a short description of IC final assembly and IC packaging.
3. State the four functions of IC packaging.
4. For each of the following, give two design constraints: performance, size/weight/form, materials, cost and assembly.
5. List and explain the two different packaging levels.
6. What is a printed circuit board (PCB)?
7. List the four steps of traditional assembly.
8. Explain the backgrind operation.
9. Describe how the die is separated from the wafer.
10. What is die attach?
11. List and discuss the three types of chip bonding.
12. What is a hermetic seal?
13. State and discuss the three methods for wirebonding.
14. What are the two primary methods for the quality control of wirebonding?
15. List the two most widely-used types of traditional IC packaging materials.
16. Explain the process for plastic packaging.
17. List and describe six different plastic packages.
18. When is ceramic packaging typically used?
19. What are the two main methods for ceramic packaging?
20. Describe the refractory ceramic process. What is HTCC and LTCC?
21. What is a pin grid array (PGA) package?
22. What is a CERDIP?
23. Describe a contactor for testing.
24. List seven advanced packaging techniques.
25. What is flip chip?
26. What are the advantages of a bump process for chip packaging?

27. Why is epoxy underfill used for flip chip?
28. What is the advantage of area array over perimeter array?
29. Describe the BGA package. Can it be placed on the circuit board with a standard surface-mount process and integrated with standard components?
30. Describe COB.
31. What is tape automated bonding (TAB)?
32. What is the benefit from a multichip module?
33. State the general definition of a chip scale package (CSP). What are the benefits of a CSP?
34. Describe wafer-level packaging.
35. For each of the following parameters of wafer-level packaging, describe its benefits: (1) package size, (2) mounted package height, (3) component reliability, (4) solder joint reliability, (5) electrical performance, (6) integration with existing SMT infrastructure, (7) alpha-particle protection, and (8) cost.

ASSEMBLY AND PACKAGING SUPPLIERS' WEB SITES

Supplier	Web Site
3M	http://www.3m.com/US/electronics_mfg/microflex/
Chip Pac	http://www.chippac.com/
Cypress Semiconductor	http://www.cypress.com/
Entegris Inc.	http://www.entegris.com/
Fujitsu	http://www.fujitsu.com/
Hitachi	http://www.semiconductor.hitachi.com/
IBM	http://www.chips.ibm.com/
IC Master	http://www.icmaster.com/mfrsearch.asp
International Microelectronics and Packaging Society, IMAPS	http://www.imaps.org/
International Micro Industries, Inc.	http://www.imi-corp.com/
International SEMATECH	http://www.sematech.org/
Karl Suss Inc.	http://www.suss.com/
Kulicke & Soffa Industries, Inc.	http://www.kns.com/
Matsushita Electronic Corp.	http://www.mec.panasonic.co.jp/e-index.html
Mitsubishi	http://www.mmc-sil.com/
Motorola	http://www.mot-sps.com/
National Semiconductor	http://www.national.com/
NEC Semiconductor	http://www.nec.com/semiconductors/
Oki Electric	http://www.oki.co.jp/semi/
Pacific Coast Technologies	http://www.pcth.com/
Philips Semiconductor	http://www.us2.semiconductors.philips.com/
SEMI	http://www.semi.org/
Silicon Sensors LLC	http://www.siliconsensors.com/
SISA, Semiconductor Industry Suppliers' Association	http://www.sisa.org/
Soldering Technology Centre	http://www.solderworld.com/
Sony	http://www.sony.com/
Tessera	http://www.tessera.com/
Texas Instruments	http://www.ti.com/
TSK America Inc.	http://www.tsk-wms.com/
Toshiba	http://www.toshiba.com/

REFERENCES

1. P. Burggraaf, "Chip Scale and Flip Chip: Attractive Solutions," *Solid State Technology* (July 1998): p. 239.
2. R. DeJule, "High Pincount Packaging," *Semiconductor International* (July 1997): p. 139.
3. J. Baliga, "Package Styles Drive Advancements in Die Bonding," *Semiconductor International* (June 1997): p. 101.
4. H. Hinzen and B. Ripper, "Precision Grinding of Semiconductor Wafers," *Solid State Technology* (August 1993): p. 53.
5. H. Blech and D. Dang, "Silicon Wafer Deformation After Backside Grinding," *Solid State Technology* (August 1994): p. 74.
6. T. Tachikawa, "Assembly and Packaging," *ULSI Technology,* ed. C. Chang and S. Sze (New York: McGraw-Hill, 1996), p. 541.
7. L. Oboler, "Wire Bonding Still at the Head of the Class," *Chip Scale Review* (July/August 1999): p. 40.
8. R. DeJule, "High Pincount Packaging," p. 142.
9. Industry News, *Chip Scale Review* (May/June 1999): p. 23.
10. P. Burggraaf, "Chip Scale and Flip Chip: Attractive Solutions," p. 239.
11. T. Tachikawa, "Assembly and Packaging," p. 552.
12. J. Sergent, "Materials for Multichip Modules," *Semiconductor International* (October 1996): p. 212.
13. T. Tachikawa, "Assembly and Packaging," p. 552.
14. R. Estes, "Flip-Chip Packaging with Polymer Bumps," *Semiconductor International* (February, 1997): p. 103.
15. A. Babiarz, "Key Process Controls for Underfilling Flip Chips," *Solid State Technology* (April 1997): p. 77.
16. J. Sergent, "Materials for Multichip Modules," p. 209.
17. T. DiStefano and J. Fjelstad, "Chip-Scale Packaging Meets Future Design Needs," *Solid State Technology* (April 1996): p. 82.
18. P. Elenius, "Wafer-Level Packaging Gains Momentum," *Solid State Technology* (April 1999): p. 46.

APPENDIX A
CHEMICALS AND SAFETY

As in any manufacturing process, there are a wide range of safety hazards in semiconductor manufacturing. Many of the liquid and gas chemicals used in semiconductor manufacturing are hazardous, toxic, flammable, and pyrophoric. However, the semiconductor industry is one of the best manufacturing industries with regards to the health and safety of its workers.[1] The definitions of the following safety terms are:

Hazardous: Any chemical or substance that has adverse effects on the health or safety of people. Some examples include acids and bases.

Toxic: Any chemical or substance that seriously damages biological tissue. Examples are phosphine and arsine.

Flammable: Any liquid or gas that is capable of igniting into fire. Examples are alcohol and acetone.

Pyrophoric: Any material that ignites spontaneously in air below 55°C (130°F). An example is silane.

To use these chemicals safely, it is important to follow safe practices. The same principle applies in the work environment: understand the chemical, its properties, and how to use it safely. Sample health hazards in semiconductor manufacturing are:

- Process chemicals
- Highly flammable gases
- Toxic gases
- Pyrophoric gases
- Corrosive gases
- Toxic or caustic liquids
- High voltages
- Mechanical hazards
- High temperatures
- Radiation (UV, laser, X-ray, etc.)
- Freezing temperatures

CHEMICAL LABELS

All chemicals must be properly labeled to convey chemical hazards. Before using any chemical, stop and read the hazard label. Make sure you know how to interpret labels (see Figure A.1 on page 602).

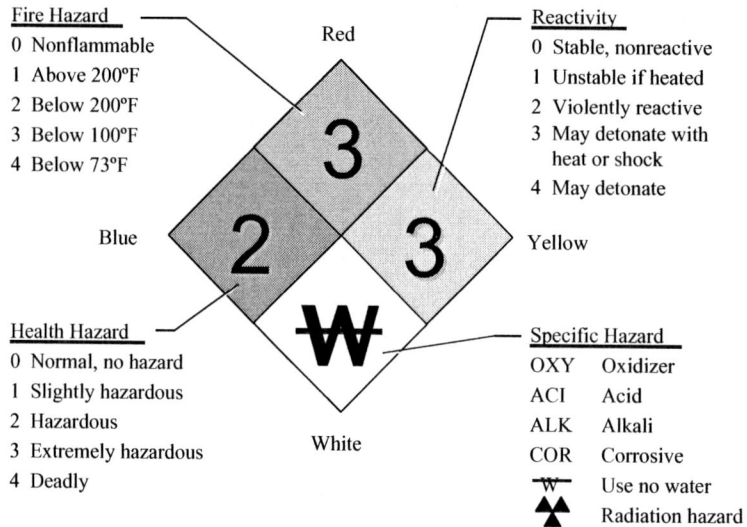

FIGURE A.1 Hazard Information Label

Health Hazard Classifications

The classification numbers for the different health hazards are:

4   Materials that could cause death or major injury with very short exposure, even though prompt medical treatment is given. This includes those materials that are too dangerous to be approached without specialized protective equipment.

3   Materials that could cause serious temporary or permanent injury with very short exposure time, even though prompt medical treatment is given. This includes those materials that require protection from all bodily contact.

2   Materials that cause temporary incapacitation or possible injury after intense or continued exposure unless prompt medical treatment is given. This includes those materials that require use of respiratory protective equipment with independent air supply.

1   Materials that cause irritation or only minor injury on exposure, even if no treatment is given. This includes those materials that require use of an approved canister-type gas mask.

0   Materials that on exposure would offer no hazard beyond that of ordinary combustible material.

EXPOSURE LIMITS

Chemicals have maximum exposure limits defined:

TLV-TWA: Threshold limit values—time-weighted average. This is a time-weighted average of substances for a normal 8-hour workday or 40-hour workweek. Nearly all workers could be repeatedly exposed, day after day, without an adverse affect.

TLV-STEL: Threshold limit values—short-term exposure limit. This is a 15-minute time-weighted average exposure, which should not be exceeded at any time during the workday, even if the 8-hour time-weighted average is within the TLV. Exposures at the STEL should not be longer than 15 minutes, and should not be repeated more than four times per day.

IDLH: Immediately dangerous to life and health. This concentration represents a maximum level from which one could escape within 30 minutes without any escape-impairing symptoms or any irreversible health effects.

PEL: Permissible exposure limit is a legal standard for exposure issued by the Occupational Safety and Health Act (commonly referred to as OSHA). The PEL is often the same value as the TLV.

Chemicals may enter the body through three primary means:

1. Contact with skin or eyes.
   - Wear safety glasses and no contact lenses.
   - Use goggles to protect normal eyewear.
   - Wear the appropriate glove type for the job. Chemicals absorbed through the pores of the skin can enter the body and cause damage to vital organs.
   - Use full face shield when pouring or mixing chemicals.

2. Ingestion (swallowing).
   - Certain toxic chemicals can be fatal when even a minute amount is ingested.
   - Never bring food or drink into areas where chemicals are being used. It is good practice to wash hands with soap and water when leaving the workplace.

3. Inhalation.
   - Breathing toxic gases may result in burns or damage to lung tissue and can pass into the bloodstream, damaging other organs.
   - The workplace must be well-ventilated. If unusual odors are detected, notify someone in charge and leave the area. Sound an alarm if appropriate.

MSDS

Safety information about chemicals is available through the *Material Safety Data Sheet* (*MSDS*). An MSDS must be available for every hazardous chemical at the workplace. Some information given on an MSDS includes:

Chemical name:	This is the trade, common, or Chemical Abstract Service name.
Date prepared:	Date of preparation is listed.
Composition of mixtures:	Including all hazardous materials over 1% and all carcinogens over 0.1%.
PEL & TLV:	Usually given in parts per million (ppm) or in milligrams per cubic meter (mg/m^3).
Health effects:	Identification of organs or systems adversely affected by overexposure.
Physical/Chemical Characteristics:	This includes boiling point, melting point, vapor pressure, specific gravity, and solubility in water.
Fire/Explosion Data:	This includes flash point (lowest temperature a chemical can be ignited with a flame) and autoignition temperature (lowest temperature a chemical will ignite spontaneously in air).
Reactivity Hazard Data:	This describes whether the material is unstable and, if so, under what conditions.
Health Hazard Data:	This gives information as to what limit of exposure will kill 50% of a test animal population (usually specified as LD50 for lethal dose).

MSDS Terminology

Some common terms used in an MSDS and how to interpret these terms are shown in Table A.1 on page 604.

TABLE A.1 Common Terms Used in an MSDS

Terminology	Definition	Precautionary Action
Avoid Contact	General rule for all chemicals, even if they are considered nonhazardous.	Do not breathe vapors and avoid contact with skin, eyes, and clothing for all chemicals.
Carcinogen	Substances that are suspected or known to cause cancer. Some may have threshold limits of exposure. Multiple exposure to suspected carcinogenic materials for even a low dose may be worse than a single massive exposure.	Exercise extreme care when handling. Do not breathe vapors and avoid all contact with skin, eyes, and clothing by wearing suitable protective equipment.
Corrosive	Living tissue as well as equipment is destroyed on contact with these chemicals.	Do not breathe vapors and avoid contact with skin, eyes, and clothing. Use suitable protective equipment.
Skin	A notation on the MSDS for substances that can be absorbed sufficiently through the skin as to cause toxic effects.	Do not allow contact with skin, eyes, or clothing.
Danger	Substances that can have serious harmful effects.	Considered dangerous chemicals. There are serious hazards associated with these chemicals.
Explosive	Substances known to explode under some conditions.	Avoid shock (dropping), friction, sparks, and heat. Isolate from other chemicals that are hazardous when spilled.
Flammable or Combustible	Substances that give off vapors that can readily ignite under usual working conditions.	Keep chemicals away from heat, sparks, flames, and other sources of ignition.
Irritant	Substances that have an irritant effect on skin, eyes, and so on.	Do not breathe vapors and avoid contact with skin and eyes.
Lachrymator	Substances that have an irritant or burning effect on skin, eyes, or respiratory tract. These are dangerous in very small quantities.	Only open in a fume hood. Do not breathe vapors. Avoid contact with skin and eyes. Avoid heating.
Mutagen	Chemical or physical agents that cause genetic alterations.	Handle with extreme care. Do not breathe vapors and avoid contact with skin, eyes, and clothing.
Peroxide Former	Substances that form peroxides or hydroperoxides upon standing or when in contact in air.	Many peroxides are explosive.
Poison	Substances that have very serious and often irreversible effects on the body. These substances are hazardous when breathed, swallowed, or in contact with the skin.	Avoid all contact with the body and use suitable protective equipment.
Stench	Substances that have or generate foul odors.	Open only in a fume hood.
Teratogen	Substances that cause the production of physical defects in a developing fetus or embryo.	Handle with extreme care. Do not breathe vapors and avoid contact with skin, eyes, and clothing. Use protective clothing.
Toxic	Substances that are hazardous to health when breathed, swallowed, or in contact with the skin. There is danger of serious damage to health by short or prolonged exposure.	Avoid all contact with the body. Do not breath vapors, dust, or mist. Use suitable protective equipment.

WET CHEMICAL SAFETY

Some liquid chemicals can be corrosive and hazardous. Technicians and engineers working around chemicals must be trained and follow all safety precautions.

Corrosive Materials

Corrosives can alter or destroy human tissues. Depending on their pH, they are acids (<7) or bases (>7). When working with corrosives:

- Clearly identify all chemicals before use (e.g., HF looks like H_2O). Do not mix incompatible chemicals (see Table A.2).
- Wear eye protection and a face shield at all times.
- Wear body and arm protection, including acid-resistant apron and sleeve guards.
- Wear gloves and boots suitable for the type of chemical.
- Do not breathe vapors. Use only under a fume hood.
- Store and use HF only in plastic containers—HF attacks glass.
- Know the location of eye wash and chemical shower.

TABLE A.2 Incompatible Chemicals

Chemical	DO NOT MIX WITH:
Acetone	Bromine, chlorine, nitric acid, and sulfuric acid
Ammonium Fluoride	Acid solutions
Antimony Trioxide	Metals and reducing agents
Arsine	Oxidizing compounds
Boron Trichloride	Moisture in air or water
Flammable Liquids	Ammonium nitrate, chromic acid, hydrogen peroxide, nitric acid, sodium peroxide, and halogens
Hydrofluoric Acid	Ammonia solutions
Hydrogen Peroxide	Copper, chromium, iron, most metals or their salts, alcohols, acetone, organic materials, aniline, nitromethane, flammable liquids, and combustible materials
Nitric Acid	Acetic acid, aniline, chromic acid, hydrocyanic acid, hydrogen sulfide, flammable liquids, and flammable gases
Oxygen	Flammable gases, liquids, or solids such as acetone, acetylene, grease, hydrogen oils, and phosphorus
Sulfuric Acid	Potassium chlorate, potassium perchlorate, potassium permanganate, and compounds with light metals such as sodium and lithium

Solvents

Many solvents are flammable. Most have harmful vapors and irritate or damage the skin. Never mix solvents with acids—a violent reaction occurs. When working with solvents:

- Wear eye protection (face mask), appropriate gloves, and protective clothing.
- Avoid breathing vapors. Use only under a fume hood or in a well-ventilated area.
- Keep solvents away from heat, sparks, and open flame. Know the nearest fire extinguisher location.
- Do not pour solvents into acid sinks or drains. Pour solvents into waste solvent containers.
- Keep solvents in a flammable materials storage cabinet.
- Dispose solvent-contaminated materials only in waste containers labeled "solvent waste only."
- Do not mix acid waste with solvent waste—could produce dangerous exothermic reaction.

Hydrofluoric Acid (HF)

HF has unique safety attributes because it does not cause pain when contacted by skin. But a human can be burnt by HF from exposure because HF penetrates the skin and flesh and then reacts with calcium in the bone. This can cause severe damage if not quickly treated in a proper manner. It is important that exposure to HF (or suspicion of being exposed to HF) be treated immediately by rinsing the area thoroughly with water and then seeking medical help. There are certain types of creams that can neutralize the HF, but they must be used with care by a trained physician.

Sulfuric Acid (H_2SO_4)

If H_2SO_4 is mixed with water or chemicals containing water (e.g., hydrogen peroxide [H_2O_2]), a large amount of heat is liberated. The general rule is to always add acid to water. Maintaining a piranha cleaning solution (H_2SO_4 mixed with H_2O_2) violates this rule because the H_2O_2 is added to the piranha solution. If this is done, be very careful with exposure due to splattering.

Chemical Hazards

All chemicals must be handled and used with proper safety precautions. In general, the procedures for handling hazardous chemicals are defined in manufacturing specifications, local fire codes, and OSHA (Occupational Safety and Health Agency) requirements. A summary of the hazards of some common wafer fab chemicals is shown in Table A.3.[2]

GAS DETECTION AND MONITORING

Gas leaks can occur for several reasons. One cause of gas leaks is the inadvertent release of gas at the gas cabinet, such as during cylinder replacement. Another source of gas is deposits on the inside of a process chamber that emit gas when exposed to air and moisture during routine maintenance. An additional gas leak concern is the gas delivery system itself, since it contains a large number of valves, fittings, and welds that are always potential leak points due to problems such as vibration, shock, or the degradation of seals over time.

Technicians must be knowledgeable of safety procedures for hazardous gas. Some recommended procedures are:[3]

- Conduct formal safety reviews and inspections
- Implement regular gas safety training programs
- Limit the number of cylinders stored on-site through just-in-time replacement programs

Good gas system design is essential for safety. Important design features for hazardous gas systems are:[4]

- Select components and materials suitable for reactive gases
- Double containment for gas lines, where appropriate
- Good ventilation around piping
- Leak testing prior to use
- Appropriate use of check valves and flow limiting orifices
- Automatic shutoff valves
- Pressure and vacuum-cycle purge on process stations
- Backup power for fire protection and exhaust systems
- Gas detection and alarm system appropriately placed, as defined in the Uniform Fire Code and local ordinances
- Steel gas cabinets with locks and external emergency shut-off valves

TABLE A.3 Commonly Used Fab Chemicals and Their Safety Hazards

Chemical Name	Symbol	Combustible or Explosive	Health hazard class	TLV—TWA (ppm)	TLV—STEL (ppm)	IDHL (ppm)	Process Applications (see note below)
Ammonia	NH_3	X	2	25	35	500	CVD
Argon	Ar		0	—	—	—	A, CVD, CG, Di, E/C, I, P/B, S, TO
Arsine	AsH_3	X	4	0.05	—	6	CVD, CG, Di, Do, I
Boron trichloride	BCl_3		3	1	—	100	Di, Do, E/C, I
Boron trifluoride	BF_3		3	1	—	100	Di, Do, I
Chlorine	Cl_2		3	0.5	1	30	E/C, TO
Carbon dioxide	CO_2		1	5000	30000	50000	P/B
Diborane	B_2H_6		3	0.1	0.3	40	CVD, Di, Do
Dichlorosilane	SiH_2Cl_2		3	5	—	100	CVD
Helium	He		0	—	—	—	A, CVD, CG, E/C, I, P/B
Hydrogen	H_2	X	0	—	—	—	A, CVD, CG, Di, E/C, I, P/B, TO
Hydrogen bromide	HBr		3	3	—	50	E/C
Hydrogen chloride	HCl		3	5	—	100	TO, C
Nitrogen	N_2		0	—	—	—	A, CVD, E/C, I, P/B, TO, Di
Nitrogen trifluoride	NF_3		3	10	15	2000	E/C, TO
Nitrous oxide	N_2O	X	2	50	—	—	E/C, TO
Oxygen	O_2	X	0	None	None	None	CVD, Di, E/C, S, TO
Phosphine	PH_3	X	4	0.3	1	200	CVD, CG, Di, Do, I
Silane	SiH_4	X	4	5	—	—	CVD, Di
Silicon tetrachloride	$SiCl_4$		3	5	—	100	CVD, E/C
Sulfur hexafluoride	SF_6		3	100	1250	—	E/C
Tetrafluoro-methane	CF_4	X	3				E/C
Tungsten hexafluoride	WF_6		3	3	6	—	CVD
Tetraortho-Silicate (TEOS)	$(C_2H_5)_4SiO_4$	X	2	10	—	1000	CVD

Note: Process applications are listed here only for reference and are described in the appropriate chapters.
A: annealing
CVD: chemical vapor deposition
CG: crystal growth
Di: diffusion
Do: doping
E/C: etch/clean
I: ion implant
P/B: purge/blanket
S: sputtering
TO: thermal oxidation

TLV-TWA: Threshold limit values—time-weighted average. Nearly all workers could be repeatedly exposed, day after day, without an adverse affect.
TLV-STEL: Threshold limit values—short-term exposure limit. Exposures at the STEL should not be longer than 15 minutes, and should not be repeated more than four times per day.
IDLH: Immediately dangerous to life and health.

PHOTO LIGHT SOURCE SAFETY

Basic rules for working with the photolithography light sources are:

1. Never look directly into a light source, regardless of the type of source.
2. Always wear eye protection when working around photolithography light sources, especially lasers. There are usually several lasers pointed in different directions.
3. Mercury arc lamps contain mercury. Mercury is a hazardous material and should be treated with care. Dispose of all mercury products according to company safety procedures.
4. Keep high voltage power supplies clean and free of dust and moisture. Do not handle high voltage parts with bare hands, even if power is shut off. Skin oils may create low resistance paths around existing circuit components. Photolithography light sources operate at high power. Be careful of high voltages.
5. Gases used for excimer lasers are hazardous. Follow company safety procedures when working around or handling these gases.
6. Wear appropriate gloves when handling mercury arc lamps and any other optical element. Oils from the skin can damage the glass surface and cause optical aberrations that distort the lithography process and destroy the lamp or optical element. More importantly the oil on the quartz of high-pressure mercury lamps may cause devitrification of the quartz, which may weaken it and cause the lamp to explode.
7. Always remove the product wafers before working on the stepper or scanner. Wafers can be damaged or contaminated and resist can be activated by inadvertent light reflections.
8. UV light causes material deterioration. Inspect all wiring, cables, connectors, and other materials in close proximity to the light source to verify acceptable condition.

ION IMPLANTATION SAFETY

There are many safety precautions designed into the implanter. The source gases used in ion implantation are toxic (see Table A.4). Dopant gas bottles are stored in an isolated gas cabinet with an exhaust system to evacuate any escaped gasses to the scrubber system. When exposure to toxic gases is possible, technicians should always work in pairs and wear a self-contained breathing apparatus (SCBA) for proper breathing protection. All high-voltage sections of the implanter are isolated with interlocked doors (the high voltage and often the entire implanter equipment will not operate if the doors are open). Special grounding sticks are used by maintenance personnel to ensure the high-voltage parts of the implanter are completely discharged before entering the implanter. Small amounts of X-rays are generated in the implanter (just as in televisions that also use high voltages) and lead shielding is used to prevent any radiation from escaping the equipment. The equipment has emergency power off buttons that will shut down the entire machine (electrical, gas, and so on) if activated by the technician.

TABLE A.4* Toxicity of Common Implant Dopant Gases (Set by National Institute of Safety and Health)

Dopant Gas	Exposure Limit (TLVA)
Arsine (AsH_3)	50 ppbv[B]
Phosphine (PH_3)	300 ppbv
Boron trifluoride (BF_3)	1 ppmv

A: TLV is threshold exposure limit

B: ppbv is part per billion by volume

*A. Kulkarni, D. Jukherjee and W. Gill, "Membrane Reprocessing of Hydrofluoric Acid Solutions," *Semiconductor International*, (July 1995).

CHEMICAL RECYCLING

Chemicals are frequently replaced in ULSI production to minimize the problems associated with contamination. This procedure generates large amounts of chemical wastes. Strict environmental regulations make it difficult to dispose of these chemicals. An important option is to recycle chemicals for reuse in the fab.

Hydrofluoric Acid. Hydrofluoric acid (HF) represents about 40% of the total hazardous waste generated by the U.S. semiconductor industry, with recent regulations prohibiting landfill disposal. HF's main use in semiconductor processing is as an etchant of oxide on the silicon surface or as a cleaner for quartz wafer carriers used in furnaces and process chambers. After use, the HF contains particles and metallic and ionic impurities.

A recycling process for HF must be capable of removing the impurities. This can be done through distillation and ion exchange of the HF to remove the impurities. Distillation is a process of boiling the liquid and then condensing the steam on a cool surface. Most contaminants do not vaporize and therefore do not pass to the final HF (referred to as distillate). Ion exchange is a method to remove ionic impurities by using resins that attract the ions.

Another way to purify HF is with membrane technology. The membrane permits HF to pass through the membrane while filtering out the contaminants.

Perfluorocarbon Compounds. Perfluorocarbon (PFC) gases are used during etch processes and for cleaning the inside walls of process chambers after usage. This cleaning process takes place under vacuum and with the introduction of intense RF (radio frequency) energy, which breaks down the gas molecules into highly reactive species that efficiently remove film deposits.

Many PFC gases will leave the reaction chamber unreacted and were previously released into the atmosphere. PFC's are generally nontoxic (except for NF_3), but their long atmospheric lifetime and strong infrared adsorption characteristics make them greenhouse gases. For this reason, there is a strong emphasis on recycling these gases after usage.

REFERENCES

1. P. Singer, "Handling Hazardous Materials: What You Should Know," *Semiconductor International* (December 1996): p. 63.
2. Adapted from E. Zdankiewicz, "Avoid False Alarms with Proper Gas Detection Equipment," *Solid State Technology,* p. 82.
3. Ibid., p. 81.
4. Ibid.

APPENDIX B
CONTAMINATION CONTROLS IN CLEANROOMS

The overall trends in wafer device dimensions and reduction in contamination are shown in Table B.1.[1]

TABLE B.1 Evolution of Chip Feature Sizes and Contamination Control

Year of Mass Production	1980	1984	1987	1990	1993	1995	1997
Wafer Diameter (mm)	75	100	125	150	200	200	200
DRAM Memory Technology (increasing value has more memory)	64K	256K	1M	4M	16M	64M	256M
Chip Size (cm^2)	0.3	0.4	0.5	0.9	1.4	2.0	3.0
Minimum Feature Size on Chip (μm)	2.0	1.5	1.0	0.8	0.5	0.35	0.25
Number of Process Steps	100	150	200	300	400	450	500
Class of Cleanroom (smaller value is cleaner)	1,000-100	100	10	1	0.1	0.1	0.1 and minienvironment
Chemical Impurity (ppb)	1,000	500	100	50	5	1	0.1

HUMAN CONTAMINATION

Normal talking can project undesirable saliva particles up to two to three feet, while sneezing can project saliva and lung particles up to 15 feet. Saliva contains dissolved minerals and salts that can contaminate wafers and cause killer defects. Some typical contaminants in saliva are: sodium (Na), calcium (Ca), iron (Fe), magnesium (Mg), chlorine (Cl), aluminum (Al), sulfur (S), potassium (K), and phosphorus (P).

DEVELOPMENT OF CLEANROOM STANDARDS

The first definition for cleanroom operation was Federal Standard 209, developed in 1963. This standard developed into a definition for a class of cleanroom based on the number of particles in a defined volume above a certain diameter. Table B.2 shows the evolution of this standard and the major criteria for each revision up to the most recent standard in effect, Federal Standard 209E.[2] In this last revision, the title of the standard was adopted as, "Airborne Particulate Cleanliness Classes in Cleanrooms and Clean Zones."

TABLE B.2 Evolution of Federal Standard 209 Specification for Cleanliness of Air

Federal Date	Standard	Highlights of Original and Revised Contents
Dec. 1963	209	• Cleanroom operation principles.
Aug. 1966	209A	Cleanroom design and testing methods: • Defined air cleanliness classifications as class 100, 10,000, and 100,000, specified as the number of particles at sizes larger than 0.5 micron per cubic foot. • Defined air flow pattern of laminar flow and turbulent flow. • Specified air velocity at 90 +/- 20 ft/min. • Specified pressure, temperature, humidity, and vibration. • Specified audio frequency noise and air exchange rate.
Apr. 1973	209B	• Changed air velocity from 90 +/- 20 ft/min to 90 +/- 20% ft/min and changed humidity from 45% to 40 +/- 5%.
May 1977	209B (amend)	• Added cleanliness class 1,000.
Oct. 1987	209C	Major revision of cleanroom classification and testing method: • Added air cleanliness classes 1 and 10. • Extended the particles measurements from 5 micron and 0.5 micron down to 0.3 micron and 0.2 micron for class 100, and down to 0.3, 0.2, and 0.1 micron for class 10 and class 1. • Clearly defined particulate samples locations and numbers of samples and measuring time.
June 1988	209D	• Corrected several typographical errors found in 209C.
Sep. 1992	209E	• Adapted the metric system. • Added descriptor to specify the maximum allowable number of ultrafine particles per cubic meter. • Added sequential airborne particle sampling plan to the single air sampling plan specified in 209D.

Metric Definition of Cleanroom Classes

The metric version of cleanroom classes is shown in Table B.3.[3] To interpret this table, "Class M2 at 0.3 µm" represents air with not more than 309 particles per cubic meter with a particle size of 0.3 µm and larger.

TABLE B.3 Metric Definition of Airborne Particulate Cleanliness Classes Per Federal Standard 209E

	Particles/m^3				
Class	0.1 µm	0.2 µm	0.3 µm	0.5 µm	5 µm
M1	3.50×10^2	7.57×10^1	3.09×10^1	1.00×10^1	
M1.5	1.24×10^3	2.65×10^2	1.06×10^2	3.53×10^1	
M2	3.50×10^3	7.57×10^2	3.09×10^2	1.00×10^2	
M2.5	1.24×10^4	2.65×10^3	1.06×10^3	3.53×10^2	
M3	3.50×10^4	7.57×10^3	3.09×10^3	1.00×10^3	
M3.5		2.65×10^4	1.06×10^4	3.53×10^3	
M4		7.57×10^4	3.09×10^4	1.00×10^4	
M4.5				3.53×10^4	2.47×10^2
M5				1.00×10^5	6.18×10^2
M5.5				3.53×10^5	2.47×10^3
M6				1.00×10^6	6.18×10^3
M6.5				3.53×10^6	2.47×10^4
M7				1.00×10^7	6.18×10^4

Cleanroom Gloves

Fab workers wear gloves as part of the cleanroom garment protocol. Cleanroom gloves are only for normal work (no chemical exposure), and may be latex or vinyl depending on the company policy. Cleanroom gloves and chemical glove types are summarized in Table B.4.[4]

TABLE B.4 Glove Characteristics

Glove	Desirable Characteristics	Undesirable Characteristics
PVC (vinyl) cleanroom glove	• Barrier to skin contaminants • Flexible; inexpensive • Low level of contaminants • Low particle levels	• Excessive sweating • Tears/splits easily • Not acid or solvent resistant
Latex cleanroom glove	• Inexpensive	• Often irritates skin
Orange latex acid glove	• Excellent acid protection • Low particle levels	• Slippery • Too warm • Chemical extractables too high (e.g., chloride)
Green nitrile solvent glove	• Adequate solvent protection for many solvents	• Not resistant to all solvents • Chemical extractables too high (e.g., sulfur)
Silver mulitlayered PVA solvent glove for special solvents (e.g., dimethyl acetamide)	• Excellent solvent protection • Low level of extractable contaminants and particles	• Lacks dexterity • Expensive

SPECIFICATIONS FOR DI WATER

There are two primary specifications for electronic grade DI water, one from the American Society for Testing and Materials (ASTM) and the other from Semiconductor Equipment and Materials International (SEMI). The ASTM standard is ASTM D-19 Standard Guide for Electronic Grade Water D5127-90 (1990), and the SEMI standard is SEMI Suggested Guidelines for Pure Water for Semiconductor Processing (1989).

ELECTROSTATIC DISCHARGE

Materials can have a likelihood of a positive surface charge or a negative charge. Some common materials used in semiconductor manufacturing and their electrostatic charge potential are shown in Figure B.1.[5] The farther apart the materials are in the series, the higher the generated voltage when the materials separate. Note that silicon is electronegative, while materials such as glass, quartz, human skin, and air are strong electropositive.

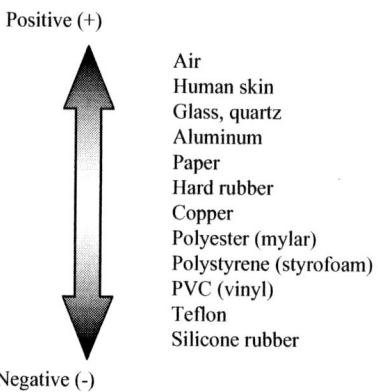

FIGURE B.1 Charge Generation Capability of Common Materials

The semiconductor manufacturing environment is especially prone to developing static electrical charges because it is maintained at a low humidity, typically 40% +/− 5% relative humidity (RH). This condition is conducive to the generation of high levels of static charge (see Table B.5).[6]

TABLE B.5 Electrostatic Voltages at Different Relative Humidity Levels

Means of Static Generation	10% to 20% Relative Humidity	50% to 90% Relative Humidity
Walking across carpet	35,000 V	1,500 V
Walking over vinyl floor	12,000 V	250 V
Worker at bench	6,000 V	100 V
Work chair padded with polyurethane foam	18,000 V	1,500 V

A subtle problem from ESD is the interruption of process equipment due to electromagnetic interference (EMI). An electrostatic discharge that occurs in a few nanoseconds can create peak currents over one amp, causing EMI over a wide band and disrupting electronic circuitry in automated equipment. This can cause the equipment to occasionally lock-up and stop, with the resulting disruption to production.

REFERENCES

1. H. Tseng and R. Jansen, "Cleanroom Technology," *ULSI Technology*, ed. by C. Chang and S. Sze (New York: McGraw-Hill, 1996), p. 2.
2. Ibid., p. 4.
3. Ibid., p. 5.
4. R. Iscoff, "Cleanroom Apparel: A Question of Tradeoffs," *Semiconductor International* (March 1994): p. 66.
5. D. Tolliver, editor, *Handbook of Contamination Control in Microelectronics* (Park Ridge, NJ: Noyes Publications, 1998), p. 157.
6. Ibid., p. 175.

APPENDIX C
UNITS

The *International System* of units (SI) is widely accepted in industry. The SI units are listed in Table C.1. SI units are based on the kilogram (kg), meter (m), second (s), and ampere (A). These, along with a few others, are the base units. All physical quantities may be derived from the base units.

TABLE C.1 International System of Units

Quantity	Unit	Abbreviation	Definition of Units
Length	meter	m	
Mass	kilogram	Kg	
Time	second	s	
Temperature	kelvin	K	
Current	ampere	A	
Frequency	hertz	Hz	1/s
Force	newton	N	Kg-m/s^2
Pressure	pascal	Pa	N/m^2
Energy	joule	J	N-m
Power	watt	W	J/s
Electric charge	coulomb	C	A-s
Potential	volt	V	J/C
Conductance	siemens	S	A/V
Resistance	ohm	Ω	V/A
Capacitance	farad	F	C/V
Magnetic flux	weber	Wb	V-s
Magnetic induction	tesla	T	Wb/m^2
Inductance	henry	H	Wb/A

It may help to have an everyday interpretation of some physical quantities.[1] *Force* is expressed in newtons (N). The scientific definition for a newton is a force of 1 newton accelerates a mass of 1 kilogram to a speed of 1 meter/second in 1 second. To appreciate a newton, realize that to hold a book that weighs 1 pound (454 grams) requires you to exert a force of about 5 N.

Energy is expressed in joules (J). By definition, an energy of 1 joule is expended in moving an object through 1 meter when it is opposed by a force of 1 newton. Formally, $1 J = 1 N m = 1$ kg-m^2/s^2. To illustrate a joule, consider that lifting a 1 pound book up 1 meter requires about 5 J of energy. Another example is each pulse of the human heart requires about 1 J of energy. As a final example, the energy to heat 1 quart (0.946 liter) of water from room temperature to its boiling point is about 18 kJ.

Power is expressed in watts (W). A 1-watt source supplies energy at the rate of 1 joule per second. Formally, $1 W = 1 J s^{-1}$. In this manner, a 100-watt lamp consumes energy at the rate of 100 joules per second. The human body involved in normal activity is rated at about 100 W, with a large amount of power used to drive the brain. Power is often expressed in kilowatts (1kW = 1,000 W) and megawatts (1MW = 10^6 W). It is interesting to note the sun radiates energy at the rate of about 70 MW per square meter of its surface. On earth, there is a mean annual solar energy flux of about 1.4 kW per square meter when measured at the equator.

SI UNIT PREFIXES

The standard prefixes for SI units are provided in Table C.2.

TABLE C.2 SI Prefixes

Prefix	Symbol	Value
femto-	f	10^{-15}
pico-	p	10^{-12}
nano-	n	10^{-9}
micro-	μ	10^{-6}
milli-	m	10^{-3}
centi-	c	10^{-2}
deci-	d	10^{-1}
deka-	da	10
hecto-	h	10^2
kilo-	k	10^3
mega-	M	10^6
giga-	G	10^9
tera-	T	10^{12}

UNIT CONVERSIONS

A *meter* is the basis for metric units of length:

1 Å = 10^{-10} m
1 nm = 10^{-9} m
1 μm = 10^{-6} m
1 mm = 10^{-3} m
1 cm = 10^{-2} m

An *angstrom* is a common unit of measure for thickness in wafer fabrication. An angstrom's relationship to other metric length units is:

$$1 \text{ Å} = 10^{-1} \text{ nm} = 10^{-4} \mu\text{m} = 10^{-8} \text{ cm} = 10^{-10} \text{ m}$$

Table C.3 demonstrates the conversion between some common units and SI units.

TABLE C.3 Conversion Between Common and SI Units

Physical Property	Common Unit	SI Unit
Length	1 in.	2.54 cm
	0.001 in. (1 mil)	25.4 microns (μm)≈25 μm
	0.0039 in. (about 4 mils)	0.1 mm
	39.3 microinches (μ inches)	1 μm
Mass	2.205 pound (lb)	1.000 kg
	1.000 lb	453.6 g
	1 ounce (oz)	28.35 g
Volume	1.000 gallon (gal)	3.785 liter (L)
	1.00 quart (qt)	0.946 L
Energy	1 eV	1.6022×10^{-19} J
	1 kWh	3.600×10^3 kJ

REFERENCES

1. Interpretation of physical quantities from P. Atkins, *The 2nd Law,* (New York: W. H. Freeman, 1994), p. 201.

APPENDIX D
COLOR AS FUNCTION OF OXIDE THICKNESS

A color chart is shown in Table D.1 for thermally grown SiO_2 films when viewed in white light. Once the film color is known, its actual thickness can be estimated from this table. A visual assessment of the oxide color gives a rough estimate of oxide thickness.

TABLE D.1* Color Chart for Thermally Grown Oxide Films

Film Thickness (μm)	Color and Comments	Film Thickness (μm)	Color and Comments
0.05	Tan	0.63	Violet-red
0.07	Brown	0.68	Bluish
0.10	Dark violet to red-violet	0.72	Blue-green to green
0.12	Royal blue	0.77	Yellowish
0.15	Light blue to metallic blue		
0.17	Metallic to very light yellow-green		
0.20	Light gold or yellow	0.80	Orange
0.22	Gold with slight yellow-orange	0.82	Salmon
0.25	Orange to melon	0.85	Dull, light red-violet
0.27	Red-violet	0.86	Violet
		0.87	Blue-violet
		0.89	Blue
0.30	Blue to violet-blue	0.92	Blue-green
0.31	Blue	0.95	Dull yellow-green
0.32	Blue to blue green	0.97	Yellow to yellowish
0.34	Light green	0.99	Orange
0.35	Green to yellow-green		
0.36	Yellow-green		
0.37	Green-yellow		
0.39	Yellow		
0.41	Light orange	1.00	Carnation pink
0.42	Carnation pink	1.02	Violet-red
0.44	Violet-red	1.05	Red-violet
0.46	Red-violet	1.06	Violet
0.47	Violet	1.07	Blue-violet
0.48	Blue-violet		
0.49	Blue		
0.50	Blue-green	1.10	Green
0.52	Green (broad)	1.11	Yellow-green
0.54	Yellow-green	1.12	Green
0.56	Green-yellow	1.18	Violet
0.57	Yellow to yellowish	1.19	Red-violet
0.58	Light orange or yellow		
0.6	Carnation pink		

*S. Ghandhi, *VLSI Fabrication Principles: Silicon and Gallium Arsenide*, 2d ed., (New York: Wiley, 1994), p. 502.

APPENDIX E

OVERVIEW OF PHOTORESIST CHEMISTRY

ORGANIC MATERIALS

Many photoresists are carbon based organic molecules. This section will introduce the basic chemistry of photoresist to students who may not be familiar with organic chemistry.[1]

Carbon, like silicon, has four electrons in its valence shell, requiring four additional electrons to complete its valence shell. Carbon readily bonds with other carbon atoms, hydrogen, and elements to the right of it on the periodic table to form complex chains of long repetitive molecules. In organic chemistry, an aromatic ring consists of six carbon atoms arranged in a planar hexagonal structure. The simplest aromatic ring compound is benzene, with a single hydrogen atom and attached to each carbon atom. A diagram of the simple benzene aromatic ring with its chemical symbol is shown in Figure E.1. In organic chemistry, the hydrogen atoms are left off the chemical symbol for simplicity.

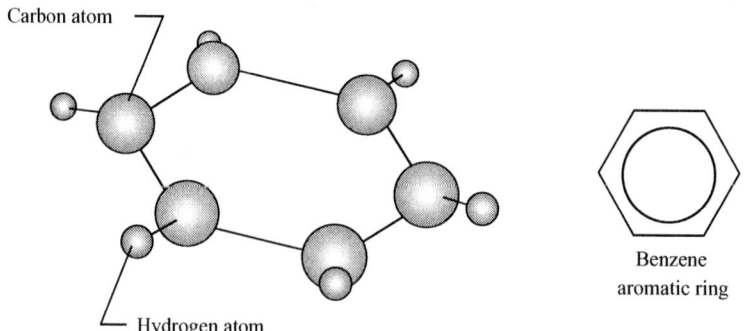

FIGURE E.1 Diagram and Symbol of Simple Benzene Aromatic Ring

Every carbon atom in the aromatic ring is covalently bonded to the adjacent carbon atoms, which acquires two electrons. Furthermore, a carbon atom gains one electron by bonding to the hydrogen. There remains one unpaired electron in each carbon's valence shell. The unpaired electrons from the six carbon atoms form a delocalized bond that takes the shape of a donut around the benzene molecule. This highly mobile delocalized bond is referred to as pi-electron and it gives the aromatic rings their unique properties.

A variety of organic compounds are formed by making simple alterations to the benzene ring. Tuolene, a common solvent, is formed by replacing one of the hydrogen atoms by a methyl (CH_3) group (see Figure E.2). An organic acid called carboxylic acid is formed by adding a carboxyl group (COOH). Aromatic rings can also attach to one another, such as napthalene.

Polymers are large molecules formed by linking many smaller repeating units called monomers. A polymer contains at least five monomers and can contain thousands of monomers. Examples of polymers are plastics, rubbers, and resins—often these are carbon based. The simplest polymer is polyethylene (see Figure E.3), consisting of a long chain of carbon atoms with two hydrogen atoms bonded to each carbon. A polymer can have branches and can crosslink, which is bonding with itself or other polymers. Crosslinking increases the polymer strength and reduces the ability of the polymer to dissolve in typical solvents, which is important for resists. At the same time, if polymers are broken into shorter chains, then the molecules dissolve more readily in solvents.

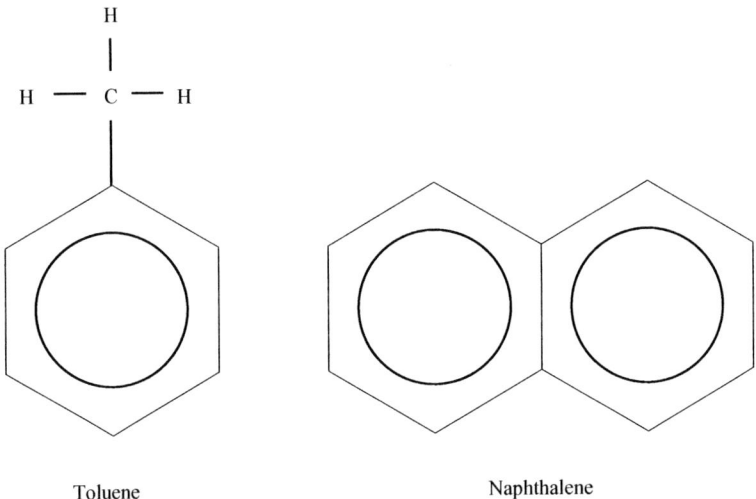

FIGURE E.2 Aromatic Based Compounds

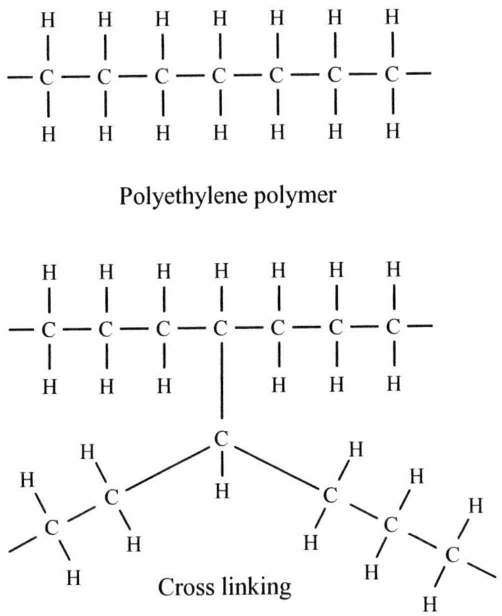

FIGURE E.3 Polyethylene Polymer and Crosslinking

DNQ RESIST

DNQ resist for i-line lithography is composed of the photoactive compound (PAC) diazonaptho-quinone (DNQ) and novolak (N) matrix resin material.[2] The chemical structure for the DNQ PAC is shown in Figure E.4 on page 620. Usually the part of the DNQ molecule below SO_2 is specific to each resist manufacturer and is represented as a generic R. Also note that the right-hand ring is not an aromatic but instead has a double bond. Novolak is a polymer whose monomer is an aromatic ring with two methyl groups and an OH group (see Figure E.5 on page 620). Solvents are added to the resist to control its viscosity for forming a film while spin coating. Most solvents for positive DNQ resists are combinations of aromatic compounds such as xylene.

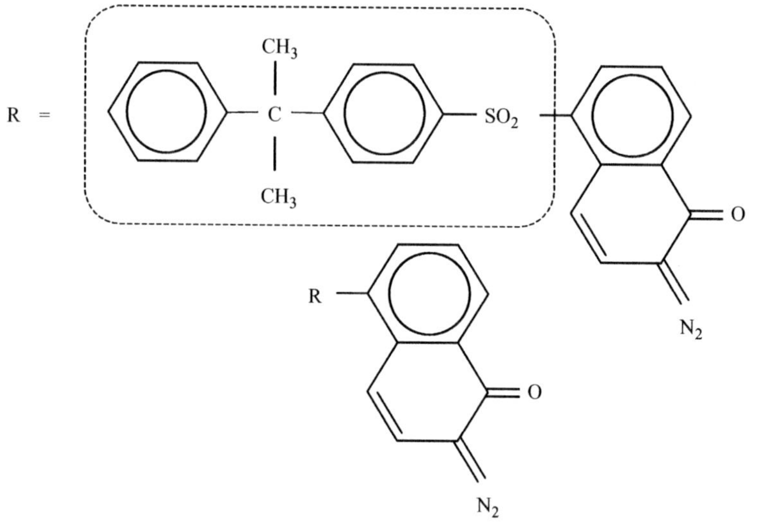

FIGURE E.4 Common Photoactive Compound of Diazonaphthoquinone (DNQ)

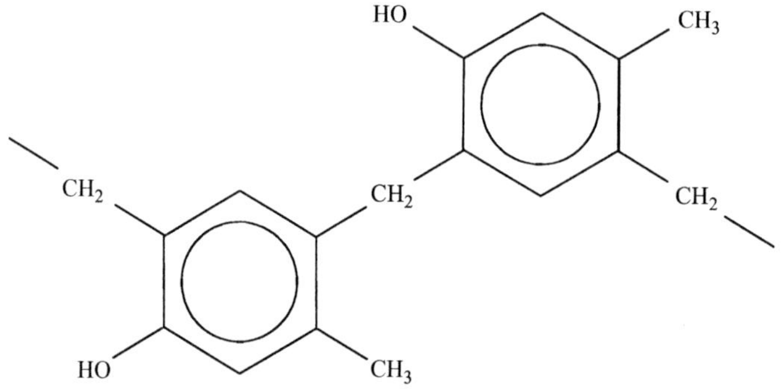

FIGURE E.5 Basic Aromatic Ring for Novolak Polymer

DNQ resist is transformed into a material readily dissolved in developer through a multistep process (see Figure E.6). First, the exposure to UV light will free the nitrogen molecule from the carbon ring, creating a highly unstable carbon site. This structure is stabilized by moving one carbon outside the ring and covalently bonding the oxygen atom to this external carbon atom. This is referred to as the Wolff rearrangement and results in ketene. In the presence of water, a new structure occurs where the double bond to the external carbon atom is replaced with a single bond and an OH group. This final product is carboxylic acid, which readily reacts with and dissolves in base developer solutions. Novolak resin is already water soluble and dissolves easily. Typical developer solutions are KOH or NaOH diluted with water.

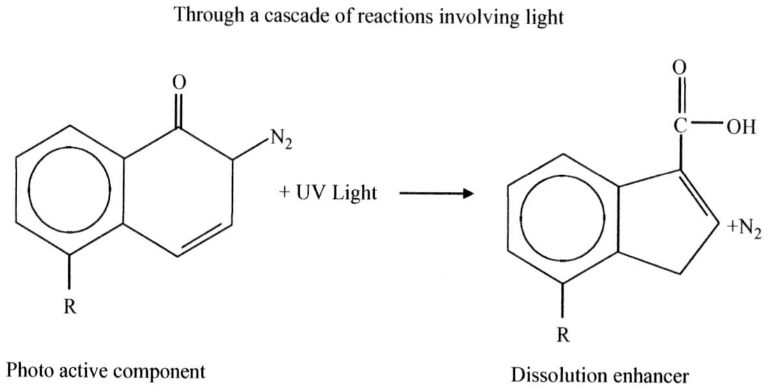

Through a cascade of reactions involving light

Photo active component   Dissolution enhancer

FIGURE E.6 Reactions of DNQ After Exposure to UV

CHEMICALLY AMPLIFIED RESIST

Chemically amplified (CA) resists have high sensitivity for deep UV lithography due to the unique catalytic reaction occurring in the resist. The resists are formulated with a t-BOC resin and an onium salt as a photoacid generator (PAG).[3] The t-BOC is the protecting group that renders insoluble the resist resin known as PHS, or poly(4-hydroxystyrene). The PAG produces a strong acid when exposed to deep UV light (see Figure E.7). The photo-generated acid is used to remove the t-BOC protecting group during a post-exposure bake (referred to as an acid-catalyzed t-BOC deprotection reaction, and shown in the dotted line). Once deprotected, the PHS resin becomes soluble in developer.

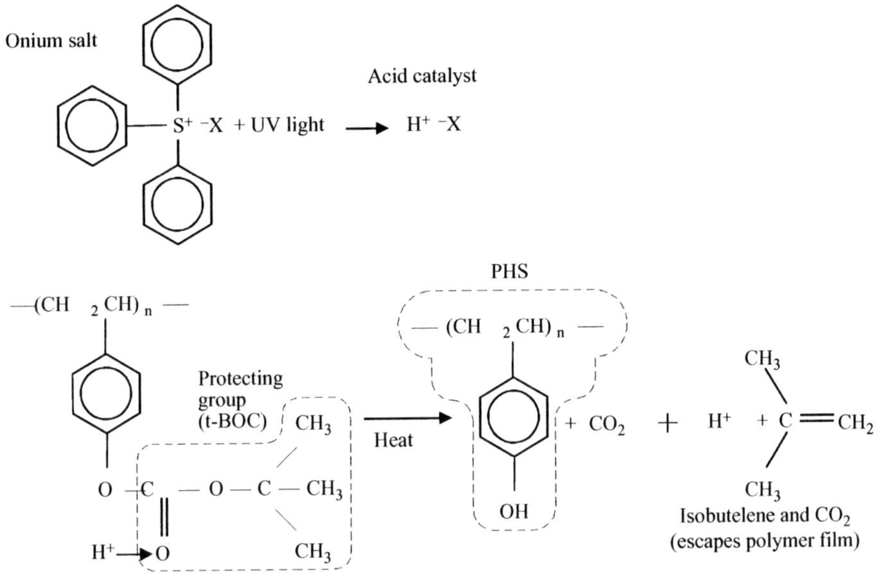

FIGURE E.7 Chemical Amplification of Photoresist with t-BOC Deprotection Reaction
Redrawn from C. G. Willson, "Organic Resist Materials," *Introduction to Microlithography*, ed. L. Thompson, C. G. Willson, and M. Bowden, 2d ed., (Washington, DC: American Chemical Society, 1994), p. 219.

REFERENCES

1. Review of basic organic chemistry from S. Campbell, *The Science and Engineering of Microelectronic Fabrication*, (New York: Oxford University Press, 1996), pp. 182–84.

2. Information for DNQ resist summarized from S. Campbell, *The Science and Engineering of Microelectronic Fabrication*, pp. 184–85.

3. CA resist summarized from T. Ueno, "Chemistry of Photoresist Materials," *Microlithography, Science and Technology*, ed. J. Sheats and B. Smith (New York: Marcel Dekker, 1998), p. 465.

APPENDIX F
ETCH CHEMISTRY

PLASMA ETCH GAS TRENDS

Table F.1 lists the typical gas chemistries used for plasma etching. Dielectric and silicon materials are usually etched with fluorine-based gas etchants such as CF_4. Aluminum etching typically occurs with chlorine-based gas chemistries such as BCl_3.

TABLE F.1* Etch Chemistries of Different Etch Processes

Etched Material	Conventional Chemistry	New Chemistry	Benefits
PolySi (polysilicon)	Cl_2 or BCl_3/CCl_4 Cl_2 or BCl_3/CF_4 Cl_2 or $BCl_3/CHCl_3$ Cl_2 or BCl_3/CHF_3 (the bottom three are sidewall passivating gases)	$SiCl_4/Cl_2$ BCl_3/Cl_2 $HBr/Cl_2/O_2$ Br_2/SF_6 SF_6 CF_4	No carbon contamination. Selectivity to SiO_2 & resist. No carbon contamination. Higher etch rate.
Al	Cl_2 BCl_3 + sidewall passivating gases $SiCl_4$	$SiCl_4/Cl_2$ BCl_3/Cl_2 HBr/Cl_2	Improved profile control. No carbon contamination.
Al with 1% Si and 0.5% Cu	Same as Al	$BCl_3/Cl_2 + N_2$	N_2 accelerates Cu etch rate.
Al with 2% Cu	$BCl_3/Cl_2/CCl_4$	SF_6 only	
WSi_2, $TiSi_2$, $CoSi_2$	CCl_2F_2	CCl_2F_2/NF_3 CF_4/Cl_2	Controlled etch profile. No carbon contamination.
Single-crystal Si	Cl_2 or BCl_3 + sidewall passivating gases	CF_3Br HBr/NF_3	Higher selectivity for trench etch.
SiO_2 (BPSG)	CCl_2F2 CF_4 C_2F_6 C_3F_8	CCl_2F_2 CHF_3/CF_6 CHF_3/O_2 CH_3CHF_2	Environmentally improved alternatives.
Si_3N_4	CCl_2F_2 CHF_3	CF_4/O_2 CF_4/H_2 CHF_3 $CH3CHF2$	Environmentally improved alternatives.

*Y. Lii, "Etching," *ULSI Technology*, ed. C. Chang and S. Sze (New York: McGraw-Hill, 1996), p. 354.

WET ETCH

Wet etch is not widely used in submicron fabrication. The following information explains how silicon and aluminum are wet etched.

Silicon

Single-crystal and polycrystalline silicon can be wet etched in a solution of hydrofluoric acid and nitric acid (HF/HNO_3) buffered with acetic acid or DI water. The buffer is used to reduce the etch rate. The etch process occurs in two steps. First, the nitric acid dissociates to produce nitrogen dioxide (NO_2), which attacks and oxidizes silicon, making it into SiO_2. Second, the HF reacts with and dissolves the oxide. Wet etching of silicon (single-crystal and polysilicon) is generally not done for submicron IC fabrication because superior results are obtained with dry plasma etching.

Aluminum

Aluminum can be etched in a mixture of phosphoric acid (H_3PO_4), nitric acid (HNO_3), acetic acid (CH_3COOH), and DI water. The etch mechanism is similar to that for silicon—the material is first oxidized (by the HNO_3) and then the oxide is dissolved by the H_3PO_4 and water. Agitation is required for aluminum wet etching to dislodge from the wafer surface small hydrogen (H_2) bubbles produced during the etch process. Aluminum etching in submicron processing is always done with plasma dry etching because of superior results.

GLOSSARY

1st level packaging The back-end processes of the wafer fab, commonly referred to as chip assembly and packaging.

2nd level packaging The assembly of the packaged IC component into the next level of assembly, such as a PC board.

aberrations For optics, deviations from the ideal lens behavior that result from design, fabrication, or usage flaws.

absorption Taking up of a liquid or gas into the bulk of another material.

acceleration column The subsystem in an ion implanter where positive ions are accelerated in an electric field to gain energy.

acceptors The trivalent dopants (Group IIIA) with three valence shell electrons that make p-type semiconductors, the most common being boron.

acid In common terms, a corrosive substance that produces H^+ ions when dissolved in water and has a pH of <7.

acoustic streaming The steady flow of liquid induced in the megasonic tank from the ultrasonic energy.

active region An area of the substrate that has been doped p-type or n-type and is part of an electrode in a semiconductor device.

activate For semiconductor technology, bringing a dopant atom from an interstitial site to a regular crystal site by exposure to high temperature.

activated dopants Dopants that are part of the silicon lattice structure and can act as a donor or acceptor of electrons.

active components Semiconductor devices that can control the direction of current flow, such as integrated circuits and transistors.

adsorption The chemical binding that occurs during chemical vapor deposition, causing the gaseous atoms and molecules to condense and chemically attach to the solid wafer surface.

aerosols Airborne particles suspended in the air.

air ionizers Special ion emitters located on the ceiling in the cleanroom that ionize the air molecules and make them conductive by gaining or losing an electron.

alignment The arranging of a reticle or photomask and wafer in correct positions with respect to each other during photolithography.

alignment marks Visible patterns placed on the reticle and the wafer to determine their position and orientation during the alignment process.

alloy For semiconductor fabrication, the alloy process step refers to a heat treatment to improve the metallurgical bond between the silicon substrate and the contact metal. For metallurgy, alloy means to blend together two or more metals to form a compound.

aluminum (Al) The metal traditionally used in semiconductor technology to form the interconnects between devices on a chip. It is usually deposited by PVD.

aluminum subtractive process The traditional metallization process where aluminum is deposited as a blanket film and then etched away (subtracted) to form the circuitry.

ambient A term describing the atmosphere (e.g., pressure and temperature) surrounding the wafers during processing.

ammonia (NH_3) A toxic combustible gas used in CVD process to create a silicon nitride layer.

ammonium fluoride (NH_4F) A salt used as a buffer in a buffered oxide etch solution.

ammonium hydroxide (NH_4OH) The base formed by dissolving ammonia in water.

amorphous Solid materials that lack a repetitive structure and demonstrate structural disorder at the atomic level.

analog The operating signals of an electronic circuit that take on a continuous range of values instead of discrete values.

angstrom (Å) Unit of measure that equals one-ten billionth (1/10,000,000,000) of a meter or 1/10,000 of a micron and often used for thin film thickness.

anion An ion that is negatively charged by the gain of one or more electrons.

anisotropic etch profile An etch sidewall profile for submicron feature sizes where the rate of etching is only in one direction perpendicular to the wafer surface.

anneal A high-temperature operation to achieve specific goals, such as relieving stress in silicon, activating ion-implanted dopants, and reducing structural defects.

antimony (Sb) Crystalline metal used for n-type doping during ion implantation to form a buried layer in a bipolar device.

antireflective coating (ARC) The reduction of standing wave effects in photoresist by applying an antireflective layer that suppresses unintended light reflection.

APCVD *See* atmospheric pressure CVD.

application specific IC (ASIC) Fully custom-designed and manufactured integrated circuits to meet an individual customer's needs.

aqueous solution Any solution dissolved in water, meaning water is the solvent.

area array Solder bumps are placed in an X-Y grid across the entire chip surface, efficiently using chip surface area for I/Os.

argon (Ar) A chemically inert, colorless, nontoxic gas used for process chamber purging, annealing and sputtering applications. Argon is monatomic (its molecules have only one atom) and is noted for its chemical inactivity.

arsenic (As) A semimetallic solid used for n-type doping in ion implantation.

arsine (AsH₃) A source gas for arsenic during ion implantation and is highly toxic and flammable.

ashing Removal of photoresist by a high-temperature oxidation process.

aspect ratio The ratio of the depth to width for a small gap, trench, or hole.

aspect ratio dependent etching (ARDE) A problem in etch uniformity that occurs because etching rates and profiles depend on feature size and pattern density.

asphyxiant A gas that by itself is not toxic to the human body, but when in an excess supply in the air, can deprive the person of the necessary oxygen for breathing.

assay number The percentage of a particular chemical in a container.

atmospheric pressure The pressure exerted at sea level by the earth's atmosphere, equivalent to 760 mm of mercury or 14.7 pounds per square inch.

atmospheric pressure CVD (APCVD) A chemical vapor deposition reactor that operates at atmospheric pressure with a simple design and high deposition rates.

atomic force microscopy (AFM) A surface profiler that scans a small, counterbalanced tip probe over the wafer to create a 3-D surface map.

atomic number A number assigned to each element that identifies the number of protons in the nucleus of an atom.

auger electron spectroscopy (AES) The identification of a material by measuring the auger electron energy emitted by the sample surface when struck by an incident electron beam.

autodoping During epitaxial growth, dopant impurities evaporate from the wafer and move into the gas stream to cause unintentional doping of the epilayer.

automatic defect classification (ADC) Software to identify and classify semiconductor manufacturing defects based on software recognition patterns.

back-end of line (BEOL) The process steps from contact metal through completion of the wafer prior to electrical test.

backgrind An operation using an abrasive on the backside of a wafer to achieve the correct wafer thickness for scribing and packaging.

backing film A compliant material sandwiched between the wafer and chuck of a CMP tool to provide a compliant support that compensates for the wafer's nonflatness.

ball grid array (BGA) A high-density IC surface-mount package with an area array of solder balls for interconnection to a substrate.

ballroom layout An approach to cleanroom layout with one large fabrication room that contains all production equipment.

barrel reactor An older batch-type cylindrical reactor design with almost pure chemical etching and isotropic sidewalls.

barrier metal A thin layer of deposited metal or metals used to make ohmic contact to silicon and isolate the silicon substrate from the primary metal used for interconnecting the circuit elements.

barrier voltage The net effect of the difference in charges across the depletion region is to create a potential difference across the junction, which results in a barrier voltage that has to be overcome before the diode can be operated.

base One of the three electrodes of the bipolar junction transistor (BJT)—usually serves as the input to the BJT.

batch A number of wafers processed as a group in a manufacturing process as opposed to single-wafer processing.

bay and chase layout A cleanroom layout in which a common corridor separates the production area (bay) from the service area (referred to as service chase or gray area).

beam blow-up In ion implantation, this describes how the diameter of the moving ion beam increases due to like charges (+) repelling each other.

beam current The flux of positive dopant ions in an ion implanter that is capable of being measured in terms of milliamperes.

beam deceleration An ion implant method to achieve a high beam current in a low-energy beam. Also referred to as decel.

beam energy The energy acquired when the ions in an ion implanter are accelerated to a high velocity.

BEOL *See* back-end of line.

best focus With regard to photolithography, refers to the location along the plane of an optical system where the best imaging can be attained.

BGA *See* ball grid array.

biasing When referring to semiconductor devices, it is the term used to describe the power supply and how it is connected to the device for different operations.

BiCMOS A device having both bipolar and CMOS transistors.

bincode number A procedure at wafer electrical test that assigns bincode numbers to categorize wafers according to test results.

bin map An electrical test wafer map that highlights the bincode number by location on a wafer.

bipolar junction transistor (BJT) A solid state device made of single-crystal semiconductor having three electrodes and makes use of holes and electrons during its normal operation.

bipolar technology Discrete and integrated devices based on the bipolar junction transistor known for fast speeds, durability, and power controlling abilities, as well as high power consumption.

bird's beak effect The structural effect caused by the growth of SiO_2 under the SiN_3 mask during the LOCOS process.

blanket deposition A film deposited over the entire surface of the wafer whether masked or not. Most films are deposited in this manner.

blower A widely used mechanical vacuum pump that provides for high throughput of gas and uses no fluids to pump. Also called a booster pump.

boat A wafer holder made from quartz or polycrystalline silicon for use in furnace operations during semiconductor fabrication. Also made of Teflon for transporting wafers between processing locations.

BOE *See* buffered oxide etch.

bonding pads Relatively large metallized pads traditionally placed around the perimeter of the IC die to provide the areas where bonding wires are connected to terminate the leads of the package.

bonding wire For traditional assembly, fine diameter wires, usually aluminum or gold, used to connect the metal bonding pads on an IC to the internal terminations of the package.

boron (B) A nonmetal element used for p-type doping in ion implantation.

boron trichloride (BCl₃) Toxic, nonflammable, corrosive gas used to etch metals and silicides and also used as a source of boron for p-type doping.

boron trifluoride (BF₃) Toxic gas used as a source of boron for p-type doping in ion implantation.

borophosphosilicate glass (BPSG) Boron and phosphorous incorporated into silicon dioxide deposition for good step coverage and gettering of mobile ionic contaminants.

borosilicate glass (BSG) Boron added to silicon dioxide for improved step coverage, but it provides poor gettering for mobile ionic contaminants.

bottom antireflective coating (BARC) An organic or inorganic dielectric material that is applied to the wafer before the photoresist is applied to suppress unintended light reflection.

boule A term that describes the cylindrical, single-crystal ingot after the crystal growing process. Also called ingot.

breakthrough step Often the initial dry etching step that removes the native oxide and surface contaminants to achieve a uniform etch.

brightfield detection In optical inspection or photolithography tools it refers to the illumination and detection technique, which has the incident light and reflected light in parallel to each other. The illuminated object appears on a bright background.

brush scrubbing An effective method using a rotating scrub brush for removing particles from the wafer surface, especially following chemical mechanical planarization.

bubbler A container holding a liquid through which some type of inert carrier gas is passed to carry some partial pressure of the liquid into a process tube or reaction chamber.

buffered oxide etch (BOE) A wet etching solution of hydrofluoric acid (HF) that is buffered with ammonium fluoride (NH₄F) for removal of an oxide layer without removing silicon.

bulk chemical distribution A system consisting of a chemical source, such as a storage vessel, a chemical delivery module, and a piping system.

bulk gases The group of gases typically consisting of oxygen (O₂), nitrogen (N₂), hydrogen (H₂), helium (He), and argon (Ar) and stored on site in large quantities for delivery to the fab area.

bulkhead equipment layout A cleanroom layout where equipment is located behind the process bay wall with access provided to the operator for controls and wafer load/unload.

bumped chip A chip from a wafer that has been specifically processed with solder or gold "bumps" to provide reflow for 2nd level attachment in lieu of wire bonding.

buried layer A heavily-doped layer between two lightly-doped layers.

burn-box A special chamber associated with CVD equipment and furnaces for the combustion of hazardous gases before removal through the main exhaust duct.

burn-in A stress test that applies voltage to a semiconductor device at an elevated temperature for a given period of time to cause devices with marginal reliability to fail, thus reducing potential failures by customers.

CA *See* chemical amplification.

cap oxide A thin oxide layer grown on the wafer surface to prevent dopant atoms from diffusing out of the silicon.

capacitance Capability of storing electrical charge. Unit of measure is the Farad (F).

capacitance-voltage test (C-V test) A test which provides information on the quality of the gate oxide structure, primarily to ensure the oxide is contamination-free, or used to characterize the dopant profile of a pn junction.

capacitive coupled plasma Refers to the way that RF is fed into a plasma process chamber. An external capacitor is used to make an electrical connection to one of two electrodes of a plasma chamber.

capacitor A discrete device with two conductors separated by a dielectric that stores electrical charge.

carbon tetrafluoride (CF₄) Colorless, odorless, nonflammable gas that is inert at room temperature. It dissociates in an RF field to form reactive fluoride ions that etch tungsten and tungsten silicide films. Also referred to as tetrafluoromethane.

caro's acid A cleaning solution prepared by mixing 380 parts of concentrated H₂SO₄ with 17 parts of 30% H₂O₂ and one part of UHP water.

carrier A charged particle such as an electron or a hole that provides the electrical current flow in semiconductors.

carrier-depletion region Used to describe the recombination area, because it has been depleted of electrons.

carrier gas An inert gas used to transport other elements to a process chamber or tube.

cassette A plastic wafer carrier used in the wafer fab to transport and store wafers.

cation An ion that has a positive charge.

caustic A substance capable of eroding or destroying by chemical action.

cavitation The energy in ultrasonic waves that causes the liquid to separate into tiny cavities that immediately collapse.

CD *See* critical dimension (CD).

CD-SEM A SEM specifically used to measure critical dimensions of structures on a wafer.

Celsius The most common temperature scale in scientific work where 100 degrees separate the freezing and boiling points of water.

center of focus (COF) In photolithography, the point from the center of the lens to the wafer being patterned where the best imaging occurs.

center slow For CMP, rotational forces give the edge of the wafer a higher polishing rate than the center, making the center of the wafer polish slower than the edge.

central processing units (CPU) Also known as microprocessor units, they are complex logic ICs capable of executing instructions programmed into a separate or internal ROM.

ceramic packaging An IC package that uses a ceramic substrate for high-performance or high-power applications.

ceramic substrate An IC packaging substrate using a high-temperature ceramic material that is inorganic, nonmetallic, and crystalline.

CERDIP Ceramic DIP package with two ceramic layers pressed together (after chip wirebonding) with a leadframe positioned between them.

channel The lightly-doped region between the source and drain of an MOS transistor that becomes conductive when the gate, source, and drain electrodes are properly biased.

channel length The length of the lightly-doped region separating the source and drain in an MOS transistor.

channeling Occurs when implanted ions are not slowed by collisions with silicon atoms and instead pass through the interstitial areas of the crystal.

charge carrier *See* carrier.

chase An area in a wafer fab between bays where the bulk of the process equipment and their supporting supplies and facilities are located. Also referred to as the "gray area."

chelating agent An agent that binds and removes metallic ions.

chemical amplification (CA) A means to increase the sensitivity of the deep UV resist substantially over that of the conventional i-line resists by reacting at an accelerated rate to the deep UV light source based on an acid-catalyzed reaction.

chemical etch mechanism An etching process that chemically reacts with the materials on the wafer surface to produce an isotropic etch profile.

chemical mechanical planarization (CMP) A manufacturing process that simultaneously uses both chemical reactions and mechanical forces to remove material and planarize a wafer surface. Also referred to as polish.

chemical properties Properties such as flammability and corrosiveness, which result from an interaction or transformation with another substance.

chemical reaction A reaction that occurs when a substance is converted into a different material with its unique composition and properties.

chemical solution A chemical mixture mixed so well that the molecules or ions of the solution are the same throughout.

chemical vapor deposition (CVD) A process of depositing a thin, solid film on a wafer surface through a chemical reaction of a gas mixture.

chip An integrated circuit (IC) or discrete device. Also referred to as die.

chip on board (COB) A process to mount IC chips by wirebond directly onto a substrate with other surface mount components. *See also* direct chip attach (DCA).

chip scale package (CSP) An IC package that has a footprint (surface area) that is <1.2 times the footprint of the die. Many advanced packaging methods are CSP.

circuit geometries Is used to refer to the relative sizes and shapes of structures on a die.

class number A number that designates the air quality inside a cleanroom by defining the particle size and density in the cleanroom air.

cleanroom An environment where humidity, temperature, and contamination are precisely controlled and suitable for wafer fabrication.

cleanroom protocol A strict operating procedure that all people, materials, and equipment must follow to minimize contamination in the cleanroom.

clearfield mask A photomask (or reticle) in lithography that has an opaque pattern in a mostly transparent mask.

cluster tool Fabrication equipment having integrated process chambers located around a central wafer handling unit for single-wafer processing.

CMOS Complementary metal oxide semiconductor. A fabrication process that incorporates p-channel and n-channel CMOS transistors within the same silicon substrate.

CMP *See* chemical mechanical planarization (CMP).

coater/developer track A cluster tool that integrates many of the operations in photolithography.

cobalt silicide The bonding of cobalt with silicon to form a low-resistance compound of cobalt silicide when raised to a temperature of >550°C.

coefficient of thermal expansion (CTE) The amount a material expands due to heating.

coherence probe microscope The primary method used to measure registration of overlaying targets on a wafer following alignment and exposure.

coherent light Light that has waves in phase and of one wavelength.

cold wall When a furnace or reactor transfers energy between a heat source and a wafer while not heating the surrounding reactor wall.

collector Along with the emitter and base, one of the three electrodes of the bipolar junction transistor that collects the majority current carriers and then passes them through a conductor to complete the electrical circuit.

collimated light Light in which the rays are parallel.

collimated sputtering A method of achieving improved directional enhancement during sputtering for increased coverage on the bottom and sides of a contact or via.

compound semiconductors Used as alternative semiconductors, compound semiconductors are formed from Group IIIA and Group VA of the periodic table.

concentration The amount of dopant in a specified area when compared to adjacent regions. Measured in percentage, parts per million, and sometimes referred to with symbols, e.g., p^-, n^-, p^+, n^+, n^{++}, p^{++}.

condensation The process of a gas changing into a liquid.

conductor A substance through which electricity can readily flow.

confocal microscope An automated microscope that scans a viewing area on a wafer in the z-axis, then reconstructs the scanned images to create 3-D renderings with improved contrast and resolution.

conformal step coverage Uniform material thickness in all regions of a wafer surface step, including the sidewalls and corners.

contact An electrical connection at the silicon surface between the devices in the silicon wafer and the first metal layer.

contact alignment The primary method of photolithography imaging during the SSI era until the early 1970s; it is only suitable for large feature sizes (>5 μm).

contact angle meters Used to measure adhesion of liquids to the wafer surface and to calculate surface energies or adhesion tension.

contamination Any undesirable substance introduced to a semiconductor wafer that affects the production or electrical performance of the microchip.

continuous spray develop Dissolving exposed resist by spraying a liquid developer on a rotating wafer, usually after completion of a post-exposure bake.

contour maps Graphic plots of measured data displayed on a wafer image on a CRT depicting deviations in ohms/square or thickness above and below the nominal value.

contrast The sharpness of the transition from exposure to non-exposure in photoresists.

conventional i-line photoresists Those resists used in optical lithography that are suitable for i-line UV wavelengths (365 nm) to pattern layer CDs down to 0.35-μm.

Cook's theory The theory that explains the basic mechanism for oxide CMP.

copper CVD For copper metallization, used to deposit a thin copper seed layer necessary for bulk copper fill with electroplating.

copper interconnect For copper metallization, copper interconnect wiring will replace aluminum metallization to achieve significant benefits in chip performance.

cost of ownership (COO) The total cost of owning and operating equipment for production, including up-front capital expenditures plus the hidden costs from areas such as consumables, scheduled downtime, equipment performance, and maintenance.

covalent bond A chemical bond between atoms in which electrons are shared.

critical dimension The width and space of critical (smallest) circuit elements in an IC. Also referred to as CD, minimum geometry, or feature size.

cryogenic aerosol cleaning The cooling of a gas (argon) to form solid ice particles that are injected onto a wafer surface to knock off particulate contamination.

cryogenic pump (cryopump) A capture pump that removes gases from a process chamber by freezing and capturing them, in effect reducing the pressure and improving the vacuum in the chamber.

crystal A solid material that has an ordered, repeatable three-dimensional pattern over a long range of many atoms.

crystal activation For ion implantation, the movement of dopants into the regular crystal sites by thermal energy.

crystal defect Any interruption in the repetitive nature of the unit cell crystal structure. Also called microdefect.

crystal growth The process of converting the polysilicon chunks of semiconductor grade silicon into a large monocrystal of silicon.

crystal lattice The repeatable order within a crystal at the atomic level of the internal structure (atomic order).

crystal orientation The relationship of the wafer surface to the crystal facets at which the crystal is sliced.

CTE *See* coefficient of thermal expansion.

current-driven current amplifier The principle of operation for the bipolar junction transistor where the input (base) current has control over the output (collector) current.

CVD *See* chemical vapor deposition.

cycle time The amount of time for a product to complete a single process operation or a series of operations.

CZ crystal puller Equipment used to permit thin films of silicon to accurately replicate the seed crystal structure and grow into a large silicon ingot.

Czochralski (CZ) method The most common technique today for growing a silicon monocrystal ingot, the Czochralski method transforms molten silicon liquid into a solid silicon ingot that has the correct crystal orientation and is doped either as n-type or p-type.

damascene An IC process that embeds a metal conductor pattern in an etched dielectric film on the silicon substrate, resulting in a planar interconnection layer.

darkfield detection In optical inspection or photolithography tools, it refers to the illumination and detection technique, which has the incident light and reflected at angles to each other, and the illuminated object appears on a dark field.

darkfield mask A photomask with a clear pattern within an opaque field.

DC bias A DC volt that develops across a plasma process chamber when an RF voltage is applied to the chamber's electrodes.

deep UV (DUV) UV wavelengths of <300 nanometers (248 nm and 193 nm) generated by excimer lasers and used as an exposure light source to pattern features at 0.25 μm and below.

defect density The number of defects per unit area (usually cm^2) of wafer surface.

defects Some characteristic of the wafer or a result of the wafer fabrication process that causes nonconformance to the specified wafer requirement.

deglaze The removal of poor quality native oxide on the wafer surface, usually through a wet etch.

degree of planarity (DP) For CMP, indicates how flat the wafer surface is at a particular step height location after CMP, relative to the initial step height prior to CMP.

dehydration bake An elevated temperature bake to attain a dry wafer surface prior to priming and resist spin coating for optimum resist adhesion.

density A physical property of a substance defined as the ratio of its mass (or weight) to its volume.

depletion mode A depletion mode transistor has the channel region doped to have normally-on operation.

depth of focus The focal length of a lens system to maintain a precise image size.

deposition A heat or physical (sputtering) process whereby a thin film of material is deposited over the surface of the wafer. The process of changing a gas into a solid to form a thin film.

deposited oxide layer A layer generated by using an external silicon source and O_2 and reacting these materials in a chamber to form a thin film on the wafer surface.

depth of focus (DOF) For optical lithography, the range around the focal point (center of focus) over which the image is continuously in focus.

descum A plasma etch operation following development that removes residual resist from uncovered areas of the wafer.

design for test (DFT) Design strategies used in early chip design to take into account the testability of the chip.

desorption The removal or release of by-products from wafer surface reactions.

develop inspect The inspection a wafer receives after the photoresist development process.

development (also develop) A chemical process that removes the photoresist from areas not defined by the mask in a lithographic procedure.

developer The liquid solution used to resolve an image after exposure.

device isolation The electrical isolation between MOS devices for wafer fabrication to reduce or eliminate parasitic field transistors.

device technology A term used to categorize the technology for building discrete or integrated devices based on a particular technology, such as TTL or CMOS.

DI water Deionized water. Water that has been specially processed to remove all ionized impurities and filtered to remove particle contamination.

diameter grinding A method used to create the precise diameter of the silicon ingot.

diborane (B_2H_6) A toxic, pyrophoric gas used as a source of boron for p-type doping in ion implant and epitaxial growth.

dichlorosilane (H_2SiCl_2) A flammable, corrosive liquid used as one of the reactant gases to deposit silicon dioxide and silicon nitride in CVD and promote the growth of epitaxial silicon.

die (singular or plural) One individual chip cut from a wafer before it is packaged.

die array The grid pattern that appears on the wafer surface resulting from the reproduction of many die patterns on the wafer. Also die matrix.

die attach The physical attachment of the die to the leadframe or substrate, primarily through epoxy attach.

die-by-die alignment *See* field-by-field alignment.

dielectric A material that is a nonconductor of current (same as insulator).

dielectric constant (k) A numerical valve representing the effectiveness of a nonconductive material to store electrical potential energy under the influence of an electric field.

die matrix *See* die array.

die separation A process to cut each individual die from the wafer using a diamond-blade dicing saw. Also known as die singulation.

diffraction The phenomenon of light passing through a narrow opening and creating interference patterns at the edge of the opening rather than the expected sharp edge.

diffraction-limited optics A term used to describe the inability of the optics in lithography tools to capture all of the diffracted light exiting from small openings in reticles.

diffusion A high-temperature process in which dopants on a wafer are redistributed due to the movement of one material through another from a region of high concentration to a region of lower concentration.

diffusion controlled The parabolic stage of oxide growth where the reaction is limited by the rate at which the oxygen diffuses through the oxide.

digital/analog An integrated circuit that converts digital data at its input terminals to an analog voltage at its output terminals.

digital circuit Digital circuits have operating signals that vary about two distinct voltage levels—a high and a low, representing the binary bits 1 and 0.

diluent An inert gas, such as nitrogen or argon, introduced into the reaction chamber, along with the reactant gases to maintain the desired reaction rate.

direct chip attach (DCA) Attaching a chip directly to a circuit board, usually through epoxy attach and wirebonding. See chip on board (COB).

directionality For dry etching, the plasma ions are directional (perpendicular to the wafer surface) which drives the plasma ions into high aspect ratio openings.

discrete As opposed to an integrated circuit, a device or a component that has a single function, such as a resistor, diode, or transistor.

dishing A CMP defect that has a reduction in the thickness of the polished material toward the center of the feature.

dislocation A crystal defect that is a discontinuity in the repetitive structure of the crystal lattice.

dissolution rate The rate at which a developer dissolves the resist.

dissolution rate monitor (DRM) A method for real-time monitoring of the development process using interferometric signal data collected by an optical sensor.

DNQ-novolak A general name for i-line photoresist representing the mixture of diazonapthoquinone (DNQ) and novolak resin.

donors An element with five electrons in its outer shell used to make n-type semiconductor. Examples include arsenic, antimony, and phosphorus.

dopant concentration (dose) The number of atoms of the dopant present in the silicon.

dopant profile A characterization of the amount of dopant actually added in a wafer doped region as a function of depth.

dopant (or impurity) An element that alters the conductivity of a semiconductor material by adding a hole or electron. Silicon is typically doped with impurities from Group IIIA or VA of the periodic table.

doped region The region where dopants reside within the silicon crystal structure.

doping The process of adding small amounts of a dopant to the silicon crystal structure in order to improve its electrical conductivity.

dose monitor The control of resist exposure in photolithography by measuring the intensity of the UV light source at the surface of the wafer.

dose, Q The number of implanted ions per unit area of wafer surface, with units of atoms/cm^2 (also stated as ions/cm^2).

downstream reactor A method of reducing damage to the wafer surface and heat buildup due to ion bombardment during plasma processing (e.g., etch) by locating the wafer away from the plasma.

drain One of the three electrodes (along with source and gate) of a metal oxide semiconductor transistor.

drive-in The 2nd part of a high-temperature diffusion process used to move dopants through the silicon crystal to the desired junction depth in the wafer.

dry etch The process of using RF energy and gas phase chemicals to remove a specific layer during semiconductor processing.

dry mechanical pump A type of mechanical pump that requires no oils for lubricating the rotating parts of the pump's compression chamber.

dry oxidation Thermal oxide growth within oxygen only and no moisture.

dry plasma etch An etch that uses a high-powered radio frequency (RF) energy to ionize a gas and create a plasma to remove surface material. *See* plasma etch, dry etch.

dry-in/dry-out A CMP tool approach that integrates polishing and cleaning.

dual damascene process A process for copper metallization that creates both the vias and interconnect wiring for each metal layer by etching holes and trenches in the ILD, depositing copper in the etched features, and using chemical mechanical planarization to remove excess copper.

dual plasma source (DPS) A high-density plasma source with two sources of RF power.

dual-polysilicon gate structure A separate p^+ doped gate for the p-channel devices to complement the n^+ dopant for the n-channel gate, which improves device performance.

dynamic RAM (DRAM) Dynamic random access memory—a device that temporarily stores digital information, but requires refresh voltage pulses to ensure data is not lost.

economies of scale An approach to efficiency in which the sheer scale of an operation makes it economical.

edge bead removal An operation after photoresist coat that removes the thick photoresist accumulated at the bottom and top edge of a wafer.

edge die The two or three rows of dice along the outer circumference of the wafer.

edge exclusion The distance from the wafer edge where no usable die can be produced.

electrically erasable PROM Programmable read-only memory—an IC that can be electrically erased and reprogrammed with user-specified information.

electrode When referring to active semiconductor devices, electrode refers to a functioning element of the device. When referring to process chambers, electrode refers to the electrical conductors that couple the electrical potential from power supplies into the chamber.

electromigration Movement of atoms (e.g., aluminum) in the conductor due to momentum transfer from the electrons carrying the current that can cause a depletion or accumulation of metal atoms.

electron beam (e-beam) lithography A direct-write method of lithography using a beam of electrons that is capable of achieving a high-resolution pattern transfer to the reticle surface.

electron cyclotron resonance (ECR) A high-density plasma reactor that uses a magnetic field to give directionality to the plasma ions.

electron shower A technique in an ion implanter that uses electrons to neutralize positive ions before they impact the wafer. This is required to prevent wafer charge build-up.

electronic stopping For ion implant, the stopping of dopant atoms is caused by interactions with the target electrons.

electronic wafer map A computer image of the good and bad die on a wafer.

electroplating A method of depositing metal atoms (e.g., copper) on a substrate. Also called electrochemical deposition (ECD).

electropolishing A chemical process to create a clean, smooth surface.

electrostatic chucks A wafer holding mechanism common in many process chambers that holds wafers using electrostatic attraction.

electrostatic discharge (ESD) Uncontrolled transfer of static charge from one object to another.

ellipsometry A nondestructive, noncontact optical film-thickness measurement technique primarily used to measure thin, transparent films.

emitter One of the three electrodes of a bipolar junction transistor (BJT).

endpoint detection A general term describing when a fabrication process has finished. An example is determining when the CMP has polished to the correct thickness.

energy dispersive spectrometer (EDX) The most widely used X-ray detection method to identify elements; often used with an SEM.

enhancement mode Enhancement mode channels are doped opposite in polarity to the source and drain regions to have normally-off MOSFET operation.

epi Short for epitaxy, the controlled growth of a layer of crystalline semiconductor material on a suitable substrate.

epitaxial layer A layer of a single-crystal semiconducting material grown on a single-crystal substrate having the same crystallographic characteristics as the substrate material.

epoxy attach The most common method for die attach in traditional IC packaging.

epoxy underfill Used in flip chip packaging to flow epoxy between the chip and the substrate to reduce stress and encapsulate the bumps.

erasable PROM Programmable read-only memory IC that can be erased by exposing to UV light.

erosion The thinning of oxide and metal in a patterned area during chemical mechanical planarization (CMP).

etch The process of removing material (such as oxides or other thin film) by chemical, electrolytic, or plasma (ion bombardment). A process which selectively removes material by chemical or physical means to form a permanent pattern on the wafer in areas not protected by the photoresist.

etch bias A measure of the reduction in feature size of a critical dimension (CD) after an etch process.

etch profile A term that describes the shape of the sidewall of an etched feature.

etch rate The speed at which material is removed from the wafer surface during etching.

etch residue The unwanted material remaining on the wafer features after etch.

etch uniformity A measure of the capability of an etch process to evenly etch across the entire surface area of the wafer, across the entire wafer lot, and from lot to lot.

etchant An acid or base in gaseous or liquid state that selectively removes a layer of material from a substrate.

etchback planarization The etching of a sacrificial dielectric material on a wafer surface to fill voids and low spots (topography) on the surface.

eutectic attach An older method of attaching a chip to a leadframe depositing gold on the chip backside to alloy to the substrate.

eutectic temperature The lowest possible melting point for a particular mixture of two elements.

evaporation An early method for depositing metal on wafers that works by heating a material inside a vacuum chamber until it vaporizes and deposits on the wafers.

excimer laser A laser (coherent) light source used for deep UV exposure in photolithography and based on krypton fluoride, krypton chloride, argon fluoride, or fluoride species.

exposure Subjecting a sensitive chemical or material to light or other radiant energy.

exposure dose The amount of light energy supplied to a photoresist.

extraction electrode In an ion implanter a high-negatively charged electrode is mounted at the outlet of the ion source to extract positive ions from the source chamber.

extreme UV (EUV) A next-generation lithography method that uses a laser-produced plasma source to produce UV wavelengths of ~13 nm.

extrinsic silicon Also known as doped silicon, extrinsic silicon is created through the process of doping.

fabless A semiconductor company that does not have their own wafer manufacturing facility but subcontracts the wafer manufacturing to other companies.

fabrication In semiconductor manufacturing, fabrication usually refers to the front-end process of making the devices in semiconductor wafers, and usually does not include the back-end processes of final assembly and packaging.

facilities Refers to the department in a manufacturing site that provides essential environmental and chemical needs (e.g., electricity, air conditioning, bulk chemicals, and DI water) to sustain fab operations.

fast ramp furnaces Vertical diffusion furnaces which are able to quickly raise and lower the temperature of a batch of wafers.

fault models Test software algorithms for detecting many different types of IC failures during electrical test.

FCC diamond For a silicon crystal, the unit cell is a variation of the face-centered cubic (FCC) structure.

feature size *See* critical dimension (CD).

FEOL *See* front-end of line.

Fick's laws A mathematical description of the diffusion process that describes how particles move through a material.

field-effect transistor (FET) A transistor consisting of a source, gate, and drain, the action of which depends on the flow of majority carriers past the gate from the source to the drain. The flow is controlled by an electric field under the gate structure.

field oxide A relatively thick oxide layer grown on a silicon wafer that protects the inactive portion of the silicon surface and prevents induced charge in the wafer surface.

field-by-field alignment A fine alignment method used by photolithography steppers to align each field on the wafer to the reticle one field at a time.

field-programmable PROM Programmable read-only memory that can be reprogrammed in the field. The two types are the EPROM and the EEPROM.

film A term used when referring to a layer of material on the surface of the substrate.

film stress Stress in a layer that may deform the substrate and create reliability concerns.

final assembly and packaging A general term used to describe the 1st level packaging process. Also referred to as back-end.

final test The electrical evaluation of the packaged device.

first interlayer dielectric (ILD-1) The dielectric film between the silicon wafer and the first metal layer.

fixed oxide charge A charge originating during thermal oxidation from the incomplete oxidation of silicon less than 2 nm away from the Si/SiO_2 interface.

flats A portion of the periphery of a circular wafer that has been removed to identify the crystal orientation.

flip chip A high-density packaging technique that mounts the active side of a chip (with the surface bonding pads) toward the substrate.

float zone An alternative crystal growth method that produces a silicon monocrystal ingot with significantly lower oxygen content.

fluorosilicate glass (FSG) A fluorinated oxide that is used as a dielectric layer in multilevel metallization.

focal length In optical lithography, the distance from the center of the lens to the point of best optical imaging or focal point.

focal plane The plane perpendicular to the axis of a lens or optical system that contains the focal point of the lens or optical system.

focal point In optical lithography, the principal focus where the light rays are refracted and converge to form a sharp image by the light.

focus A condition where most of the individual rays of radiated energy are made to converge together into a single point.

focused ion beam (FIB) A destructive technique that can be used to deposit or etch materials on wafers and is often used in the etch mode to prepare wafer structures for failure analysis.

footprint The area of floorspace occupied by equipment.

forward bias The polarity of the connections and amplitude of a voltage source needed to force a pn junction into the conduction mode.

four-point probe A measurement method with four evenly spaced probes in a line that is a widely used method to measure sheet resistance.

Frenkel defect Otherwise known as a vacancy-interstitial pair, it occurs when an atom leaves its lattice site and takes up position in a void.

front-end of line All processes including wafer start through contact etch.

front-opening unified pod (FOUP) A handling system for wafers that integrates the wafer cassette and tool load/unload functions with a standard mechanical interface (SMIF).

functional tests Electrical tests that verify a chip's functionality for meeting the data sheet specification.

furnace flat zone A zone in the furnace where temperatures can be controlled to less than 0.5°C even at temperatures of over 1000°C.

g-line The wavelength (436 nm) of a particular color of light contained in the light spectrum of ionized mercury.

gallium (Ga) A soft metal element used for p-type doping in ion implantation.

gallium arsenide (GaAs) The most common compound semiconductor material. It produces higher speed devices than those made from silicon.

gap fill The filling of narrow spaces with a material without creating voids or other defects.

gas One of three states of matter in which the particles fill a container regardless of the shape.

gas cabinet The enclosure where the high-pressure compressed gas cylinders are stored. It contains the valves, gauges, and safety devices needed to handle and store the compressed gases.

gas manifold An arrangement of pipes that connects several gas cylinders to a common gas supply line.

gas phase nucleation A process where reactant molecules attach to each other to form larger molecules. In CVD, if this process occurs in the gas phase, undesired particles are generated.

gas purge A method of flushing undesirable residual gases, atmospheric gases, and water vapor from a process chamber and the gas delivery system.

gas throughput The mass quantity of gas flow through a system at standard conditions.

gate An electrode that controls current flow in a metal oxide semiconductor transistor.

gate oxide The thin layer of thermal oxide that separates the gate electrode (terminal) from the channel region of the semiconductor substrate.

gate oxide integrity Relates to the degree of gate oxide quality needed to maintain the functionality of the MOSFET.

germanium (Ge) A brittle, grayish-white metallic element having semiconductor properties. Widely used in crystal diodes and early transistors.

getter Any chemical or process that immobilizes or ties up impurities.

glass A deposited film of silicon dioxide doped with elements (e.g., boron and/or phosphorus) to achieve improved properties.

glazing A flattened chemical mechanical planarization polishing pad condition that occurs after repeated use.

global alignment A fine die alignment scheme used with steppers in photolithography where the orientation of the wafer relative to the reticle is obtained by making a series of alignment measurements on several die around the wafer.

global planarization Surface planarization across the entire wafer, such as chemical mechanical planarization.

glow discharge An alternate term for plasma since it glows in a vacuum.

gray area Also called service chase or equipment bay, is the service support area outside a cleanroom.

gross defect Related to the structure of the crystal and typically occurring during crystal growth. Two examples are twin crystal planes and crystal slip.

grown oxide layer An oxide layer grows on a silicon wafer by providing externally supplied high-purity oxygen in an elevated-temperature environment to react with the substrate.

halogen In the periodic table, the elements under the Group VIIB column. They are fluorine (F_2), chlorine (Cl_2), bromine (Br_2), iodine (I_2), and astatine (At_2). Halogens are extremely reactive, have strong odors, and they are poisonous to breathe.

hardbake A post-development thermal bake used to drive out any residual solvent and harden the resist for follow-on processing.

HEPA filter High-efficiency particulate air filter. A specially constructed filter membrane that allows a high volume of air flow and stops small particles of some predetermined size from passing through.

hermetic sealing The sealing of the final ceramic package to protect the device from the external environment, specifically moisture and contamination.

heteroepitaxy A type of epitaxy where the film and substrate are two different materials (e.g., Si on Al_2O_3).

heterogeneous reaction A CVD chemical reaction that takes place on (or very close to) the wafer surface to deposit the solid film.

hexamethyldisilazane (HMDS) A solvent used in a dehydration bake oven to prime the wafer surface prior to coating the wafer with photoresist.

high-density plasma CVD (HDPCVD) A plasma used in chemical vapor deposition with a high-density mixture of gases at low pressure that is directed toward the wafer surface in the reaction chamber to deposit films that can fill high aspect ratio gaps.

high-density plasma etch The dry etching method suitable for high aspect ratio gaps, with a low reactor pressure and a high density of active species in the plasma.

high-pressure oxidation A special pressure vessel contained inside a furnace that allows thermal oxidation at elevated pressure levels of 5 to 25 atmospheres. Also referred to as HiPOx.

high-temperature diffusion furnace Furnaces that operate at temperatures near 1200°C and can run a variety of processes, including oxidation, diffusion, and deposition.

high vacuum A vacuum condition characterized by having few collisions between gas molecules (molecular flow), resulting in very clean process chambers.

high vacuum pumps Pumps used for achieving high and ultra-high vacuum from 10^{-3} torr to 10^{-9} torr for process chambers.

hillock A protrusion on the surface of metal films due to electromigration that can lead to shorting between adjacent conductors.

homoepitaxy In epitaxy, the condition where the film and substrate are the same material.

homogeneous reaction A chemical vapor deposition reaction that occurs in the gas phase above the wafer surface and results in poorly adhering, low-density films.

horizontal furnace A furnace that derives its name from the horizontal position of the quartz tube where wafers are located and heated during batch furnace processing.

hot electrons Electrons accelerated by large lateral electric fields from the source and drain (nMOSFET), leading to collisions with other carriers. The collisions cause the hot electrons to scatter resulting in damage to the transistor.

hot wall The condition where furnace or reactor walls are heated during processing.

hydrochloric acid (HCl) A poisonous solution of hydrogen chloride gas in water.

hydrofluoric acid (HF) An extremely hazardous acid used to etch silicon dioxide.

hydrogen (H_2) A non-toxic flammable gas that is both a reactant used in wafer processing, and sometimes a by-product in other processes.

hydrogen chloride (HCl) A corrosive gas used in etching, epitaxy, oxidation, and passivation.

hydrogen peroxide (H_2O_2) An unstable and strong oxidizer that is soluble in water and alcohol and is used as a catalyst in etch formulations for piranha solution.

hydrophilic A wafer surface with affinity toward water, allowing water to spread across it in large puddles. Also called hydrated.

hydrophobic A dry wafer surface (aversion to water). Also called dehydrated.

hyperfiltration A filtration process that can separate impurities as small as 0.005 microns. See also reverse osmosis.

i-line The wavelength (365 nm) of a particular color of light contained in the light spectrum of ionized mercury.

IC final assembly The back-end process of separating each good die from the wafer and attaching the die to a metal leadframe or substrate.

IC packaging Enclosing the die in a protective package for assembly to the next higher level.

IC reliability The probability that IC devices will perform properly in their usage environment over their intended life.

I_{DDQ} testing A wafer sort electrical test that measures the steady state I_{DDQ} current from the source to the drain (MOSFET) to verify there is no device defect.

impurity Also referred to as dopant, an element added to a semiconductor to improve its conductivity.

in situ measurements Process parameter measurements such as film thickness and resistivity made in a process chamber while processing wafers.

index of refraction Represents how much a light ray bends as it passes through the interface of the two media based on its change in velocity.

indium A soft metal element used for p-type doping in ion implantation.

inductively coupled plasma (ICP) As opposed to capacitive coupled plasma which creates damaging electric fields, RF is applied to a coil outside the process chamber which creates magnetic fields in the gas and accelerates electrons and excites the plasma.

inert gas Gases, such as He, Ar, and N_2, that do not readily form compounds and are widely used in semiconductor manufacturing because they do not react with other chemicals.

infrared interference A nondestructive and relatively fast technique for measuring dopant concentrations of lightly doped layers based on optical reflectance.

ingot A cylindrical single-crystal silicon mass formed after the crystal growing process. Also called boule.

ink mark The traditional wafer sort procedure that puts a drop of ink on each unacceptable die to highlight the die for rejection during chip assembly.

in-line parametric test A wafer electrical test performed on test pattern structures located in the scribe region on wafers. It is done as early as possible after the completion of the front-end processes. Also referred to as wafer electrical test or DC test.

input/output (I/O) pins The external connections to an IC through which the input and output signals are transferred to and from the chip.

insulator A material with poor conductance of electricity that is used to isolate conductors from one another to allow for safe current flow without arcing.

integrated circuit (IC) A microminiature electronic circuit built on a die of single-crystal semiconductor. Also referred to as a chip, microchip, or device.

integrated measurement tool A tool which has sensors that permit it to function as a part of the process and deliver in situ (real-time) measurement data.

interconnect The conductor materials, such as aluminum, doped polysilicon, or copper, that form the metal wiring circuitry to carry electrical signals to different parts of the chip.

interconnect delay The delay in an electrical signal due to the resistance and capacitance time delay of the interconnect wiring.

interface-trapped charge Positive or negative charges trapped at the gate oxide/silicon interface that result from structural defects, oxidation-induced defects, or metal impurities.

interferometer A precision optical instrument that measures interference patterns due to movements of one surface relative to a reference plane. Applications include in situ position detection step height and film thickness measurements.

interlayer dielectric (ILD) An insulator material between each metal layer or between the first metal layer and silicon. Same as interlevel dielectric.

interstitial A point defect where an atom becomes located in a void within the crystal structure.

intrinsic silicon Pure silicon, containing no contaminants or impurities.

ion An atom that has either gained or lost electrons making it a charged particle (either negative or positive, respectively).

ion analyzer The subsystem in an ion implanter that separates the desired dopant ions from the main body of ion species.

ion beam milling or ion beam etching (IBE) A directional plasma reactor capable of anisotropic etching of small features.

ion implantation A process for introducing dopants into the wafer through a high-voltage ion bombardment.

ion implanter A tool for doping wafers in which gases carrying the desired dopant, e.g. As, P, B, BF_2, and are ionized inside the implanter and accelerated by high voltages to give the dopant the energy needed to penetrate into the wafers.

ion projection lithography (IPL) A next-generation lithography technique that uses ion beams to expose resists, either through a mask or by serially writing on the resist with a finely focused beam.

ionization Removal of an electron from an atom to create a positively charged atom or molecule.

ionized metal plasma PVD A sputtering process where metal is ionized in an RF plasma and forced to travel in a highly directional path toward the wafer for conformal coverage of high aspect ratio gaps.

IPA vapor dry Wafer drying through the displacement of water by heated solvent vapors of isopropyl alcohol.

isolation regions Areas on the wafer with dielectric isolation between active transistors on a substrate.

isotropic etch profile An etch process that removes material in the same rate in all directions (laterally and vertically).

JFET One of two types of FET's, the junction field-effect transistor (JFET) is so named because the gate actually forms a physical pn junction with the other electrodes of the transistor.

junction (pn) The line of demarcation in a wafer where the number of p- and n-type carriers are exactly equal with a surplus of p-type on one side and n-type on the other side.

junction depth, x_j The exact depth in the wafer where the p-type doped region meets the n-type doped region.

junction spiking Microalloying by diffusion of the contact metal into the silicon.

Kelvin Also called the absolute scale, it is the base unit of temperature for the metric unit system.

killer defects Defects that cause the chip to fail during electrical test.

kinetically controlled reaction A chemical vapor deposition reaction limited by the kinetic energy of the reactants at the surface. Also referred to as reaction rate limited.

laminar air flow Air that flows uniformly in parallel streams of uniform velocity with a minimum amount of turbulence (the type of air flow used in cleanrooms).

lapping A mechanical operation to remove damage left by wafer slicing and achieve a high degree of parallelism and flatness of the wafer.

latchup An effect observed when the pn junctions in CMOS devices produce parasitic transistors, which can unintentionally turn on the CMOS device.

lateral diffusion The diffusion of dopant that's parallel to the substrate surface.

law of reflection Describes the relationship between an incident light ray and the corresponding reflected ray, stating that the angle of incidence is equal to the angle of reflection.

LDD *See* lightly doped drain.

leadframe A formed or etched copper strip, sometimes plated with silver or palladium, that provides external electrical connections with inner leads for die attachment and outer leads for attachment to the substrate.

leakage current Undesirable mobility of charge carriers in a semiconductor device which consumes battery power and disrupts normal functions of the device.

lens For optical lithography, an optical element used to refract light from an object that passes through it in order to form an image of the object.

lens compaction In optical lithography, lens damage caused by a laser beam that is a densification (structural rearrangement) of the lens material.

light An electromagnetic wave that is radiant energy and visible to the eye.

light intensity The power per unit area (mW/cm^2) for the photolithography light source that is measured on the surface of the resist. Also referred to as brightness.

light scattering (laser scattering or scatterometry) A particle inspection method using light that reflects from particles on the wafer surface.

lightly doped drain (LDD) A low energy implant of either p-type or n-type dopant to form shallow source and drain junctions sometimes referred to as source/drain extension.

linear An alternate term for analog circuits, i.e. linear circuits.

linear accelerator An ion implanter subsystem used to accelerate ions to high velocity and obtain beam energies >200 keV.

linear stage The initial growth stage of thermal oxidation.

linewidth The width of interconnect wiring; often refers to the minimum linewidth. *See also* Critical dimension (CD).

liquid Either a pure substance, such as pure water, or a chemical mixture.

lithography *See* Photolithography.

loaded brush A CMP cleaning condition of particles embedded in the cleaning brush.

loading effects The change in etch rate for wafers with significant surface area (slow etch) versus wafers with small surface areas that require etch (fast etch).

loadlock Where wafers are placed prior to entering the process chamber, isolating the inner regions of the tool from the atmospheric workplace environment.

Local Interconnect (LI) The method used to form metal connecting lines between transistors and contacts at the substrate level.

local oxidation of silicon (LOCOS) The traditional method of selectively oxidizing silicon through openings in a silicon-nitride mask that covers the substrate. Not suitable for device technologies below 0.35 μm.

local planarization Surface planarization across a small area of the wafer.

LOCOS *See* Local oxidation of silicon.

logic A collection of circuit elements that perform a function.

lot A batch of wafers processed at the same time.

low-pressure chemical vapor deposition (LPCVD) A widely used chemical vapor deposition tool that has superior film properties, operates at a medium vacuum, and is reaction rate limited.

low vacuum A vacuum above about 1 torr where gas flow is primarily by collisions between molecules (also known as viscous flow).

LPCVD *See* low-pressure chemical vapor deposition.

LSI Large scale integration. Semiconductor fabrication with a device density between 5,000 to 100,000 devices on an individual die.

magnetic CZ (MCZ) A technique to achieve homogeneous, large diameter crystals, using a magnetic field around the silicon melt to stabilize the crystal during its growth.

magnetically enhanced RIE (MERIE) A high-density plasma etch reactor that uses a magnetic field to increase the electron and ion concentration near the wafer surface.

magnetron sputtering A physical vapor deposition process that uses magnets to increase the ion bombardment rate on the target, which increases the ionization rate.

majority carrier The type of charge carrier that makes up more than one-half of the total number of charge carriers in a region of a substrate. One example would be electrons in an n-type region.

make-up loop The part of the deionized (DI) water installation which removes particles, including total organic carbons (TOCs), bacteria, microorganisms, and ionic impurities.

mask A transparent quartz plate covered with chrome patterns used in photolithography for making integrated circuits by exposing photoresist. The patterns are opaque areas that prevent light passage. Sometimes called photomask.

mask-programmable gate array A chip customized to meet the functional needs of individual customers through differences in the interconnect layers of the manufacturing process.

mass flow controller (MFC) An instrument that uses the heat transfer property of a gas to directly measure and control the mass flow rate of the gas into a process chamber.

mass spectrometer An instrument used to identify molecules based on the mass of the molecule.

mass-transport limited reaction Describes how the rate of a chemical vapor deposition reaction cannot proceed more rapidly than the rate at which reactant gases are supplied from the main gas stream to the wafer substrate, regardless of temperature.

mean free path (MFP) The average distance a gas molecule moves before it strikes another molecule.

medium vacuum A vacuum from about 1 torr to 10^{-3} torr that is a transition between low and high vacuum.

megasonic cleaning Cleaning with ultrasonic energy with frequencies near 1 MHz to effectively remove particles at lower bath temperatures.

melt A term describing the liquid silicon used during ingot fabrication.

membrane contactor A microporous filter made of hydrophobic polypropylene hollow fibers.

membrane filter A filter that uses a thin membrane of polymer or ceramic with small penetrating pores as the filter medium.

mercury arc lamp The UV illumination source in all conventional i-line steppers.

MESFET Metal gates used in gallium arsenide JFETs.

metal contacts *See* contacts.

metal impurities Metallic impurities, such as alkali metals, that are found in many common chemicals and processes and are strictly controlled in all materials used in a wafer fab.

metal stack A metal film structure composed of different layers.

metallization The use of sputtering or CVD processes to deposit a thin film of metal and pattern it to form the desired interconnection arrangements on the wafer surface.

metalorganic CVD (MOCVD) A vapor phase epitaxy process that uses halides and metalorganic sources to deposit polycrystalline and amorphous films of metals and oxides.

metrology The science of measurement to determine dimensions, quantity, or capacity.

microchip *See* integrated circuit.

microdefect *See* crystal defect.

microlithography A term describing the process to pattern the ultraminiature features necessary for submicron wafer fabrication. Also referred to as photolithography.

microloading An etch phenomenon where the etch rate is slower in small openings, to the point that etching can actually stop in geometries with high aspect ratios. *See also* aspect ratio dependent etching (ARDE).

micron (μm) A term for micrometer that is a unit of linear measure that equals one-millionth (1/1,000,000) of a meter. There are 25.4 microns in one-thousandth of an inch.

microprocessor unit *See* CPU.

microroughness Small-scale deviation of the actual wafer surface from a nominal plane surface, with many small, closely spaced peaks and valleys.

Miller indices A set of numbers used to describe atomic orientation in a crystal lattice.

minienvironment A process tool localized environment created by an enclosure that isolates wafers from the cleanroom and potential contamination.

minimum geometry The smallest linewidth or spacing between lines or features on a semiconductor die. Also referred to as CD, critical dimension, or feature size.

minority carrier The type of charge carrier that makes up less than one-half the total number of charge carriers in a region of a substrate. One example would be holes in an n-type region, where electrons are the majority carriers.

mix and match To optimize the photolithography process, the type of photoresist and equipment technology are matched to the criticality of the wafer layer.

mobile ionic contaminants (MICs) Metallic ions that are highly mobile in semiconductor materials.

mobile oxide charge A trapped charge in the gate oxide due to mobile ionic contaminants (MICs).

molecular beam epitaxy (MBE) An evaporative deposition process using high vacuum and capable of excellent uniformity and thickness control.

molecular flow A type of gas flow that occurs when the distance between molecules is so great they have little influence on each other and movement is completely random.

monitor wafers (test wafers) Blank or unpatterned wafers used to monitor process performance.

monocrystal *See* single crystal.

monolithic device A device whose circuitry is completely contained on a single die or chip.

Moore's law Named after Gordon Moore, predicts that the number of transistors on a chip would double every eighteen months.

MOS Metal oxide semiconductor. Refers to field-effect transistor (FET) in which current flows through a channel of semiconductor material that is controlled by the electric field of a gate structure.

MOSFET One of two types of FET's, the Metal Oxide Semiconductor Field Effect Transistor is the most popular. It has a metal (conductor) gate over thermal oxide dielectric.

motor current endpoint An endpoint detection method for CMP that monitors the magnitude of the current supplied to the motor used to rotate the wafer carrier. The magnitude of the current is related to the type of material polished.

MSI Medium-scale integration. Semiconductor fabrication technology with a device density of 50 to 5,000 devices on a single die.

multichip module (MCM) Package that mounts several die onto one substrate to achieve a higher density of silicon chips.

multilevel metallization The parts of the wafer fab process that involve the deposition, patterning, planarization, and dielectric isolation of layers of metal interconnects.

Murphy's model A widely used yield prediction model for wafer fabrication that assumes the defect density varies across the wafer and from wafer to wafer.

nanometer (nm) A unit of measure equaling one-billionth (1/1,000,000,000) of a meter.

native oxide Thin oxide contamination layer that forms on a silicon wafer if exposed either to air at room temperature or DI water.

n-channel MOSFET A MOSFET device doped to have predominantly a negative charge during conduction while the source and drain are n-type. Also called nMOSFET.

negative lithography In photolithography, printing a pattern on the wafer surface that is opposite the pattern in the mask.

negative resist A resist that becomes insoluble when exposed to light and cannot be washed away by solvents. Negative resist is not suitable for submicron lithography due to swelling.

negative resist development A solvent development wash to remove unexposed negative resist, which is not crosslinked and therefore soft and soluble.

neutral beam trap A method of trapping neutral ions in an ion implanter to remove them from the positive-charged ions in the beamline.

next-generation lithography The new lithography methods under evaluation to eventually replace optical lithography when it is no longer feasible.

nitric acid (HNO_3) A liquid corrosive oxidizer prepared from sulfuric acid, nitrates, and ammonia that is used to clean silicon wafers and wet etch metals.

nitrogen (N_2) A nontoxic, nonflammable gas used for chamber purging, blanketing, and cleaning and used as a CVD carrier gas.

nitrogen trifluoride (NF_3) A toxic gas used as a fluoride source in plasma processing, such as for etching polysilicon, silicon nitride, tungsten silicide, and tungsten films and for cleaning of CVD chambers.

nitrous oxide (N_2O) A nontoxic, nonflammable gas used in combination with silane for CVD of silicon nitride layers.

NMOS *See* n-channel MOSFET.

noncritical layers In multilevel metallization, usually the upper metal levels with wider linewidths, often 0.5 μm and larger, and less sensitivity to particulate contamination.

nonvolatile memory A semiconductor memory device designed to store digital data in the form of an electrical charge, even after power is turned off.

normality A chemical concentration unit that refers to the quantity of substances (stoichiometry) in a solution.

notch An intentionally fabricated indent on the wafer edge oriented such that the diameter passing through the center of the notch is parallel with a specified crystal direction.

novolak The phenol-formaldehyde polymer resin material in positive i-line photoresists.

npn An npn bipolar transistor has a base of p-type silicon sandwiched between an emitter and a collector of n-type silicon.

n-type silicon When a pentavalent (Group VA) element (e.g., phosphorus or arsenic) is added to pure silicon, the resulting n-type material has electron majority carriers and is therefore negative. Extra negative electrons find no unoccupied bonds to bind them, so they are free to wander and constitute electric current.

nuclear stopping During ion implant, ions are stopped in the silicon crystal by collisions between atoms, causing a displacement of silicon atoms.

nucleation For thin film growth, the first stage where clusters of stable nuclei are formed.

nuclei coalescence For thin film growth, the second stage where the stable nuclei grow into larger island clusters based on surface mobility and density. Also referred to as island growth.

numerical aperture (NA) In photolithography, the ability of a lens to collect diffracted light; this is also a measure of how much light the lens can accept and image by converging the diffracted light to a single point.

n-well A localized n-type doped region where pMOSFETs are constructed.

objective The main system of lenses in an optical microscope that magnifies the object being viewed.

off-axis illumination (OAI) A technique in optical lithography that has the incident light strike the mask at an angle in order to align the diffraction fringes with the lens, which serves to reduce the resolution limit of the optics.

ohmic contact A low-resistance contact to the silicon that has resistance inversely proportional to the area of the contact.

op amp A high-gain and high-input impedance amplifier that is adapted to a variety of electronic control applications.

optical interferometry endpoint A CMP endpoint detection method based on light reflectance of film layers depending on the type of film and its thickness.

optical lithography In photolithography, the use of optics and light to accurately project and expose a mask pattern onto a resist-coated wafer.

optical microscope (light microscope) Traditionally one of the most widely used methods of inspecting the wafer surface for defects such as particles and scratches.

optical proximity correction (OPC) In photolithography, the introduction of selective image size biases (alterations) into the reticle pattern to compensate for optical proximity effects due to light diffraction and interference between closely spaced features.

optical pyrometer A noncontact temperature measuring device based on measuring the infrared radiation given off by the heated object.

optics The study of the physical properties and composition of light that is important for optical lithography.

organic compound A chemical compound of carbon, hydrogen, and one or more other elements, such as oxygen or nitrogen.

out-diffusion A form of irregular doping during epitaxial growth where the substrate acts as a source of dopant impurities that diffuse into the epilayer.

outgassing The release of absorbed gases or vapors from materials usually accelerated by heating or by reducing the pressure (as in a vacuum) surrounding it.

overdrive A condition during electrical test where probe needles pierce (or "scrub") the surface of the probed aluminum pad to ensure good continuity.

overetch step An additional etch step during plasma etch to remove remaining residues.

overflow rinser The flowing of DI water through and around wafers, sometimes with a nitrogen bubbler to aid flow at the wafer surface.

overlay accuracy In photolithography, a measure of the alignment tool's ability to align the reticle pattern onto the existing wafer pattern. Also referred to as registration, or overlay registration.

overlay budget For photolithography alignment, the maximum relative displacement between a patterned layer and the previously defined layer.

overlay registration The ability to align mask patterns during photolithography to the existing wafer pattern that was previously etched into the wafer.

oxidation The chemical process of joining oxygen with another element. For semiconductors, the production of silicon dioxide by the joining of oxygen and silicon.

oxidation-induced stacking faults (OISFs) A form of unit cell dislocation due to layer stacking errors caused by incomplete oxidation at the Si/SiO_2 interface.

oxide Silicon dioxide. An insulating film applied to the surface of the wafer.

oxidizer A substance that unites with oxygen and removes electrons from atoms or ions. Also, used to classify gases that support combustion.

oxide-trapped charge A positive or negative charge trapped in the bulk gate oxide away from the interface.

ozone (O_3) A triatomic oxygen molecule, denoted by the symbol O_3. It is much more reactive than oxygen.

package The protective container for an electronic component, with terminals to provide electrical access to the chip(s) inside.

pad conditioning The mechanical regeneration of CMP polishing pads to slow down the effects of wear.

pad oxide A thin layer of thermal oxide used to reduce stress between the nitride mask and silicon.

paddle A structure in a horizontal diffusion furnace made of silicon carbine that supports the wafer-bearing quartz sled.

parabolic stage The second, slower phase of oxidation growth that starts after about 150Å of oxide thickness.

parallel-plate (planar) reactor A dry plasma reactor with two parallel plates that are symmetrical in size and positioned in the reactor.

parallel testing Electrical test procedures that permit simultaneous testing of different sections of a chip to increase tester throughput.

parameter A numerical measurement describing some characteristic of a manufacturing process.

parametric test *See* wafer electrical test.

parasitic An undesirable stray capacitance, inductive coupling or resistance leakage, as well as undesired transistor actions.

parasitic capacitance Unwanted capacitance created unintentionally by the layout of the materials on the wafer.

parasitic resistance Unwanted resistance created as a by-product of the IC design (size, shape, material type, and so on).

parasitic transistors pn junctions in CMOS devices that produce parasitic transistors that can create a latchup condition in CMOS ICs that causes transistors to unintentionally turn on.

partial pressure The partial pressure of each gas in the system is the pressure the system would have if all of the other gases were completely removed from the system.

particle density The number of particles in a given wafer surface area.

particles per wafer per pass (PWP) The number of particles above a certain critical size that are added to a wafer at an operation.

passivation A layer of material put over a wafer to stabilize and protect the surface. Silicon dioxide or silicon nitride are often used for IC passivation.

passivation layer The top layer of silicon nitride that protects the microchip from chemical action, moisture, contamination, and handling.

passive components Devices such as resistors and capacitors that conduct current without polarity and are constructed on silicon substrates but have parasitic losses that are detrimental to IC performance.

pattern sensitivity For CMP, the polishing rate of a feature is influenced by the topography of the surrounding features.

patterned etching An etch process that uses a masking layer to define a pattern on the wafer surface to remove only selected portions of the wafer surface material.

pattern wafers Wafers that have an image of a die matrix formed into a layer of photoresist or a permanent pattern that has been etched into the substrate.

patterning The transferring of an image from a reticle or mask to the photoresist. See photolithography.

pc board Printed circuit board. A common low-cost substrate for 2nd level packaging.

p-channel MOSFET A type of MOSFET device doped to have a channel with predominantly a positive charge during conduction and p-type source and drain. Also called PMOSFET.

PCM See process control monitor.

PEB See post-exposure bake.

PECVD See plasma enhanced chemical vapor deposition.

PEL Permissible exposure limit. A time-weighted average exposure limit (typically 8 hours) or a ceiling exposure limit for employees.

pellicle A means to minimize particulate contamination on reticles used in photolithography by protecting the reticle surface with a thin, optically transparent membrane.

pentavalent Elements from Group VA of the periodic table having five valence electrons.

perimeter array Traditional packaging with wirebonding that has bonding pads around the die perimeter, thus limiting the number of I/O (input/output) pins in the package.

pH scale Ranging from 0 to 14, this scale indicates the strength or weakness of an acid (0 to 7) or base (7 to 14).

phase-shift mask (PSM) In photolithography, an advanced method to overcome problems associated with light diffraction through small openings in the reticle by shifting the phase of the exposing light.

phosphine (PH_3) A toxic, pyrophoric gas used as a source of phosphorus in CVD, ion implant, and epitaxial growth.

phosphoric acid (H_3PO_4) A phosphorous-based oxygen acid used to etch silicon nitride.

phosphorus (P) A poisonous solid used for n-type doping in ion implantation.

phosphorus oxychloride ($POCL_3$) A chlorinated phosphorus compound used as an n-type dopant.

phosphosilicate glass (PSG) Phosphorous is incorporated into silicon dioxide during deposition to achieve a low deposition temperature, relatively planar surface, and excellent gap fill characteristics.

photoacid generator (PAG) A sensitizer used in chemically amplified deep UV (DUV) resist that boosts the resist's sensitivity to DUV light.

photoacoustics A noncontact technology based on light-induced sound pulses to generate an acoustic pulse that is directed toward a film stack on a wafer.

photoactive compound (PAC) A sensitizer used in positive i-line photoresist that promotes dissolution in the developer after exposure to UV light.

photolithography The process to produce a three-dimensional pattern on the surface of the wafer using a light-sensitive photoresist material and controlled exposure to light. Also referred to as photo, photomasking, masking, litho, lithography, or patterning.

photomask In photolithography, a quartz plate that contains the circuit patterns etched on a chrome layer needed to define one process layer for the entire wafer. Sometimes used interchangeably with the term reticle (pattern for a single field). Also referred to as a mask.

photoplate The term used for a mask before images have been formed on its chrome-coated surface.

photoresist In photolithography, an organic compound that experiences a change in solubility in a solvent (called a developer) when exposed to ultraviolet (UV) light.

photoresist asher The removal of a photoresist layer by reacting atomic oxygen atoms with resist material, often with microwave plasma generation in a downstream reactor.

photoresist stripping A general term that describes removal of the resist with either wet stripping or dry methods.

physical etch mechanism Dry etching with plasma to provide energetic species (bombarding ions) that are accelerated toward the wafer surface through the use of an electrical bias.

physical vapor deposition (PVD) Primarily a physical rather than chemical process to deposit a material by physically dislodging atoms from the surface of a solid target. Also referred to as sputtering.

pigtail The tubing which leads from the gas cylinder to the gas panel and is typically shaped into two or more coils. This allows the cylinder change-out to be performed without unduly stressing and damaging the tubing.

pin grid array (PGA) Conventional pin-in-hole package suitable for high I/O (input/output) and high-performance ceramic package.

pinhole A small undesired hole in an oxide, opaque region of a mask or reticle, or in a photoresist layer.

piranha A strong cleaning solution that combines sulfuric acid (H_2SO_4) and hydrogen peroxide (H_2O_2) to remove organic contaminants.

pitch In the case of repetitive features on a mask or reticle, pitch is the distance separating the features from center-to-center.

planar Existing essentially in a single plane. A process in which all pn junctions intersect the top surface of the semiconductor material, such that these intersections are permanently protected by the masking oxide and all contacts to the device can be made at the top surface.

planar capacitors Capacitors built laterally on the wafer substrate.

planar process The planar process of forming integrated circuit and semiconductor components is based on the use of a single surface for referencing each successive operation.

planarity Describes thickness variations for wafer distances ranging from microns to millimeters.

planarization The process of creating a smooth surface over topographic features on a wafer. Most commonly done on advanced ICs by chemical mechanical planarization.

plasma Referred to as the fourth state of matter. An electrically conductive gas composed of ionized particles and energized in an RF field.

plasma-based dry cleaning A cleaning method using plasma to etch the surface of the wafer and remove contaminants.

plasma electron flood A method of controlling wafer surface charging during ion implantation by immersing the ion beam and wafer in a high-density plasma.

plasma enhanced CVD (PECVD) In chemical vapor deposition, the use of plasma energy to create and sustain the CVD reaction.

plasma etch A dry etch process using reactive gases energized by a plasma field and may produce isotropic or anisotropic etch profiles depending on the placement relative to the electric field creating the plasma.

plasma-induced damage Damage to sensitive IC devices on the wafer from the energized ions, electrons, and excited molecules of plasma. An example is trapped charges on the gate electrode, causing a breakdown of the thin gate oxide.

plasma potential distribution The electric potential distribution of a plasma glow discharge that has a strong effect on the etching capability of the system.

plastic dual in-line package (DIP) Traditional pin-in-hole plastic package where the pins are located on the two long sides of the rectangular package.

plastic leaded chip carrier (PLCC) Early surface mount plastic package.

plastic packaging Most common packaging material for traditional packaging.

plug A metal that fills the vias to form the electrical connection between the two metal layers.

pMOS (p-channel) *See* p-channel MOSFET.

pn junction The region on a doped silicon crystal in which the conductivity changes from p-type to n-type. This junction creates silicon's useful characteristics as a semiconductor.

pn junction diode A pn junction diode with specific current flow characteristics is formed when a region of n-type semiconductor is adjacent to a region of p-type semiconductor.

pnp A bipolar junction transistor with a p-type emitter and collector terminals and an n-type base.

point defect A defect at a particular location in the crystal lattice, the most basic type of which is a vacancy.

Poisson's model An early yield model that assumes uniform defect density across a wafer, which is rarely true for large diameter wafers. Also called exponential model.

polarization The process of changing a nonpolarized form of radiation into one that's polarized. This can occur, for example, when nonpolarized light reflects off of surfaces.

polarized light Given that light is a form of electromagnetic energy, polarized light has all of its electric field components aligned in parallel.

polish *See* chemical mechanical planarization (CMP).

polish rate Represents the speed at which the material is removed for CMP.

polished wafer edge (edge grind) A finishing operation to contour a smooth radius on the edge of the wafer.

polishing loop The final part of a water purification system that removes contaminants remaining after the make-up loop.

polishing pad A pad that attaches to the top surface of the CMP platen to help transport the slurry and promote uniform polishing of wafers.

polycide A refractory metal reacted with polysilicon to achieve a lower resistance.

polycrystal Crystal unit cells in a random orientation. Also referred to as poly. Doped poly is an electrical conductor and used as such for the MOS gate electrode.

polymer formation Polymers that are sometimes intentionally deposited on the sidewalls of an etch feature to inhibit lateral etching by acting as sidewall passivation.

polymerization For negative photoresist, the resist becomes insoluble and hardens when exposed to UV light.

polysilicon A polycrystalline silicon deposited on the wafer during IC fabrication. In its doped condition, it is the most popular material used for the MOSFET gate electrode. Also referred to as poly.

polysilicon gate Doped polysilicon used as the gate electrode. Also referred to as poly gate.

positive lithography In photolithography, the printing of a pattern on the wafer that is the same pattern as on the mask.

positive resist A photoresist where regions of the resist exposed to light are removed in a developer solution, whereas resist not exposed to light underneath the opaque mask pattern remains on the wafer.

positive resist development The dissolution of positive resist involves a chemical reaction between the developing solution and the exposed resist.

post-develop inspection An inspection to identify and remove wafers with defective resist coatings before they continue to the following processes.

post-exposure bake (PEB) A thermal bake following resist exposure to catalyze critical resist chemicals in chemically amplified DUV resists or improve adhesion and reduce standing waves for conventional i-line resists.

ppb Part per billion. The presence of one part in a billion parts.

ppm Part per million. The presence of one part in a million parts.

ppt Part per trillion. The presence of one part in a trillion parts.

preamorphization Implantation of a wafer with an electrically inactive species, such as Si, to amorphize a thin layer of the single-crystal structure to reduce channeling during a subsequent implant operation.

precursors In gas phase reactions, the atoms and molecules that form intermediate reactions and are transported to the wafer surface for adsorption and reaction.

predeposition For a diffusion process, dopants are introduced into a thin layer of the wafer surface at high temperature.

premetal dielectric (PMD) The dielectric layer between the silicon and the first metal layer. Also referred to as local interconnect (LI) oxide.

Preston equation An equation that describes the chemical mechanical planarization (CMP) removal rate as a function of pressure and velocity.

primary orientation flat The longest flat edge on the wafer orients the wafer to be parallel to the wafer to be parallel to the desired crystal plane. Also referred to as the major flat.

print bias In photolithography, the difference between a feature dimension on a reticle and the same printed dimension on the wafer.

printed circuit board (PCB) Epoxy laminated material with circuitry etched into layers of copper and commonly used as the substrate for 2nd level packaging of ICs. Also called substrate or carrier.

probe The functional testing of the electrical parameters of each die on a wafer. Also referred to as wafer sort.

probe card The interface between the automatic tester and the device under test (DUT). It is often a printed circuit board with probe needles to make physical and electrical contact to the DUT.

prober This highly automated tool positions the wafer for contact with the probe card interface needles during electrical test.

process All the factors and events that are predefined and executed with monitors and controls to produce a desired effect on the product being manufactured.

process chamber A controlled vacuum environment where intended chemical reactions occur under controlled conditions. *See also* reactor.

process chemicals Chemicals used during wafer fabrication.

process control monitor (PCM) *See* test structure.

process latitude The capability of a process (e.g., photolithography) to consistently produce product that meet the specified requirements, with an objective to achieve the largest process latitude possible for production.

process recipe A set of software instructions that contains the specified process conditions followed during wafer processing at a workstation.

programmable array logic (PAL) A logic IC containing a network of programmable logic gates used to create a custom logic circuit.

programmable logic array Available in both mask-programmed and field-programmed versions, the programmable logic array differs from the PAL in that both the input AND gates and the output OR gates are programmable.

programmable logic devices ICs that use a variety of logic components.

programmable read-only memory An IC that can be reprogrammed in the field and is less expensive than the mask-programmable ROM. There are two types, the EPROM and the EEPROM.

projected range, R_p The depth traveled by implanted ions into the wafer, depending on the ion mass and energy, target mass, and beam direction.

proximity aligner An alignment and exposure tool that evolved from the contact aligner and was common in the early 1970s for linewidth dimensions of 2 to 4 µm.

p-type silicon When trivalent dopant atoms (Group IIIA) are added to silicon, the majority carriers are holes and are therefore referred to as positive.

puddle develop A resist development method that applies a small puddle of resist to form a meniscus over the entire wafer, thus minimizing developer volume.

pump speed An indication of how effectively a vacuum pump can remove gases, expressed in units of volume per unit time (e.g., liters/second).

punchthrough A condition where the channel is shorted (punched through) by undesirable leakage current, causing device failure. Reduced channel length and high electric fields at the drain end of the channel make punchthrough more likely.

purge A routine cleaning procedure where an inert gas is used to displace undesirable gases from a process chamber or gas panel.

purge cycle This consists of two steps: a pressurization step followed by a depressurization step.

PVD *See* physical vapor deposition.

p-well A localized p-type doped region where nMOS transistors are constructed.

pyrogenic steam High purity water vapor for wet oxidation.

pyrogens Fragments shed by bacteria.

pyrolytic A chemical reaction caused by the action of heat. For example, TEOS breaks down into silicon dioxide and other compounds when it is heated to a particular temperature.

pyrophoric Materials that ignite spontaneously in air at or below 54.4°C.

quad flatpack (QFP) Popular surface mount plastic package with a high number of leads on all four sides of the package.

quadrupole mass analyzer (QMA) A mass analyzer consisting of four cylindrical rods that have a constant DC potential and RF potential.

quality measures Measures which define the requirements for specific aspects of wafer fabrication to ensure acceptable device performance as measured by electrical tests and device reliability.

quartz Silicon dioxide.

quartz tube A quartz cylinder used to hold wafers in a high-temperature furnace.

quartz wafer boat A quartz boat that holds the wafers horizontal in the furnace tube.

queue time The maximum amount of time allowed to lapse between two sequential operations.

radiation damage The crystal damage caused during ion implantation as the high-energy ions enter the crystal and collide with substrate atoms.

radical A chemical fragment created when an energetic electron bombards a neutral gas molecule.

random access memory (RAM) A memory device that can be accessed to read its stored data or can be erased and new data rewritten into it.

range The total distance an ion travels in the silicon during ion implantation.

rapid thermal anneal (RTA) High-temperature exposure of the wafer with an extremely fast ramp and short dwell time at the target temperature.

rapid thermal processor (RTP) A single-wafer high temperature processing tool used for RTA processes.

RCA clean The industry standard wet clean process consisting of multiple cleaning steps using a combination of DI water, peroxide, and ammonium hydroxide, followed by DI water, hydrogen peroxide, and hydrochloric acid. *See* standard clean 1 and standard clean 2.

reaction rate limited For chemical vapor deposition, at lower pressures the arrival rate of reactants exceeds the rate at which they are consumed by the chemical reaction process at the wafer surface.

reactive ion etch (RIE) An etch process similar to the parallel-plate plasma reactor with wafers placed on an RF-powered electrode that is much smaller in size than the grounded electrode size to achieve both chemical and physical etching.

reactivity The tendency of a chemical to undergo chemical reactions and its ability to release energy.

reactor A term used to describe a process chamber for chemical reactions in semiconductor manufacturing, such as a reactor for depositing film layers.

read-only memory (ROM) A nonvolatile memory IC used for permanent storage of unalterable data.

recombination The process in which a free electron in the conduction band gives up its energy and falls into a hole in the covalent bond of the valence band.

redistribution During diffusion, a modification of dopant concentration due to silicon surface oxidation.

reflection spectroscopy (reflectometry) One of the three general types of reflection microscopes, with the other two being optical microscopy and ellipsometry.

reflective notching An undesirable condition in photolithography where vertical surfaces on the sides of etched structures reflect light into the resist where exposure is not intended.

reflow A high-temperature process that rounds the corners of a doped oxide insulating layer and slopes the sidewalls to improve step coverage.

refraction The property of a transparent substance that addresses how much light bends as it travels through it.

refractory metal A metal that has high temperature stability and is commonly used as a barrier metal in multilevel metallization, and commonly found in the Periodic Groups IVA, VA, and VIA.

regeneration The process by which frozen gases are removed from a cryopump.

registration *See* overlay accuracy.

relative index of refraction, n Represents how much the light ray bends as it passes through the interface of the two media based on its change in velocity.

residual gas analyzer (RGA) A process chamber instrument used to identify the types of gas molecules remaining in an evacuated system.

resist *See* photoresist.

resist development A general term describing the chemical dissolution of the soluble resist in the developer solvent.

resistance The opposition of current flow, accompanied by the dissipation of heat.

resistivity The measure of a material's resistance to current flow.

resolution In photolithography, the smallest isolated feature that can be printed on a resist-coated wafer.

reticle In photolithography, a quartz plate that contains a 4X, 5X, or 10X pattern to be reproduced on the wafer with a field size of a single die or several dice, and commonly used with a step-and-repeat stepper or step-and-scan system.

retrograde well Implanted wafer regions with peak dopant profile buried at a certain depth (e.g., several microns below the wafer surface). Also referred to as vertically modulated well.

reverse bias The polarity of the connections and amplitude of the voltage source needed to cut off the flow of current through a pn junction.

reverse osmosis (RO) A filtering technique by flowing water under pressure across a membrane filter to separate ionized salts, colloids, and organic materials.

RF Radio frequency—typically refers to alternating current having frequencies between low kHz to the low GHz. For industrial use, 13.56 MHz is the most common.

RF sputtering A diode sputtering system with a cathode and anode configuration and RF frequency source that can sputter nonconducting target materials.

rinse The use of DI water to neutralize, clean, and remove another liquid from a wafer.

RO *See* reverse osmosis.

roots blower A dry type vacuum pump that uses the action of two lobes or impellers rotating on parallel shafts to move the gas molecules. Also referred to as lobe pump.

roughing pumps Pumps which have several purposes: to achieve a rough vacuum (pressure down to 10^{-3} torr) in a chamber, to evacuate the entry area for wafers into a cluster tool (known as the loadlock), and to exhaust a high vacuum pump.

RTA *See* rapid thermal anneal.

RTP *See* rapid thermal processor.

salicide A method used to create silicides that are properly aligned with the exposed silicon of the source, drain, and polysilicon gate. Taken from the expression "self-aligned silicide."

scaling The shrinking of the overall dimensions and operating voltages of existing IC devices while maintaining electrical properties to attain improved operating performance as well as compactness.

SCALPEL A next-generation lithography method using an electron beam imaging approach (SCattering with Angular Limitation Projection Electron Beam Lithography).

scanner *See* scanning projection aligner.

scanning electron microscope (SEM) A microscope that produces an image by using reflected electron beams that scan the surface of the object and magnify the image up to 100,000X and greater.

scanning projection aligner An alignment and exposure system where the wafer and mask are mounted parallel to each other on a carriage while scanning through a narrow beam of light to project the image of the mask onto the wafer.

Schottky diode A high-speed diode formed when metal is brought in contact with lightly doped n-type semiconductor material.

screen oxide layer A thin oxide layer that reduces channeling effects by randomizing the direction of implanted ions as they enter the silicon lattice. Also called a sacrificial layer.

scribe line The separation between adjacent die on the wafer. This path is used as the cutting area in sawing a wafer into the individual die. Also referred to as saw lane.

scribe line monitors (SLMs) Test structures located in the scribe line region between the individual die.

scumming A residue film left on the wafer surface after resist development if the wafer has been underdeveloped (such as from inadequate developer volume) or improperly rinsed.

secondary electrons Electrons that are released from substances as a result of impact from high energy electrons or ions.

secondary electron flood A method to control wafer charging during ion implantation by flooding the surface of the wafer with low-energy electrons using an electron shower.

secondary ion mass spectrometry (SIMS) A method of eroding a wafer surface with ions accelerated in a magnetic field to analyze the surface material composition.

Seed's model A yield prediction model that assumes varying defect density across the wafer and from wafer to wafer and suitable for VLSI/ULSI technology.

selective etching Etching that is done so that certain material is removed but other materials or areas of the material are not affected by the etchant.

selective oxidation Silicon is selectively masked with silicon nitride to grow oxide only in the unmasked areas, since oxygen diffuses very slowly through the nitride mask.

selectivity A general term used in material removal that describes that rate of removal for one material versus the removal rate for the underlying or masking material. An example is CMP selectivity during polish.

semiconductor grade silicon The highly refined silicon used for ingot fabrication.

semiconductors An element such as silicon or germanium that has conductivity in the range between an insulator or a conductor. Integrated circuits are fabricated from semiconductors.

sensitivity For photolithography, the minimum amount of light energy (measured in mJ/cm^2) at a certain wavelength that is needed to produce a good pattern in the photoresist.

shallow trench isolation (STI) A device isolation technology based on an oxide-filled trench.

sheet resistance, R_s The end-to-end resistance of a thin film on a wafer that is proportional to the film resistivity and inversely proportional to its thickness. It is measured in the units of ohms/square.

sheet resistivity, ρ_s Term often used interchangeably with sheet resistance.

shot size The amount of resist dispensed per wafer.

shrinking A term describing the coordinated reduction in device feature size to improve IC performance as well as compactness. *See* scaling.

SI units International system of units based on the metric system.

sidewall spacers Used alongside the poly gates to prevent the higher S/D implant from penetrating too close to the channels where S/D punchthrough could occur.

silane (SiH$_4$) A colorless, flammable, and pyrophoric gas used as a silicon source for chemical vapor deposition epitaxial growth, polysilicon deposition, and low-temperature silicon dioxide and silicon nitride CVD.

silicide A metal compound used as an interface between silicon and other metals that is thermally stable and provides uniform electrical properties with low resistance.

silicon (Si) The element used in most semiconductor devices. Used to create wafers for semiconductor fabrication with a single-crystal structure. Silicon with randomly oriented crystals is called polycrystalline silicon, polysilicon, or poly.

silicon dioxide (SiO$_2$) High-quality stable electrical insulator material that is thermally grown or deposited on silicon wafers.

silicon nitride (Si$_3$N$_4$) A film deposited with chemical vapor deposition used as a passivation, masking, and insulating layer.

silicon on sapphire *See* SOS.

SOI, silicon on insulator A technique for producing a substrate with a deep oxide insulator layer that provides improved performance of low-power and high-speed CMOS devices.

silicon tetrachloride (SiCl$_4$) A colorless, corrosive, nonflammable liquid that forms hydrogen chloride when combined with water. It is used for epitaxial growth and high-temperature CVD deposition of silicon dioxide. Also called tetrachlorosilane.

silicon tetrafluoride (SiF$_4$) A colorless, corrosive gas used in its pure state and in oxygen mixtures to etch silicides, silicon, and polysilicon films.

single crystal silicon Silicon with a repeatable crystal structure having long-range order. Also referred to as monocrystal.

silicon on insulator (SOI) *See* SOI.

silicon tetrachloride (SiCl$_4$) A corrosive, colorless liquid used for high-temperature chemical vapor deposition of silicon dioxide and epitaxial deposition of single-crystal silicon.

silylation To place a thin layer of silicon on a surface, such as on resist for top surface imaging.

SIMOX Separation by implanted oxygen. A well-defined horizontal oxide layer in a wafer that serves as vertical isolation for devices in a silicon wafer.

single crystal A material with a continuous, regular crystalline lattice with no internal grain boundaries. Contrast with polysilicon.

slip A gross defect in a crystal, occurring when there is slippage of the crystal along one or more crystal planes.

slurry A mixture of abrasive particles and chemicals that is used to remove unwanted material from the wafer surface during CMP.

SMIF *See* standard mechanical interface.

sodium hydroxide (NaOH) A caustic base in the form of white crystals used as an etchant for polysilicon and silicon nitride.

soft bake In photolithography, the resist-coated wafer is subjected to an elevated temperature step to drive off solvent and improve adhesion. Also called pre-bake.

solid A state of matter that has its own fixed shape and will not conform to the container shape.

solvent The substance in which the solute dissolves.

SOS Silicon on sapphire. A fast MOS technology in which silicon is epitaxially grown on a sapphire wafer and etched away between transistors. Each device is thus isolated by air or oxide from other devices.

source One of the three electrodes of a field-effect transistor, along with the gate and drain.

source/drain (S/D) implants Highly doped regions (10^{20}–10^{21} ions/cm^3) that interface with the lightly doped active channel and well regions (10^{16}–10^{17} ions/cm^3) of a MOS transistor. S/D regions are doped to have the opposite conductivity type as the well that surrounds them.

spatial coherence The degree to which all the wavefronts emitted by an illumination source are in phase.

spatial signature analysis A defect analysis tool used after electrical test that recognizes unique bin map failure distributions to pinpoint the source of the wafer defects.

specialty gases The unique cylinder gases provided to the semiconductor industry for wafer fabrication in small quantities.

species In chemical reactions, a particular chemical substance that is an atom, ion, or molecule. In ion implantation it is the specific ion of an element that is separated from other elements for the purpose of implanting only one specific substance into the wafer.

specific gravity The ratio of the density of a substance to the density of water, measured at 4°C.

specific heat The physical property that specifies the amount of heat (energy) required to change the temperature of a given mass of a material; it is different for each gas.

speckle Interference patterns formed in the lithography image of the photoresist due to poor control of spatial coherence.

spectroscopic ellipsometry Analyzes the polarization change of laser light to measure thin film thickness.

spin coating (spin) An operation in photolithography where a controlled amount of photoresist is dispensed on a wafer while it is spinning to attain a uniform thickness of the resist.

spin dryer Wafer drying by high-speed rotation to remove moisture while wafers are sprayed with heated nitrogen.

spin-on-dielectric (SOD) A dielectric material that is applied to the wafer by spin coating with applications for interlayer dielectric.

spin-on-glass (SOG) A solvent-based liquid applied to a wafer surface by spinning and forms a thin, solid film with the properties of silicon dioxide. This was previously used for gap fill and planarization of the ILD before the wide acceptance of chemical mechanical planarization.

spray cleaning Wet cleaning chemicals are sprayed onto wafers placed in a cassette and rotated inside a sealed chamber.

spray rinser Using the physical force of the flowing water to dislodge residual chemicals on the wafer surface, a spray rinser is typically used in conjunction with wafer cleaning and drying.

spreading resistance probe A metrology tool used to measure both dopant concentration depth profiles and resistivity.

sputtering A method of depositing a film of material on an IC wafer. A target of the desired material is bombarded with excited ions that knock atoms from the target, which are then deposited on the wafer. *See* physical vapor deposition (PVD).

sputter etch A dry cleaning step before deposition in a processing chamber that removes native oxides and oxide residues from reverse sputtering on the wafer surface.

sputtered aluminum The primary process for interconnect wiring metallization since the early 1970s, with the aluminum alloy blanket deposited as a "slab" interconnect in a plane, followed by patterning.

sputtering yield In sputtering, the number of atoms ejected by the target (cathode) per incident ion that strikes it to determine the rate of sputter deposition.

SSI Small scale integration. Semiconductor fabrication technology that has a device density of between 2 to 50 devices on a single die.

stacking faults A form of dislocation fault due to layer stacking errors.

standard clean 1 (SC-1) The RCA cleaning mixture of $NH_4OH/H_2O_2/H_2O$ (ammonium hydroxide/hydrogen peroxide/DI water) mixed with a ratio range of 1:1:5 to 1:2:7.

standard clean 2 (SC-2) The RCA cleaning mixture of $HCl/H_2O_2/H_2O$ (hydrochloric acid/hydrogen peroxide/DI water) mixed in a ratio range of 1:1:6 to 1:2:8.

standard mechanical interface (SMIF) A handling interface system between the process tool and the wafer enclosure (pod).

standing waves In photolithography, interference between the incident light waves and reflected light waves that causes nonuniform exposure of the photoresist.

static RAM Static random access memory uses flip-flops for fast read-write memory cells based on transistors.

statistical process control (SPC) A method that uses statistics to identify and document process variables for process characterization and control.

step coverage The thickness of a deposited film over steps or features (surface topography) relative to the thickness on the top surface.

step height The depth of an etched opening or feature on the wafer surface.

step-and-repeat aligner A common alignment and exposure tool used for photolithography that projects only one exposure field (which may be one or more chips on the wafer) at a time, then has to repeat the process for all other fields. Also referred to as a stepper or aligner.

step-and-scan system The optical lithography step-and-scan is a hybrid tool that combines the technology from scanning projection aligners and step-and-repeat steppers by using a reduction lens to scan the image of a large exposure field onto a portion of the wafer.

stepper *See* step-and-repeat aligner.

stepping motor driver A bipolar IC used for controlling stepping motors. Sometimes referred to as stepper motor.

stepper A photolithography tool that aligns and exposes one die or a few die at a time. The tool "steps" to each subsequent die on the wafer.

stoichiometry Indicates how closely the deposited film approaches the ratio of the elements given by its nominal chemical formula.

straggle, ΔR_p The distribution of distances traveled by the dopant atoms implanted in the wafer.

stress A measure of force divided by area, measured in units of either pounds per square inch (Psi) or Pascal (Pa).

striations Defects appearing as lateral grooves on photoresist due to standing waves.

stripping The process of completely removing a coating such as photoresist.

structure Represents how the layers and materials for a particular device or feature are formed in silicon.

subatmospheric CVD (SACVD) *See* low-pressure chemical vapor deposition.

submicron Wafer feature sizes below 1 μm.

sub-quarter micron Wafer feature sizes below 0.25 μm.

substrate Another term for wafer. The material on which a microelectronic device is built. Such material may be active, like silicon, or passive, like alumina ceramic.

sublimation A process by which a solid can change directly to a gas.

substitutional atom A foreign atom displacing an atom of the host crystal and taking up its position in the crystal lattice.

subwavelength lithography The imaging of wafer critical dimensions that is significantly less than the light source wavelength during resist exposure.

sulfur hexafluoride (SF_6) A colorless, odorless, nontoxic, nonflammable gas used for dry plasma etching prior to CVD, including oxide and nitride etching and wafer cleaning. SF_6 is inert and disassociates in the presence of an RF field to form reactive fluoride ions.

sulfuric acid (H_2SO_4) A poisonous, corrosive liquid used to clean wafers and remove photoresist. Concentrated sulfuric acid can cause severe burns.

surface profiler Surface metrology tool which uses a stylus that comes in contact with the wafer surface with a low force capable of profiling soft production wafer films.

surface tension The energy required to increase the surface area of contact.

susceptor A general term for a wafer holder usually used when there is a means to heat the wafer holder because the wafers need to be heated.

target chamber The subsystem where ion implantation of the ion beam into the wafer surface takes place. Also referred to as an end station.

target In physical vapor deposition, a solid source material composed of the exact material composition to be deposited.

temperature ramp rates The rate of temperature change that the wafers are exposed to during heating and cooling, and measured in °C/min or °C/sec.

temperature A measure of how hot or cold a substance is relative to another substance.

TEOS A chemical source (tetraethylorthosilicate) for the CVD deposition of silicon dioxide.

test algorithm A computer program written by engineers that controls the test instrumentation to perform electrical tests.

test coverage Describes the percent of circuit nodes actually tested in wafer sort.

test structures Special structures located on wafers and used to test specific parameters for in-line parametric test. Also referred to as process control monitors.

test vectors During electrical test, the binary test patterns (string of 0s and 1s) that are input to the devices to verify correct output.

thermal budget The amount of thermal exposure (temperature multiplied by time) for a wafer during processing.

thermal oxide An oxide layer grown on silicon in a high-temperature furnace.

thermocompression bonding A wirebond method that uses thermal energy and pressure to bond an Au wire to the chip pad and the leadframe inner lead.

thermocouple A sensor that converts heat into millivolts that can be measured by sensitive instrumentation calibrated in °C or °F.

thermogravimetric analysis (TGA) A measurement of the solvent content of the resist as a function of temperature.

thermosonic bonding A wirebonding technique combining ultrasonic vibration, heat, and pressure to form a bond referred to as a ball bond.

thin films A thin layer of a dielectric or conductor material on the wafer surface.

thin small outline package (TSOP) A popular plastic package with a low height profile.

III-V compound Refers to semiconductors formed by combinations of materials in groups IIIA and VA of the periodic table of elements. Commonly refers to compounds of gallium, such as GaAs.

threshold The input voltage at which the output logic level changes state.

threshold voltage The lowest attainable voltage value (V_{GS}) that will turn on a FET.

threshold voltage adjustment implant A low energy implant used in CMOS processes to slightly-dope the well region of the MOSFETs with an exact amount of dopant (same as the well) to achieve a specified threshold voltage requirement for each FET.

throughput Refers to the number of wafers processed during a given time period.

time of flight SIMS (TOF-SIMS) Identifies ions of different mass by measuring the time it takes for an ion to travel the length of a fixed path, since the charged particle velocity is a function of mass.

titanium silicide The bonding of titanium with silicon to form a low-resistance compound of titanium silicide when raised to a temperature >700 °C. Also referred to as tisilicide.

TLV Threshold limit value. An exposure level under which most people can work consistently for 8 hours a day, day after day, with no harmful effect.

top antireflective coating (TARC) A method to reduce reflection at the interface between the resist surface and air.

top surface imaging Photoresist is imaged only at its top surface to provides a wider process latitude for focus.

topography Three-dimensional shapes on the surface of a wafer, usually referring to an irregular, bumpy surface containing IC structures in any state of completion. It is caused by processing steps such as patterning and etching.

torr The most common vacuum unit in the United States for vacuum; it is equivalent to the height of one mm of mercury (Hg) in a barometer.

toxic The quality of being dangerous to human health and life.

track system (also tracks) An integrated tool for processing many of the steps of photolithography (from surface preparation to develop) without operator intervention.

transient enhanced diffusion (TED) Dopant diffusion that occurs in the anneal step after implantation that is caused by extra interstitial atoms from the dopant ion implanted in the silicon.

transistor An active semiconductor device with three electrodes that may be used either as an amplifier or a switch.

transmission electron microscope (TEM) Similar to the SEM, differing in that the beam of electrons is transmitted through an ultrathin slice of the sample.

trench An etched region on a silicon substrate which is used for forming isolation regions between devices or for fabricating storage capacitors.

trench capacitor A three-dimensional device formed by dry etching a trench into silicon; it has replaced the planar storage capacitor for DRAM because of the shrinking memory cell size.

trichlorosilane (TCS or $SiHCl_3$) A flammable, corrosive liquid used as a silicon source in epitaxial growth and in the production of polycrystalline silicon.

triode planar reactor An RF dry plasma etch reactor that adds a third electrode to attain control over the amount of ion bombardment.

triple well For CMOS devices, a buried implanted well layer beneath the standard n-type and p-type retrograde wells for improved device performance and packing density.

trivalent Elements from Group IIIA of the periodic table, having three valence electrons.

tungsten (W) A refractory metal deposited to fill the via between adjacent metal levels and provide an electrical connection (plug).

tungsten etchback An etch process used to fill vias by depositing blanket tungsten into via openings followed by a dry plasma etch step to remove the excess tungsten layer.

tungsten hexafluoride (WF_6) A toxic, corrosive, nonflammable liquid used in tungsten silicide CVD.

tungsten plugs The most common metal currently used to fill vias, which is why plugs are often referred to as tungsten plugs.

turbomolecular pump (turbo pump) A versatile, reliable, and clean pump that uses high-speed turbine blades for pumping action and is widely used for fab equipment.

twin planes (twinning) A gross defect in which the crystal grows in two different directions from the same plane.

twin-well (twin-tub) An approach to define the active regions of the nMOS and pMOS transistors, consisting of a p-well and an n-well.

ULSI Ultra large scale integration. Semiconductor fabrication technology that creates a device density of >1 million devices on a single die.

ultralow penetration air (ULPA) Filters that have an efficiency of 99.9995% or better for particulate diameters greater than 0.12 micron.

ultrafiltration A filter that uses pressure and flow through a membrane with pore sizes ranging from 10 angstroms to 0.2 micron.

ultrafine particles Particles smaller than 0.1 μm in diameter, down to the smallest diameter detectable with a discrete particle counter.

ultrahigh purity (UHP) Required of chemicals in semiconductor manufacturing, contaminants are controlled to the parts per billion (ppb) to parts per trillion (ppt) range.

ultrahigh vacuum A continuation of high vacuum with stringent control of the vacuum chamber design and materials to minimize undesirable gas contaminants.

ultrashallow junctions A junction depth of tens of nanometers to scale the source and drain (S/D) junctions with the short channel length necessary for high-performance ICs.

ultrasonic bonding A wirebonding technique based on ultrasonic energy and pressure to form a wedge bond between the wire and pad.

ultraviolet *See* UV.

undercut The lateral etching that occurs as the etching proceeds vertically.

uniformity A term widely used in wafer fabrication to assess the entire wafer.

unit cell The simplest arrangement of atoms that, when repeated in three directions, gives the crystal structure.

unpatterned etching (stripping) The entire wafer surface is etched with no mask present. *See also* wet chemical strips.

unpatterned wafers Bare silicon wafers or wafers with blanket films used as monitor wafers in process runs to provide information about process conditions.

UV Ultraviolet. That portion of the electromagnetic spectrum with wavelengths between 4 nm and about 480 nm.

vacancy A crystal point defect occurring when an atom is removed from its lattice site.

vacuum Removal of gas molecules (e.g., air, moisture, and gas residues) in a closed container to achieve a pressure less than atmosphere.

vacuum wand A hand-held device that uses suction to pick up the wafer.

Van der Pauw method A method of measuring sheet resistivity with four probes on the periphery of an arbitrarily shaped sample.

vapor phase epitaxy (VPE) An epitaxial deposition process that can combine several source gases to deposit compound semiconductors.

vapor pressure Pressure exerted by vapor in a closed container at equilibrium.

vapor prime A common method in photolithography for applying the HMDS primer to the wafer surface with no contact of the wafer.

vaporization The process of changing from a liquid into a gas.

variable angle spectroscopic ellipsometry (VASE) A method of varying the incident light angle to accurately assess a film's thickness.

vertical furnace A conventional hot wall furnace, oriented vertically, having better temperature control and uniformity than a horizontal furnace.

vias Openings that pass through the various dielectric layers and are filled with a metal (i.e., tungsten) to form an electrical pathway from one metal layer to the metal layer either above or below it.

viscous flow Flow occurring when the gas behaves much like a fluid, usually at pressures greater than about 1×10^{-2} torr. Each molecule is strongly influenced by its neighbors.

VLSI Very large scale integration. Semiconductor fabrication technology that has a device density of between 100K to 1 million devices on a single die.

volatile memory A memory device that allows data to be stored and changed as desired, however, the data is lost when the power is turned off.

volatile The tendency of a substance to become a gas. High vapor pressure materials are volatile.

voltage regulator A single-chip device that regulates voltage delivered to a load.

wafer A round slice of silicon used to manufacture semiconductor devices in wafer fabrication, usually anywhere from 75 mm (3") to 300 mm (12") in diameter. During manufacture, a wafer may contain several hundred die. Also referred to as substrate.

wafer cassettes An open wafer carrier to hold one or more wafers.

wafer charging During ion implantation, the ion beam strikes the wafer and causes some positive ions to accumulate in the masking layer with an electric charge buildup.

wafer electrical test (WET) A test procedure that is performed on the wafer, usually after metal contact etch, to verify certain electrical characteristics (e.g., resistance and capacitance) meet specifications.

wafer etch Process for the chemical removal of a thin layer of the silicon surface during wafer preparation.

wafer flat or notch A region ground into the ingot to identify crystal orientation, dopant type, and mask/reticle location reference. For 200mm and larger wafers, an edge flat has been replaced by a notch.

wafer flatness Linear thickness variation across the wafer.

wafer-level reliability (WLR) A specialized form of parametric testing that uses wafer test structures to assess the reliability of the devices on the wafer.

wafer slicing The creation of thin silicon wafers from the monocrystal ingot.

wafer sort yield Percentage of acceptable die that pass the wafer sort test.

wafer sort The primary wafer level electrical test to verify each die functions properly at the completion of wafer fabrication. Also called electrical sort, wafer probe, or probe.

wafer tests The electrical tests (i.e., parametric and sort) done on wafers during fabrication.

wafer tilt The most widely used method to reduce channeling during ion implantation by tilting the wafer (e.g., 7°) relative to the incident ion beam.

wafer to wafer non-uniformity (WTWNU) Measure of the amount of film thickness variation across multiple wafers and often used to describe global planarization.

wafer-level packaging Creation of the 1st level interconnections and the package I/O (input/output) terminals on the wafer before it is diced.

water deionization Process of removing the electrically active salt ions using specially manufactured ion exchange resins. Referred to as DI water.

wavelength dispersive spectrometer (WDX) A spectrometer for identifying materials which operates based on a diffracting crystal and photo counter.

well A localized implanted or diffused region on a wafer, where a MOSFET can be constructed that is of opposite dopant type as the host well.

WET *See* wafer electrical test.

wet chemical strips Wet chemicals are sometimes used to remove wafer surface layers, often to achieve the high selectivity of wet etch.

wet cleaning station Cleaning tools used in the diffusion and thin films area to remove contamination and native-grown oxide before inserting the wafers into the process chambers.

wet etch Liquid chemicals such as acids, bases and solvents are used to chemically remove wafer surface material and is generally applicable only for larger geometries.

wet oxidation Water vapor introduced into thermal oxidation for an increased growth rate.

wet sinks A series of acid and rinse tanks housed in fume hoods. Also referred to as wet bench.

wirebonding The most common method for electrically connecting the aluminum bonding pads on the chip surface to the package inner lead terminals.

wiring A term that describes the electrical circuitry on a wafer layer. *See also* interconnect.

within-wafer nonuniformity (WIWNU) Measure of the amount of film thickness variation across a wafer surface and often used to describe global planarization.

X-ray Radiations of short wavelengths between ultraviolet and gamma rays in the electromagnetic spectrum.

X-ray fluorescence (XRF) X-ray beams focused on a surface and used for making film thickness measurements.

X-ray lithography An established next-generation lithography technology using an X-ray source to pattern an image on a resist-coated wafer with critical dimensions below 100 nm.

X-ray photoelectron spectroscopy (XPS) A spectroscope used to identify chemical species on a sample surface, analyzing to a sample depth of about 2 nm.

yield The ratio of the number of good units (e.g., die, wafers, etc.) to the total number of units produced. Usually expressed as a percent.

yield management systems An effort to improve fab yield by linking defect and parametric data back to the individual workstations in the fab process.

zeta potential A positive or negative electrical charge that can build up in colloids (very fine suspended particles in a liquid).

SOURCES

1. Official dictionary, SEMATECH, Inc., 1995. (http://www.sematech.org)
2. The Semiconductor International Manufacturing Process, *Glossary of Semiconductor Terms,* (Fullman Company, 1998).
3. *Microelectronics Glossary of Terms,* Integrated Circuit Engineering Corporation.
4. M. Madou, *Fundamentals of Microfabrication,* (New York: CRC Press, 1997).
5. Specialty gas information from Web page of Solkatronic Chemicals, Inc., Fairfield, NJ.

INDEX

193 nm DUV resists, 356
Aberration, 379
Absolute index of refraction, 377
Absorption, 95
AC tests. *See* Wafer sort tests
Acceleration column, 491–94
Accelerator, linear. *See* Linear accelerator
Acid gases, 107
Acid, hydrofluoric. *See* Hydrofluoric acid (HF)
Acids, 100
Acoustic streaming, 139
Action, etching. *See* Etching action
Activated dopants. *See* Dopants, activated
Activation, 479
Activation, crystal. *See* Crystal activation
Active components, 46–59
Additives, 352
Adhesion, 301, 350
Adhesion, film. *See* Film adhesion
Adjustment, threshold voltage implant. *See* Threshold voltage implant adjustment
Adjustment, threshold voltage. *See* Threshold voltage adjustment
Adsorption, 95, 266
Advanced assembly and packaging, 586–95
Advanced lithography, 414, 424–29
Advanced resist processing, 428–29
Advanced temperature control, 246
Advanced technology, 16
Advantages of CMP, 522–23
Advantages of ion implant, 484
Aerosols, 115
Air, 120
 filtering, 124
 ionization, 126
Airflow principle, 123
Alignment and exposure, 344
 quality measures, 407–08
 system, 206
 troubleshooting, 408–09
Alignment, 400–04
 coarse. *See* Global alignment
 definition, 400
 global. *See* Global alignment
Alignment mark deterioration, 403
Alignment marks, 401–03
 definition, 401
Alkali, 100
Alkaline substance. *See* Alkali
Alternatives to RCA clean, 142
Aluminum and metal stacks, 462–63
Aluminum, 4, 303–05
 metallization, 259
 subtractive process, 326
Aluminum-copper alloys, 305
Ambient, 242
Amorphous, 211
 materials, 69

Amorphous quartz. *See* Silica, fused
Analog circuits. *See* Circuits, analog
Analytical equipment, 171–177
Analyzer magnet, mass. *See* Mass analyzer magnet
Analyzer, ion. *See* Ion analyzer
Anion, 24
Anisotropic
 plasma etcher, 210
 profile, 438
Anneal, 206, 303
 furnace. *See* Furnace anneal
 rapid thermal. *See* Rapid thermal anneal (RTA)
Annealing, 499–501. *See also* Thermal treatments
 definition, 499
Anode, 315
Antireflective coating (ARC), 210, 272–75
Application specific slurry, 526. *See also* CMP slurry and pad
Applications
 CMP. *See* CMP applications
 dry etch. *See* Dry etch applications
Aqueous solutions, 99
Area array, 589
Argon-flouride (ArF), 375
Array
 area. *See* Area array
 perimeter. *See* Perimeter array
Asher overview, 467
Asher, photoresist. *See* Photoresist asher
Ashing, 142
 plasma. *See* Plasma ashing
Aspect ratio, 261
 dependent etching (ARDE), 333
Assay number, 130
Assembly and packaging
 advanced. *See* Advanced assembly and packaging
 quality measures, 596
Assembly
 extraction. *See* Extraction assembly
 traditional. *See* Traditional assembly
Atmospheric pressure, 404
 chemical vapor deposition (APCVD), 519
 CVD (APCVD), 270, 271
Atom, 22
Atomic force microscope (AFM), 173–74
Atomic mass unit, 26
Auger electron spectroscopy (AES), 174
Auger, Pierre, 174
Autodoping, 289
Automated
 test equipment (ATE), 554, 586. *See also* Final test
Automatic defect classification (ADC), 152

Back-end-of-line (BEOL), 532
Back-end. *See* Integrated circuit (IC) manufacturing process, back-end
Backgrind, 466. *See also* Wafer preparation, final assembly
Backing film, 531
Bacterial control, 129

INDEX

Bake
    DUV post-exposure. *See* DUV post-exposure bake (PEB)
    hard. *See* Hard bake
    post-exposure. *See* Post-exposure bake (PEB)
Ball grid array (BGA), 589–90
Ballroom layout, 123
Bardeen, John, 3
Barrel
    plasma etcher, 446
    reactor, 446
Barrier
    dopant. *See* Dopant barrier
    layer metallurgy (BLM), 588. *See also* Flip chip
    metals, 307–08
Barrier voltage, 47
Base, 100–01
Baseline compensation (BLC), 290
Basic sputtering steps, 315–16
Batch furnace, hot wall. *See* Hot wall batch furnace
Bay and chase layout, 123
Beam
    blow-up, 493
    current, 485
    deceleration, 493
    etching, ion. *See* Ion beam milling
    high-current and high-energy. *See* High-current and high-energy beam
    high-current and low-energy. *See* High-current and low-energy beam
    milling, ion. *See* Ion beam milling
    trap, neutral. *See* Neutral beam trap
Bias, 399
    etch. *See* Etch bias
BiCMOS technology, 57–58
Bin, 554
    map, 556
Bincode number, 556
Bipolar
    junction transistor (BJT), 49
    technology, 52
Bird's beak effect, 237
Bladder carrier, 532
Blanket tungsten CVD deposition, 321–23
Blower/booster pump, 186
Blow-up, beam. *See* Beam blow-up
Boat, wafer. *See* Wafer boat
Borophosphosilicate glass (BPSG), 269, 519
Borosilicate glass (BSG), 269
Bottom antireflective coating (BARC), 383
Boule. *See* Single-crystal ingot
Boundary layer, 267
Box, burn. *See* Burn box
Brattain, Walter, 3
Breakthrough step, 460
Brightfield detection, 161
Bromine gas chemistry, 354
Brush
    loaded. *See* Loaded brush
    scrubbing, 139
Budget, thermal. *See* Thermal budget
Buffered
    HF (BHF), 465
    oxide etch (BOE), 465
Bulk
    chemical distribution (BCD) system, 102
    gas distribution (BGD) system, 103
    gases, 103–04
    resistivity, 87

Bulkhead
    equipment layout, 131
    installation, 131
Buried oxide (BOX), 507
Burn box, 245
Burn-in test, 552
By-product removal, 266

Calculating resolution, 386
Calibration, 554
Cap oxide, 479
Capacitance, 31
Capacitance-voltage (C-V) test, 168–70
Capacitor, trench. *See* Trench capacitor
Capillary tip, 578
Captive chip producer, 8
Carbon, 79
Caro's acid, 137
Carrier
    bladder. *See* Bladder carrier
    wafer. *See* Wafer carrier
Carrier-depletion region, 47
Carrier. *See* printed circuit board (PCB)
Carriers, hot. *See* Hot carriers
Catalyzed, surface. *See* Heterogeneous reaction
Cathode, 315
Cation, 24
Cavitation, 139
CD-SEM, 166
Celsius, 93
Cell, tetrahedron. *See* Tetrahedron cell
Center of focus (COF), 276
Center slow [tool], 528
Central processing units. *See* Microprocessing unit
Ceramic
    DIP (CERDIP), 584
    packaging, 584–86
CERDIP. *See* Ceramic DIP
CGA (Compressed Gas Association) connector, 105
Challenges for in-line parametric testing, 555
Chamber, process. *See* Process chamber
Channel length, 211
Channeling, 501–02
    definition, 501
Characteristics, thin-film. *See* Thin-film characteristics
Charge neutralization, space. *See* Space charge neutralization
Charge
    fixed oxide. *See* Fixed oxide charge
    interface-trapped. *See* Interface-trapped charge
    mobile oxide. *See* Mobile oxide charge
    oxide-trapped. *See* Oxide-trapped charge
Charging, wafer. *See* Wafer charging
Chelating agents, 142
Chemical
    amplification (CA), 355
    for DUV resists, 355–56
    distribution, 102
    impurities, 79
    mechanical
        planarization (CMP), 83, 139, 204, 209, 269, 273, 321, 322, 388, 403, 517, 520–21, 520–34
        polish, 517
    mechanism, 444
    processes, CVD. *See* CVD chemical processes
    properties, 92
    reaction, 92
    reaction for oxidation, 231
    strips, wet. *See* Wet chemical strips

vapor cleaning, 137
vapor deposition (CVD), 203, 264, 265–69, 320
Chemically amplified DUV resists, 310
Chemistry, bromine gas. *See* Bromine gas chemistry
Chip. *See also* Microchip
  bonding, 576. *See also* Die attach
    epoxy attach, 576
    eutectic attach, 576
    glass frit attach, 577
  design, 81
  functionality. *See* Wafer sort
  performance, 284
  scale package. *See* Chip scale packaging
  sorting. *See* Wafer sort
  speed, increasing, 9–10
  fabrication. *See* Wafer fabrication
Chip on board (COB), 590
Chip scale packaging (CSP), 592
Chloride, hydrogen. *See* Hydrogen chloride (HCl)
Chlorinated agents, use of in oxidation. *See* Use of chlorinated agents in oxidation
Chlorine trifluoride (ClF_3), 274
Chuck, wafer. *See* Wafer chuck
Circuit geometry, 10. *See also* Feature size
Circuit integration, 4–6
  history of, 4
Circuits
  analog, 44
  digital, 44–45
Class number, 120
Classifying gases, 107
Clean
  CMP. *See* CMP clean
  post-CMP. *See* Post-CMP clean
  post-copper CMP. *See* Post-copper CMP clean
Cleaning, 84
  wafer. *See* Wafer cleaning
Cleanliness, 81
Cleanroom, 114
  garments, 121–22
  layout, 123
  protocol, 121–22
Clear-field mask, 340
Cluster tool, 182, 273
CMOS
  manufacturing steps, 205–222
  process flow, 199–204
CMP
  advantages of. *See* Advantages of CMP
  applications, 535–38
  clean, 532–34
  equipment manufacturers, 534
  equipment, 529–32
  mechanisms, 523–25
  metal. *See* Metal CMP
  planarity, 521–22. *See also* Chemical mechanical planarization
  quality measures, 538–39
  slurry and pad, 526–29
  troubleshooting, 540
  variables, 529
Coalescence, nuclei. *See* Nuclei coalescence
Coarse alignment. *See* Global alignment
Coater/developer track system, 201, 342
Coefficient
  diffusion. *See* diffusion coefficient, D
  of thermal expansion (CTE), 98, 588
Coherence probe microscopy (CPM), 168
Cold wall, 248
Cold-wall reactor, 270
Collimated sputtering, 320
Color interference contrast, 163
Column, acceleration. *See* Acceleration column
Combined physical and chemical mechanism, 444
Comparison of photo tools, 405
Complementary metal-oxide semiconductor (CMOS), 52–58
Compound, 22
  semiconductors, 39–40
Computer as host or server/network, 554–55
Condensation and vaporization, 95
Conditioning, 527. *See also* Polishing pad
Conductivity, 29, 301
Conductor, 23, 29
Confocal contrast microscope, 163
Conformal step coverage, 167
Consumables, 526. *See also* CMP slurry and pad
Contact, 300
  aligner, 278–79
    definition, 278
  angle, 171
    meters, 171
  formation, 214
  metal etch, 464
Contactor,. *See also* final test, 586
Contaminants, 75
  and particles, 351
  mobile ionic. *See* Mobile ionic contaminants (MICs)
Contamination, 114
  particle. *See* Particle contamination
Continuous film, 263
Continuous spray development, 419–20
Contour maps, 155
Contrast, 163, 349
Control
  advanced temperature. *See* Advanced temperature control dose.
  model-based temperature. *See* Model-based temperature control
  system, furnace. *See* Furnace control system
  temperature. *See* Temperature control
  thermocouple, 244
Controlled
  diffusion. *See* Diffusion controlled
  reaction-rate. *See* Reaction-rate controlled
Controller, dose. *See* Dose controller
Conventional i-line
  PEB, 416
  photoresists, 351–54
  resists, 419
Converter chips, 57–58
Cook's theory, 523
Cooling
  elastomeric. *See* Elastomeric cooling
  gas. *See* Gas cooling
  wafer. *See* Wafer cooling
  rate, 79
Copper, 305–07
  barrier metals, 308–09
  challenges, 306–07
  CVD, 323
  damascene structure, 326–28
  electroplate, 323–25
  interconnect, 305
  metal slurry, 526. *See also* CMP slurry and pad
  overburden, 326
  polish, dual-damascene. *See* Dual-damascene copper polish
Correction, optical proximity. *See* Optical proximity correction (OPC)
Correlation microscope, 168
Corrosion, 301

INDEX

Cost of ownership (COO), 406
Covalent bond, 28
Coverage, film-step. *See* Film-step coverage
Creation, ion. *See* Ion creation
Critical dimension (CD), 165–66, 206, 439
    generations, 337
Critical layers, 259
Crucible, 73–79
    holder, 73
Cryoarrays, 187
Cryogenic aerosol cleaning, 142
Cryopump, 187
Crystal, 33, 69
    activation, 480
    defect, 78, 79, 87
        dislocation, 78, 79
        gross, 78
        point, 78
    growth, 73–81
    ingot, 74
    lattice, 69, 78
    orientation, 71–73, 74, 235–36
    plane, 72, 80
    puller, 74, 81
    single, 75
    solid, 79
    structure, 68–71, 78, 79, 80, 81
Crystal defects, 563. *See also* Wafer sort yield
    in silicon, 78–80
Crystal dislocations, 83
Crystal. Seed, 73
Crystalline defect formation, 79
Crystal-melt interface, 79
Crystals, silicon 73, 74, 75
Cu precursors, 323
Cup, Faraday. *See* Faraday cup
CurpraSelect, 323
Current, beam. *See* Beam current
Current-driven current amplifier, 51
CVD
    chemical processes, 265
    deposition systems, 269–82
    equipment design, 270
    gas-flow dynamics, 267
    high-density plasma. *See* high-density plasma CVD (HDPCVD)
    quality measures, 292
    reaction, 266–69
        steps, 266
    reactor
        configuration, 270
        heating, 270
        summary, 270–71
    troubleshooting, 292–94
Cyclotron resonance, electron. *See* Electron cyclotron resonance
Cylinder change-out, 106
CZ crystal puller, 72–73
Czochralski (CZ) method, 72, 73, 74, 75, 79. *See also* CZ process

Damage
    plasma. *See* Plasma damage
    plasma-induced. *See* Plasma-induced damage
    radiation. *See* Radiation damage
Damascene, 214
Dark space, 195
Darkfield detection, 161, 163
Dark-field mask, 340
Data
    management, 152
    trends, 551. *See also* Interpreting parametric data
DC tests, 449–50. *See also* In-line parametric test
De Forest, Lee, 2
Deal and Grove linear parabolic model, 236
Decel. *See* Beam deceleration
Deceleration, beam. *See* Beam deceleration
Deep buried layers, 503
Deep UV (DUV), 354
    photoresists, 354–57
Defect
    density, 78, 149
    generation, 79
    reduction. *See* Yield management systems
Defect inspection (DI), 208
Defects, 149
Deglaze, 481
Degree of planarization (DP), 413. *See also* CMP planarity
Dehydrated. *See* Hydrophobic
Dehydration bake, 346
Deionization, 127
Deionized water, 102
Delay, PEB. *See* PEB delay
Density, 97–98
Dep/etch. *See* Deposit/etch
Dep:etch (D:E) ratio, 281
Depletion mode, 58
Deposit/etch, 325
Deposited oxide layer, 225
Deposition, 97, 301
    atmospheric pressure chemical vapor. *See* Atmospheric pressure chemical vapor deposition
    chemical vapor. *See* Chemical vapor deposition (CVD) film. *See* Film deposition
    techniques, film. *See* Film deposition techniques
Depressions, 79
Deprotection, 355
Depth
    junction. *See* Junction depth
    of field. *See* Depth of focus (DOF)
    of focus (DOF), 386–87
Design
    for test (DFT), 559
    RTP. *See* RTP design
DESIRE (Diffusion–Enhanced SIlylated Resist), 319
    process, 319–20
Desorption, 266
Detection
    endpoint. *See* Endpoint detection
    motor current endpoint. *See* Motor current endpoint detection
    optical endpoint. *See* Optical endpoint detection
Develop, 344, 416–22
    continuous spray. *See* Continuous spray develop
    inspect, 345, 422–23
    puddle. *See* Puddle develop
    quality measures, 320–30
    troubleshooting, 431–32
Developer, 418
    selectivity, 418
    speed. *See* Dissolution rate
    temperature, 421
    time, 421
    volume, 421
Development
    methods, 419–21
    negative resist. *See* Negative resist development
    parameters, resist. *See* Resist development parameters

652 INDEX

positive resist. *See* Positive resist development
puddle. *See* Puddle development
resist. *See* Resist development
Device
    fabrication, 79
    geometries, 78
    isolation, 286–87
    protection and isolation, 227
    technology, 79
    yield, 78
Device under test (DUT), 553
Dewetted, 141
DI water. *See* Deionized water
Diameter. *See also* Wafer geometry
    control, 81
    grinding, 81
Diameter Index Safety System (DISS) cylinder valve, 105
Diazonapthoquinone (DNQ), 353
Dichloroethylene (DCE), 234
Die
    attach, 575–76. *See also* Wafer preparation, final assembly
    separation, 575. *See also* Wafer preparation, final assembly
    singulation. *See* Die separation
Dielectric, 30
    and performance, 282–87
    between metal layers, 229–30
    breakdown, 78
    constant (k), 31, 32, 282
    dry etch, 457
    etch, 436
    layers, 259
    first interlayer. *See* First interlayer dielectric (ILD-1)
    gate oxide. *See* Gate oxide dielectric
    interlayer. *See* Interlayer dielectric (ILD)
    premetal. *See* Premetal dielectric (PMD)
    spin-on. *See* Spin-on dielectric
Differential lens, 493
Diffraction, 379–80
    orders, 380
Diffusion, 200, 232, 479–82
    bay, 200
    coefficient, D, 479
    controlled, 235
    definition, 479
    furnace, vertical. *See* Vertical diffusion furnace (VDF)
    lateral. *See* Lateral diffusion
    principles, 479–81
    process, 481–82
    thermal. *See* Thermal diffusion
    transient enhanced. *See* Transient enhanced diffusion (TED)
Diffusivity, 479
Digital circuits, *See* Circuits, digital
Dilute cleaning chemistries, 137
Dimension, critical. *See* Critical dimension
Dimensional
    control, 81
    specifications, 77
Dioxide
    stress in silicon. *See* Stress in silicon dioxide
    thermal silicon. *See* Thermal oxide
Direct chip attach (DCA). *See* Chip on board (COB)
Directionality, 438
Disadvantages of ion implant, 484
Dishing, 525. *See also* CMP mechanisms
Dislocations, 79–80
    edge, 79
Dispense, 357

Dissolution, 79
    rate, 309
    rate monitor (DRM), 421
    system, gas. *See* Gas distribution system
    plasma potential. *See* Plasma potential distribution
    potential. *See* Potential distribution
DI-water
    filtration, 128
    installation, 127
    rinse, 140
DNQ-novolak, 353
Donors, 35–36
Dopant
    barrier, 228
    concentration, 159–61
    effects, 235
    movement, 479–80
    profile, 477
    sources, 481–82
    uniformity, 81
Dopant
    pentavalent phosphorus, 74
    trivalent boron, 74
Dopants, 35, 73, 74
    activated, 480
Doped, 74
    oxide, 272
    region, 477
    regions, 477–79
Doping, 34–35, 475
    during CVD, 268
Dose (Q), 484–85
    control, 499
    controller, 499
    monitor, 376
Double-sided polishing (DSP), 83
Downstream
    etch systems, 448
    reactor, 448
Drain, lightly doped. *See* Lightly doped drain (LDD)
Drive-in [step], 479
Dry
    cleaning, 142
    etch, 436, 443–45
        applications, 456–64
        definition, 443
        dielectric. *See* Dielectric dry etch
        metal. *See* Metal dry etch
        silicon. *See* Silicon dry etch
        troubleshooting, 470–71
    in/dry out [process], 426
    mechanical pump, 185
    oxygen, 231
    plasma etcher, 208
    thermal oxidation, 250–51
Dual
    in-line package (DIP), 582
    plasma source (DPS), 453
Dual-damascene
    copper polish, 537–38
    process, 326
Dual-polysilicon gate structure, 506
Dump
    rinse, 141
    rinser, 141

DUV
    post-exposure bake (PEB), 306
    process requirements, 356–57
    resists, chemically amplified. See Chemically amplified DUV resists
Dynamic
    dispense, 359
    SIMS, 172
Dynamic random access memory (DRAM), 61

E–beam lithography, 286–87
E–beam. See Electron beam
Economies of scale, 76
Edge
    contour, 82–83
    exclusion, 423
    grind. See Polished wafer edge
Edge-bead removal (EBR), 360
Edison, Thomas, 2
Effect
    bird's beak. See Bird's beak effect
    pressure. See Pressure effect
Effects
    dopant. See Dopant effects
    loading. See Loading effects
    pattern density. See Pattern density effects
Eight basic steps of photolithography, 342–45
Elastomeric cooling, 497
Electrical
    current, 28
    parameters, 79
    sort. See Wafer sort
Electrical test and sort department, 222
Electrochemical deposition (ECD). See Copper electroplate
Electrode, 315
    drain, 53
    source, 53
    suppression. See Suppression electrode
Electrofill. See Copper electroplate
Electromagnetic spectrum, 261–62
Electromigration, 305
Electron, 22, 29
    beam, 396
    cyclotron resonance (ECR), 280, 451
    flood, plasma. See Plasma electron flood
    flood, secondary. See Secondary electron flood
    volt (ev), 22
Electron-hole pair, 29
Electronic Industries Association of Japan (EIAJ), 587
Electronic Numeric Integrator and Calculator (ENIAC), 3
Electronic
    stopping, 486
    wafer map. See Wafer sort, performing
Electrons, hot. See Hot carriers
Electroplating. See Copper electroplate
    fundamentals, 324
Electropolishing, 105
Electrostatic
    chuck (ESC), 133, 281
    discharge (ESD), 119, 125, 397
    scanning, 495
Element, 22
    heating. See Heating element
Elements,
    pentavalent, 35
    trivalent, 35
Ellipsometry, 156
Emission spectroscopy, optical. See Optical emission spectroscopy

End-of-line
    back. See Back-end-of-line (BEOL)
    front. See Front-end-of-line (FEOL)
End removal, 81
End station, 498
Endpoint
    detection, definition, 455
    motor current. See Motor current endpoint
    optical interferometry. See Optical interferometry endpoint
Energy- and wavelength-dispersive spectrometer (EDX and WDX), 176
Energy-band theory, 23
Energy-dispersive spectrometer (EDX), 176
Enhanced diffusion, transient. See Transient enhanced diffusion (TED)
Enhancement
    mode, 58
    plasma. See Plasma enhancement
Environmental conditions, 404
Epitaxial
    growth, 289, 290
    layer, 87–88
    silicon, 88, 206
Epitaxy, 289–91
    definition, 289
    growth methods, 290–91
Epoxy
    attach. See Chip bonding
    underfill, 588–89
Equation, Preston's. See Preston's equation
Equipment
    CMP. See CMP equipment
    furnace. See Furnace equipment
    manufacturers, CMP. See CMP equipment manufacturers
Erosion, 525. See also CMP mechanisms
E–sort. See Electrical test and sort department
Etch, 202, 435
    applications, dry. See Dry etch applications
    bias, 439–40
        definition, 439
    buffered oxide. See Buffered oxide etch (BOE)
    chambers, vacuum for. See Vacuum for etch chambers
    contact metal. See Contact metal etch
    dielectric. See Dielectric etch
    dielectric dry. See Dielectric dry etch
    dry. See Dry etch
        definition. See Dry etch, definition
    high-density plasma. See High-density plasma etch
    inspection, 469
        quality measures, 469–70
    metal. See Metal etch
    metal dry. See Metal dry etch
    parameters, 437–43
    poly gate. See Poly gate etch
    processes, 436–37
    profile, 438–39
        definition, 438
        isotropic. See Isotropic etch profile
    rate, 437–38
        definition, 437
    reactive ion. See Reactive ion etch (RIE)
    reactors, plasma. See Plasma etch reactors
    residue, 441–42
    resistance, 350
    silicon. See Silicon etch
    silicon dry. See Silicon dry etch
    single-crystal silicon. See Single-crystal silicon etch
    system review, 453–55

654 INDEX

systems, downstream. *See* Downstream etch systems
uniformity, 441
wet. *See* Wet etch
    oxide. *See* Wet oxide etch
    types of. *See* Types of wet etch
Etchback, 209, 328, 518–19
    planarization. 518–19
    tungsten. *See* Tungsten etchback
Etcher, barrel plasma. *See* Barrel plasma etcher
Etchers, high-density plasma. *See* High-density plasma etchers
Etching, 83
    action, 444
    aspect ratio dependent. *See* Aspect ratio dependent etching (ARDE)
    ion beam. *See* Ion beam milling
    patterned. *See* Patterned etching
    silicon trench. *See* Silicon trench etching
    unpatterned. *See* Unpatterned etching
Ethylene glycol monomethyl ether acetate (EGMEA), 360
Ethylenediamine-tetra-acetic acid (EDTA), 142
Eutectic
    attach. *See* Chip bonding
    temperature, 303
Evaporation, 313–14
Excimer laser, 372, 374–76
    definition, 374
Exclusion
    edge. *See* Edge exclusion
    zone, 81
Exhaust flow, 422
Exponential model. *See* Poisson's model
Exposure
    control, 376
    dose, 350
    sources, 372–76
Extraction
    and ion analyzer, 490–91
    assembly, 490
Extreme UV, 424

Fab yield response. *See* Wafer sort
Fabless company, 8
Fabrication
    cost, 77
    process, 75
        semiconductor, 82
    requirements, 81
Facility, 123–26
Factors affecting oxide growth, 235
Faraday cup, 499
Faraday's law of electrolysis, 325
Fast ramp
    furnace, 245
    vertical furnace, 245
Fault models, 559–60
Faults, stacking, oxidation-induced. *See* Oxidation-induced stacking faults (OISFs)
Feature size, 10, 15
Fick's laws, 233, 479
Field oxide layer, 228
Field-by-field alignment, 294
Field-effect transistor (FET), 52–55
    JFET, 52,
    MOSFET, 52, 53–55
        nMOS (n-channel), 53
        pMOS (p-channel), 53

Film
    adhesion, 263
    backing. *See* Backing film
    continuous. *See* Continuous film
    deposition, 260
        techniques, 264
    formation, 277
        of silicide. *See* Formation of silicide films
    growth, 263
    layering terminology, 258–60
    oxide. *See* Oxide film
        nature of. *See* Oxide film, nature of
        uses of. *See* Oxide film, uses of
    precursors
        at wafer, 266
        formation of. *See* Formation of film precursors
    purity and density, 262
    stress, 158
    structure, 263
    thickness, 153–57
        metrologies, 153
    thin. *See* Thin film
Film-step coverage, 261
Filter efficiency, 130
Filters, 130
Final
    assembly and packaging, 571
    test, 586
Fine alignment. *See* Field-by-field alignment
First interlayer dielectric (ILD–1), 259, 268, 321
First mask exposure, 403
Fixed oxide charge, 233
Fixed-quality area (FQA), 81, 86
Flat zone, 242
Flatness, 81, 82, 86
Flats, 81
Fleming, John, 2
Flip chip, 587–88
Float-zone method, 75
Flood
    plasma electron. *See* Plasma electron flood
    secondary electron. *See* Secondary electron flood
Flow, exhaust. *See* Exhaust flow
Fluorinated compounds, 107
Fluorosilicate glass (FSG), 269
Focal
    length, 378
    point, 378
Focused ion beam (FIB), 176–77
Formation
    of film precursors, 266
    of silicide films, 239
    polymer. *See* Polymer formation
Forward bias, 48
Foundry, 9
Four-point probe, 154
    resistivity check, 81
Freeman ion source, 489
Front-end-of-line (FEOL), 532
Front-opening unified pod (FOUP), 134
Functional tests, 558–59
Furnace
    anneal, 500
    control system, 245
    equipment, 238–40
    fast ramp. *See* Fast ramp furnace
    horizontal. *See* Horizontal furnace

hot wall batch. *See* Hot wall batch furnace
tube. *See* Process chamber
    materials, 242
vertical. *See* Vertical furnace
    diffusion. *See* Vertical diffusion furnace (VDF)
    fast ramp. *See* Fast ramp vertical furnace
Furnaces, horizontal versus vertical. *See* Horizontal versus vertical furnaces

Gain, 51
Gallium arsenide (GaAs), 40, 291
Gap fill, 217, 279
Gaps, high aspect ratio. *See* High aspect ratio gaps
Gas, 91–92
    cooling, 497
    distribution system, 244–45
    line connections, 105
    piping, 105
    purge, 105
    stick, 106
    transport to deposition zone, 266
Gases, 103–08
Gate oxide, 52, 210
    dielectric, 228
    integrity (GOI), 168, 228
    structure, 228
Gate
    poly. *See* Polysilicon gate
    polysilicon. *See* Polysilicon gate
    structure, dual-polysilicon. *See* Dual-polysilicon gate structure
Generator, photoacid. *See* Photoacid generator (PAG)
Germanium, 4, 33
Getter, 75
Gettering, 234
Glass, 226
    frit attach. *See* Chip bonding
    reflow, 239, 411
Glazing, 419
Global
    alignment, 404
    planarization. *See* Planarization, global
Glow discharge, 193–95
Grid, 291
Gross defects, 80
Grove, Sir William Robert, 314
Grown oxide layer, 225
Growth
    factors affecting oxide. *See* Factors affecting oxide growth
    film. *See* Film growth
    island. *See* Island growth
    model, oxidation. *See* Oxidation growth model
    phase, initial. *See* Initial growth phase
    process, 80
    rate, 79
    rate of oxide. *See* Rate of oxide growth
    thermal oxidation. *See* Thermal oxidation growth
Guardbanding, 559, 560

Handling, 132–33
Hard bake, 344, 422
Head, polishing. *See* Polishing head
Heat, 93
    zones, 242
Heating element, 243
Height, step. *See* Step height
Helicon, 280
    wave, 452
Hermetic sealing, 577

Heteroepitaxy, 289
Heterogeneous reaction, 266
Hexamethyldisilazane (HMDS), 320, 343, 346
HF
    buffered. *See* Buffered HF (BHF)
    last step, 137
High
    temperature processing, 80
    vacuum, 184
    vacuum pumps, 184, 186–88
High aspect ratio gaps, 261
High-current
    and high-energy beam, 493
    and low-energy beam, 493
High-density plasma
    CVD (HDPCVD), 193, 217, 268, 269, 270, 280, 531
    etch, 450
    etchers, 450–52
High-efficiency particulate air (HEPA), 114, 125
High-k dielectric constant, 285
High-temperature
    cofired ceramic (HTCC), 585
    diffusion furnace, 200
High-vacancy densities, 79
Hillocks, 305
Historical perspective [etch], 466
Hollerith
    Herman, 3
    mechanical tabulating machine, 3
Homoepitaxy, 289
Homogeneous reaction, 266
Horizontal
    furnace, 239, 240
    versus vertical furnaces, 240–48
Hot
    carrier, 505
    DI-water rinsing, 141
    electron. *See* Hot carrier
    wall batch furnace, 239–40
Hot-neutral CVD, 281
Hot-wall reactor, 270
Humans, 121–22
Humidity, 404
Hybrid scanning, 496
Hydrated. *See* Hydrophilic
Hydrides, 107
Hydrofluoric acid (HF), 137, 141, 304, 357
Hydrogen
    chloride (HCl), 234
    silsesquioxane (HSQ), 288
Hydrophilic, 346
    surface, 141
Hydrophobic, 346
    surface, 141
Hydroxide, tetramethyl-ammonium. *See* Tetramethyl-ammonium hydroxide (TMAH)
Hyperfiltration. *See* Reverse osmosis (RO)

IC
    capacitor structures, 46
    electrical tests, 546–47
    metrology, 150–52
    products, 60–62
    reliability, 552
    resistor structures, 45
I_{DDQ} testing, 559, 560
I-line PEB, conventional. *See* Conventional i-line PEB
I-line resists, conventional. *See* Conventional i-line resists

Imaging, top-surface. *See* Top-surface imaging
Implant
  adjustment, threshold voltage. *See* Threshold voltage implant adjustment
  parameters, ion. *See* Ion implant parameters
  trends in process integration, ion. *See* Ion implant trends in process integration
Implantation
  ion. *See* Ion implantation
  troubleshooting, ion. *See* Ion implantation troubleshooting
Implanter, ion. *See* Ion implanter
Implants, source/drain. *See* Source/drain implants
Importance of alignment and exposure, 368–70
Impurities. *See* Dopants
Impurity control, 75
In situ sputter etch, 314
Increase in number of processing steps, 562
Increased die size, 562
Inductively-coupled plasma (ICP), 280, 451–52
  definition, 451
Inert gas, 92, 103
Infrared (IR), 360
Ingot, 72–73, 75, 81, 82
  growth, 76, 79, 81, 82
  monocrystal, 75
Initial growth phase, 236
Ink mark. *See* Wafer sort, performing
In-line parametric test, 547–55
  equipment, 552–55
Inner lead bond region (ILB), 591
Inspect, develop. *See* Develop inspect
Inspection
  etch. *See* Etch inspection
  post-develop. *See* Post-develop inspection
Insulator, 23
Integrated
  circuit (IC), 4
    advanced, 78
    fabrication, 7
    final assembly, 571
    manufacturing process, back-end. *See* Final assembly and packaging
    packaging, 571
  measurement tool, 151
Integration
  ion implant trends in process. *See* Ion implant trends in process integration
Integrity, gate oxide. *See* Gate oxide integrity
Interconnect, 300
  delay, 284
  local. *See* Local interconnect (LI)
Interface, oxide-silicon. *See* Oxide-silicon interface
Interface-trapped charge, 233
Interference of light waves, 370–71
Interlayer dielectric (ILD), 215, 283, 300, 370, 371, 457, 519, 532, 536
  first. *See* First interlayer dielectric (ILD-1)
Internal diameter saw, 82
International 300 mm Initiative (I300I), 77
Interpreting parametric data, 551
Interstitial, 79
  impurities, 79
Interstitial movement, 479–80
Ion, 23–24
  analyzer, 382
  beam etching (IBE). *See* Ion beam milling
  beam milling, 448–49
    definition, 448

  creation, 381
  etch, reactive. *See* Reactive ion etch (RIE)
  implant, 202, 203
    advantages of. *See* Advantages of ion implant
    disadvantages of. *See* Disadvantages of ion implant
    parameters, 376–79
    quality measures, 400–01
    trends in process integration, 395–400
  implantation, 117, 238, 367–69, 374–79
    overview, 375
    troubleshooting, 401–02
  implanter, 203, 206, 375–76
  implanters, 380–94
  projection lithography (IPL), 426
  sheath. *See* Dark space
  source, 380–81
  source, Freeman. *See* Freeman ion source
Ionic
  bond, 26
  contaminants, mobile. *See* Mobile ionic contaminants (MICs)
Ion-induced deposition, 281
Ionization, 24
Ionized
  metal plasma (IMP), 318, 320
  metal plasma PVD (IMPPVD or ionized PVD), 320
IPA vapor dry, 142
Island growth, 263
Isolation
  regions, 208
  shallow trench. *See* Shallow trench isolation (STI)
Isopropyl alcohol (IPA), 137
Isotropic etch profile, 438

Joint Electronic Device Council (JEDEC), 587
Junction
  depth, 206, 478
  spiking, 304
Junctions, ultrashallow. *See* Ultrashallow junctions

Kelvin, 93
Kerf, 82
Kilby, Jack, 4
Killer defects, 115
Kinetically controlled, 267
Krypton-fluoride (KrF), 375

Laminar airflow, 123
Laminated ceramic, 586
Langmuir, 314
Lapping, 82
Larger wafer diameter, 561–62
Laser scattering. *See* Light scattering
Laser-scribing, 81
Latchup, 59–60, 206
Lateral diffusion, 480–81
Lattice, 79
  site, 79
Law of reflection, 376
Laws, Fick's. *See* Fick's laws
Layer
  stacking errors, 79
  thickness, polish. *See* Polish time and layer thickness
Layering terminology, film. *See* Film layering terminology
Leadless chip carrier (LCC), 583
Leakage current failure, 78
Lens, 378–79
  compaction, 378
  definition, 378

differential. *See* Differential lens
    material, 378–79
LI
    oxide polish, 535–36
    tungsten polish, 536
Lifetime, 29
Light, 370–72
    intensity, 373
    microscope. *See* Optical microscope
    scattering, 163. *See also* Surface perfection
        defect detection, 163–64
        on patterned wafers, 165
    spectrum, 337
Light-emitting diode (LED), 291
Lightly doped drain (LDD), 505
    implant, 211
    implants process, 211
Limit, solid solubility. *See* Solid solubility limit
Linear
    accelerator, 493
    stage, 234
Linewidth, 206
Liquids, 91, 99–103
Lithography. *See* Photolithography
    advanced. *See* Advanced lithography
    ion projection. *See* Ion projection lithography (IPL)
    next-generation. *See* Next-generation lithography
    subwavelength. *See* Subwavelength lithography
    X–ray. *See* X–ray lithography
Loaded brush, 533
Loading effects, 330
Loadlock, 188
Lobe pump. *See* Blower/booster pump
Local interconnect (LI), 214–15, 535, 536
    process, 214–15
Local
    oxidation of silicon (LOCOS), 237, 286, 357, 535
    planarization. *See* Planarization, local
Low vacuum, 183
Low-k dielectric requirements, 284
Low-pressure chemical vapor deposition (LPCVD), 208, 270, 273–77, 321
Low-temperature cofired ceramic (HTCC), 585

Magnetic
    CZ (MCZ), 74. *See also* Czochralski method
    field, 74
Magnetically enhanced RIE (MERIE), 453
Magnetron. *See* Magnetically enhanced RIE (MERIE)
    sputtering, 318, 319–20
Main-etch step, 460
Makeup loop, 127
Mask. *See* Photomask
Masking. *See* Photolithography
Masks, phase-shift. *See* Phase-shift masks (PSM)
Mass
    analyzer magnet, 490
    flow controllers, 189
    flow controller (MFC), 189
Mass-transport limited, 267
Material selectivity, underlying. *See* Underlying material selectivity
Materials, furnace tube. *See* Furnace tube materials
Matter, 22
Mean free path (MFP), 184
Measurement equipment, 151
Measurements, quality. *See* Quality measurements [oxidation]
Measures
    etch inspection quality. *See* Etch inspection quality measures
    ion implant quality. *See* Ion implant quality measures
Mechanical
    damage, 82
    scanning, 496
    shock, 80
    stress, 79, 83
Mechanisms, CMP. *See* CMP mechanisms
Medium vacuum, 184
Megasonic cleaning, 138
Megasonics, 138–39
Melt, 74, 76, 79
Membrane
    contactor, 128
    filter, 130
Memory
    nonvolatile, 61–62
    volatile, 60–61
Merchant chip supplier, 8
Mercury arc lamp, 372–73
Metal
    CMP, 523–24
    contacts, 214
    CVD, 320–23
    deposition systems, 313–25
    dry etch, 462–64
    etch, 436
        contact. *See* Contact metal etch
    impurities, 116
    layers, 259
        dielectric between. *See* Dielectric between metal layers
    oxide, 53
    plugs, 312–13
    polish, 523–25
    sputtering tools-physical vapor deposition (PVD), 203
    stack, 217
Metal-1 interconnect formation, 217
Metal-2 interconnect formation, 219–20
Metal-3 to pad etch and alloy, 220–21
Metallic
    contaminants, 75
    impurities, 116–17
Metallization
    aluminum. *See* Aluminum metallization
    multilevel. *See* Multilevel metallization
    quality measures, 329
    schemes, 325–28
    troubleshooting, 330
Metalorganic CVD (MOCVD), 291
Methods, development. *See* Development methods
Metrology, 149
Microchip, 2, 76, 77, 78
    fabrication. *See* Wafer fabrication
Microchips, price of, 12–13
Microdefect. *See* Crystal defect
Microlithography, 336
Microloading. *See* Aspect ratio dependent etching (ARDE)
Microprocessor unit (MPU), 61
Microroughness, 86
Miller indices, 72
Milling, ion beam. *See* Ion beam milling
Minienvironment, 133–34
Mix and match, 296
Mobile
    ionic contaminants (MICs), 117, 127, 233, 234, 249, 262, 269, 418
    oxide charge, 233

Model, oxidation growth. *See* Oxidation growth model
Model-based temperature control, 246
Modifications to RCA clean, 137
Modulated well, vertically. *See* Vertically modulated well
Molecular-beam epitaxy (MBE), 291
Molecule, 22
Momentum transfer, 316
Monitor
    dissolution rate. *See* Dissolution rate monitor (DRM)
    wafers, 150
Monocrystal
    interface, 75
    seed, 75
    silicon, 75
        growth, 73–75
Moore, Gordon, 10, 14
Moore's law, 10–11, 15, 556
Motor current endpoint, 531
    detection, 531
Movement
    dopant. *See* Dopant movement
    interstitional. *See* Interstitional movement
    substitutional. *See* Substitutional movement
Multichip module (MCM), 591
Multilevel metallization, 258
Murphy's model, 564

n+ S/D implant, 213
Native oxides, 118, 206
Nature of oxide film. *See* Oxide film, nature of
Neck-down procedure, 79
Negative
    i-line photoresists, 352–53
    lithography, 339–40
    resist, 339, 417
        development, 417
    versus positive resists, 348
Neutral beam trap, 386
Neutralization, space charge. *See* Space charge neutralization
Next-generation lithography, 424
Nine nines efficiency, 130
Nitride, silicon. *See* Silicon nitride
n-LDD implant, 211
Noble gas. *See* Inert gas
Noncritical layers, 259
Nonuniformity
    wafer-to-wafer. *See* Wafer-to-wafer nonuniformity
    within wafer. *See* Within-wafer nonuniformity
Normality, 418–19, 421
Notch, 81
Notched wafer, 81
Novolak, 353
Noyce, Robert, 4, 14, 15
npn transistor, 49–50
Nuclear stopping, 486
Nucleation, 263
Nuclei coalescence, 263
Number of I/O leads, 589
Numerical aperture (NA), 380–81
n-well, 53
    formation, 206

Off-axis illumination (OAI), 399
Ohmic contact, 303
Ohm's law, 303
Opaque films, 154–55
Operational amplifier (op-amp), 60
Optical
    emission spectroscopy, 347
    endpoint detection, 531
    enhancement techniques, 398–400
    exposure, 370
    filters, 371
    interferometry endpoint, 531
    lithography, 336, 367, 370–88
    microscope, 161–62
    microscopy, 156, 161–63
    proximity correction (OPC), 305, 398–99
    pyrometer, 248
    system, 162
Optics, 376–85
    definition, 376
Organic
    contaminants, 117
    contamination, 117
Orientation, 81
    crystal. *See* Crystal orientation
Out-diffusion, 289
Outer lead bond region (OLB), 591
Output checks, 558
Overburden, 529. *See also* Polish time and layer thickness
Overdrive. *See* Wafer positioning
Overetch step, 460
Overflow rinsers, 140
Overlay
    accuracy, 338, 400
    budget, 338, 400
    registration (OL), 208
    registration, 167–68
Over-temperature thermocouple, 244
Overview
    [of ion implantation]. *See* Ion implantation, overview
    of areas in a wafer fab, 200–04
    asher. *See* Asher overview
Oxidation growth
    model, 231–38
    thermal. *See* Thermal oxidation growth
Oxidation, 265
    chemical reaction for. *See* Chemical reaction for oxidation
    dry thermal. *See* Dry thermal oxidation
    local, of silicon. *See* Local oxidation of silicon (LOCOS)
    process, 248–51
        recipe, 250
    selective. *See* Selective oxidation
    troubleshooting, 253–54
    wet. *See* Wet oxidation
Oxidation-induced stacking faults (OISFs), 238
Oxide, 457–59
    buried. *See* Buried oxide (BOX)
    cap. *See* Cap oxide
    charge
        fixed. *See* Fixed oxide charge
        mobile. *See* Mobile oxide charge
    dielectric, gate. *See* Gate oxide dielectric
    etch
        buffered. *See* Buffered oxide etch (BOE)
        wet. *See* Wet oxide etch
    film, 226–29
        nature of, 226–27
        uses of, 227
    growth
        factors affecting. *See* Factors affecting oxide growth
        rate of. *See* Rate of oxide growth
    integrity, gate. *See* Gate oxide integrity

layer
    deposited. *See* Deposited oxide layer
    field. *See* Field oxide layer
    grown. *See* Grown oxide layer
    screen. *See* Screen oxide layer
    pad. *See* Pad oxide
    polish, 523. *See also* CMP mechanisms
        LI. *See* LI oxide polish
        STI. *See* STI oxide polish
    sacrificial. *See* Sacrificial oxide
    slurry, 526. *See also* CMP slurry and pad
    structure, gate. *See* Gate oxide structure
    thermal. *See* Thermal oxide
    versus silicon doping, 269
Oxide-silicon interface, 233
Oxide-trapped charge, 233
Oxidizing gas, 103
Oxygen, 75, 79
    content, 86
    dry. *See* Dry oxygen
    impurities, 79
Oxygen-induced stacking faults (OISF), 79
Ozone (O_3), 142

p+ S/D implant, 213
Packaging, 84
    ceramic. *See* Ceramic packaging
    levels, 573
        1st level, 573
        2nd level, 573
    plastic. *See* Plastic packaging
    traditional. *See* Traditional packaging
    wafer-level. *See* Wafer-level packaging
Pad
    oxide, 237
    polishing. *See* Polishing pad
Parabolic stage, 235
Parallel
    plate (planar) reactor, 447
    scanning, 496
    testing, 559
Parameters
    etch. *See* Etch parameters
    ion implant. *See* Ion implant parameters
    resist development. *See* Resist development parameters
Parametric testing, 220
Parasitic
    capacitance structures, 46
    resistance structures, 45
    transistors, 59
Particle
    contamination, 294, 443
    density, 116
Particles, 87, 115
    in the ion implant process, 502
    per wafer per pass (PWP), 116, 164
Parts
    per billion (ppb), 100
    per billion-atomic (ppba), 172
    per million (ppm), 100
    per trillion (ppt), 100
Passivation
    layer, 221
    surface. *See* Surface passivation
Passive component structures, 45–46
Pattern
    density effects, 524–25
    sensitivity, 524–25. *See also* CMP mechanisms

Patterned
    etching, 328
    surface defects, 164–65
    wafers, 150, 164. *See also* Production wafers
Patterning. *See* Photolithography
Patterning/planarization, 301
PEB
    conventional i-line. *See* Conventional i-line PEB
    delay, 415
Pellicle, 397
Performing wafer sort. *See* Wafer sort, performing
Perimeter array, 589
Periodic table of the elements, 24–26
Perspective, historical [etch]. *See* Historical perspective [etch]
PETEOS, 279
pH, 101
    scale, 101
Phase, initial growth. *See* Initial growth phase
Phase-shift mask (PSM), 305, 398
Phosphosilicate glass (PSG), 268, 272
Photo. *See* Photolithography
Photoacid generator (PAG), 306, 355
Photoacoustics, 157
    technology, 157
Photoactive compound (PAC), 353, 383
Photolithography, 201–02, 276, 336
    concepts, 336–39
    equipment, 278–96
    processes, 339–41
Photolysis, 265
Photomask, 336, 394
Photoresist, 348
    asher, 359
    dispensing methods, 357–60
    physical properties, 349–51
    quality measures, 362
    removal, 358–360
    selectivity, 351
    stripping, 358
    troubleshooting, 363
Physical
    dimensions, 85
    mechanism, 444
    properties, 92
    vapor deposition (PVD), 313, 314
Physics of sputtering, 316–18
Pin grid array (PGA), 585, 589. *See also* Ceramic packaging
Pin-in-hole (PIH), 581
Piranha mixture, 137
Planar
    reactor, triode. *See* Triode planar reactor
    technology, 4
Planarity, 521. *See also* CMP planarity
Planarization
    chemical mechanical. *See* Chemical mechanical planarization
    degree of. *See* Degree of planarization (DP)
    etchback. *See* Etchback planarization
    global, 516
    local, 516
    traditional. *See* Traditional planarization
Planarized wafer. *See* Wafer, planarized
Plasma, 91, 92, 192–95
    ashing, 359
    CVD, high-density. *See* High-density plasma CVD (HDPCVD)
    damage, 467
    definition, 192
    electron flood, 497
    enhancement, 236

etch reactors, 446–47
etch, high-density. *See* High-density plasma etch
etcher, 202
    barrel. *See* Barrel plasma etcher
etchers, high-density. *See* High-density plasma etchers
etching, 193
    inductively-coupled. *See* inductively-coupled plasma
    potential distribution, 444
    resist stripper, 202
    source, dual. *See* Dual plasma source
Plasma-assisted CVD, 277–81
Plasma-based dry cleaning, 142
Plasma-enhanced chemical vapor deposition (PECVD), 192, 218, 270, 278, 282, 320
Plasma-induced damage, 334
Plastic
    leaded chip carrier (PLCC), 583. *See also* Types of plastic packages
    packages, types of. *See* Types of plastic packages
    packaging, 581–84
    packaging. *See* Packaging, plastic and Traditional packaging
p-LDD implant, 211–12
Plug, 300
    polish, tungsten. *See* Tungsten plug polish
pn junction, 39, 46–47, 79
    diode, 46–48
pnp transistor, 50
Point defects, 78–79
Point-of-use (POU), 102, 128
Poisson's model, 564. *See also* Wafer sort yield models
Polish, 204, 209, 409. *See* also Chemical mechanical planarization
    chemical mechanical. *See* Chemical mechanical polish
    dual-damascene copper. *See* Dual-damascene copper polish
    LI oxide. *See* LI oxide polish
    LI tungsten. *See* LI tungsten polish
    metal. *See* Metal polish
    oxide. *See* Oxide polish
    rate and uniformity, 528–29. *See also* CMP slurry and pad
    STI oxide. *See* STI oxide polish
    time and layer thickness, 529. *See also* CMP slurry and pad
    tungsten plug. *See* Tungsten plug polish
    wafer edge, 83. *See also* Edge grind
Polisher, 520–21. *See also* Chemical mechanical planarization
Polishing, 83
    head. *See* Wafer carrier
    loop, 127
    pad, 527. *See also* CMP slurry and pad
Poly gate. *See* Polysilicon gate
    etch, 460
    structural process, 210
Poly. *See* Polysilicon film
Polycide, 309
    structure, 276
Polymer formation, 442
Poly-si. *See* Polysilicon film
Polysilicon, 53
    bar, 75
    doped, 75
    film, 276
    gate, 210, 506
    granular, 72
Polyvinyl alcohol (PVA), 140
Positive
    i-line photoresists, 353–54
    lithography, 339, 340–41
Positive resist, 340, 418
    development, 418
Post-accelerator, 493

Post-CMP clean, 533–34
Post-copper CMP clean, 534
Post-develop inspection, 423
Post-exposure bake (PEB), 344, 383, 415
    definition, 414
    DUV. *See* DUV post-exposure bake (PEB)
Posts. *See* Wirebonding
Potential
    distribution, 444–46
    plasma. *See* Plasma potential distribution
    zeta. *See* Zeta potential
Preamorphization, 502
Pre-bake. *See* Soft bake
Precursor
    adsorption, 266
    diffusion, 266
Precursors, 266
Predep. *See* Predeposition
Predeposition, 479
Premetal dielectric (PMD), 259
Pressure
    and vacuum of a gas, 93–95
    effect, 236
    in CVD, 267–68
    of a gas, 93
Preston's equation, 523
Primary flat, 81
Priming techniques, 346
Principles, diffusion. *See* Diffusion principles
Print bias, 399
Printed circuit board (PCB), 573
Probe. *See* Wafer sort
    card, 553
    interface, 553
Prober. *See* Wafer positioning
Process
    chamber, 181, 242, 498–99
        contamination, 195
        gas flow, 189
    characterization [soft bake], 361
    chemicals, 99–107, 130
    control monitors (pcms), 548–49
    DESIRE. *See* DESIRE process
    diffusion. *See* Diffusion process
    flow. *See* Wafer preparation
    gases, 130
    latitude, 339
    maturity, 563. *See also* Wafer sort yield
    oxidation. *See* Oxidation process
        recipe, 250
Processes
    CVD chemical. *See* CVD chemical processes
    etch. *See* Etch processes
Processing, advanced resist. *See* Advanced resist processing
Processor, rapid thermal. *See* Rapid thermal processor (RTP)
Product, 92
Production
    equipment, 131
    wafers, 150
Profile
    anisotropic. *See* Anisotropic profile
    dopant. *See* Dopant profile
    etch. *See* Etch profile
    isotropic etch. *See* Isotropic etch profile
    sidewall. *See* Sidewall profile
    thermocouple, 244
Projected range, R_p, 485
Projection lithography, ion. *See* Ion projection lithography (IPL)

Properties, 92
    of materials, 92–99
Propylene glycol monomethyl ether acetate (PGMEA), 360
Protection and isolation, device. *See* Device protection and isolation
Proximity
    aligner, 389–90
        definition, 389
    correction, optical. *See* Optical proximity correction (OPC)
Puddle
    develop, 420–21
    dispense and spin, 346
Pump speed, 189
Punchthrough, 211, 504
    stoppers, 505
Purity and density, film. *See* Film purity and density
p-well, 53, 59
    formation, 207
Pyrogenic steam, 231
Pyrogens, 127
Pyrolysis, 265, 276
Pyrometer, optical. *See* Optical pyrometer

Quad flatpack (QFP), 582. *See also* Types of plastic packages
Quadrupole mass analyzer (QMA), 191
Quality measurements [oxidation], 251–54
Quality measures, 84–87, 152–71. *See also* Wafer preparation
    alignment and exposure. *See* Alignment and exposure quality measures
    assembly and packaging. *See* Assembly and packaging quality measures
    CMP. *See* CMP quality measures
    CVD. *See* CVD quality measures
    definition, 152
    develop. *See* Develop quality measures
    etch inspection. *See* Etch inspection quality measures
    ion implant. *See* Ion implant quality measures
    metallization. *See* Metallization quality measures
    photoresist. *See* Photoresist quality measures
    test. *See* Test quality measures
    wirebond, 579
Quartzware, 242

Radiation damage, 484
Radical chemicals, 208
Radicals, 194
Radio frequency (RF), 193
Raised source/drain (S/D) structures, 289
Range, 485
    projected. *See* Projected range, Rp
Rapid thermal anneal (RTA), 240, 310, 500–01
Rapid thermal process (RTP), 213
Rapid thermal processor (RTP), 240, 246–47, 253, 312
Rate
    dissolution. *See* Dissolution rate
    etch. *See* Etch rate
    limiting step, 267
    monitor, dissolution. *See* Dissolution rate monitor (DRM)
    of oxide growth, 234
    polish. *See* Polish rate
Rates, temperature ramp. *See* Temperature ramp rates
Ratio
    aspect. *See* Aspect ratio
    gaps, high aspect. *See* High aspect ratio gaps
RCA clean, 135–37
Reactant, 92
Reaction
    for oxidation, chemical. *See* Chemical reaction for oxidation
    heterogeneous. *See* Heterogeneous reaction
    homogeneous. *See* Homogeneous reaction
Reaction-rate
    controlled, 235
    limited, 267
Reactions, surface. *See* Surface reactions
Reactive ion etch (RIE), 450, 466
Reactor, 181, 247, 265
    barrel. *See* Barrel reactor
    downstream. *See* Downstream reactor
    parallel plate (planar). *See* Parallel plate (planar) reactor
    triode planar. *See* Triode planar reactor
Reactors, plasma etch. *See* Plasma etch reactors
Rectifiers, silicon-controlled. *See* Silicon-controlled rectifiers (SCRs)
Redeposition, 281
Redistribution, 479
Reducing gas, 103
Reduction, 265
Reduction-oxidation (redox), 265
Reflection, 281
    of light, 376
    spectroscopy, 156–57
Reflective notching, 382
Reflectometry. *See* Reflection spectroscopy
Refraction, 158, 376
Refractive index, 158
Refractory
    ceramic, 585. *See also* Ceramic packaging
    metals, 308
Regeneration, 188
Registration, 400
Relative index of refraction, 376
Reliability, 301
Removal
    photoresist. *See* Photoresist removal
    residue. *See* Residue removal
Residual gas analyzer (RGA), 190–92
Residue
    etch. *See* Etch residue
    removal, 468
Residues, 441–42
Resin, 351
Resist
    adhesion, 350
    development, 413
        parameters, 421–22
        negative. *See* Negative resist development
        positive. *See* Positive resist development
    negative. *See* Negative resist
    positive. *See* Positive resist
    processing, advanced. *See* Advanced resist processing
    viscosity, 350
Resistance, 30
    heaters, 73
Resistivity (r), 29–30, 74, 153
    measurement procedure, 81
Resists
    chemically amplified DUV. *See* Chemically amplified DUV resists
    conventional i-line. *See* Conventional i-line resists
Resolution, 385–88, 338, 349
    photolithography definition, 385
    versus depth of focus, 388
Resonance, electron cyclotron. *See* Electron cyclotron resonance
Reticles, 336, 394–97
    definition, 394
    fabrication, 396–97

materials, 393
  reduction and size, 395–96
Retrograde
  implant, 205
  wells, 503
Reverse
  bias, 47–48
  osmosis (RO), 128, 130
  sputter. *See* Sputter etch
Review, etch system. *See* Etch system review
RF (radio frequency)
  energy, 194
  heating coils, 73
  sputtering, 318–19
RGA as real-time monitor, 191–92
RIE, magnetically enhanced. *See* Magnetically enhanced RIE (MERIE)
Rinse, 313
Robotic
  material handling, 81
  wafer handlers, 132
Roots blower, 186
Roots-type blower, 186
Rough vacuum. *See* Low vacuum
Roughing pumps, 184, 185–86
Roughness. *See* Surface perfection
Roundness, 81
RTP design, 247
Runout, 339

Sacrificial oxide, 502
Salicide, 311
Sandwich. *See* Metal stack
Scaling, 10, 56
SCALPEL (Scattering with Angular Limitation Projection Electron Beam Lithography), 425–26
Scanner. *See* Scanning projection aligner
Scanning
  electron microscope, 164, 165–66
  projection aligner, 388, 390
    definition, 390
    system, 494–95
  electrostatic. *See* Electrostatic scanning
  hybrid. *See* Hybrid scanning
  mechanical. *See* Mechanical scanning
  parallel. *See* Parallel scanning
Scatterometry. *See* Light scattering
Schottky diode, 51
Screen oxide layer, 502
Scribe line monitors (SLMs). *See* Process control monitors
Scrubbers, 139–40, 245
Scumming, 421
Secondary
  electron flood, 497
  flat, 81
Secondary-ion mass spectrometry (SIMS), 171–72
Seed end, 81
Seed's model, 564. *See also* Wafer sort yield models
Selection of aluminum, 303
Selective oxidation, 236
Selectivity, 528–29. *See also* Polish rate and uniformity
  developer. *See* Developer selectivity
  in etch, 440–41
  photoresist. *See* Photoresist selectivity
  underlying material. *See* Underlying material selectivity
Self-aligned salicide, 311
SEMATECH, 15
Semiconductor, 4, 23, 32–33, 78
  industry, 2, 81
    and Asia, 14–15
    careers in, 16–18
    changes in, 14–16, 76
    development of, 2–4
  manufacturing, 80, 81
    chemical properties for, 93–99
Semiconductor Leading Edge Technology (Selete), 77
Sensitivity, 350
  pattern. *See* Pattern sensitivity
Sensitizer, 352
Shallow trench isolation (STI), 237–38, 261, 286, 465, 535
  definition, 207
  process, 207–09
Shaping operations, 81, 82
Sheet
  resistance, 153–54
  resistivity, 154
Shock
  mechanical, 80, 82
  thermal, 80, 82
Shockley, William, 3
Shrinking feature sizes, 56, 563. *See also* Wafer sort yield
Sidewall
  profile, 459
  spacer, 212
    formation, 212
Siemens process, 68
Silane (SiH_4), 210, 271
Silane-based borophosphosilicate, 272
Silica, 68, 127
  fused, 72. *See also* Amorphous quartz
Silicides, 309–12
  films, formation of. *See* Formation of silicide films
  structure. *See* Polycide structure
Silicon, 3, 4, 7, 33–39, 68
  dioxide (SiO_2), 34, 238, 271, 274–75
    PECVD, 278
    stress in. *See* Stress in silicon dioxide
    thermal. *See* Thermal oxide
  doped, 37
  etch, 328
    dry, 459–62
    single-crystal. *See* Single-crystal silicon etch
  extrinsic, 35
  high-purity 75
  ingot, 76
  monocrystal 81
  intrinsic, 33
  local oxidation of. *See* Local oxidation of silicon (LOCOS)
  metallurgical-grade (MGS) 68
  nitride (Si_3N_4), 275, 459
    PECVD, 279
  n-type 35–36, 72–74
  polycrystalline, 71. *See also* Polysilicon
  p-type, 35, 36–37, 72, 74
  semi conductor grade (SGS), 68, 76, 79. *See also* Electronic-grade silicon
  oxynitride (SiOxNy), 276, 279
  trench etching, 353–54
Silicon-controlled rectifiers (SCRs), 243
Silicon-on-insulator (SOI), 507
silylation, 319
SIMOX (Separation by IMplanted OXygen), 507–08
Simultaneous deposition and etching, 281
Single in-line
  package (SIP), 582. *See also* Types of plastic packages

Single-crystal
    ingot, 80
    silicon etch, 461
Sinter, 303
Sintering, 307
SiO$_2$ deposition
    with silane, 271
    with TEOS-ozone, 271
    with SiH$_4$, 275
    with TEOS, 275
Slip, 80. See Gross defect
Slippage, 80
Slurry and pad, CMP. See CMP slurry and pad
Slurry, 526. See also CMP slurry and pad
    application-specific. See Application-specific slurry
    copper metal. See Copper metal slurry
    oxide. See Oxide slurry
    tungsten metal. See Tungsten metal slurry
Small-outline IC (SOIC), 582
Socket. See Contactor
Soft bake, 344, 360–61
    equipment, 360
Solid, 91
    solubility, 480
        limit, 480
Solid-state transistor, 3
Solutes, 99
Solvents, 99, 101–02, 352
    evaporation, 357
Source
    dual plasma. See Dual plasma source
    Freeman ion. See Freeman ion source
    ion. See Ion source
Source/drain implant (S/D), 210–11, 505–06
    definition, 505
    processes, 212
Sources
    dopant. See Dopant sources
    of particles. See Cleanliness
    of reticle damage, 397
Space charge neutralization, 493–94
Spatial
    coherence, 375–76
    signature analysis (SSA), 556
Specialty gases, 103, 104–07
Species, 266
Specific gravity (SG), 98
Specifications, 81
Spectroscopic ellipsometry, 156
Spectroscopy, optical emission. See Optical emission spectroscopy
Speed, developer. See Dissolution rate
Spike thermocouple. See Control thermocouple
Spin
    coat, 343, 348–60
    coating, 357
        equipment, 357–58
        parameters, 358–59
    dryer, 141
Spin-off, 357
Spin-on
    dielectric (SOD), 264, 287, 288
    dielectrics, 287–88
    films, 412. See also Traditional planarization
Spin-on-glass (SOG), 287–88
    system, 203, 209
Spin-up, 357

Spray
    cleaning, 139
    develop, continuous. See Continuous spray develop
    dispense and spin, 347
    rinse, 141
    rinser, 141
Spreading resistance probe (SRP), 160–61
Sputter etch, 281, 308, 318
Sputtered aluminum, 312
Sputtering, 314–20
    mechanism, 316
    yield, 317
Stacking faults. See Dislocation
    oxidation-induced. See Oxidation-induced stacking faults (OISFs)
Stacks, aluminum and metal. See Aluminum and metal stacks
Stage
    linear. See Linear stage
    parabolic. See Parabolic stage
Stagnant layer, 267
Stand-alone measurement tool, 151
Standard
    clean 1 (SC–1), 135–36, 137
    clean 2 (SC–2), 135, 136–37
    mechanical interface (SMIF), 134
    spin coaters, 288
Standing waves, 382–83
States of matter, 91–92
Static
    dispense, 358
    dissipative, 126
    SIMS, 172
Station, end. See End station
Statistical process control (SPC), 152, 565. See also Yield management systems
Steam, pyrogenic. See Pyrogenic steam
Step
    breakthrough. See Breakthrough step
    coverage, 167
    height, 329
    main-etch. See Main-etch step
    overetch. See Overetch step
Step-and-repeat aligner, 278, 280–82
    definition, 280
Step-and-scan
    alignment marks, 403
    system, 368, 388, 393–94
        definition, 393
Stepper, 201, 278. See also Step-and-repeat aligner
Stepper motor driver, 60
Steps, CVD reaction. See CVD reaction steps
STI
    oxide
        fill, 209
        polish, 427
        polish-nitride strip, 209–10
        trench etch, 207–08
Stoichiometry, 262, 312
Stoppers, punchthrough. See Punchthrough stoppers
Stopping
    electronic. See Electronic stopping
    nuclear. See Nuclear stopping
Storage and handling, 350
Straggle, Δr_p, 485
Stress, 98–99, 301
    in silicon dioxide, 238
Stripping, 328
    photoresist. See Photoresist stripping
Strips, wet chemical. See Wet chemical strips

Structure
- dual-polysilicon gate. *See* Dual-polysilicon gate structure
- film. *See* Film structure
- gate oxide. *See* Gate oxide structure
- monocrystal, 70–71. *See also* Single crystal
- polycrystal, 70–71

Structures, test. *See* Process control monitors
Sub-fab area, 123
Sublimation, 96
- and deposition, 96–97

Substitutional
- impurities, 79
- movement, 372

Substrate, 4, 33, 573. *See also* Printed circuit board (PCB)
Subwavelength lithography, 314, 398
Suppression electrode, 490
Surface
- catalyzed. *See* Heterogeneous reaction
- defects, 78
- mount technology (SMT), 581
- passivation, 227
- planarity, 278
- polishing. *See* Wafer preparation
- profiler, 167
- reactions, 266
- tension, 98, 350
- topography. *See* Topography

Susceptor. *See* Wafer holder
Synchrotron, 427
System
- furnace control. *See* Furnace control system
- gas distribution. *See* Gas distribution system
- scanning. *See* Scanning system
- wafer transfer. *See* Wafer transfer system

Systems, downstream etch. *See* Downstream etch systems

Tang end, 81
Tape automated bonding (TAB), 591. *See also* Advanced assembly and packaging
Target, 315
Techniques, film deposition. *See* Film deposition techniques
Technology nodes, 10
Temperature, 93, 404
- and humidity, 125
- control, 244
  - advanced. *See* Advanced temperature control
  - model-based. *See* Model-based temperature control
- developer. *See* Developer temperature
- ramp rates, 245
- uniformity, 415

TEOS oxide, 280
Terminology, film layering. *See* Film layering terminology
Test
- AC. *See* Wafer sort tests
- algorithm, 554
- coverage, 558–59. *See also* Functional tests. *See also* Wafer sort
- DC. *See* DC tests
- functional. *See* Functional tests
- issues at wafer sort. *See* Wafer sort, test issues at
- quality measures, 565–66
- structures. *See* Process control monitors
- troubleshooting, 567
- vectors, 558. *See also* Functional tests
- wafer sort. *See* Wafer sort tests
- wafers. *See* Monitor wafers

Tester instrumentation, 554
Tetraethylorthosilicate (TEOS), 271
Tetrahedron cell, 226

Tetramethyl-ammonium hydroxide (TMAH), 418–19
Thermal
- budget, 226
- coefficient of expansion (TCE), 581
- decomposition. *See* Pyrolysis
- diffusion, 475
- energy. *See* Heat
- expansion, 98
- gradient, 79
- oxidation, 79
  - dry. *See* Dry thermal oxidation
  - growth, 231–38
- oxide, 226
- process steps, 83
- shock, 80

Thermal-wave system, 159
Thermocouple (TC), 244
- control. *See* Control thermocouple
- over-temperature. *See* Over-temperature thermocouple
- profile. *See* Profile thermocouple
- spike. *See* Control thermocouple

Thermogravimetric analysis (TGA), 361
Thickness uniformity, 262
Thin films, 203–04, 225, 260
- characteristics, 260–63

Thin small outline package (TSOP), 582. *See also* Types of plastic packages
Threshold
- voltage, 55
- voltage adjustment, 504
- voltage implant adjustment, 504

Throughput, 189
Tilt, wafer. *See* Wafer tilt
Time
- developer. *See* Developer time
- polish. *See* Polish time and layer thickness
- of flight secondary-ion mass spectrometry (TOF-SIMS), 173

Tisilicide. *See* Titanium silicide
Titanium silicide, 213
Tool repeatability, 77
Top antireflective coating (TARC), 384–85
Topography, 226, 515
Top-surface imaging, 428
Torr, 95, 182
Total organic carbon (TOC), 127
Total test time, 451. *See also* Wafer sort, test issues at
Total-reflection XRF (TRXRF), 157
Tracks. *See* Coater/developer track system
Traditional
- aluminum structure, 325–26
- assembly, 574–80
- packaging, 580–86
  - plastic, 581–84
- planarization, 410–12

Transfer system, wafer. *See* Wafer transfer system
Transient enhanced diffusion (TED), 501
Transistor, 3, 78
Transmission electron microscope (TEM), 86, 174
Transport to deposition zone, gas. *See* Gas transport to deposition zone
Trap, neutral beam. *See* Neutral beam trap
Trench
- capacitor, 506
- etching, silicon. *See* Silicon trench etching

Trends in process integration, ion implant. *See* Ion implant trends in process integration
Triboelectricity, 119
Trichloroethane (TCA), 234

INDEX **665**

Trichloroethylene (TCE), 234
Trichlorosilane, 68
Triode planar reactor, 340
Triple well, 395
Tube, furnace. *See* Process chamber
Tungsten, 463
    CVD, 321
    etchback, 463–64
    metal slurry, 526. *See also* CMP slurry and pad
    plug, 216, 312
        polish, 536
    polish, LI. *See* LI tungsten polish
Tunneling current, 286
Turbo pump. *See* Turbomolecular pump
Turbomolecular pump, 186
Twin planes. *See* Gross defect
Twinning. *See* Gross defect
Twin-tub. *See* Twin-well
Twin-well, 205
    process, 205–07
Types
    of alignment, 403–04
    of contamination, 114–19
    of metals, 301–13
    of parametric tests, 550
    of photoresists, 348
    of plastic packages, 582–83. *See also* Plastic packaging
    of wet etch, 465–66

Ultrafiltration, 128
Ultrafine particles, 120
Ultrahigh
    purity (UHP), 102
    vacuum, 184
Ultra-low penetration air (ULPA), 125
Ultrapure deionized (DI) water, 126
Ultrashallow junctions, 506–07
Ultraviolet (UV) lamps, 129
Underlying material selectivity, 458–59
Undoped
    oxide (UDOX), 272
    silicate glass (USG), 272
Uniformity, 333, 521. *See also* CMP planarity
    etch. *See* Etch uniformity
    polish. *See* Polish rate and uniformity
    temperature. *See* Temperature uniformity
    thickness. *See* Thickness uniformity
Unit cell, 69–70, 78, 79
    face-centered cubic (FCC), 70. *See also* FCC diamond structure
Unpatterned
    etching, 436
    surface defects, 161–64
    wafers, 161
UPW. *See* Ultrapure deionized (DI) water
Use of chlorinated agents in oxidation, 234
Usefulness of yield models, 564
Uses of oxide film. *See* Oxide film, uses of
UV, extreme. *See* Extreme UV

Vacancy, 78. *See* Crystal defect, point defect. *See also* Void
Vacancy-interstitial pair, 79. *See also* Crystal defect, Frenkel defect, Point defect.
Vacuum, 182–84
    for etch chambers, 456
    in integrated tools, 188
    of a gas, 95
    pumps, 184–88
    ranges, 182–84
    tube, 2–3
Valence shell, 22
Van der Pauw method, 155
Vapor
    phase epitaxy (VPE), 290
    pressure, 95–96
    prime, 343, 345–47
        prime and dehydration bake, 347
        prime coating, 347
Variable-angle spectroscopic ellipsometry (VASE), 156
Variables, CMP. *See* CMP variables
Vertical
    diffusion furnace (VDF), 239
    furnace, 239, 241–42
        fast ramp. *See* Fast ramp vertical furnace
Vertically modulated well, 503
Via-1 and Plug-1 formation, 216
Via-2 and Plug-1 formation, 217
Vias, 216, 258, 300
Vibration, 294
Viscosity, 350
Viscous flow, 183
Visual inspection (VI), 208
VLSI device fabrication, 79
Voids, 79
Volatile, 96
Voltage amplifying device, 52
Volume, developer. *See* Developer volume

Wafer, 2, 6, 72, 75, 77, 81
    bias and heat load, 280–81
    boat, 242
    carrier, 531–32. *See also* CMP equipment
    cassettes, 132
    charging, 497
    chuck, 422
    cleaning, 345–46, 481
        steps, 137
    cooling, 496–97
    diameter, 76, 77, 79, 82
    drying, 141–42
    edge chipping, 80
    electrical test (WET), 221. *See also* In-line parametric test
    etch, 83
    evaluation, 84
    fabrication, 8–9, 72–75, 78–83
        factories, 2, 5
    fabs, 2, 81
    flatness. *See* Flatness
    flats, 81
    geometry, 81
    holder, 270
    isolation technology, 133
    lapping, 82–83
    level reliability (WLR), 552
    notched, 81
    planarized, 408
    positioning, 553–54
    preparation, 80–84
        requirements, 81
        final assembly. *See* Traditional assembly
    priming, 346–47
    probe. *See* Wafer sort
    rinse, 140–41

scrubbers, 204
slicing, 82
sort, 555–60
    definition, 555
    performing, 556–57
    test issues at, 559–60
    tests, 557–59
    yield, 560–61
    yield models, 563–565
surface, 75, 78, 81
test, 545, 547–65
test structures, 548–49
tilt, 501–02
transfer system, 244
wet cleaning, 135–142
yield, 78
Wafer-level packaging, 592–95. *See also* Advanced assembly and packaging
Wafer-to-wafer nonuniformity (WTWNU), 522
Wall, cold. *See* Cold wall
Warpage. *See* Wafer geometry
Water, 126–29
    deionization, 127
Wave, helicon. *See* Helicon wave
Wavelength-dispersive spectrometer (WDX), 176
Wedge bond. *See* Thermocompression bonding
Wet
    chemical strips, 465–66
    chemistry, 135
    cleaning station, 200
    etch, 436, 464–65
    etch, types of. *See* Types of wet etch
    oxidation, 231
    oxide etch, 465
    sinks, 138
Wet-clean equipment, 138–142
Wet-cleaning overview, 135–37
Wire
    saw, 82
Wirebond quality measures, 579
Wirebonding, 577–80
    thermocompression bonding 577, 578
    thermosonic ball bonding 577, 579
    ultrasonic bonding 577, 578
Wiring. *See* Interconnect
Within-wafer nonuniformity (WIWNU), 522
Workstation design, 131–34

X–ray
    analysis, 79
    film thickness, 157
    fluorescence (XRF), 157
    lithography, 318
    photoelectron spectroscopy (XPS), 174

Yield, 151–52, 560–63
Yield management systems, 564–65

Zeta potential, 129, 533
Zone, flat. *See* Flat zone
Zones, heat. *See* Heat zones